A BASIC COURSE IN
STATISTICS

FIFTH EDITION

A BASIC COURSE IN
STATISTICS

G. M. CLARKE AND D. COOKE

John Wiley & Sons, Ltd

First published in Great Britain in 1978
Fifth edition published in 2004 by
Arnold, a member of the Hodder Headline Group,
338 Euston Road, London NW1 3BH

John Wiley & Sons Ltd, The Atrium, Southern Gate, Chichester, West
Sussex, PO19 8SQ, United Kingdom

For details of our global editorial offices, for customer services and
for information about how to apply for permission to reuse the copyright
material in this book please see our website at www.wiley.com.

British Library Cataloguing in Publication Data
A catalogue record for this book is available from the British Library

Library of Congress Cataloging-in-Publication Data
A catalog record for this book is available from the Library of Congress

ISBN 978-0-470-97387-5

2 3 4 5 6 7 8 9 10
2007 2006 2005

Typeset in 11/13 Times Ten Roman by Charon Tec Pvt. Ltd, Chennai, India

CONTENTS

CONTENTS

Bibliography 684

Note to Reader: Please note the CD has been converted to URL. Go to the following website http://www.wiley.com/WileyCDA/WileyTitle/productCd-0470973870.html

PREFACE TO THE FIFTH EDITION

The fourth edition contained two new chapters and much additional new material. We have not made any changes of that nature in this fifth edition. Instead we have substantially altered, and increased, the material on computing. Previously we included hints on how to carry out the computing exercises using MINITAB, though the computing exercises, to be found at the end of most chapters, are not tied to any particular statistical package. We have now added sections throughout giving detailed instructions on how to use MINITAB to carry out the statistical procedures introduced in the book. The hints have been completely revised. We have taken advantage of the substantially improved version of the package, MINITAB 14, introduced in December 2003, though users of other packages will still find better ways of handling some of the exercises.

We thank all of those who have helped us during the production of previous editions, Dr E. E. Bassett for valuable discussions on this edition and Mr J. M. Bremner for allowing us to use his extension of 'Anscombe's data' in Exercise 23.13(3). We are grateful to acknowledge the permission of the National Birthday Trust Fund and Heinemann Medical Books Ltd to use material from the survey reported in 'British Births 1970'. We thank Dr R. Chamberlain for helping us with this material.

We are also grateful to acknowledge the permission of the following boards to use questions from their A-level papers: Associated Examining Board, University of Cambridge Local Examinations Syndicate, Joint Matriculation Board, Oxford Delegacy of Local Examinations, the University of London Examinations and Assessment Council, Cambridge University Press (Diploma question), Oxford and Cambridge Schools Examination Board (who were also responsible for the Mathematics in Education and Industry (MEI) and Schools Mathematics Project (SMP) syllabuses), Southern Universities' Joint Board, and Welsh Joint Education Committee. The questions we have used are labelled with the appropriate board in the text. We also thank the Royal Statistical Society for permission to take questions from their examination papers, which are labelled (RSS), and to continue using material from the examinations of the Institute of Statisticians (IOS) which has now merged with the Royal Statistical Society.

We are indebted for permission to reprint parts of tables to the Biometrika Trustees, to the Literary Executors of the late Sir Ronald A. Fisher, FRS, and Dr Frank Yates, FRS, and to Oliver and Boyd Ltd, Edinburgh. The tables have been drawn from E. S. Pearson and H. O. Hartley, *Biometrika Tables for Statisticians*, Volume I, third edition (1966), and from R. A. Fisher and F. Yates, *Statistical Tables for Biological, Agricultural and Medical Research*, sixth edition (1963). The data on cuckoo eggs, used in Chapter 5, are taken from *Biometrika* **1** (1901), and Table A9 is based on one originally published in the *Journal of the American Statistical Association*.

Finally we thank our publishers for encouraging us to write this book in the first place, and for giving us the opportunity to publish new editions at appropriate times. An innovation for this edition, suggested by our publishers, has been to put some material on a website (see Introduction, page xxi). We shall welcome readers' and users' comments at any time; those on the previous editions have been very useful and have helped us to clarify or improve the text where it seemed desirable.

G. M. Clarke

D. Cooke

2004

INTRODUCTION

The number of people studying statistics increased steadily after its vital importance was realized in and immediately after the Second World War. First it was studied only at graduate level in universities, but then it spread through into undergraduate courses and finally to all age-groups in schools. As syllabuses have changed regularly since the first edition of this book was written, so we have made changes in the text, introducing new topics and emphasizing computing and the use of statistical packages. For this fifth edition we have made more changes, reflecting the recent rapid advances in statistical computer packages. Not many UK students now study pure mathematics and statistics as a double subject for A level, but those who do should find all the topics they need. First-year university courses in statistics cover similar ground, and the text is at a suitable level for these.

The mathematical level is roughly that required for the algebra and calculus sections of A-level pure mathematics, or that to be studied along with statistics in a Part I university or college course. We also expect this text to go on being useful to people working for early professional examinations, such as the Ordinary and Higher Certificates of the Royal Statistical Society.

PRACTICAL APPLICATIONS

The importance of statistics stems from its usefulness. Statistics is concerned with the collection and analysis of data in order to obtain a better understanding of phenomena, and its methods have proved relevant to a very wide range of subjects. We believe strongly that one should begin statistics by considering the questions asked in a typical investigation that uses statistics: how are we to collect data that can be used in a meaningful way, and how far can we go in explaining the result by using only simple descriptive statistics? From this, we come to appreciate the limitations of simple statistical procedures, and to see the power of the approach of the applied mathematician, who sets up a model of a situation in order to explore its implications. The basis of models used in statistics is the mathematics of probability, which we introduce in Chapters 6 and 7. Probability theory must hold a central place in the development. But our experience, from teaching and examining students of a wide range of attainments and backgrounds, has been that the majority find it difficult to master the level of abstraction required in probability theory. Thus, they cannot hope to use probability theory as a tool unless they first become well acquainted with practical examples of random phenomena. We do not therefore *begin* with probability theory, as some texts do; and throughout the book we introduce new topics with examples and then show how to set up appropriate models.

DISCRETE AND CONTINUOUS DATA

We pass from the fundamentals of probability and develop the basic ideas of statistical theory with *discrete* random variables only. We have found that many of the important ideas of statistical inference can be understood in this way without mathematical complication. It is therefore over half-way through the book before the ideas of calculus (Chapter 14) and of the exponential function (Chapter 15) are required in order to deal with continuous random variables; for the student who is studying the necessary mathematics simultaneously with statistics, this gives ample time to meet these ideas. Moreover, any development of the statistical theory of continuous random variables must centre upon the normal distribution, and the mathematical complexity of the normal distribution may easily deter a student if it is met too early in a course.

PROJECTS

Right from the first edition, we included projects at the end of each chapter. The case for practical work in statistics is just as strong as in the experimental sciences, but schools and school examining boards were slow to accept this, particularly when the subject was seen as part of mathematics, not requiring laboratory time. Practical work may on occasion take a lot of time to achieve what can seem rather little, though the rapidly increasing use of computers in schools and colleges has reduced the load of arithmetic involved in analysing data collected in projects. We are aware of some imaginative courses, particularly at college and university level, where large sets of data can be explored in a wide variety of ways; even when it is difficult to find space in a timetable we must still stress how important it is to do practical work at the same time as studying the theory.

Our aim has been to make the projects as real as possible, without demanding very special equipment or an excessive amount of time in collecting data. We have included some artificial projects, based only on dice, coins and playing cards, because they exhibit some ideas in a particularly simple form, but we believe the proportion of these should be kept small because they do not give an opportunity to meet the real-life problems of collecting data. So many attempts to use statistics are founded upon sand because data have been carelessly collected. Half the value of a project is in thinking how to collect data that do represent some population; then, and only then, can we generalize from the results we obtain. Chapters 3 and 12, in particular, discuss the collection of data, by sample surveys, experiments and simulation, in more detail than has usually been found in a text of this type. In the fourth edition we added notes on presenting data in simple graphs and diagrams that can help considerably in interpreting and understanding them. Modern computer packages have a wide range of graphics tools of this type, which should be used whenever they will help to clarify a written report on the results of an analysis.

We have found that students enjoy and gain from project work; but an essential part of it is to write a report at once and to make plain what has been achieved, before the detail is forgotten. We therefore hope that all students will keep practical (loose-leaf) notebooks which will include data record sheets (we discuss the planning of these in Chapter 1) and reports on projects. (The parts of a report might have the headings: Title – Summary – Introduction – Method of data collection – Analysis of results – Discussion – Conclusion.) Wherever possible, at least one project should be done with each chapter. It is often possible to divide projects among a class or group of students. If each student does a different project and reports the results to the rest of the class, all students will become acquainted with a wide range of projects. On the other hand, in projects where data are difficult to collect it may be best for all students to do the same project and pool the data for analysis. When students are working alone, it may not be possible to carry out projects, even using simulated data as discussed in the computing sections; so we should finally point out that the text is self-sufficient without the projects.

CONTENTS

The text is arranged in 24 chapters. Not all are of the same length, and some will prove more difficult than others, though we believe all are very important in developing a student's knowledge of statistical theory and methods. Chapter 20, on the Poisson distribution, serves to revise many earlier topics, and Chapters 21 to 24, on correlation, analysis of variance and regression, are somewhat detached from the development of the main thread of thought through the text. Some material from these chapters, therefore, may be taken earlier in a course if desired. Chapter 22 follows on directly from material near the ends of Chapters 17 and 18. Otherwise, our order of development is entirely deliberate. We include a brief appendix of mathematical results which are very directly relevant to the statistical theory; this serves as an easily accessible reference and will be of particular use if topics have not been completely covered in mathematics before they are required in statistics.

VARIATES AND RANDOM VARIABLES

In any application of mathematics, it is vital to distinguish between the observations we make – the *data* – and the mathematical explanation we give of these – the *model*. We have tried to emphasize this distinction by differentiating between two words, variate and random variable, which are often regarded as synonyms, though applied statisticians talk of the data they handle as being values of variates while theoretical statisticians talk of random variables. We therefore use *variate* to denote a quantity we can observe and measure, and reserve the term *random variable* for a quantity we propose as a model. A related, but more common, distinction is that we usually denote a parameter in a population or in a probability function by a Greek letter, while the estimate of this parameter that we make from the sample is denoted by a Latin letter. We have used the bar notation (e.g. $\bar{x}$, $\bar{y}$) to denote a sample mean. We considered using both the bar notation and m, corresponding to the population mean μ, but rejected m to avoid the confusion resulting from the use of two symbols. We would have liked to use ν, π for the parameters of the binomial distribution, but this seemed such a drastic change from the commonly used n, p that we compromised on n, π.

COMPUTING

We have used methods which students who take the subject at a higher level will not have to 'unlearn'. One particular example is the formula for a variance calculated from data, $\sum x_i^2 / n - \bar{x}^2$. This is attractive because it can be neatly expressed in the phrase 'mean of the squares minus the mean squared'. Unfortunately, as we have found with university students, particularly those in other subjects who need to study statistics, it leads to confusion later when the appropriate formula for an estimated variance is $[\sum x_i^2 - (\sum x_i)^2 / n](n - 1)$. We give in most chapters an indication of how to use MINITAB 14 to obtain results and information explained in the chapter. For certain procedures MINITAB has sophisticated options that are not required at this level, and we have often not mentioned these in our explanation. There are also some default settings in MINITAB which give figures whose application the novice may not appreciate. Although they appear automatically they will not be appropriate in every analysis, and should only be included in a report after taking advice from a statistician. We have given a warning in these cases.

Statistical packages may be used without any knowledge of the underlying programming. Readers who are interested in programming statistical computations may refer to our book (with Dr A. H. Craven) *Basic Statistical Computing* (second edition, Arnold, London, 1990) which uses BASIC, or to *Statistical Computing in Pascal* by the same authors. We have included some references to the first text using *CCC* as a shorthand for its title, and although these texts are now out of print copies are likely to be found in many libraries. Much can be done very quickly with modern statistical packages, and more complicated analyses can be carried out. But for small sets of data the small pocket calculators now widely available at low cost offer an immediate way of doing many standard analyses without the need to load data into a powerful package. It remains our view that the computer is best used as a *supplement* to the traditional tools of pen, paper and hand calculator rather than as a complete replacement of them.

EXAMPLES AND EXERCISES

There are many worked examples and a large number of exercises to give the student practice. Simple exercises follow immediately on the introduction of a new concept, leading up to questions of examination standard at the ends of chapters and in the revision exercises. There are six sets (A–F) of revision exercises following Chapters 5, 9, 13, 16, 19 and 24. Set A covers the first five chapters. Set B is mainly on the material of Chapters 6 to 9 but also contains some questions on earlier material, and the remaining sets are made up in a similar way. There are computing exercises at the end of most chapters. Hints for doing these, using MINITAB, are included near the end of the book (p. 706). Throughout the book other exercises for which a computer might be used are labelled .

WEBSITE

There is a website which accompanies this edition; it may be found at www.basicstatistics.co.uk. Several of the larger data sets from exercises and computing exercises in this book (those marked) will be found there, together with the 'Student Data Set' originally published in *CCC*. There are also additional hints on the computing exercises. The website also contains a selection of multiple-choice questions (each with five possible answers); they are grouped according to the material in each chapter and at a similar level to our treatment. Other sources of such questions can be located by an internet search.

LIST OF PROJECTS

LIST OF DATA SETS ON THE WEB

Each set is marked in the text. The title in capitals gives the name of the web file.

NOTATION

r	particular value of a discrete variate or random variable, 33, 169; sample correlation coefficient, 546
$r_1, r_2, \ldots$	discrete variate values, 33
s^2	estimated variance, 255
$S, S_B, S_G, S_R,$	
S_T, S_{LR}, S_{MR}	corrected sums of squares in analyses of variance, 576, 587, 622, 655
S_{pp}, S_{pq}	corrected sums of squares and products, 651
T	Wilcoxon rank-sum statistic, 285
T_+, T_-	Wilcoxon signed-rank statistics, 291
T_i	total of all observations on treatment i, 576
t	Student's t distribution, 426
$t_{(n-1, 0.05)}$	95% point of t distribution with $n - 1$ degrees of freedom, 451
U	Mann–Whitney statistic, 287
$\mathrm{Var}[R]$	variance of the random variable R, 208
X	name of a random variable, 169
x	particular value of a continuous variate or random variable, 30, 169
$\bar{x}$	mean of a set of observations, 30
$\hat{Y}, \hat{y}$	predicted values of Y, y in a linear model, 625, 654
z	standardized normal score, 355
α, β	parameters of a regression line, 606; parameters of (continuous) uniform distribution, 325
β_i	regression coefficient of Y on X_i, 649
λ	parameter of the exponential distribution, 327; parameter of the Poisson distribution, 513
$\hat{\lambda}$	maximum likelihood estimate of parameter λ, 462
$\tilde{\lambda}$	method of moments estimate of parameter λ, 458
μ	population mean, 52; mean of a random variable, 198, 322
π	proportion in a population, 52; one of the parameters of the binomial distribution, 174; parameter of the geometric distribution, 177
ρ	population correlation coefficient, correlation coefficient of two random variables, 548
σ	standard deviation of a random variable, 208
σ^2	variance of a random variable, 208; variance of a population, 322
$\sum_{i=1}^{N} x_i$	sum of a set of N observations, i.e. $x_1 + x_2 + \cdots + x_N$, 30
$\Phi(b)$	cumulative distribution function of normal distribution, 349
χ^2	chi-squared distribution, 475
X^2	statistic used in a χ^2 test, 474
$\ln$	is always used for $\log_e$ in this text

1 POPULATIONS AND VARIATES

OVERVIEW

An example of a statistical investigation, the Survey of British Births, *is used in order to introduce the study of statistics. Tables, graphs and diagrams are shown to provide convenient initial summaries of the information in data.*

1.1 A STATISTICAL INVESTIGATION

Statistics is the science that studies the collection and interpretation of numerical data.

An example of an investigation using statistics was the *Survey of British Births, 1970.* The aim of that investigation was to improve the survival rate and the care of British babies at and soon after birth, by collecting and analysing data on new-born babies.

Data collected on one or two babies would be of little use, because the items of information being measured would vary substantially from one baby to another. For example, their weights at birth would vary considerably. The aim was to gain information about a *population*, namely 'all British births'.

Definition 1.1

A **population** is the collection of items under discussion. It may be finite, as in this example, or infinite; it may be real, as here, or hypothetical (as in some of the models we set up later in the book).

It was not practicable to collect full information on every birth in Britain, even for a single year. A smaller population was specified which was expected to be very similar to the whole population of British births. This smaller population consisted of all babies who were born (alive or dead) after the 24th week of gestation, between 0001 hours on Sunday 5 April and 2400 hours on Saturday 11 April 1970.

A large amount of information was collected about each baby, its mother, and the circumstances of the birth. Data recorded included the birth-weight and sex of the baby, the place of birth (home, hospital or elsewhere), and whether the mother smoked or not. Some of these data are measured (e.g. weight), whereas others are classified (e.g. sex must be male or female). We call these classified records *attributes.* Each of the quantities or attributes recorded is called a *variate.*

Definition 1.2

A **variate** is any quantity or attribute whose value varies from one unit of investigation to another.

The units in this example were the individual babies in the population. Note that 'home', 'hospital' and 'elsewhere' are the three possible values of the variate 'place of birth'. It is convenient to stretch the meaning of the word 'value' to include cases like this, rather than invent a new word for variates that are not numerical.

Definition 1.3

An **observation** is the value taken by a variate for a particular unit of investigation.

The number of variates used exceeded 100, and the number of babies in the population investigated was over 17 000. There were therefore about 17 000 observations on each of 100-plus variates. However, as is usual in any large or complex statistical investigation, there were 'missing data': not every variate was recorded on every baby. Mishaps led to some of these measurements not being available, though in fact there were few such mishaps in this study. Some variates were simply not applicable to all babies, e.g. the cause of death applied only to the dead babies. Even so, the total number of observations – the mass of data – was enormous.

    The original 'raw' data were collected by filling in questionnaire forms. These data were then transferred to punched cards so that a computer could be used to deal with them. The prime problem with such an amount of data is to summarize them so that they can be interpreted.

    Variates differ in nature, and the methods of analysis appropriate to a variate will obviously depend on its nature. We can distinguish between *quantitative variates* (like the birth-weight of the baby, or the daily number of cigarettes smoked by the mother at a specified time during pregnancy) and *qualitative variates*, or attributes (such as the sex of the baby or the place of birth).

Definition 1.4

A **quantitative variate** is a variate whose values are numerical.

Definition 1.5

A **qualitative variate** or **attribute** is a variate whose values are not numerical.

Quantitative variates can also be divided into two types: they may be *continuous*, if they can take any value we care to specify within some range, or *discrete* if their values change by steps or jumps. Thus birth-weight is continuous, because there is no reason why a baby should not have a weight of 6.943762 lb – even if no scales could measure it this accurately! However, a variate like the number of children previously born to the mother must be a whole number $0, 1, 2, \ldots$, going up in steps; decimal values are certainly not allowed here.

Definition 1.6

A **continuous variate** is a variate which may take all values within a given range.

Definition 1.7

A **discrete variate** is a variate whose values change by steps.

The choice of which variates to record is an important decision to be made in any investigation. It is a decision which needs careful thought beforehand, as we hope everyone will remember in carrying out any of the projects suggested in this book. In the present example the most important variate is birth-weight, because it was already known that the ability of a baby to survive is closely associated with birth-weight. Babies with low or particularly high birth-weights have relatively less chance of survival.

A question of particular interest was whether smoking by the mother had any influence on the weight of the baby. This could be investigated by comparing the weights of babies born to mothers who smoked with the weights of babies whose mothers did not smoke.

Let us first look at how many of the mothers smoked. We consider the qualitative variate *smoking habit* that can take only the following 'values': 'mother smokes now', 'mother has given up smoking', 'mother has never smoked'. This is a very simple variate, and the *frequency table* (Table 1.1) summarizes the information on it and demonstrates the *frequency distribution* of the possible 'values'.

Definition 1.8

The **frequency distribution** of a (discrete) variate is the set of possible values of the variate, together with the associated frequencies.

Relative frequencies (i.e. the frequencies in the groups expressed as a proportion of the total frequency) have been worked out also, because they help in making comparisons.

We look at the birth-weights separately for the two clearly defined populations: (1) babies whose mothers smoke now, and (2) babies whose mothers have never smoked. Each of these two populations of babies gives us a population of measurements (of the birth-weights recorded for individual babies); we need convenient summaries of these populations of measurements. We shall again

Table 1.1 *Smoking habits of mothers*

	Smokes now	Given up smoking	Never smoked	Total
Frequency	7149	2818	6563	16 530
Relative frequency	0.4325	0.1705	0.3970	1.000

draw up frequency tables, one for each population. This time the variate being summarized is weight, a continuous measurement. Instead of looking at the frequency of each variate-value that occurs we first group the values into intervals, that is, subdivisions of the total range of possible values of the variate. In this example the variate is conveniently collected into the class-intervals 1–500, 501–1000, . . . , 4501–5000, 5001–5500 grams.

Definition 1.9

A **class-interval** is a subdivision of the total range of values which a (continuous) variate may take.

Definition 1.10

The **class-frequency** is the number of observations of the variate which fall in a given interval.

Definition 1.11

The **frequency distribution** of a (continuous) variate is the set of class-intervals for the variate, together with the associated class-frequencies.

Table 1.2 shows the frequency distribution of birth-weight in the two populations. We cannot make a direct comparison of these two sets of frequencies because the total frequencies in the two populations differ. In order to obtain two sets of figures expressed in the same units, which *can* be compared, we calculate the relative frequencies as shown.

Comparison is helped by drawing a graph. The *frequency polygon* illustrating a set of frequencies, or, as here, a set of relative frequencies, is obtained by plotting

Table 1.2 *Frequency distribution of birth-weight of babies according to whether mother smokes or has never smoked*

Interval (g)	1–	501–	1001–	1501–	2001–	2501–
Frequency S†	4	38	72	123	479	1661
Frequency N†	4	20	25	73	236	1089
Relative frequency S†	0.001	0.005	0.010	0.017	0.067	0.232
Relative frequency N†	0.001	0.003	0.004	0.011	0.036	0.166

Interval (g)	3001–	3501–	4001–	4501–	5001–
Frequency S†	2831	1560	343	36	2
Frequency N†	2530	1927	551	101	7
Relative frequency S†	0.396	0.218	0.048	0.005	0.000
Relative frequency N†	0.385	0.294	0.084	0.015	0.001

† S refers to the population of babies whose mothers smoke; N refers to the population of babies whose mothers have never smoked.

class-frequencies or relative frequencies as *y*-values against the centre-points of class-intervals as *x*-values. Then the plotted points are joined by straight lines.

In Fig. 1.1, the two frequency polygons for the populations are shown on the same graph. One can see that although the two distributions have a similar shape, the distribution of birth-weight for babies whose mothers smoke is to the left of the corresponding distribution for those whose mothers have never smoked. Considered as a whole, the birth-weights for babies whose mothers smoke are rather smaller than those for babies whose mothers have never smoked. There is a great deal of overlapping of the two populations: many *individual* babies in the 'smokers' population have weights greater than individuals in the 'non-smokers' population. Nevertheless the difference between the two populations viewed in their entirety is quite distinctive.

The apparent evidence about smoking from this survey is not conclusive. The women chose for themselves whether or not to smoke. There is a possibility, which cannot be excluded and should not be overlooked, that the women who did smoke might be different in other ways (e.g. richer or more nervous) from those who did not. The difference in birth-weight of the babies might be a result of some other unknown factor, not of smoking. Before using results to make claims about possible causes, an investigator should always check up, so far as possible, whether other factors may be influencing the results.

This example has illustrated how simple tables and graphs help to summarize a great deal of information. In the examples which now follow we consider the practical processes of constructing tables and graphs in more detail.

Figure 1.1 *Frequency polygons of birth-weights of babies whose mothers never smoked (o---o), and whose mothers smoked at the time of the survey (•—•).*

1.2 CONSTRUCTING TABLES AND GRAPHS

Example 1.1 Frequency table and bar diagram of a discrete variate

A man kept count of the number of letters he received each day over a period of 100 days (excluding Sundays). The observations were:

```
02111    20010    11000    31201
10010    11020    00101    02120
02010    10103    12000    01000
10101    02012    12010    22101
00005    01120    02102    00210
```

Construct a frequency table and give a suitable graphical representation of the data.

The first step is to make a *tally count* of the data. A manual way of doing this is shown below. The line totals give the frequency table.

Value	Tally	Frequency
0	⊤⊦⊦ ⊤⊦⊦ ⊤⊦⊦ ⊤⊦⊦ ⊤⊦⊦ ⊤⊦⊦ ⊤⊦⊦ ⊤⊦⊦ ⊤⊦⊦ ⏐⏐⏐	48
1	⊤⊦⊦ ⊤⊦⊦ ⊤⊦⊦ ⊤⊦⊦ ⊤⊦⊦ ⊤⊦⊦ ⏐⏐	32
2	⊤⊦⊦ ⊤⊦⊦ ⊤⊦⊦ ⏐⏐	17
3	⏐⏐	2
4		0
5	⏐	1
	Total	100

The grand total provides a check that no observation has been missed. A good overall check is to repeat the tally count taking the observations in the reverse order.

The frequency distribution of a discrete variate is best represented by a *bar diagram* (Fig. 1.2).

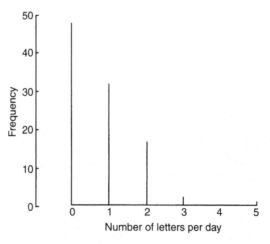

Figure 1.2 *Bar diagram of number of letters received each day.*

The height of the bars is proportional to the frequency of each variate-value. The bars may be thickened, though still centred on the variate-values, to make the bars more obvious. The thickness has no significance. The bars must be kept distinct (cf. the histogram in the next example), to show that the variate-values are distinct.

Example 1.2 Frequency table and histogram of a continuous variate

Measurement of the lengths of 100 eggs of cuckoos gave the following results (all readings in millimetres):

22.5, 20.1, 23.3, 22.9, 23.1, 22.0, 22.3, 23.6, 24.7, 23.7,
24.0, 20.4, 21.3, 22.0, 24.2, 21.7, 21.0, 20.1, 21.9, 21.9,
21.7, 22.6, 20.9, 21.6, 22.2, 22.5, 22.2, 24.3, 22.3, 22.6,
20.1, 22.0, 22.8, 22.0, 22.4, 22.3, 20.6, 22.1, 21.9, 23.0,
22.0, 22.0, 22.1, 22.0, 19.6, 22.8, 22.0, 23.4, 23.8, 23.3,
22.5, 22.3, 21.9, 22.0, 21.7, 23.3, 22.2, 22.3, 22.8, 22.9,
23.7, 22.0, 21.9, 22.2, 24.4, 22.7, 23.3, 24.0, 23.6, 22.1,
21.8, 21.1, 23.4, 23.8, 23.3, 24.0, 23.5, 23.2, 24.0, 22.4,
23.9, 22.0, 23.9, 20.9, 23.8, 25.0, 24.0, 21.7, 23.8, 22.8,
23.1, 23.1, 23.5, 23.0, 23.0, 21.8, 23.0, 23.3, 22.4, 22.4.

Construct a frequency table. Illustrate the data graphically by means of: (a) a stem-and-leaf diagram; (b) a histogram; (c) a cumulative frequency diagram.

With data such as these the practice is to record to the nearest value for the number of decimal places chosen. Thus a recorded value of 22.5 represents a number in the interval 22.45–22.55 mm.

A suitable class-interval must be chosen. The maximum and minimum observations are 25.0 and 19.6. The range of the observations is therefore $25.0 - 19.6 = 5.4$. (More precisely it is $25.05 - 19.55 = 5.50$, but this is an unnecessary refinement here.) About 10 intervals is usually suitable; division of the range by 10 gives

Interval (mm)	Centre (mm)	Tally count	Frequency
19.5–19.9	19.7	I	1
20.0–20.4	20.2	I I I I	4
20.5–20.9	20.7	I I I	3
21.0–21.4	21.2	I I I	3
21.5–21.9	21.7	THL THL I I	12
22.0–22.4	22.2	THL THL THL THL THL I I	27
22.5–22.9	22.7	THL THL I I	12
23.0–23.4	23.2	THL THL THL I	16
23.5–23.9	23.7	THL THL I I	12
24.0–24.4	24.2	THL I I I	8
24.5–24.9	24.7	I	1
25.0–25.4	25.2	I	1
		Total	100

a value of 0.54, which suggests a class-interval of 0.5 mm. The end-points of the interval should be chosen so that no observation can fall on them. They will thus be expressed to one place more of decimals than the actual observations. Suitable intervals are 19.45–19.95, 19.95–20.45, ... The *class-centres* are 19.7, 20.2, ... An alternative form of specification of the intervals is 19.5–19.9, 20.0–20.4, ...

A tally count for the cuckoo-egg measurements is given in the table on p. 7.

(a) An alternative to a tally count table is a *stem-and-leaf* diagram (or *stem plot*) invented by J. W. Tukey (American, 1915–2000). It has the virtue of retaining all the information on individual values; we may think of it as a concise way of writing down a set of numbers. A stem-and-leaf diagram for the cuckoo-egg data takes the form:

	Length of cuckoo eggs (in tenths of mm)	Depth
19		0
19	6	1
20	1 4 1 1	5
20	9 6 9	8
21	3 0 1	11
21	7 9 9 7 6 9 9 7 9 8 7 8	23
22	0 3 0 2 2 3 0 0 4 3 1 0 0 1 0 0 3 0 2 3 0 2 1 4 0 4 4	50
22	5 9 6 5 6 8 8 5 8 9 7 8	50
23	3 1 0 4 3 3 3 4 3 2 1 1 0 0 0 3	38
23	6 7 8 7 6 8 5 9 9 8 8 5	22
24	0 2 3 4 0 0 0 0	10
24	7	2
25	0	1

Stem unit = 1 mm;   Leaf unit = 0.1 mm

Tukey calls each row of the diagram a *stem*; the *stem labels* such as 19 or 20 are put to the left of the ruled line while the digits to the right of the line are the *leaves* on the stems. A measurement 20.1 is represented on the third row, i.e. the stem with label '20', by the digit, or leaf, '1'. The remaining three leaves on the same stem in this example represent the measurements 20.4, 20.1 and 20.1. The intervals used are identical with those in the tally count. The diagram above was constructed by reading along the rows of the data and adding the leaves one by one. Intervals should be of 5 units, as for the cuckoo eggs, or of 2 or 10 units.

The column on the right might be used for a frequency count, as in the tally count above. Tukey recommends using it, as we have done, for *depth*, which is a measure of how deep any observation is in the distribution. The column gives cumulative depths counting from each end of the distribution. Thus, on the '21' stem the cumulative depth is $11 = 1 + 4 + 3 + 3$, the sum of the frequency counts of the earlier stems. For the cuckoo-egg data the *median* (see p. 24) happens to fall on an interval boundary, because it has depth $\frac{1}{2}(100 + 1) = 50.5$. If this does not happen, there will be a central interval that cannot sensibly be cumulated with the totals from each end. The convention is to give the frequency count for that interval in brackets (see the example on p. 90).

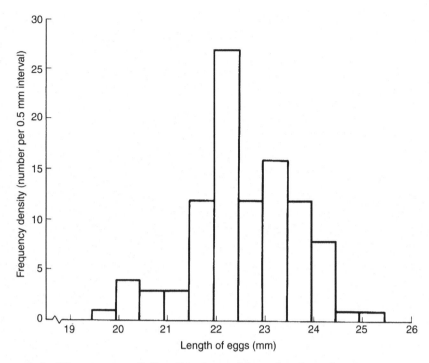

Figure 1.3 *Histogram of distribution of length of cuckoo eggs.*

It may be useful to put the leaves on each stem in increasing numerical order, e.g. the leaves on the stem labelled '20', which run '1 4 1 1', would be ordered as '1 1 1 4'. In this way all the numbers can be put in order; this will be helpful when finding a median.

(b) *Histogram.* We found frequency polygons to be useful diagrams for comparing the two frequency distributions in the British births example. An alternative diagram for representing the frequency distribution of a continuous variate is a histogram.

In a histogram, a rectangle is drawn above each class-interval (see Fig. 1.3) such that the *area* of each rectangle is proportional to the frequency of the observations falling in the corresponding interval. If intervals are of the same width then the height of the rectangles is proportional to the frequency, but it is not always practicable to make all intervals the same width. The vertical axis is labelled 'frequency density'; since area represents frequency it follows that the dimension of the height of a rectangle is number per unit class-interval of egg length, which should be described as 'frequency density' rather than 'frequency'.

(c) *Cumulative frequency diagram.* We first prepare a *cumulative frequency table.* We form the cumulative frequencies from the frequency table by adding in the frequencies, one by one, beginning from the left. The cumulative frequency corresponding to each class-interval gives the total frequency in that interval and in all intervals to the left of it; it therefore gives us the total frequency of observations that fall below the upper end-point of the class-interval. Hence it is the upper end-point that we put in the table and use in the graph.

Upper end-point	19.45	19.95	20.45	20.95	21.45	21.95	22.45	22.95
Cumulative frequency	0	1	5	8	11	23	50	62

Upper end-point	23.45	23.95	24.45	24.95	25.45
Cumulative frequency	78	90	98	99	100

Figure 1.4 shows the cumulative frequencies plotted against the corresponding end-points. We must always make the first cumulative frequency in our table 0 to give a starting point for our diagram. This 0 will correspond to the upper end-point of the interval below the one that contains the first observation. The diagram that we have drawn in Fig. 1.4 may be called a cumulative frequency polygon, since it consists entirely of straight lines. When a statistician wants to make inferences about a population on the basis of a sample from it, the first step is very often to smooth this polygon into a curve. This will be illustrated in Fig. 4.2 (p. 77).

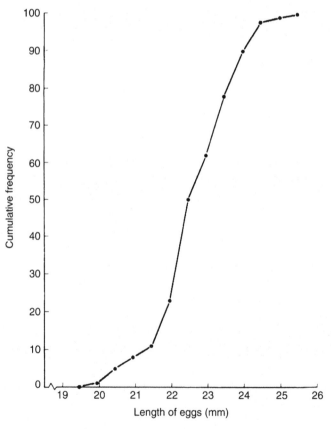

Figure 1.4 *Cumulative frequency diagram of distribution of length of cuckoo eggs.*

Example 1.3 Histogram of a continuous variate, with unequal grouping intervals

A certain disease affects children in their early years, and sometimes kills them. The frequency table of the age at death, in years, of 95 children dying from this disease is:

Age at death (years)	0–	1–	2–	3–	4–	5–10	Total
Frequency	10	40	20	10	5	10	95

Draw a histogram to represent the data.

Note that the convention with age is to record not to the nearest number of years but to the number of *completed* years. A child said to be aged 3 years has an age in the interval 3 to 4 years, including the lower end-point but excluding the upper end-point; in symbols, if the age is x, $3 \leqslant x < 4$. This interval is most conveniently denoted 3–, and this convention has been used in the preceding table. The class-centre of the class-interval 3– is 3.5 years.

The histogram (Fig. 1.5) is drawn in a similar manner to that in Example 1.2 except that we must take care with the final class-interval 5–10, since it is longer than the other class-intervals. We do *not* draw a rectangle with *height* equal to the frequency, 10, over the final interval, because that does not represent the data fairly. The incidence of the disease wanes with age after age 1; we see that there were 10 cases per year of age at age 3, 5 cases per year of age at age 4 and 2

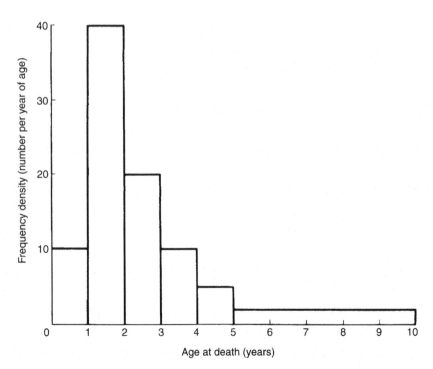

Figure 1.5 *Histogram of distribution of age at death of diseased children, demonstrating the use of unequal class-intervals.*

(=10/5) cases per year of age over ages 5–10. We make the *area* of the rectangle for the last age-group equal to the frequency 10, rather than the height of the rectangle. This illustrates the general rule for histograms: it is the *area* of each rectangle that is proportional to the frequency.

Note also that the frequency distribution of the age at death is not symmetrical about a centre but rises quickly to a peak and falls gradually away. Frequency distributions which are not symmetric are said to be *skew*.

1.3 TABLES AND DIAGRAMS

We have already seen the convenience of tables and diagrams in summarizing data. We now make some general comments about them.

Data tables are used widely in the collection, manipulation and summary of data. Tables allow the relations between numbers to be exhibited clearly and reduce mistakes when collecting and handling data; they aid the calculation of derived quantities and the interpretation of data. Three main types of table may be recognized: (1) raw data tables, or record sheets, (2) calculation tables, (3) summary tables.

Many points about table construction will appear so obvious to the reader that it does not seem profitable to discuss them. Nonetheless, tables are often poorly constructed so we shall now give a check-list of points to heed. The points under 'summary tables' owe a great deal to the recommendations of A. S. C. Ehrenberg.

1.3.1 Points to note when constructing tables

General
1. Make the table self-explanatory: provide a title, a brief description of the source of the data, state in what units the figures are expressed, label rows and columns where appropriate.

Raw data tables
2. Plan the record sheet so that: (a) recording is straightforward and unambiguous; (b) derived quantities may be worked out conveniently on the sheet (e.g. allow room for totals).

3. Keep the table as simple as possible, subject to (2).

4. When recording numbers, carry enough figures so that information is not thrown away but do not record many meaningless digits. When in doubt, a rule of thumb is to record to three significant figures.

5. Make alterations clearly; strike out a wrong number and write the correct value beside it, rather than attempting to alter figures.

6. Distinguish between zero values and missing observations.

Calculation tables

7. When possible carry out calculations on the record sheet. (Transfer of data from one sheet to another introduces errors.)

8. When data are transferred from the record sheet, double-check the numbers transferred. Check, for example, that the total of columns of transferred numbers equals the total of the original numbers on the record sheet, as well as reading the numbers over again.

9. Give the calculation a logical pattern on the sheet.

10. Choose units that will simplify the calculation. Consider a suitable coding (see p. 33); for example, decimal numbers may be multiplied by a power of 10 to convert them to integers.

Summary tables

11. Make the table clear and simple, with the main numbers for comparison close to each other. It is usually easier to compare numbers in the same column than those in the same row.

12. If the table is becoming complicated break it down into smaller tables.

13. Arrange rows and columns in some natural order or in size order. Where possible, put big numbers at the top of the table.

14. Give row and column averages or totals, where they are meaningful.

15. Choose units that the reader will understand and that will keep the table simple, e.g. a unit of millions of £ may be convenient. State the unit of measurement.

16. Give a clear title and a verbal summary of the main points of the table.

17. Round all numbers to two effective digits. Effective digits are those which vary within a set of data. Each of the following sets of data is quoted to two effective digits: (a) 129, 136, 119, 151; (b) 6.2, 7.3, 4.2, 3.8; (c) 290, 530, 640, 310. Note that this does not contradict (4) above; that remains an appropriate rule when collecting and processing data (as opposed to presenting a summary). To fix the final digit, round to the nearest number. Thus 1.32 becomes 1.3 and 27.9 becomes 28. If the value before rounding is exactly in the middle of the two possible rounded numbers, then round to the even digit. Thus 2.45 becomes 2.4 and 2.75 becomes 2.8. It is desirable not to consistently round in one direction, say up, because that would yield misleading averages.

Diagrams

In a table use is made of the geometrical ordering of rows and columns to exhibit relations between the numbers in the table. In diagrams (which we take to include graphs), the magnitude of numbers is represented geometrically in order to aid comparisons.

In this chapter, frequency diagrams have been considered. A simple plot of the values of the observations on a line is often useful when assessing the nature of data and detecting wild observations (see Fig. 5.1, p. 90). Plots of pairs of variates (see scatter diagrams, p. 93) are also useful.

These diagrams are adequate for the projects suggested in this book but the reader will find elaborations of these diagrams, and other types, in the presentation of statistical data in newspapers and magazines. The reader should inspect examples critically and decide whether they are helpful or whether they mislead.

As with tables, we give a check-list of points to heed when constructing graphs.

1.3.2 Points to note when constructing graphs

1. Make the graph self-explanatory: provide a title and a brief description of the source, label the axes, state the units, mark in the scales and give a key if it is needed.

2. Choose a scale that: (a) is convenient; (b) ensures that most of the graph-paper is used.

3. Beware of misleading if the origin is not included.

1.4 THE USE OF COMPUTERS

Over the past fifty years the greatest change in the study and practice of statistics, as in many human activities, has been the increasing use of computers. Computation has always been important in statistics. In the early 1950s statisticians carried out their analyses by literally turning the handles of calculating machines. Large investigations used punched card machinery, as was done with the *1970 Survey of British Births*, which is discussed earlier in this chapter. Over time, desk calculating machines became electric, then electronic. Simultaneously, the immediate postwar experimental computers were the inspiration for huge mainframe computers occupying a whole room. These decreased in size as they used, in turn, valves, transistors and silicon chips. The two streams came together in the personal computer that we use today.

We see three main uses for computers in the study of statistics: (1) to facilitate the analysis of data sets which are too large to be investigated by hand; (2) to aid the drawing of graphs and diagrams, which allow critical study of data before analysis, and improve presentation of results; (3) to simulate probability models in order to illustrate probability and statistical theory.

As students become more familiar with computers the size of a set of data that can be analysed more conveniently by computer, rather than by hand, becomes smaller and smaller. Yet there is still a place for pencil and paper analysis. The teacher, having just marked thirty scripts, can gain an immediate impression of the overall performance of his pupils by setting out the marks in a stem-and-leaf diagram such as that we describe on p. 8. Scratching down the numbers, to use John Tukey's phrase, gives an acquaintance with every item of information. Manipulating the data on a computer is more impersonal; odd, and therefore important, items can be missed.

Drawing graphs and diagrams is extremely important in statistics but producing them by hand can be very laborious. Here there is no doubt of the usefulness of a personal computer. The variety of graphs, and the speed of their production by the modern computer, is amazing.

Mimicking the rolling of a die by generating a sequence of digits on a computer is a very simple example of the simulation of a physical process. To roll the die hundreds of times would be laborious and time-consuming; to simulate it hundreds or even thousands of times on a computer is quick and easy. Simulations of this sort allow deeper insight into probability theory and, in many cases, it is only by using a computer that sufficient repetitions can be made to be informative.

In the early days of computing, users had to produce their own programs. An example, to calculate simple statistics, is given on p. 67. Nowadays, computer packages are available that carry out complex calculations and operations on a one-line command or on choice of an item in a menu. We have picked out one widely used statistical package, MINITAB, and describe, through the book, its leading features.

1.5 INTRODUCTION TO MINITAB

When MINITAB is started on the computer two windows appear. The *Data window* contains a *worksheet* which is similar in appearance to a spreadsheet. Data for analysis may be entered in the columns of the worksheet. The *Session window* displays the results of analyses and may also be used to enter commands.

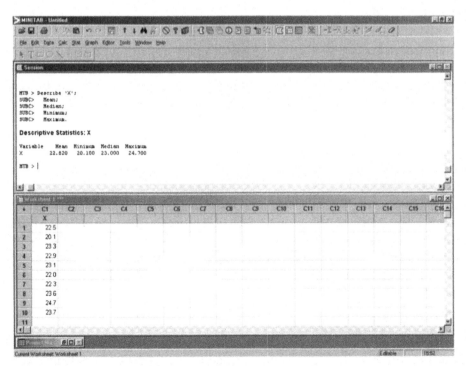

Figure 1.6 MINITAB initial display.

Instructions may be given to MINITAB in two ways. As with Windows, items on a menu bar may be clicked and dialogue boxes completed. We shall call this the *menu method*. Alternatively, commands may be typed in the Session window. We shall call this the *session command method*.

Suppose data are entered into Column 1. The column of data is automatically labelled C1. To produce a histogram from the data click **Graph** on the menu bar. A menu appears; from this click on **Histogram**. In the dialogue box that appears on the screen select **Simple** (it is the default setting and probably already selected) and click **OK**. In the next dialogue box enter C1 under **Graph variables**, ignoring other choices, and click **OK**. (MINITAB uses 'variable' as an alternative term to 'column'.) The graph will be displayed. We might express the procedure more concisely as

> **Graph>Histogram/**
> Select **Simple/**
> Enter **Graph variable/**

The symbol '>' represents movement of the cursor from one word to another and the symbol '/' represents a click on a menu item or on an OK box. Occasionally two successive clicks will be necessary, and this is denoted by '//'. The relevant words on the computer screen are shown here in bold type.

The alternative method is to use session commands. Click anywhere in the Session window and then click **Editor** on the menu bar. Click on **Enable Session Commands**. A prompt 'MTB>' appears on the left-hand side of the window. To produce a histogram type

> HISTOGRAM C1

after the prompt and press 'Return'. Again a graph appears, in its own window. For clarity, we shall write the command words in capitals, but this is not necessary when using MINITAB. Either upper or lower case letters may be used for commands and arguments; only the first four letters of a command word are necessary.

We recommend that readers enable the session commands even if they intend using only the menu approach. If this is done, the session commands equivalent to the menu commands appear in the Session window and provide a record of the analysis.

To enter data into a worksheet, choose say the first column and click the cell below 'C1'. The name of the variate may be entered. This is usually desirable but is not necessary. Use the arrow down key (↓) to move the cursor to the cell below and enter the first value. Repeat this procedure to enter the whole set. Data may also be entered through the Session window. For data that are to go into column 1 use

> SET C1

A prompt 'DATA>' appears. Type in the data, separated by spaces or commas, over one or more lines and signal completion by typing END. For data in several columns use READ, which also provokes the DATA prompt.

Of the forms of data that MINITAB offers we shall make use of columns and stored constants. A column may contain numbers or text and is referred to by C plus a number, as in C1 or C23, or by name. A name may be typed at the head of a column or assigned by, for example, the command

NAME C1 'Length'

Individual entries in a column may be referred to as C1(1), C3(2), etc. A stored constant is a single number or a text string. It is referred to by K plus a number, as in K15. It can be given a name by the NAME command. Its value may be displayed by use of commands such as

PRINT K1

or

PRINT 'density'

To discover the names of columns and constants in use, type the session command

INFO

To carry out arithmetic, MINITAB offers a calculator obtained by choosing

Calc>Calculator/

Typical operations are adding two columns and putting the result in a third column, or transforming data by, for example, constructing a new column of values which are the logarithms of an original column of data. Equivalent operations can be carried out by using LET in session commands such as

LET C3 = C1 + C2
LET C2 = LOGE(C1)

1.6 MINITAB: GRAPHS, DIAGRAMS AND TABLES

1.6.1 Graphs and diagrams

MINITAB can be used to produce the bar charts, histograms and stem-and-leaf diagrams that we have described in this chapter. With data in a column of the worksheet, a bar chart is produced by choosing

Graph>Bar Chart/
Select **Simple/**
Enter **Categorical variable**, i.e. column name/

To obtain a histogram choose

Graph>Histogram/
Select **Simple/**
Enter **Graph variable/**

MINITAB uses the term 'cutpoints' for the endpoints of intervals. It uses a set of rules to choose the cutpoints but the user may wish to choose another set of intervals. We find it easier to do this using session commands, rather than the menus. To construct a histogram from the cuckoo egg data, as it was done in Example 1.2, assume the 100 values are in a column labelled 'Length' and use the commands

HISTOGRAM 'Length';
CUTPOINTS 19.45:25.45/0.1.

This produces a histogram with extreme cutpoints 19.45 and 25.45, and class-interval 0.1. The specification of the cutpoints is an example of MINITAB's notation for patterned data (see Section 7.11). Alternatively, the number of intervals K may be specified. The subcommand is

NINTERVAL K

MINITAB attempts to choose convenient cutpoints subject to the stated number of intervals, or 'bins', to use the language of MINITAB.

MINITAB offers many options for comparing histograms of more than one set of data. Histograms can be overlaid on the same graph, with many display options, or separate graphs may be set side-by-side by using **Editor>Layout Tool**.

For stem-and-leaf plots, choice of

Graph>Stem-and-Leaf/
Enter **Graph variable/**

gives the diagram in the Session window with the leaves in increasing order. The size of the stem interval, called **Increment**, may be chosen or left to MINITAB. Diagrams based on data from many columns appear successively in the Session window and therefore are not as easy to compare as histograms.

1.6.2 Tables

MINITAB may also be used to carry out tally counts. For discrete variates choose

Stat>Tables>Tally Individual Values/
Enter **Variable**, select for **Display** one or more of **Counts, Percents, Cumulative Counts** or **Cumulative percents/**

This would be appropriate for the letter data in Example 1.1. Continuous variates, such as egg length in Example 1.2, are more awkward to tally in MINITAB. See Section 7.11.

If it is wished to cumulate the frequencies in a column, choose

Calc>Calculator/
Select **Partial sums** for **Expression**, enter input column in **PARS( )**, and **Store result** in output column/

The sum of the first i rows of the input column is put in the ith row of the storage column.

1.7 EXERCISES ON CHAPTER 1

1. The variates below were recorded for individual people. State whether each variate is qualitative or quantitative and, if the latter, whether discrete or continuous: (a) age, (b) year of birth, (c) sex, (d) height, (e) colour of hair, (f) number of pairs of shoes possessed, (g) favourite vegetable, (h) age rank in household (i.e. position in order of age among people in the same household), (i) surname.

2. Choose a population of people (e.g. teachers in your school or college, Members of Parliament, inhabitants of your town or village) and guess appropriate relative frequency distributions for the variates listed in Question 1. Sketch diagrams to represent the frequency distributions.

3. (a) People's weights are recorded to the nearest kilogram, and presented in a table. The classes in this table are labelled 60–64, 65–69 and so on. What are the exact upper and lower limits covered by each class, and what is the mid-point of each class?
   (b) An insurance company asks new customers to fill in on a form 'age last birthday'. When processing its records, it lists the number of new customers and their ages in a table, typical classes being 30–34, 35–39 and so on. What are the upper and lower limits and the mid-point of each class?
   (c) A quiz contains 30 questions. One hundred people write down their answers, and these are marked right or wrong. The number of correct answers is recorded and summarized in a table whose class-intervals are 0–8, 9–12, 13–16, 17–20, 21–24, 25–30. What are the mid-points of these classes?
   (d) People attending a doctor's surgery by appointment may have to wait some time before being seen. This waiting time is recorded on a particular day in Dr Smith's surgery, to the nearest minute. The records of waiting time are summarized in a table whose classes are 'less than 5', '5 to 9', '10 to 14', '15 or more' minutes. What are the upper and lower limits of each class, and the mid-point of each class?

4. In 3(c) above, the numbers in the classes were 10, 8, 22, 29, 20, 11 respectively. Suggest how these could be shown in a diagram.

5. In 3(d) above, the numbers in the classes were 6, 9, 9, 4 respectively. Show these in a suitable diagram.

6. The numbers of people travelling from a particular village on the daily bus to town were recorded for a period of 30 days, as follows: 6, 3, 2, 7, 4, 0, 5, 1, 3, 2, 6, 2, 4, 4, 3, 0, 5, 2, 1, 2, 4, 3, 5, 2, 4, 6, 3, 0, 7, 1. Make a frequency table to summarize these records, and show them in a suitable diagram.

7. The times, measured to the nearest second, taken by 30 students to complete an algebraic problem are given below:

$$47, \quad 61, \quad 53, \quad 43, \quad 46, \quad 46, \quad 68, \quad 48, \quad 72, \quad 57,$$
$$48, \quad 54, \quad 41, \quad 63, \quad 49, \quad 42, \quad 58, \quad 65, \quad 45, \quad 44,$$
$$43, \quad 51, \quad 45, \quad 38, \quad 48, \quad 46, \quad 44, \quad 52, \quad 43, \quad 47.$$

Group these times into a frequency table using eight equal class-intervals, the first of which contains measured times in the range 35–39 seconds.

Draw a histogram of the grouped frequency distribution. Which is the modal class? [*Note.* See Definition 2.8.] (*Welsh*)

8. The following figures show the attendance (in thousands, to the nearest thousand) at 32 matches of a national football team:

$$42, \ 51, \ 34, \ 35, \ 29, \ 10, \ 16, \ 51,$$
$$47, \ 51, \ 35, \ 31, \ 17, \ 67, \ 43, \ 23,$$
$$25, \ 36, \ 32, \ 51, \ 46, \ 12, \ 21, \ 29,$$
$$14, \ 47, \ 31, \ 35, \ 29, \ 23, \ 10, \ 34.$$

(a) Construct a stem-and-leaf diagram for these data.
(b) What are the advantages of a stem-and-leaf display, when compared with a histogram, for illustrating these data? (*IOS*)

9. The number of males, in litters of rats which contain 5 rats altogether, was observed with the following results over 100 different litters. Summarize these results and illustrate them:

2, 5, 3, 1, 3, 4, 2, 2, 0, 3, 2, 4, 1, 3, 3, 2, 2, 2, 1, 2, 3, 4, 3, 0, 4,
5, 3, 3, 3, 2, 3, 2, 2, 1, 1, 4, 1, 3, 2, 2, 4, 3, 4, 3, 2, 3, 2, 2, 3, 4,
2, 3, 0, 1, 1, 3, 5, 2, 2, 4, 3, 2, 5, 4, 3, 2, 4, 3, 3, 4, 2, 1, 3, 2, 1,
2, 4, 3, 1, 3, 3, 4, 2, 3, 1, 2, 3, 4, 3, 2, 2, 4, 3, 3, 4, 1, 2, 3, 2, 3.

10. A standard pack of playing cards contains 52 cards in all, 13 of each of the four suits hearts, diamonds, spades and clubs. A hand of cards consists of 13 out of the 52, and before dealing out a hand the pack must be well shuffled. One hundred hands are obtained in this way from different packs, and the number of spades in each hand is recorded. Two of these hands contain no spades, there are 6 hands with 1 spade, 20 hands with 2 spades, 28 hands with 3, 23 with 4, 12 with 5, 4 with 6, 2 with 7, and one each with 8, 10, and 11 spades respectively.

Summarize this information in a suitable table and in a suitable diagram.

11. The heights of 187 plants of a species were recorded to the nearest millimetre, as shown in the table below. Summarize them in class-intervals of width 5 mm, and show the results in a suitable diagram. Also summarize the height records in class-intervals of *other* widths (e.g. 4 mm, 2 mm, 10 mm), and show the results in a diagram. Comment on the similarities and/or differences in these summaries:

Height	26	27	28	29	30	31	32	33	34	35	36	37	38
Frequency	1	0	2	1	0	1	3	0	1	0	4	1	2

Height	39	40	41	42	43	44	45	46	47	48	49	50	51
Frequency	6	10	12	8	6	15	17	20	13	9	12	7	8

Height	52	53	54	55	56	57	58	59	60	61	62	63	64	65
Frequency	8	6	4	4	0	1	1	0	1	1	1	1	0	0

12. Railway trains due to arrive at a main terminus in London are not always on time; records are kept of how late each is. Some trains have to be cancelled due to shortage of staff or to engine failure. A month's figures for each of four termini A, B, C, D are presented in a report as follows.

    Trains due into A: 5400. Of these 205 were cancelled, 35 were twenty minutes late or more, 65 were between fifteen and twenty minutes late, 160 between ten and fifteen minutes, 840 between five and ten minutes, 2725 between two and five minutes and the rest were less than two minutes late arriving.

    Trains due into B: 3660. Of these 40 were cancelled, 30 were twenty minutes late or more, 215 were between fifteen and twenty minutes late, 205 between ten and fifteen minutes, 825 between five and ten minutes, 2230 between two and five minutes, and the rest were less than two minutes late arriving.

    Trains due into C: 6250. Of these 38 were cancelled, and the numbers late, in the same time-intervals as for A and B, were respectively 112, 335, 968, 1465, 2482.

    Trains due into D: 2830. Of these 2 were cancelled, and the numbers late, in the same time-intervals as for A and B, were respectively 12, 76, 190, 775, 1075.

    (i) Present this information in a table, labelled as clearly as possible for quick understanding. Design the table to allow 'secondary information', such as percentages, totals, ratios to be calculated and entered in suitable places.

    (ii) Show in a diagram the distribution of degrees of lateness for *one* of the four termini.

    (iii) Compare in a diagram the percentages of trains cancelled for the four termini.

13. Two examination centres, one in Africa and one in Asia, have entered students for the same external examination. Their marks (%) are:

African Centre:	56,	67,	42,	48,	55,	61,	52,	39,	47,
	58,	50,	40,	59,	62,	44,	57;
Asian Centre:	78,	34,	37,	72,	52,	68,	27,
	55,	65,	40,	75,	33,	66.

    Use stem-and-leaf diagrams to compare these two sets of data.

1.8 COMPUTING EXERCISES

1. Produce histograms or bar charts and stem-and-leaf diagrams in order to investigate the following sets of data:
   (a) weights of students in Exercise 23.13(1);

(b) reaction times in Exercise 13.10(9);

(c) number of defectives in Example 16.9;

(d) word data in Revision Exercises A(4).

Notice particularly whether the distributions:

(i) are symmetric or skew;

(ii) have a single peak or more than one;

(iii) are compact or widely spread.

2. Compare the two sets of data in Exercise 5.6(1) by using histograms and stem-and-leaf diagrams.

3. Investigate the effect on the appearance of the histogram of the cuckoo data (Example 1.2) of varying the class-interval. One investigation might begin with an interval size of 1.5 mm which is decreased by steps of 0.1 to 0.1 mm.

4. Repeat Question 3 with other sets of data and investigate the rule that for a data set consisting of n values, a satisfactory histogram is obtained by using $\sqrt{n}$ intervals.

1.9 PROJECTS

1 Distribution of digits in relative atomic masses

Aim. To investigate the frequency distributions of the first and third significant digits in relative atomic masses.

Equipment. Table of relative atomic masses.

Preliminary work. Consider what you would expect these distributions to be. (Remember the first significant digit can never be 0.)

Collection of data. Make a tally count directly from the table for: (a) the first significant digit; (b) the third significant digit.

Analysis. Make a frequency table for each distribution. Represent the distributions by bar diagrams.

Note. Alternative sources of data which give similar 'first digit distributions' are other tables of physical constants, populations of towns and areas of countries.

2 Traffic distributions

Aim. To investigate the distribution of the number of vehicles passing a particular point in one-minute intervals.

Equipment. Watch with second hand, or digital count of seconds.

Collection of data. Choose a fairly busy road on which you might expect a uniform flow of traffic. Decide what type of vehicles should be included in the count and where the passing point is to be. Count the number of vehicles passing in one-minute intervals, in one direction, over a period of one hour.

Repeat, if time allows, at a point where traffic flow is interrupted, e.g. near traffic lights, or junctions. (In a class project a variety of sites could be chosen.) *Analysis.* Make a frequency table and bar diagram for each distribution.

SUMMARY

Statistics is the science that studies the collection and interpretation of data. Observations are records of **variates** which are measured on the units of investigation. Variates may be **quantitative** or **qualitative**; quantitative variates may be **discrete** or **continuous**.

In many statistical investigations, observations of a particular variate are used to gain information about an underlying **population**, or to compare populations. Preliminary investigation usually centres on the **frequency distribution** of the observations.

Tabular and diagrammatic methods are used to investigate frequency distributions. Tables include **tally counts** and **frequency tables**. Diagrams include bar diagrams for discrete variates, **histograms** and **frequency polygons** for continuous variates, **stem-and-leaf diagrams** and **cumulative frequency diagrams**.

2 MEASURES OF THE CENTRE OF A SET OF OBSERVATIONS

OVERVIEW

How do we choose a single value to represent a set of observations? We look at possible choices in this chapter.

2.1 INTRODUCTION

How long did it take you to travel from home to school, or college, today? If you measure this time carefully, to the nearest minute, you will find it varies from day to day: it is a *variate*. Suppose that an investigation is being made of the journey times of the pupils in a school. Questions that might be asked are, for example, 'do journey times change from winter to summer?', 'are the younger pupils slower than the older ones?'.

Data might be collected from all the school, or from selected groups in it, several times during a year. In one group of nine people, on one day, these were the journey times:

17, 30, 14, 16, 26, 15, 27, 18, 26 minutes.

(Journey time is a continuous variate, of course, but we should require special equipment to record it more accurately than to the nearest minute.) This set of observations contains a lot of information, which cannot be appreciated just by looking at the figures. We must find methods of extracting the most important bits of information, so that we can use them in comparing these journey times with those from other groups or those collected at other times.

2.2 THE MEDIAN

As written down (in the haphazard order in which the nine people in the group were asked) the observations show no particular pattern. If we *rank* the observations, that is write them in order of size beginning with the smallest, we obtain: 14, 15, 16, 17, 18, 26, 26, 27, 30. Another way of considering the observations is to plot them on a linear scale (Fig. 2.1). This is called a *dot diagram* or *dot plot*. This begins to put ideas into our heads, such as whether the first five came from

Figure 2.1 *Dot diagram for data of Section 2.1.*

one direction and the remaining four from another. But let us try to keep things simple for the present, and look for *one* figure to represent this whole population of journey times. Bearing in mind some of the patterns for frequency distributions that we have seen in Chapter 1, we shall not want a 'representative figure' to be near the extremes, i.e. near to the largest or smallest observations. Instead we look for something in the middle. We might choose the observation that has as many observations above it as below it in the ranked order. The value taken by this chosen observation is called the *median*. We give a temporary working definition: the median M of a set of observations is the middle one of the ranked observations.

For our nine journey times, the median is equal to 18, because in the ranked order

$$14, \quad 15, \quad 16, \quad 17, \quad \boldsymbol{18}, \quad 26, \quad 26, \quad 27, \quad 30$$

the value 18 has four observations below it and also four above it. So we summarize these journey times by saying that their median value is 18 minutes.

It is clear from what we have just done that when the number N of observations is odd, there is a unique 'middle' observation, the $\frac{1}{2}(N+1)$th in rank order. But for N even, we need a slightly modified rule. Suppose we had a tenth person in our group, with a journey time of 35 minutes. In the rank order of all ten observations, 18 now has four below and five above, while the next one (26) has five below and four above. Neither of these can therefore be M; the accepted rule when N is even is to place M midway between the two middle observations (the $\frac{1}{2}N$th and the $[\frac{1}{2}N+1]$th), i.e. between the fifth and sixth for this example. The median journey time for the group of ten is thus $\frac{1}{2}(18+26)=22$ minutes. We can now give a complete definition for M.

Definition 2.1

The **median** M of a set of N observations which have been ranked in order of size is equal to the value taken by the middle (the $\frac{1}{2}[N+1]$th) observation when N is odd, and is half the sum of the values of the two middle observations (the $\frac{1}{2}N$th and $[\frac{1}{2}N+1]$th) when N is even.

For sets of data containing more than 10 observations a stem-and-leaf diagram (see p. 8) can be a useful aid when finding the median. The stem which contains the median may be located by cumulating the frequencies and, by putting the leaves on that stem in rank order, the median may easily be found.

2.2.1 Exercise

Find the medians of the following populations of observations. Plot the values of the observations in each population on a linear scale (a dot plot) and mark the median:

(a) 21, 15, 33, 24, 12.
(b) 21, 6, 62, 47, 9.
(c) 17, 8, 14, 12, 9, 3, 21, 5, 12, 16.
(d) 34, −12, 16, 0, 11, −2, 20, 1, 17, 12.

You may have spotted one of the special properties of M as a measure: it ignores completely the actual sizes of the observations, except for those in the middle of the rank order. Thus if the maximum journey time in our first group of nine people had been 90 minutes instead of 30, M would stay unaltered at 18. This may be an advantage or a disadvantage according to the way in which the journey times varied.

If there were just a few pupils in the whole school who had very long journey times (or very short ones) we might think it an advantage to use a central measure that was not influenced by these very few people. On the other hand, if, say, the journey times of older and younger pupils were being compared, and we found that the times for the two age-groups differed little, we might prefer a measure that was influenced by all the observations. Such a measure would be sensitive to small changes or differences in the time, whereas the median might change very little or not at all.

2.3 THE ARITHMETIC MEAN

Another measure of the centre of a set of observations is the arithmetic mean (very often simply called 'the mean').

Definition 2.2

The (arithmetic) **mean** of a set of observations is the sum of the observations divided by the number of observations.

The mean of the nine journey times on p. 24 is

$$(17 + 30 + 14 + 16 + 26 + 15 + 27 + 18 + 26)/9 = 189/9 = 21 \text{ minutes.}$$

It happens to be a whole number in this example, but a moment's thought tells us that there is no reason at all why it should be so in general, even when all the observations are whole numbers.

The mean uses the actual values of all the observations, and is therefore particularly useful in detecting small differences between sets of observations. It has been found a convenient measure to use in the theory of statistics, whereas the median is usually not easy to deal with theoretically. For these reasons, the mean is the usual choice for a single measure to summarize a set of observations. It would be the natural measure to use in summarizing the observations of journey time.

An exception to this choice might be in the case where the longest journey time had been 90 minutes rather than 30, for then the mean would be $27\frac{2}{3}$, and might be thought (with justice) to be no longer in the middle of the observations – it is unduly influenced by the one very extreme value, often called an *outlier* – whereas the median, at 18, is a more representative measure in this case. In general, the median is a better measure than the mean when the population of data has a skew distribution; the mean can be misleading if the distribution is not symmetric or almost symmetric.

Another case in which the median is useful is when dealing with a measurement that takes a long time to obtain. Electric light bulbs sometimes fail very quickly, and sometimes last a very long while. Once we have observed the lifetimes of just over half of a batch of bulbs, we know the median lifetime. We would have to wait until all the bulbs in the batch had failed before we could calculate the mean.

2.3.1 Exercise

Calculate the medians and the means of the populations of observations (a)–(c). Plot each set of observations on a linear scale (a dot plot) and mark in the mean and the median.

(a) There are 12 similar houses in a block, and the total incomes of the 12 households in the same week are £ 74, 59, 125, 62, 46, 53, 108, 60, 92, 126, 72, 147.

(b) The heights of 10 male students aged 18 are 178, 173, 175, 181, 179, 171, 169, 175, 173, 178 cm.

(c) Twelve children in a class caught an infectious disease (chicken-pox). Counting the day when the first child became ill as day 0, the days on which the 12 became ill were 0, 22, 18, 14, 1, 15, 17, 1, 17, 19, 3, 20.

Example 2.1 Median and mean of grouped discrete data
Calculate: (a) the median; (b) the mean; of the letter data in Example 1.1, p. 6.

We copy the frequency table from Example 1.1, and add a row of cumulative frequencies:

No. of letters per day	0	1	2	3	4	5
Frequency	48	32	17	2	0	1
Cumulative frequency	48	80	97	99	99	100

(a) *Median.* There are 100 observations. The median is half the sum of the 50th and 51st observations in the ranked order. We see from the cumulative frequencies that both these observations equal 1. Hence the median number of letters per day is 1.

(b) *Mean.* Of the 100 observations 48 are 0s, 32 are 1s, etc. Hence the mean equals $(48 \times 0 + 32 \times 1 + 17 \times 2 + 2 \times 3 + 0 \times 4 + 1 \times 5)/100 = 77/100 = 0.77$.

Example 2.2 Median and mean of grouped continuous data

Calculate: (a) the median; (b) the mean, of the following data on the height in centimetres of 10 plants in pots. The data have been grouped.

Class-interval	11.5–16.5	16.5–21.5	21.5–26.5	26.5–31.5
Class-centre	14.0	19.0	24.0	29.0
Frequency	2	5	2	1
Cumulative frequency	2	7	9	10

(a) *Median.* The median is obviously inside the interval (16.5, 21.5). If we assume that the five observations which lie in this interval are equally spread out within it, and put each at the centre of its own small interval, we obtain the diagram:

From the definition above, the median is half the sum of the $\frac{10}{2} = 5$th and $(\frac{10}{2} + 1) = 6$th observations. This value is at the end of the third of the five equal intervals into which the interval $(16.5, 21.5)$ is divided, and is $16.5 + \frac{3}{5}(5.0) = 19.5$. (To illustrate the case when the total number is odd, this example may be reworked omitting the single observation in the class interval $(26.5, 31.5)$, to give a set of 9 observations. The median is then the fifth observation and is $16.5 + \frac{2.5}{5}(5.0) = 19.0$.) Note that if records of the ungrouped data had been preserved the median could also have been calculated from them. It would usually have a slightly different value from the median calculated using grouped data, because we would know exactly where each observation lay instead of having to assume them equally spread out within the groups.

(b) *Mean.* When calculating the mean from grouped continuous data we act as if all the observations in a given interval are equal in value to the class-centre of that interval. The argument is that we would expect some observations to be above the centre and some below it, and it is therefore reasonable to replace the observations in an interval by the same number of observations placed at the centre. We then proceed as with grouped discrete data. The mean is therefore

$$(2 \times 14.0 + 5 \times 19.0 + 2 \times 24.0 + 1 \times 29.0)/10 = 200/10 = 20.$$

This value is an approximation to the true mean of the set of observations, which could be obtained from the individual observations. With a large number of observations the two values are not likely to differ much. (The method has been illustrated with only 10 observations for simplicity; with so few observations one is not likely to group the data.) With a large number of observations a substantial reduction in computing time is obtained by grouping the data to calculate the mean. Note also that the choice of grouping intervals will influence the calculated value of the mean. Two people who use different grouping intervals for the same set of data are likely to obtain different values for the mean.

2.3.2 Exercises

1. Define the mean and the median of a frequency distribution. Calculate the median of the frequency distribution given below, and explain why you might prefer it to the mean as a measure of central tendency. Represent the distribution by a histogram, assuming the last class extends to £15 000.

Annual income of 184 employees in a factory

Annual income (£)	Frequency
1 000–	9
1 500–	25
2 000–	47
2 500–	43
3 000–	30
4 000–	17
5 000–	8
6 000–	4
10 000–	1

(*Oxford*)

2. The distribution of households according to daily protein intake (grams per head) is given in the following table. Plot a histogram of the data, and *without* calculation say whether you would expect the mean or the median to have the larger value. Then calculate the values of the mean and the median.

Daily protein intake (g)	Percentage of households
Less than 20	2
20–	4
25–	14
30–	16
35–	17
40–	14
45–	19
55–	8
65 and over	6

(*IOS*)

2.4 Σ-NOTATION

We may express the definition of the arithmetic mean in a simple formula if we introduce a convenient piece of notation. Individual observations are often represented by a subscript notation. Thus the journey times x for the nine people on p. 24 would be called x_1, x_2, x_3, x_4, x_5, x_6, x_7, x_8, x_9 identified with the order in which they are recorded. That is, $x_1 = 14$, $x_2 = 17$, $x_3 = 16$, and so on until $x_8 = 18$ and $x_9 = 26$. A typical one of these may be labelled x_i, and in order to list them all, the subscript i must take the values 1 to 9 in turn. The whole set of observations is $\{x_i : i = 1, 2, \ldots, 9\}$.

The sum of a set of N observations, i.e. $x_1 + x_2 + \cdots + x_N$, may be written $\sum_{i=1}^{N} x_i$, by introducing the $\sum$ sign which stands for 'the sum of' and is the Greek letter capital sigma, the equivalent of the Latin S. The expression is read 'sigma x_i, i equals one to N', and is defined as follows.

Definition 2.3

$\sum_{i=1}^{N} x_i$ is the **sum** of the quantities x_i as i takes successively the values 1, 2, \ldots, N$.

The summation is said to be over i, which is the *index of summation*, and the *range of summation* is from 1 to N.

The arithmetic mean of the set of observations x_i $(i = 1, 2, \ldots, N)$ may be represented by $\bar{x}$ (x-bar). Hence we are led to the following definition.

Definition 2.4

The (arithmetic) **mean** $\bar{x}$ of the observations $x_1, x_2, \ldots, x_N$ is $\sum_{i=1}^{N} x_i / N$.

We shall find this $\sum$-notation very useful throughout the book and will therefore explore it further.

When the range of summation is obvious, the summation sign is sometimes simply written $\sum$ instead of, for example, $\sum_{i=1}^{N}$. Note that the resulting sum does not include i in any form. Hence the result is unchanged whatever letter we use for the index of summation. For example,

$$\sum_{r=1}^{N} x_r = \sum_{i=1}^{N} x_i.$$

Sometimes the way that x_i depends on i is known. For example, if $x_i = i$ we have

$$\sum_{i=1}^{4} x_i = \sum_{i=1}^{4} i = 1 + 2 + 3 + 4 = 10.$$

Similarly,

$$\sum_{i=3}^{5} i(i+1) = 3 \times 4 + 4 \times 5 + 5 \times 6 = 12 + 20 + 30 = 62.$$

The lower end of the range can be any positive or negative integer, or zero. The summation is over the values of the index of summation as it increases by steps of one from the lower end of the range to the upper end.

2.4.1 Exercises

1. Show that

   (i) $\displaystyle\sum_{i=1}^{4} i^2 = 30,$ (ii) $\displaystyle\sum_{r=1}^{3} r^3 = 36,$ (iii) $\displaystyle\sum_{i=2}^{4} \frac{1}{i} = \frac{13}{12},$

   (iv) $\displaystyle\sum_{i=1}^{5} 1 = 5,$ (v) $\displaystyle\sum_{i=-2}^{2} i = 0,$ (vi) $\displaystyle\sum_{i=1}^{1} x_i = x_1.$

2. If x_i $(i = 1, 2, 3)$ and y_i $(i = 1, 2, 3)$ take the values shown in the following table:

x_1	x_2	x_3	y_1	y_2	y_3
6	1	2	5	3	4

   confirm the relations

   (i) $\displaystyle\sum_{i=1}^{3} y_i = 12,$ (ii) $\bar{y} = 4,$ (iii) $\displaystyle\sum_{i=1}^{3} x_i = 9,$

   (iv) $\displaystyle\sum_{i=1}^{3} (x_i + y_i) = 21 = \sum_{i=1}^{3} x_i + \sum_{i=1}^{3} y_i,$ (v) $\displaystyle\sum_{i=1}^{3} 2x_i = 18 = 2 \sum_{i=1}^{3} x_i.$

Rules for operation with $\sum$

We can formulate rules that the symbol, or operator, $\sum$ obeys. In many cases it is easier to evaluate an expression by applying the rules than by using the basic definition. The rules are

(1) $\displaystyle\sum_{i=1}^{N} (x_i + y_i) = \sum_{i=1}^{N} x_i + \sum_{i=1}^{N} y_i;$

(2) $\displaystyle\sum_{i=1}^{N} cx_i = c \sum_{i=1}^{N} x_i,$ where c is a constant;

(3) $\displaystyle\sum_{i=1}^{N} c = Nc,$ where c is a constant;

(4) $\displaystyle\sum_{i=1}^{N} x_i = \sum_{i=1}^{K} x_i + \sum_{i=K+1}^{N} x_i,$ provided $1 \leqslant K < N.$

Rules (1), (2) and (4) follow at once from the basic definition, and (3) follows from (2) by putting x_i equal to 1 for all i. In (4), we may split a sum into two parts by taking K anywhere between the upper and lower limits allowed for N.

Example 2.3 Sum of deviations about the mean

Show that $\sum_{i=1}^{N} (x_i - \bar{x}) = 0$, i.e. that the sum of deviations of a set of observations from their mean is zero.

$$\sum_{i=1}^{N}(x_i - \bar{x}) = \sum_{i=1}^{N}x_i - \sum_{i=1}^{N}\bar{x}, \quad \text{by rule (1),}$$

$$= \sum_{i=1}^{N}x_i - N\bar{x}, \quad \text{by rule (3),}$$

$$= N\bar{x} - N\bar{x} = 0, \quad \text{using the definition of } \bar{x}.$$

Example 2.4 Sum of squares of deviations about the mean

Show that

$$\sum_{i=1}^{N}(x_i - \bar{x})^2 = \sum_{i=1}^{N}x_i^2 - N\bar{x}^2 = \sum_{i=1}^{N}x_i^2 - \left(\sum_{i=1}^{N}x_i\right)^2 \bigg/ N.$$

Since $\sum$ is a 'linear' operation, one which involves addition and not multiplication, we must first square out $(x_i - \bar{x})^2$ and then deal with it term by term.

$$\sum_{i=1}^{N}(x_i - \bar{x})^2 = \sum_{i=1}^{N}(x_i^2 - 2\bar{x}x_i + \bar{x}^2)$$

$$= \sum x_i^2 - 2\bar{x}\sum x_i + \sum \bar{x}^2, \quad \text{by rules (1) and (2).}$$

$$= \sum x_i^2 - 2N\bar{x}^2 + N\bar{x}^2, \quad \text{using definition of } \bar{x} \text{ and rule (3),}$$

$$= \sum x_i^2 - N\bar{x}^2$$

$$= \sum x_i^2 - \left(\sum x_i\right)^2 \bigg/ N, \quad \text{using definition of } \bar{x}.$$

These identities are useful later. $\sum_{i=1}^{N}x_i^2$ is called the *raw sum of squares*. $\sum_{i=1}^{N}(x_i - \bar{x})^2$ is called the *corrected sum of squares* since it may be obtained by subtracting the *correction term*

$$\left(\sum_{i=1}^{N}x_i\right)^2 \bigg/ N$$

from the raw sum of squares. Thus the corrected sum of squares is an alternative, and more convenient, name for the sum of squares of deviations about the mean.

2.5 THE MEAN OF GROUPED DATA

We have seen, in the examples above (2.1 and 2.2), that when a set of observations has been summarized in a frequency table, the labour of calculating the mean is reduced. We now give a formal statement of the process.

A set of N observations of a discrete variate r, which have been grouped according to the k distinct values that it is possible for r to take, may be expressed in a frequency table:

Variate value	r_1	r_2	r_3	$\ldots$	r_k	Total
Frequency	f_1	f_2	f_3	$\ldots$	f_k	$\sum_{i=1}^{k} f_i = N$

The total of all the observations is simply the sum of each possible r_i times the frequency with which that r_i occurs, that is $\sum_{i=1}^{k} f_i r_i$. Hence we have the following definition.

Definition 2.5

The **mean** of a set of N observations of a discrete variate, grouped so that the value r_i $(i=1,2,\ldots,k)$ occurs with frequency f_i, is $\bar{r} = \sum_{i=1}^{k} f_i r_i / N$.

Given a set of N observations of a continuous variate x, we first group them by choosing convenient intervals and then count the number of observations that fall in each interval. If there are k intervals, and the *centres* of these intervals are at x_i $(i=1,2,\ldots,k)$ we may summarize the observations in the grouped frequency table:

Centre of interval	x_1	x_2	x_3	$\ldots$	x_k	Total
Frequency	f_1	f_2	f_3	$\ldots$	f_k	$\sum_{i=1}^{k} f_i = N$

To calculate the mean we approximate the values of the observations in an interval by taking them all to be equal to the value of the centre of that interval, x_i. Hence we have the following definition.

Definition 2.6

The **mean** of a set of N grouped observations of a continuous variate, in which f_i observations fall in the interval whose centre is x_i $(i=1,2,\ldots,k)$, is $\bar{x} = \sum_{i=1}^{k} f_i x_i / N$.

The ratio f_i/N denotes the *relative frequency* with which an observation falls in the interval whose class-centre is x_i. The mean may therefore also be found by multiplying each class-centre x_i by the relative frequency, f_i/N, with which an observation falls in the corresponding class-interval, and summing. The mean of a discrete variate may be defined in a similar way, in terms of the relative frequency with which each of the possible values of the variate occurs.

2.6 CODING

Many of us are nowadays quite happy to add together a set of N quite large numbers, and then divide them by N, because we have pocket calculators to help us! But if we are finding the mean of the observations 1054, 1066, 1025, 1070, 1033, we are not being very clever to add them up just as they stand. It is pretty clear that the mean is also going to begin 10 ... , and it is the last two figures we really have to determine. We use a process called *coding*.

Given a set of N observations $\{x_i: i = 1, 2, \ldots, N\}$, we may wish to write each observation in the form $(u_i + a)$, where a is a constant. The sum of the observations, $\sum_{i=1}^{N} x_i$, is then

$$\sum_{i=1}^{N} (u_i + a) = \sum_{i=1}^{N} u_i + Na.$$

Therefore the mean of the set $\{x_i: i = 1, 2, \ldots, N\}$ is

$$\bar{x} = \frac{1}{N} \left(\sum_{i=1}^{N} u_i + Na \right) = \bar{u} + a.$$

If we add a to each observation in the set $\{u_i\}$, we also add a to the mean.

Hence in our example a may be taken as 1000; then $u_1 = 54$, $u_2 = 66$, $u_3 = 25$, $u_4 = 70$, $u_5 = 33$. The mean $\bar{u} = (54 + 66 + 25 + 70 + 33)/5 = 49.6$, and so the mean of the original observations is $\bar{x} = \bar{u} + 1000 = 1049.6$. We can of course check this directly. From now on, we shall call u the *coded variate* and x the *original variate*. The value chosen for a is called the *working zero* for the original observations. (We also may call the original observations the *raw data*; after coding they become the *coded data*.) With modern computing facilities, coding is not necessary, although when using a pocket calculator instead of a computer the simpler arithmetic is still very welcome.

2.6.1 Exercises

1. Repeat the calculation of the mean of 1054, 1066, 1025, 1070, 1033 taking: (i) $a = 1025$; (ii) $a = 1050$.

2. Calculate the mean amount of nitrogen per gram in eight samples of dried strawberry leaves where the individual measurements were 2.032, 2.016, 2.022, 2.057, 2.034, 2.040, 2.019, 2.045.

Example 2.5 Determining a mean using coding

There are 50 breeding pairs of a particular species of bird in a wood. The frequency distribution of the number of eggs in 50 nests is given below. Calculate the mean number of eggs per nest: (a) without coding; (b) with coding.

(a) *Without coding.* The first two columns in Table 2.1 give the frequency distribution, and the third column is derived from them. Hence, denoting the number of eggs by r, we have (using the formula on p. 33)

$$\bar{r} = \frac{\sum f_i r_i}{\sum f_i} = \frac{127}{50} = 2.54.$$

Table 2.1

r	f	fr	u	uf
1	9	9	−2	−18
2	17	34	−1	−17
3	15	45	0	−35
4	6	24	1	6
5	3	15	2	6
	50	127		+12
				−35
				−23

(b) *With coding.* The coded variate $u = r - 3$ is used, i.e. a working zero is taken at the value 3 in the original data. The values of u appear in the fourth column of Table 2.1, and the fifth column is derived from the second and fourth columns. In setting out the calculation, it is convenient to use the space in the uf column opposite 0 in the u column to sum the negative uf values; this sum is carried below and added to the sum of the positive uf values. From Table 2.1 we have

$$\bar{u} = \frac{\sum u_i f_i}{\sum f_i} = \frac{-23}{50} = -0.46.$$

We took $r - 3 = u$, that is $r = u + 3$, so that $a = 3$. Hence $\bar{r} = \bar{u} + 3 = 3 - 0.46 = 2.54$.

The choice of working zero is arbitrary; it is often convenient to make $u = 0$ correspond to the greatest frequency f. In this example, there would be little to choose between that coding and the one we have used. Remember that the point of coding is to simplify the arithmetic.

Decimal points may be troublesome in calculations. Hence if we had a set of observations $\{x_i\} = \{0.12, 0.37, 0.41, 0.24\}$ we might prefer to work with the coded set $\{u_i\} = \{12, 37, 41, 24\}$. The coded variate now is $u = 100x$. The most general relation that we can employ in coding is to combine change of scale (i.e. multiplication) with the use of a working zero: $x = a + bu$. For example the set $\{x_i\} = \{7.23, 7.18, 7.20, 7.29\}$ might in this way be coded $\{u_i\} = \{23, 18, 20, 29\}$, so that $u = 100(x - 7)$, that is $u = 100x - 700$, or $x = 7 + 0.01u$. This last form has $a = 7, b = 0.01$.

When $x = a + bu$, it follows at once that $\bar{x} = a + b\bar{u}$.

In scientific work, we sometimes find it convenient to measure observations on a scale (e.g. the Celsius scale) which gives a value x when we really wish to use a scale (e.g. the Fahrenheit scale) which gives a value y, with $y = a + bx$. (In the Celsius–Fahrenheit example, $a = 32$, and $b = \frac{9}{5}$.)

*Example 2.6 Calculation of the mean of a continuous variate
using coding*

Calculate the mean of the frequency distribution described by the first three columns of Table 2.2.

Table 2.2

Interval	Centre, x_i	f_i	u_i	$f_i u_i$
6.1–6.5	6.3	3	−2	−6
6.6–7.0	6.8	16	−1	−16
7.1–7.5	7.3	32	0	−22
7.6–8.0	7.8	20	1	20
8.1–8.5	8.3	9	2	18
		80		38
				−22
				16

The class-centres are denoted by x. Choose a working zero of 7.3 and change the scale by doubling the distance between centres, i.e. use the coded variate $u = 2(x - 7.3)$. Hence the original variate $x = 0.5u + 7.3$. The computation proceeds as in Example 2.5, and we find that

$$\bar{u} = \frac{\sum f_i u_i}{\sum f_i} = \frac{16}{80} = 0.2.$$

$$\bar{x} = 0.5\bar{u} + 7.3 = 0.1 + 7.3 = 7.4.$$

2.7 THE MODE

Definition 2.7

The **mode** of a discrete variate is that value of the variate which occurs most frequently.

For the letter data of Example 1.1, p. 6, it is 0, that being the number of letters which most often arrived.

The mode is not directly appropriate to a continuous variate, since if we measure to sufficient accuracy it is doubtful whether even two observations would take exactly the same value. So it is not sensible to try and find a 'most frequent value'. Instead we define a modal class.

Definition 2.8

The **modal class** of a continuous variate, grouped in equal class intervals, is the class that contains the highest frequency.

For the plant data of Example 2.2 the modal class is 16.5–21.5. The modal class depends on what class-intervals are used, and is thus somewhat arbitrary. For this reason (among others) it is not a very useful summary measure. However, the basic idea of the mode is used to describe the shape of a frequency distribution. Distributions which rise to a peak and fall are said to be *unimodal*. Those which have more than one peak are said to be *multimodal*, and one particular case of this, with just two peaks, is a *bimodal* distribution.

2.8 THE GEOMETRIC MEAN

Definition 2.9

The **geometric mean** of the N positive observations $x_1, x_2, \ldots, x_N$ is the Nth root of the product $(x_1 x_2 x_3 \ldots x_N)$.

Although we very rarely use this directly, we sometimes do so in disguise by considering the set $\{\log x_i\}$ instead of the $\{x_i\}$ actually observed. The mean of the logarithms of a set of numbers is equal to the logarithm of the *geometric* mean of the numbers. This is often applied to skew distributions since the distribution after taking logarithms may be more symmetrical and its mean therefore a good measure of its centre. Thus the geometric mean is often regarded as a better central measure than the arithmetic mean for a skew distribution.

2.9 WEIGHTED AVERAGES AND INDEX NUMBERS

The following data give the basic weekly wages in 1976 of the three categories of worker in a factory:

Type of worker	Number of workers	Basic weekly wage
Skilled	110	£62.50
Semiskilled	140	£56.50
Unskilled	170	£48.00

If we want to know the average wage per person in the whole factory we must calculate the total wage bill and divide it by the number of workers. The total weekly wage bill is $£(110 \times 62.50) + (140 \times 56.50) + (170 \times 48.00) = £22\,945$. The total number of workers is $110 + 140 + 170 = 420$, and the average wage is thus $£22\,945/420 = £54.63$.

We can regard this as a *weighted average* (or *weighted mean*) of the wages for the three categories, weighted according to the numbers in each category. Looking at it this way, we multiply £62.50 by the weight 110, £56.50 by 140 and £48.00 by 170, then add these three products and divide by the sum of the weights.

Definition 2.10

A **weighted average** of the quantities x_i $(i = 1, 2, \ldots, k)$, using weights w_i, is

$$\sum_{i=1}^{k} w_i x_i \bigg/ \sum_{i=1}^{k} w_i.$$

Weighted averages are also used in working out index numbers. The purpose of an index number is to show how a quantity that cannot be measured directly varies over time. Examples are indices of industrial production, business activity and the cost of living. Suppose, for illustration, that we wish to calculate a food expenditure index in Great Britain. First we must list all the items (or at least all the important items) of food that people buy: bread, meat, milk, cheese, butter, fruit, vegetables, and so on through the list of items it is possible to buy in the shops. The required index will be a weighted average of the prices of these items, using as weights the proportions of total expenditure spent on each item.

A family budget survey carried out by the government statistical service provides estimates of these proportions, which are needed before an index number can be set up initially, and which form the $\{w_i\}$ in Definition 2.10. The $\{x_i\}$ are the prices of the items; prices vary, of course, from shop to shop and in different parts of the country, but an average retail price at a specific type of shop can be found. The index number is calculated as the weighted average using these $\{w_i\}$ and $\{x_i\}$. When an index number is first constructed, the $\{w_i\}$ and $\{x_i\}$ used are thus those existing at a particular time, called the *base period* (often the *base year*) of the index.

The prices $\{x_i\}$ included in the index will change continuously and often quite rapidly; it is, however, fairly easy to collect a new set of $\{x_i\}$ – every month, say, or every quarter – and to work out a new weighted average using them. We relate this new value to the original by expressing the new value as a percentage of the original, that is by taking the original index number in the base period as 100 and stating the new index as, for example, 105 if it shows a 5% increase over the original or 88 if it shows a 12% decrease. All subsequent calculations are related to the value in the base period as 100, and *not* to the value most recently calculated.

The $\{w_i\}$ are kept the same for quite a long period of time, but of course people's buying habits do change for many reasons: so at relatively long intervals of time a new set of $\{w_i\}$ from a new survey will be used to construct a completely new index. This in its turn will be called 100 and used as a basis for index calculations in the future.

Example 2.7

An index is made up by combining five items A, B, C, D, E whose relative weights are 40, 20, 15, 15, 10. The index was set up in the year 1975. Price relatives in 1978 and 1981 for the five items are respectively 120, 105, 110, 102, 96 and 150, 126, 121, 153, 132. Using 1975 as base, calculate indices for 1978 and 1981. Calculate also an index for 1981 using 1978 as base year.

The term *price relative* is often used to express the percentage that a current price is of the price in a base period: if the price of A was 100 in 1975, then in 1978 it was 120 (see above) and, in 1981, 150; the price relatives can thus be used as $\{x_i\}$ when calculating indices for 1978 and 1981. Thus for 1978 we have

Item	A	B	C	D	E
Weight w_i	40	20	15	15	10
Relative weight $w_i / \sum_i w_i$	0.40	0.20	0.15	0.15	0.10
Price relative x_i	120	105	110	102	96

giving

$$\sum_i w_i x_i \Big/ \sum_i w_i = (0.40 \times 120) + (0.20 \times 105) + (0.15 \times 110)$$
$$+ (0.15 \times 102) + (0.10 \times 96) = 110.4.$$

This is the index number for 1978 relative to 1975 as base year. Similarly, using the same weights $\{w_i\}$ with the price relatives for 1981 as $\{x_i\}$, the index number for 1981 is 139.5.

In order to calculate an index for 1981 relative to 1978, we must first obtain price relatives for 1981 by comparison with 1978.

$$\frac{1981\,\text{price}}{1978\,\text{price}} = \frac{1981\,\text{price}}{1975\,\text{price}} \Big/ \frac{1978\,\text{price}}{1975\,\text{price}},$$

i.e. the ratio of the two sets of price relatives already given. So for A we have $150/120 = 125$; for B, $126/105 = 120$; for C, $121/110 = 110$; for D, $153/102 = 150$; and for E, $132/96 = 137.5$, all expressed as percentages. Using again the original set of weights, with these new price relatives, we find that the index number for 1981 taking 1978 as base year would be 126.75.

There is a developed theory of index numbers (see Bibliography, p. 684) and readers who need to follow up this topic will find that there are alternative ways of defining an index number. Indices are calculated for a wide variety of quantities: share prices, industrial production, retail prices of all goods, and so on. The choice of the list of items to be included in an index is sometimes controversial; economists will often have a major influence on the decision.

2.10 MINITAB: CALCULATION OF MEAN AND MEDIAN; DOT PLOTS

There are many ways of calculating the mean of a column of data in a worksheet. The most direct is to use the command

MEAN C

The equivalent menu approach is to choose

    Calc>Column Statistics/
    Select **Mean**, Enter **Input variable/**

'Mean' may be replaced by 'Median' in both methods.
    Choice of

    Stat>Basic Statistics>Display Descriptive Statistics/
    Enter **Variable/**

gives both mean and median as well as the maximum, minimum, number of data and some statistics that we shall meet later. If more than one column is named the results are presented side-by-side. Choice of **Statistics** in the dialogue box allows one to choose exactly which statistics are displayed.

To calculate the mean when the data are given as a frequency table with values in C1 and frequencies in C2, use the commands

    TABLE C1;
    FREQUENCIES C2;
    MEAN C1.

This will produce a list of the values with the mean at the base. The mean may also be obtained using **Stat>Tables>Descriptive Statistics/** but how to complete the dialogue box is not clear.

Qualitative variates, or attributes, may be considered as MINITAB variables. Thus their values, e.g. 'smoked' or 'never smoked', or A, B, C, D, may be entered in a column. Put another way, the data in a column may be numerical values or text. Numerical data from two groups, labelled A and B, might be entered in separate columns, or all the numerical data might be put in C1 and corresponding labels put in C2. In the latter case, the data are said to be 'stacked', to use MINITAB terminology, while in the former case they are said to be 'unstacked'. To move the data from one form to the other choose **Data>Stack** or **Data>Unstack Columns**.

The menu choice that produces the basic statistics for the stacked data, set out under each other for easy comparison, is

    Stat>Basic Statistics>Display Descriptive Statistics/
    Variable C1 and **By Variable** C2/

Dot plots are produced by choosing

    Graph>Dotplot/
    Select **Simple/**
    Enter **Graph variable/**

This choice will give a separate graph for each column. To put the plots on the same graph choose

    Graph>Dotplot/
    Select **Simple** under **Multiple Y's/**
    Enter **Graph variables/**

To put all the plots on one graph with stacked data, where variate values are in C1 and labels in C2, choose

>**Graph>Dotplot/**
>Select **With Groups/**
>Enter C1 as **Graph variable** and C2 as **Categorical variable/**

MINITAB offers a large number of data sets for practice. They are described under 'Sample Data Sets' in MINITAB Help. To obtain a particular set of data choose

>**File>Open Worksheet**

Ensure that the listing is for '*Files of type* *.mtw', and select the chosen file.

2.11 EXERCISES ON CHAPTER 2

1. 'Over 60% of the people in this country earn less than the average wage', said the Opposition spokesman. Is this nonsense?

2. There are 30 children in a class. Eight were aged 9 at their last birthday, and three were aged 11; the remainder were aged 10. What is (i) the median age; (ii) the modal age?

3. Tiny Tim had a birthday party yesterday. The ages of the people present were 3, 4, 4, 4, 4, 5, 5, 5, 33, 38, 40. (i) Find the mean of these data. (ii) Divide the data into two natural groups, describe what people are put into each group and calculate the mean of each group. (iii) How would you summarize the data?

4. In a new factory eleven light bulbs of each of type A and type B were fitted. When any of the bulbs failed the length of life in months was recorded. After 18 months eight of type A had failed at ages 4, 9, 11, 13, 13, 14, 17, 18 and six of type B at ages 7, 12, 15, 15, 16, 18. Suggest, and calculate for each type, a measure that might be used to compare the lifetimes of the two types of bulbs.

5. The weekly wages of ten carpenters in a factory are £151, 153, 153, 154, 155, 157, 160, 161, 161, 165. (i) Find the mean wage of the carpenters. (ii) If all those who earn less than the mean wage have their wage increased to that value, find the new mean wage. (iii) If this procedure of increasing wages is repeated many times, what is the ultimate distribution of the wages?

6. Calculate the mean, median and mode of the sets of observations that follow. Plot each set of observations on a linear scale and mark in these averages. Comment on the relative merits of each average for each set of data:
   (a) Weekly wages, in £, of ten workers, including the manager, in a factory: 184, 184, 184, 185, 185, 185, 185, 186, 188, 300.

(b) Price, in pence per kilogram, of potatoes in five shops: 42, 41, 42, 43, 40.

(c) Price, in pence, of a can of baked beans (all of the same brand) in five shops: 34, 35, 34, 34, 34.

(d) Age, in years, of four children in a family: 6, 8, 12, 14.

Questions 7–12 are exercises on $\sum$-notation

7. The set of observations $x_1 = 3$, $x_2 = 2$, $x_3 = 1$, $x_4 = 0$, $x_5 = -3$ has the total $\sum_{i=1}^{5} x_i = 3$, and $\sum_{i=1}^{5} x_i^2 = 23$. Evaluate the expressions that follow: (i) by using the definition of $\sum$ (Definition 2.3); and (ii) by simplifying the expressions, using the rules for $\sum$ (p. 31), and then evaluating:

(a) $\displaystyle\sum_{i=1}^{5} (x_i + 10)$, (b) $\displaystyle\sum_{i=1}^{5} (2x_i + 3)$, (c) $\displaystyle\sum_{i=1}^{4} (x_i + x_{i+1})$,

(d) $\displaystyle\sum_{i=1}^{4} (x_i - x_{i+1})$, (e) $\displaystyle\sum_{i=1}^{5} (x_i - 1)(x_i + 1)$.

8. Write the following expressions more concisely by using $\sum$-notation:

(a) $p_0 + p_1 + p_2 + p_3 + p_4$, (b) $x_1 y_1 + x_2 y_2 + x_3 y_3 + x_4 y_4$,

(c) $1^2 + 2^2 + 3^2 + 4^2 + 5^2$, (d) $x_1 + 2x_2 + 3x_3 + 4x_4 + 5x_5$,

(e) $2 + 4 + 6 + 8 + 10$, (f) $1 + 3r + 5r^2 + 7r^3$.

9. (i) Show that

$$\sum_{r=0}^{n} [(r+1)^2 - r^2] = (n+1)^2.$$

(Write out the first two and last two terms and note the cancelling.)

(ii) Show that

$$\sum_{r=0}^{n} [(r+1)^2 - r^2] = 2\sum_{r=0}^{n} r + n.$$

(Use the rules for $\sum$.)

(iii) From (i) and (ii) show that $\sum_{r=1}^{n} r = \frac{1}{2}n(n+1)$

10. A discrete variate takes the value x_i ($i = 1, 2, \ldots, k$) with frequency f_i; the total number of observations $\sum_{i=1}^{k} f_i = N$ and the mean value is $\bar{x}$. Show that

(i) $\displaystyle\sum_{i=1}^{k} f_i(x_i - \bar{x}) = 0$

(ii) $\displaystyle\sum_{i=1}^{k} f_i(x_i - \bar{x})^2 = \sum_{i=1}^{k} f_i x_i^2 - \left(\sum_{i=1}^{k} f_i x_i\right)^2 \bigg/ N.$

11. Show that

$$\sum_{i=1}^{N} (x_i - d)^2 = \sum_{i=1}^{N} (x_i - \bar{x})^2 + N(\bar{x} - d)^2.$$

12. One set of n_1 observations has mean $\bar{x}_1$ while another set of n_2 observations has mean $\bar{x}_2$. Show that if the two sets are combined the overall mean is $(n_1\bar{x}_1 + n_2\bar{x}_2)/(n_1 + n_2)$.

13. Find the means of the following sets of data by using a suitable coding:

(a)	x_i	6.5	8.5	10.5	12.5	14.5	Total
	f_i	2	8	22	13	5	50

(b)	x_i	1020	1040	1060	1080	Total
	f_i	12	4	7	17	40

(c)	x_i	0.135	0.145	0.155	0.165	0.175	Total
	f_i	40	21	9	7	3	80

14. Find the mean and median of the following sets of grouped data:

(a) Length of life, in years, of a species of bird

x	0–	1–	2–	3–	4–	5–	6–	7–	8–	Total
f	49	26	11	7	3	2	1	0	1	100

(b) Height of plants recorded to nearest millimetre

x	26–30	31–35	36–40	41–45	46–50	51–55	56–60	61–65	Total
f	4	5	23	58	61	30	3	3	187

(c) Length of rally (i.e. number of strokes per point) in a game of tennis

r_i	1	2	3	4	5	6	7	8	9	10	11	12	Total
f_i	8	60	80	50	41	20	6	9	3	0	1	1	279

15. The numbers of moths collected in a light trap on five successive nights were 210, 4, 376, 12, 50. (i) Calculate the mean of the data. (ii) Find the logarithms of the observations. Find the mean of the data after they have been transformed to logarithms, and take the antilogarithm of this mean to obtain the geometric mean of the observations. (iii) Compare the arithmetic and geometric means.

16. Define the terms: (a) skew; (b) unimodal; (c) median. Give an example of an observed variable whose distribution is likely to satisfy the criteria for each of the four cases below:

(i) continuous, skew, unimodal, mean greater than median;
(ii) discrete, multimodal;

(iii) discrete, symmetrical, unimodal;

(iv) continuous, symmetrical, multimodal. *(Oxford)*

17. Explain the relative advantages and disadvantages of the mean, median, mode and geometric mean as measures of average. Find the mean, median, mode and geometric mean for the following set of observations: 17, 19, 4, 12, 19, 23, 9, 27, 14. Also illustrate these data in a suitable graph.

18. Explain concisely under what conditions you would prefer: (i) the (arithmetic) mean; (ii) the median; as a measure of central tendency in a set of data.

The following table (from the Registrar General's annual return) shows the age at marriage of females in 1969. Display this information in a suitable diagram, and calculate the median age of marriage.

Without carrying out any more calculations, state which of the two measures, mean and median, you expect will be the larger for these figures. Justify your choice.

Age in years at marriage	Number of females (thousands)
Under 18	27
18 and under 19	35
19 and under 20	50
20 and under 21	58
21 and under 25	173
25 and under 30	48
30 and under 35	17
35 and under 45	18
45 and under 55	13
55 and over	12

19. The following data are used to form an index of prices. All prices were taken as 100 at the time chosen as base-period, and the 'price index' column shows the change for each item since then. The 'weight' column shows the relative amounts of expenditure on each item, the weights being coded to sum to 1000. Calculate an index of prices: (a) using only the first four items, (b) using all eight items; to show how the present costs compare with the base-period.

Item	Price index	Weight
Food	149.7	348
Rent and rates	104.2	88
Clothing	147.1	97
Fuel and light	140.1	65
Household durable goods	136.6	71
Miscellaneous goods	137.3	35
Services	123.9	79
Drink and tobacco	108.5	217

20. A cost of living index is calculated using the weights shown, with base-year 1972:

Food	Housing	Clothes and fuel	Other items
4	3	2	1

It is known that the value of the index for 1973 is 110 and for 1974 is 121. In 1973 the cost of food had risen by 9% and the cost of housing had risen by 8% above their 1972 values. In 1974 the costs of food and housing had risen by 16 and 20% respectively above their 1972 values. Clothes and fuel increased by a% and other items by b% each year above the previous year's values.

Calculate the possible values of a and b. (*JMB*)

2.12 COMPUTING EXERCISES

1. Two aptitude tests, A and B, are each claimed to be successful in assessing a candidate's potential as a computer programmer. The maximum possible score for each test is 50. Nine candidates took aptitude test A both before and after a two-week preliminary training period. Eight candidates took aptitude test B before and after the same two-week training period. The following results were obtained.

Test A			Test B		
Candidate	Before training	After training	Candidate	Before training	After training
1	16	13	10	19	21
2	5	22	11	47	50
3	23	29	12	35	41
4	15	28	13	18	20
5	40	48	14	37	36
6	31	30	15	7	12
7	39	46	16	34	38
8	6	17	17	27	25
9	24	33			

(i) Denote the scores before and after training by X and Y respectively. Calculate $Z = Y - X$ for each candidate.
(ii) Compare the X values for tests A and B by calculating means and medians and constructing dot plots for the two sets of data.
(iii) Repeat, as in (ii), for the variates Y and Z.

(iv) Review your results and hence describe how performance compares on the two tests.

2. Produce dot plots and calculate means and medians of the following sets of data:
    (a) weights of students in Exercise 23.13(1);
    (b) reaction times in Exercise 13.10(9);
    (c) number of defectives in Example 16.9;
    (d) word data in Revision Exercises A(4).
    Compare the values of the mean and the median for each set, and relate the similarity or difference to the symmetry or skewness of the distribution.

3. Calculate the mean and median, and construct a dot plot, for each of the two sets of data in Exercise 5.6(1). Hence compare the two sets of results.

2.13 PROJECTS

1 Age of pennies in circulation

Aim. To investigate the distribution of the age of pennies in circulation. (Any convenient denomination of coin may be used, provided it has been in circulation for some years.)

Collection of data. Ask people to show you the date on a penny chosen at random from the coins they have. Note down the last two digits of the date. Collect dates from at least 60 coins.

Analysis. Calculate the age in years of each coin. Make a stem-and-leaf diagram, a frequency table and a histogram of the distribution. Calculate the mean, median and mode of the data.

2 Throwing sixes with a die

Aim. To estimate the average number of throws of a die required for a six to appear.

Equipment. Die and shaker.

Preliminary work. Guess what the average number will be.

Collection of data. Throw the die until a 6 appears. Count the number of throws. (*Include* the throw giving the 6.) Repeat 50 times.

Analysis. Make a frequency table from the data. Calculate the mean and median.

3 Index of personal expenditure on leisure items

Aim. To construct an index showing the expenditure of a group of people on leisure items and to follow its change over time.

Collection of data. Ask a group of people, e.g. in the same school or college class, to record their expenditure on all forms of leisure for an extended period, preferably at least a month. This should include expenses on sport, music, theatre, cinema, etc. and contain costs of material or equipment bought as well as costs of attending or taking part in events. The proportion of the whole group's expenditure spent on each different leisure category will be the weight to be given to that category. Categories need to be chosen in such a way that it is possible to define a 'price' (perhaps itself a weighted average price) for each category.

Analysis. Construct an index number as a weighted average, using these weights and prices. At suitable intervals of time, update the prices and calculate a new index.

SUMMARY

The commonly used measures of the centre of a set of observations are the arithmetic mean and the median. The **arithmetic mean** is the sum of the observations divided by the number of observations. The **median** is the middle observation when the observations are put in rank order.

$\sum$-notation is useful for describing, manipulating and making calculations with observations.

Calculation of a mean may be done from data grouped in a frequency table. Calculation of the mean, and other statistical quantities, may be simplified by **coding** data, i.e. transforming the data to a more amenable scale.

Less commonly used measures of the centre of a set of observations are the mode and the geometric mean. The **mode** of a discrete variate is that value of the variate that occurs most frequently. The **geometric mean** of the N positive observations $x_1, x_2, \ldots, x_N$ is the Nth root of the product $(x_1 x_2 \ldots x_N)$.

A **weighted average**, or weighted mean, of quantities x_i ($i = 1, 2, \ldots, k$) takes the form $\sum w_i x_i / \sum w_i$. **Index numbers**, often used in economics, are examples of weighted averages.

3 SAMPLES AND POPULATIONS

OVERVIEW

If data are to be informative, they must be collected properly. We consider how to choose a sample to gain reliable information about a population.

3.1 A SAMPLING INVESTIGATION

A group of pupils in a school plan to investigate how long it takes to travel between home and school. They would like to discover the journey times for all the 2000 pupils in their school, but realize that they do not have the time to collect and analyse such a large amount of data. They argue that it should not be necessary to collect information from the whole school, but that information from some of the pupils should give them what they want provided these pupils are chosen properly. So they decide to collect data from only a part of the complete school population. We call this part a *sample* of the school population.

Definition 3.1

A **sample** is any subset of a population.

An investigation of this type is said to be a *survey* of a population. In the particular example we are discussing, information is collected by sampling: such an investigation is called a *sample survey*. If a survey plans to collect information from every member of a population, it is called a *census* of that population. In most advanced countries, regular censuses of population are taken; the United Kingdom census of population takes place every ten years, the most recent having been in 2001.

How should a sample be chosen? We would like the sample to be an image in miniature of the whole population, and to reproduce the characteristics of the population. In our example, the most important summary characteristic is the mean journey time for the population of schoolchildren. We would therefore like the mean journey time in the sample to be close to the mean in the whole population.

3.2 SOME METHODS OF CHOOSING A SAMPLE

The investigators might choose to ask only their friends. This certainly has the merit of convenience, but how good a sample are they likely to obtain? It is easy to see that the mean journey time for the members of this sample might be very different from that for the whole population. There is a good chance that many of the friends would live close together, but little chance that their mean distance from school is close to the mean distance for the whole population.

A method that superficially appears much better is to choose a 'representative' sample. The investigators would choose pupils they considered were 'typical' of the various categories of pupils in the whole school. This could be a very good or a very bad representation of the whole population of pupils, depending on how much the investigators knew about the population. Experiment suggests that people tend to be bad at choosing samples in this way. Dr F. Yates carried out an experiment in which he spread 1200 stones of varying sizes on a table, and asked twelve observers to choose three samples each. The samples were to consist of 20 stones each, to represent as nearly as possible the size distribution of the whole collection. It was found that the observers tended to pick samples whose mean weight was above the mean weight of the population; in fact, this was so for 30 of the 36 samples. The mean weight in the population was 1.91 oz, and for the samples it was 2.34 oz. (Some data from a similar experiment will be found in Exercises 3.11(1).)

Even if the mean weight for the samples had happened to come near to the population value, the result of choosing a sample in this way would be unconvincing to a person who had no special reason to trust the judgement of the chooser. People who disagreed with the results of a sample survey based on this type of sample might claim that the chooser had very bad judgement. It would be very difficult to reply to these objections, because there is no yardstick which allows someone to judge how well the sample has been chosen.

3.2.1 Random sampling

Criticism that a sample has been chosen in a subjective way may be avoided by choosing a *random sample* from the population. There are many sampling schemes that may correctly be called random. We shall only define a *simple* random sample, which is by far the most straightforward of these. Other, more complex, random sampling schemes are of particular use in certain special types of problem.

Definition 3.2

A (simple) **random sample** is a sample which is chosen so that every member of the population is equally likely to be a member of the sample, independently of which other members of the population are chosen.

We allow some chance procedure, such as drawing names from a hat, or tossing a coin, to determine whether any particular member of the population enters the sample. The choice of the sample is most conveniently done from a list. The school in our example would certainly have a list of its 2000 pupils. We would number the names in the list 1 to 2000. Then we would obtain 2000 tickets (all the same size and colour), number these 1 to 2000, put them into a hat and mix them well. If we required a random sample of 200 names we should draw 200 numbered tickets from the hat. After drawing each ticket, we should put it aside and mix the remainder thoroughly before drawing another. The pupils whose names in the list corresponded to the numbers on the tickets we have drawn would form the sample. This method of choosing the numbers is obviously laborious; more convenient methods using tables or a computer are described later in the chapter (see p. 53).

A model of simple random sampling

We can investigate some of the properties of random sampling by taking samples from a small model population consisting of one each of the ten digits 0, 1, 2, 3, 4, 5, 6, 7, 8, 9. Suppose we choose samples of five units. We could list all 252 possible samples, though it would be tedious to do so; examples are (0, 1, 2, 3, 4), (2, 3, 7, 8, 9), (0, 2, 4, 6, 8). Each of the 252 possible samples is equally likely to occur if we choose a simple random sample.

Let us look at the relation between the means of the samples and the mean of the population. The mean of the population is 4.5. The mean values of the three examples of samples given above are 2.0, 5.8, 4.0, respectively. Note that the digit after the decimal point in all the sample means in this example will be even, since the means are whole numbers divided by 5. The frequency distribution of the means of all possible 252 samples is summarized in a frequency table:

Interval	1.8–2.4	2.6–3.2	3.4–4.0	4.2–4.8	5.0–5.6	5.8–6.4	6.6–7.2	Total
Frequency	4	24	59	78	59	24	4	252
Relative frequency	0.016	0.095	0.234	0.310	0.234	0.095	0.016	1.000

Each interval has been specified (e.g. 2.6–3.2) to make clear which values fall in the interval. In general, and with larger samples, the intervals will not be so well separated, and in the limit they would actually be continuous. When constructing a histogram (Fig. 3.1) to illustrate the information in this table, we shall therefore want to make the intervals continuous. The end-points are thus 0.1 beyond the values quoted in the table, e.g. the end-points of the interval 2.6–3.2 are 2.5 and 3.3.

A high proportion of the means of the random samples are close to the mean of the population. If we choose a random sample then, especially if the sample size is not too small, there is a high chance that the mean of the sample will be close to the mean of the population. However, it must be admitted that there is also a chance, though a small one, that the mean of the sample could be appreciably different from the mean of the population; for example, there is one sample with a mean of 2.0.

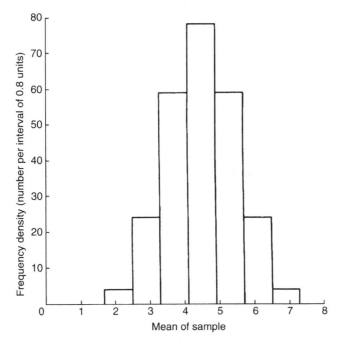

Figure 3.1 *Histogram of distribution of mean of sample of five digits chosen at random from the ten digits 0 to 9.*

Another desirable property possessed by random samples is that the mean of the means of all possible random samples is equal to the population mean. We see that the mean of the frequency distribution of sample means above is equal to the population mean of 4.5.

The reason for this can be seen more clearly from a simpler example. Consider the ten possible samples of size two that can be chosen from a model population $(1, 2, 4, 5, 8)$ which has a mean of 4.

The possible samples are:

	Sample number									
Unit	1	2	3	4	5	6	7	8	9	10
1	×	×	×	×						
2	×				×	×	×			
4		×			×			×	×	
5			×			×		×		×
8				×			×		×	×
Total	$(1+2)$	$(1+4)$	$(1+5)$	$(1+8)$	$(2+4)$	$(2+5)$	$(2+8)$	$(4+5)$	$(4+8)$	$(5+8)$

Call the sample totals $T_1(=1+2), T_2(=1+4), \ldots, T_{10}(=5+8)$; the sample means are $T_i/2$, where $i = 1, 2, \ldots, 10$. The mean of the sample means is

$$\frac{1}{10}\left(\sum_{i=1}^{10} \frac{T_i}{2}\right) = \frac{T_1 + T_2 + T_3 + T_4 + T_5 + T_6 + T_7 + T_8 + T_9 + T_{10}}{20}$$

$$= \frac{\left[\begin{array}{l} (1+2)+(1+4)+(1+5)+(1+8)+(2+4) \\ +(2+5)+(2+8)+(4+5)+(4+8)+(5+8) \end{array} \right]}{20}$$

$$= \frac{4(1+2+4+5+8)}{20} = \frac{1+2+4+5+8}{5} = 4,$$

i.e. the mean of the population.

To summarize: the merits of choosing a random sample as opposed to a non-random sample are that: (1) the method is objective; (2) the mean of a random sample has a high chance of being close to the mean of the population; (3) the mean of the means of all possible random samples equals the population mean.

3.3 COMPLETING THE INVESTIGATION

We assume that a random sample of 200 pupils has been chosen. The variate to be measured on each member of the sample must be specified precisely. A reasonable measure of the length of journey time could be obtained by asking each member of the sample to record the times of the two journeys, to and from school, made each day for a week. The mean journey time for each person for the week would then be used as the variate to be studied, x. This continuous variate x could be summarized in a frequency distribution, and its mean value determined, as described in Chapter 2.

We really want to know the mean value of the journey time in the whole population; but since we do not know it we must be content with an *estimate* of it. The sample mean $\bar{x}$ provides this estimate of the true (but unknown) mean value of the whole population of journey times. Our aim in statistics is to find estimates that are good ones.

Sample and population values

A major distinction in statistics is that between the sample and population values of quantities such as the mean, or the proportion having a certain attribute. A useful convention is to use Latin letters for the sample values and Greek letters for the population values. So we shall use p as a symbol for the proportion in a sample, and π (the Greek letter corresponding to p) for the proportion in the population from which the sample is drawn. We have already introduced the bar-notation (e.g. $\bar{x}$, $\bar{r}$) to represent a mean; we shall use μ to represent the corresponding population mean.

3.3.1 Sampling to estimate a proportion

We may be interested to know what proportion of pupils make use of public transport when travelling to and from school. If a random sample of pupils is chosen, the true (but unknown) value of the proportion π for the whole population may be estimated by the value of the proportion p in the sample. Thus, if a number a, out of a random sample of size n, make use of public transport, the estimate of the true proportion π in the population who use public transport is the sample proportion $p = a/n$.

In fact this procedure is equivalent to estimating a mean. If we define a discrete variate r which takes the value 1 for a pupil who travels by public transport and 0 for one who does not, the sample is a collection of observations $r_1, r_2, \ldots, r_n$ each of which takes the value 0 or 1. There will be a ones and $(n-a)$ zeros among the r_i. The mean of the sample is $\bar{r} = \sum_{i=1}^{n} r_i/n = a/n = p$, the proportion in the sample who do use public transport. Similarly, the mean value of r in the whole population equals π, the proportion who travel by public transport in the whole population. Thus once again we are estimating a population mean by the corresponding sample mean.

3.3.2 Sampling to estimate a total

If we estimate the mean acreage of a population of N farms to be $\bar{x}$ $(=\sum_{i=1}^{n} x_i/n)$ from a sample of size n, then we can estimate the total acreage of the population of farms by $N\bar{x}$. Similarly, if we estimate the proportion of pupils who travel by public transport to be p $(=a/n)$, then we estimate the total number of pupils who travel by public transport by Np (N being the total number of pupils in the school and n the number of pupils in the sample).

3.4 RANDOM NUMBERS

Drawing numbers from a hat or throwing dice can be used to choose random samples but, for greater convenience, statisticians prefer to use random numbers that come from tables or have been generated by computer. Use of tables is discussed in this section. The computer generation of random numbers is discussed in Section 3.5.

Random number tables contain lists of the digits 0–9 in random order. An example of a line in such a table, in which the digits are grouped in pairs, is:

$$10 \quad 43 \quad 67 \quad 89 \quad 70 \quad 80 \quad 62 \quad 80 \quad 03 \quad 42 \quad 71 \quad 35.$$

(There is a table of random numbers on p. 713 of this book.)

Suppose we wish to draw four days at random from the days of a year numbered 1 to 365. We begin to read digits from some point in the table. (It is not necessary to choose the starting point at random but one should vary the starting point each time the table is used.) Suppose we obtain the line above. We may read the digits in threes:

$$104 \quad 367 \quad 897 \quad 080 \quad 628 \quad 003 \quad 427 \quad 135.$$

We accept numbers in the range 001–365, and reject all numbers outside this range and numbers that have already been chosen. This gives us as a random sample the days numbered 104, 080, 003, 135. The rejections may be reduced by using the equivalence $001 \equiv 366, 002 \equiv 367, \ldots, 365 \equiv 730$; we then reject only 000 and numbers greater than 730. This method gives us the random sample

$$104, \quad 367 - 365 = 002, \quad 080, \quad 628 - 365 = 263.$$

We cannot use the table numbers greater than 730 by subtracting 730 ($= 2 \times 365$) from them, because the numbers 731 to 999 produce by this method only the numbers 1 to 269. Thus, for example, three of the table numbers (001, 366, 731) generate 1 whereas only two (300, 665) generate 300. This would not give all the numbers 1–365 an equal chance of being in the sample, as in the case with the two methods illustrated and as is required for random sampling.

It is interesting to know how some of the published tables of random numbers were compiled. The 100 000 random digits for Kendall and Babington Smith's tables were obtained from a machine specially constructed for the purpose. 'An observer was placed before a machine consisting of a circular disc divided into ten equal sections in which were inscribed the digits 0 to 9. The disc rotated at high speed and every now and then a flash occurred from a nearby electric lamp of such short duration that the disc appeared at rest. The observer had to watch the disc and write down the number occurring in the division indicated by a fixed pointer.'

Fisher and Yates' table of 15 000 digits was obtained by choosing the 15th–19th digits from a table of logarithms to 20 places. The digits were adjusted because they were found to contain too many sixes.

We all have an intuitive idea what random numbers are but it is impossible to give a precise definition. The best we can do is to think of the properties that random numbers ought to have, and check whether any particular list has these properties. Obviously the ten digits ought to occur with approximately equal frequency, and there should not be long runs of one digit. See also Chapter 19 where we consider how to test whether sequences of digits are 'random'.

3.5 PSEUDO-RANDOM NUMBERS

How can a computer be made to produce random numbers? One might put a large table of random numbers in the computer store, but this would either be wasteful of valuable easy-access store or difficult to access through being put in a peripheral store. One requires a method, which must work according to a fixed rule and therefore be deterministic, which will produce numbers that have the properties of true random numbers. We shall give examples of such methods.

Let us take a four-digit number (e.g. 6729), square it (45 279 441), and record the middle two digits (79); then pick out the middle four digits (2794) and repeat the process using them. The next square is 07 806 436; we record 06 and use 8064 as the next four-digit number, and so on. The 'random numbers' produced are 79, 06, ... Since there are exactly 10 000 four-digit numbers, at some stage our first four-digit number will be repeated; from that point on the sequence of 'random numbers' we have obtained will be repeated exactly, i.e. the method is cyclic. The cycle length turns out to be much less than 10 000 in practice, and depends on the starting number. A more serious objection to this method is that all the possible numbers in the range 00–99 do not occur equally frequently.

A better method is to generate the numbers by a recurrence relation. We call the pseudo-random numbers x_0, x_1, x_2, ... Thus, we might begin with the number $x_0 = 7$. We find $(21x_0 + 3)$, which is $(147 + 3)$ or 150. To keep the numbers in the range 00–99, we divide the result by 100 and keep only the remainder. Thus $x_1 = 150 - 100 = 50$. Then we form $(21x_1 + 3)$ or $(1050 + 3)$ which equals 1053, and put $x_2 = 1053 - 1000 = 53$. In general we generate numbers by the relation

$$x_{n+1} = ax_n + b \pmod{m}.$$

In our example, $a = 21, b = 3, m = 100$. The mod m is an abbreviation for modulus m; it means that we use the remainder after dividing by m. These methods are devised for computers, so the recommended values of a and m are large. For example, one recommended choice is $a = 16\,807$, $b = 0$, $m = 2^{31} - 1$. Division by m of the integers produced gives random numbers in the range 0 to 1. These can be used to give random numbers in any required range. Computer generation of pseudo-random numbers is discussed at greater length in CCC, Chapter 7.

3.6 SYSTEMATIC SAMPLES

A sample which has been obtained by choosing names at regular intervals (say, every 10th) from a list or by choosing every sixth house, say, on a street, is said to be a systematic sample. A list of names in alphabetical order can often be regarded as being in random order with respect to the variate of interest. In these cases a systematic sample from an alphabetical list can be treated as if it were a random sample. This can be very convenient in practice. A direct element of randomness is usually introduced by choosing the first name at random, e.g. choosing one of the first ten names on the list at random and then taking every successive tenth name on the list.

But there are dangers in using systematic sampling. Suppose we decide to take every fourth house in a street, in an enquiry on heating costs. If the street consists of terraces of four houses (two corner houses and two central ones) then the sample would consist entirely of either the more exposed corner houses or the more sheltered central houses, and would lead to a very poor estimate of the average heating cost per house in the whole street.

We should also note a possible difficulty in using an alphabetical list of people's names, such as that in a telephone directory: it may not be in a completely random order. The reason is that members of the same nation or origin may well come together through having similar surnames, e.g. Scotsmen whose surnames begin with Mac. However, this is unlikely to spoil a sample unless the proportion of adjacent related names in the whole list is substantial.

3.7 SOME USEFUL TERMS

The aim of a sample survey is to obtain information about a certain population, e.g. all the children in a school or all the voters in a country. Because of practical difficulties it is often not possible to take a sample from the population we would like to sample. Some children will be off school because of illness or holidays; some voters will live in inaccessible places. Gaining information from these members of the population may be very expensive in terms of time or money, or even quite impossible. For practical reasons the investigator often has to settle for obtaining information about a population which he believes has similar properties to the population he really wishes to sample, though not being identical with it. It is convenient to distinguish these two populations by giving them separate names.

Definition 3.3

The **target population** is the population about which we want information.

Definition 3.4

The **study population** is the population about which we can obtain information.

In order to take a sample we must specify the members of the population from which we choose the sample. These might be individual children, as in the survey above, or the families in a town if the survey was concerned with family expenditure. The child or family is said to be the *sampling unit*.

Definition 3.5

A **sampling unit** is a potential member of the sample.

The most convenient way of choosing a sample is usually to draw up, or obtain, a list of the sampling units and choose the sample from the list by the use of random numbers. In the school survey example, a list of the pupils in the school was used. Another form of list which is often used for sampling is the electoral roll for a particular district; this is a list of all the people who may vote at national and local elections and includes all adults over 17, with minor exceptions. It is revised each year, but clearly becomes less accurate as the year goes on, because people move in or out and some die. However, it is easily available and probably provides the best local list available.

    In some cases it may not be necessary to list explicitly all the sampling units. For example, a biologist when sampling a field to investigate the plant life may choose a 1-metre square as the sampling unit and divide the whole field into 1-metre squares by the use of a grid; any particular square may be specified by two coordinates representing, for example, distances east and north of some convenient starting point. He chooses each member of the sample by taking a pair of random numbers, of which the first indicates distance east, and the second distance north, of the starting point.

3.8 SAMPLING WITH AND WITHOUT REPLACEMENT

If we choose a sample by pulling numbered tickets out of a hat, we have a choice after each draw of either returning the ticket to the hat for the next draw, or leaving it out of the hat. The first method is called sampling (the population of numbered tickets) *with replacement*; the second, sampling *without replacement*. When choosing a sample to estimate the mean journey time, for example, it does not seem sensible to include a particular member of the population more than once in the sample; we obtain more information if all the sample members are different. We therefore sample without replacement. But in other circumstances when we choose samples, and particularly in theoretical work, we often sample with replacement. For example, a table of random numbers is a sample with replacement from the population of the ten digits 0–9.

3.9 MINITAB: RANDOM NUMBERS; SAMPLING

MINITAB is a source of pseudo-random numbers. To generate random digits in the range 0–9, similar to the tables of random digits (see Table A1, p. 713), and store them in a number of columns, choose

Calc>Random Data>Integer/
Enter number of **Rows of data**, and **Columns** in which to be **stored**,
Enter **minimum** of 0 and **maximum** of 9/

Expressed another way, we are choosing a random sample of values from the integers between chosen maximum and minimum values. We are sampling what MINITAB calls the *integer* distribution and we call the *discrete uniform distribution* (Section 8.4).

The basic form in which pseudo-random numbers appear is as values in an interval with endpoints 0 and 1. A typical value given by MINITAB is 0.659 843. The general form of the menu choice is

Calc>Random>Uniform/
Enter number of **Rows of data** and name of **Columns** in which to be **stored**,
give **lower** and **upper endpoints**/

The default values of the endpoints are 0 and 1 but any numerical values may be chosen. The values may be regarded as a random sample from the *continuous uniform distribution* (Section 14.5). MINITAB calls it simply the *uniform* distribution.

As will be seen from the menus displayed when using this approach, random samples may be generated from a wide range of theoretical distributions. Two only interest us at this stage: the continuous uniform distribution (**Uniform**) and the discrete uniform distribution (**Integer**).

MINITAB will choose random samples from sets of data. The choice of a sample of size two from the population of five digits $(1, 2, 4, 5, 8)$, assumed placed in C1, as was demonstrated on page 51, would be carried out by choosing

Calc>Random Data>Sample from Columns/
Enter **Sample** size 2, **column** C1 to be sampled, **Store** sample in C2/

Sampling is *without replacement*, by default; sampling *with replacement* may be selected. One way of repeating the sampling is to select (using the Windows procedure) the session command that appears, provided session commands are enabled:

MTB>SAMPLE 2 C1 C2

Choose **Edit>Command Line Editor**, replace C2 by C3 in the dialogue box and click **Submit Commands**.

Having chosen the samples, one would be interested in, e.g. the means or medians, which would be obtained by operating on the columns. But, continuing the example of samples of size two from five digits, if one wanted 100 samples it would be very clumsy to spread the result over 100 columns each containing just two observations. A neater method would be to choose samples of size 100 in two columns and regard the rows, containing two observations, as the samples. (The sampling, of course, has to be with replacement.) In this case we would be interested in the row statistics. To calculate the row means, choose

Calc>Row Statistics/
Select **Mean**, enter **Input variables** and where to **Store result/**

Many other statistics, e.g. median, sum, max, may be chosen instead of the mean.

3.10 EXERCISES ON CHAPTER 3

1. A population consists of the ten digits 0, 1, 2, 3, 4, 5, 6, 7, 8, 9. List the 45 possible distinct samples of two digits that would be equally likely choices if random samples were drawn from the population without replacement (so that no sample can contain identical digits). Find the mean of each sample and construct the frequency distribution of these means. Show that the mean of the distribution equals the mean of the population.

2. Assume that, for each of the quantities (a)–(d) listed below, you have to carry out a sample survey to estimate the quantity. In each case state: (i) the sampling unit; (ii) the target population; (iii) the study population;

(iv) the variate; (v) how you would calculate the estimate. (*Example.* Estimate the number of children in the families that live in a block of high-rise flats. The sampling unit is a flat; the target population is all the flats in the block; the study population is all the flats from which a response can be obtained; the variate is the number r_i of children who live in flat i. If the study population was substantially smaller than the target population the non-responders should be investigated to check if they were different from the responders in respect of number of children. If population size is N and the sample size is n, the required estimate is $N\bar{r}$ where $\bar{r} = \sum_{i=1}^{n} r_i/n$.)

(a) The total number of chairs in a school;
(b) the proportion of households in a town with microwave ovens;
(c) the number of worms in a field;
(d) the mean height of children who live on a particular road.

In Questions 3–6 use the table of random digits on p. 713 when choosing the random samples.

3. Select, without replacement, five words at random from the ten words: 1, a; 2, the; 3, cat; 4, dog; 5, bit; 6, red; 7, blue; 8, hen; 9, ate; 10, mad. (Interpret the random digit 0 to mean 10.)

4. Assume that Members of Parliament are numbered 1 to 650 on a list. Choose a random sample, without replacement, of the numbers 1 to 650, which would determine a random sample of Members of Parliament.

5. Choose a random sample, with replacement, of ten from the two letters H, T. (This would produce a run of letters that might represent ten tosses of a coin that landed heads or tails.)

6. Choose a random sample, without replacement, of four days from the 30 days of November.

7. Repeat question 6 but instead of using a table of random digits to choose the sample use one or more coins to generate the random numbers.

8. The methods described below were intended to select, without replacement, four letters at random from the 26 letters of the alphabet, which we assume are put in correspondence with the numbers 1 to 26 by associating A with 1, B with 2, ..., Z with 26. State whether the method used is valid or invalid; if the latter, explain your reasons.
    (a) From tables we obtained a line of random digits beginning 3, 2, 7, 2, 5, ... From the correspondence of numbers with letters, 3 gave C, 2 gave B, 7 gave G, we rejected '2' since it had already been used, 5 gave E. Hence the sample chosen was C, B, G, E.
    (b) We read pairs of digits from a table of random digits: 87, 63, 16, 40, 12, 89, 88, 50, 14, 81, 06. We rejected 00 and all numbers over 26 and thus used 16, 12, 14, 06, which gave the sample P, L, N, F.
    (c) We used the pairs of digits given in (b) but subtracted 26, or a multiple of 26, if necessary, to obtain a number in the range 1 to 26. We obtained

$87 - 78 = 9$, $63 - 52 = 11$, 16, $40 - 26 = 14$. This yielded the sample I, K, P, N.

(d) We proceeded as in (c) but rejected, from the initial set of numbers, 00 and all that exceeded 78. We therefore obtained $63 - 52 = 11, 16, 40 - 26 = 14, 12$. This gave the sample K, P, N, L.

9. Choose a random sample of 25 pages of this book using random digits. Find the proportion of pages in the sample that have exercises on them, and hence estimate the proportion of pages in the book that have exercises on them.

10. A telephone directory contains 720 pages of names and there are *approximately* 300 names to a page. A name is chosen from the directory by choosing a number at random in the range 1 to 720 to determine the page. Then a three-digit random number is used to determine the name on that page; the names are numbered from the top left-hand corner. The name corresponding to the appropriate number is chosen, unless the number is too large in which case a new random number is chosen and the selection of the name from the page repeated.

(a) Does this give a random sample of the names in the directory?

(b) If not, is the method likely to be satisfactory for practical purposes?

11. Comment on the following suggestions for obtaining a random sample. Give a better method if you can.

(a) A random sample of all the adults in a town is required: stand outside the largest supermarket in the town and stop every 10th person who leaves.

(b) A random sample of the children cared for by one doctor is required: choose ten cards at random from the doctor's file of cards (one per family) and pick a child at random from those in each family; if there is no child in the family, select another card.

(c) A random sample of the children in a school is required for an investigation: ask children to sign up if they would like to take part in a 'useful investigation' and promise to pay each chosen £1; choose at random from the volunteers until the required number is obtained.

3.11 COMPUTING EXERCISES

1. *Pebble sampling.* In a repeat of Dr Yates' experiment (Section 3.2), 100 pebbles were spread out on a table and participants were asked to select:

(a) ten pebbles which appeared to be representative of the population of pebbles, and

(b) a random sample of ten pebbles from the population.

The mean weights (g) of the samples chosen by 50 participants are given below:

(a) *'Representative' (or 'judgement') sample*

39.85, 27.31, 50.04, 58.09, 44.00, 40.07, 68.29, 61.71, 47.91, 32.13, 53.27, 53.96, 45.40, 45.07, 41.90, 41.46, 36.47, 39.40, 57.69, 41.41, 34.34, 53.18, 34.60, 48.24, 32.78, 33.26, 37.45, 36.78, 35.28, 34.35, 51.50, 49.69, 48.15, 44.03, 64.97, 47.86, 56.64, 37.72, 35.15, 40.84, 46.57, 40.93, 27.28, 28.10, 64.18, 56.64, 63.30, 55.32, 64.02, 47.66.

(b) *Random sample*

25.58, 37.44, 27.04, 34.35, 50.60, 42.16, 37.41, 56.48, 33.47, 42.86, 44.80, 48.96, 20.90, 37.43, 44.04, 43.27, 38.59, 26.69, 54.78, 41.00, 33.31, 28.90, 40.32, 49.21, 24.49, 52.50, 42.86, 40.57, 57.79, 26.07, 35.00, 34.65, 27.11, 40.22, 29.09, 40.58, 29.41, 44.60, 47.66, 32.37, 28.27, 34.37, 40.90, 32.78, 34.46, 34.82, 52.73, 48.30, 24.60, 26.73.

Compare the two sets of data by constructing diagrams (e.g. histograms, dot plots) and calculating descriptive statistics (e.g. means, medians). Decide which is the more satisfactory method of sampling the pebbles. (The true mean weight of the 100 pebbles was 37.659 g.)

2. Generate 100 pseudo-random digits in the range 0–9. Construct a bar diagram of the data and assess whether the data appear to have a random distribution.

3. Generate 100 pseudo-random numbers in the range 0–1. Construct a histogram of the data and assess whether the data appear to have a random distribution.

4. Choose 100 random samples of size 4, with replacement, from the weights of 20 female students listed in Exercise 5.6(1). Construct a histogram of the 100 sample means and compare the mean of the distribution of sample means with the mean of the 20 weights.

3.12 PROJECTS

1 Mean length of pencils

Aim. To compare the estimates of the mean length of a population of pencils derived by two methods of sampling.

Equipment. A collection of 40 pencils of varying sizes.

Collection of data. Choose a sample of 5 pencils by: (a) selecting pencils that are representative of the whole population; (b) random sampling using random digits. Record the lengths of the pencils in each sample to the nearest millimetre. Repeat, preferably allowing no person to choose a representative sample more than once, until at least 30 samples of each type are obtained.

Analysis. Find the mean of each sample of 5 pencils. Compile frequency tables for the means obtained by each method. Draw a histogram and calculate the mean of each distribution. Compare the distribution of the sample estimates with the true mean.

2 Sampling from a map

Aim. To estimate: (a) the number of churches; (b) the total length of classified road; over the area covered by an Ordnance Survey map.

Equipment. 1:50 000 Ordnance Survey map.

Collection of data. Let the sampling unit be a 2 × 2 kilometre square (1 × 1 kilometre squares are marked on the map) and let the sample size be 10% of the total number of 2 × 2 squares. Use random number tables to choose a pair of coordinates which will specify a square to be included in the sample. Count the number of churches a_i, and measure the total length of classified road x_i (the classified roads are coloured brown or red or blue) for square i ($i = 1, \ldots, n$) in the sample.

Analysis. Let the number of squares in the population be N. The estimate of the number of churches is $(\sum_{i=1}^{n} a_i)N/n$. The estimate of the total length of classified road is $(\sum_{i=1}^{n} x_i)N/n$.

SUMMARY

Populations of interest are usually investigated by taking samples from them. To gain reliable information a random sample should be chosen.

A **simple random sample** is a sample which is chosen so that every member of the population is equally likely to be a member of the sample, independently of which other members of the population are chosen.

When a sampling unit is drawn from a *finite* population and returned to the population before the next unit is drawn, the sampling is said to be **with replacement**. Otherwise the sampling is said to be **without replacement**.

In order to choose a random sample a numbered list is made of the units in the population under investigation. A random sample of the units is chosen with the help of either a table of random numbers, or a list of pseudo-random numbers generated by a computer.

Systematic samples are often more convenient than completely random samples. It must be shown that the order in the list which is systematically sampled is effectively random and the first unit of the systematic sample must be chosen in a random way.

4 THE MEASUREMENT OF VARIABILITY

OVERVIEW

Statistics may be defined as the study of variability. If there were no variability there would be no need for the science of statistics. We look at ways of measuring the variability of populations and sets of observations.

4.1 INTRODUCTION

In Chapter 2, we looked at ways of measuring the centre of a set of observations. That is only part of the story which the observations can tell us. We accept that they *will* vary, from one observation to another in the same set; we now ask how this variability can be measured.

A doctor investigated the effect of lack of sleep on the ability of people to do simple tasks. One of the tasks he chose for his subjects was to press a button as quickly as possible after a red light came on; the response-time was recorded, in hundredths of a second, by automatic equipment wired into the light circuit. For each subject, he measured five response-times when the subject was fully rested; a typical set for one subject was 61, 52, 55, 58, 54 hundredths of a second. He also measured five response-times when the subject had been awake for 24 hours; the set for the same subject was 60, 87, 53, 73, 82 hundredths of a second. The mean of the first set (when rested) is 56, and of the second set (awake 24 hours) is 71 hundredths of a second. The mean of the second set is larger than that of the first, though the two sets of times do overlap. What struck the doctor as interesting was that the second set of response-times differed more from one another than did the first. It seems that the subjects may become more erratic when they are tired. There may be an increase in response-times too as the subjects become tired, but it is the increase in variability which is even more noticeable.

4.2 RANGE

One way to measure this variability is simply to look at the highest and the lowest of the five observations in a set, and calculate the difference between them.

Definition 4.1

The **range** of a set of observations is the difference in values between the largest and smallest observations in the set.

For the first set of response-times recorded above, the range is $61 - 52 = 9$ hundredths of a second, while for the second set it is $87 - 53 = 34$ hundredths of a second. The difference between these ranges is very large. In small sets of observations such as these, the range can be a useful measure of variability. It uses only two of the observations, however, the lowest and highest in the set. It ignores the pattern of distribution of the observations in between, and can thus be relatively uninformative in large sets of data. Since it is usually difficult to develop mathematical properties of the range, we shall not be able to use it much in the later chapters that deal with statistical theory. However, range is used as a measure of variability in the application of *control charts*, where samples are usually very small; this is discussed in Section 16.6.

4.3 MEAN DEVIATION

A better approach, having found the mean $\bar{x}$ of a set of observations, is to measure variability by seeing how closely the individual observations cluster round $\bar{x}$. We can look at the average (i.e. the arithmetic mean) distance of an x_i from $\bar{x}$. For the five records in the first set above, $\bar{x} = 56$, and so the deviations $(x_i - \bar{x})$ are $+5, -4, -1, +2, -2$. The sum of these deviations is zero; this must always be so in any set of data since it follows from the definition of $\bar{x}$ that $\sum_{i=1}^{N}(x_i - \bar{x}) = 0$ (Example 2.3, p. 31). We shall therefore use the *absolute* value (i.e. the value ignoring the sign) of each deviation to form an average. The mean deviation of the set of observations is $\frac{1}{5}(5 + 4 + 1 + 2 + 2) = \frac{14}{5} = 2.8$. For the second set of response-times, $\bar{x} = 71$ and the deviations $(x_i - \bar{x})$ are $-11, +16, -18, +2, +11$; the mean deviation is thus $\frac{1}{5}(11 + 16 + 18 + 2 + 11) = \frac{58}{5} = 11.6$.

In the algebraic formula for mean deviation, we need a symbol to represent the operation 'take the absolute value of ...'. The appropriate symbol is the modulus sign; thus $|x|$ is called 'the modulus of x', and its value is the absolute value of x. For example, $|-2| = 2, |3| = 3$.

Definition 4.2

The **mean deviation** of a set of observations $x_1, x_2, \ldots, x_N$ is the mean of the absolute deviations from the mean and equals

$$\frac{1}{N}\sum_{i=1}^{N}|x_i - \bar{x}|.$$

This is a measure of variability that has been used in the past, but is little used now. Its virtues are that it uses all the observations and it is relatively easy to calculate. Unfortunately, it is even more difficult to develop any mathematical theory for the mean deviation than it is for the range, and nowadays there is no premium on easy computations.

An alternative definition of mean deviation uses deviations about the median, $x_i - M$, instead of about the mean.

4.3.1 Exercises

1. (a) The blood pressures of a group of students were measured. They were 118, 135, 128, 132, 124, 137. Find the range and the mean deviation of these figures.
   (b) A group of managers also had their blood pressures measured. The records for these were 129, 138, 135, 158. Calculate the range and the mean deviation in this group.

2. Find the range of the observations 5, 8, 3, 2, 7, 4, 1, 6, and also of the observations 10, 10, 17, 17, 11, 16, 17, 10. Find the mean deviations in the two sets of observations. Which of the two measures (range or mean deviation) gives the better idea of variability? (It may help to draw a dot diagram.)

4.4 VARIANCE AND STANDARD DEVIATION

Pursuing the idea of measuring how closely a set of observations cluster round their mean, let us square each deviation $(x_i - \bar{x})$ instead of taking its absolute value. This, again, gives us a set of positive quantities. For the first set of response-times in Section 4.3, we have:

Deviation $(x_i - \bar{x})$:	+5	−4	−1	+2	−2
Squared deviation $(x_i - \bar{x})^2$:	+25	+16	+1	+4	+4

The next measure of variability that we are going to define, the *variance*, is the most commonly used one: it is the mean of the squared deviations. For the first set of response-times, then, the variance is $\frac{1}{5}(25 + 16 + 1 + 4 + 4) = \frac{50}{5} = 10$. A similar calculation for the second set of response-times gives their variance as 165.2. Again, using this measure of variability, there is a large difference between the two sets of records.

Definition 4.3

The **variance** of a set of observations $x_1, x_2, \ldots, x_N$ is the average of the squared deviations from their mean and equals

$$\frac{1}{N} \sum_{i=1}^{N} (x_i - \bar{x})^2.$$

The first step in working out the variance is to calculate the corrected sum of squares $\sum_{i=1}^{N} (x_i - \bar{x})^2$, either directly or from the formula

$$\sum_{i=1}^{N} x_i^2 - \frac{(\sum x_i)^2}{N}$$

(see p. 32). In words, this is 'the sum of the squares *minus* the total squared and divided by N'. This corrected sum of squares is then divided by the number of observations.

As is clear from the definition, the variance is measured in units of x^2 rather than x. This is sometimes a nuisance as well as making it harder to understand what the size of the variance is telling us, and to compare variability in two sets of observations. We therefore define a measure of variability which is closely related to the variance but is expressed in the same units as the observations.

Definition 4.4

The **standard deviation** is the positive square root of the variance, and equals

$$\sqrt{\frac{1}{N} \sum_{i=1}^{N} (x_i - \bar{x})^2}.$$

The standard deviations of the first and second sets of response-times are 3.16 and 12.85 hundredths of a second respectively. A small standard deviation tells us that the observations cluster closely round their mean, while a large standard deviation says that the observations are much more scattered. The standard deviation is a very commonly used measure of variability, even though it takes a little effort to compute. It does use all the observations, and because the variance can be studied mathematically, it is possible to develop theoretical results involving the standard deviation.

Example 4.1 Mean, variance and standard deviation of data

In a city there are six professional football clubs. Last season they had 25, 30, 18, 27, 28 and 22 players respectively on their full-time paid staffs. Find the mean, variance and standard deviation of the number of full-time paid staff.

Let us call the number of full-time paid staff r. (We use r since we are dealing with a discrete variate. Because the formulae developed earlier in the chapter are quite general, we wrote them in terms of x.) It is easiest to lay out the calculation in the form of a table such as Table 4.1.

Table 4.1 *Calculation of mean, variance and standard deviation of data*

Club	r_i	$r_i - \bar{r}$	$(r_i - \bar{r})^2$	r_i^2
A	25	0	0	625
B	30	5	25	900
C	18	−7	49	324
D	27	2	4	729
E	28	3	9	784
F	22	−3	9	484
	6⌊150		6⌊96	3846
	Mean $\bar{r} = 25$		16 = variance	

Hence the standard deviation is 4.

The mean will not usually work out to be a whole number, and if it does not then the values of $(x_i - \bar{x})$ are awkward to deal with. For hand calculation it is usually more convenient to use the identity that the corrected sum of squares equals the raw sum of squares minus the correction term (p. 32). In Table 4.1, $\sum_{i=1}^{6} r_i^2 = 3846$, $N = 6$, $\sum_{i=1}^{6} r_i = 150$, so the variance is computed as

$$\frac{1}{6} \sum_{i=1}^{6} (r_i - \bar{r})^2 = \frac{1}{6} \left[\sum_{i=1}^{6} r_i^2 - \left(\sum_{i=1}^{6} r_i \right)^2 \bigg/ 6 \right]$$

$$= \frac{1}{6}(3846 - 150^2/6) = \frac{1}{6}(3846 - 3750) = 16.$$

Unfortunately this is not a good method when using a computer. We consider the calculation of variance by computer in Example 4.2; the other examples in this chapter assume that we are calculating variance by hand.

Example 4.2 Calculation of mean, variance and standard deviation by computer

Give a flow diagram for the calculation of mean, variance and standard deviation by computer.

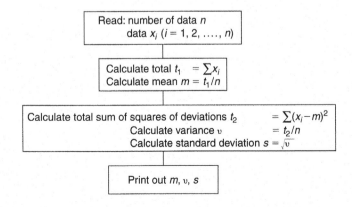

Loops would be used to read in the data in block 1 and to calculate the totals in blocks 2 and 3.

This diagram describes the best simple method of calculating a variance by computer. The method which depends on subtracting a correction term (often the best method for hand calculation) is a bad method for computer calculation. Subtracting the correction term from the uncorrected sum of squares $\sum x_i^2$ may lead to a substantial loss of significant figures by taking the difference between two numbers, both of which are large. In hand calculation this error would usually be recognized and avoided (though be careful with hand-held calculators which have a key for variance calculation). In computer calculation the error can easily go unnoticed. A fuller discussion of calculating the variance by computer is given in *CCC*, Chapter 6.

4.4.1 Exercises

1. For each of the following sets of observations, calculate the variance: (i) by using the definition $(1/N) \sum_{i=1}^{N} (x_i - \bar{x})^2$; (ii) from the formula

$$\frac{1}{N} \left\{ \sum_{i=1}^{N} x_i^2 - \frac{1}{N} \left(\sum_{i=1}^{N} x_i \right)^2 \right\}$$

(a) $1, 1, 4$;   (b) $3, 6, 9, 12, 15$;   (c) $3, 0, -3, 6, -6$;

(d) $2, 2, 2, 8, 8, 8$;   (e) $-1, -1, -4$;   (f) $\frac{1}{4}, \frac{1}{2}, \frac{3}{4}$.

Calculate also the standard deviation for each set.

2. The heights of six people are 172, 169, 174, 162, 161, 164 cm. Find the mean and the variance of these observations. The first three were males, the others females. Find the mean and the variance for the males alone, and for the females alone.

3. Two different strategies in playing darts are: (i) to aim always for the bull (the centre of the board); (ii) to aim always for the treble twenty (the top of the board). Fifteen scores of a player using strategy (i) were 56, 36, 52, 23, 32, 23, 25, 19, 49, 60, 36, 36, 56, 20, 20, and fifteen scores of the same player using strategy (ii) were 33, 67, 2, 24, 28, 13, 11, 33, 8, 20, 25, 65, 92, 36, 49.

Draw two parallel lines on a sheet of graph-paper; mark the scores on strategy (i) on one line and those on strategy (ii) on the other, and make a visual comparison of the scatter in the two groups. Calculate the range, the mean, the variance and the standard deviation for each group.

4.5 VARIANCE AND STANDARD DEVIATION OF A GROUPED FREQUENCY DISTRIBUTION

We can rewrite the variance formula for use with a frequency table, in the same way as we did for the mean in Chapter 2. Let the table of values of the discrete variate be:

Variate value:	r_1	r_2	r_3	...	r_k	Total
Frequency:	f_1	f_2	f_3	...	f_k	$\sum_{i=1}^{k} f_i = N$

The corrected sum of squares is $\sum_{i=1}^{k} f_i(r_i - \bar{r})^2$.

Definition 4.5

The **variance** of a set of N observations of a discrete variate, grouped so that the value r_i ($i = 1, 2, \ldots, k$) occurs with frequency f_i, is

$$\frac{1}{N} \sum_{i=1}^{k} f_i(r_i - \bar{r})^2 \quad \text{or} \quad \frac{1}{N}\left(\sum_{i=1}^{k} f_i r_i^2 - \frac{\left(\sum_{i=1}^{k} f_i r_i\right)^2}{N} \right).$$

The expression in the large brackets may be stated in words as 'the sum of the squares of *all* the observations *minus* the total squared and divided by N'.

As the following example shows, the calculation can be set out in the form of a table quite simply.

Example 4.3 Mean, variance and standard deviation of a grouped discrete frequency distribution

In a population of 30 work-people in the same establishment, sickness and absence records were kept daily for a quarter (three months totalling 91 days). The numbers of workers absent each day are shown in the first two columns of Table 4.2. Find the mean, variance and standard deviation of the number absent per day.

The mean is

$$\frac{(\sum f_i r_i)}{N} = \frac{106}{91} = 1.165.$$

The variance is

$$\frac{1}{N}\left(\sum f_i r_i^2 - \frac{(\sum f_i r_i)^2}{N} \right) = \frac{1}{91}\left(318 - \frac{106^2}{91} \right)$$

$$= \frac{194.53}{91} = 2.1377.$$

Table 4.2

Number of people absent, r_i	Number of days, f_i	$f_i r_i$	$f_i r_i^2$
0	44	0	0
1	19	19	19
2	10	20	40
3	8	24	72
4	7	28	112
5	3	15	75
6 or more	0	0	0
$\sum_{i=1}^{k} f_i = N = 91$		106	318

The standard deviation is $\sqrt{2.1377} = 1.46$.

The case where continuous data have been grouped into a frequency table is dealt with in basically the same way as the case of a discrete variate which we have just discussed.

Definition 4.6

The **variance** of a set of N grouped observations of a continuous variate, in which f_i observations fall in the interval whose centre is x_i $(i = 1, 2, \ldots, k)$, is

$$\frac{1}{N} \sum_{i=1}^{k} f_i (x_i - \bar{x})^2 \quad \text{or} \quad \frac{1}{N} \left(\sum_{i=1}^{k} f_i x_i^2 - \frac{\left(\sum_{i=1}^{k} f_i x_i \right)^2}{N} \right).$$

4.5.1 Relation between variance of coded and original data

If we use the coding $x_i = a + b u_i$, so that the original observation x_i is coded into u_i, we have seen (Section 2.6) that $\bar{x} = a + b\bar{u}$.

Now

$$x_i - \bar{x} = (a + b u_i) - (a + b\bar{u})$$
$$= b(u_i - \bar{u}).$$

Hence the variance of x is

$$\frac{1}{N} \sum_{i=1}^{k} f_i (x_i - \bar{x})^2 = \frac{1}{N} \sum_{i=1}^{k} b^2 f_i (u_i - \bar{u})^2$$

$$= \frac{b^2}{N} \sum_{i=1}^{k} f_i (u_i - \bar{u})^2.$$

Thus

$$(\text{variance of } x) = b^2 (\text{variance of } u).$$

Therefore, after finding the variance of the coded variate u, we multiply by b^2 to obtain the variance of the original variate x. Similarly, the standard deviation of x is equal to b times the standard deviation of u.

A similar relation between the variance of the coded and original variates holds when the data are grouped.

4.5.2 Exercises

1. The number of goals scored per game by a football player during 1975–76 were as follows:

Number of goals	0	1	2	3	4 or more
Number of games	23	14	3	2	0

Calculate the mean, variance and standard deviation of the number of goals per game.

2. A grass lawn is marked out in squares, and 100 of these squares are examined; the number of clumps of daisies in each square is recorded. The results are summarized in the following table:

Number of clumps	0	1	2	3	4
Number of squares	56	29	7	5	3

Find the variance and standard deviation of the number of clumps per square. Find also the range. Which measure gives more information about the scatter among the records?

Find also the mean number of clumps per square, the median and the mode. Which of these is the most useful summary measure of central tendency in the records?

3. The heights of 100 men are measured in centimetres as follows:

Height	Frequency
Less than 168	5
168–	10
173–	24
178–	29
183–	19
188–	10
193 or more	3

Find the variance and the standard deviation of height. If height had been measured in inches instead of centimetres, what would have been the variance and standard deviation? (1 inch = 2.54 cm.)

4. Two dice were thrown 100 times. The number of times each total score $2, 3, \ldots, 12$ occurred is shown in the following table:

Score	2	3	4	5	6	7	8	9	10	11	12
Frequency	2	5	8	11	14	16	15	10	9	6	4

Find the variance and the standard deviation of the total score.

5. A hundred samples of a certain machine part were taken from a large consignment, and the lifetime of each part was measured. The results obtained are summarized in the following grouped frequency table:

Lifetime (hours)	0–	5–	10–	15–	20–	25–30
Number of parts	37	20	17	13	8	5

Plot a histogram and calculate the mean, variance and standard deviation of the lifetime. Find the lifetime values equal to: (a) mean plus one standard deviation; (b) mean plus two standard deviations. Estimate graphically the proportions of parts having lifetimes in excess of each of these values.

4.6 THE INTER-QUARTILE RANGE (OR MID-SPREAD)

The variance, the standard deviation and the mean deviation go naturally with the arithmetic mean. They are based on deviations from the arithmetic mean, and the averaging process used is the same as that for calculating the mean. In this section we look at a measure of variability that is based on the rank order of a set of observations, and is therefore related to the median. (The range, which we mentioned briefly on p. 63, is also based on the rank order of a set of observations.)

An airline flies a daily service between two cities, using a 100-seat aircraft for the flight. The numbers of seats sold on the first fifteen days of September are $87, 67, 98, 57, 74, 100, 83, 60, 99, 88, 54, 72, 78, 75, 93$ in date order. Let us arrange these observations in rank order:

Rank order	1	2	3	4	5	6	7	8	9	10	11	12	13	14	15
Number of seats sold	54	57	60	67	72	74	75	78	83	87	88	93	98	99	100
	.	.	.	Q_1	.	.	.	M	.	.	.	Q_3	.	.	.

The median divides the rank order into two equal parts: it lies at the eighth observation and so has the value 78. We now find the middle value of each of the two parts of the rank order. There are seven observations smaller than the median, so the middle of the lower part falls at the fourth observation, having the value $Q_1 = 67$. Similarly, the middle of the upper part is at the twelfth observation, having the value $Q_3 = 93$. Q_1 is called the *first* or *lower quartile* and Q_3

is the *third* or *upper quartile*. We could (though we hardly ever do!) call the median M the second quartile and label it Q_2. The point is that Q_1, M, Q_3 divide the distribution of ranked observations into quarters, and this is the reason for calling them quartiles. The difference between the upper and lower quartiles is $Q_3 - Q_1 = 93 - 67 = 26$. This is the *inter-quartile range*. (It is sometimes called the *mid-spread*: see p. 92.) It is a measure of the spread in the ranked distribution because it is the length of the interval which contains the central 50% of the observations. Often it is $\frac{1}{2}(Q_3 - Q_1)$, the *semi-inter-quartile range*, which is used as the measure of variability; in our example this has the value 13.

Definition 4.7

The **quartiles** of a set of observations are the values below which fall 25%, 50% and 75% of the observations as arranged in rank order. These are called respectively the first, second and third quartiles and are denoted by Q_1, Q_2 or M, and Q_3.

Definition 4.8

The **inter-quartile range** of a set of observations equals $Q_3 - Q_1$. An alternative name is *mid-spread*.

4.6.1 Exercise

In the first fifteen days of December in the same year, on the same air service as described above, the numbers of seats sold were $38, 75, 50, 84, 62, 46, 96, 67, 33, 55, 65, 42, 83, 70, 49$ in date order. Find the median, quartiles and semi-inter-quartile range for these observations. Plot September and December observations on a sheet of graph-paper, mark in the measures you have calculated, and compare them. Work out also the range, mean deviation and standard deviation for each month's data.

4.6.2 Calculation of quartiles using depths

A different approach to the calculation of quartiles comes from using Tukey's concept of 'depth' introduced on p. 8: we recall that 'depth' is a measure of how deep an observation is in a set of observations that have been arranged in rank order. The two extreme values both have depth 1. The median of a set of N observations has depth $\frac{1}{2}(N+1)$. Thus for a set of 9 observations the median has depth $5 = \frac{1}{2}(9+1)$; for a set of 10 observations it has depth $5\frac{1}{2} = \frac{1}{2}(10+1)$, where we interpret depth $5\frac{1}{2}$ to mean the average of the two observations with depths 5 and 6. This agrees with our definition above.

Tukey defines the 'hinges' of a set of data as those observations that have depth $\frac{1}{2}(1 + \text{depth of median})$, where 'depth of median' is quoted as an integer; if the depth of the median includes $\frac{1}{2}$ the fraction is ignored. These hinges are the first and third quartiles, though the values obtained will be slightly different from those found by the method we have outlined above, and which we make precise in Example 4.4. For the airline data, with 15 observations, the median has depth $\frac{1}{2}(15+1)=8$ and the hinges have depth $\frac{1}{2}(1+8)=4\frac{1}{2}$. This would give values $\frac{1}{2}(67+72)=69.5$ and $\frac{1}{2}(88+93)=90.5$, by contrast with the values 67 and 93 which we calculated. Neither method can be claimed 'more correct' than the other; there is an element of arbitrariness in calculating quartiles.

4.6.3 Box-and-whisker plot

J. W. Tukey introduced the box-and-whisker plot as well as the stem-and-leaf plot (p. 8). Consider the 'seats sold' data in Section 4.6. On a linear scale (Fig. 4.1) vertical marks are made corresponding to the minimum, the first quartile, the median, the third quartile and the maximum; namely at the values 54, 67, 78, 93 and 100.

The lines for the first and third quartiles are then joined to make a box corresponding to the central 50% of the observations, and whiskers are taken out from the box to the maximum and minimum values. This is often a convenient way of representing a distribution and is particularly useful for comparing distributions.

Figure 4.1 *Box-and-whisker plot for data of Section 4.6.*

4.7 QUANTILES

If we have a very large mass of data, we might want to divide it into more than four parts to make a summary of it. It is often helpful to draw a cumulative frequency curve, i.e. to plot the cumulative frequency (p. 9) in the vertical direction against increasing values of the variate plotted horizontally. (Such a curve is often called an *ogive* to describe its general shape.) From this, required results can be read off.

The median and quartiles divide a set of ranked observations into four parts. Two other sets of measures, *deciles* and *percentiles*, are useful.

Definition 4.9

The **deciles** of a set of ranked observations on a variate are the variate-values which divide the set into 10 equal parts.

Definition 4.10

The **percentiles** of a set of ranked observations on a variate are the variate-values which divide the set into 100 equal parts.

More generally, we can consider measures which divide a set of ranked observations into q parts; one of these measures will be the pth q-tile, where p takes one of the values $1, 2, \ldots, q-1$. Thus for the quartiles $q = 4$ and $p = 1, 2, 3$. The general term for these measures is *quantiles*; the quartiles, deciles and percentiles are examples.

Example 4.4 Calculation of quantiles of a set of observations

Find a general formula for the pth q-tile of a set of n observations. Calculate (i) the median; (ii) the first quartile; (iii) the seventh decile, of the 13 observations: 15, 34, 7, 12, 18, 9, 1, 42, 56, 28, 13, 24, 35.

First of all put the observations in rank order. Two cases must be considered, depending on whether np/q is, or is not, an integer.

(a) np/q is an integer. Let $r = np/q$; consider the rth observation in rank order, x_r, and the next observation x_{r+1}. For a value between x_r and x_{r+1}, the frequency to the left equals np/q and that to the right equals $n - np/q$ or $n(q-p)/q$. The pth q-tile may be taken to be anywhere between x_r and x_{r+1}. The convention is to take it equal to $(x_r + x_{r+1})/2$.

(b) np/q is not an integer. The pth q-tile will fall on an observation in this case; let it be x_r. We require that r satisfies the inequalities

$$r - 1 < np/q \quad \text{and} \quad r > np/q.$$

If we choose

$$r = \text{INT}(np/q) + 1,$$

where $\text{INT}(x)$ denotes the integer part of x, we find the inequalities are satisfied.

We may derive a general formula to cover the two cases. Using the formula in (b) to calculate the pth q-tile from the left, we obtain x_{r_1}, where

$$r_1 = \text{INT}(np/q) + 1.$$

Applying the same formula to calculate the pth q-tile from the right, we find that it is at x_{r_2}, where

$$r_2 = n - \text{INT}(n(q-p)/q).$$

If np/q is not an integer then $r_1 = r_2$ and the two calculations provide a useful check on each other. If np/q is an integer then the pth q-tile is between x_{r_1} and x_{r_2}. Thus a general formula covering both cases is that the pth q-tile equals $(x_{r_1} + x_{r_2})/2$, where r_1 and r_2 are defined as above.

We first rank the data in this example, for which $n = 13$;

Rank	1	2	3	4	5	6	7	8	9	10	11	12	13
Observation	1	7	9	12	13	15	18	24	28	34	35	42	56

(i) $p=1$, $q=2$. $np/q=6\frac{1}{2}$, so $r=\text{INT}(6\frac{1}{2})+1=7$ gives the median: it is the seventh value from the beginning (and also from the end), having the value 18.
(ii) $p=1$, $q=4$. $np/q=3\frac{1}{4}$, so $r=\text{INT}(3\frac{1}{4})+1=4$ gives the first quartile as the fourth value from the beginning, namely 12.
(iii) $p=7$, $q=10$. $np/q=9\frac{1}{10}$, so $r=10$ gives the seventh decile as the tenth observation from the left, which is 34. The calculations in (ii) and (iii) check from the right also.

The general formula above gives a convenient method for a computer program to calculate a pth q-tile (see *CCC*, Chapter 3).

Example 4.5 Deciles and quartiles of a grouped continuous variate

The incomes of married couples over retiring age in 1973 (as reported in the UK government publication *Social Trends*, 1975) are shown in columns 1 and 2 of Table 4.3. Draw a cumulative frequency curve for the data, and from it estimate the lowest decile, the median, the lower and upper quartiles of income. Check the result for the median by direct calculation. Use the curve to estimate the proportion of married couples who had a gross weekly income between £22 and £28.

The first step is to add the third column to the table. Then these values of the cumulative frequency are plotted against the upper end-points of the class-intervals of the income distribution. The end of the first interval is £13.99, so that the cumulative frequency for income £13.99 is 144. There were no incomes below £12, so we set the cumulative frequency at £11.99 equal to zero. We add the remaining points from the table, as shown by the circles in Fig. 4.2. We could, as in Fig. 1.4 on p. 10, join these by straight lines. But we are going to use the resulting diagram to make inferences about a large population of figures, and there seems no reason to suppose that income varies in a discontinuous way through this population. It is much more intuitively reasonable to assume that the income variation is continuous, and so to put the best smooth curve we can through the circles on the diagram. This is the *cumulative frequency curve* that the question asked for.

Table 4.3 *Gross weekly income of married couples (persons over retiring age in 1973)*

Income (£)	Number of married couples (thousands)	Cumulative frequency (thousands)
12–	144	144
14–	330	474
16–	329	803
18–	247	1050
20–	412	1462
25–	206	1668
30 or over	391	2059

There is a problem when we reach the last interval: no upper limit is quoted for it. Remembering what the data are, it is quite likely that there will be a few very large values in them. In the present problem, as we shall see, this does not affect any of the measures we want to calculate. But in order to complete the cumulative frequency curve we will take a working upper limit of £50; the shape of the curve then appears reasonable.

The total frequency, in thousands, is 2059. One-tenth of the total frequency must lie below the first decile. Thus, from the graph we need to find the income corresponding to 205.9 on the vertical scale: it is £14.40 as accurately as we can read it. The first decile is thus £14.40.

From the graph, the median has to have half the total frequency, i.e. 1029.5, below it. The income corresponding to this is £19.80. To check this by calculation, we find, in exactly the same way as for Example 2.2, that

$$M = 17.99 + \frac{226.5}{247} \times 2 = £19.82.$$

The quartiles correspond to cumulative frequencies of $\frac{1}{4} \times 2059 = 514.75$ and $\frac{3}{4} \times 2059 = 1544.25$; they are therefore, from the graph, $Q_1 = £16.20$ and $Q_3 = £26.20$.

Finally, from the graph the cumulative frequency up to £22 is 1230 thousands, and up to £28 is 1590 thousands, so that the number of married couples having incomes between £22 and £28 is 360 thousands. This is a proportion $360/2059 = 0.175$ (i.e. $17\frac{1}{2}\%$) of the whole.

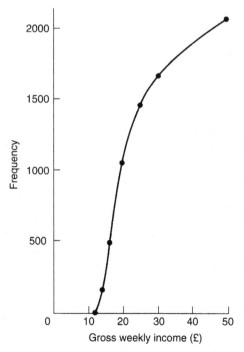

Figure 4.2 *Cumulative frequency curve for income of married couples.*

We rarely calculate percentiles in this way, though it is useful to have a word which includes deciles, quartiles and median or any other desired subdivision of the whole: Q_1, M and Q_3 could be called the 25th, 50th and 75th percentiles, while the first, second, ..., ninth deciles could be called the 10th, 20th, ..., 90th percentiles. In statistical inference, as we shall see later, percentiles such as the 95th or 99th can be useful.

4.7.1 Exercises

1. Calculate the median and the semi-inter-quartile range of the following data, on the duration of unemployment. Do this first by drawing a cumulative frequency curve and then check by direct calculations. Explain why the median and the semi-inter-quartile range would be preferred to the mean and standard deviation for this distribution.

Unemployment in Great Britain, July 1974

Duration in weeks	Number of unemployed (thousands)
Up to 1	53
Over 1 and up to 2	41
Over 2 and up to 3	28
Over 3 and up to 4	21
Over 4 and up to 8	57
Over 8 and up to 13	43
Over 13 and up to 26	65
Over 26 and up to 39	37
Over 39 and up to 52	24
Over 52	112

Suggest why this particular set of class-intervals has been used.

2. The times spent waiting for a mechanic to attend to a machine breakdown, in a factory containing a large number of machines, were as shown in the following table. Calculate the mean and standard deviation of the time spent waiting. Would you expect the median and semi-inter-quartile range to give similar information to the mean and standard deviation? Why?

Waiting time, t (minutes)	Number of breakdowns	Waiting time, t (minutes)	Number of breakdowns
0	8	$2 < t \leqslant 3$	24
$0 < t \leqslant \frac{1}{2}$	15	$3 < t \leqslant 4$	14
$\frac{1}{2} < t \leqslant 1$	20	$4 < t \leqslant 5$	8
$1 < t \leqslant 1\frac{1}{2}$	36	$5 < t \leqslant 7$	3
$1\frac{1}{2} < t \leqslant 2$	32		

4.8 MINITAB: THE TWO VARIANCES; BOX-AND-WHISKER PLOTS

If MINITAB is used to calculate the 'descriptive statistics' of the football staff data on page 66, it gives a value of 19.2 for variance and not 16 as we calculated in Table 4.1. The quantity we have introduced in this chapter is more fully described as the 'population variance' whereas MINITAB calculates the 'estimated (population) variance' or 'sample variance'. The root of the distinction is that, in this chapter, we assume the set of data under study is the complete population, whereas the MINITAB 'variance' assumes the set of data is a sample from a larger population and calculates a value that is a good estimate of the variance of the underlying population. The arithmetical difference between the two quantities is that we use a divisor N while MINITAB is using a divisor $N - 1$. Multiplication of the MINITAB value by $(N - 1)/N$, i.e. 5/6 in this case, gives the value 16 that we calculated.

The distinction between the two 'variances' is important and understanding it is a key step in understanding statistics, but the reader should not worry about it at this early stage. When MINITAB is being used for the exercises in this chapter, remember that an adjustment, using a hand-calculator or the computer's calculator, may need making. If the variability of two sets of data is being compared it is often the relative size of the measures that matters and not the absolute values.

The reader may be relieved to be told that when, later in the book, we wish to calculate a variance for a set of data it is usually the sample variance that is appropriate and the unadjusted MINITAB value gives us the quantity we require. Note also that since standard deviation is the square root of variance, there is a population and a sample standard deviation. The MINITAB value must be multiplied by the square root of $(N - 1)/N$ to give the population standard deviation.

MINITAB offers an easy method of obtaining a box-and-whisker plot from data in a column. Choose

> **Graph>Boxplot/**
> Under **One Y** choose **Simple/**
> Enter **Graph Variable/**

If the cursor is made to hover over the centre of the box plot, MINITAB displays the values of $Q_1, M, Q_3, Q_3 - Q_1$ and N.

Sets of data in different columns may be compared by choosing

> **Graph>Boxplot/**
> Under **Multiple Y's** choose **Simple/**
> Enter **Graph Variables/**

If the observations from all groups are all in column C1 with group labels in column C2, an appropriate choice is

> **Graph>Boxplot/**
> Under **One Y** choose **With Groups/**
> Enter **Graph Variable** C1 and **Categorical variable** C2/

MINITAB picks out extreme observations, or *outliers*. If any value is so large that it is further than 1.5 times the mid-spread beyond the third quartile then it is marked separately, by an asterisk. The whisker is carried out to the largest value which is not an outlier. Low values are treated similarly.

4.9 EXERCISES ON CHAPTER 4

1. Calculate the mean and standard deviation of the number of letters in the words in this sentence. (*SMP*)

2. A set of observations has mean 20 and standard deviation 4. What is the effect on the mean and standard deviation of: (a) adding 3 to each observation; (b) multiplying each observation by 10; (c) multiplying each observation by 100 and then adding 4; (d) subtracting 12 from each observation and then dividing by 2?

3. Six dice were thrown, and the number of dice showing even scores noted. The process was repeated 100 times, with the following results:

Number of even scores	Number of throws
0	1
1	19
2	36
3	25
4	13
5	4
6	2

Calculate the mean, variance and standard deviation of the number of even scores per throw.

4. The Much Booking village football team, in 200 matches, had scores given in the following frequency table. The number of goals scored by the team in one match is x, and the frequency of each value of x is f.

x	0	1	2	3	4	5	6	7	8	9	10	11	12
f	38	35	28	21	18	19	17	15	5	1	2	0	1

Write down the modal score and find the arithmetic mean, median, variance and standard deviation of x. (*AEB*)

5. For the data of each part (a), (b), (c) of Exercise 2.11(14), calculate the variance and the standard deviation.

6. For the data of each part (a), (b), (c) of Exercise 2.11(13), calculate the variance and the standard deviation, using the same coding as in calculating the mean.

7. Find the range, the mean deviation, the standard deviation and the mid-spread of the observations

$$16, \ 5, \ 7, \ 13, \ 2, \ 9, \ 3, \ 20, \ 13, \ 6, \ 5.$$

Illustrate these data by (i) a dot plot, (ii) a box-and-whisker plot.

8. Derive a formula for the standard deviation of the population of the first N natural numbers, $1, 2, 3, \ldots, N$.

9. Construct a box-and-whisker plot for the data on cuckoo egg lengths given in Example 1.2.

4.10 COMPUTING EXERCISES

1. *Railway ticket office data.* Customers were observed buying tickets at a railway booking office during the morning rush hour. The times taken, in seconds, are tabulated below, classified according to whether cash, credit card or cheque was used. Compare the three distributions by constructing histograms and calculating descriptive statistics. (When using 'cash' data, the reader may wish to work with only the first 100 items.)

Credit card customers

55, 43, 52, 50, 57, 54, 59, 54, 67, 65, 60, 60, 60, 66, 55,
59, 65, 55, 67, 67, 44, 52, 63, 63, 61, 60, 61, 61, 45, 51,
63, 58, 64, 61, 63, 41, 46, 56, 60, 52, 56, 48, 53, 68, 60,
64, 44, 64, 61, 59, 67, 53, 58, 55, 73, 49, 57, 64, 55, 64,
57, 51, 48, 42, 70, 53, 57, 58, 57, 56, 43, 58, 50, 55, 55,
62, 72, 66, 41, 46, 60, 56, 53, 57, 66, 57, 58.

Cheque customers

64, 51, 57, 59, 60, 70, 58, 56, 50, 55, 46, 61, 62, 65, 37,
50, 55, 53, 55, 54, 65, 57.

Cash customers

16, 38, 21, 18, 13, 23, 11, 40, 16, 19, 38, 15, 30, 30, 24,
16, 18, 17, 20, 36, 28, 30, 38, 37, 13, 18, 21, 12, 34, 18,
25, 17, 32, 15, 42, 19, 47, 26, 14, 32, 37, 39, 15, 22, 29,
20, 18, 15, 22, 13, 17, 16, 28, 23, 23, 17, 11, 30, 44, 13,
23, 19, 18, 33, 16, 30, 16, 17, 18, 15, 26, 24, 33, 26, 13,
17, 23, 21, 28, 20, 17, 15, 36, 44, 16, 19, 42, 37, 11, 11,
34, 14, 12, 37, 25, 18, 15, 19, 16, 33, 13, 26, 15, 28, 24,
18, 35, 34, 14, 24, 13, 17, 11, 11, 32, 13, 13, 13, 51, 37,
20, 16, 25, 30, 11, 15, 18, 24, 23, 18, 23, 29, 15, 31, 17,
12, 20, 36, 16, 17, 20, 12, 15, 51, 17, 19, 18, 76, 18, 22,
40, 16, 37, 21, 39, 16, 17, 36, 13, 29, 28, 11, 12, 17, 12,

19, 14, 14, 27, 22, 13, 20, 13, 26, 58, 19, 13, 18, 34, 20,
15, 30, 11, 10, 33, 19, 24, 14, 13, 37, 14, 14, 21, 15, 10,
22, 15, 40, 16, 70, 22, 13, 28, 17, 40, 47, 14, 12, 11, 20,
32, 31, 26, 12, 24, 15, 15, 17, 39, 18, 41, 18, 28, 26, 13,
22, 14, 31, 21, 23, 30, 18, 44, 12, 22, 25, 33, 27, 23, 22,
14, 22, 12, 16, 36, 12, 15, 39, 41, 27, 25, 22, 12, 13, 11,
24, 19, 17, 22, 39, 16, 15, 18, 15, 28, 16, 23, 30, 21, 19,
14, 15, 19, 30, 19, 40, 30, 14, 18, 31, 25, 16, 19, 13, 16,
16, 16, 24, 24, 11, 46, 28, 37, 15, 26, 13, 13, 13, 22, 30,
30, 18, 26, 21, 13, 21, 28, 49, 21, 21, 16, 36, 12, 15, 25,
33, 16, 19, 22, 17, 23, 15, 31, 24, 15, 13, 22, 15, 22, 13,
27, 36, 24, 15, 12, 26, 20, 22, 17, 40, 20, 17, 17, 11, 19,
17, 36, 22, 16, 49, 11, 27, 26, 12, 29, 24, 18, 15, 22, 27,
16, 13, 19, 42, 10, 16, 13, 16, 18, 18, 20, 30, 11, 35, 24,
10, 14, 12, 37, 13, 13, 21, 27, 30, 21, 18, 13, 16, 25, 27,
33, 21, 13, 20, 20, 31, 15, 22, 28, 27, 77, 11, 13, 13, 22,
14, 25, 19, 16, 19, 29, 25, 16, 28, 26, 32, 14, 30, 39, 19,
15, 15, 27, 20, 34, 10, 32, 13, 17, 20, 24, 12, 22, 11, 33,
12, 35, 13, 18, 24, 36, 18, 30, 28, 34, 19, 11, 32, 11, 29,
50, 17, 24, 11, 15, 18, 15, 22, 19, 21, 11, 42, 26, 23, 13,
16, 13, 27.

2. A certain college conducted a sample survey of recent graduates in order to determine how well they had progressed in their careers after leaving the college. Questionnaires were sent by post to a random sample of 50 recent graduates, and about half of these responded. Amongst other things, they were asked to state their current salary. These results, together with the sex of the respondent (1 = male, 2 = female) are given below. Compare the distributions of salaries of men and women.

Respondent number	Sex	Present salary (US $000)	Respondent number	Sex	Present salary (US $000)
1	1	26	12	1	28
2	1	20	13	2	17
3	2	20	14	2	25
4	2	14	15	2	18
5	1	28	16	2	22
6	1	18	17	2	25
7	2	24	18	1	21
8	2	18	19	2	29
9	2	19	20	1	29
10	1	22	21	2	21
11	2	19	22	2	25

3. Choose a random sample (*with* replacement) of 10 values from the data on number of words in the first 100 sentences of the *Origin of Species* (Revision Exercises A(4)) (p. 100). Repeat this 100 times and compare the distribution of sample means with the distribution of the original values. In particular, compare the means, variances and shapes of the two distributions.

4.11 PROJECTS

1 Strategy at darts

Aim. To compare the mean and variance of the scores obtained at darts by aiming at: (a) the top of the board; (b) the middle of the board.

Equipment. Dart board and three darts.

Collection of data. The two strategies used are: (A) aiming at the treble 20; (B) aiming at the bull. Use random digits to construct a list of the pairs of letters AB occurring five times in random order, e.g. ABBABAABAB. This gives the order in which the strategies will be used; the random element is introduced so that neither strategy is favoured. Using the first strategy in the list, throw three darts, record the total score, and do this three times. Repeat using the next strategy, and so on through the list. This will give a total of 15 scores (each of them the total of throws with three darts) for each strategy.

Analysis. Mark the 15 values for each strategy on a line, setting the line for one strategy below that for the other, in order to aid comparison. Find: (i) the mean; (ii) the variance of the scores on each strategy; and see if one strategy leads to higher or lower values for each of these measures.

2 Word length comparisons

Aim. To examine whether the length of words used differs from one author to another.

Equipment. Two books of each of two authors of similar date (e.g. Jane Austen and Walter Scott, Lewis Carroll and Charles Darwin).

Collection of data. From each work take samples of at least 250 words. Count the number of letters in each word. Samples should represent the whole text, but need not be completely random. Consider, and make some rules about, how to deal with abbreviations, hyphenated words and so on.

Analysis. Summarize the four sets of data by suitable tables and diagrams. Find the mean and variance of word length in each set.

Notes. Many variants on this are possible. The variate recorded might be sentence length or frequency of common words such as 'the' or 'a'. Newspapers or magazines might be used instead of books.

If two variates are plotted against each other for a number of samples from each book, then a region is obtained for each author. This often allows easy visual discrimination of authors.

SUMMARY

The most important measure of the variability of a population is the variance. The **variance of a population** of N observations $x_1, x_2, \ldots, x_N$ is defined as

$$\frac{1}{N} \sum (x_i - \bar{x})^2 = \frac{1}{N} \left\{ \sum x_i^2 - \frac{(\sum x_i)^2}{N} \right\}.$$

The **standard deviation** is the square root of the variance.

Quantiles are statistics which divide a population, or a set of observations, into a chosen number of equal parts. The most commonly used quantiles are the three **quartiles** which divide a set into four equal parts: in increasing order the dividers are the first quartile, the median and the third quartile. Deciles divide into 10 equal parts, **percentiles** divide into 100 equal parts.

Other measures of variability are:

(i) the **range**, which is the difference between the largest and smallest observations in a set;

(ii) the **mean deviation**, which equals

$$\frac{1}{N} \sum |x_i - \bar{x}|;$$

(iii) the **mid-spread**, or **inter-quartile range**, which equals the difference between the first and third quartiles. The semi-inter-quartile range, or quartile deviation, which is half of this, may be used instead.

5 LOOKING AT DATA

OVERVIEW

At the beginning of any statistical analysis the data should be carefully scrutinized to check their quality. The next step is to draw simple diagrams, make tables and calculate simple statistics. In this chapter we look at some of the available methods.

5.1 INTRODUCTION

In the early years of the twentieth century a biologist, O. H. Latter, investigated whether the sizes of eggs laid by cuckoos in other birds' nests were related to the species of bird whose nest was chosen. Data on length and breadth of the eggs, together with the species of bird in whose nest the egg was found, were collected and published in 1901 in the first volume of the statistical journal *Biometrika*. We shall use these data to illustrate the way data are looked at by statisticians and research scientists. The data are being used solely to illustrate methods of investigation; we are not making a serious analysis of them.

The ideas presented in this chapter owe a great deal to the American statistician, J. W. Tukey, who developed his thoughts under the rubric *exploratory data analysis* (EDA), and the British statistician, C. Chatfield, who although concerned with closely related ideas referred to *initial data analysis* (IDA). Their work is summarized in Tukey (1977) and Chatfield (1995). Tukey and Chatfield both set out to make explicit, and extend, the practices developed by experienced statisticians when they were beginning an analysis: scanning data, making diagrams and working out simple statistics. This process might be described as 'getting a feel for the data'.

Chatfield picks out three main stages in IDA:

1. data preparation and scrutiny,

2. data description,

3. model formulation.

In this chapter we consider the first two stages.

5.2 DATA PREPARATION AND SCRUTINY

5.2.1 Structure of the cuckoo egg data

A total of 243 eggs were recorded. For each egg, two quantitative continuous variates (length and breadth, in millimetres) were recorded as well as the qualitative variate 'species of bird in whose nest the egg was found'. We do not think of the three variates in the same way. The set of eggs from nests of one species may, on the average, be different in size from the set of eggs from nests of another species. We could say that the difference in size is *explained* by the species difference. Thus nest species may be called an **explanatory variate**. Length and breadth are called **response variates**. The words derive from usage in experiments. If we put some plants in a warm environment and similar plants in a cold environment and record their growth, we could use the height of the plants after a time as a measure of the *response* of plants to growth in different temperatures. The word 'response' is also applied to situations where we have not deliberately imposed experimental treatments, but expect to find differences that can be explained by nest species, variety of plant or some other physical or biological factor.

We should note that this classification of variates depends on the questions being asked of the data in a particular investigation; different questions and investigations may logically require different classifications of variates.

In his paper, Latter presented two tables, one of lengths and the other of breadths, each having the eggs numbered 1 to 243 and arranged in the same order by species. For 12 of the 42 species named, there was only one egg measured, while the most common species had 45 eggs measured. If we assume that each egg came from a different nest then each table offers a simple **hierarchical classification** whose levels may be described as (i) 'species' and (ii) 'nests within species', which is the same as 'eggs within species' because there is one egg per nest. If, on the other hand, a number of eggs had been measured in any one nest then we would have a more complicated hierarchical classification with levels (i) 'species', (ii) 'nests within species' and (iii) 'eggs within nests'. If it were possible to assign each nest to one of a limited number of locations, and only one egg was measured in each nest, then we would have a **cross-classification** specified by 'species' × 'location' with an egg occupying each of the cells so formed.

To appreciate the structure of a set of data we need to know what the response variates are and to know the classification imposed by the explanatory variates on each response variate. Before analysis we also need to know the statistical structure of the data. Most commonly, data are independent of each other and chosen groups may be treated as random samples from some population. This is the case with the cuckoo egg data if only one egg is recorded from each nest. But if more than one egg is recorded from some nests then caution is required. Measurements of eggs in the same nest, when mixed in with measurements of eggs from other nests, cannot be regarded as independent. Measurements from eggs in the same nest are likely to be more similar than those from eggs in different nests, even if the nests are of the same species of bird. The technical term is that they are correlated (see p. 94). If there were only a few of these correlated measurements it would be best to discard values, at random, to leave only one observation per nest. If there were so many that rejecting observations was not sensible then the analysis becomes more complicated.

5.2.2 Quality

How accurate are the data? The cuckoo egg records (examples are 22.5 and 20.1 for length, with corresponding breadths 17.0 and 14.9) appear to be expressed to the nearest tenth of a millimetre. Can we believe this? A frequency distribution of the final recorded digit is informative. The distribution for lengths is:

Final digit	0	1	2	3	4	5	6	7	8	9	Total
Frequency	58	27	23	30	15	13	12	12	19	34	243

With accurate measurement, the distribution would be a uniform one with a frequency of approximately 24 for each digit. (Later in the book, Example 19.1, p. 474, we show how to make a strict test in similar situations when the differences are not as clear-cut as here.) The actual distribution is far from uniform. The digit '0' is particularly popular with a frequency of 58, more than double what might be expected. The range of digits '4' to '8' are chosen much less frequently than the remaining digits. The evidence shows that the final digit is not reliable. One can treat these data as being reliable only to the nearest millimetre, after rounding the data.

The distribution of final digits should always be inspected. A common finding is that '0' and '5' are more popular than the remaining digits. If we had found that was so with the egg data we would infer that the data were correct to the nearest half-millimetre. Another common finding is for the even digits to be chosen more frequently than odd digits. The apparent pattern of preferences for the cuckoo data is a most unusual one.

Data may be measured to a greater accuracy than their variability justifies. Recording, or copying, too many digits can lead to errors so it is wise to use no more than are necessary. In Chapter 1 we suggested, as a rule of thumb, that data might be recorded to three significant figures. We can offer a more precise rule now. If the data are recorded, or rounded, to one-third or less of a standard deviation of the data, then as a consequence of this the calculated variance will increase by less than 1%. Therefore we might choose to record to the largest convenient interval that is less than one-third of a standard deviation. For example, the lengths of the cuckoo eggs found in the nests of robins had a standard deviation of 0.78. We should therefore choose a recording interval of $0.78/3 = 0.26$ or less. The largest convenient interval that satisfies this requirement is 0.1. It was therefore desirable to record the eggs to one-tenth of a millimetre, but this does not seem to have been effectively achieved.

The above considerations are relevant whether the data are to be operated on by hand or by computer. If the analysis is to be done by hand, serious consideration should be given to possible coding of the data (see p. 33). Intermediate calculations involving squares or products of the raw data should be carried out to twice the number of decimal places in the raw data. (If the data have been coded to whole numbers one can stay in whole numbers.) Means should usually be recorded to one more decimal place than the raw data, variances to between four and six significant figures, standard deviations to two or three significant figures.

If calculations are done by computer then far more digits will be used than would be the case for hand calculation. If data are abstracted from computer output for further calculation by hand then the considerations above should apply. One should beware copying data to many decimal places simply because they are there on the output. With summary tables, derived from hand or computer calculation, the rule given in Chapter 1 (p. 13) should apply: round all numbers to two effective digits. It is a mistake to suppose that presenting more decimal places, or more digits, somehow makes results more 'accurate'; this is a spurious accuracy and only makes it harder to appreciate data.

Missing information should be noted. The nest species is unknown for 20 eggs. The measurement data for these eggs provide limited information, though they do not create any difficulties when using a hierarchical classification by species. Missing data in cross-classification data can create serious difficulties in analysis; if, for example, we wished to classify by species and by location, the absence of some species from some locations would be very inconvenient.

Sets of data should also be inspected for wild observations, or *outliers*: values which are much too big or too small. A rule of thumb is to pick out observations which deviate from the mean by more than 2.5 standard deviations. A related rule picks out observations which are more than 1.5 times the *mid-spread* (p. 80) beyond the appropriate quartile. A misplaced decimal point is a common slip. Observations should only be rejected if there is very strong evidence that a mistake has been made. If in doubt, the data should be analysed with, and again without, the suspect observation. This can be helpful where, for example, we suspect that two digits have been transposed, such as 32 instead of 23.

5.2.3 Rough estimates of standard deviation

It is frequently useful to be able to find a rough estimate of the standard deviation of a set of data. The method that follows is based on the range. Let the range of a random sample of n observations be denoted by w_n. If the sample is from a particular distribution (the normal distribution which is described in Chapter 15) then the relation between the average value of the range w_n and the standard deviation σ is approximately $\sigma = w_n/\sqrt{n}$, when $3 \leqslant n \leqslant 12$ (see also p. 388).

We can illustrate the method with data from the random sample of stones (p. 61):

$$
\begin{array}{ccccccccc}
25.6 & 37.4 & 27.0 & 34.4 & 50.6 & 42.2 & 37.4 & 56.5 & 33.5 \\
42.9 & 44.8 & 49.0 & 20.9 & 37.4 & 44.0 & 43.3 & 38.6 & 26.7
\end{array}
$$

The range of the first nine observations is $(56.5 - 25.6) = 30.9$. Dividing by $\sqrt{9} = 3$ gives an estimate of 10.3 for σ. The second row of observations yields an estimate of $(49.0 - 20.9)/3 = 28.1/3 = 9.4$. Taking the average we estimate σ to be $(10.3 + 9.4)/2 = 19.7/2 = 9.8$. Calculation of σ gives an actual value of 9.1.

It is a good idea to introduce averaging into the procedure. Thus with eight observations, instead of finding the range of the eight observations, it would be better to split the data into two sets of four; obtain the range of each subset, average and divide by the appropriate factor. The data should always be in random order if they are to be split up.

Although the relation is only strictly true for data that follow a normal distribution the method seems to work quite well for most data.

5.3 DATA DESCRIPTION

We have met, earlier in the book, most of the tools we need when viewing and describing data. For graphical representation we have introduced histograms, frequency diagrams, dot plots, stem plots and box plots, while for numerical measures we have the mean and median, quartiles, the range and mid-spread, the standard deviation and variance. So far we have been concerned almost entirely with a single sample or population. More commonly we wish to compare samples, and that is what we shall be looking at in most of the remainder of the chapter.

5.3.1 Comparison of two groups

The cuckoo egg data allow us to compare the lengths of cuckoo eggs that were found in the nests of different bird species. We choose to make the comparison between eggs found in robin and wren nests. There are almost the same number of observations in each group; comparisons between groups are easier to make when this is so. We shall use the lengths given below. Recall that we found the final digit unreliable. We have redistributed the final digit for some of those observations ending in .0 in order to avoid many repeat observations.

Robin nest
21.8, 23.4, 23.3, 22.4, 22.4, 23.7, 23.6, 23.5, 23.9, 22.3, 22.6, 22.6, 22.5, 22.1, 21.1, 23.4.
Wren nest
19.8, 22.1, 21.5, 20.9, 22.4, 21.0, 22.3, 21.5, 20.3, 20.9, 22.0, 20.0, 20.8, 21.2, 21.6.

5.3.2 Diagrams

One of the simplest and most useful diagrams for comparing samples is the dot plot (Fig. 5.1). Although there is quite a lot of overlap between the values in the two samples, the robin data have a similar spread to the wren data and are shifted a little over a millimetre, on the egg-length scale, to the right. The data

Figure 5.1 *Dot plot for comparison of cuckoo eggs from robins' and wrens' nests.*

suggest that bigger eggs are laid in the nests of robins than in those of wrens. This seems reasonable since robins are bigger birds and have bigger eggs than wrens. In a dot plot such as this the groups can obviously be seen to be different if there is no overlap, or overlap of just one or two points. With greater overlap it becomes a matter of experience to judge if there is a real difference. When the number of values becomes large, histograms or stem plots are more informative than dot plots. Small crosses are often more convenient than dots to show the values of measurements on a diagram.

Another useful diagram for comparing two samples is the back-to-back stem plot. It is an obvious extension of the simple stem plot (p. 8).

Robin nest				Wren nest	
			19		
			19	8	1
			20	30	3
			20	998	6
1		1	21	02	(2)
2		8	21	556	7
6		1344	22	1430	4
(3)		566	22		
7		434	23		
4		9567	23		

Stem unit = 1 mm   Leaf unit = 0.1 mm

Again the difference in average size of the two groups shows up. The leaves have been left unordered. The depths of the leaves of the two plots are shown on the outside.

5.3.3 Summary statistics

To summarize a set of observations we need, as a minimum, a measure of the centre of the observations and a measure of spread. An obvious choice is to use the mean and the standard deviation. Thus we might compare the eggs in the robin and wren nests by the table:

	Mean (mm)	Standard deviation (mm)
Robin nests	22.8	0.78
Wren nests	21.2	0.80

Comparison of two groups is most straightforward when the variability of the two groups is about the same, as in this case. There is a difference in means of 1.6 but, on the material we have covered so far, there is not sufficient information in the table for us to decide whether a real difference between groups is indicated. Using methods developed later (p. 431), a back-of-an-envelope calculation with the means and standard deviations would indicate a real difference between the eggs in the robin and wren nests.

In EDA, the stress is on keeping calculation to a minimum so the preferred pair of measures would be the median and the range. Both can be obtained conveniently from the stem plot. For the robin data, the depth of the median is (no. of observations $+ 1)/2 = (16 + 1)/2 = 8.5$; the median is therefore $(22.6 + 22.6)/2 = 22.6$. The range is $23.9 - 21.1 = 2.8$. We derive the table:

	Median (mm)	Range (mm)
Robin nests	22.6	2.8
Wren nests	21.2	2.6

Again we see the measures of variability (the ranges) are almost the same; the medians are almost identical with the means.

Tukey also suggested a more elaborate summary which he called the five-number summary. It consists of the median M, the quartiles Q_1 and Q_3, the minimum (or lowest value) L and the maximum (or highest value) H, arranged in an open upturned box:

Consider the robin data. We have seen already that the median has depth 8.5 and its value is 22.6. The first quartile Q_1 is the middle observation of the ranked data from L to M. Its depth is therefore $(8 + 1)/2 = 4.5$; we ignore the half in the depth of the median when making the calculation. The depth of Q_3 is also 4.5. From the stem plot we find $Q_1 = (22.3 + 22.4)/2 = 22.35$, which we round to 22.4; $Q_3 = (23.4 + 23.5)/2 = 23.45$, which we round to 23.4. Making similar calculations for the wren nest data, we obtain the summaries:

Robin nests ($n = 16$)					Wren nests ($n = 15$)			
M 8.5		22.6			M 8		21.2	
Q 4.5	22.4		23.4	1.0	Q 4.5	20.8	21.8	1.0
1	21.1		23.9		1	19.8	22.4	

Sample size is particularly likely to influence the extreme values so we denote it by n and insert the value in the title. The labels to the left of each summary show the depths of the median M and the quartiles Q; the maximum and minimum values always have depth 1. On the right-hand side of each summary is shown the mid-spread, $Q_3 - Q_1$, which can be a useful measure of variability.

5.3.4 Comparison of many groups

Provided there are at least about ten observations for each group, box plots are a good way of comparing many groups. Five-number summaries provide the data needed to draw the diagrams. Figure 5.2 illustrates the comparison of the lengths of cuckoo eggs found in the nests of three species: robin, wren and tree-pipit.

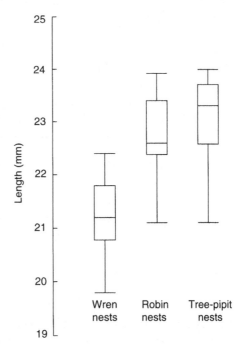

Figure 5.2 *Box plot for comparison of lengths of cuckoo eggs from nests of wrens, robins and tree-pipits.*

It can be seen that eggs from the nests of robins and tree-pipits do not differ much in size but that nests of both these species contain larger eggs than are found in wrens' nests. A refinement in the drawing of box plots is worth mentioning here. It is to prevent comparisons being misled because of outliers. If any value is so large that it is further than 1.5 times the mid-spread beyond the third quartile then it is marked in separately, by a dot, and labelled. If outliers are picked out in this way the whisker is carried out to the largest value which is not an outlier. Low values may be treated similarly.

Dot plots are also useful provided the number of observations in each group is not large. Stem plots and histograms, arranged side by side, are rather clumsy.

5.3.5 Bivariate data

Another type of question we might ask is, do the longer eggs have bigger breadths? We may investigate this in a *scatter plot* (biplot is an alternative term) in which we plot the length of each egg against its breadth. We do this in Fig. 5.3 for the tree-pipit data. To help judge if there is an association between the two variates it is useful to insert lines at the median values of the two variates, parallel to the axes (see the dotted lines in Fig. 5.3). If there is no association the points will be randomly scattered over the plane and approximately equal numbers will fall in each of the four quadrants formed. If most of the points fall in the first (top-right) and third (bottom-left) quadrants, and few in the other two quadrants, a positive association is indicated. This appears to be the case for the tree-pipit

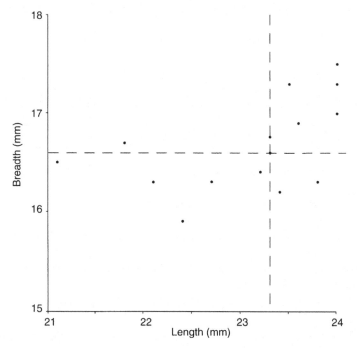

Figure 5.3 *Scatter plot of length and breadth of cuckoo eggs found in nests of tree-pipits.*

data; as might be expected, the longer eggs tend to be broader. If most points fall in the second (top-left) and fourth (bottom-right) quadrants, this indicates a negative association. See also Figs 21.1 and 21.2.

Two variates that are associated are said to be correlated and we discuss this thoroughly in Chapter 21. In our diagram (Fig. 5.3) we marked in median values because these could be put in easily, with little calculation; they are also little affected if the distributions should be skew. In Chapter 21 we use similar diagrams but insert lines at the *mean* values because the theory can be developed best if we work with deviations of the observations from their means.

5.4 CONCLUSION

A great deal of information can be extracted from data using the simple methods we have described in this chapter. The methods are useful not only for beginners but also for experienced statisticians. Large effects will exhibit themselves at this stage; being aware of them will often prevent silly mistakes in the later detailed analyses. Moreover, the methods allow a check to be made on, for example, equality of variability or nature of distribution, which determine the assumptions made at the next stage of analysis.

Further information can be extracted from data by setting up a mathematical model and manipulating that. The remainder of the book is devoted to developing these more sophisticated methods.

Note. Most of the exercises which follow in Section 5.6 can be solved using a computer. No special computer exercises are provided for this chapter.

5.5 MINITAB: SCATTER AND TIME SERIES PLOTS; FIVE-NUMBER SUMMARY

A scatterplot is obtained by the command

PLOT C*C

where the Y variate is in the first column and the X variate is in the second; or, by

Graph>Scatterplot/
Select **Simple/**
Enter **Y variable** and **X variable/**

Lines may be inserted, to judge association, by the command

Editor>Crosshairs/

Moving the cursor over the graph changes the position of crossed lines which may be placed to correspond with the median values. There is a readout of the values where the lines cross the axes in the left-hand corner of the window. To fix the lines on the graph, session commands can be used. The PLOT command is followed by the subcommands

REFERENCE 1 K;
REFERENCE 2 K.

where the constants take the values of the X and Y medians, respectively.

How a variate changes over time is often of interest. Suppose there is a column of data which represent the mean daily temperature, on successive days, at a particular site. Requesting MINITAB to make a time series plot of the data will produce a graph in which the Y coordinates of the points are the temperatures, and the successive X coordinates are at equally spaced intervals of time. Successive points are joined by lines. The menu choice is

Graph>Time Series Plot/
Select **Simple/**
Enter **Series** variable, i.e. column/

The session command is

TSPLOT C.

Choice of

Stat>Basic Statistics>Graphical Summary/
Enter **Variable/**

produces a five-number summary, box plot and histogram, among other information.

5.6 EXERCISES ON CHAPTER 5

1. The weights (kg) of 20 male (M) and 20 female (F) students were:

M 72, 58, 60, 68, 59, 72, 85, 73, 57, 76, 69, 74, 63, 50, 64, 71, 66, 67, 65, 61;
F 68, 63, 60, 58, 49, 63, 58, 57, 55, 57, 65, 47, 59, 67, 55, 52, 33, 53, 53, 54.

Compare the two distributions by using (i) stem plots, (ii) numerical summaries, (iii) box plots. Write a brief report on the results.

2. The fuel consumption of a motor vehicle is measured under controlled conditions with four different mixtures A, B, C, D. The results (km/L) are:

A	13.31	14.04	13.68	13.75	13.12	14.11	13.96
B	14.28	14.47	14.03	15.62	15.10		
C	15.04	14.77	15.13	15.45	14.98	15.51	
D	14.66	13.93	15.05	14.21	14.42	14.30	14.25

(i) Make a rough estimate of the standard deviation and hence decide whether the data might be rounded. (*Note*: the standard deviation should be estimated separately for each mixture and an average taken.)

(ii) Compare the four mixtures by using dot plots.

(iii) Summarize the data in a table of the mean (or, median) and the range for each mixture. Write a brief report on how the mixtures compare.

3. A student observed customers buying tickets at a railway booking office. The times taken, in seconds, for customers paying by cash, cheque or credit card are tabulated as follows:

Cash customers
16, 38, 21, 18, 13, 23, 11, 40, 16, 19, 38, 15, 30, 30, 24, 16, 18, 17, 20, 36.
Cheque customers
64, 51, 57, 59, 60, 70, 58, 56, 50, 55, 46, 61, 62, 65, 37, 50, 55, 53, 55, 54.
Credit card customers
55, 43, 52, 50, 57, 54, 59, 54, 67, 65, 60, 60, 60, 66, 55, 59, 65, 55, 67, 67.

(i) Look at the distribution of final digit to see if there is evidence of some digits being preferred to others.

(ii) Make a comparison between the three methods of payment by using stem plots and numerical summaries.

4. At the beginning of an investigation involving house bricks, the length, in centimetres, of ten bricks was recorded:

21.7, 21.1, 21.6, 22.1, 21.2, 21.6, 22.0, 21.5, 21.9, 21.3.

Make a rough estimate of the standard deviation and hence recommend how many decimal places should be used in the records.
Carry out the same procedure assuming the original records were:

22, 21, 22, 22, 21, 22, 22, 22, 22, 21.

Would your recommendation change?

5. The lengths (in metres) of 50 rolls of cotton were measured and are recorded below. Comment on these data.

18.288	18.314	18.250	18.288	18.237
18.237	18.288	18.262	18.314	18.288
18.275	18.314	18.288	18.275	18.262
18.288	18.262	18.301	18.288	18.326
18.301	18.288	18.262	18.339	18.288
18.288	18.339	18.288	18.262	18.301
18.542	18.237	18.275	18.288	18.262
18.262	18.262	18.288	18.314	18.288
18.288	18.301	18.314	18.288	18.314
18.275	18.288	18.224	18.275	18.288

(1 metre = 39.37 in)

(*IOS*)

6. In a study of different treatments for the management of diabetes there arose some concern about the loss of weight of patients during treatment. It was of interest to discover whether or not weight loss after one year of treatment had begun was related to a patient's initial weight. Sixteen newly diagnosed patients were weighed at diagnosis and after one year. The values for initial weight X in pounds and weight loss Y at one year, also in pounds, are given below. Investigate whether X and Y are related by drawing a scatter plot.

X	225	235	173	223	200	199	129	242
Y	15	44	31	39	6	16	21	44
X	140	156	146	195	155	185	150	149
Y	5	12	−3	19	10	24	−3	10

7. During a trial of a new drug a patient was given a placebo (i.e. the patient believed he was receiving a drug but in fact he was given an inert substance) for 28 days, no treatment for 7 days and the new drug for 28 days. Respiratory capacity (in cubic centimetres) was measured three times per day and the maximum recorded. The results are given below.

   Compare the new drug with the placebo by making (a) stem plots, (b) five-number summaries and (c) box plots. Assume there are no effects of time. Are any of the observations outliers? If there are any, mark them separately on the box plot and take the whisker out to the observation before.

   Placebo   420, 400, 410, 410, 380, 460, 450, 480, 400, 460, 490, 400, 420, 490, 480, 500, 470, 430, 450, 490, 480, 490, 490, 530, 540, 530, 530, 540.

   New drug   400, 420, 380, 390, 450, 310, 420, 470, 400, 310, 410, 510, 410, 440, 490, 420, 590, 410, 400, 450, 470, 440, 600, 490, 410, 440, 410, 470.

8. As part of an investigation into the growth of healthy children 36 children of different ages were asked to sit still for five minutes, after which their heart rates were measured. The values for age A, in years, and heart rate R, in beats per minute, are given below. Draw up a table giving the heart rates of children for each of the ages represented. Find the mean heart rate for each age and draw a graph of heart rate against age.

A	4	8	10	10	6	4	18	12	2	18	16	6
R	88	73	70	65	67	97	60	66	104	61	68	81
A	2	10	16	8	4	14	10	14	16	2	8	12
R	88	65	65	69	90	59	69	64	64	114	64	64
A	16	18	8	6	6	14	2	18	12	4	14	12
R	58	62	69	72	78	63	109	68	72	79	69	62

9. The heights H, in centimetres, and the weights W, in kilograms, of 30 women students are given below. Make a scatter diagram to investigate the relation between height and weight. Are there any outliers?

H	157	170	168	159	157	160	162	161	168	155
W	49	57	54	53	49	50	52	47	53	49
H	166	161	160	168	176	162	161	170	175	156
W	58	64	51	52	70	49	49	63	58	57
H	170	173	177	161	168	164	160	154	148	161
W	55	57	65	47	59	67	55	52	33	53

5.7 PROJECT

1 Prices of goods in shops

Aim. To compare the prices of goods in shops.

Collection of data. Prices of about 10 items from at least three shops should be collected. A possibility is to consider common foods and a list including bread, milk, potatoes, tea, coffee, butter, sugar and baked beans might be drawn up. Great care will be needed in specifying each item in order that comparable items are being sought in each shop. Interpretation is easier if the quantities chosen are such that the average prices of items (e.g. of tea and of coffee) do not differ greatly. There is obviously great scope in the choice of items.

Analysis. Compile a two-way table of items by shops; calculate the total cost per shop and the average price per item. Analysis should not stop with this but should then proceed to make comparisons within the table, numerically and by diagram. For example, parallel dot plots might be made for the price of the items (one line for each item) using a common scale, and the items from a given shop linked together. This might be refined by expressing the prices of each item as a deviation from the average price of the item over all the shops.

SUMMARY

Three main stages may be picked out in the initial analysis of data:

> **data preparation and scrutiny**;
> **data description**;
> **model formulation**.

We have looked at the first two stages.

Data should be scrutinized to assess their **accuracy** and to detect **outliers**. Investigation of the distribution of the final digits of observations is often informative.

Many variates may be recorded in an investigation. For each statistical analysis we need to be clear which variates are **response variates** and which **explanatory variates**. The explanatory variates may be interrelated by a **hierarchical classification** or a **cross-classification**.

Two, or more, groups may usefully be compared using:

(i) dot plots, provided the number of observations is not large;
(ii) stem-and-leaf plots;
(iii) means and standard deviations;
(iv) medians and ranges;
(v) five-number summaries, i.e. minimum, first quartile, median, third quartile, maximum;
(iv) box-and-whisker plots.

The **association** of pairs of variates may be investigated by a **scatter plot**.

REVISION EXERCISES A

1. A market gardener sowed 20 seeds of a particular plant in each of 100 specially prepared seed trays. The number of seeds that germinated per tray had the following frequency distribution:

Number germinating	20	19	18	17	16	15	14 or fewer
Number of trays	73	15	5	4	2	1	0

(i) Exhibit the distribution diagrammatically.
(ii) Calculate the arithmetic mean and the standard deviation of the distribution.
(iii) Calculate the overall proportion of seeds that germinated. *(Welsh)*

2. The weekly earnings of a random sample of women workers in 1974 were:

Weekly earnings (£)	No. of women
Less than 15	1
15–	4
20–	28
25–	42
30–	33
35–	18
40–	13
45–	9
50 and over	2

(i) Calculate the mean and standard deviation of this distribution.
(ii) What will happen to the values of the mean and standard deviation if every woman has an increase of £4.40 per week?

(iii) What will the mean and standard deviation be if every woman has an increase of 10% of previous earnings?

(*Note.* Treat end classes as of the same width as other classes.) (*IOS*)

3. Weekly incomes of heads of households in a particular town in the UK in 1971, based on data from the Family Income Survey, were as follows:

Weekly income (£)	Number of households	Weekly income (£)	Number of households
Less than 10	124	35–	63
10–	74	40–	42
15–	68	45–	25
20–	85	50–	29
25–	96	60–	21
30–	82	80 or over	16

Illustrate these figures by: (i) a histogram; (ii) a cumulative frequency graph.

From the second graph, estimate the quartiles, the median and the top decile. Also estimate the percentage of heads of households earning between £24 and £46 per week.

What difficulties arise in estimating the lowest decile, and how might they be overcome?

4. The words in the first 100 sentences of Darwin's *Origin of Species* were counted and the following numbers found:

52, 73, 25, 44, 15, 24, 37, 27, 33, 53
34, 21, 22, 38, 72, 18, 95, 37, 58, 38
12, 37, 63, 20, 92, 49, 14, 46, 47, 21
69, 38, 48, 20, 13, 16, 26, 13, 33, 23
44, 32, 18, 53, 22, 10, 22, 17, 28, 83
22, 22, 28, 80, 34, 36, 41, 56, 16, 64
36, 21, 22, 30, 25, 26, 83, 28, 65, 49
41, 41, 70, 56, 45, 19, 19, 32, 30, 36
77, 41, 27, 46, 30, 64, 34, 53, 38, 64
39, 11, 51, 32, 39, 33, 24, 21, 31, 18.

Construct the frequency distribution table by choosing a suitable grouping interval.

(a) Calculate the mean of the distribution (i) from the original data, (ii) from the grouped data.

(b) Construct a stem-and-leaf diagram for these data.

(c) Calculate the median and quartiles, and draw a box-and-whisker plot.

5. The arithmetic mean of 17 numbers is 8, and that of 26 other numbers is 13. Their standard deviations are 8 and 3 respectively. Find the mean and the standard deviation of the 43 numbers combined.

6. The number of rooms per dwelling was recorded in the UK Censuses in 1961 and 1971, with the following results:

Number of rooms	Percentage of dwellings	
	1961	1971
1	1	2
2	5	4
3	12	9
4	27	23
5	35	30
6	13	23
7	3	5
8 or more	4	4

For each distribution, calculate the mean (μ) and the standard deviation (σ), and also the percentage coefficient of variation ($= 100\ \sigma/\mu$). Comment on the results. (*IOS*)

7. The following stem-and-leaf plots summarize the data obtained in an experiment relating the adult weights of *Drosophila persimilis* insects to the densities at which they were reared as larvae. Draw the corresponding box-and-whisker plots for each density, and use these to compare the main characteristics of the four sets of data.

Larval density = 5 dm^{-3} (25 insects)

```
1.0    3
1.1    3 5 9
1.2    1 2 2 6 6 7 8 8 9 9
1.3    1 3 3 4 6 8 9
1.4    0 2 5
1.5    3
```

Note: 1.0   3 represents 1.03 mg

Larval density = 10 dm^{-3} (40 insects)

```
0.7    1 6 9
0.8    3 4 4 6 7 9 9
0.9    0 2 2 2 4 5 5 6 8 9 9 9
1.0    0 1 1 2 2 2 3 6 6 7 8 9
1.1    2 4 7
1.2    0 8
1.3    8
```

Larval density = 20 dm^{-3} (50 insects)

0.3	4 8
0.4	2 4 9
0.5	1 7
0.6	0 1 4 4 5 6 6 8 9 9
0.7	0 0 1 2 2 2 3 3 4 4 5 6 8 8 8 9 9
0.8	1 1 2 4 4 6 7 7 8
0.9	0 3 7 8
1.0	2 5 7
1.1	6

Larval density = 40 dm^{-3} (18 insects)

0.2	7
0.3	3 7
0.4	0 1 1 5 7 8 9
0.5	0 3 4 7 7
0.6	1 5
0.7	6

8. Ten chemical analyses carried out to estimate the percentage of carbon in an organic compound gave the results 32, 27, 33, 29, 26, 35, 34, 28, 30, 32. Find the mean, mode, median and geometric mean of the percentage of carbon.

   Five more analyses gave the results 29, 31, 31, 36, 29. Calculate the same four measures for all 15 analyses.

9. A man travels on the 8.05 train every morning. It is not always on time, and he notes each day the actual time when it leaves his station. Unfortunately, his watch gains 6 minutes a day, and he does not always remember to correct it in the morning.

   His records of the number of minutes his train is late each day are: 3, 0, 7, 2, 4, 8, 6, 10, 4, 5, 2, 9, 7, 1, 2, 0, 8, 2, 6, 1, over a period of 20 days. Make a dot plot of the data. Calculate the mean and the median of these figures and comment on their use as summaries of the data.

10. Two groups of observations of weights of a particular plant, in two habitats A, B, were as follows:

    Group A: 56.7, 14.8, 8.8, 108.3, 7.5, 61.4, 103.1, 64.1 grams
    Group B: 11.9, 24.4, 55.0, 18.8, 2.1, 1.6, 43.9, 33.4, 44.8, 12.8 grams.

    Make adjacent dot plots of the two groups of observations.
       Calculate the arithmetic mean, the geometric mean and the median weight in each group.
       If you wished to compare the two groups, which measure would you think it best to use, and why?

11. Samples of eggs are taken from the nests of a species of bird in two localities, and the weights of the eggs are recorded. The following are the frequency distributions. Represent these data graphically, and, without extensive calculations, judge whether the localities differ in their average egg weight or the scatter of the weights.

	Weight (g) – class-centres							
Locality	2.5	3.5	4.5	5.5	6.5	7.5	8.5	Total
A	3	61	84	196	80	33	0	457
B	1	72	88	174	102	46	5	488

12. A batch of 50 ball bearings had the following distribution of diameters:

Diameter (cm)	Number of ball bearings
0.160–	1
0.162–	3
0.164–	9
0.166–	20
0.168–	14
0.170–	2
0.172–	1
0.174–	0

Calculate the mean, median and standard deviation of the diameter.
   A second batch of 100 ball bearings has mean diameter 0.165 cm and the standard deviation of diameter is 0.003 cm. Calculate the mean and standard deviation of the diameter for all 150 ball bearings.

13. The frequency distribution of a variable x is classified in equal intervals of size c. The frequency in a class is denoted by f and the total frequency is N. If the data are coded into a variable u by means of the relation $x = a + cu$, where x takes the central values of the class-intervals, show that the mean $\bar{x}$ and the standard deviation s of the distribution are given by

$$\bar{x} = a + c\bar{u}$$

$$s^2 = c^2 \left[\frac{\sum fu^2}{N} - \left(\frac{\sum fu}{N} \right)^2 \right].$$

   The following table gives the frequency distribution of the intelligence quotients x of 500 scholars. Find the mean and standard deviation of the distribution.

x	82–85	86–89	90–93	94–97	98–101	102–105	106–109
f	5	19	32	49	71	92	75

x	110–113	114–117	118–121	122–125	126–129	130–133
f	56	39	28	18	10	6

Total $N = 500$ *(AEB)*

14. In a certain industry, the numbers of employees in 1970 were as below, by age-groups:

Age last birthday	15–19	20–24	25–29	30–34	35–39	40–44	45–49	50–54	55–59	60–64
Number (thousands)	66	65	56	50	42	37	35	30	24	22

Calculate the arithmetic mean, median, variance and standard deviation of the ages of employees in the industry.

Estimate the percentage of employees whose ages lie within one standard deviation of the arithmetic mean.

15. The following table shows the distribution of the ages in years on their last birthdays of 120 students who registered for a particular degree at a certain college on 1 October 1974:

Age in years on 1 October 1974	18	19	20	21	22	23
Number of students	50	32	22	8	5	3

(a) Estimate the mean age (in years and months) of these students on 1 October 1974.

(b) Estimate to the nearest month in each case: (i) the median; and (ii) the semi-inter-quartile range; of the ages of these students on 1 July 1977.

(Welsh)

16. Numbers of books borrowed by each of 200 schoolboys in one year are shown in the table. Arrange these data in suitable class-intervals and display in a diagram.

Number of books	0	6	10	12	15	16	17	18	19	20	21	22	23	24	25	26	27	28	29	30	31	32	33	34	35	36	37	38	40	44
Number of boys	34	1	3	4	6	4	3	5	2	11	3	6	8	13	20	5	6	9	5	12	4	5	2	3	10	8	2	1	4	1

Calculate the mean, variance and standard deviation of these observations.

17. (a) In an arithmetic test given to a class consisting of n girls and n boys, the n boys scored a mean mark of m_1 with a standard deviation of σ_1, and the n

girls scored a mean mark of m_2 with a standard deviation of σ_2. Show that the standard deviation, σ, of the marks of all $2n$ pupils is given by

$$\sigma^2 = \frac{1}{2}(\sigma_1^2 + \sigma_2^2) + \frac{1}{4}(m_1 - m_2)^2.$$

(b) With 1963 as base-year, the index numbers of production in three engineering industries in 1971 were a_1, a_2, a_3 respectively. Determine the corresponding index numbers for 1963 if 1971 is taken as base-year. Calculate: (i) the arithmetic mean; and (ii) the geometric mean; of each of the two sets of index numbers. By studying the answers you obtain, explain why the geometric mean is more appropriate than the arithmetic mean when averaging index numbers. (*Welsh*)

18. A manufacturer of a new soap powder wishes to predict the likely volume of sales in a town. Four schemes, as below, are proposed for selecting people for a questionnaire. Discuss the merits of each and choose one, explaining why you think it the best.
   (i) Take every 20th name on the electoral register of the town.
   (ii) Choose people entering a supermarket, ensuring that the numbers in each sex, age and social class category are proportional to the number in the population.
   (iii) Select houses at random from a town plan and interview one person from each.
   (iv) Choose at random one name from each page of the telephone directory and ring them up. (*Oxford*)

19. In an examination, there are five papers, the marks for each being expressed as percentages. However, they are not all given equal importance but are combined with the weightings shown in the table. The optional subject may vary from one candidate to another, and some candidates may not offer one at all. The percentage marks of eight candidates A–H are shown. Calculate their weighted average percentage marks and rank them in order of overall performance.

Subject	Weight	A	B	C	D	E	F	G	H
Mathematics	27.5	68	53	29	47	61	32	48	58
English	27.5	54	60	47	43	55	40	66	61
French	17.5	48	58	50	33	60	25	49	42
Science	17.5	67	55	24	44	52	33	40	56
Optional subject	10.0	50	37	22	–	46	58	–	45

The board who review the examination wonder what would be the effect of omitting the optional subject altogether, while keeping the same relative weighting for the other four subjects. Recompute the weighted average percentage marks and find the new ranking.

Comment on the use of the two different schemes if candidates are required to achieve (a) 35%, (b) 40% in order to pass this examination.

6 PROBABILITY

OVERVIEW

Theoretical statistics is based on probability. We define probability and show how to calculate probabilities.

6.1 INTRODUCTION

So far we have been looking at populations of observations, and at samples drawn from these populations. We have developed ways of summarizing real-life data, so that we can make inferences from them. In order to devise mathematical ways of making inferences, we shall need to set up mathematical models of the real-life situations which we think are underlying our data. These models make use of probability theory.

Probability theory was developed by mathematicians to solve problems that arose in gambling games with cards and dice. The theory did not begin to take form until the late seventeenth century. The essential idea was that probability was the *relative frequency* of possible occurrence of some chosen outcome.

Suppose that a supermarket employs 40 people, of whom 15 work full-time and the remaining 25 part-time. The manager will have a list of the employees' names, and it would be easy to select one employee at random as described in Chapter 3 (using random digits). Will the one selected be a full-time or a part-time employee? *After* selection, we shall *know*; *before* selection we do not know, but we do have the information that there are 15 who work full-time and 25 part-time. Can we use this information to *predict* what our selection will be? Of course such a prediction is in the nature of a guess, but, as we shall see, not a wild one: we can attach numbers to it.

By making a random selection of one from the 40, we are equally likely to choose any one of the employees; no group or person is favoured in any way in the selection process. The *relative frequency* of full-timers among the 40 is $\frac{15}{40}$ and of part-timers $\frac{25}{40}$. So, since every one of the 40 is equally likely to be chosen, we shall say that the *probability* of choosing a full-timer is $\frac{15}{40}(=\frac{3}{8})$ and the probability of choosing a part-timer is $\frac{25}{40}(=\frac{5}{8})$. We have set the probability of selection of one type of worker, on a single occasion, equal to the relative frequency of that type of worker in the whole population.

If we go on sampling with replacement, several times, choosing one employee at random each time from the whole population, we may calculate the proportion of samples in which the employee chosen was a full-timer. This proportion, i.e. the relative frequency of full-timers in the samples, would be close to $\frac{15}{40}(=\frac{3}{8})$ provided we repeated the sampling a large number of times.

This does not give an *exact* prediction of what will happen, but it is a very useful way of comparing the possibilities numerically. The probabilities $\frac{3}{8}$ and $\frac{5}{8}$ are easy to compare for size. Consider another population, of total number 60, in which 30 are full-time and 30 part-time; in this situation the probability of selecting a full-time worker, when one worker is picked at random from the whole population, would be $\frac{30}{60} = \frac{1}{2}$, and of course the probability of selecting a part-time worker is also $\frac{1}{2}$. Finally, in selecting one at random from 50, of which 5 are full-time and 45 part-time, the probability that a full-timer is selected is $\frac{5}{50} = \frac{1}{10}$, and for a part-timer it is $\frac{45}{50} = \frac{9}{10}$. All these probabilities are easy to understand and compare as fractions, but usually with less simple numbers it is best to write probabilities as decimals.

There is another way of looking at the selection process. Going back to the first illustration above, in the population of 40 workers, we ask first of all how many different ways there are of selecting one person from 40. The answer of course is 40: any one of the population could be selected. Now we ask how many of these ways give a full-time employee as the selection: this is 15. The ratio

$$\frac{\text{number of ways of selecting one full-time employee}}{\text{total number of ways of selecting one employee}}$$

is the *probability* of selecting a full-time employee. It is $\frac{15}{40} = \frac{3}{8}$. This ratio rule works so long as we are considering the number of *equally likely* ways of selection, and is sufficiently useful to be used as a working definition until a more mathematical one is presented.

Working definition of probability

In a process which can be carried out in n equally likely ways, r of these ways lead to a particular result. The probability of obtaining this result in one run of the process is r/n.

Let us call the probability of selecting a full-timer, in the first example, p. Since it has been defined as a ratio, it is clearly a positive number between 0 and 1. If in our population there had been no part-timers, then p would have been $\frac{40}{40} = 1$, and we would have been bound to pick a full-timer as our selection. In the same way, if all had been part-timers, p would have been $\frac{0}{40} = 0$, and it would have been impossible to pick a full-timer. Something that is bound to happen has probability 1; something that cannot happen has probability 0.

Further, it is clear that if the probability of selecting a full-timer is p, then the probability of not doing so is $1 - p$: either we do or we do not, and these two possibilities between them account for all the ways of selecting one employee. We shall return to these ideas in a little while.

We have interpreted probability as a relative frequency. This usually allows probabilities to be determined in an objective way. Another interpretation of probability is as a *degree of belief*, and this often lies behind statements made in everyday life. For example, someone interested in horse racing may say 'I believe, with probability 0.8, that Smoky Joe will win the Derby tomorrow.' This interpretation uses subjective probability and it can be difficult to determine numerical values. It has been seriously suggested that the same type of approach may be used in assessing the truth of a scientific theory, determining a subjective probability by asking what odds the proposer would accept in a bet. In this book we restrict ourselves to the relative frequency interpretation.

6.2 WINNING THE TOSS

Team captains often have to toss a coin to determine, for example, who shall bat first in cricket or who shall kick off or choose ends in football. One captain tosses the coin, and the other calls. Provided the coin is 'fair' (properly balanced) and is tossed properly, there is no reason why one of the two possible results, heads and tails, should occur any more often than the other. Heads (H) and tails (T) are equally likely results for a toss because the tossing process does not favour one more than the other. If we neglect the possibility of losing the coin or having it stand on its rim against some obstacle, the probability of a head, which we call P(H), and that of a tail, P(T), must both be $\frac{1}{2}$ at any toss. This is because the two possible results H, T are equally likely, and between them they account for all possible results.

Let us suppose that Horace, whenever he is captain and calls, calls heads. In a series of three games in which Horace calls each time, we can show the possible results of the tosses in what is called a *tree diagram* (Fig. 6.1), and then use this to see how many times he has won. Each of the eight possible paths through the tree is equally likely.

After the third game, Horace may have won all three tosses, if the coin came down heads every time. There is only one path through the tree leading to the end-point 3. But there are three possible paths through the tree to reach the end-point '2 tosses won', and similarly three paths to the end-point 1. There is only one path leading to the end-point 0. The results H and T at each toss are equally likely; but the results 0, 1, 2, 3 wins in three tosses are *not* equally likely. We use the listing of *equally likely* events to find probabilities, and we must make sure we have this listing before using the ratio definition of probability. A tree diagram can be a useful way of obtaining it. It is then easy to write down the

Figure 6.1 *Tree diagram showing possible results of three tosses of a coin.*

probabilities of Horace winning 0, 1, 2 or 3 tosses, using our working definition (p. 107) for equally likely results.

Number of tosses won	0	1	2	3
Probability	$\frac{1}{8}$	$\frac{3}{8}$	$\frac{3}{8}$	$\frac{1}{8}$

Note again that, having listed all the possible results, we have a set of probabilities that add to 1.

Example 6.1

A card is drawn at random from a conventional pack of 52 cards. What is the probability of drawing: (a) a red card; (b) an ace; (c) a court card (i.e. a King, Queen or Jack)?

(a) There are 26 red cards in the pack. The probability of drawing a red card is $\frac{26}{52} = \frac{1}{2}$.
(b) There are 4 aces in the pack. The probability of drawing an ace is $\frac{4}{52} = \frac{1}{13}$.
(c) There are 12 court cards in the pack. The probability of drawing a court card is $\frac{12}{52} = \frac{3}{13}$.

Example 6.2

Two fair dice are thrown. What is the probability that: (i) the total score is 2; (ii) the total score is 4?

The first die can show any of 6 numbers, and so can the second die. Hence there are 36 distinct equally likely possible results. (A tree diagram might be drawn to illustrate them.)
(i) A total of 2 arises only if both dice show a 1. This is just one of the 36 equally likely results. Hence the probability is $\frac{1}{36}$.
(ii) A total of 4 arises if the pairs of scores on the dice are $(1, 3)$, $(2, 2)$ or $(3, 1)$. (Note that $(1, 3)$ and $(3, 1)$ are distinct results.) These are three of the 36 equally likely results. The probability is therefore $\frac{3}{36} = \frac{1}{12}$.

6.2.1 Exercises

1. In a particular country in 1993, of the people aged 65 and over, 4.7 million were females and 2.9 million were males. If a person in this age-group had been chosen at random what would the probability be that the person was female?

2. In a small country, there are vehicles of only three types: 20% of all vehicles are Japanese-made estate cars, 45% are Japanese-made saloon cars, and 35% are German-made saloon cars. If a vehicle is selected at random, what is the probability that
   (a) it is Japanese-made,
   (b) it is a saloon car?

3. What is the probability that two successively chosen random digits are identical?

4. Assume that boys and girls are equally likely to be born. What is the probability of there being no boys in a family of three children? (Compare with Horace's tosses.)

5. If a single die is thrown, what is the probability that:
   (a) the face marked '6' will be uppermost;
   (b) an even numbered face will be uppermost;
   (c) a face containing a score which is a prime number will be uppermost?

6. If two dice are thrown together, what is the probability that:
   (a) both faces marked '6' will be uppermost;
   (b) the total score on the two dice will be an even number;
   (c) the total score will be divisible by 3?

7. Two digits are picked from a table of random digits. What is the probability that:
   (a) both will be odd numbers;
   (b) their sum will be twice their difference;
   (c) their product will be 4?

8. What is the probability that a person, chosen at random from the whole population, will have a birthday on:
   (a) the same day of the week as you;
   (b) the same day of the year as you?

6.3 TERMINOLOGY AND NOTATION

Before developing further ideas, we need to define some standard words used in probability theory.

Definition 6.1

A **trial** is any process which, when repeated, generates a set of results or observations.

Thus, selecting one supermarket employee from a list is a trial; so is measuring a journey time; so is tossing a coin (or two coins one after the other); so is weighing a person or an animal. The word 'experiment' is quite often used instead of 'trial', but we avoid it because it has a more limited technical meaning to a scientist.

Definition 6.2

An **outcome** is the result of carrying out a trial.

Thus, an outcome is to pick one particular employee, say number 23 in the list; to obtain a measurement, say 19 minutes, for a journey time; to note H or T when tossing a coin (or if tossing twice to note, for example, HH); to obtain a weight in kilograms or grams.

We shall consider a basic set of equally likely possible outcomes of a trial, and we shall want to look at these outcomes either individually or in some convenient combination. For example, in tossing three coins, we may either wish to look at any of the eight individual results shown in the tree diagram (Fig. 6.1) or we may be interested in the set of results (i.e. HHT, HTH, THH) which led to Horace winning two of the tosses.

Definition 6.3

An **event** is a set which consists of one or more of the possible outcomes of a trial.

An event is thus a subset of the set of all possible outcomes. For example, when Horace calls heads in three tosses, any of the following are events:

A Horace wins three times;
B Horace wins the middle one of the three;
C Horace wins exactly one of the three.

We can attach probabilities to these (and many other) events by extending our previous definition, as we show below.

Frequently in mathematics we can appreciate more easily what a piece of analysis means if we can illustrate the situation by a diagram. The language of sets is useful in probability, and so also is the type of diagram (a Venn diagram) that is used to illustrate sets. We may draw a diagram in which the full set of equally likely possible outcomes is represented by a set of points, and events by subsets of these.

Figure 6.2 shows eight points, which represent the eight equally likely outcomes of the three tosses of a coin with which captain Horace has been concerned (p. 108). We label the complete set of possible outcomes S. It is equivalent to the universe or universal set, in conventional set language. It is also called the *possibility space*, a term which we shall use, or the *sample space*. Each outcome is represented by a point. The precise arrangement of the points is not important, so long as they are laid out in an orderly way which allows events to be illustrated.

The event A, 'Horace wins three times', is represented by one point only {HHH}; the event B, 'Horace wins the middle one of the three', is represented by the set of points {HHH, HHT, THH, THT}, the event C, 'Horace wins exactly one of the three', is represented by the set {THT, HTT, TTH}. We make the distinction that an *elementary event* is just one of the possible outcomes, while

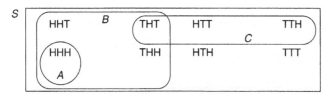

Figure 6.2 *Venn diagram showing the possibility space for the results of three tosses of a coin.*

a *compound event* includes several separate outcomes. Thus A is an elementary event while B and C are each compound events.

The probability of event A is the number of outcomes in event A (i.e. one, HHH) divided by the total number of possible outcomes (i.e. eight). But this equals the number of points in subset A in Fig. 6.2, divided by the number of points in the set S. If we use the notation $N(S)$ to represent the number of outcomes, or points, in S, and similarly write $N(A)$ for the number of outcomes, or points, in A, we see that the probability of A equals $N(A)/N(S)$. This leads us to the following definition. (It is sometimes called the definition of classical probability because it was the first satisfactory mathematical definition of probability. Later we extend the definition to situations where all outcomes are not equally likely.)

Definition 6.4

If a trial has a set of equally likely possible outcomes S, then the **probability** of the event E is $N(E)/N(S)$.

We see that $N(B) = N(\{\text{HHH, HHT, THH, THT}\}) = 4$. Hence $P(B) = N(B)/N(S) = \frac{4}{8} = \frac{1}{2}$. Similarly, $P(C) = N(C)/N(S) = \frac{3}{8}$.

The complete set of possible outcomes S is an event and $P(S) = N(S)/N(S) = 1$, as we would expect. We will also call the empty set $\emptyset$ an event (so extending the definition of p. 111). For example,

$$Z \quad \text{Horace wins four times}$$

is an event, containing none of the outcomes of S. Obviously, since $N(Z) = 0$, we must have $P(Z) = 0$.

The event S is *certain*: we *must* obtain one of the possible outcomes. Likewise the event Z is *impossible* in three tosses. The probability of a certain event is 1, and of an impossible event 0. Probabilities in general must lie in the range $[0, 1]$, as is clear from the definition.

Let us consider further examples.

(a) *Trial*: throw of a die.
A set of possible outcomes $S = \{1, 2, 3, 4, 5, 6\}$ is the possibility space. 'Throwing a six' is the event $\{6\}$. Its probability is

$$\frac{N(\{6\})}{N(S)} = \frac{1}{6}.$$

'Obtaining an odd number' is the event $\{1, 3, 5\}$. Its probability is $\frac{3}{6} = \frac{1}{2}$.

There may be many ways of choosing a possibility space S; if so, we choose the one that is most convenient for our purpose. With the throw of a die we might alternatively classify the outcomes as even (0) or odd (1). Then $S = \{0, 1\}$. In this case 'obtaining an odd number' is the event $\{1\}$. Its probability is

$$\frac{N(\{1\})}{N(S)} = \frac{1}{2},$$

as above.

(b) *Trial*: tossing two coins.

An outcome may be specified by two letters, each of which may be H (head) or T (tail). The first letter denotes the result for a specific coin (which might, for example, be the first tossed), and the second letter denotes the result for the other coin.

$$\text{The possibility space } S = \{HH, HT, TH, TT\}.$$

'Obtaining two heads' is the event $\{HH\}$. Its probability is $\frac{1}{4}$.

'Obtaining at least one head' is the event $\{HH, HT, TH\}$. Its probability is $\frac{3}{4}$.

(c) *Trial*: choice of a domino at random from a standard set of 28 dominoes.

An outcome may be specified by an unordered number pair, e.g. (1, 3). The elementary events of S may be listed:

$$
\begin{array}{ccccccc}
(0,0) & (1,0) & (2,0) & (3,0) & (4,0) & (5,0) & (6,0) \\
 & (1,1) & (2,1) & (3,1) & (4,1) & (5,1) & (6,1) \\
 & & (2,2) & (3,2) & (4,2) & (5,2) & (6,2) \\
 & & & (3,3) & (4,3) & (5,3) & (6,3) \\
 & & & & (4,4) & (5,4) & (6,4) \\
 & & & & & (5,5) & (6,5) \\
 & & & & & & (6,6)
\end{array}
$$

Thus $N(S) = 28$.

The event A, 'at least one of the numbers on the domino is a 6', is the set $\{(6,0),$ $(6,1), (6,2), (6,3), (6,4), (6,5), (6,6)\}$. $N(A) = 7$. Hence

$$P(A) = \frac{N(A)}{N(S)} = \frac{7}{28} = \frac{1}{4}.$$

The event B, 'the domino is a double', is the set $\{(0,0), (1,1), (2,2), (3,3), (4,4),$ $(5,5), (6,6)\}$. $N(B) = 7$.

$$P(B) = \frac{N(B)}{N(S)} = \frac{7}{28} = \frac{1}{4}.$$

The event C, 'the sum of the numbers is a multiple of 3', is the set $\{(2,1), (3,0),$ $(3,3), (4,2), (5,1), (5,4), (6,0), (6,3), (6,6)\}$. $N(C) = 9$.

$$P(C) = \frac{N(C)}{N(S)} = \frac{9}{28}.$$

6.3.1 Exercises

1. List the outcomes (elementary events) in a possibility space S for each of the following trials. Also list the outcomes included in each of the events specified, and hence find the probability of each event.
   (a) *Trial*: a throw of a die.
   Events: (A) the result is an even number; (B) the result is a multiple of 3.

(b) *Trial*: Amanda holds a small party for three guests. She cannot remember which coat each guest brought so she hands the three coats back at random.

Events: (*A*) each guest receives the correct coat; (*B*) exactly one of the three guests receives the correct coat (note that the lucky person is not specified).

(c) *Trial*: Jim always has a drink after his lunch each day and chooses at random whether to drink tea, coffee or water.

Event: Jim has a drink after lunch.

2. A psychological test contains two items, each of which consists of a statement to which the subject expresses his response on a five-point scale. The statements are 'pop music is noisy' and 'motor-bikes are a bigger noise problem than aircraft'. If the subject agrees with a statement very strongly, a mark of $+2$ is allocated; agrees mildly, $+1$; is neutral, 0; disagrees mildly, -1; disagrees very strongly, -2. List a set of possible outcomes of the test. Assuming that every pair of possible scores for the two items is equally likely, find the probabilities of the following events: (*A*) the total score on the two items is 0; (*B*) the score for the pop-music item is lower than for the other item; (*C*) the difference between the scores on the two items is not more than 1.

3. A box contains three dice: one red die whose faces are labelled $1, 2, 3, 4, 5, 6$ and two green dice whose faces are labelled $0, 0, 1, 1, 2, 2$. Two of the three dice are selected at random and thrown. What is the probability that:
   (a) both green dice are selected;
   (b) a red and a green die are selected;
   (c) one of the dice shows 6;
   (d) exactly one of the dice shows 0;
   (e) both the dice show 2;
   (f) the total score on both dice is 4;
   (g) the product of the scores on the two dice is 0;
   (h) the difference between the scores on the two dice is 3?

4. Five people, A, B, C, D, E, dine at a round table. They take their places at random.
   (a) What is the probability that A and B sit next to each other?
   (b) A is married to B and C to D while E is single. What is the probability that no members of a married couple sit next to each other?
   (c) Three of the people are men, Tom, Dick and Harry, and two women, Jill and Joan. If it is *not* known which person has been labelled A, B, C, D or E, what is the probability that Tom is married to Jill?
   (d) What is the probability that Dick does not sit next to Joan?

5. There are six copies of a certain textbook in a school library, all of which were purchased at different times. Three are of the first edition and the other three of the second edition. When they are returned after use, no attention is given to putting them in order. Assuming they are all on the shelf, what is the

probability that:
(a) they are arranged in order of age (i.e. time of purchase);
(b) the three books of the first edition are all together;
(c) the three books of the first edition are all together, and so are those of the second;
(d) the newest of the first edition is next to the oldest of the second edition?

6. A box containing 100 apples is sent to market. Eight of the apples have a skin blemish. An inspector takes a random sample of five apples from the box. What is the probability that he will pick: (a) none, (b) one, (c) two with a blemish?

6.4 PROBABILITY WHEN OUTCOMES ARE NOT EQUALLY LIKELY

We have developed the basic ideas of probability using outcomes that were equally likely. But we often wish to apply probability in situations where the outcomes are *not* equally likely. We may be interested in the consequences of tossing a biased coin, for which the probability of obtaining a head is 0.75 instead of 0.5. Alternatively, there may be situations where we can specify equally likely outcomes, but it is more convenient to work with outcomes that are not equally likely. For example, in the coin-tossing exploits of captain Horace, we were really interested in the number of times he won out of three. As we saw from the tree diagram (Fig. 6.1), there were probabilities of $\frac{1}{8}, \frac{3}{8}, \frac{3}{8}, \frac{1}{8}$ of winning $0, 1, 2, 3$ times. We may still set up a possibility space (Fig. 6.3) with points to represent the four outcomes $0, 1, 2, 3$, but now we must attach a specific probability to each point since the points are no longer equally probable. The probability of an event in this new possibility space is defined as the sum of the probabilities attached to all the points that fall within the event. Thus, P(Horace wins two or three times) = $\frac{3}{8} + \frac{1}{8} = \frac{1}{2}$. We may think of probability as a small mass attached to each point. We evaluate the probability of an event E by summing the masses attached to the points that are included in the event and call the total the probability mass $M(E)$ of the event. The probability mass of the whole possibility space $M(S)$ is, of course, equal to 1. With a possibility space of n equally likely outcomes the probability mass attached to each point is $1/N(S) = 1/n$. Since there are $N(E)$ points in E the total probability mass of E equals $N(E)/n = N(E)/N(S)$, which gives the same definition of probability for equally likely outcomes as before.

Figure 6.3 *Possibility space for number of heads when tossing three coins.*

The more general definition of the probability of E may be written $M(E)/M(S)$, by analogy with $N(E)/N(S)$, but this simplifies to $M(E)$ since $M(S)$ equals unity.

Let us consider more examples of possibility spaces.

(a) *Trial*: two throws of a die.

The outcomes are ordered number pairs which may be conveniently represented in a table. The reason for this particular form of table will become clear later.

$$
\begin{array}{llllll}
(1,1) & (1,2) & (1,3) & (1,4) & (1,5) & (1,6) \\
& (2,1) & (2,2) & (2,3) & (2,4) & (2,5) & (2,6) \\
& & (3,1) & (3,2) & (3,3) & (3,4) & (3,5) & (3,6) \\
& & & (4,1) & (4,2) & (4,3) & (4,4) & (4,5) & (4,6) \\
& & & & (5,1) & (5,2) & (5,3) & (5,4) & (5,5) & (5,6) \\
& & & & & (6,1) & (6,2) & (6,3) & (6,4) & (6,5) & (6,6)
\end{array}
$$

The 36 number pairs may be regarded as the points in a possibility space S in which each point is equally likely and of probability mass $\frac{1}{36}$. If we are interested in the total score after the two throws, the possibility space in which each point is a total score is more useful. From the table above, by counting how many number pairs there are in each column, we can see that $P(\text{Total} = 2) = \frac{1}{36}$, $P(\text{Total} = 3) = \frac{2}{36}$, and so on.

The possibility space in which the points represent totals may be shown as follows.

$\frac{1}{36}$	$\frac{2}{36}$	$\frac{3}{36}$	$\frac{4}{36}$	$\frac{5}{36}$	$\frac{6}{36}$	$\frac{5}{36}$	$\frac{4}{36}$	$\frac{3}{36}$	$\frac{2}{36}$	$\frac{1}{36}$
•	•	•	•	•	•	•	•	•	•	•
2	3	4	5	6	7	8	9	10	11	12

The elementary events are $\{2\}, \{3\}, \ldots, \{12\}$. The probability masses are as shown in the diagram. The event (Total score ≥ 10) consists of the elementary events $\{10\}, \{11\}, \{12\}$. It therefore has probability mass $\frac{3}{36} + \frac{2}{36} + \frac{1}{36} = \frac{6}{36} = \frac{1}{6}$. Hence $P(\text{Total score} \geq 10) = \frac{1}{6}$.

(b) *Trial*: blood group of an individual selected at random from the white population of the USA.

Human blood is commonly classified into four types labelled O, A, B, AB. Among American whites the relative percentages have been estimated to be 45, 41, 10 and 4, respectively. Using O, A, B and AB as elementary events, we may represent the possibility space with assigned probabilities as follows:

O	A	B	AB
•	•	•	•
0.45	0.41	0.10	0.04

A person of blood group A can receive blood from persons of types O and A only; blood of other types leads to clotting. Hence if a patient is of type A the probability that a person chosen at random will be able to supply appropriate blood is equal to the probability of the event {O, A}. The total probability mass of this equals $0.45 + 0.41 = 0.86$. A table of similar probabilities for the four types of blood is as follows:

Blood group of patient	Can receive blood from	Probability
O	O	0.45
A	O, A	$0.45 + 0.41 = 0.86$
B	O, B	$0.45 + 0.10 = 0.55$
AB	O, A, B, AB	1

6.4.1 Exercises

1. A man has noticed that when he sings his dog will (A) howl, (B) ignore him or (C) hide in a corner. (A) is twice as likely as (B) which is twice as likely as (C). Find the probabilities associated with the elementary events A, B and C in the possibility space made up of them.

2. Two dice are thrown. Find the probability masses in the possibility space in which the elementary events are (A) both dice show even numbers, (B) both dice show odd numbers and (C) one die shows an odd and one an even number.

3. A possibility space is made up of four elementary events A, B, C, D where $P(A) = p^2$, $P(B) = pq$, $P(C) = pq$ and $P(D) = q^2$. Express, in the simplest terms, $P(A \text{ or } C)$.

6.5 DEFINITION OF PROBABILITY

We have developed a general procedure for evaluating probabilities. It has three stages:

(i) *Specify the points in a possibility space.* For the simple case of tossing two fair coins where the outcome is the number of heads, we might specify three points corresponding to 'no heads', 'one head' and 'two heads'.

(ii) *Assign a probability to each point in the possibility space such that the probabilities sum to* 1. For our simple example the probabilities $\frac{1}{4}$, $\frac{1}{2}$ and $\frac{1}{4}$ are assigned to 0, 1 and 2 heads respectively.

(iii) *Calculate the probability of an event E by summing the probabilities assigned to the points which are included in the event E.* In the example, the event E 'at least one head' is made up of the events 'one head' and 'two heads'. Hence $P(E) = \frac{1}{2} + \frac{1}{4} = \frac{3}{4}$. We may think of the process as a summation of probability masses.

We will re-express the procedure in a more formal way. These statements, or similar ones which are equivalent, are often referred to as the *axioms of probability theory*. It is possible to derive the whole of the theory from them.

We are considering only possibility spaces in which there is a finite number of outcomes. These spaces are called *discrete*, by contrast with spaces in which outcomes may take any of an infinite number of values within a range, which are called *continuous spaces*.

Definition 6.5

A **discrete possibility space** S consists of the n elementary events $s_1, s_2, \ldots, s_n$.

Definition 6.6

A **probability function** is defined on the possibility space S such that

(i) $P(s_i) > 0$ for $i = 1, \ldots, n$.

(ii) $\sum_{i=1}^{n} P(s_i) = 1$.

(iii) For any event E in S,

$$P(E) = \sum_{S_i \in E} P(s_i)$$

where the summation is over all points in E.

We may write $\sum_{S_i \in E} P(s_i)$ as $M(E)$, the probability mass of E.

We have seen above that this definition includes our earlier one (Definition 6.4, p. 112) for the special case when all outcomes are equally likely. To distinguish, we shall call Definition 6.4 the definition of *classical probability*.

6.6 SETTING UP PROBABILITY MODELS

To use the theory of probability in practical situations we need to specify a set of outcomes together with associated probabilities. Description of the set of outcomes is usually straightforward; let us consider the determination of the values of the probabilities. We may be able to choose outcomes that are equally likely. There may be natural symmetry present, as in trials such as tossing a coin or throwing a die, or the symmetry may be imposed by, for example, choosing a pupil at random from a class. In a limited number of cases we may be able to find appropriate values for unequally likely outcomes by derivation from equally likely outcomes, as in the example on the total score shown by two dice.

We often want to set up models in which outcomes are not equally likely, and the probability of each outcome is unknown. For example, we may be concerned with the probability that a baby dies within the first year of its life. It is not obvious what this probability is, and if we want to use it in a model we shall assign it a theoretical value π. The concern of the statistician is then to *estimate* π. This could be done by taking a sample of n babies, and noting the number a that die within the first year of life. The relative frequency in the sample, a/n, is then taken as the estimate of π, and can be used for π in developing any models needed. In fact, in such a situation we can *define* π as the limit of this relative frequency as n increases.

6.7 COUNTING METHODS

In calculating probabilities, it is essential to be able to work out $N(S)$ and $N(E)$ as straightforwardly as possible. That branch of algebra called 'permutations and combinations' is very helpful here. We develop the ideas by stating methods or theorems and giving examples.

A multiplication principle

If two operations A, B are carried out, and there are m different ways of carrying out A and k different ways of carrying out B, then the combined operations A and B may be carried out in mk different ways.

Example 6.3

A toy manufacturer makes a wooden toy in two parts; the top part may be coloured red, white or blue and the bottom part brown, orange, yellow or green. How many differently coloured toys can he produce?

A red top part may be combined with a bottom part of any of the four possible colours. Similarly, either a white or a blue top part may be combined with each of the four different coloured bottom parts. Hence the number of differently coloured toys is $3 \times 4 = 12$. Readers who are doubtful about the multiplication principle should illustrate this example by a tree diagram.

Each of the orders in which n distinct objects can be arranged is said to be a *permutation* (or an *arrangement*) of the n objects. Thus ABC, ACB, BAC, BCA, CAB, CBA are all the possible permutations of the three letters, A, B, C.

Theorem 6.1 Permutations of n objects

The number of permutations of n distinct objects, taken all together, is $n \times (n-1) \times (n-2) \times \cdots \times 2 \times 1$.

Example 6.4

Another toy includes a set of five wooden rings of the same size, in order from top to bottom, forming one section of the toy. There is one ring of each of the colours black, white, red, yellow and green. How many different toys may be produced? (The rest of the toy, apart from this section, is standard.)

The number of different colour orders possible for this section of the toy is $5 \times 4 \times 3 \times 2 \times 1$: any of the five colours may go in first place, then any of the four not yet used may go in second place, any of the three so far not used may go in third place, either of the remaining two in fourth place, and the last place is occupied by the one colour still unused. (This is an extension of the argument in the previous example.) The total is therefore 120.

Because $n \times (n-1) \times \cdots \times 2 \times 1$ is often used, it has a special name and symbol *n-factorial, n!*. The number of possible colour orders in the example above is 5!.

It follows from the previous two examples that if we have two distinct sets of objects, m in the first set and k in the second, and each set may be arranged in any order, then the total number of arrangements of the first set in any order followed by the second set in any order is $m! \times k!$.

Example 6.5

A company codes its customers by giving to each customer an eight-character code. The first three characters in this code are the letters A, B, C in any order and the remaining five are the digits 1 to 5 in any order. Each letter and each digit can occur only once. How many customers can the company code in this way?

The first three positions can be filled in 3! ways and the next five in 5! ways, so that the total is $3! \times 5! = (3 \times 2 \times 1) \times (5 \times 4 \times 3 \times 2 \times 1) = 6 \times 120 = 720$.

Similar forms of argument can be used when there is some restriction on the permitted orders.

Example 6.6

There are 10 people employed in the service department of a company. A particular job requires 3 people, assigned to specific parts (*a, b, c*) of the job, together with a supervisor. Only 3 of the 10 employees are qualified to supervise this job. In how many ways can a team be selected, and allocated to the specific parts of the job?

Note that we must have a supervisor, and that this place can be filled in only 3 ways. Now any of the remaining 9 employees can be assigned to part *a* of the job, then any of the remaining 8 to *b* and finally any of the remaining 7 to *c*. This gives $3 \times 9 \times 8 \times 7 = 1512$ ways altogether of choosing and allocating a team.

We have assumed that those qualified to supervise may also be included as ordinary members of the team if picked. If this is not so, there is more restriction on the choice. First, the supervisor is chosen (as before) from the 3 available, so there are still 3 ways of filling that position. The other 2 people qualified to supervise cannot be chosen for other parts of the job, so the remaining choices must be made from the 7 employees not qualified to supervise. Thus the employee assigned to part *a* of the job may be chosen in 7 ways, then the one for *b* in 6 ways and finally the one for *c* in 5 ways. Thus the number of ways of choosing and allocating the team now is $3 \times 7 \times 6 \times 5 = 630$.

We may be interested in the number of permutations, or arrangements, of two digits when the two have been chosen from the ten digits $0, 1, 2, \ldots, 9$. Some of the permutations are $01, 10, 12, 21$. The total number of permutations is said to be the number of permutations of ten things taken two at a time.

Theorem 6.2 Permutations of n objects taken r at a time

The number of permutations of n distinct objects taken r at a time is $n \times (n-1) \times (n-2) \times \cdots \times (n-r+1)$.

The first place in the permutation may be filled in n ways, the second in $(n-1)$ ways from among all the objects which remain, the third similarly in $(n-2)$ ways; we continue in this manner until the final choice is made from the remaining $(n-r+1)$ objects.

This number is often given the symbol ^nP_r. In factorial notation

$$^nP_r = \frac{n!}{(n-r)!}.$$

Example 6.7

What is the number of permutations of ten distinct digits taken two at a time?

The first digit can be chosen in 10 ways and, having filled the first place, the second digit can be chosen in 9 ways. Hence there are $10 \times 9 = 90$ permutations.

Suppose that we do not care in what order the r objects are chosen in the previous example, but are only interested in *which* r are chosen from the n. Then all the $r!$ ways in which we may arrange those r objects that are chosen are equivalent. So the number of ways in which r objects may be chosen from n distinct objects, with *no* attention paid to the order of choice, is

$$\frac{^nP_r}{r!} = \frac{n(n-1)\ldots(n-r+1)}{1,2\ldots r}.$$

This is the number of different *selections* or *combinations* of r objects which may be made from n distinct objects. The symbol for this is ^nC_r or $\binom{n}{r}$, and in factorial notation

$$\binom{n}{r} = \frac{n!}{r!(n-r)!}.$$

Theorem 6.3 Combinations of n things taken r at a time

The number of combinations of n things taken r at a time is

$$\frac{n!}{r!(n-r)!} \quad \text{or} \quad \binom{n}{r}.$$

The expression $\binom{n}{r}$ arises in another, slightly different, way. Suppose that we have n objects altogether, and that they fall into two types: for illustration we will use capital and small letters. Of these n objects there are r of one type (say, capitals) and $n-r$ of the other (small letters). In how many distinct ways can we arrange these two types of letter?

It is only the order of *types* which matters. For present purposes, taking a specific example with $n=6$ and $r=3$, the orders ABCabc and CABbca are the same; in fact, using one of the rules above, there are $3! \times 3!$ orders in all of which the types go capital–capital–capital–small–small–small. Obviously this is also true for any other scheme of arranging the types capital and small. The whole set of six letters A, B, C, a, b, c can be arranged in 6! ways. But the number of distinct orders of types is only $6!/(3! \times 3!)$. Try writing down all the distinct possible orders of types in a systematic way and then enumerating them: this is much more tedious, even with small numbers, as in this example, where the answer

$$\frac{6 \times 5 \times 4 \times 3 \times 2 \times 1}{3 \times 2 \times 1 \times 3 \times 2 \times 1}$$

is equal to 20.

Theorem 6.4 *Permutations of two types of object*

The number of distinct ways of arranging n objects, of which r are of one type and $n - r$ of another, and where objects of the same type are indistinguishable, is

$$\frac{n!}{r!(n-r)!} = \binom{n}{r}.$$

This can easily be seen by using exactly the same argument as we have used above in the special case.

The symbol $\binom{n}{r}$ is defined by the expression $n!/r!(n-r)!$ quite clearly and unambiguously for all values of r from 1 to $n-1$; note in passing that because of this definition $\binom{n}{r}$ must equal $\binom{n}{n-r}$. But what, if any, meaning can $\binom{n}{r}$ have if $r = 0$ or n? Clearly there is only one distinct way of arranging n objects which are all of the same type; also there is only one possible selection of n objects from n, namely all of them. So $\binom{n}{n}$ must be 1. Thus $n!/n!0! = 1$, which can only be true if we define $0! = 1$. (Our previous definition of $r!$ does not cover the case $r = 0$, and so we are perfectly at liberty to add this extra convention to extend the definition to this case.) Obviously $\binom{n}{0} = 1$ also.

6.7.1 Exercises

1. If everyone had only two names, a forename and a surname, how many distinct pairs of initials would be possible?

2. A manufacturer supplies a model of a car in six colours and with three optional extras: hi-fi, leather seats and metallic paint. How many distinct types of car may be ordered?

3. A four-volume dictionary is replaced on a shelf. In how many different orders may the volumes be replaced?

4. How many distinct vehicle licence plates can be formed if the first three symbols on a licence plate are letters and the last three symbols are digits? (Assume *all* selections possible.)

5. Assuming that any arrangement of letters forms a word, how many four-letter words can be formed from: (a) KATE; (b) JILL?

6. When Mother is serving meals she always serves Tiny Tim first, Jane second, William third and then the adults (Uncle Jack, Aunt Jill and Father) in any order, and herself last. In how many ways can she do this? However, Father, when serving drinks, always serves first the adults, excluding himself, in any order and then the children, in any order, and finally himself. In how many ways can he do this?

7. A man places four cards, from left to right, on a table from a standard pack of 52 playing cards. How many distinct arrangements can he produce?

8. A football league contains 8 teams. How many matches have to be played in order that each team plays every other team?

9. Assuming that any arrangement of one or more letters forms a word, how many words can be formed from: (a) RAT; (b) TRAVEL?

10. A class contains 9 boys and 11 girls. In how many ways can a committee of two boys and two girls be formed from the class?

11. A nursery offers a collection of roses of assorted colours, which is to contain one red, one yellow, one white and two pinks. It has two different named red roses available, three different named yellows, two different named whites and five different named pinks. In how many different ways can the collection be made up?

6.8 USE OF COUNTING METHODS IN PROBABILITY

We are now able to return to the calculation of probabilities using the relative-frequency definition.

Example 6.8

Three fair cubical dice are thrown. Find the probability that: (i) the sum of the scores is 18; (ii) the sum of the scores is 5; (iii) none of the three dice shows a 6; (iv) the product of the scores is 90. (*AEB*)

Let an outcome be defined by an ordered triple, e.g. $(2, 1, 6)$, where the first digit denotes the number appearing on die 1, the second digit that on die 2 and the third digit that on die 3. S therefore contains $6^3 = 216$ outcomes.

(i) A score of 18 (event E) occurs only when each die shows 6. Hence E contains only one outcome and its probability equals

$$\frac{N(E)}{N(S)} = \frac{1}{216}.$$

(ii) This event occurs with the outcomes $(1, 1, 3), (1, 3, 1), (3, 1, 1)$ and $(1, 2, 2)$, $(2, 1, 2), (2, 2, 1)$. So the probability of the event is

$$\frac{6}{6^3} = \frac{1}{36}.$$

(iii) If none of the three dice shows a 6, each may show any of the numbers $1, 2$, $3, 4, 5$. The number of outcomes in which this may happen is 5^3. Hence the probability of the event is

$$\frac{5^3}{6^3} = \frac{125}{216}.$$

(iv) The prime factors of 90 are $2, 3, 3, 5$. The only group of three feasible numbers that may be derived from these is 3, 5, 6. We must count the outcomes as these numbers occur in the different dice, e.g. $(3, 5, 6), (3, 6, 5), (5, 6, 3) \ldots$. We want the number of permutations of three things, which is 3! or 6. Hence the probability of the event is

$$\frac{6}{6^3} = \frac{1}{36}.$$

Example 6.9

A batch of 20 items, taken from an industrial production line, contains 3 which are faulty. An inspector, not knowing this, selects 2 at random from the 20. What is the probability that: (i) both selected items are faulty; (ii) at least one selected item is faulty?

If the 20 items were labelled A, B, C, ..., T then the outcome of a trial might be denoted by an unordered pair of letters, e.g. AB, EP. The total number of outcomes is the number of ways of selecting any 2 items from 20; so

$$N(S) = \binom{20}{2} = \frac{20 \times 19}{1 \times 2} = 190.$$

(i) Let E be the event that both items are faulty. That is, we select two from the three faulty and none from the 17 perfect items available. The number of ways of doing this is $\binom{3}{2} \times \binom{17}{0}$, which is $N(E)$. So the required probability is

$$\frac{N(E)}{N(S)} = \binom{3}{2} \times \binom{17}{0} \Big/ \binom{20}{2} = 3 \times 1 \Big/ \binom{20}{2}$$

$$= 3 \times 1 \times \left(\frac{2 \times 1}{20 \times 19} \right) = \frac{3}{190} = 0.0158.$$

(ii) The required event F includes outcomes in which: (a) one item is faulty; (b) both items are faulty. The number for case (a) is $\binom{3}{1}\binom{17}{1}$ and for case (b) is $\binom{3}{2}\binom{17}{0}$. The number $N(F)$ is the sum of these and equals

$$\frac{3!}{2!1!} \times \frac{17!}{16!1!} + \frac{3!}{2!1!} \times 1 = 3 \times 17 + 3 = 54.$$

Hence

$$P(F) = \frac{N(F)}{N(S)} = \frac{54}{190} = \frac{27}{95} = 0.2842.$$

An alternative approach is to use the fact that the number of outcomes with at least one faulty is equal to the total number of outcomes minus the number of outcomes with none faulty. Now the number of outcomes with none faulty is

$$\binom{17}{2}\binom{3}{0} = \frac{17!}{15!2!} \times 1 = \frac{17 \times 16}{1 \times 2} = 17 \times 8 = 136.$$

Hence $N(F) = 190 - 136 = 54$. We then proceed as before.

6.9 SIMULATION

How is it possible to check that probability theory really works? At the simplest level we might toss a coin many times and verify that the proportion of heads, say, settles close to a fixed value, 0.5 if the coin is fair. There are records of historical interest. The eighteenth-century French naturalist Buffon tossed a coin 4040 times and observed 2048 heads and 1992 tails. A more elaborate test was made, at the end of the nineteenth century, by the English biologist Weldon who tossed 12 dice 26 306 times and each time observed the number of dice which showed 5 or 6. Making tests such as these by hand is extremely laborious and, for that reason, only the simplest are attempted. Doing similar tests by computer is relatively easy and the verification of predictions for complicated situations becomes possible. These ideas are taken up in Section 6.12.

By using the computer's ability to produce many pseudo-random numbers, it is possible to generate chance events and investigate them. For example, we can generate the results of 100 trials in which there are two possible outcomes of equal probability. The computer is programmed to generate pseudo-random numbers in the range $(0, 1)$. A value less than 0.5 is recorded as a head while a value greater than 0.5 is recorded as a tail. This allows us to mimic tossing a coin and counting the numbers of heads and tails. Investigations of this sort are called *simulations*; to *simulate* is to imitate or mimic. In the next section we show how to use MINITAB to make simple simulations.

6.10 MINITAB: GENERATING RANDOM SAMPLES

Simulations using MINITAB are based on the menu choice **Calc>Random Data**. We met this in Section 3.9 and saw how to generate random integers or random numbers in any chosen interval. Expressed another way, we were choosing random samples from the **Uniform** and **Integer** distributions, using MINITAB's terminology. We now introduce another theoretical distribution that MINITAB makes available.

Independent trials which have two possible outcomes, often denoted by 0 and 1, with fixed probabilities, are called Bernoulli trials (see Section 8.5). They provide a model for the results of coin tossing, for example. The menu choice that generates an appropriate sequence of 0s and 1s is

> **Calc>Random Data>Bernoulli/**
> Specify no. of **rows of data**, enter **Storage columns** and state
> **probability of success/**

6.11 EXERCISES ON CHAPTER 6

1. How many four-digit numbers can be formed from the digits 1, 2, 3, 4 using each digit once only? How many of these numbers are less than 3000 and how many are divisible by 4?

2. Four dice are thrown at the same time. What is the probability: (i) that all four will show different scores; and (ii) that the four will show consecutive scores (irrespective of the order in which the dice are observed)?

3. Neglecting those born on 29 February, what is the probability that none of a group of 20 randomly chosen people has a birthday on the same day as anyone else in the group? If the group consists of n people, what would this probability be? What should n be so that this probability is: (a) less than $\frac{1}{2}$; (b) less than $\frac{1}{3}$?

4. Six blue balls, four red balls and two white balls are placed in a straight line. In how many of the possible arrangements will both end balls be of the same colour? The balls are split into two identical sets, each of which is placed in a line. In what proportion of the possible arrangements will:
   (a) all four end balls be of the same colour;
   (b) at least two of the four end balls be of the same colour;
   (c) not more than two of the four end balls be of the same colour?   (*IOS*)

5. There are three bus services between a suburb and the town centre. Number 1 goes into town at 15 and 45 minutes past the hour, number 101 at half past the hour, and number 111 on the hour. At random times during the day, four people go to the bus stop to wait for a bus into town. What is the probability that:
   (a) all four catch a number 1;
   (b) all three different-numbered services will be used;
   (c) no one will catch a number 1?
   Illustrate this problem by constructing a possibility space.

6. The following table gives the population of a small village, classified by age (in years) at last birthday.

Age (years)	Males	Females
0–4	18	18
5–14	38	36
15–29	67	65
30–44	56	56
45–59	46	47
60–64	15	17
65–74	21	27
75–84	10	19
85 and over	2	5

One person is selected at random from the village. What is the probability that this person is:
(i) female;
(ii) aged 30 years or over;
(iii) a male between 30 and 59 years old;
(iv) a female under 60 years old?

7. (a) Sixteen football clubs enter a competition. They are divided at random into two groups, each club playing all the others in its group, and the final is between the winners of the two groups. What is the probability that the second best club will not appear in the final? (Assume that all results of matches reflect the true abilities of the clubs involved.)

(b) If the sixteen clubs are divided into four groups, with the winner of each group entering the final stage of the competition, what is the probability that the second best club will not appear in the final stage? What is the probability that: (i) the third best club; (ii) the fifth best club; will appear in the final stage?

6.12 COMPUTING EXERCISES

1. Simulate tossing a coin, and calculate the proportion of heads at regular intervals. Observe whether this proportion tends to $\frac{1}{2}$. (Try 1500 trials and calculate the proportion after every 50 trials.)

2. Simulate the situation in Section 6.3, in which the quantity of interest is the number of times that Horace wins the toss in a series of three games. Find the relative frequencies of the four possible outcomes $(0, 1, 2, 3)$.

6.13 PROJECTS

1 Relative frequency of sixes when throwing a die

Aim. To find the relative frequency of sixes obtained when throwing a die, and compare it with the theoretical probability.

Equipment. Die and shaker.

Collection of data. Throw a die 100 times and note the number obtained on each throw. It is convenient to record in blocks of five results, perhaps one block per line.

Analysis. Count the number of sixes in each block and derive the cumulative total of sixes for each block. Express these cumulative totals as a proportion of the total number of throws. Plot these proportions against the total number of throws, i.e. $5, 10, 15, 20, \ldots$. Join the successive points by straight lines. Draw in the line for the proportion $\frac{1}{6}$ and note if the observed proportions tend to settle close to this line.

2 Distribution of telephone subscribers

Aim. To estimate the proportion of telephone subscribers within an area that are on a particular exchange.

Equipment. A telephone directory (white pages).

Collection of data. From the exchanges included in the area, choose one that has an appreciable number of subscribers. Use random digits to choose a subscriber, by first choosing a page at random, and then choosing a subscriber at random from that page. (It may help to choose a column on the page as the second step, and then finally an entry in that column.) Note the name of the subscriber's exchange.

Repeat to obtain a sample of at least 50 subscribers.

Analysis. Find the proportion of subscribers in the sample that are on the chosen exchange. This is the estimate of the required proportion.

Notes

1. Suppose our estimate is p. Using the definition at the beginning of this chapter, we shall then say that the probability of any further, randomly chosen, entry being on the chosen exchange is p. So we have also estimated a *probability*.

2. An alternative record, instead of noting the exchange, is to take the last digit of the telephone number. The records of last digits can then be examined to see whether there is an excess of any particular digits (e.g. in some areas subscribers renting several lines may have numbers ending in 1 or in 0).

3. It is possible that yellow-pages directories will give different answers from white pages for the same area. Comparisons could be made.

SUMMARY

Classical probability is concerned with trials that have *equally likely* outcomes. A situation is represented by a **possibility space** S made up of **elementary events** which represent the equally likely outcomes. Let $N(E)$ equal the number of elementary events in the event E. Then the **probability** of E equals $N(E)/N(S)$.

The definition of probability is extended to cases where all outcomes are not necessarily equally likely by assigning *probability masses* to the elementary events such that the total probability mass of the events in S is one. The **probability** of the event E equals the total mass of the elementary events in E, $M(E)$.

A formal approach is to define a probability function on a discrete possibility space S consisting of the n elementary events $s_1, s_2, \ldots, s_n$. A **probability function** P satisfies the following axioms:

(i) $P(s_i) > 0$ for $i = 1, 2, \ldots, n$;

(ii) $\sum_{i=1}^{n} P(s_i) = 1$;

(iii) For any event E in S,

$$P(E) = \sum_{S_i \in E} P(s_i).$$

Counting methods are an important aid in evaluating probabilities. Some basic rules are:

(i) the number of **permutations**, or **arrangements**, of n distinct objects, *taken all together*, is

$$n \times (n-1) \times (n-2) \times \cdots \times 2 \times 1, \quad \text{i.e.} \quad n!;$$

(ii) the number of **permutations**, or **arrangements**, of n distinct objects *taken r at a time*, is

$$n \times (n-1) \times (n-2) \times \cdots \times (n-r+1), \quad \text{i.e.} \quad n!/(n-r)!;$$

(iii) the number of **combinations**, or **selections**, of n distinct objects taken r at a time is

$$n!/r!(n-r)! \quad \text{or} \quad \binom{n}{r}.$$

7 PROBABILITIES OF COMPOUND EVENTS

OVERVIEW

We study how the probability of compound events, derived from the union and intersection of events, is related to the probability of simple events.

7.1 INTRODUCTION

We can sometimes express one event E, in which we are particularly interested, in terms of other events using the operations possible with sets. We show in this chapter that the probability of E may then be related to the probabilities of the other events. This can often simplify calculating the probability of E.

7.2 THE INTERSECTION OF TWO EVENTS

Let us look again at our friend Horace's coin tossing. In Fig. 7.1, each possible outcome of a sequence of three tosses is represented by a point, and the particular events that we are interested in are illustrated.

We shall adopt the term *possibility space* for the set of all possible outcomes S, as explained on p. 111. The events illustrated are:

      A Horace wins three times;
      B Horace wins the middle one of the three;
      C Horace wins exactly one of the three.

Remember, Horace always calls heads.

It is clear that some of the points can appear in more than one event. For example, THT appears in B and C. These two events can occur together: if one of B or C occurs it does not render the other impossible. On the other hand, A and C cannot both happen together; they have no points in common in Fig. 7.1.

Figure 7.1 *Diagram illustrating some events in the possibility space for three tosses of a coin.*

In set theory, the elements that are common to two sets form the *intersection* of the sets. We adopt the same language for events. Look again at events B and C as represented in Fig. 7.1; they have just one point in common, that point which represents THT. In other words, the outcome THT is part of both events B and C: it is the intersection of the events B, C. This intersection, being a subset of the possibility space S, is itself an event. We may label it as the event $B \cap C$, which we read 'B intersection C', or 'B and C'. This use of 'and' follows normal English usage, since the compound event $B \cap C$ would be expressed in words as the event in which: 'Horace wins the middle one of the three tosses' and 'Horace wins exactly one of the tosses'. That is, $B \cap C$ is an event where *both* the component events B, C hold, and it is satisfied when 'Horace wins exactly one of the tosses *and* that toss is the middle one'; so it is the single outcome THT.

Definition 7.1

The **intersection** of the events Q, R is the event consisting of those points in the possibility space which appear in both the events Q and R.

The number of points of S common to Q and R is $N(Q \cap R)$.

Returning to Fig. 7.1 (and Horace), $N(B \cap C) = 1$; $N(A \cap B) = 1$, since $A \cap B$ consists of HHH alone; $N(A \cap C) = 0$, so that $A \cap C$ is the empty set and we could write $A \cap C = \emptyset$. A and C are said to be *mutually exclusive* events.

Definition 7.2

Two events are **mutually exclusive** if they have no points in common. (Sometimes the same term is used for events as for sets, so that mutually exclusive events are also said to be *disjoint*.)

Referring to our definition of classical probability, that $P(E) = N(E)/N(S)$, we can immediately write down $P(A \cap B) = \frac{1}{8}$, $P(A \cap C) = 0$, $P(B \cap C) = \frac{1}{8}$. We would, of course, obtain the same answers if we assigned a probability mass of $\frac{1}{8}$ to each of the eight outcomes and used the general definition of probability (Definition 6.6(iii), p. 118).

7.3 THE UNION OF TWO EVENTS

Those points of S which occur in either B or C, or both, are called the *union* of the events B and C, again using the language of set theory; this union is clearly a subset of S and so is itself an event. We label it $B \cup C$, which we read 'B union C' or 'B or C'. When expressing the compound event $B \cup C$ in words we must be careful to remember that it means event B or event C or *both*. Thus, in words 'B or C' is the event: 'Horace wins the middle one of the three tosses' or 'Horace wins exactly one of the three tosses' or Horace does both of these things. By considering the verbal expression of the event one *may* manage to decide that it is equivalent to the event 'Horace wins just the first toss or just the last toss or the middle toss, no matter what happens at the first or last toss'. But any attempt at this type of analysis in words quickly shows the value of a diagram such as Fig. 7.1. From the figure we see that $B \cup C$ is {HHH, HHT, THH, THT, HTT, TTH}. In the same way we could examine unions of other events: for example, we see from Fig. 7.1 that $A \cup B$ is {HHT, THT, HHH, THH} and is identical with B.

Definition 7.3
The **union** of the events Q, R is the event consisting of those points in the possibility space S which appear in Q or R or both.

We see that $N(A \cup B) = 4$ and $N(B \cup C) = 6$. Thus $P(A \cup B) = N(A \cup B)/N(S) = \frac{4}{8} = \frac{1}{2}$ and $P(B \cup C) = \frac{6}{8} = \frac{3}{4}$.

Example 7.1 Coin tossing with a biased coin
Suppose that Horace's coin is biased, so that the probability is as shown in brackets in the following table (note that the probabilities sum to one).

$$\text{HHT}(\tfrac{2}{27}) \quad \text{THT}(\tfrac{4}{27}) \quad \text{HTT}(\tfrac{4}{27}) \quad \text{TTH}(\tfrac{4}{27})$$

$$\text{HHH}(\tfrac{1}{27}) \quad \text{THH}(\tfrac{2}{27}) \quad \text{HTH}(\tfrac{2}{27}) \quad \text{TTT}(\tfrac{8}{27}).$$

Define the events A, B and C as on p. 130, and as illustrated in Fig. 7.1.
Find $P(A \cap B), P(B \cap C), P(A \cup B), P(B \cup C)$.

$$P(A \cap B) = P(A) = P\{\text{HHH}\} = \tfrac{1}{27}.$$
$$P(B \cap C) = P\{\text{THT}\} = \tfrac{4}{27}.$$
$$P(A \cup B) = P(B) = P\{\text{HHT, THT, HHH, THH}\}$$
$$= M\{\text{HHT, THT, HHH, THH}\}$$
$$= \tfrac{2}{27} + \tfrac{4}{27} + \tfrac{1}{27} + \tfrac{2}{27} = \tfrac{9}{27}$$
$$= \tfrac{1}{3}.$$
$$P(B \cup C) = M\{\text{HHT, THT, HHH, THH, HTT, TTH}\}$$
$$= \tfrac{2}{27} + \tfrac{4}{27} + \tfrac{1}{27} + \tfrac{2}{27} + \tfrac{4}{27} + \tfrac{4}{27} = \tfrac{17}{27}.$$

A quicker method of obtaining the final result is to say that

$$P(B \cup C) = 1 - M\{\text{HTH, TTT}\} = 1 - (\tfrac{2}{27} + \tfrac{8}{27})$$
$$= 1 - \tfrac{10}{27} = \tfrac{17}{27}.$$

7.3.1 Exercises

1. A possibility space for two successive tosses of a fair coin is $S = \{$TT, TH, HT, HH$\}$. Draw a diagram of the space and mark in sets corresponding to

the events:

> A both tosses have the same result;
> B there is at least one head;
> C both tosses are heads.

List the points in the events $A \cup B, A \cup C, B \cup C, A \cap B, A \cap C, B \cap C$, and express these events in words.
  Show that $P(A \cup B) + P(A \cap B) = P(A) + P(B)$.

2. State whether the pair of events listed for each trial below are mutually exclusive.
   (a) Choice of a child at random from a class of boys and girls: (i) the child is a girl; (ii) the child is left-handed.
   (b) Throw of two fair dice: (i) total of 10; (ii) total of 11.
   (c) Deal of a card from an ordinary pack of 52 playing cards: (i) the ace of spades; (ii) a black card.
   (d) Deal of a hand of two cards: (i) the hand includes at least one red card; (ii) the hand includes at least one black card.

3. Find the probability of the union of the two events in each of 2(b), 2(c), 2(d) above, by finding the probabilities of the individual events and their intersection.

4. Let R be an event in the possibility space S. Show that (i) $R \cup S = S$, (ii) $R \cap S = R$, (iii) $R \cap \emptyset = \emptyset$.

7.4 THEOREMS ON THE PROBABILITY OF THE UNION OF TWO EVENTS

If two events, Q and R, are mutually exclusive, the probability of their union takes a particularly simple form. Since each elementary event belongs either to Q or to R, but not to both, the probability mass of $Q \cup R$ equals the probability mass of Q plus the probability mass of R. In symbols we write

$$P(Q \cup R) = M(Q \cup R), \quad \text{from Axiom (iii), p. 118,}$$
$$= M(Q) + M(R), \quad \text{since } Q \text{ and } R \text{ are mutually exclusive,}$$
$$= P(Q) + P(R).$$

Theorem 7.1 Addition rule for mutually exclusive events
If Q and R are mutually exclusive events of a discrete possibility space S then

$$P(Q \cup R) = P(Q) + P(R).$$

Example 7.2

Find the probability of obtaining at least two aces in a hand of three playing cards.

An outcome is specified by the three cards, in any order, that might make a hand. Hence

$$N(S) = \binom{52}{3} = \frac{52 \cdot 51 \cdot 50}{1 \cdot 2 \cdot 3} = 22\,100.$$

Let A_1 be the event that exactly two of the cards are aces. Thus

$$N(A_1) = \binom{4}{2}\binom{48}{1} = \frac{4 \cdot 3}{1 \cdot 2} \cdot 48 = 288.$$

and

$$P(A_1) = \frac{288}{22\,100} = 0.0130.$$

Let A_2 be the event that exactly three of the cards are aces. Then

$$N(A_2) = \binom{4}{3}\binom{48}{0} = \frac{4 \cdot 3 \cdot 2}{1 \cdot 2 \cdot 3} \cdot 4.$$

and

$$P(A_2) = \frac{4}{22\,100} = 0.0002.$$

The required event is $A_1 \cup A_2$, and since A_1 and A_2 are mutually exclusive

$$P(A_1 \cup A_2) = P(A_1) + P(A_2) = 0.0130 + 0.0002 = 0.0132.$$

7.4.1 General case

We now state and prove the more general form of the theorem when Q and R are not necessarily mutually exclusive.

Theorem 7.2 Probability of union of two events

If Q and R are events of a discrete possibility space S then

$$P(Q \cup R) = P(Q) + P(R) - P(Q \cap R).$$

The theorem may be explained in terms of probability masses. If Q and R intersect, the probability mass of Q plus the probability mass of R exceeds the probability mass of $Q \cup R$ since the probability mass of $Q \cap R$ is counted twice. Hence

$$M(Q \cup R) = M(Q) + M(R) - M(Q \cap R).$$

A formal proof is best understood from a Venn diagram (Fig. 7.2). Assume that Q and R intersect. (When Q and R do not intersect the theorem is true from Theorem 7.1.) The three mutually exclusive events produced are labelled A, B

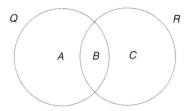

Figure 7.2 *Venn diagram to illustrate proof of Theorem 7.2.*

and C where $A \cup B = Q$, $B \cup C = R$ and $B = Q \cap R$, as shown in the diagram. We have

$$
\begin{aligned}
P(Q \cup R) &= P(A \cup B \cup C) \\
&= P(A \cup B) + P(C), \quad \text{using Theorem 7.1, since } A \cup B \text{ and } C \text{ are} \\
&\qquad\qquad\qquad\qquad\quad \text{mutually exclusive,} \\
&= P(A) + P(B) + P(C), \quad \text{again using Theorem 7.1} \\
&= [P(A) + P(B)] + [P(B) + P(C)] - P(B), \quad \text{by rearrangement} \\
&= P(A \cup B) + P(B \cup C) - P(B), \quad \text{using Theorem 7.1} \\
&= P(Q) + P(R) - P(Q \cap R).
\end{aligned}
$$

Example 7.3

An urn contains three balls: one black, one white and one green. A ball is chosen at random and its colour noted; it is then replaced. One more ball is chosen at random and its colour noted. Let Q be the event that at least one of the chosen balls is black, and R be the event that the balls are of the same colour. Specify a possibility space for the problem. List the points of the space in the events Q, R, $Q \cup R$, $Q \cap R$ and hence find the probabilities of each of these events.

For any two events C, D express $P(C \cup D)$ in terms of $P(C), P(D), P(C \cap D)$, and show that the relation holds for the two particular events Q and R above.

Use B, W, G to denote the black, white and green balls respectively. The possibility space $S = \{BB, BW, BG, WB, WW, WG, GB, GW, GG\}$, and each outcome is equally likely.

The event that at least one of the balls is black,

$$Q = \{BB, BW, BG, WB, GB\}.$$

The event that both the balls are of the same colour,

$$R = \{BB, WW, GG\}.$$

$$Q \cup R = \{BB, BW, BG, WB, GB, WW, GG\}.$$

This event cannot be given a concise description in words. Finally,

$$Q \cap R = \{BB\}$$

and is the event that both balls are black.

$$P(Q) = \frac{N(Q)}{N(S)} = \frac{5}{9}, \quad P(R) = \frac{3}{9} = \frac{1}{3},$$

$$P(Q \cup R) = \frac{7}{9}, \qquad P(Q \cap R) = \frac{1}{9}.$$

Now, by Theorem 7.2

$$P(C \cup D) = P(C) + P(D) - P(C \cap D).$$

If we replace C by Q and D by R, the value of the right-hand side of this expression is

$$\frac{5}{9} + \frac{1}{3} - \frac{1}{9} = \frac{5 + 3 - 1}{9} = \frac{7}{9},$$

which equals the value of the left-hand side, and hence the relation holds.

7.5 COMPLEMENTARY EVENTS

Definition 7.4

The **complementary event** Q^c to the event Q is the set consisting of all those points of the possibility space S that are not in Q.

Thus Q and Q^c are mutually exclusive and $Q \cup Q^c = S$. If we use Theorem 7.1 with the complementary events Q and Q^c, we obtain

$$P(Q \cup Q^c) = P(Q) + P(Q^c).$$

But the left-hand side is $P(S)$, which is equal to 1. Hence we obtain the following theorem.

Theorem 7.3

If Q and Q^c are complementary events then

$$P(Q) = 1 - P(Q^c)$$

Example 7.4

Find the probability of obtaining at least one ace in a hand of four cards.

An outcome is specified by four cards in any order. We have

$$N(S) = \binom{52}{4} = \frac{52 \cdot 51 \cdot 50 \cdot 49}{1 \cdot 2 \cdot 3 \cdot 4} = 270\,725.$$

Let Q be the event 'at least one ace in a hand of four cards'. Then Q^c is the event 'no aces in a hand of four cards'. So for Q^c the four cards must be selected from the 48 non-aces, and thus

$$N(Q^c) = \binom{48}{4} = \frac{48 \cdot 47 \cdot 46 \cdot 45}{1 \cdot 2 \cdot 3 \cdot 4} = 194\,580.$$

and

$$P(Q^c) = 0.7187.$$

Therefore

$$P(Q) = 1 - P(Q^c) = 1 - 0.7187 = 0.2813.$$

In this example it is obviously easier to compute $P(Q^c)$ than to find $P(Q)$ directly, since Q is 'one ace or two aces or three aces or four aces' and would be quite complicated to deal with. The previous example could also be worked this way, and as will be seen there is little to choose between finding $P(Q^c)$ and finding $P(Q)$ directly in that case.

Example 7.2 (repeated)

Find the probability of obtaining at least two aces in a hand of three playing cards.

The required probability is the complement of the probability that no aces or one ace will be obtained. Let B_1 be 'no aces', B_2 be 'one ace'. Thus

$$N(B_1) = \binom{48}{3} = \frac{48 \cdot 47 \cdot 46}{3 \cdot 2 \cdot 1} = 17\,296.$$

We have already found $N(S) = 22\,100$, so

$$P(B_1) = \frac{17\,296}{22\,100} = 0.7826.$$

Similarly

$$N(B_2) = \binom{4}{1}\binom{48}{2} = 4 \cdot \frac{48 \cdot 47}{2 \cdot 1} = 4512,$$

and so

$$P(B_2) = \frac{4512}{22\,100} = 0.2042.$$

The events B_1, B_2 are mutually exclusive, and so

$$P(B_1 \cup B_2) = P(B_1) + P(B_2) = 0.9868.$$

This is the probability of 0 or 1 aces, and the probability of 2 or 3 aces is thus $1 - 0.9868 = 0.0132$, as we obtained directly in the original example.

7.5.1 Exercises

1. Two fair dice are thrown. Find the probability of: (i) a score greater than 3; (ii) a score of 7 or 11; (iii) a score that is not divisible by 4.

2. If A and B are two events in the same possibility space, show that:
   (i) $P(A) = P(A \cap B) + P(A \cap B^c)$,
   (ii) $P(A \cap B^c) = P(A) - P(B) + P(A^c \cap B)$.

3. A certain type of component used in manufacturing transistor radios is sent out in boxes of 50. One such box contains 5 that are faulty. What is the probability that a manufacturer using 10 components selected at random from this box will pick 2 or more of these faulty ones?

7.6 THE INTERSECTION AND UNION OF THREE OR MORE EVENTS

Intersections and unions of any number of events may be defined by extending the methods developed for two events. Let us consider, as an example, an international committee of nine people. The chairman is chosen at random from the members of the committee, and we are interested in the probability that different types of people become chairman. The composition of the committee is illustrated in Fig. 7.3, each person being represented by a point. There are six nations represented: China (C), Russia (R) and America (A) each having two members, and there is one member from each of Egypt (E), Britain (B) and Germany (G). There are two females (F) on the committee, the Egyptian and one of the Americans, and two young (Y) people, the Egyptian and one of the Russians; 'young' here means under 40. Each person is equally likely to be chosen as chairman and Fig. 7.3 represents a possibility space S for this choice. The events in this space are shown by italic letters in an obvious notation.

The three events E, F, Y intersect and their intersection contains a single point. We write the compound event $E \cap F \cap Y$, 'E intersection F intersection Y'. So $N(E \cap F \cap Y) = 1$, and from our definition of classical probability we see that

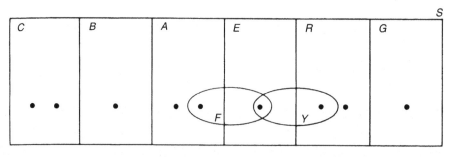

Figure 7.3 *Possibility space for choice of chairman of committee.*

$P(E \cap F \cap Y) = \frac{1}{9}$. Thus the probability that the chairman is female, young and Egyptian is $\frac{1}{9}$.

In a similar way we can form the union of the three events E, F, Y, which we write $E \cup F \cup Y$ and call 'E union F union Y'. It consists of all points in any of E, F and Y, and we see from Fig. 7.3 that $N(E \cup F \cup Y) = 3$. Hence $P(E \cup F \cup Y) = \frac{3}{9} = \frac{1}{3}$. The general rule for expressing the probability of the union of many events in terms of the probabilities of the component events is complex and we shall not give it in this book. But there is a simple rule for specially related events, which we will illustrate by evaluating $P(C \cup B \cup A)$. We see from Fig. 7.3 that

$$N(C \cup B \cup A) = N(C) + N(B) + N(A),$$

and

$$M(C \cup B \cup A) = M(C) + M(B) + M(A)$$

when a probability mass of $\frac{1}{9}$ is attached to each outcome, and hence that

$$P(C \cup B \cup A) = P(C) + P(B) + P(A) = \frac{2}{9} + \frac{1}{9} + \frac{2}{9} = \frac{5}{9}.$$

The events C, B, A are *mutually exclusive*.

Definition 7.5

Two or more events are **mutually exclusive** if no two of them have points in common.

For a general set $E_1, E_2, \ldots, E_k$ of *mutually exclusive* events defined in S,

$$M(E_1 \cup E_2 \cup \cdots \cup E_k) = M(E_1) + M(E_2) + \cdots + M(E_k).$$

is obviously true if each event contains a finite number of points. Hence we have the following theorem.

Theorem 7.4

If $E_1, E_2, \ldots, E_k$ are mutually exclusive events of a discrete possibility space S, then

$$P(E_1 \cup E_2 \cup \cdots \cup E_k) = P(E_1) + P(E_2) + \cdots + P(E_k).$$

We also note that the events, defined by the various nationalities, between them account for every person. We say that these events are *exhaustive*.

Definition 7.6

Two or more events are said to be **exhaustive** if their union equals the whole of the possibility space S.

We see that for a set of mutually exclusive and exhaustive events $E_1, E_2, \ldots, E_k$

$$\sum_{i=1}^{k} P(E_i) = 1.$$

7.6.1 Exercises

1. Three fair coins are tossed. For each of the set of events listed below state whether the events are: (1) mutually exclusive; (2) exhaustive.
   (a) (i) No heads; (ii) no tails.
   (b) (i) No heads; (ii) at least one head.
   (c) (i) At least two heads; (ii) at least two tails.
   (d) (i) One tail; (ii) two tails; (iii) three tails.

2. Find the probabilities of the events in 1(d) above. Hence find the probability of the event 'at least one tail'.

7.7 CONDITIONAL PROBABILITY

A group of ten first-year university students may be classified by sex and their major subject, according to Table 7.1. We suppose the students take part in a small lottery. Tickets will be bought by the students; the winner will receive the combined takings and be able to celebrate. In fact each economics student decides to buy three tickets while the physics students each buy one ticket. There are therefore 20 tickets in the draw.

If we are interested in the probability of a particular type of student winning the lottery then an appropriate possibility space S is shown in Fig. 7.4. Each point in P represents one ticket, with a probability of 0.05 of being chosen; each point in E represents three tickets, with a probability of 0.15 of being chosen. We may check that the total probability is

$$5 \times 0.15 + 5 \times 0.05 = 0.75 + 0.25 = 1.$$

Consider the events:

 E the winning ticket belongs to a student who takes economics,
 F the winning ticket belongs to a female student.

The probabilities of these events are $P(E) = 5 \times 0.15 = 0.75$ and $P(F) = 3 \times 0.15 + 0.05 = 0.50$. The probabilities of related compound events are $P(E \cap F) = 3 \times 0.15 = 0.45$ and $P(E \cup F) = 5 \times 0.15 + 0.05 = 0.80$.

Suppose a ticket is chosen at random and it can be seen that it belongs to a female student: event F has therefore occurred. *Before* we check whether she is taking physics or economics, can we use the information that the ticket belongs to a female student to predict what the answer will be? Clearly we can do so by

Table 7.1

	Male (M)	Female (F)
Physics (P)	4	1
Economics (E)	2	3

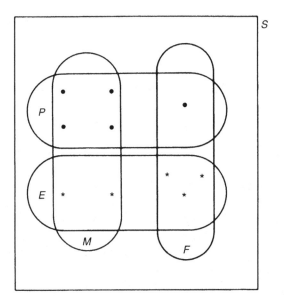

Figure 7.4 *Possibility space for lottery outcomes, classified by sex and subject of study of ticket holder. Key to probability masses:* • *0.05,* ∗ *0.15.*

restricting ourselves to that part of S in Fig. 7.4 representing F. We *know* F has happened, so the rest of S (that is F^c) is irrelevant. We have a new possibility space F. In order that the probabilities in this new space sum to 1 we must divide each of the current probabilities by $P(F) = 0.50$. Hence the probability of the point that represents a physics student is $0.05/0.50 = 0.10$, while the probability of each of the remaining three points that represent economics students is $0.15/0.50 = 0.30$. Knowing that the ticket of a female student has been chosen, we can say that the probability that she is a physics student is 0.10 while the probability that she is an economics student is 0.90.

Working definition

The *conditional probability* of R given Q is the probability that event R will occur given that event Q has already occurred. It is written $P(R \mid Q)$; note that $R \mid Q$ does *not* stand for a set.

We confine our attention to that part of S in which Q has occurred. The set of possible outcomes therefore becomes the set Q; this is the new possibility space. We rescale all the probabilities in the new space by dividing by $P(Q)$. Within the new space we look for the event R to occur also, i.e. the event of interest is $Q \cap R$. The probability is therefore $P(Q \cap R)/P(Q)$. This derivation allows us to frame a more precise definition of conditional probability.

Definition 7.7

The **conditional probability** of R given Q is

$$P(R \mid Q) = \frac{P(Q \cap R)}{P(Q)}.$$

In our example,

$$P(E \mid F) = \frac{P(E \cap F)}{P(F)} = \frac{0.45}{0.50} = 0.90.$$

Of course, $P(F \mid E)$ is not the same as this because the events now occur in a different order; the expression is not symmetrical at all:

$$P(F \mid E) = \frac{P(E \cap F)}{P(E)} = \frac{0.45}{0.75} = 0.6.$$

Of those taking economics, 60% are female.

In defining $P(R \mid Q)$ and $P(Q \mid R)$ we divide $P(Q \cap R)$ by $P(Q)$ and $P(R)$ respectively, so that the result for $P(R \mid Q)$ appears to hold only if $P(Q) \neq 0$, and that for $P(Q \mid R)$ only if $P(R) \neq 0$. However, if there are no outcomes in Q, or no outcomes in R, then certainly there are no outcomes in $Q \cap R$ either! So the conditional probability only has a well-defined meaning so long as we are not trying to condition on an impossible event.

The definition above can be rewritten

$$P(Q \cap R) = P(Q)P(R \mid Q) \tag{7.1}$$
$$= P(R)P(Q \mid R).$$

This shows that another way of finding the probability that two events Q, R have occurred is to multiply the probability of the first by the probability of the second, given that the first has happened. Intuitively this is to be expected. The factor 'the probability of the first' gets us into the right subspace, and then the other factor gives us the right proportion of that subspace.

The two ways of writing $P(Q \cap R)$ are often called the *multiplication rules* for probabilities.

Another way of illustrating these probabilities is to draw a tree diagram (Fig. 7.5) in which the first branch distinguishes between tickets held by students reading physics and by those reading economics. We mark the probability of a

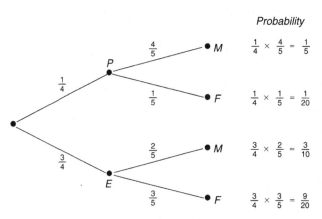

Figure 7.5 *Tree diagram for finding probabilities of winning ticket by sex–subject combinations.*

ticket being held by a physics student (P), which is $5 \times \frac{1}{20} = \frac{1}{4}$, on the appropriate branch, and similarly mark the probability of a ticket being held by an economics student (E), which is $5 \times \frac{3}{20} = \frac{3}{4}$. The second branch then distinguishes between tickets held by male (M) and female (F) students. The appropriate probabilities are conditional probabilities. From P, we proceed to M with conditional probability $\frac{4}{5}$ and to F with conditional probability $\frac{1}{5}$; from E we proceed to M and F with conditional probabilities $\frac{2}{5}$ and $\frac{3}{5}$ respectively. At each end-point of the tree we find the probability of reaching that end-point, by multiplying the probabilities on the sections of the tree which lead to it. The end-points themselves thus represent the intersections of pairs of events as shown in Fig. 7.5, and between them they account for all possible results.

Example 7.5

A card is drawn at random from a standard pack of playing cards. We are told it is an ace. What is the probability it is the ace of diamonds?

The most direct approach is to say that the ace is equally likely to be any of the four aces and hence the probability that it is the ace of diamonds is $\frac{1}{4}$. But it is instructive to consider it as a conditional probability.

Consider a possibility space that has the 52 distinct cards as equally likely outcomes. Let A be the event 'drawing an ace' and D be the event 'drawing the ace of diamonds'. We require the conditional probability $P(D \mid A)$.

Now

$$P(A) = \frac{4}{52} = \frac{1}{13} \quad \text{and} \quad P(A \cap D) = P(D) = \frac{1}{52}.$$

Hence

$$P(D \mid A) = \frac{P(A \cap D)}{P(A)} = \frac{\frac{1}{52}}{\frac{1}{13}} = \frac{1}{4}.$$

Example 7.6

There are two children in a family. What is the conditional probability that both are boys, given that at least one is a boy?

Consider a possibility space {BB, BG, GB, GG}. Let T be the event 'both children are boys', i.e. {BB}, and let L be the event that 'at least one is a boy', i.e. {BB, BG, GB}.

Hence

$$P(L) = \frac{3}{4} \quad \text{and} \quad P(L \cap T) = P(\{BB\}) = \frac{1}{4}.$$

Therefore

$$P(T \mid L) = \frac{P(L \cap T)}{P(L)} = \frac{\frac{1}{4}}{\frac{3}{4}} = \frac{1}{3}.$$

7.7.1 Exercises

1. Two fair dice are thrown. (a) If it is known that the total score was 6, what is the probability that the difference between the scores on the two dice was 2? (b) If it is known that the difference between the scores was 4, what is the probability that the total score was 8?

2. Explain what is meant by the conditional probability of an event A given event B. A standard pack of 52 cards of 4 suits, each with 13 denominations, is well shuffled and dealt out to four players, N, S, E and W, who each receive 13 cards. If N and S have exactly ten cards of a specified suit between them, show that the probability that the three remaining cards of the suit are in one player's hand (either E or W) is 0.22. (*Oxford*)

3. The probability that a blue-eyed person is left-handed is $\frac{1}{7}$. The probability that a left-handed person is blue-eyed is $\frac{1}{3}$. The probability that a person has neither of these attributes is $\frac{4}{5}$. What is the probability that a person has both attributes? (*SMP*)

4. Three girls, two of whom are sisters, and five boys, two of whom are brothers, meet to play tennis. They draw lots to determine how they should split up into two groups of four to play doubles.
   (a) Calculate the probabilities that one of the two groups will consist of: (i) boys only; (ii) two boys and two girls; (iii) the two brothers and the two sisters.
   (b) If the lottery is organized so as to ensure that one of the two groups consists of two boys and two girls, calculate the probability that the two brothers and the two sisters will be in the same group. Given that the two brothers are in the same group, calculate the probability that the two sisters are also in that group. (*Welsh*)

5. For the population described in Exercise 6.11(6), what is the probability that one randomly selected person will be:
   (i) 65 years old or more, given that she is female;
   (ii) male, given that he is between 15 and 44 years old (inclusive);
   (iii) of working age, 15–64 years old (inclusive), given that the person is under 65;
   (iv) below school age, which is 5 years old, given that the person is not of working age?

7.8 INDEPENDENCE

In discussing conditional probability, we considered a pair of events Q, R, which could be interrelated; one of them could (and did) give information about the other. It is perfectly possible to carry out trials and define two events in such a way that one event gives no information at all about the other. Suppose our trial is a game in which a fair die is thrown once and a fair coin is tossed twice. Define two events as follows:

> A the score with the die is 3;
> B both tosses of the coin give tails.

Since A and B both refer to the same trial, they may clearly occur together. But information on what happened to the die is no help at all in predicting what may have happened with the coin. It is intuitively obvious in a case like this that $P(B \mid A)$ must simply be equal to $P(B)$, and similarly $P(A \mid B) = P(A)$. From this, we see that $P(A \cap B) = P(A)P(B)$ (from Equation (7.1), p. 142).

Definition 7.8

The events Q, R are **independent** if and only if

$$P(Q \cap R) = P(Q)P(R).$$

It may be thought that the intuitive definition $P(Q \mid R) = P(Q)$ would be the better one to adopt for a starting point. However, it gives slight difficulty if $P(R) = 0$, for we have already warned against trying to condition on an impossible event. (If $P(Q) = 0$, there is no difficulty, because the probability of Q conditional on any other event will then also be 0 and so the equality holds.) Also, since (Equation (7.1))

$$P(Q \cap R) = P(Q)P(R \mid Q) = P(R)P(Q \mid R)$$

the definition we have chosen immediately implies $P(R \cap Q) = P(R)$ and $P(Q \mid R) = P(Q)$. Conversely, if $P(R \mid Q) = P(R)$ or $P(Q \mid R) = P(Q)$, Equation (7.1) implies the equation in Definition 7.8.

Notice that independence is a symmetrical relation: if Q is independent of R then R is independent of Q.

To prove this, assume $P(Q \mid R) = P(Q)$, i.e. Q is independent of R. Then

$$P(Q \cap R) = P(R)P(Q \mid R)$$
$$= P(R)P(Q).$$

But also $P(Q \cap R) = P(Q)P(R \mid Q)$, from Equation (7.1).

Therefore (provided $P(Q)$ is not zero) we must have $P(R \mid Q) = P(R)$, and R is independent of Q. So we may simply speak of 'Q and R being independent'.

Let us return to our example of throwing a die and two coins. The possibility space for this example is shown in Fig. 7.6. From this we see that $N(A) = 4, N(B) = 6, N(S) = 24$ and $N(A \cap B) = 1$. It then follows that

$$P(A) = \tfrac{4}{24} = \tfrac{1}{6}; \ P(B) = \tfrac{6}{24} = \tfrac{1}{4}; \ P(A \cap B) = \tfrac{1}{24}.$$

We can work out the conditional probabilities directly from the diagram, or we can use the formula. From the diagram,

$$P(B \mid A) = \frac{N(A \cap B)}{N(B)} = \frac{1}{4},$$

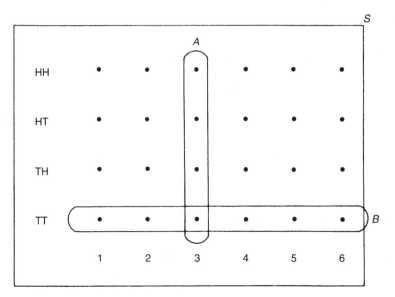

Figure 7.6 *Possibility space for throwing a die and two coins.*

while from the formula,

$$P(A \mid B) = \frac{P(A \cap B)}{P(B)} = \frac{\frac{1}{24}}{\frac{1}{4}} = \frac{1}{6}.$$

So $P(A \mid B) = P(A)$ and $P(B \mid A) = P(B)$; also $P(A)P(B) = \frac{1}{6} \times \frac{1}{4} = \frac{1}{24} = P(A \cap B)$, and this shows formally that A and B are independent.

Example 7.7

A salesman for a company sells two products, A and B. During a morning he makes three calls on customers. Suppose the chance that on any call he makes a sale of product A is 1 in 3 and the chance that he makes a sale of product B is 1 in 4. Suppose also that the sale of product A on any call is independent of the sale of product B, and that the results of the three calls are independent of one another. Calculate the probability that the salesman will:

(a) sell both products, A and B, at the first call;
(b) sell one product at the first call;
(c) make no sales of product A during the morning;
(d) make at least one sale of product B during the morning. *(IOS)*

Figure 7.7 illustrates the possible outcomes of the first call. Because sales of A and B are independent,

$$P(A \cap B) = P(A)P(B) = \frac{1}{3} \times \frac{1}{4} = \frac{1}{12}.$$

(We adopt the obvious notation that event A is 'product A is sold' and event B is 'product B is sold'.) So the answer to (a) is $\frac{1}{12}$. If just one product is sold, this may be either A or B, and the sum of the probabilities $\frac{1}{4} + \frac{1}{6} = \frac{5}{12}$ is thus the answer to (b). (In terms of events, this is $P(A \cap B^c) + P(A^c \cap B)$.)

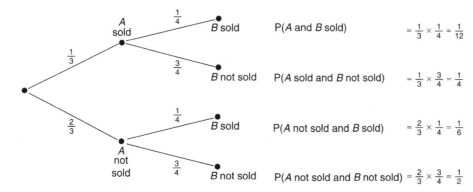

Figure 7.7 *Tree diagram showing possible sales of two products.*

At the first call, the probability of not selling A is $\frac{2}{3}$, and at the other two calls it is again $\frac{2}{3}$ each time; the results of the three calls are independent of one another. Thus the probability of not selling A at all is $(\frac{2}{3})^3 = \frac{8}{27}$.

Finally, the probability of not selling B at all is $(\frac{3}{4})^3$, and so the probability of selling B at least once, which is the complement of this, is $1 - (\frac{3}{4})^3 = 1 - \frac{27}{64} = \frac{37}{64}$.

In decimal form, the answers are: (a) 0.083; (b) 0.417; (c) 0.296; (d) 0.578.

Example 7.8

In the example illustrated in Fig. 7.4 check whether or not the events (a) the winning ticket belongs to a male student, (b) the winning ticket belongs to a student of physics, are independent.

Write M for the event (a) above and P for the event (b) above. From Fig. 7.4, $P(M) = 0.50$, $P(P) = 0.25$ and $P(M \cap P) = 0.20$.

Thus $P(P)P(M) = 0.25 \times 0.50 = 0.125$. This is not equal to $P(P \cap M)$, so P and M are not independent.

Beware of confusing mutually exclusive events and independent events. If two events Q, R are mutually exclusive, the sets Q and R are disjoint, and $Q \cap R = \emptyset$. However, if two events Q, R are independent, then apart from the exceptional case when $P(Q)$ or $P(R)$ is zero the sets Q and R are not disjoint, so $Q \cap R \neq \emptyset$. Independent events usually have at least one outcome in common.

7.8.1 Independence of many events

We need a formal definition of independence to ensure that we have a satisfactory theory, but in practice we usually introduce independence as an assumption of our model. Thus, when setting up a model for throwing a real die we *assume* that the outcome of the second throw is independent of the outcome of the first throw, and provided we throw the die properly, and it has no corners chipped off, the assumption is a reasonable one. This allows us to write down almost immediately the probability of a joint event, e.g.

P('6 on first throw' $\cap$ 'even number on second throw')
  = P('6 on first throw') $\times$ P('even number on second throw')
  = P('6 on any throw') $\times$ P('even number on any throw') = $\frac{1}{6} \times \frac{1}{2} = \frac{1}{12}$.

We can extend the idea of independence from two to many events. A formal definition becomes complex and we shall not attempt it. We accept it as an intuitive notion and shall often make the assumption in models that many events are independent. We are simply extending the idea that one event gives us *no information* about another. Using this assumption for three events A, B, C we can write

$$P(A \cap B \cap C) = P(A \cap (B \cap C)) = P(A) \times P(B \cap C) = P(A) \times P(B) \times P(C).$$

We have, in general, the following theorem.

Theorem 7.5
If the events $E_1, E_2, \ldots, E_k$ are independent then

$$P(E_1 \cap E_2 \cap \cdots \cap E_k) = P(E_1) \times P(E_2) \times \cdots \times P(E_k).$$

Example 7.9
Ten independent throws of a fair die are made. What is the probability that it shows a six every time?

The required probability is the product of the probabilities that the die shows a six each time, and is $(\frac{1}{6})^{10}$.

Example 7.10 Binomial probabilities
Independent trials are made, in which the probability of success at any trial is π, and of failure is $1 - \pi$. What is the probability of r successes in n trials?

This provides a reasonable model for many real situations. A trial may be to toss a coin, in which *success* is to obtain a head and *failure* is to obtain a tail. Each toss is independent of every other unless the tossing is done in a very half-hearted or careless way. Or a trial may be to throw a die, in which *success* is to score a 6 and *failure* is to obtain a score other than 6. Each throw should again be unaffected by what happened at previous throws, i.e. independent of them. In both these cases, there is no reason for the probability π of success to change from one trial to another.

In industrial production, a trial may be to produce one object such as a nut, a screw, a motor car axle, a transistor component. Even when the machinery producing them is working well, there will be an occasional faulty nut, screw, axle or component coming off a continuous production line. So there is a small, constant, probability $(1 - \pi)$ of *failure*, i.e. a faulty product, and a constant probability π of *success*, i.e. a sound product. If the machinery is working well, each product is sound or faulty independently of what happens to its neighbours on the production line. (If, however, a machine is not working properly, this simple idea of constant π and independent trials will not be good enough.)

The sexes of individual children within a family of n children can often be considered independent. Each *trial* is a new birth, *success* is a girl and *failure* a boy (success and failure are merely convenient technical words!), and the probability π of success is constant. The value of π in this case may not be exactly $\frac{1}{2}$, but as a first approximation could be taken so.

Suppose there are two successes (S) and one failure (F) in three trials, i.e. $n=3$ and $r=2$. One possible sequence of results for the three trials is then SFS, and, since the results of each trial are independent of those at other trials, the probability for this sequence is $\pi \times (1-\pi) \times \pi = \pi^2(1-\pi)$. The two other possible sequences are FSS, SSF, and each of these has probability $\pi^2(1-\pi)$ also. The possible sequences of results are mutually exclusive events, so the probability of $r=2$ is the sum of the probabilities of the three sequences, which is $3\pi^2(1-\pi)$.

In the general case, the probability of any one particular sequence that contains r successes and $n-r$ failures is $\pi^r(1-\pi)^{n-r}$. The number of different possible sequences of this type is $\binom{n}{r}$ (see Theorem 6.4, p. 122), and so the total probability of r successes in n trials is

$$\binom{n}{r} \pi^r(1-\pi)^{n-r}.$$

7.8.2 Exercises

1. For each of the following sets of events state whether the events are: (i) mutually exclusive; (ii) independent.
   (a) In picking two digits from a random number table, the events are: (1) the first digit is 5; and (2) the digits total 12.
   (b) In a test of roadworthiness of a motor car, the events are: (1) the car fails the test because of defective steering; (2) it fails because of defective brakes; (3) it fails because of defective tyres.
   (c) In throwing two fair dice, each of which may score 1, 2, 3, 4, 5 or 6, the events are: (1) the first die scores 3; (2) the total score is 7.
   (d) As in (c), except that the events are: (1) the first die scores 4; (2) the total score is 9.
   (e) In drawing two playing cards from a standard pack of 52, the events are: (1) both cards are red; (2) the ace of clubs is drawn.

2. Let A and B be two events such that $P(A)$ and $P(B)$ are both greater than 0. By considering the two ways of writing $P(A \cap B)$, show that

$$P(A \mid B) = \frac{P(A)P(B \mid A)}{P(B)}.$$

3. If p_1 and p_2 are the probabilities of two independent events, show that the probability of the simultaneous occurrence of these two events is $p_1 p_2$.

In 18 games of chess between A and B, A has won 8, B has won 6, and 4 have been drawn. Now A and B are to play a tournament of 3 games. On the basis of the above data, estimate the probability that:

(a) A will win all three games;   (b) A and B will win alternately;
(c) two games will be drawn;   (d) A will win at least one game.   (*AEB*)

4. Show that if two events A and B are independent, then: (i) A and B^c, (ii) A^c and B^c are also independent.

5. Show that if A and B are independent events, such that A is a subset of B, then either $P(A) = 0$ or $P(B) = 1$.

6. A game consists of tossing a coin and throwing a die. A random variable X is given the value $+1$ if the coin showed a head and -1 for a tail; also Y is the score on the die, whose faces are numbered 1, 2, 3, 4, 5, 6. The random variable U is defined as $U = XY$, the product of X and Y. Write down the list of possible values U may take, and their probabilities.

7.9 INVERSE PROBABILITY

If you are answering a 'multiple-choice' question, where you are given several possible answers to choose from and merely have to tick the right one, life is fine if you really know the answer. But if you do not know, what do you do? You may decide to guess – at least there is some chance of hitting the right one! Let us assume that a candidate in an examination who does not know the correct answer will always guess. When there are m possible answers to choose from, we will further assume that guessing is purely 'random', not 'intelligent', so that every answer is *equally likely* to be picked.

We shall go on to consider how the probability of obtaining a correct answer is modified when an element of guessing is involved. But a more interesting question, which we consider later in the section, is 'Given that the candidate produced the correct answer, what is the probability that he really knew it?' This may be called *inverse probability*. There are two routes to the correct answer with a multiple-choice question: *either* the candidate knows it, and marks it correctly on the answer sheet, *or* the candidate does not know, but happens to guess correctly. There are two events involved in reaching the answer by the first route: event A is 'the candidate knows the answer' and event B is 'the candidate marks the correct answer on the sheet'. The second route begins with the event A^c, the complement of A, i.e. the answer is not known; the candidate therefore guesses. One of A or A^c *must* happen, and these two form a mutually exclusive and exhaustive pair. Event B can occur with either A or A^c; $P(B|A)$ is the probability that the candidate both knew the answer and marked it correctly, while $P(B|A^c)$ is the probability that the candidate did not know but happened to guess correctly. This last probability is $1/m$ when there are m possible answers and guessing is random.

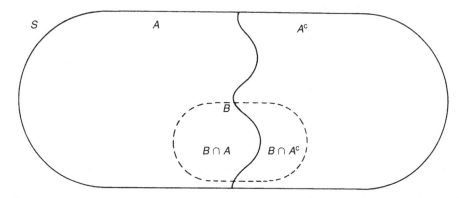

Figure 7.8 *Possibility space for answering multiple-choice question. Event A is 'candidate knows answer', and B is 'candidate gives correct answer'.*

Figure 7.8 illustrates the situation: S, the possibility space, is divided into A and A^c, and event B may occur with either A or A^c. We can write $M(B) = M(B \cap A) + M(B \cap A^c)$, to obtain $P(B) = P(B \cap A) + P(B \cap A^c)$. Using this together with Equation (7.1), we find

$$P(B) = P(B \mid A)P(A) + P(B \mid A^c)P(A^c).$$

In the multiple-choice test, suppose that you know the answer to three-quarters of the questions, and guess the remaining quarter. When you know the answer you have probability 0.95 of marking the answer sheet correctly. Each question has five possible answers, so your probability of making a correct guess is $\frac{1}{5}$. Event B is 'obtain the correct answer' and A is 'the answer is known'; in symbols the information about the results is $P(A) = 0.75$, $P(A^c) = 0.25$, $P(B \mid A) = 0.95$, $P(B \mid A^c) = 0.20$.

Hence the probability of obtaining a correct answer to a question is

$$\begin{aligned} P(B) &= P(B \mid A)P(A) + P(B \mid A^c)P(A^c) \\ &= (0.95 \times 0.75) + (0.20 \times 0.25) = 0.7125 + 0.0500 \\ &= 0.7625. \end{aligned}$$

We now look at the more interesting question. If a candidate obtains a correct answer (event B), what is the probability that he really knew it (event A), rather than having to guess? We want to know $P(A \mid B)$; in other words, we now have the information that his answer was correct, and we want to find out how he got there. From Equation (7.1),

$$P(A \mid B)P(B) = P(B \mid A)P(A).$$

Also, as we have just shown,

$$P(B) = P(B \mid A)P(A) + P(B \mid A^c)P(A^c).$$

Thus

$$P(A \mid B) = \frac{P(B \mid A)P(A)}{P(B)} = \frac{P(B \mid A)P(A)}{P(B \mid A)P(A) + P(B \mid A^c)P(A^c)}. \qquad (7.2)$$

In the example, therefore,

$$P(A \mid B) = \frac{0.95 \times 0.75}{0.7625} = 0.9344.$$

In other words, given that a candidate obtained the right answer there is a probability of 0.9344 that he actually knew it (as opposed to having guessed).

A situation where a similar problem arises is in medical tests to see if a patient has a particular disease. A diagnostic test can be carried out, but its results are not absolutely certain: sometimes the test misses detecting the disease and sometimes it 'detects' when the disease is not really there. The following example illustrates this.

Example 7.11 Probability that a person tested positive has disease

A test for a certain disease has a probability of 0.8 of detecting the disease, when it is present, but also a probability of 0.1 of falsely 'detecting' it when it is not present. If a person selected at random from the population gives a positive result to the test, what is the probability that the person really has the disease, given that the disease is actually present in (i) 1/1000, (ii) 1/5 of the whole population?

Label the underlying events.

D the person has the disease,
T the test gives a positive result.

We are given numerical values for $P(T \mid D)$, $P(T \mid D^c)$ and $P(D)$. We seek the value of $P(D \mid T)$.

(i) $P(T \mid D) = 0.8$, $P(T \mid D^c) = 0.1$, $P(D) = 0.0001$.

Hence

$$P(D \mid T) = \frac{P(T \mid D)P(D)}{P(T \mid D)P(D) + P(T \mid D^c)P(D^c)}$$

$$= \frac{0.8 \times 0.0001}{0.8 \times 0.0001 + 0.1 \times 0.9999} = \frac{8}{10\,007} = 0.0008.$$

It is disappointing to see that, even though we have what would appear to be quite a reliable test, it is of very little use for detecting a rare disease. Compare with part (ii).

(ii) $P(T \mid D) = 0.8$, $P(T \mid D^c) = 0.1$, $P(D) = 0.2$.

$$P(D \mid T) = \frac{0.8 \times 0.2}{0.8 \times 0.2 + 0.1 \times 0.8} = \frac{16}{24} = \frac{2}{3} = 0.6667.$$

There is another way of arguing. Using (ii) as the example, consider testing 100 people. On average $16\,(= 100 \times 0.8 \times 0.2)$ will test positive and have the disease, while $8 (= 100 \times 0.1 \times 0.8)$ will test positive and not have the disease. Hence of those testing positive the proportion who have the disease is, on average, $16/24 = 0.6667$. This argument is based on rewriting $P(D \mid T)$ as $NP(T \mid D) P(D)$ divided by $[NP(T \mid D) P(D) + NP(T \mid D^c) P(D^c)]$, where N is the number of tests, and interpreting the components as mean values of 'successes' in binomial trials (see p. 148).

7.9.1 Exercise

1. In Example 7.11, let the true proportion of the whole population having the disease be π. If the probabilities of true and false detection remain 0.8 and 0.1, find an expression in terms of π for the probability that a person giving a positive test result really has the disease. Plot a graph of this probability against π. Repeat with the detection probabilities changed to 0.95 and 0.02.

7.9.2 More than two mutually exclusive exhaustive events

If S can be divided not just into A and A^c but into a larger set of mutually exclusive, exhaustive events $A_1, A_2, \ldots, A_k$ it is easy to show that the probability of another event B is related to this set $\{A_i\}$ by the rule

$$P(B) = P(B \mid A_1)P(A_1) + P(B \mid A_2)P(A_2) + \cdots + P(B \mid A_k)P(A_k)$$

$$= \sum_{i=1}^{k} P(B \mid A_i)P(A_i).$$

This is sometimes called the *total probability rule* for finding $P(B)$.

Example 7.12

In a factory, a certain brand of chocolates is packed into boxes on four different production lines A_1, A_2, A_3, A_4. Records show that a small percentage of boxes are not packed properly for sale: 1% from A_1, 3% from A_2, $2\frac{1}{2}$% from A_3 and 2% from A_4. If the percentages of total output that have come from the production lines are 35% from A_1, 20% from A_2, 24% from A_3 and 21% from A_4, what is the probability that a box chosen at random from the whole output is faulty?

Take event A_i as 'box produced on line A_i', $i = 1$ to 4; event B is 'box is faulty'.

$$P(B) = P(B \mid A_1)P(A_1) + P(B \mid A_2)P(A_2) + P(B \mid A_3)P(A_3) + P(B \mid A_4)P(A_4)$$

$$= (0.01 \times 0.35) + (0.03 \times 0.20) + (0.025 \times 0.24) + (0.02 \times 0.21)$$

$$= 0.0197.$$

The result in Equation (7.2) can be extended too. If we find a box that is faulty, what is the probability that it came from production line A_2? We require

$P(A_2 \mid B)$. From Equation (7.1), this is equal to

$$\frac{P(B \mid A_2)P(A_2)}{P(B)}.$$

Writing $P(B)$ in 'total probability' form we obtain

$$P(A_2|B) = \frac{P(B|A_2)P(A_2)}{\sum_{i=1}^{4} P(B|A_i)P(A_i)}.$$

The general form of this result is usually called *Bayes' theorem*.

Theorem 7.6 (Bayes' theorem)

If $A_1, \ldots, A_k$ are a set of mutually exclusive, exhaustive events in a possibility space S, and B is any other event in S such that $P(B) > 0$, then

$$P(A_i|B) = \frac{P(B \mid A_i)P(A_i)}{\sum_{i=1}^{k} P(B \mid A_i)P(A_i)} \quad i = 1, 2, \ldots, k.$$

The result in Equation (7.2) – which is the special case for $k = 2$ – is often called Bayes' theorem too.

Example 7.12 (continued)

What is the probability that a faulty box comes from each of the production lines?

$$P(A_1|B) = \frac{P(B \mid A_1)P(A_1)}{P(B)} = \frac{P(B \mid A_1)P(A_1)}{\sum_{i=1}^{4} P(B \mid A_i)P(A_i)}$$

$$= \frac{P(B \mid A_1)P(A_1)}{0.0197} = \frac{0.01 \times 0.35}{0.0197} = 0.1777.$$

Similarly,

$$P(A_2 \mid B) = \frac{0.03 \times 0.2}{0.0197} = 0.3046,$$

$$P(A_3 \mid B) = \frac{0.025 \times 0.24}{0.0197} = 0.3046,$$

$$P(A_4 \mid B) = \frac{0.02 \times 0.21}{0.0197} = 0.2132.$$

Another way of saying this is that about 18% of faulty boxes come from line A_1, 30% each from A_2 and A_3, 21% from line A_4.

7.9.3 Exercises

1. There are four urns: U_1 contains 4 black and 1 white balls; U_2 and U_3 each contain 2 black and 3 white balls; U_4 contains 1 black and 4 white balls. An

urn is chosen at random and a ball taken at random from it. The ball is white. What is the probability it came from U_4?

2. A large college buys red and green pens in the ratio 1:2. The college has 3 suppliers for these pens. Supplier A provides 70% of the red pens and 40% of the green pens. The remainder of the red pens come from supplier B and the remainder of the green pens from supplier C. The college knows that 20% of the pens supplied by A, 15% of those supplied by B and 10% of those supplied by C are defective. The pens are all stored together in a box. A member of staff grabs a pen at random.
(a) Find the probability that the pen will be defective.
The member of staff found that the pen was defective and immediately complained to the college management about the quality of pens from supplier A.
(b) Find the probability that the defective pen was supplied by A.

(London)

3. Three balls are drawn at random *without* replacement from a box containing 8 white, 4 black and 4 red balls. Calculate the probabilities that there will be
   (i) at least one white ball;
   (ii) two white balls and one black ball;
   (iii) two balls of one colour and the other of a different colour;
   (iv) one ball of each colour.
If, instead, each ball *is* replaced in the box before the next ball is drawn, calculate the probability that the three balls drawn will consist of one of each colour. *(Welsh)*

4. Two teams, A and B, play a football match against each other. The number of goals each scores in the match is independently distributed with the distributions shown below. Calculate the probability of A winning, of B winning and of a draw. If a series of three matches is played, what is the probability of A winning the series?

| Number of goals | Probability | |
	A	B
0	0.3	0.2
1	0.3	0.4
2	0.3	0.3
3	0.1	0.1

(Oxford)

5. (a) In the *Who, What or Where* game, three contestants each choose one of the three categories of question. Assuming that the contestants choose independently and that each is equally likely to select any of the categories, find the probability that: (i) all will choose the same category; (ii) all will choose different categories; (iii) two will be alike and the third different.

(b) From a pack of 52 cards, seven are taken at random, examined and replaced. The cards are shuffled and another seven are drawn at random. Find the probability that at least one card will be drawn twice.   (*AEB*)

6. On a certain course, students may study one, two or three optional subjects. Sixteen study sociology only, 21 study politics only, 20 study history only, 8 study history and politics, 7 study sociology and politics, 5 study history and sociology and 3 study history, politics and sociology.

    How many students are there on the course? Find the probability that a student selected at random studies one optional subject only.   (*IOS*)

7.10 SIMULATION

Mathematical models of real-life situations quickly become very complex and the theoretical analysis of even relatively simple models involving probabilities is often difficult. In these situations, simulation can be a very powerful method of investigating the system. It is used a great deal in scientific research and in the analysis of industrial processes. Let us illustrate the idea by considering how we might simulate a game of tennis.

Let us suppose that A has a weak serve and has a chance of only 0.01 of winning from a serve, without a rally taking place (see Fig. 7.9). B is not much better at serving and has a chance of 0.10 of winning directly from a serve. If the game goes to a rally, A has a chance of 0.55 of winning and B a chance of 0.45.

At each stage of the game we decide the next step by using random numbers. For example, at the beginning of the first game when A is serving, the event that

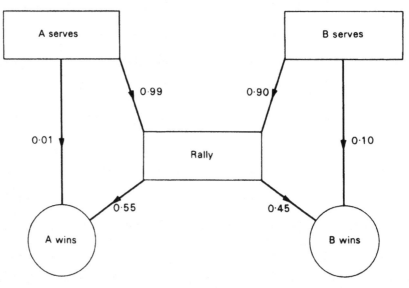

Figure 7.9 *Flow diagram for model of game of tennis.*

A wins directly has probability 0.01. We therefore choose a pair of random digits; if we obtain 01 then A wins directly, otherwise the game moves to a rally. A wins a rally with probability 0.55; hence for this step we choose a two-digit random number, and if it is any of 01–55 inclusive then A wins, otherwise B wins. The process is best set out as a table, like Table 7.2. We use r to denote a random number, R denotes a rally and A or B signifies that the player of that name has won.

B has done well in these two games, but we could be confident that B would usually win only by simulating a large number of games. For this example, it is possible to work out the probabilities that A or B will win, and in fact A has the advantage. The simulation method may easily be developed to become more realistic. We could attach separate probabilities to first serve and second serve, and even of winning on each stroke, but we must beware of making our model too complex. Data from a simulation method, such as we have illustrated, could be analysed statistically when the method is repeated many times.

We also use the simulation method in Chapter 11 to investigate how the proportion of people with telephones varies from sample to sample, when we take samples of 50 from a population of households in a town. The possibilities of the method are endless. We might simulate the movement of queues at the check-out desk of a supermarket, the transport of articles from place to place, or the distribution of patients' waiting-time in a doctor's waiting room.

The results of simulations are often very variable. If a clear picture is to emerge we need to do a large number of simulations and therefore it is usually wise to use a computer. Examples of computer simulations are given in the computing exercises throughout this book, and the use of computers in simulations is discussed in *CCC*, Chapters 7 and 9.

Table 7.2 *Simulation of a game of tennis*

	Serve		Rally		Score	
	r	Outcome	r	Outcome	A	B
A serves	13	R	84	B	–	15
	40	R	39	A	15	15
	25	R	03	A	30	15
	93	R	74	B	30	30
	03	R	58	B	30	40
	96	R	58	B	B wins	
B serves	30	R	07	A	15	–
	42	R	73	B	15	15
	09	B	–	–	15	30
	45	R	61	B	15	40
	84	R	50	A	30	40
	79	R	88	B	B wins	

7.11 MINITAB: ENTERING PATTERNED DATA AND CODING

7.11.1 Entering patterned data

MINITAB offers many shortcut ways of entering patterned data into a column. Often the numbers act as labels although they might be successive intervals of time. For example,

```
SET C1
4(1) 3(2)
END
```

will put the values 1 1 1 1 2 2 2 into C1. Other examples of specifications and what they yield are:

1:6	1 2 3 4 5 6
6:1	6 5 4 3 2 1
3:9/2	3 5 7 9 i.e. 3 to 9 in steps of 2
1.3:2.5/0.3	1.3 1.6 1.9 2.2 2.5
2(1:4)	1 2 3 4 1 2 3 4
(1:3)2	1 1 3 3

Similar results may be obtained by using the menu

Calc>Make Patterned Data/
Select **Simple Set of Numbers** for numbers at equal steps; otherwise select **Arbitrary Set of Numbers/**

7.11.2 Coding

Data in a column might represent the numbers that appeared in successive throws of a die. Each datum could be classified as being 'six' or 'less than six' and the data, transformed in this way, entered into another column. Making changes of this sort is called 'coding' by MINITAB. The coded values may be determined and stored in another column by use of the menu

Data>Code

As in the example, one may change numeric data to text or use many other options. For example, one might replace 6 by 1 and numbers less than 6 by 0, making a numeric to numeric change. The process may also be carried out by using the session command CODE.

Coding may be used when making a tally count of a continuous variate. Suppose, in Example 1.2, that the 100 lengths of cuckoo eggs are stored in C1 and we choose to code each value to the centre point of the interval in which it falls. We choose

Data>Code>Numeric to Numeric/
Code data from column C1 **Into column** C2,
Original (19.5:19.9) **New** 19.7,
Original (20.0:20.4) **New** 20.2, etc./

Finally, using

Stat>Tables>Tally Individual Values/
For **Variable** C2 **Display Counts/**

will give the required frequencies opposite the class centres, appearing in the Session window. If one wants the results stored in columns use

TALLY C2;
STORE C3 C4.

7.12 EXERCISES ON CHAPTER 7

1. A trial has eight possible outcomes, each equally likely, numbered $1, 2, \ldots, 8$. Events A, B, C are defined by $A = \{1, 2, 3, 4\}$, $B = \{2, 3, 5, 6\}$, $C = \{2, 4, 5, 6\}$. Show that: (i) A and B are independent; (ii) A and C are independent; (iii) $P(D) = P(A)P(B)P(C)$, where D is the event that A, B, C all occur.   (*IOS*)

2. You are given the following probabilities relating to two events A and B: $\Pr(A) = 0.5$, $\Pr(B) = 0.7$, $\Pr(A \text{ or } B) = 0.8$. Calculate: (i) $\Pr(A \text{ and } B)$; (ii) $\Pr(A \text{ and not-}B)$; (iii) the conditional probability $(B \mid A)$.   (*SMP*)
(*Note*: $\Pr(E)$ is the same as $P(E)$, and not-B is the same as B^c.)

3. The probability that a candidate attempts this question is $\frac{9}{10}$ and, having done so, the probability of success is $\frac{2}{3}$. Find the probability that the examiner will find at least one correct solution in the first three scripts which he marks.
    (*SMP*)

4. A census showed that 20 per cent of all married couples living in a certain district had no child under 16 years of age, 50 per cent had one child under 16, and 30 per cent had two or more children under 16. The census also showed that both the husband and the wife were in employment in 70 per cent of couples having no child under 16, in 30 per cent of couples having one child under 16, and in 10 per cent of couples having more than one child under 16.
    (i) If a married couple is chosen at random, calculate the probability that both the husband and the wife are in employment.
    (ii) If a married couple is chosen at random from those couples where both the husband and the wife are in employment, calculate the probability that the couple has at least one child under 16.

159

(iii) For a particular purpose a sociologist assigns two scores to each married couple. The first of these is based on the number of children under 16, being 0 for a couple having no such child and being 1 otherwise. The second score is based on whether or not both the husband and the wife are in employment, being 0 if they are and 1 otherwise. Calculate the mean of the sum of the two scores for all married couples in the district.

(*Welsh*)

5. Suppose that 5 men out of every 100 and 25 women out of every 10 000 in a population are colourblind. A colourblind person is chosen at random. What is the probability of the person being male? (Assume males and females to be in equal numbers.)

A random sample of the males of the above population is to be made. How big should the sample be to ensure that the probability of it containing at least one colourblind person is 90%?

6. In a routine test for early warning of an unsuspected disease, 8% of those tested react (but do not all have the disease), while 7% in fact have the disease (but do not all react). The probability that a person neither has the disease nor reacts is 0.9. What is the probability that a person with the disease will be detected?

7. (a) Express $P(A \cap B)$ in terms of $P(A)$ and $P(B)$ in each of the cases when A and B are (i) mutually exclusive, (ii) independent.

(b) An enquiry sent (in writing) to company C has a probability 0.7 of being dealt with the next day, probability 0.2 of waiting until the second day after it was sent, and probability 0.1 of waiting until the third day. An enquiry made to company K has corresponding probabilities of 0.3, 0.4 and 0.3.

Two enquiries are sent, by the same person at the same time, one to each company. Assuming that the answering times of C and K are independent, calculate the probabilities that (i) at least one enquiry will be dealt with the next day, (ii) the enquiry made to company C will be dealt with one day earlier than that made to K.

(c) If three people all make an enquiry simultaneously to company K, find the probability that all three will be dealt with at the same time.

8. Three men, A, B and C, share an office with a single telephone. Calls come in at random in the proportions $\frac{2}{5}$ for A, $\frac{2}{5}$ for B, $\frac{1}{5}$ for C. Their work requires the men to leave the office at random times, so that A is out for half his working time and B and C each for a quarter of theirs.

For calls arriving in working hours, find the probabilities that:
 (i) no one is in to answer the telephone;
 (ii) a call can be answered by the person being called;
 (iii) three successive calls are for the same man;
 (iv) three successive calls are for three different men;
 (v) a caller who wants B has to try more than three times to get him.

(*JMB*)

9. A bank wishes to compare the relative merits of having, over the very busy period 12 noon–2 p.m.: (a) three cashiers available for all business; (b) one cashier available for cashing cheques only and two cashiers available for all business. The bank knows that approximately half the customers wish to cash cheques only. Explain how to carry out a simulation to compare the two methods, paying particular attention to:
   (a) the specification of the model,
   (b) the quantities to be analysed.
   (You are *not asked* to carry out the simulation.)

7.13 COMPUTING EXERCISES

1. Simulate throwing a die, and count the number of throws required to obtain a six. (Include the throw that produces the six.) Repeat many times (e.g. 500) and hence find the distribution of 'number of throws to obtain a six.' Plot a bar diagram and find the mean of the distribution.

2. Some cigarette manufacturers used to offer cards in packets of cigarettes, and these cards could be collected to make a set. Simulate this situation, assuming there are 10 cards in a set and that any of the 10 cards is equally likely to be found in a packet. Find the number of packets that must be bought to obtain a complete set. Repeat many times and so find the distribution of 'number of packets bought to produce a complete set.' Plot a bar diagram and find the mean of the distribution.

7.14 PROJECTS

1 Matching cards

Aim. To investigate the relative frequency of no matches in two sets of cards.
Equipment. A standard pack of playing cards.
Collection of data. Pick out the Jack, Queen, King and Ace of one suit, and put them down in a line on a table. Pick out a second set, consisting of similar cards from another suit. Shuffle the second set and place them below the cards on the table. Count the number of matches. Repeat at least 50 times, shuffling well between each placing. (Shuffling is difficult with only four cards. It is made easier if the remaining nine cards of the suit are included for shuffling. When dealing the cards keep dealing to the same position until one of the court cards or the ace appears, then deal to the next place.)
Analysis. Compile a frequency table for the number of matches. Find the relative frequency of no matches and compare with the calculated probability of $\frac{3}{8}$. (Verify this probability by writing down all the ways in which no match occurs.)

2 Simulation of use of beds in a hospital

Introduction and aim. A health authority is considering setting up a 'cottage hospital', for minor illnesses, in a small town. At present all patients have to travel a long distance to a major hospital. The aim of the simulation is to see whether the proposed number of beds will be adequate.

Equipment. A table of random numbers.

Collection of data. There are four beds in the hospital. From previous records, the distribution of the number of patients becoming ill each day is:

Number	0	1	2	Total
Frequency	30	50	20	100

The length of stay of patients has the distribution:

Length of stay (days)	1	2	3	4	5	6	7	Total
Frequency	6	10	20	40	15	6	3	100

Assume that on day 0, two beds are occupied and, counting day 0 as the first day of these patients' stay, they are in bed for 3 and 5 days. It is recommended that a table be drawn up as shown below. An example of the progress of the simulation over the first few days is given to illustrate the procedure.

	New patients		Length of stay, l Patient 1		Patient 2		Bed occupancy (+ = occupied)				Patients not accepted	No. of empty beds
Day	r	No.	r	l	r	l	1	2	3	4		
0							+	+	−	−		2
1	27	0					+	+	−	−	0	2
2	92	2	43	4	86	5	±	+	+	+	0	0
3	95	2	89	5			+	+	+	+	1	0
4	04	0					+	±	+	+	0	0
5	16	0					+	−	±	+	0	1
6	01	0					+	−	−	±	0	2
							±					

Continue the simulation for about 40 days. Repeat the simulation with a different number of beds.

Analysis. For each simulation run find the percentage of patients who are not accepted, and the mean number of empty beds per day. Find the distribution of these quantities over the various runs of the simulation.

Compare the results for four beds with results using a different number of beds.

3 Probability of a car being a hatchback

Aim. To estimate the conditional probability of a saloon car being a hatchback, i.e. having a rear door instead of a luggage boot, given its age (year of registration).

Collection of data. Make the study population the cars in a large car park. Choose a random sample of the cars, note the type of car and the 'year' letter and/or numbers in the registration number (in the UK; or other age identification, if available, in other countries). Take care to choose a sample which is as random as possible – this may not be easy. Aim to include at least 100 cars.

Analysis. Classify the age information into about five categories. Draw up a table with 'age' as the column headings and 'saloon' or 'hatchback' for the two rows. Count the number of cars that come into each category. Make sure that there are at least 15 cars in each column by combining neighbouring columns if necessary. Find the proportion of hatchback cars in each column. These are estimates of the required conditional probabilities. Consider critically how valid these estimates are.

SUMMARY

Events are sets of points in a possibility space S.

The **union** of the events Q and R is the event consisting of the points which appear in Q or R or both.

The **intersection** of the events Q and R is the event consisting of those points which appear in both the events Q and R.

Two events are **mutually exclusive** if they have no points in common.

Two or more events are said to be **exhaustive** if their union equals the whole of the possibility space S.

The **complementary event** Q^c to the event Q is the set consisting of all those points in the possibility space S that are not in Q.

Probability of union of two events. If Q and R are events in a possibility space then

$$P(Q \cup R) = P(Q) + P(R) - P(Q \cap R).$$

Addition rule for mutually exclusive events. If Q and R are mutually exclusive events then

$$P(Q \cup R) = P(Q) + P(R).$$

More generally, if $E_1, E_2, \ldots, E_k$ are mutually exclusive events then

$$P(E_1 \cup E_2 \cup E_k) = P(E_1) + P(E_2) + \cdots + P(E_k).$$

The **conditional probability** of R given Q, $P(R \mid Q)$, is the probability that the event R will occur given that the event Q has already occurred.

$$P(R \mid Q) = P(Q \cap R)/P(Q).$$

Two events are **independent** if knowing the outcome of one event does not affect the probability of the other event. Formally, the events Q and R are independent if and only if

$$P(Q \cap R) = P(Q)P(R).$$

Bayes' theorem. If $A_1, A_2, \ldots, A_k$ are a set of mutually exclusive, exhaustive events in a possibility space S, and B is any event in S such that $P(B) > 0$, then

$$P(A_i \mid B) = \frac{P(B \mid A_i)P(A_i)}{\sum_{i=1}^{k} P(B \mid A_i)P(A_i)}, \quad i = 1, 2, \ldots, k.$$

8 DISCRETE RANDOM VARIABLES

OVERVIEW

Models for variates such as the number of left-handers in samples of 10 people, or the number of tosses of a coin required for a head to appear, are provided by discrete random variables. We show how these random variables are defined and begin to investigate their properties and uses.

8.1 PREDICTING THE NUMBER OF LEFT-HANDED CHILDREN IN A CLASS

A discrete variate is one whose values are distinct from each other; the values are usually (but not necessarily always) integers. For example, in a class of N students, the variate might be the number who write with their left hands; it could take values 0, or N, or any whole number in between, but not $1\frac{1}{2}$, or 0.36, or any negative number.

Suppose that a teacher has been running a dressmaking class with groups of 10 pupils at a time. For those who are left-handed she has provided left-handed scissors; she has three pairs, and has always found this number sufficient. Now the school has a new building, so that dressmaking classes can contain 30 pupils at the same time. How many pairs of left-handed scissors will she need? Of course, if she provided 30 she would be *certain* of always having enough. But it is extremely unlikely that everyone in a class of 30 would be left-handed, and it would be unnecessarily extravagant to have many pairs of special scissors that were never going to be used. The actual number of left-handed pupils in a class of 30 will vary from class to class: it is a *variate*. In order to make a rational decision about how many more special scissors to buy, the teacher will want to estimate the minimum number that will suffice in all except extreme cases. She will not want to be *certain* of *always* having enough, but will make do with being *pretty sure* that she has enough for the types of group of 30 which she expects to meet.

To begin making her decision she needs to know the frequency distribution of the number of left-handed pupils in classes of 30. But she cannot observe this variate until after she has moved into the new building, and then it will be too late. Suppose the teacher asks us to help: what can we do?

First, we ask what information she already has on left-handers. Let us suppose that she has a record of the number of left-handed pupils in each class of 10 that she has taught in the past, and finds that there have been 30 left-handed pupils out of a total of 300 pupils. The valuable information for us is that the *proportion* of left-handers in her pupils to date has been 0.1.

We approach the problem by constructing a model of the real-life situation, and using the model to predict what will happen in the future.

8.1.1 A physical model

We might use coloured beads to make a model. We obtained a large number of beads, of which one-tenth were blue and the rest white, and mixed them up very thoroughly. The blue beads formed one-tenth of the population and represented the left-handed pupils; the white beads represented those who were right-handed. We mimicked what happens in classes of 30 by picking 30 beads at random and counting how many blue ones there were among these. This sample of 30 was then returned to the whole population. All the beads were mixed thoroughly, another sample of 30 beads was drawn and the blue ones counted. By taking 500 samples in this way we generated the frequency distribution of the number of blue beads in random samples of 30 (Table 8.1). This number is a variate: what relation does it have to the variate our dressmaking teacher is really interested in?

We argue that this sampling process mirrors what happens when classes are set up; therefore the frequency distribution which we obtain predicts the distribution we would find if we actually observed the number of left-handers in groups of 30. In doing this, we make two major assumptions. Let us look at these more closely. (1) We assume that the proportion of left-handers in the whole population (from which our groups of 30 will be drawn) stays *constant*, and that it will be the same for future pupils as it has been for past pupils. There is no reason to suppose that the proportion of left-handers among the children may be changing, so this assumption appears reasonable. (In contrast, if we were interested in the number of pupils having computers at home, we could not reasonably assume that the proportion stayed constant over time.) (2) We treat the new classes as if they were *random samples*, with respect to left-handedness, from some population. If, for example, left-handers tended to stick together as a group, so that if one left-hander chose a particular class there was a high chance that the others would do so too, this assumption would be false. But we do not think that in practice this, or anything like it, happens; so we can accept this assumption too.

Table 8.1 *Frequency distribution of the number of blue beads in 500 random samples of 30 beads*

Number of blue beads	0	1	2	3	4	5	6	7	8	≥9	Total
Frequency	21	59	108	118	97	54	24	15	4	0	500
Relative frequency	0.04	0.12	0.22	0.24	0.19	0.11	0.05	0.03	0.01	0	

The basic assumptions of the bead model therefore seem reasonable, making it an instructive model to consider. Its failings are more of a practical nature: choosing a random sample from a large pile of beads is not easy. Without special apparatus, adequate mixing of the beads is also difficult. The blue beads may well differ in size and weight from the white ones, and this would influence how frequently they were chosen, perhaps making their probability of choice noticeably different from the intended 0.1. Also, the choice of enough samples to generate an adequate frequency distribution takes a long time. An alternative approach is to construct a theoretical model.

8.1.2 A theoretical model

We now consider a population of hypothetical pupils whose only interesting attribute is whether they are labelled L, corresponding to left-handed, or R, corresponding to right-handed. Among the real pupils, the *relative frequency* of left-handers in the past has been 0.1. We therefore specify that in our hypothetical population there will be a *probability* of 0.1 that a pupil chosen at random shall be labelled L. As we saw in Chapter 6, probability is an idealization of relative frequency. Corresponding to the variate which takes the values 'left-handed' or 'right-handed' in the population of pupils who take the dressmaking classes, we have devised an attribute which takes the values L or R in a hypothetical population. We call this new concept a *random variable*.

Working definition

A **random variable** is any quantity or attribute whose value varies from one unit of a hypothetical population to another.

We may compare this with the definition of a variate, given earlier: a variate is any quantity or attribute whose value varies from one unit to another of the population being investigated. As with variates, we recognize qualitative, discrete and continuous random variales.

By observing the variate we construct a *frequency distribution*. Note that we call the two forms, 'left-handed' and 'right-handed', that the variate may take, the 'values' of the variate:

Variate value	Left-handed	Right-handed
Frequency	30	270
Relative frequency	0.1	0.9

For the random variable, we specify a *probability distribution*:

Value	L	R
Probability	0.1	0.9

Table 8.2 *Probability distribution of number of hypothetical pupils labelled L in random samples of 30, using a binomial probability model*

Number labelled L	0	1	2	3	4	5	6	7	8	$\geqslant 9$
Probability	0.04	0.14	0.23	0.24	0.18	0.10	0.05	0.02	0.01	0

This random variable is a qualitative one. But qualitative random variables (attributes) take only a finite number of values, so we consider them with discrete random variables in this chapter.

Now that we have specified this basic random variable we can derive the probability distributions of other random variables by using the theory of probability. We will then be able to make predictions about the variates corresponding to these random variables.

We wish to predict the relative frequency distribution of the variate 'the number of left-handed pupils in classes of 30'. The corresponding random variable is the number of hypothetical pupils labelled L in random samples of size 30. We must calculate the probability distribution of this random variable. Its possible values are $0, 1, 2, \ldots, 30$. Calculating the probabilities for these values is identical with the calculation that $0, 1, 2, \ldots, 30$, out of 30 independent trials are successful. We considered a similar problem earlier (Example 7.10, p. 148) and saw that the probability that r trials were successful was

$$P(r) = \binom{30}{r}(0.1)^r(0.9)^{30-r}, \quad r = 0, 1, 2, \ldots, 30.$$

(Note that the formula for $P(r)$ by itself is not enough: to specify a probability distribution completely we must also give the list of values that the variable r may take.) These probabilities are shown in Table 8.2, to two decimal places.

These probabilities give the teacher who asked our advice some useful information to help her make her decision. If she has six pairs of left-handed scissors, then in only three classes out of 100 is she likely to need more, and the probability of needing more than seven pairs is only 0.01.

Notice that the teacher still has her own decision to make; the statistician does not do that for her, but simply gives her extra information upon which to base the decision.

8.2 RANDOM VARIABLES

Random variables may be (and widely are) studied for their intrinsic mathematical interest. But they are also useful to us because they provide models for the variates that we can observe in real life. To specify a discrete random variable we need to know its set of possible values and the probability with which it takes each of these values. This may be done by the use of a table, e.g.

r	0	1	2
P(r)	0.3	0.6	0.1

or by the use of a formula, e.g.

$$P(r) = \binom{30}{r}(0.1)^r(0.9)^{30-r}, \quad r = 0, 1, 2, \ldots, 30.$$

In both cases, a function of r is specified. For a discrete random variable we call this function the *probability mass function* of the variable. (Note that a qualitative random variable can be specified only by a table, such as that near the foot of p. 167.)

When we have collected data on a variate, and made an observed frequency distribution, we often want to specify a random variable in order to model this variate. We did so above for the variate that takes the values 'left-handed' and 'right-handed'. In such cases it is natural to specify the random variable in the form of a table. The probability with which each possible value of the random variable occurs in the table is made equal to the relative frequency of the corresponding value of the observed variate.

In other situations we may know the mechanism by which a variate is generated. To model such a variate, we begin by specifying a simple random variable, or giving the probabilities of simple events, using our knowledge of the mechanism. Then we can derive the distribution of a random variable which models the variate we want to study. In the scissors example we began with the simple 'L/R' random variable and derived from it a random variable to model the variate we were interested in. This was 'the number of left-handed children in classes of 30'. In these latter cases we shall often obtain a formula to describe the probability mass function of the random variable.

The set of values that a random variable can take is, of course, a possibility space. The values form the outcomes and we assign probabilities to them. The sum of these probabilities must be unity. In the majority of examples the set of values is finite but models in which there are an infinite number of values can be useful (see the geometric distribution, Section 8.7).

From another viewpoint we may regard the set of values that a random variable *can* take as a hypothetical infinite population. The particular values that *do* occur in actual sets of observations are samples from this population.

It is useful to distinguish between the name of a random variable and the particular values that the random variable can take. We shall use a capital letter, e.g. X, for the name, and the corresponding small letter, e.g. x, to denote a particular value. For discrete random variables we shall frequently use the name R, and denote particular values by the corresponding small letter r.

8.2.1 Exercises

1. A survey of 200 households in Notown of the number of TV sets per household gave the frequency distribution:

Number of sets	0	1	2	3
Frequency	8	180	10	2

   Specify a random variable to model the variate 'number of TV sets per household in Notown'.

2. Specify a random variable to model the variate 'the number of heads in two tosses of a coin', by assuming that the probability of heads at any toss is: (a) $\frac{1}{2}$; (b) π (which is an unknown value between 0 and 1).

3. For each of the following variates specify a random variable that might serve as a model for the variate. Give the set of values that the variate takes, together with the probabilities of these values, expressed either as numerical values or by a formula. Check that the probabilities sum to one.
   (i) The number obtained by throwing a fair die.
   (ii) The number of cups of tea (or coffee) that you drink per day.
   (iii) The number of people wearing glasses in a random sample of four from a population in which the proportion of people with glasses is: (a) 0.2; (b) π (any value between 0 and 1).
   (iv) The hair colour of people in your school (or college, or at your place of work).

8.3 FAMILIES OF RANDOM VARIABLES

If the probability of choosing a pupil labelled L is π, and the size of the random sample is n, the set of probabilities for choosing r pupils labelled L in the sample is

$$\binom{n}{r} \pi^r (1 - \pi)^{n-r}, \quad r = 0, 1, 2, \ldots, n.$$

These probabilities define a *family* of random variables, in fact the binomial family which we return to discuss later (Section 8.6). A particular individual member of the family is specified by giving numerical values to n and π; these are called the *parameters* of the distribution. We shall often use small Greek letters to denote parameters of distributions (though here the Latin n is so commonly used for sample size that in this case we bend our rule). Since random variables specify hypothetical populations the Greek notation is consistent with the distinction we made earlier (p. 52) between population values, for which we use Greek letters, and sample values, usually indicated by the corresponding Latin letters.

8.4 THE DISCRETE UNIFORM DISTRIBUTION

We have already seen examples of this distribution. It is the distribution in which all the possible values have equal probabilities; examples have been the result of tossing a fair coin and the distribution of random digits.

If we toss a coin and record which face is shown when it lands, we are observing a variate which can take the value 'heads' or 'tails'. If we state, in addition, that we are treating the coin as a 'fair coin', we are setting up a mathematical model in which we are defining a random variable that has two outcomes; we may call these outcomes H and T, and each has probability $\frac{1}{2}$. Previously, we defined a fair coin simply as one that was 'properly balanced'; but the only reasonable way of *testing* whether a coin is fair is to see how well the results of tossing it several times agree with the mathematical model that says H and T occur with equal probability. In fact, real coins usually do agree very well with this model; try making a long sequence of tosses for yourself.

In a similar way, we may model what happens when we throw a six-sided die. We suppose that we have a properly balanced die, which when thrown will show any of the scores $1, 2, 3, 4, 5, 6$ with equal probability $\frac{1}{6}$ each. The construction of an ordinary cubic die is such that there is no reason to expect one face to be favoured over any other, and therefore the model is a reasonable one. We describe the random variable mathematically by writing $P(R = r) = \frac{1}{6}$ ($r = 1, 2, 3, 4, 5, 6$), where r represents the score at a particular throw. Very often we omit to mention R, the name of the random variable, and simply write $P(r) = \frac{1}{6}$.

Clearly $P(r)$ is zero for any score we care to mention for r except the integers 1 to 6; and also, since one or other of these six outcomes is bound to occur, we have $\sum_{r=1}^{6} P(r) = 1$. An obvious way to represent the possibility space for this random variable is to draw a part of the real line (x-axis) and place spots at each of the six points representing $1, 2, 3, 4, 5, 6$.

When a table of random digits is constructed (see, for example, Table A1 on p. 713), we suppose that each position in it is equally likely to be occupied by any one of the digits $0, 1, 2, \ldots, 9$. So a model for this is a random variable R with $P(r) = \frac{1}{10}$ ($r = 0, 1, \ldots, 9$), where r is the value of the digit occupying any given position in the table. We may test whether particular sets of digits (e.g. those generated by a computer program) can be regarded as random by checking if they agree with this model. (There are various ways of doing this, and we shall mention some later, on pp. 473, 480.)

Definition 8.1

A discrete random variable R, taking values $1, 2, 3, \ldots, \kappa$ such that

$$
P(R = r) = \begin{cases} \dfrac{1}{\kappa} & r = 1, 2, 3, \ldots, \kappa \\ 0 & \text{otherwise} \end{cases}
$$

follows the **discrete uniform distribution**.

We have chosen to define the discrete uniform distribution on the set of integers $1, 2, 3, \ldots, \kappa$. But, by relabelling, the distribution is equivalent to any distribution defined on any consecutive set of integers, including positive and negative integers and zero, when all values are taken with equal probability. Hence we shall call any of these discrete uniform distributions.

One way to write the model for the result of tossing a fair coin is $P(H) = P(T) = \frac{1}{2}$. But we can make our random variable look exactly like a discrete uniform distribution in the following way. We associate a value of the random variable R with each possible outcome of the toss: take $R = 1$ whenever a head results and $R = 0$ for a tail. Then the model is written

$$P(R = r) = \begin{cases} \frac{1}{2} & r = 0, 1 \\ 0 & \text{otherwise.} \end{cases}$$

A random variable R used in this way, *associated* with the results rather than *equal* to them, can be a very useful concept. It is called a *dummy variable* or an *indicator variable*.

8.5 THE BERNOULLI DISTRIBUTION

Suppose that we now toss a coin that may *not* be 'fair', but has some probability π $(0 < \pi < 1)$ of showing heads on any toss. This idea generalizes to any trial with only two outcomes that are not necessarily equally likely. For example, we may observe trains arriving at a railway station which may be on time (or early) with probability π, or late with probability $1 - \pi$.

In each of these examples, we may associate the two possible outcomes of the trial with the values $1, 0$ of a dummy variable:

$$R = 0 \text{ when a tail is shown,} \qquad R = 1 \text{ for a head;}$$

and

$$R = 0 \text{ when a train is late,} \qquad R = 1 \text{ otherwise.}$$

The probability distribution of R in these situations is

$$P(R = 1) = \pi,$$
$$P(R = 0) = 1 - \pi.$$

It is very often called Bernoulli's distribution, after Jacob Bernoulli (Swiss, 1654–1705), one of a family of eminent mathematicians who lived in Europe in the seventeenth and eighteenth centuries.

8.6 THE BINOMIAL DISTRIBUTION

This is the most important discrete probability distribution, and one of the most important probability distributions in the whole of statistics. Let us imagine a population that contains only two types of member, say sound and faulty items coming from a continuously operating production line in a factory. The whole population of items contains an unknown proportion of sound ones. There is a scheme of inspection: a random sample consisting of n items is taken from the end of the production line, and in this sample the variate 'number of sound items' is observed. At various times this random sampling process is repeated. Thus we obtain a set of observations of the variate.

In order to set up a random variable R that models this number of sound items, we need to make assumptions very similar to those required by our teacher of dressmaking, described on p. 165. Now we require that every individual item produced has the same probability, π, of being sound, and that items are selected singly, independently of one another. This is what is meant by the sample being random, since in a random sample we cannot predict what future sample members will be selected even if we know which have been picked so far. Therefore knowledge of existing members gives no information about the probabilities of values which will be observed in future selections; in other words, the probabilities of these values are not conditional on any previous values, and we can simply multiply the probabilities.

We can specify the probability distribution of R if we know the probability that a random sample of size n, drawn from a population in which a special type of item occurs with probability π, will contain exactly r items of the special type. As we saw in Example 7.10, p. 148, this is

$$\binom{n}{r} \pi^r (1 - \pi)^{n-r}, \quad r = 0, 1, \ldots, n.$$

We have already remarked that this represents a whole family of distributions, in which n, π are parameters. We shall usually be told the sample size n; this will be determined partly by how quickly items can be inspected. We may then want to use our model, with a particular value inserted for π, to compare with the observations actually made on the variate.

This family of probability mass functions is called the *binomial family*. As is shown in the Appendix (p. 674), the expression $(a + b)^n$, in which n is a positive whole number, as sample size obviously must be, can be expanded in the *binomial series*.

The successive terms in the expansion are

$$a^n; \binom{n}{1} a^{n-1} b; \binom{n}{2} a^{n-2} b^2; \ldots; \binom{n}{n-1} ab^{n-1}; b^n.$$

If we now put b equal to π and a equal to $(1-\pi)$, we see that these terms are exactly $P(R = 0); P(R = 1); P(R = 2); \ldots; P(R = n-1); P(R = n)$. The sum $\sum_{r=0}^{n} P(R = r)$ should be 1, because R *must* equal one of the integers $0, 1, \ldots, n$, which are mutually exclusive and exhaustive. We have in fact expanded $[(1 - \pi) + \pi]^n$, or 1^n, which is 1; so the sum of the terms in the series expansion must be 1, and this is just the sum of all probabilities $P(r)$.

Definition 8.2

A discrete random variable R is said to follow the **binomial distribution** if

$$P(R = r) = \binom{n}{r} \pi^r (1 - \pi)^{n-r}, \qquad r = 0, 1, 2, \ldots, n,$$

where $0 < \pi < 1$.

The conditions that give rise to a binomial distribution are:

   (i) there is a *fixed* number n of trials;
  (ii) only *two* outcomes, 'success' and 'failure', are possible at each trial;
 (iii) the trials are *independent*;
 (iv) there is a *constant probability* π of success at each trial;
  (v) the variable is the *total number of successes* in n trials.

    Let us give two more examples of variates for which the binomial distribution provides a reasonable model.

Number of girls in families of a given size

We must consider only families in which there are a given number of children in order to satisfy condition (i). Only two possible outcomes, boy and girl, can occur at each birth; condition (ii) is satisfied by equating 'girl' with 'success'. The evidence is that the sex of a baby is independent of the sex of the children already in the family, so that condition (iii) is satisfied. The probability of a girl being born appears to be constant and is, in fact, slightly less than one-half; a more precise value from recent data is 0.49 (though for many practical purposes it is satisfactory to use $\frac{1}{2}$). The variate, the number of girls in families of, say, 4 children is therefore well modelled by the binomial random variable with $n = 4$ and $\pi = 0.49$. This random variable has the probability distribution:

r	0	1	2	3	4
$P(r)$	$(0.51)^4$	$4(0.51)^3(0.49)$	$6(0.51)^2(0.49)^2$	$4(0.51)(0.49)^3$	$(0.49)^4$
	0.068	0.260	0.375	0.240	0.058

(Owing to rounding off in the third place of decimals, we find these probabilities add to 1.001, but this cannot always be avoided.)

Number of observations ending in 0 or 5

A scientist asked his assistant to count the number of diseased plants on 60 field plots. When looking at the results he was struck by the high proportion of observations that ended in 0 or 5, and wondered if his assistant had been careless in making the records. Because he remembered once before telling the assistant to record some very large numbers to the nearest 5, the scientist wondered if the assistant had done so again without being told. So he worked out what distribution one would expect the number of observations ending in 0 or 5 to follow. He argued that if the last digit in each recorded observation was equally likely to be any one of $0, 1, 2, \ldots, 9$, then there was a constant probability of 0.2 that the last digit of any observation would be a 0 or a 5, if the recording was done conscientiously. Also, the final digit of each observation would be independent of the final digits of the other observations. In a total of 60 observations, the number of observations ending in 0 or 5 should follow a binomial distribution with $n = 60$ and $\pi = 0.2$. Hence the probability distribution of the number of observations ending in 0 or 5, assuming accurate recording, could be calculated. This would provide a basis against which to compare the 60 observations he was looking at.

Example 8.1

Forty per cent of a sweet assortment are 'soft centres' and the remainder 'hard centres'. A handful of four sweets may be regarded as a random sample from a huge pile of the sweets. What is the probability that there are three or four 'soft centres' in a handful?

The number of 'soft centres' in a handful has a binomial distribution with $n = 4$, $\pi = 0.4$.

We do not need to list the whole distribution, but only those probabilities we are interested in. The probability we require is

$$P(3 \text{ or } 4) = P(3) + P(4) = \binom{4}{3}(0.4)^3(0.6) + (0.4)^4 = 4(0.4)^3(0.6) + (0.4)^4$$

$$= 0.1536 + 0.0256 = 0.1792 \cong 0.18.$$

Example 8.2 Binomial probability

What is the probability that in 10 tosses of a fair coin, exactly 5 come down heads?

The distribution of the number of heads is binomial with $n = 10$, $\pi = 0.5$. The probability required is

$$\binom{10}{5}(0.5)^5(0.5)^5 = \frac{10 \cdot 9 \cdot 8 \cdot 7 \cdot 6}{1 \cdot 2 \cdot 3 \cdot 4 \cdot 5} \cdot \frac{1}{2^{10}} = \frac{252}{1024} = 0.246.$$

8.6.1 Exercises

1. For each variate below, state whether the binomial random variable would provide a satisfactory model, giving your reasons. If it would, state the values of n and π.
   (i) The number of sixes obtained in three successive throws of a die.
   (ii) The number of sixes obtained when three dice are thrown simultaneously.
   (iii) The number of girls in the families of British prime ministers.
   (iv) The number of aces in a hand of four cards dealt from a standard pack of cards.
   (v) The number of children, in a class of 30, whose birthday anniversary falls on a Sunday this year.
   (vi) The number of tosses required when a coin is tossed until the first head appears.

(vii) The number of men in random groups of four successive people chosen as people leave a theatre. The theatre audience was large and there were approximately equal numbers of men and women.

2. For the binomial distribution with $n = 5$ and $\pi = 0.2$, calculate the probability of:
   (i) two successes;
   (ii) at least four successes;
   (iii) at least one success.

3. The probability that Jim can hit the bull on a target is $\frac{1}{3}$. Is he certain to score a bull if he takes 3 shots? Find the probability of his scoring at least one bull in 3 shots.

4. Individual cards, chosen at random from a set of 20, are given away with bars of chocolate. David needs one card to complete the set he is collecting. If he buys five bars, what is the probability that: (i) none of them contains the required card; (ii) at least one contains the required card (make use of (i))?

5. Show, for the binomial variable R with $n = 3$, that $P(R = 1 \text{ or } 2) = 3\pi(1 - \pi)$.

6. I threw three dice 216 times and obtained the following frequency distribution of the number of sixes per throw:

Number of sixes	0	1	2	3	Total
Frequency	110	85	20	1	216

Calculate the frequencies you would expect for a binomial random variable with $n = 3, \pi = \frac{1}{6}$ and comment whether this model appears a reasonable one for the data.

8.7 THE GEOMETRIC DISTRIBUTION

Suppose that we continue tossing a coin until a head first appears. We record the number of tosses required for this, which may be any integer from 1 upwards, and then begin tossing again. We can build up a frequency distribution for the variate 'number of tosses until the first head'. Similarly, if we are looking at traffic passing an observation point we may record the number of cars until the first estate car passes (or any other chosen type that we can recognize easily). As soon as an estate car passes, we make our record and the process begins again.

In both of these examples, the sample size (number of tosses or number of cars) is *not fixed*: instead sample size is the variate being observed. A random variable to model it thus has to be of a slightly different type from those met so far in this chapter. Again (as in the binomial) we imagine a population consisting of two types of item only, one type (heads or estate cars) occurring with probability π; we also imagine that each member drawn from this population is independent of every other. In coin-tossing that is a safe assumption; with makes of car we should take elementary precautions like making sure we were not near a local car dealer's showroom at the time a delivery of new vehicles is being made. Further, π must be constant for all observations made: it must not change during the time the experiment or survey is going on.

At each toss of the coin we obtain either a head, with probability π, or a tail with probability $1 - \pi$. Thus, the random variable R, 'number of tosses until first head', cannot be less than 1; with probability π it will actually be 1. If $R = 2$, we must have obtained tail–head in that order, and since individual tosses are independent of one another the probability for this is $(1 - \pi)\pi$. In the same way, $R = 3$ requires tail–tail–head, with probability $(1 - \pi)^2\pi$. It is easy to see that $P(R = r)$ is $(1 - \pi)^{r-1}\pi$. Nothing like a binomial coefficient comes in because there is only one possible sequence of results leading to each value of R. Note, however, that we can give no firm upper limit to the values R may take, because we can never specify a point where we *must* have tossed a head. R is therefore a random variable that takes integer values $1, 2, 3, \ldots$ without upper limit. This is called being 'countably infinite'.

Definition 8.3

A discrete random variable R is said to follow the **geometric distribution** if

$$P(R = r) = (1 - \pi)^{r-1}\pi, \quad r = 1, 2, 3, \ldots,$$

where $0 < \pi < 1$.

The conditions that give rise to the geometric distribution are:

  (i) there is a *sequence* of trials;
 (ii) only *two* outcomes, 'success' and 'failure', are possible at each trial;
(iii) the trials are *independent*;
 (iv) there is a *constant probability* π of success at each trial;
  (v) the variable is the *number of trials* taken for the first success to appear (the successful trial is included in the count).

The possibility space contains an infinite number of points corresponding to the possible values $r = 1, 2, 3, \ldots$ of the variable. The probabilities associated with the points form the successive terms in a geometric series. Their sum is a 'sum to infinity' since we specify no upper limit on r, and is equal to $\pi(1 + w + w^2 + w^3 + \cdots)$, where $w = 1 - \pi$. The series is convergent since $w < 1$ and the sum is

$$\pi \cdot \frac{1}{1 - w} = \frac{\pi}{\pi} = 1.$$

Because the sum of all the probabilities is one, we have a properly defined distribution.

This is our first example of a possibility space which contains an infinite number of points. We still call the variable a discrete random variable because the values change by jumps and not continuously. We shall consider only two types of discrete random variable. The first type, like the binomial distribution, takes only a finite set of values. The second type, like the geometric distribution, takes values in the set of natural numbers $1, 2, 3, \ldots$ or in a set that can be put in one–one correspondence with the set of natural numbers. The probabilities must, of course, sum to one.

A sampling inspection scheme

In a continuously operating production line producing bolts which should be 3 cm long, 5% of the output has to be rejected as too short. An inspection scheme consists of picking bolts at random from the production, and noting how many are picked up until a faulty one is found. If R denotes the total number picked up (including the faulty one), its probability distribution is geometric with $\pi = 0.05$:

r	1	2	3	4	5	6	7
P(r)	0.05	$(0.95)(0.05)$	$(0.95)^2(0.05)$	$(0.95)^3(0.05)$	$(0.95)^4(0.05)$	$(0.95)^5(0.05)$	$(0.95)^6(0.05)\ldots$
	0.05	0.0475	0.0451	0.0429	0.0407	0.0387	0.0368 $\quad\ldots$

The values of P($R = r$) drop very gradually; for $r = 20$ we still have a probability of 0.0189.

We are assuming all the time that the production line stays in proper adjustment, so that π remains constant.

8.8 ACCEPTANCE SAMPLING IN QUALITY CONTROL

An important application of the binomial distribution is in studying the proportion of 'faulty' items found in samples from a large population of identical items. (The word 'defective' is often used instead of 'faulty', and modern jargon even has 'non-conforming' (to specification) for 'faulty' – we shall use 'faulty'!) The ideas were developed in industry, where an item or component may be mass-produced on a production line, for subsequent use in a product such as a radio or television, a motor vehicle or a piece of household equipment. Batches of the item may be available, or the production line may be continuous and random samples can be taken at convenient intervals. By using rules based on the number of faulty items in a sample we wish to decide whether the batch sampled, or the current production, should be accepted as satisfactory or whether it should be rejected. The sample from a batch or from the production line contains n items, and we assume that the conditions for the binomial distribution (Section 8.6, p. 173) apply. The probability of an item in the whole production being faulty is π; thus π can be thought of as the *proportion* of faulty items in the whole output, and clearly we want the probability of acceptance of a batch to be high when π is small and to be low when π is large.

These ideas can be extended to many other fields of study, where we can set up a criterion for classifying each item observed as 'sound' or 'faulty'. Standards of performance in an administrative or service system can be assessed this way, such as the proportion of medical patients treated within a specified time, or the record of punctuality of a transport system. Consumer protection laws specify what minimum weight should be attained in a food product that has a weight stated on it, and if more than a very small proportion of items of the product fall below this minimum weight the producer will be committing an offence. A packing station may be making up packets of four apples for sale in shops and supermarkets; it will only classify a pack as 'sound' if none of the four apples shows any sign of surface damage or blemish.

Example 8.3 A single sampling scheme

A large-scale process is producing screws to a specified diameter. If the diameter of a screw is not within 0.5 mm of specification, the screw is classified as 'faulty'. The process gives a faulty screw with probability π. At random intervals of time, a random selection of $n = 8$ screws is made from the output.
(a) Find, in terms of π, the probability that the number r of faulty screws in the sample will be 0 or 1.
(b) The following *decision rule* is proposed: if the number r is 0 or 1, the batch is accepted and the process is allowed to continue uninterrupted, but otherwise the output is monitored closely for a while. Find, for values of π from 0.01 to 0.20, the probability that a batch will be accepted.
(c) Comment on the result of using this decision rule.

(a) For $n = 8$, the probability of $r = 0$ or 1 is $P(0) + P(1)$ in the binomial $(8, \pi)$, which is

$$(1 - \pi)^8 + 8\pi(1 - \pi)^7 = (1 - \pi)^7(1 + 7\pi).$$

(b) Our decision rule accepts a sample with a probability $(1 - \pi)^7(1 + 7\pi)$ when the output population contains a proportion π of faulty individuals. Table 8.3 shows, in the second column, these probabilities of acceptance for values of π between 0.01 and 0.20.

(c) There is a high probability of acceptance if π is anything up to 0.04 (i.e. 4% faulty in the population). Up to $\pi = 0.10$ (10% faulty) this decision rule will often not cause any extra monitoring of the system. Even when $\pi = 0.20$ (20% faulty) a batch is still roughly equally likely to be accepted (probability 0.503) as rejected (probability 0.497). This does not seem to be a good scheme.

Altering the decision rule. Instead of allowing $r = 0$ or 1 faulty items among the random sample of $n = 8$ items, let us not allow any faulty: we will accept a sample only if it has $r = 0$ faulty, and otherwise will monitor the process more closely. Now the value of $P(0) = (1 - \pi)^8$, and from the third column of Table 8.3 we see that this decreases quickly as π gets above even 0.01. In this way the scheme is improved considerably.

Altering the sample size. An alternative to insisting on having no faulty items in a sample is to keep the original decision rule, $r = 0$ or 1, but take a larger random

Table 8.3 *Probabilities of acceptance in the one-stage sampling scheme of Example 8.3*

Sample size, n	8	8	12	12
Decision rule, $r =$	0 or 1	0	0 or 1	0
π				
0.01	0.9973	0.9227	0.9938	0.8864
0.02	0.9897	0.8508	0.9769	0.7847
0.03	0.9777	0.7837	0.9514	0.6938
0.04	0.9619	0.7214	0.9191	0.6127
0.05	0.9428	0.6634	0.8816	0.5404
0.06	0.9208	0.6096	0.8405	0.4759
0.07	0.8965	0.5596	0.7967	0.4186
0.08	0.8702	0.5132	0.7513	0.3677
0.09	0.8423	0.4703	0.7052	0.3225
0.10	0.8131	0.4305	0.6590	0.2824
0.12	0.7520	0.3596	0.5686	0.2157
0.14	0.6889	0.2992	0.4834	0.1637
0.16	0.6256	0.2479	0.4055	0.1234
0.18	0.5634	0.2044	0.3359	0.0924
0.20	0.5033	0.1678	0.2749	0.0687

sample, say $n = 12$. The fourth column of Table 8.3 shows the probabilities of acceptance for the same set of values of π; for $n = 12$, $P(0) + P(1)$ in the binomial $(12, \pi)$ is

$$(1 - \pi)^{12} + 12\pi(1 - \pi)^{11} = (1 - \pi)^{11}(1 + 11\pi).$$

Altering the decision rule and the sample size. The final column of Table 8.3 uses $r = 0$ only as the decision rule and takes samples of $n = 12$, so that if there are any faulty items in the sample of 12 the process is immediately monitored more closely. We see that this will happen quite often even when π is very small. Which of these combinations of sample size and decision rule we adopt will depend on whether it is important to avoid values of π less than about 4–5%. The combination $n = 12$, $r = 0$ could perhaps be over-cautious; if we are to use $r = 0$ then $n = 8$ (the third column of Table 8.3) may be considered satisfactory in situations where a fairly small proportion of faulty items can still be tolerated – for example, they are not a total loss but can be sold at a reduced price. But if the decision rule uses $r = 0$ or 1, then $n = 8$ is not enough to avoid letting through too many samples with an appreciable value of π; for this rule, $n = 12$ looks about right. The manufacturer does not want to incur extra cost in monitoring the process when π is very small, but wants to avoid passing batches, and selling them to customers, when π is higher, say 10–20%. Exercise 8.8.1(1) shows a possible scheme with n much higher (40) and r correspondingly relaxed a little.

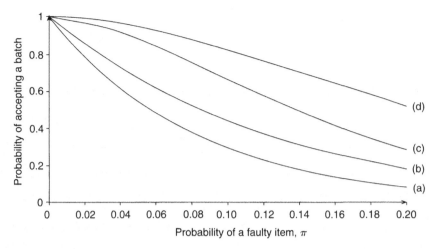

Figure 8.1 *Operating characteristic curves for the sampling schemes of Table 8.3: (a) sample size n = 12, decision rule r = 0; (b) n = 8, r = 0; (c) n = 12, r = 0 or 1; (d) n = 8, r = 0 or 1.*

Operating characteristic curves. The choice between schemes discussed above is often made easier if a graph is drawn of the probability of acceptance plotted against the value of π. When $\pi = 0$, this probability is 1; and we want the probability to reduce quickly as π increases. Figure 8.1(a) shows an acceptable shape of graph and Fig 8.1(d) a very undesirable one.

These curves have always been known by industrial statisticians as *operating characteristic curves*, or *OC curves*: they indicate the important characteristics of a sampling scheme by showing how quickly the probability of acceptance decreases as the quality of a batch decreases. Figure 8.1 shows the operating characteristic curves for all four decision rules summarized in Table 8.3, plotted on the same graph. In Fig. 8.1(a) there is a reasonably rapid drop in probability of acceptance, although perhaps not as rapid as the manufacturer would really like if he is working in a very competitive market. Figure 8.1(b) is similar in shape, but the scheme does not seem sensitive enough for values of π which could lead to serious complaints from customers, such as $\pi = 0.10$ as we have already remarked. Figures 8.1(c) and (d) show schemes that are most unlikely to be acceptable.

By using larger sample sizes, the OC curve can be made to drop off sharply for very small values of π; this will be needed if it is important to avoid any appreciable amount of faulty items.

In assessing industrial sampling schemes, two probabilities called *producer's risk and consumer's risk* have been found very useful; they are related to type I and II errors in statistical inference, and are discussed in Section 13.3.

Example 8.4 A double sampling scheme

A food product is pre-packaged, for safe storage before and after sale. The packaging must be airtight if the product is to be acceptable to the customer. Random samples, each of n items, are taken at intervals, and the probability of an item being found faulty is π. The following decision rule is proposed: a batch will be accepted if none of the n items in the sample is faulty, and if the number, r, of faulty items is greater than a specified number the batch will be rejected; if r is greater than 0 but not greater than the specified number, a second sample of n items will be taken.

(a) If $n = 5$ items are examined in each sample, and the batch is rejected immediately if r is 2 or more, find the probabilities of immediate acceptance and rejection of a batch.

(b) If $r = 1$, a further sample of n items is taken. The decision rule for the second sampling is to accept the batch if this sample contains no faulty items. Draw a tree diagram to illustrate the possible outcomes and their probabilities, and calculate the overall probability of acceptance for values of π from 0.01 to 0.20.

(c) If the decision rule for the second sampling is changed to accepting the batch when the second sample contains $r = 0$ or 1 faulty, repeat the calculations of the overall probability of acceptance.

(d) Discuss and compare the results.

(a) The batch is accepted at the first sampling only if the number faulty, r, in the sample is 0; $P(0) = (1 - \pi)^5$, and so the batch is accepted with this probability. The rule also states that if r is 2 or more, the batch will be rejected immediately; the probability of this is $P(2) + P(3) + P(4) + P(5)$, or more simply

$$1 - P(0) - P(1) = 1 - (1 - \pi)^5 - 5\pi(1 - \pi)^4 = 1 - (1 - \pi)^4(1 + 4\pi).$$

(b) If $r = 1$ we do not make an immediate decision but instead we take a second sample, also of $n = 5$ items. We base the final decision on the combined results of the two samples, and a tree diagram is often useful for keeping track of possible outcomes and their probabilities. The proposed decision rule for the second sampling is a relatively simple one, to accept only if it contains no faulty items (so that we have only observed one faulty item in all 10 studied). Figure 8.2 illustrates the situation in a tree diagram and column 2 in Table 8.4 shows the probability of acceptance for values of π from 0.01 to 0.20: this probability is $(1 - \pi)^5 + 5\pi(1 - \pi)^9$.

(c) The third column in Table 8.4 shows the probabilities of acceptance if we agree to accept on the second sampling when $r = 0$ or 1 (instead of only for $r = 0$). Drawing the revised tree diagram helps us to see that the probability is now

$$(1 - \pi)^5 + 5\pi(1 - \pi)^8(1 + 4\pi).$$

(d) A sampling scheme having three possibilities in the decision rule may be more economical in sampling than the single sampling scheme. When there is only a small number of faulty items in the first sample, a decision is deferred until a further sample has been taken. Because we may ask for a second sample, this is often called a *double sampling* scheme.

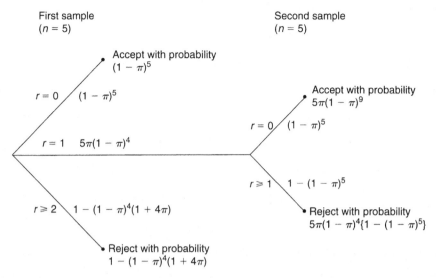

Figure 8.2 *Two-stage sampling scheme of Example 8.4 using the decision rule of Example 8.4(b).*

Table 8.4 *Probabilities of acceptance in the double sampling scheme of Example 8.4*

π	Rule for accepting at second sampling stage	
	$r = 0$	$r = 0$ or 1
0.01	0.9967	0.9990
0.02	0.9873	0.9958
0.03	0.9728	0.9904
0.04	0.9539	0.9827
0.05	0.9313	0.9728
0.06	0.9058	0.9607
0.07	0.8778	0.9464
0.08	0.8479	0.9301
0.09	0.8166	0.9118
0.10	0.7842	0.8918
0.12	0.7176	0.8471
0.14	0.6506	0.7972
0.16	0.5848	0.7434
0.18	0.5216	0.6872
0.20	0.4619	0.6297

Sample size $n = 5$. First sample accepted if $r = 0$; second sample taken if $r = 1$; accepted at second sample as shown in column headings.

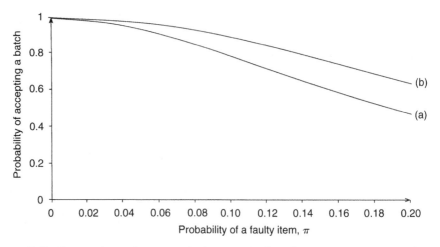

Figure 8.3 *Operating characteristic curves for the two-stage sampling schemes of Table 8.4, using the decision rule of (a) column 2, (b) column 3.*

When the more generous rule in (c) is adopted for the second sample, there is a very high probability of accepting output containing 10% faulty ($\pi = 0.10$), and even for 20% faulty ($\pi = 0.20$) the probability is as high as 0.63.

Operating characteristic curves can be drawn for double sampling schemes, and those for the two schemes of Table 8.4 are given in Fig. 8.3. Neither of these schemes seems adequate, as the probability of acceptance drops off much too slowly as π increases; larger values of n would be necessary.

The advantage of a double sampling scheme is more obvious when the sample size n is a little larger than in the examples given so far, say $n = 10$ or more. We only need to take the second sample of n (making $2n$ altogether) if the result after the first sample is inconclusive. When r is very small or quite large in the first sample the second sample is unnecessary, and the average amount of sampling needed in repeated examinations of the output is kept as low as possible. We discuss the idea of *average sample number* in Section 9.8.

8.8.1 Exercises

1. For Example 8.3, compute the probabilities of acceptance for $n = 40$ and $r \leqslant 3$. Compare these with the probabilities given in Table 8.3 for $n = 12$ and $r = 0$ or 1.

   Draw the operating characteristic curve for $n = 40$ and $r \leqslant 3$. Compare it with the other OC curves for this example and comment on which of the sampling schemes achieves good results both for the manufacturer and for the customer who uses the output.

2. In a production process, samples of 10 items are selected randomly from each batch of output, and inspected. If the number of faulty items is less than 2 the batch is accepted; otherwise the process is stopped for the machinery to be

adjusted. Find the probability that the batch is accepted (i) if $\pi = 0.04$ is the true proportion of faulty items in the whole output, (ii) if $\pi = 0.07$.

3. A sampling scheme involves selecting $n = 30$ items from each batch supplied by a manufacturer, and accepting the batch if there are no more than $c = 2$ faulty items in the sample.
   (a) Find the probability of accepting a batch when the manufacturer is supplying items of which the proportion faulty is (i) $\pi = 0.05$; (ii) $\pi = 0.1$; (iii) $\pi = 0.2$.
   (b) Repeat the calculations in (a) if n is reduced to 20.
   (c) Repeat the calculations in (a) if n is increased to 50.
   (d) Repeat the calculations in (a), with $n = 30$, if c is increased to 3.

4. Chocolates are drawn randomly from a production process and placed in boxes which hold 15. At intervals a box is inspected and if more than 3 chocolates are damaged the process is adjusted. On average in normal running, 10% of chocolates are damaged.
   Find the probability that there will be sufficient damaged chocolates to cause the process to need adjustment.

5. A process produces a proportion π of faulty items. Samples of 100 items are taken from batches of the production at random. If there are less than 5 faulty items in a sample, the corresponding batch will be accepted; otherwise the batch is rejected. Find an expression, in terms of π, for the probability that a batch is accepted.
   If $\pi = 0.04$, evaluate this probability and calculate how many batches would be drawn, on average, before a batch is rejected.

8.9 CUMULATIVE DISTRIBUTION FUNCTIONS AND MEDIANS

If we know the probability mass function of a random variable R, we can sometimes find an algebraic formula for $P(R \leqslant b)$, where b stands for some particular one of the set of values that R can take. When the algebra of this defeats us, as it often does, we have to write out a cumulative table numerically, as we did for a variate in Chapter 1.

For the geometric distribution

$$P(R \leqslant b) = \pi + \pi(1 - \pi) + \pi(1 - \pi)^2 + \cdots + \pi(1 - \pi)^{b-1}.$$

Writing $w = 1 - \pi$,

$$P(R \leqslant b) = \pi(1 + w + w^2 + \cdots + w^{b-1}),$$

which is the sum of a geometric series containing b terms, and is equal to

$$\pi \cdot \frac{1 - w^b}{1 - w} = \frac{\pi(1 - w^b)}{1 - (1 - \pi)} = 1 - w^b = 1 - (1 - \pi)^b.$$

The cumulative distribution function for any possible value of b is therefore given by the function

$$F(b) = 1 - (1 - \pi)^b.$$

Definition 8.4

The (cumulative) **distribution function** of the random variable X is given by the function defined as $F(b) = P(X \leqslant b)$.

As indicated, the word 'cumulative' is often left out. A common notation is to use a small letter f when representing the probability mass function, writing this as $f(x)$, and a capital letter when representing the distribution function, $F(b)$.

The obvious definition to take for the median of a random variable is that value M of the variable such that $F(M) = \frac{1}{2}$, or $P(R \leqslant M) = \frac{1}{2}$. Unfortunately, with a discrete distribution there is no guarantee that there will actually be a value M of the random variable such that $F(M)$ equals exactly $\frac{1}{2}$, since F is not a continuous function but is a step function. Even if we can solve the equation it does not always give a sensible result. We prefer the following.

Definition 8.5

The **median** of the discrete random variable R, which has probability mass function $f(r)$ and range of values r_0 to r_k, is the value M such that

$$\sum_{r=r_0}^{M} f(r) \geqslant \frac{1}{2} \quad \text{and} \quad \sum_{r=M}^{r_k} f(r) \geqslant \frac{1}{2}.$$

This is equivalent to requiring that $P(R \leqslant M) \geqslant \frac{1}{2}$ and $P(R \geqslant M) \geqslant \frac{1}{2}$. In some cases this does not lead to a unique variable value for M but to two possible values. We then take M midway between these two possible values, as in Example 8.6(ii) below.

Example 8.5 Median of a geometric distribution

Produce a table to show the probability mass function and the distribution function of the geometric distribution with $\pi = \frac{1}{4}$. Find the median of the distribution.

r	1	2	3	...	r	...
$f(r)$	$\frac{1}{4}$	$\left(\frac{1}{4}\right)\left(\frac{3}{4}\right)$	$\left(\frac{1}{4}\right)\left(\frac{3}{4}\right)^2$	...	$\left(\frac{1}{4}\right)\left(\frac{3}{4}\right)^{r-1}$	...
$f(r)$	$\frac{1}{4}$	$\frac{3}{16}$	$\frac{9}{64}$	...	$\frac{3^{r-1}}{4^r}$	...
$F(r)$	0.2500	0.4375	0.5781	...	$1 - \left(\frac{3}{4}\right)^r$	...

This median is 3, since

$$\sum_{r=1}^{3} f(r) = 0.5781 > \frac{1}{2},$$

and

$$\sum_{r=3}^{\infty} f(r) = 1 - 0.4375 = 0.5625 > \frac{1}{2}.$$

A check with the values 2 and 4 shows that neither of them satisfies the definition of the median.

Example 8.6 *Medians of discrete uniform distributions*

Find the medians of the discrete uniform distributions (p. 171) with (i) $\kappa = 3$; (ii) $\kappa = 6$.

(i) By symmetry, we would expect the median to be 2. We see also that

$$\sum_{r=1}^{2} f(r) = \frac{2}{3} > \frac{1}{2} \quad \text{and} \quad \sum_{r=2}^{3} f(r) = \frac{2}{3} > \frac{1}{2},$$

which confirms that $M = 2$.

(ii) $F(3) = \frac{1}{2}$, so it is tempting to take 3 as the median. But by symmetry 4 has an equal claim. We therefore take the mean of 3 and 4, i.e. 3.5, as the median. We see that this satisfies the definition since $P(R \leqslant 3.5) = \frac{1}{2}$ and $P(R \geqslant 3.5) = \frac{1}{2}$.

8.9.1 The binomial distribution

The distribution function of a binomial random variable is

$$F(b) = P(R \leqslant b) = \sum_{r=0}^{b} \binom{n}{r} \pi^r (1 - \pi)^{n-r}.$$

This does not reduce to a simple expression, though tables of the cumulative binomial distribution have been constructed for a wide range of numerical values of n and π.

8.10 DISTRIBUTION OF A FUNCTION $g(R)$ GIVEN THE DISTRIBUTION OF R

Sometimes it is not the actual value, r, of R which is important or interesting, but some function of r. We could, for example, invent a game in which we throw an ordinary six-sided die and observe the score r, but receive a number of points equal to $r - 3$, or $2r$, or r^2, or $|r - 4|$, say, rather than r itself. Once given the probability mass function for r, we can easily obtain that for the number of points.

In Table 8.5, we show R, which is a discrete uniform random variable, and also the four random variables S, T, U, V obtained from it. There is only one value of

Table 8.5 *Probability distribution of a discrete random variable R, and of variables related to it*

R	1	2	3	4	5	6
$S = R-3$	-2	-1	0	1	2	3
$T = 2R$	2	4	6	8	10	12
$U = R^2$	1	4	9	16	25	36
$V = |R-4|$	3	2	1	0	1	2
Probability	$\frac{1}{6}$	$\frac{1}{6}$	$\frac{1}{6}$	$\frac{1}{6}$	$\frac{1}{6}$	$\frac{1}{6}$

$S = R - 3$ that corresponds to each particular value of R: when $R = 6$, $S = 3$ and we cannot have $S = 3$ unless $R = 6$.

The transformation from R to S is one-to-one. Therefore $P(S = 3) = P(R = 6)$, the event '$S = 3$' being equivalent to the event '$R = 6$'. The probability mass function of S is thus given by the second row and the last row in Table 8.5, ignoring the other rows. In the same way $T = 2R$ and $U = R^2$ are found from R by one-to-one transformations, so that their distributions are also easy to read off from Table 8.5.

However, the transformation leading from R to V is not one-to-one; $V = |R - 4|$ gives $V = 3$ when $R = 1$, and this event occurs with probability $\frac{1}{6}$. But $V = 2$ arises in two ways, from $R = 2$ and $R = 6$. So $P(V = 2) = P(R = 2) + P(R = 6) = \frac{1}{6} + \frac{1}{6} = \frac{1}{3}$. Similarly $V = 1$ arises from $R = 3$ and from $R = 5$, while $V = 0$ arises only from $R = 4$. A new table is needed to show the distribution of V:

V	0	1	2	3
Probability	$\frac{1}{6}$	$\frac{1}{3}$	$\frac{1}{3}$	$\frac{1}{6}$

8.10.1 Exercises

1. Find the median of each of the random variables S, T, U, V defined in this section.

2. What difficulty would arise in an attempt to find the distribution of $Q = \sqrt{R}$, given that of R?

8.11 CHOICE OF A RANDOM SAMPLE FROM A DISTRIBUTION

When carrying out a simulation that is stochastic (i.e. it includes random elements), we choose random samples from distributions. Consider the binomial distribution with parameters $n = 4, \pi = \frac{1}{2}$. The probabilities of the possible values

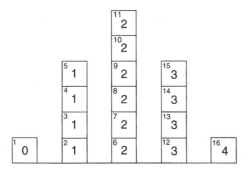

Figure 8.4 *Block diagram representing distribution from which a random sample is to be drawn.*

of r are given by:

r	0	1	2	3	4
$P(R = r)$	$\frac{1}{16}$	$\frac{4}{16}$	$\frac{6}{16}$	$\frac{4}{16}$	$\frac{1}{16}$

When choosing samples we should pick 0 with a probability $\frac{1}{16}$, 1 with a probability $\frac{4}{16}$, etc. One method would be to pick labelled discs out of a hat. We would label one disc with '0', four discs with '1', six discs with '2', four with '3' and 1 with '4'. Choose a disc out of the hat, record its number, return it to the hat and then give all the discs a good shake before making the next choice. In this way we would obtain a random sample of the distribution. It is more convenient to use random numbers to make the choice. We can illustrate the method using Fig. 8.4. Instead of discs for the numbers we have used boxes, and built them up into a block diagram. There is one box for '0', there are four for '1' and so on. We next number the boxes successively from 1 to 16. From a table of random digits we might take pairs in the range 01 to 16 (rejecting all that fell outside this). If our pair of random digits is 08, the corresponding sample value from the distribution is a 2; if the next pair of digits is 03, the next sample value is a 1, and so on. It is wasteful to ignore all pairs of digits outside the range 01 to 16. In order to be more economical, we would treat 21, 41, 61 and 81 as equivalent to 01; 22, 42, 62, 82 as equivalent to 02; and so through until 36, 56, 76 and 96, which are equivalent to 16. All pairs of digits not accounted for in this equivalence would have to be ignored; but there are now relatively few of these. A set of random digits taken from a table, and the corresponding members of the sample from this binomial distribution of r, are:

Random digits	27	92	43	86	04	16	01	95	97	73
Equivalent in range 01–16	07	12	03	06	04	16	01	15	–	13
Sample value	2	3	1	2	1	4	0	3	–	3

We can apply the same method to any discrete distribution, and this includes 'empirical' distributions, i.e. those based on observed data rather than on theoretical models. For example, we might observe the frequency distribution, over 50 days, of the number of cups of coffee a person drinks per day, with the result:

Number of cups per day	0	1	2	3	Total
Number of days	4	38	6	2	50
Cumulative frequency	4	42	48	50	
Corresponding random numbers	01–04	05–42	43–48	49–50	

By including the cumulative frequencies, it is easy to allocate which random numbers correspond to each value of the variable. In this particular example it is clear that, for economy in using random digits from a table, we should also make 51–54, 55–92, 93–98, 99–00 correspond to 0, 1, 2, 3 respectively.

This method can be extended to continuous variables, using cumulative frequencies in intervals, and after working through later chapters readers will be able to do this.

Programs for generating random samples from the common statistical distributions are given in *CCC*, Chapter 9.

8.12 MINITAB: PROBABILITY FUNCTIONS; SAMPLING A DISCRETE DISTRIBUTION

MINITAB will give the value of theoretical probability functions for particular arguments. The binomial probabilities that we calculated, on p. 174, for the number of girls in families of size 4, can be found quickly from MINITAB. Assume the numbers 0, 1, 2, 3, 4 are in C1, then choosing

Calc>Probability Distribution>Binomial/
Select **Probability**, enter **Number of trials** 4 and **Probability of success** 0.49, enter **Input column** C1/

will give the values in the table on p. 174. We call the function that is evaluated the *probability mass function*. MINITAB calls it the *probability function*.

Values of the *cumulative distribution function* are given by selecting **Cumulative probability** instead of **Probability** in the dialogue box. For the binomial distribution it evaluates the function in Section 8.9.1.

MINITAB also evaluates the *inverse cumulative distribution function*, namely the value of the variable for which the cumulative distribution function equals a chosen probability value. **Inverse cumulative probability** is selected in the dialogue box. For continuous random variables this is a unique value, but for discrete random variables, as we have shown when discussing the calculation of the median (Definition 8.5), there may not be a unique value. In this case MINITAB gives values of the cumulative distribution function in the region of an appropriate value so that the user can make a judgement.

Probability functions may be evaluated for many other distributions as well as the binomial. For example, to evaluate the probability mass function of the integer distribution (i.e. discrete uniform distribution) choose

Calc>Probability Distribution>Integer/
Select **Probability**, enter **Minimum value** and **Maximum value**, enter the **Input column**, or **Input constant**, and, if required, the **optional storage** column/

8.12.1 Sampling a discrete distribution (Section 8.11)

MINITAB will choose a random sample from a discrete distribution specified by its possible values and the associated probabilities. Choose

Calc>Random Data>Discrete/
Specify no. of **rows of data** to be generated, enter columns in which to be **stored** and columns which contain **values** and their **probabilities**/

8.13 EXERCISES ON CHAPTER 8

1. The random variable R has the probability distribution:

r	1	2	3	4
$P(R=r)$	c	$c/2$	$c/3$	$c/4$

Find: (i) c; (ii) $P(R \geqslant 2)$.

2. Two fair coins are tossed. A head is recorded as 0 and a tail as 1; the variable of interest is the product of the two records, e.g. if both coins give tails then the variable equals $1 \times 1 = 1$. Find the probability distribution of this random variable.

3. Jack and Jill play a game. They each throw a die. The one who makes the higher score is given, by the other player, the difference in pence between that score and the score of the other player; if the scores are equal then neither player receives anything. Find the probability distribution of the amount (positive or negative) that Jack receives at each play.

4. You have a die. For each of the probability distributions (a)–(d), state how you might generate a variate that you would expect to be well modelled by the distribution: (a) a discrete uniform distribution; (b) a Bernoulli distribution; (c) a binomial distribution; (d) a geometric distribution. In each case state the value(s) of the parameter(s) of the distribution.

5. Show that $P(R > b)$ for a geometric distribution with parameter π is $(1 - \pi)^b$. (*Hint*: the formula for the sum of a geometric series is useful.) Hence show that the distribution function of the geometric distribution is $F(b) = 1 - (1 - \pi)^b$.

6. Find the median of the geometric distribution with parameter $\pi = \frac{1}{5}$.

7. A fair coin is tossed twice. Let R be the number of heads. Find the distribution of: (i) R; (ii) $1/(1+R)$.

8. If I throw a peanut in the air there is a probability of $\frac{4}{5}$ that I shall catch it in my mouth. Assuming that throws are independent, what is the probability that in four throws, I catch: (i) all four; (ii) exactly two; (iii) at least one?

9. A die is thrown ten times. Find which of the following results has the higher probability: just one six, or just two sixes. (*SMP*)

10. Draw bar diagrams for binomial distributions each of which have $\pi = \frac{1}{2}$, and for which $n = 2, 4, 6, 8, 10$. Place the diagrams on one sheet, underneath each other, using similar scales. Describe the effect that increasing n has on the diagram of the distribution.

11. Repeat Question 10 with $\pi = \frac{1}{10}$.

12. Define the binomial distribution, stating the conditions under which it will arise, and find its mean.

A manufacturer of glass marbles produces equal large numbers of red and blue marbles. These are thoroughly mixed together and then packed in packets of six marbles which are random samples from the mixture. Find the probability distribution of the number of red marbles in a packet purchased by a boy.

Two boys, Fred and Tom, each buy a packet of marbles. Fred prefers the red ones and Tom the blue ones, so they agree to exchange marbles as far as possible, in order that at least one of them will have six of the colour he prefers. Find the probabilities that, after exchange: (i) they will both have a set of six of the colour they prefer; (ii) Fred will have three or more blue ones. (*JMB*)

13. (a) Let $p(k)$ denote the probability of exactly k successes in n independent trials of a random experiment in each of which the probability of a success is p. Show that

$$\frac{p(k+1)}{p(k)} = \frac{(n-k)p}{(k+1)q}, \quad k = 0, 1, 2, \ldots, n-1,$$

where $q = 1 - p$. Hence, or otherwise, find the most probable number of successes when $n = 16$ and $p = 0.2$.

(b) The probability that a typewritten page is free from error is $\frac{1}{2}$; the probability that it contains exactly n errors is inversely proportional to n for $n = 1, 2, 3, 4$; and the probability that it contains more than four errors is $\frac{1}{12}$. Calculate the probability that: (i) a page will contain exactly one error; (ii) six pages will consist of four pages containing no error and the remaining two pages containing exactly one error each. (*Welsh*)

14. A certain brand of tea has a picture card in every packet. The cards form a set of 50 different pictures and are distributed among the packets so that any packet purchased is equally likely to contain any one of the cards. A boy has

collected 47 different cards. Find the probabilities that he will get the three cards he needs to complete the set: (i) if he opens only three packets; (ii) if he opens at most four packets.

Another boy needs only one card to complete the set. Find the probability distribution of the number of packets this boy needs to open to complete the set.

Calculate to two decimal places the probability that he gets the remaining card by opening not more than ten packets. *(JMB)*

15. A fair coin is tossed repeatedly and it is decided to stop tossing as soon as three heads have been obtained. Show that the probability, p_n, that exactly n tosses are necessary is given by $p_n = (n-1)(n-2)2^{-n-1}$.

Hence show that

$$\sum_{n=1}^{\infty} (n-2)(n-3)2^{-n} = 2. \qquad (Oxford)$$

16. Let R be a binomial random variable with parameters n, π, and let $P(R = r)$ be denoted by p_r. By considering the ratio p_r/p_{r-1} show that

$$p_r = \frac{n-r+1}{r} \cdot \frac{\pi}{1-\pi} \cdot p_{r-1}.$$

Hence generate successively the probabilities $p_0, p_1, \ldots, p_4$ for the binomial distribution with $n = 4, \pi = \frac{1}{5}$.

17. Use a table of random digits in order to draw random samples of size 5 from: (a) a binomial distribution with $n = 3, \pi = \frac{1}{4}$; (b) a distribution specified by the table:

r	0	1	2	3	4	5
p_r	0.20	0.40	0.20	0.10	0.07	0.03

18. Random samples are taken from a stream to check whether the water is of acceptable quality. Six samples are taken at a time, but if the first three examined (chosen at random from the six) are all good, the other three are not studied, and no further action is taken until the next routine assessment is due. If at least two of the first three studied are poor, a more detailed inspection is carried out. If exactly one of the first three is poor, the remaining three are also studied. If these three are all good, no further action is taken until the next routine assessment is due; but if not, there is a more detailed inspection.

Write down the list of possible outcomes when a set of six samples is taken, find an expression for the probability of each of these in terms of π, the probability of a poor sample; and illustrate these results using a tree diagram. Find the probability of each outcome when (i) 10%, (ii) 20% of all samples are poor.

19. A sampling process involves taking crates of $n = 12$ bottles at random time intervals from a drinks production line and noting any that are scratched

or otherwise marked. When none are scratched or marked, the production batch is accepted, and when $r = 1$ or 2 are scratched or marked another sample of $n = 12$ is taken from the same batch. If r is 3 or more, a full inspection of the output from the process is ordered.

When two samples have been required, a batch is accepted if no more than two scratched or marked bottles are found among the whole 24; otherwise a full inspection is ordered. If a proportion $\pi = 0.05$ of scratched or marked bottles is allowable, calculate the probability
  (i)   that the second sample will be needed;
  (ii) that the batch will be accepted;
  (iii) that a full inspection of the batch will be ordered.

8.14 COMPUTING EXERCISES

1. Compute the probability of r successes, for $r = 0, 1, 2, \ldots, 8$ and $r \geqslant 9$, for the binomial distribution with parameters $n = 30$ and $\pi = 0.1$. Check your results against those in Table 8.2.

2. Carry out a computer simulation of the bead model in Section 8.1. To do this, draw a random sample of 100 observations from a binomial distribution with parameters $n = 30$, $\pi = 0.1$. Produce a frequency table similar to that in Table 8.1, and compare with the theoretical values given in Table 8.2.

   Increase the sample size and show that a better agreement of simulated with theoretical values is obtained.

3. By simulation, investigate the distribution of the number of throws of a die required until two sixes have appeared.

8.15 PROJECTS

1 Sex distribution of children in families

Aim. To find the distribution of the number of girls in families of (a) two, (b) three children, and compare with the predicted distributions.

Preliminary work. Predict what these distributions should be on the assumption that boys and girls are equally likely and that births are independent.

Collection of data. Find out the sexes of the two children, in families of size two, for as many families as you can. Repeat, using a separate recording sheet, for families that include exactly three children. Try to obtain data for at least 50 families of each size.

Analysis. Find the number of girls in each family for families including exactly two children. Form the relative frequency table and compare it with the probabilities you would predict. Repeat with the data from the three-children families.

2 Generation of random digits

Aim. To generate a table of the digits $0, 1, \ldots, 9$ in random order.
Equipment. Two dice and a shaker.
Collection of data. Roll two dice, one after the other. If the *first* die shows 6, reject the result and roll again. Otherwise record the scores on the two dice in order.
Analysis. Note the score $(1, 2, 3, 4, 5)$ on the first die. If the second die shows an even score $(2, 4, 6)$ add 5 to the score on the first die, otherwise use this first score as it stands. Make a frequency table of the scores 1–10. Now let score 10 stand for digit 0, scores 1–9 stand for digits 1–9. Check whether the set of digits appears to be a sample from a discrete uniform distribution.
Note: The digits used in the order they were generated can be used later for testing other properties of random digits.

3 Number of throws of a die to obtain a six

Aim. To examine whether the number of throws of a die required to obtain a 6 follows the geometric distribution.
Equipment. Die and shaker, as in Project 2.13(2).
Collection of data. Follow the instructions given in 2.13(2).
Analysis. Calculate the expected frequencies in a geometric distribution with $\pi = \frac{1}{6}$. Compare the observed table of frequencies with this. (See Project 19.13(1)(b) for the completion of this by a hypothesis test.)

SUMMARY

A **random variable** is any quantity or attribute whose value varies from one unit of a population to another. Formally, a random variable is a function whose domain is a possibility space.

A **discrete random variable** is one that has a finite set of values. (The definition may be extended to include variables that have a countably infinite set of values.) It is specified by a probability mass function in which a probability is assigned to each one of the set of values that the variable may take.

The (cumulative) **distribution function** of a random variable R is the function defined by

$$F(b) = P(R \leqslant b).$$

The **probability mass functions** of common discrete random variables are:

(i) discrete uniform distribution,

$$P(R = r) = 1/\kappa, \quad r = 1, 2, \ldots, \kappa;$$

(ii) binomial distribution,

$$P(R = r) = \binom{n}{r} \pi^r (1 - \pi)^{n-r}, \quad r = 0, 1, 2, \ldots, n,$$

where $0 < \pi < 1$;

(iii) geometric distribution,

$$P(R = r) = \pi(1 - \pi)^{r-1}, \quad r = 1, 2, 3, \ldots,$$

where $0 < \pi < 1$.

The conditions that give rise to a binomial distribution are: (i) fixed number n of trials; (ii) only two possible outcomes, 'success' and 'failure'; (iii) independent trials; (iv) constant probability π of success; (v) the variable is the total number of successes in n trials.

In **quality control**, acceptance of a batch of products is made to depend on the results of a suitable sampling scheme. It is an important application of sampling a binomial distribution.

9 EXPECTATION OF RANDOM VARIABLES

OVERVIEW

The phrase 'expected value' was introduced into probability from gambling because of the interest in expected winnings. We show that the concept is the same as the mean of a random variable and that the expectation operator offers us a useful technique for investigating random variables.

9.1 EXPECTED VALUES

Alan invites Bill to play a game. Alan will roll a die. If a '6' turns up, he will pay Bill 6p; if a '1' turns up he will pay him 1p; otherwise Bill is to pay Alan 2p. Should Bill accept the invitation?

Bill needs to know how much he is likely to win. Of course he cannot know exactly what he is going to win on one play of the game until he has actually played it. He will have to be satisfied with a list of the possible winnings and a probability assigned to each. Better still, he would like a summary of this information given in one figure.

An appropriate theoretical model of the game is to assume that the numbers on the die, 1 to 6, will each turn up with probability $\frac{1}{6}$, a reasonable enough assumption provided the die is not loaded. (This is of course an example of the discrete uniform distribution given in Definition 8.1, p. 171.) On this model, Bill will receive 6p or 1p, each with probability $\frac{1}{6}$, while he will pay out 2p with probability $\frac{4}{6} = \frac{2}{3}$. Bill's 'winnings' can therefore be positive or negative! The distribution of winnings may be summarized in a table:

Winnings (p)	-2	1	6
Probability	$\frac{2}{3}$	$\frac{1}{6}$	$\frac{1}{6}$

Over a long series of actual games, the *relative frequency* with which each outcome turns up should be close to its *probability* on the theoretical model. Thus, in a total of 6000 games, Bill is likely to receive 6p in about 1000, and 1p in about 1000; and he is likely to pay out 2p in about 4000. Hence, over the 6000 games Bill's winnings (in pence) are likely to equal

$$6 \times 1000 + 1 \times 1000 - 2 \times 4000.$$

If we divide by 6000, so that the result is expressed as an average per game, we obtain

$$6 \times \frac{1}{6} + 1 \times \frac{1}{6} - 2 \times \frac{2}{3} = \frac{6 + 1 - 8}{6} = -\frac{1}{6}.$$

This is called Bill's *expected winnings* per game. Since it is negative it suggests that Bill would be wise not to play, unless he can persuade Alan to be more generous. This interpretation of the word 'expected' fits in with ordinary usage; but so do other interpretations. We might, for example, have said that Bill's 'expected' winnings were −2p, since that is the value most likely to occur (i.e. it is the mode). But this is not the interpretation we will use. The word 'expected' is given a precise technical meaning: 'expected winnings' is to mean the sum of each possible value of winnings multiplied by the probability which that value has on the theoretical model of the game. Whenever the word 'expected' is applied to a random variable, it is used in this technical sense.

Finding the expected value is an averaging process. The 'expected winnings' is the value we obtain when we 'average out' all the possible winnings; it is the average winnings per game that would happen over a very long run of games, and is a similar quantity to the mean of a set of observations. If we have a set of observations $r_1, r_2, \ldots, r_k$ occurring with respective frequencies $f_1, f_2, \ldots, f_k$, and $N = \sum_{i=1}^{k} f_i$, we recall that the mean of the set is defined as $\sum_{i=1}^{k} f_i r_i / N$. Thus the mean of a set of observations of a variate equals the sum of the products of each value r_i with the relative frequency f_i / N. The expected value of R is an expression of the same form, except that each *relative frequency* is replaced by a *probability*.

Suppose that the random variable R takes the values $\{r_i\}$. In order to make expressions as neat and concise as possible, in this chapter and later, we shall often write p_i instead of $P(R = r_i)$, for the probability with which R takes the value r_i. Then the expected value of R equals $\sum_{\text{all } i} p_i r_i$. This leads us to regard the *expected value*, or *expectation*, and *mean* of R as equivalent terms. Another equivalent, and older, term which we shall not use is *mathematical expectation*.

Definition 9.1

The **mean** or **expected value** or **expectation** of the discrete random variable R, which takes the value r_i $(i = 1, 2, \ldots)$ with probability p_i, equals $\sum_{\text{all } i} p_i r_i$.

From now on we shall reserve the bar notation (e.g. $\bar{r}$ or $\bar{x}$) for the mean of a sample, and use μ for the mean of a random variable. It is convenient to introduce a symbol to represent the operation of taking the expected value, or mean, of a random variable. We write the 'expected value of the random variable R' as $E[R]$. It is a function of the whole set of possible values of R and their associated probabilities. We can think of $[R]$ as representing the set of values and associated probabilities and E denotes taking the expected value, or mean, of that population. We use *square* brackets to enclose R.

Example 9.1

The random variable R takes the values 1, 2, 3, 4, 5, 6 with equal probability. Find E[R].

$$E(R) = 1 \times \frac{1}{6} + 2 \times \frac{1}{6} + 3 \times \frac{1}{6} + 4 \times \frac{1}{6} + 5 \times \frac{1}{6} + 6 \times \frac{1}{6} = \frac{21}{6} = \frac{7}{2}.$$

Example 9.2

The random variable R takes the values 1, 2, 3 with probabilities $P(1) = 0.2$, $P(2) = 0.3, P(3) = 0.5$. What is the expected value of R?

$$E[R] = 0.2 \times 1 + 0.3 \times 2 + 0.5 \times 3 = 0.2 + 0.6 + 1.5 = 2.3.$$

Example 9.3

b is a constant. Show that $E[b] = b$.

b takes its only possible value with probability 1. Hence

$$E[b] = b \times 1 = b.$$

9.1.1 Exercises

1. A statistician pays his son pocket money each week in the following way. He tosses three 50p coins and gives to his son any that come down heads. What is the expected weekly pocket money of the boy?

2. The random variable R takes the values 2, 6, 24 with probabilities $P(2) = \frac{1}{2}$, $P(6) = \frac{1}{3}, P(24) = \frac{1}{6}$. Find the expected value, or mean, of R.

3. A box contains 3 red balls and 2 blue balls. Balls are successively drawn without replacement until a blue ball is obtained. Calculate the expected number of draws required. (*IOS*)

4. Find the mean (i.e. expected value) of the discrete uniform distribution with parameter κ.

9.2 THE MEANS OF THE BINOMIAL AND GEOMETRIC DISTRIBUTIONS

9.2.1 The binomial distribution

When R is defined by

$$P(r) = \binom{n}{r} \pi^r (1 - \pi)^{n-r}$$

for values of r from 0 to n, the mean or expected value of R is

$$\sum_{r=0}^{n} rP(r) = \sum_{r=0}^{n} r\binom{n}{r}\pi^r(1-\pi)^{n-r}$$

$$= \sum_{r=0}^{n} r\frac{n!}{r!(n-r)!}\pi^r(1-\pi)^{n-r},$$

and since the term for $r=0$ is zero, this is

$$= n\pi\sum_{r=1}^{n}\frac{(n-1)!}{(r-1)!(n-r)!}\pi^{r-1}(1-\pi)^{n-r}$$

$$= n\pi[\pi+(1-\pi)]^{n-1} \quad \text{using the binominal series (Appendix I)}$$

$$= n\pi. \quad\quad\quad\quad\quad \text{since } \pi+(1-\pi)=1.$$

Example 9.4

Find the mean number of girls in families of four children, if the probability of a female birth is 0.49.

Assuming independence of births (and survival of all children), the binomial distribution with $n=4$ and $\pi=0.49$ is the appropriate model. Hence the mean is $n\pi = 4 \times 0.49 = 1.96$.

9.2.2 The geometric distribution

The geometric random variable R is such that $p_r = \pi(1-\pi)^{r-1}$. Hence

$$E[R] = p_1 + 2p_2 + 3p_3 + \cdots + rp_r + \cdots$$

$$= \pi + 2\pi(1-\pi) + 3\pi(1-\pi)^2 + \cdots + r\pi(1-\pi)^{r-1} + \cdots.$$

The form of the series is clearer if we write w for $(1-\pi)$, since then

$$E[R] = \pi(1 + 2w + 3w^2 + \cdots + rw^{r-1} + \cdots).$$

From the binomial theorem (Appendix I),

$$(1-w)^{-2} = 1 + 2w + 3w^2 + \cdots + rw^{r-1} + \cdots.$$

(The expansion is valid since $|w| = |(1-\pi)| < 1$.) Hence

$$E[R] = \frac{\pi}{(1-w)^2} = \frac{\pi}{\pi^2} = \frac{1}{\pi}.$$

9.2.3 Exercises

1. If a quality control inspector takes 10 items at random from a production line, what is the expected value of the number of faulty items if the whole production contains: (i) 5%; (ii) 10% faulty?

2. If a die is thrown 50 times, what is the expected number of sixes scored?

3. A farmer sows 1000 seeds of a certain plant, of which the germination rate is 90%. Find the expected value of the number that germinate.

4. The probability that, when batting, Horace misses a cricket ball bowled to him by Fred is $\frac{1}{3}$. Fred bowls 18 balls to Horace. Find the expected number that Horace will hit. If Horace scores 2 from each hit, find his expected score after 18 balls.

5. The random variable R has the following probability distribution:

r	-6	-5	5	9
$P(r)$	0.30	0.25	0.25	0.20

Find $E[R]$ and comment on its use as an 'average' of the possible values of R.

6. (a) If 20% of cars are estate cars, work out the mean number of cars passing an observation point up to the first estate car noted.
   (b) A fair coin is tossed until the first head appears. Calculate the mean number of throws needed to achieve this.
   (c) In the sampling inspection scheme described on p. 178, find the mean number of items observed before a faulty one is found.

7. Show that $E[R] = 1/\pi$ for the geometric distribution by the following method. Write the infinite series $S = 1 + 2w + 3w^2 + \cdots$ (in the notation of p. 200), calculate $-2wS$, w^2S and add to obtain $(1 - 2w + w^2)S$.

9.3 EXPECTATION OF A LINEAR FUNCTION OF A RANDOM VARIABLE

Consider the random variable U which is obtained by increasing the values of a random variable R by 3, i.e. U takes the value $r + 3$ with probability $P(r)$. For example, let R have the distribution

r	2	3
$P(r)$	π	$1 - \pi$

Then U has the distribution

$u = r + 3$	5	6
$P(u)$	π	$1 - \pi$

Write $U = R + 3$. The expected value of U is

$$E[U] = 5\pi + 6(1 - \pi) = 6 - \pi.$$

We see that $E[U]$ could also be calculated more directly from R, as

$$(2 + 3)\pi + (3 + 3)(1 - \pi) = [2\pi + 3(1 - \pi)] + [3\pi + 3(1 - \pi)]$$
$$= E[R] + 3,$$

i.e. $E[R + 3] = E[R] + 3$.

We have shown, in this simple case, the result that if the values of a random variable are increased by 3 then the mean (i.e. the expected value) also increases by 3. It is intuitively obvious that, if b is a constant,

$$E[R + b] = E[R] + b.$$

Example 9.5

In a game a die is thrown and the banker pays out, in pence, twice the number of pips on the side of the die which is uppermost. What would be a fair price to pay to play the game?

We define random variables R, the number of pips that is uppermost on the die, and U, which equals $2R$. The two distributions may be shown in one table:

r	1	2	3	4	5	6
$u = 2r$	2	4	6	8	10	12
$P(r) = P(u)$	$\frac{1}{6}$	$\frac{1}{6}$	$\frac{1}{6}$	$\frac{1}{6}$	$\frac{1}{6}$	$\frac{1}{6}$

$$E[U] = 2 \times \frac{1}{6} + 4 \times \frac{1}{6} + 6 \times \frac{1}{6} + 8 \times \frac{1}{6} + 10 \times \frac{1}{6} + 12 \times \frac{1}{6} = \frac{42}{6} = 7.$$

We see, by comparison with Example 9.1, that doubling the winnings doubles the expected value of the winnings, i.e.

$$E[2R] = 2E[R].$$

The *fair price* to pay to play a game equals the expected winnings. It is therefore 7p per game.

Theorem 9.1 *Expected value of $aR + b$*

If a and b are constants and $aR + b$ denotes the random variable, derived from the discrete random variable R, that takes the value $ar_i + b$ with probability p_i, then:

(i) $E[aR + b] = \displaystyle\sum_{\text{all } i} (ar_i + b)p_i$;

(ii) $E[aR + b] = aE[R] + b$.

Note that here, instead of giving a list of all possible values of r_i, we have simply written $\sum_{\text{all } i}$, 'the sum over all i', in the expression. When it is clear from the context what values of r_i we are talking about, we shall use this notation for simplicity, or even write $\sum$ alone.

Proof. Define U to be the random variable $aR + b$. The values $ar + b$, of U are distinct provided $a \neq 0$. Consider this case first. The distribution of U is therefore:

$u = ar + b$	$ar_1 + b$	$ar_2 + b$	$\ldots$	$ar_i + b$	$\ldots$
$P(u)$	p_1	p_2	$\ldots$	p_i	$\ldots$

Hence

$$E[U] = \sum_{\text{all } i}(ar_i + b)p_i, \quad \text{which is result (i) of Theorem 9.1,}$$

$$= \sum ar_i p_i + \sum bp_i = a\sum r_i p_i + b\sum p_i$$

$$= aE[R] + b, \quad \text{since } \sum p_i = 1.$$

If $a = 0$, $U = b$, i.e. U takes the value b with probability 1.

The left-hand sides (LHS) of both (i) and (ii) in Theorem 9.1 then equal $E[b] = b$.

The right-hand side (RHS) of (i) equals

$$\sum_{\text{all } i} bp_i = b\sum_{\text{all } i} p_i = b = \text{LHS}.$$

The RHS of (ii) equals $b = \text{LHS}$.

Example 9.6

The mean of R is 5. Find the mean of the random variables: (i) $2R - 1$; (ii) $-3R$.

(i) $\qquad\qquad E[2R - 1] = 2E[R] - 1 = 2 \times 5 - 1 = 10 - 1 = 9;$

(ii) $\qquad\qquad E[-3R] = -3E[R] = -3 \times 5 = -15.$

9.4 THE EXPECTED VALUE OF A FUNCTION OF A RANDOM VARIABLE

Finding the distribution of a function of a random variable, such as R^2, or $R(R+1)$, or in general $g(R)$, may sometimes be quite difficult, but fortunately there is a relatively simple way of finding $E[g(R)]$. We saw above (Theorem 9.1) that

$$E[aR + b] = \sum_{\text{all } r}(ar + b)P(r).$$

There is a more general result:

$$E[g(R)] = \sum_{\text{all } r} g(r)P(r).$$

We shall illustrate this result in this section for $g(R) = R^2$, and prove it in Theorem 9.2.

We shall derive $E[R^2]$ for a particular random variable R in two ways: (1) by finding the distribution of $U = R^2$ and using the definition of expected value; (2) by using the formula

$$E[R^2] = \sum_{\text{all } r} r^2 P(r).$$

(1) Consider the random variable U, which is derived from the random variable R

r	2	3
$P(r)$	π	$1 - \pi$

by squaring the values, i.e. U takes the value r^2 with probability $P(r)$. The variable U has the distribution

$u = r^2$	4	9
$P(u) = P(r)$	π	$1 - \pi$

Hence

$$E[U] = 4\pi + 9(1 - \pi) = 9 - 5\pi.$$

(2) Also,

$$\sum r^2 P(r) = 2^2\pi + 3^2(1 - \pi) = 4\pi + 9(1 - \pi) = 9 - 5\pi,$$

which equals $E[U]$.

We must take a little care when the values of R^2 are not distinct.
(1) Consider the random variable:

r	-2	-1	0	1	2
$P(r)$	0.2	0.2	0.2	0.2	0.2

The values of r^2 are $(-2)^2 = 4, (-1)^2 = 1, 0^2 = 0, 1^2 = 1, 2^2 = 4$. Since $U = 4$ arises both from $(-2)^2$ and from $(+2)^2$, its probability is the total $(0.2 + 0.2)$ attached to both the values -2 and $+2$, i.e. it is 0.4. Similarly $U = 1$ arises from $(-1)^2$

and from $(+1)^2$ and its probability is $(0.2+0.2)=0.4$. The distribution of U is therefore

u	0	1	4
$P(u)$	0.2	0.4	0.4

The expected value of U is, by definition,

$$E[U] = 0 \times 0.2 + 1 \times 0.4 + 4 \times 0.4 = 0.4 + 1.6 = 2.0.$$

(2) We also see that

$$\sum_{\text{all } r} r^2 P(r) = (-2)^2 \times 0.2 + (-1)^2 \times 0.2 + 0^2 \times 0.2 + 1^2 \times 0.2 + 2^2 \times 0.2$$

$$= 0.2(4 + 1 + 0 + 1 + 4) = 10 \times 0.2 = 2.0,$$

and hence is equal to $E[U]$.

Theorem 9.2 Expected value of $g(R)$

If the discrete random variable R takes the value r_i with probability p_i, then the random variable which is denoted by $g(R)$ has expectation

$$E[g(R)] = \sum_{\text{all } i} g(r_i)p_i.$$

Proof. Write $U = g(R)$.

(i) If the values of U are distinct then U takes the value $g(r_i)$ with probability p_i. Hence

$$E[g(R)] = E[U] = \sum_{\text{all } i} g(r_i)p_i,$$

as required.

(ii) If the values of U are not all distinct, as in the example above, suppose that u_1 arises from the values $r_1, r_2, \ldots, r_k$, which have probabilities $p_1, p_2, \ldots, p_k$ respectively. That is, $u_1 = g(r_1) = g(r_2) = \cdots = g(r_k)$. The values of the variables may (if necessary) be relabelled to make this so. Hence

$$u_1 P(u_1) = u_1(p_1 + p_2 + \cdots + p_k)$$

$$= u_1 p_1 + u_1 p_2 + \cdots + u_1 p_k$$

$$= g(r_1)p_1 + g(r_2)p_2 + \cdots + g(r_k)p_k.$$

A similar analysis may be made for any other value of U that arises from more than one value of R. Hence

$$E[U] = \sum_{\text{all } j} u_j P(u_j) = \sum_{\text{all } i} g(r_i)p_i.$$

Corollary

$$E[g(R) + h(R)] = E[g(R)] + E[h(R)].$$

Proof. Replacing $g(R)$ in Theorem 9.2 by $g(R) + h(R)$, we obtain

$$E[g(R) + h(R)] = \sum_{\text{all } i} [g(r_i) + h(r_i)]p_i, \qquad \text{from Theorem 9.2,}$$

$$= \sum_{\text{all } i} g(r_i)p_i + \sum_{\text{all } i} h(r_i)p_i, \qquad \text{from the properties of } \sum (\text{Section 2.4}),$$

$$= E[g(R)] + E[h(R)], \qquad \text{from Theorem 9.2.}$$

Example 9.7

The random variable R has the distribution

r	-1	0	1
$P(r)$	0.2	0.5	0.3

Find the distributions of: (i) $R + 1$; (ii) R^2; (iii) $R(R - 1)$; (iv) R^3. In each case find the mean of the distribution: (a) directly by deriving the distribution of the function of R; (b) by the use of Theorem 9.2.
(i) The distribution of $U = R + 1$ is

u	0	1	2
$P(u)$	0.2	0.5	0.3

(a) $E[U] = 0 \times 0.2 + 1 \times 0.5 + 2 \times 0.3 = 0.5 + 0.6 = 1.1$;
(b) Also $E[R + 1] = (-1 + 1) \times 0.2 + (0 + 1) \times 0.5 + (1 + 1) \times 0.3$
$\qquad = 0.5 + 0.6 = 1.1$.
(ii) The distribution of $U = R^2$ is

u	0	1
$P(u)$	0.5	$0.5 (= 0.2 + 0.3)$

(a) $E[U] = 0 \times 0.5 + 1 \times 0.5 = 0.5$;
(b) $E[R^2] = (-1)^2 \times 0.2 + 0^2 \times 0.5 + 1^2 \times 0.3 = 0.2 + 0.3 = 0.5$.
(iii) The distribution of $U = R(R - 1)$ is

u	0	2
$P(u)$	$0.8 (= 0.5 + 0.3)$	0.2

(a) $E[U] = 0 \times 0.8 + 2 \times 0.2 = 0.4$;

(b) $E[R(R-1)] = 2 \times 0.2 + 0 \times 0.5 + 0 \times 0.3 = 0.4$.

(iv) The distribution of $U = R^3$ is

u	-1	0	1
$P(u)$	0.2	0.5	0.3

(a) $E[U] = -1 \times 0.2 + 0 \times 0.5 + 1 \times 0.3 = -0.2 + 0.3 = 0.1$;

(b) $E[R^3] = -1 \times 0.2 + 0 \times 0.5 + 1 \times 0.3 = -0.2 + 0.3 = 0.1$.

Example 9.8

The random variable R has the distribution

r	-1	0	1
$P(r)$	0.2	0.6	0.2

Calculate: (i) $E[R]$; (ii) $E[R^2]$; (iii) $E[R^3]$; and hence calculate (iv) $E[(R-1)^2]$; (v) $E[(R-1)^3]$.

(i) $E[R] = -1 \times 0.2 + 0 \times 0.6 + 1 \times 0.2 = -0.2 + 0.2 = 0$;

(ii) $E[R^2] = 1 \times 0.2 + 0 \times 0.6 + 1 \times 0.2 = 0.2 + 0.2 = 0.4$;

(iii) $E[R^3] = (-1)^3 \times 0.2 + 0 \times 0.6 + 1 \times 0.2 = -0.2 + 0.2 = 0$.

(iv) $E[(R-1)^2] = E[R^2 - 2R + 1]$

$= E[R^2] + E[-2R] + E[1]$ from Corollary of Theorem 9.2

$= E[R^2] - 2E[R] + 1$ from Theorem 9.1

$= 0.4 - 2 \times 0 + 1$ from above,

$= 1.4$.

(v) $E[(R-1)^3] = E[R^3 - 3R^2 + 3R - 1]$

$= E[R^3] - 3E[R^2] + 3E[R] - E[1]$

$= 0 - 3 \times 0.4 + 3 \times 0 - 1 = -1.2 - 1.0 = -2.2$.

9.4.1 Exercises

1. A fair coin is tossed three times. Denote the number of heads by the random variable R. Find: (i) $E[R]$; (ii) $E[R^2]$; and hence calculate (iii) $E[(R-1)^2]$.

2. A fair coin is tossed twice. Denote the number of heads by the random variable R. Find the distribution of $1/(R+1)$ and hence calculate its expectation. Show that $E[1/(R+1)] \neq 1/(E[R]+1)$.

3. (a) Show that

$$E[R^2] = E[R(R-1)] + E[R].$$

(b) Show that

$$E[(R - E[R])^2] = E[R(R-1)] + E[R] - (E[R])^2.$$

9.5 VARIANCE AS AN EXPECTED VALUE

By comparison with the definition of the variance of a set of observations (see p. 65) we define the variance of a random variable.

Definition 9.2

The **variance** of a discrete random variable R which takes the value r with probability p_r, and has mean μ, is

$$\text{Var}[R] = \sum_{\text{all } r} p_r(r - \mu)^2.$$

We often use σ^2, instead of $\text{Var}[R]$, to denote variance. From the above, we see that $\text{Var}[R]$ may be written

$$\text{E}[(R - \mu)^2] \quad \text{or} \quad \text{E}[(R - \text{E}[R])^2].$$

We may derive alternative forms for the variance. Thus

$$\text{Var}[R] = \text{E}[(R - \mu)^2] = \text{E}[R^2 - 2R\mu + \mu^2]$$
$$= \text{E}[R^2] + \text{E}[-2R\mu] + \text{E}[\mu^2] = \text{E}[R^2] - 2\mu\text{E}[R] + \mu^2$$
$$= \text{E}[R^2] - \mu^2 \quad \text{or} \quad \text{E}[R^2] - (\text{E}[R])^2, \quad \text{since } \mu = \text{E}[R].$$

The standard deviation σ is the positive square root of the variance.

Example 9.9

Find the variance and standard deviation of the random variable R which takes the values $1, 2, 3, 4, 5, 6$ with equal probability.

From Example 9.1, $\text{E}[R] = \frac{7}{2}$.

$$\text{E}[R^2] = 1^2 \times \frac{1}{6} + 2^2 \times \frac{1}{6} + 3^2 \times \frac{1}{6} + 4^2 \times \frac{1}{6} + 5^2 \times \frac{1}{6} + 6^2 \times \frac{1}{6}$$
$$= \frac{1}{6}(1 + 4 + 9 + 16 + 25 + 36) = \frac{91}{6}.$$

$$\sigma^2 = \text{Var}[R] = \text{E}[R^2] - (\text{E}[R])^2 = \frac{91}{6} - \left(\frac{7}{2}\right)^2 = \frac{35}{12} = 2.9167.$$

The standard deviation

$$\sigma = \sqrt{\frac{35}{12}} = 1.71.$$

9.6 THE VARIANCE OF THE BINOMIAL DISTRIBUTION

For a binomial random variable R,

$$p_r = \binom{n}{r} \pi^r (1 - \pi)^{n-r}.$$

Using the relation we found in Exercise 9.4.1(3(b)) we can write

$$\text{Var}[R] = \text{E}[R(R - 1)] + \text{E}[R] - (\text{E}[R])^2.$$

This relation is often more convenient for calculation than the ones developed in Section 9.5. When working with the binomial distribution, $\text{E}[R^2]$ is not easy to evaluate but $\text{E}[R(R - 1)]$ is much simpler. (This is true generally when a probability mass function has a factorial in r in the denominator: we shall see another example in Section 20.7.)

We know from p. 200 that $\text{E}[R] = n\pi$; it remains to determine $\text{E}[R(R - 1)]$:

$$\text{E}[R(R - 1)] = \sum_{r=0}^{n} r(r - 1)p_r = \sum_{r=0}^{n} \frac{r(r - 1)n!}{r!(n - r)!}\pi^r(1 - \pi)^{n-r},$$

and since the terms for $r = 0, 1$ are both zero,

$$\text{E}[R(R - 1)] = n(n - 1)\pi^2 \sum_{r=2}^{n} \frac{(n - 2)!}{(r - 2)!(n - r)!}\pi^{r-2}(1 - \pi)^{n-r}$$

$$= n(n - 1)\pi^2[\pi + (1 - \pi)]^{n-2}, \quad \text{using the binomial series,}$$

$$= n(n - 1)\pi^2, \quad \text{since } \pi + (1 - \pi) = 1.$$

Thus

$$\text{Var}[R] = \text{E}[R(R - 1)] + \text{E}[R] - (\text{E}[R])^2$$

$$= n(n - 1)\pi^2 + n\pi - n^2\pi^2$$

$$= n\pi - n\pi^2 = n\pi(1 - \pi).$$

Example 9.10

Find the mean, variance and standard deviation of the number of ones in six throws of a fair die.

The number of ones has a binomial distribution with $n = 6$, $\pi = \frac{1}{6}$. Hence the mean is $n\pi = 6 \times \frac{1}{6} = 1$, and the variance is $n\pi(1 - \pi) = 6 \times \frac{1}{6} \times \frac{5}{6} = \frac{5}{6}$. The standard deviation is $\sqrt{\frac{5}{6}} = 0.913$.

9.6.1 Exercises

1. The random variable R takes values 2, 3, 4 with probabilities $p_2 = \frac{1}{4}, p_3 = \frac{1}{2}$, $p_4 = \frac{1}{4}$. Show that $\mu = 3$. Calculate σ^2 from: (i) $\sum_{\text{all } r} p_r (r - \mu)^2$; (ii) $E[R^2] - \mu^2$.

2. Calculate the mean, variance and standard deviation of the random variables specified in (a)–(d):

(a) r	2	-2
p_r	$\frac{1}{2}$	$\frac{1}{2}$

(b) r	1	2	3
p_r	0.5	0.3	0.2

(c) r	11	12	13
p_r	0.5	0.3	0.2

(d) r	-2	-1	0	1	2
p_r	0.1	0.2	0.4	0.2	0.1

3. Calculate the variance and standard deviation of each of the random variables specified in Questions 1, 2, 3 and 4 in Exercises 9.2.3.

4. A fair coin is tossed 10 times. Find (a) the expected value, (b) the variance of the number of tails, and find also the probability that this number exceeds its expected value.

5. A packet of 20 sweets is made up by selecting at random from a pile of sweets in which 40% are 'soft centres' and the rest 'hard centres'. Find the mean (μ) and standard deviation (σ) of the number of soft centres in a packet. Find also the probability that this number is less than: (i) $\mu - \sigma$; (ii) $\mu - 2\sigma$.

6. In the bead-sampling experiment described in Section 8.1, the whole population of beads contained 10% of the blue ones. Find the mean (μ) and standard deviation (σ) of the number of blue beads in samples of 30 from this population. Find: (i) $\mu + \sigma$; (ii) $\mu - \sigma$; (iii) $\mu + 2\sigma$; (iv) $\mu - 2\sigma$. Examine the table of results (Table 8.1), and find what proportion of samples were: (a) outside the range $\mu \pm \sigma$; (b) outside the range $\mu \pm 2\sigma$.

7. Find the general formulae for the expected value and the variance of the Bernoulli distribution as defined in Section 8.5, p. 172.

9.7 VARIANCE OF A LINEAR FUNCTION OF A RANDOM VARIABLE

We have seen that $E[aR + b] = aE[R] + b$ (Theorem 9.1). The variance of $U = aR + b$ is

$$\text{Var}[U] = E[(U - E[U])^2].$$

However,

$$U - E[U] = (aR + b) - (aE[R] + b) = a(R - E[R]).$$

So $(U - E[U])^2 = a^2(R - E[R])^2$, and therefore

$$\text{Var}[U] = a^2 E[(R - E[R])^2] = a^2 \text{Var}[R].$$

Thus

$$\text{Var}[aR + b] = a^2 \text{Var}[R].$$

9.7.1 Standardized variables

One result involving a linear function of a random variable is very generally useful, and can be proved using only the E operator, so that we shall now state it in terms of a 'random variable X' (discrete or continuous).

From any random variable X, which has mean μ and variance σ^2, we can construct a new random variable U which has mean 0 and variance 1.

We write

$$U = \frac{X - \mu}{\sigma}.$$

Then

$$E[U] = E\left[\frac{X - \mu}{\sigma}\right] = \frac{1}{\sigma}(E[X - \mu]) = \frac{1}{\sigma}(E[X] - E[\mu]) = \frac{1}{\sigma}(\mu - \mu) = 0.$$

$$\text{Var}[U] = E[U^2], \quad \text{since the mean } E[U] \text{ is zero,}$$

$$= E\left[\left(\frac{X - \mu}{\sigma}\right)^2\right] = \frac{1}{\sigma^2}E[(X - \mu)^2] = \frac{\sigma^2}{\sigma^2} = 1.$$

It is often much simpler to work with standardized variables when deriving theoretical (or numerical) results, and then to transform back to the actual variables as the final step. The process is analogous to coding variates.

9.7.2 Exercises

1. The random variable R has mean μ and variance σ^2. Find the mean and variance of the random variables: (i) $5R$; (ii) $-R$; (iii) $10 - R$; (iv) $R - \mu$; (v) R/σ; (vi) $\frac{2}{3}(R - 5)$.

2. The random variable R has the binomial distribution with parameters $n = 10$ and $\pi = \frac{1}{5}$. Write down the standardized variable corresponding to R.

All the results that we have so far stated and proved for discrete random variables R in this chapter hold also for continuous random variables X. Indeed, we could have written them in terms of a 'random variable X' without specifying whether this meant discrete or continuous, and this is often done. However, the proofs for continuous variables demand replacing probability mass functions by density functions, and summation by integration. The reader will be in a position to do this after studying Chapter 14, but meanwhile may assume that all the results presented so far in this chapter are valid in general.

9.8 AVERAGE SAMPLE NUMBER IN ACCEPTANCE SAMPLING

When discussing acceptance sampling schemes in Section 8.8, we remarked (p. 184) that a double sampling scheme may have the advantage over a single sampling scheme when sample size, n, in the first sample is not very small. The second sample will be necessary only if the result after the first sampling is inconclusive. When r, the number faulty, in the first sample is very small or quite large a decision is made immediately, and the average amount of sampling needed in repeated examinations of the output is kept as low as possible.

Example 9.11

The output of a mass-produced electrical component contains on average 4% of faulty items when the process is operating under normal conditions. Random samples of $n = 20$ items can be taken from batches of this output. The decision rule for this sampling scheme is to accept a batch if there is no more than one faulty item in the sample, and to reject the batch if there are three or more faulty items in the sample. When $r = 2$ in the first sample a second random sample is taken immediately; this also contains $n = 20$ items. The whole batch is accepted if there are no faulty items in the second sample. Find the probability of accepting a batch, and the average sample number required.

Using a binomial model, with $n = 20$ and $\pi = 0.04$,

$$P(r = 0) = (0.96)^{20} = 0.4420 \quad \text{and} \quad P(r = 1) = 20(0.96)^{19}(0.04) = 0.3683.$$

The probability of accepting the batch after one random sample of 20 items is therefore $0.4420 + 0.3683 = 0.8103$. Further,

$$P(r = 2) = \frac{20 \times 19}{2}(0.96)^{18}(0.04)^2 = 0.1458;$$

and this is the probability that a second random sample will be called for before making a decision. Finally, the batch will be rejected after one random sample with probability $1 - 0.8103 - 0.1458 = 0.0439$.

If acceptance requires that there shall be no faulty items at all in the second sample ($r = 0$), it is easily found that the overall probability of accepting a batch is 0.8747 and that of rejection is 0.1253.

However, our main interest now is in the *average sample number*, that is the expected number of items that would have to be sampled to reach a decision when the sampling process is repeated a large number of times. This is

$$20\,P(\text{decision after first sample}) + 40\,P(\text{second sample required})$$

$$= 20 \times 0.8542 + 40 \times 0.1458 = 22.916.$$

Thus, on average, only about three more items are sampled than in single sampling with $n = 20$, and so this double sampling scheme would often be worthwhile.

There is no reason why the size of the second sample should be the same as the first, except that of practical simplicity. Modifications of the scheme could be studied in the same way as we considered various sample sizes and decision rules in Example 8.3, p. 179, and a scheme which led to reasonable probabilities of acceptance and rejection for the values of π likely to be met would be the one to use.

9.8.1 Proportion of batches accepted and rejected

The management responsible for operating a process may want information expressed in a different way: 'if we agree to set up a scheme which requires M samples to be taken during the working day, and a single sampling scheme is used, how many rejected samples shall we expect to have?' This is a reasonable approach if they have to make provision for rejected material, or if the costs of monitoring a process more closely are considerable.

Again the binomial model can be used, assuming that the probability of acceptance does not change systematically (which is the same as saying that π does not change). The parameters in the binomial will be M and P, where P is the probability of rejecting a sample according to whatever scheme is used. For samples taken four times per hour during an 8-hour working day, $M = 32$, and if the decision rule in 'Altering the decision rule' in Example 8.3 is used for a process which produces faulty items with probability $\pi = 0.03$, we see from Table 8.3 that the probability of rejecting a sample is $P = 1 - 0.7837 = 0.2163$.

The expected number of rejected samples in a day's work is therefore $MP = 32 \times 0.2163 = 6.92$, i.e. about 7; seeing the results expressed like this, management may well feel that the rule is too strict. Of course this information is already there in Table 8.3, but not in such an obvious way.

9.8.2 Combining both samples to make a decision

The rule at the end of a double sampling scheme where both samples have been necessary is often based on the combined number of faulty items in the total sample; where each sample has been of the same size, n, the decision is therefore based on $2n$ observations. It is tempting to think of this as a problem based on a binomial distribution with sample size $2n$, but this would be wrong because the $2n$ arose as two separate, independent sets, each of n, and the outcomes of the two steps in the sampling scheme must be kept separate.

Let us consider two alternative sampling schemes, each of which uses 24 items altogether. In a routine process for checking the weight of a packaged food product, which from experience is believed to give 10% of packets outside specified tolerance limits, the two schemes proposed for examining each batch of the product are:

> I: a single sample of $n = 24$ items is taken, and the batch is rejected if there are three or more packets outside the limits, i.e. there are $r \geqslant 3$ faulty items;
> II: a double sampling scheme based on samples of size $n = 12$ is to be used, in which a batch will be accepted at the first stage if there are no faulty items in the sample and will be rejected when there are three or more; a second sample is taken only when there are one or two faulty items in the first one. When the second sample is required, the combined number of faulty items in both samples is considered; if it is not greater than two the batch is accepted and if it is three or more the batch is rejected.

Scheme I has probability of rejection $P(r \geqslant 3)$ in the binomial distribution with $n = 24$ and $\pi = 0.1$. This is $1 - P(0) - P(1) - P(2)$; and

$$P(0) = (0.9)^{24} = 0.0798,$$

$$P(1) = 24(0.9)^{23}(0.1) = 0.2127,$$

$$P(2) = \frac{24 \times 23}{2}(0.9)^{22}(0.1)^2 = 0.2718.$$

The probability of rejection is therefore 0.4357.

In scheme II the first sample has $n = 12$ and $\pi = 0.1$, so the probability of immediate acceptance is $P(0) = (0.9)^{12} = 0.2824$. When $r = 1$ or 2, a second sample is taken:

$$P(1) = 12(0.9)^{11}(0.1) = 0.3766,$$

and

$$P(2) = \frac{12 \times 11}{2}(0.9)^{10}(0.1)^2 = 0.2301,$$

so the probability of needing the second sample is $0.3766 + 0.2301 = 0.6067$. Otherwise the batch is rejected, and the probability of this is $1 - 0.2824 - 0.6067 = 0.1109$.

Now consider the cases where a second sample is needed. The batch is to be rejected if there are altogether three or more faulty items; this can happen in two ways:

either there is 1 in the first sample, and 2 or more in the second sample
or there are 2 in the first sample, and 1 or more in the second sample.

We cannot include $r = 0$ or 3 from the first sample because we have already come to a decision about these cases. The probability of $r = 1$ followed by $r \geqslant 2$ is

$$P(1)\{1 - P(0) - P(1)\} = 0.3766\{1 - 0.2824 - 0.3766\} = 0.3766 \times 0.3410 = 0.1284.$$

Also the probability of $r = 2$ followed by $r \geqslant 1$ is

$$P(2)\{1 - P(0)\} = 0.2301 \times 0.7176 = 0.1651.$$

Thus the total probability of rejection when the second sample is needed is $0.1284 + 0.1651 = 0.2935$.

Overall, therefore, scheme II has probability of rejection $0.1109 + 0.2935 = 0.4044$. For scheme I, the probability of rejection was slightly higher, and to compare the two schemes the *average sample number* is useful. In scheme I we always sample 24 items. In scheme II the probability of needing the second sample is 0.6067, so that we use 24 items with probability 0.6067 and only 12 items with probability 0.3933. The expected sample size is

$$12 \times 0.3933 + 24 \times 0.6067 = 19.28.$$

It is likely that we would prefer scheme II, since it gives a smaller average sample number than scheme I and has similar probability of rejection for the given value of π.

9.8.3 Exercises

1. A sampling scheme consists of selecting at random six items from a production batch, making a rapid visual inspection and accepting the batch when there are no faulty items. If there are more than two faulty items the batch is rejected.

   When the number of faulty items is one or two, another sample is taken, this time of 12 items. If this contains two or less faulty items, the batch is accepted and otherwise it is rejected. Calculate the probabilities of rejection when (i) the proportion faulty in the whole production is $\pi = 0.08$; (ii) $\pi = 0.15$; and find also the average sample numbers for these values of π.

2. In a double sampling scheme, a random sample of n items is taken at the first stage and also (where necessary) at the second stage. If p is the probability of taking a decision (either to accept or to reject) at the first sample, write down an expression for the *average sample number* required.

   In a food production process, packaged items are sampled as they come off a production line; a random sample of five items from each production batch is checked to see if the contents are neatly placed and tightly packed. A batch is accepted if all five sampled items are satisfactory and rejected if there are three or more unsatisfactory items; otherwise a further sample is taken, also of five items, before making a decision. If in fact the machine is giving 80% of satisfactory items, what is the probability that this second sample will be necessary?

   When a second sample is taken, what is the probability that, out of the two samples (10 items) there are 9 satisfactory? Find also the average sample number for this scheme.

9.9 PROBABILITY GENERATING FUNCTIONS

The probabilities with which a discrete random variable takes its possible values can sometimes be conveniently summarized by a *probability generating function*. For a random variable which takes only integer values, we may construct a function of an arbitrary variable t by the equation

$$G(t) = p_0 t^0 + p_1 t^1 + p_2 t^2 + \cdots + p_r t^r + \cdots .$$

In this, the probability of each possible value r of the random variable R is multiplied by t^r and all the products are summed. (We assume that the possible values of R are $0, 1, 2, 3, \ldots$. A more general definition would be to write $G(t) = \sum_{\text{all } r} p_r t^r$, though the values of R are limited to the positive integers or zero.) For example, if we consider the random variable 'the number shown when a die is thrown', then $p_1 = p_2 = p_3 = p_4 = p_5 = p_6 = \frac{1}{6}$. Hence

$$G(t) = \frac{1}{6}t^1 + \frac{1}{6}t^2 + \frac{1}{6}t^3 + \frac{1}{6}t^4 + \frac{1}{6}t^5 + \frac{1}{6}t^6 = \frac{1}{6}(t^1 + t^2 + t^3 + t^4 + t^5 + t^6).$$

As another example, consider tossing a fair coin and let the random variable be the number of tosses required for a head to appear. Here $p_1 = \frac{1}{2}, p_2 = (\frac{1}{2})^2$, $p_3 = (\frac{1}{2})^3, \ldots$. The probability generating function

$$G(t) = \left(\frac{1}{2}\right)t + \left(\frac{1}{2}\right)^2 t^2 + \left(\frac{1}{2}\right)^3 t^3 + \cdots + \left(\frac{1}{2}\right)^r t^r + \cdots$$

$$= \left(\frac{t}{2}\right) + \left(\frac{t}{2}\right)^2 + \left(\frac{t}{2}\right)^3 + \cdots + \left(\frac{t}{2}\right)^r + \cdots .$$

But this is an infinite geometric series and, provided we restrict t so that $|t/2| < 1$, then it has the sum

$$\frac{t/2}{1 - t/2} = \frac{t}{2 - t}.$$

We may carry out the reverse operation, and expand the function $t/(2 - t)$ using the binomial theorem. This gives

$$G(t) = t(2 - t)^{-1} = \frac{t}{2}\left(1 - \frac{t}{2}\right)^{-1} = \frac{t}{2} + \left(\frac{t}{2}\right)^2 + \left(\frac{t}{2}\right)^3 + \cdots .$$

We have *generated* the probabilities with which the random variable takes the values $1, 2, 3, \ldots$; they are the coefficients of $t^1, t^2, t^3, \ldots$.

This illustrates how the probability generating function $G(t)$ may provide a compact summary of the probabilities with which a random variable takes its values. Comparing the form of the function $G(t)$ with the expression in Theorem 9.2, we see that $G(t)$ can be considered as an expectation: it is $E[t^R]$.

The probability generating function sometimes offers a convenient way of obtaining the mean and variance of a random variable. If we differentiate $G(t)$ with respect to t, then

$$\frac{dG}{dt} = G'(t) = p_1 + 2p_2 t + 3p_3 t^2 + \cdots + rp_r t^{r-1} + \cdots .$$

Putting $t = 1$ in the equation, and denoting the value of $G'(t)$ so obtained by $G'(1)$,

$$G'(1) = p_1 + 2p_2 + 3p_3 + \cdots + rp_r + \cdots .$$

But the right-hand side of this equation is the mean μ of the random variable. Hence μ equals $G'(1)$.

If we differentiate $G'(t)$ again, or differentiate $G(t)$ twice, we obtain

$$G''(t) = 2 \cdot 1p_2 + 3 \cdot 2p_3 t + 4 \cdot 3p_4 t^2 + \cdots + r(r-1)p_r t^{r-2} + \cdots .$$

Putting $t = 1$ in the equation,

$$G''(1) = 2 \cdot 1p_2 + 3 \cdot 2p_3 + 4 \cdot 3p_4 + \cdots + r(r-1)p_r + \cdots = \sum_{\text{all } r} r(r-1)p_r$$

$$= \sum_{\text{all } r} r^2 p_r - \sum_{\text{all } r} rp_r = E[R^2] - E[R].$$

Rearranging, we find

$$E[R^2] = G''(1) + E[R] = G''(1) + G'(1).$$

The variance,

$$\text{Var}[R] = E[R^2] - (E[R])^2 \quad \text{(see p. 208)},$$

so

$$\sigma^2 = G''(1) + G'(1) - [G'(1)]^2.$$

It is therefore quite straightforward to derive the mean and the variance using the generating function. Another property of the generating function is that $G(1) = \sum_{\text{all } r} p_r = 1$.

Example 9.12 Probability generating function of Bernoulli distribution

Find the probability generating function of the Bernoulli distribution with parameter π, and use it to find the mean and variance of the distribution.

For the Bernoulli distribution,

$$P(R = 1) = \pi \quad \text{and} \quad P(R = 0) = 1 - \pi.$$

Therefore

$$G(t) = (1 - \pi)t^0 + \pi t^1$$

$$= 1 - \pi + \pi t.$$

$$G'(t) = \pi, \quad \text{so } G'(1) = \pi.$$

The mean is thus π.

$$G''(t) = 0, \quad \text{so } G''(1) = 0.$$

The variance is

$$\sigma^2 = G''(1) + G'(1) - [G'(1)]^2$$

$$= 0 + \pi - \pi^2$$

$$= \pi(1 - \pi).$$

Example 9.13 Probability generating function of binomial distribution

Find the probability generating function of the binomial distribution with parameters n, π. Hence determine the mean and variance of the distribution.

For the binomial random variable R,

$$p_r = \binom{n}{r} \pi^r (1 - \pi)^{n-r}, \quad r = 0, 1, \ldots, n.$$

$$G(t) = \sum_{r=0}^{n} \binom{n}{r} \pi^r (1 - \pi)^{n-r} t^r = \sum_{r=0}^{n} \binom{n}{r} (\pi t)^r (1 - \pi)^{n-r}$$

$$= [\pi t + (1 - \pi)]^n = [1 + \pi(t - 1)]^n.$$

$$G'(t) = n\pi [1 + \pi(t - 1)]^{n-1} \quad \text{and} \quad \mu = G'(1) = n\pi.$$

$$G''(t) = n(n - 1)\pi^2 [1 + \pi(t - 1)]^{n-2} \quad \text{and} \quad G''(1) = n(n - 1)\pi^2.$$

Hence

$$\sigma^2 = G''(1) + G'(1) - [G'(1)]^2$$

$$= n(n - 1)\pi^2 + n\pi - n^2\pi^2 = n\pi(1 - \pi).$$

9.10 MINITAB: STANDARDIZATION OF A VARIATE

To standardize a variate, as discussed in Section 9.7.1, choose

Calc>Standardize/
Enter **Input column** and column to **Store results**,
Select **Subtract mean and divide by std. dev./**

This procedure uses the mean and standard deviation calculated from the data. If required, chosen values of the mean and standard deviation can be inserted in the dialogue box.

9.11 EXERCISES ON CHAPTER 9

1. A discrete random variable R has the probability distribution function

$$P(R = r) = \begin{cases} k(r + 1) & \text{for } r = 0, 1, 2, 3, 4, \\ 0 & \text{otherwise.} \end{cases}$$

Find the value of the constant k. Find also the expectation and variance of R. What is the probability that R exceeds its expected value? (*IOS*)

2. The discrete uniform distribution R with κ possible values has $P(R=r) = 1/\kappa, r = 1, 2, \ldots, \kappa$. Find a general formula for its expected (mean) value and its variance.

3. If the City football team wins its match next weekend, each player will receive a £200 bonus; if it draws, each player will receive £50; if it loses, there will be no bonus. The manager assesses their probability of winning as 0.4, of drawing 0.3 and of losing 0.3. What is the expected bonus for each player? The manager tells them that if they put in an extra training session, their probability of winning will rise to 0.5, and that of drawing will remain 0.3. What is the expected bonus now?

4. (a) When I travel up to town by train, I must first take a bus to my nearest railway station. If the bus arrives on time, I can catch a fast train which reaches town in 45 minutes. If the bus is late, I must catch a slower train, which takes 58 minutes. There is a probability of 0.6 that the bus will be late. What is my expected rail-travelling time?
(b) Following some changes in the one-way system near the railway station, my bus now has a probability 0.6 of being on time. If I travel up to town every day from Monday to Friday one week, what is my total expected rail travel time for the whole week?

5. The geometric random variable R takes the value r $(=1, 2, \ldots)$ with probability $p_r = \pi(1-\pi)^{r-1}$, where $0 < \pi < 1$. Show that

$$E[R(R-1)] = \pi[2.1(1-\pi) + 3.2(1-\pi)^2 + \cdots$$
$$+ r(r-1)(1-\pi)^{r-1} + \cdots].$$

Multiply both sides of this equation by $(1-\pi)$ and subtract the result from the equation to show that $E[R(R-1)] = 2(1-\pi)/\pi^2$. Hence show that $\text{Var}[R] = (1-\pi)/\pi^2$.

6. Prove that $E[R^2] \geqslant (E[R])^2$. When does the equality hold?

7. Two unbiased coins are tossed; for each, a score of 1 is recorded for a head and 2 for a tail. The two scores are multiplied together to give a product P. Find the mean value of P, and the standard deviation of P. (SMP)

8. A multiple-choice test paper consists of ten questions. In each question, the candidate has to choose the correct answer from a list of five answers. Consider a candidate who chooses the answer to each question at random from the listed answers. What is the probability distribution of the number of correct answers for such a candidate and what is his expected number of correct answers?

Suppose each correct answer is awarded 4 marks and each incorrect answer carries a penalty of 1 mark. Determine the expected overall mark of a candidate who randomly chooses the answer to each question.

Suppose a candidate has probability $\frac{1}{4}$ of knowing the correct answer to each question, and that when he does not know the correct answer he chooses the answer randomly. For this candidate, determine the expected number of correct answers and his expected overall mark on the paper. *(Welsh)*

9. Define the binomial distribution, stating clearly the conditions under which it applies, and derive its mean.

   A man canvasses people to join an organization. For each new recruit he receives 25p. The probability that a person he canvasses will join is 0.2. Calculate to three decimal places the probability that he will obtain three or more recruits from ten people canvassed. State the amount of money that he would be expected to obtain on average from ten canvassings. State how many people he would need to canvass each evening to average ten new recruits per evening.

   Calculate the number of people he must arrange to canvass to be 99% certain of obtaining at least one recruit. *(JMB)*

10. An event has probability p of success and $q\ (=1-p)$ of failure. Independent trials are carried out until at least one success and at least one failure have occurred. Find the probability that r trials are necessary $(r \geqslant 2)$ and show that this probability equals $(\frac{1}{2})^{r-1}$ when $p = \frac{1}{2}$.

    A couple decide that they will continue to have children until either they have both a boy and a girl in the family or they have four children. Assuming that boys and girls are equally likely to be born, what will be the expected size of their completed family? *(Oxford)*

11. A penny which has a probability π of coming down heads is tossed until the kth head appears. Let R be the number of tosses.
    (i) By considering the probability of there being $k-1$ heads in the first $r-1$ tosses and that the rth toss is a head, show that R is such that

    $$P(R = r) = \binom{r-1}{k-1} \pi^k (1-\pi)^{r-k}, \quad r = k, k+1, k+2, \ldots.$$

    (ii) By comparison of the series $p_k + p_{k+1} + p_{k+2} + \cdots$ with the binomial expansion of $(1-x)^{-k}$ (Appendix I), show that

    $$\sum_{r=k}^{\infty} \binom{r-1}{k-1} \pi^k (1-\pi)^{r-k} = 1.$$

    (Because of this correspondence of the probabilities with the terms of $(1-x)^{-k}$, the distribution is called the *negative binomial* distribution.)
    (iii) Show that $E[R] = k/\pi$.

12. In a model of the game of billiards, the probability of player A scoring a point is p and of missing a shot is $q = 1-p$. Player A continues playing until he misses, when the turn passes to his opponent. The number of points he accumulates is the length of his turn, and this can be zero. Show that the probability that player A has scored exactly k points after r turns have been completed is

$$\binom{r+k-1}{k} q^r p^k.$$

Hence show that the expected number of points he scores after r turns is rp/q. *(Oxford)*

13. A random variable R takes the values 0 and 2, each with probability $\frac{1}{2}$. Find the probability generating function $G(t)$ of R, and show $G(1) = 1$. Derive the mean and variance of R from $G(t)$.

14. Find the probability generating function of the geometric distribution with parameter π. Use it to derive the mean and variance of the distribution.

15. Small components are being mass-produced and samples of 30 items are taken at random from batches. It is desirable that the proportion of substandard components shall not exceed 4%. A double sampling scheme is proposed in which a batch will be accepted when there is at most one substandard item in the sample and rejected when there are three or more. Otherwise a second sample is examined and the batch is accepted if there are at most three substandard items in the total of 60 sampled. Find the probabilities of requiring a second sample, of accepting a batch, and of rejecting a batch.

 If in fact the proportion of substandard components is 0.06, rather than 0.04, carry out the same calculations and comment on how well the scheme is able to distinguish between the two situations.

9.12 COMPUTING EXERCISES

1. Simulate the game described in Section 9.1. Find Bill's winnings (note that they can be positive or negative) for each throw of the die, and generate 500 throws. Produce a frequency table for the winnings and find the mean value. Compare it with the theoretical value.

2. Consider the following game. Ann tosses a fair coin 5 times, unless she throws a head before the 5 tosses are completed, in which case the game ends. If Ann throws a head on the ith throw ($i = 1, 2, 3, 4, 5$) she receives i^2 pence from Betty. But if the ith toss is a tail then Ann pays 2^i pence to Betty. Calculate, theoretically, the distribution of Ann's winnings per game. Choose 100 random samples from the distribution and construct a graph of Ann's fortune over time, assuming she begins with 600 pence.

9.13 PROJECTS

1 Telephone directory search to find subscribers on a particular exchange

Aim. To estimate the mean number of subscribers that have to be scanned to find a subscriber on a particular exchange.

Equipment. Telephone directory.

Collection of data. Choose a page of a telephone directory at random. Begin scanning the names at the top left-hand corner of the page and move down the first column until a subscriber on the chosen exchange is found. Record the number of names scanned, including that of the subscriber on the chosen exchange. Repeat 50 times.

Analysis of data. Construct a frequency table and find the mean of the data. If the subscribers' names were in random order with respect to exchange, a geometric distribution would be obtained.

2 Sex distribution of children in families (Data from earlier project)

Aim. To estimate the mean and variance of the number of girls in families of (a) two, (b) three children.

Collection of data. As for Project 8.15(1).

Analysis. From the frequency table find the mean and variance of the number of girls in families of (a) two, (b) three children.

3 Distribution of number of sixes in five throws of a die (Data from earlier project)

Aim. To estimate the mean and variance of the number of sixes in five throws of a die.

Collection of data. As for Project 6.13(1).

Analysis. Work with the data in blocks of five results. Form the frequency table of the number of sixes per block of five results. Find the mean and variance of the distribution.

SUMMARY

For a random variable R which takes the value r with probability p_r:

(i) the **mean**, or **expected value**, $E[R]$ equals $\sum rp_r$;

(ii) the **expected value of a function** $g(R)$, $E[g(R)]$, equals $\sum g(r)p_r$;

(iii) the **variance** $\text{Var}[R]$ equals $E[(R-\mu)^2] = \sum (r-\mu)^2 p_r$, where $\mu = E[R]$ is the mean of R.

A **binomial distribution**, with parameters n, π, has mean $n\pi$ and variance $n\pi(1-\pi)$.

A **geometric distribution**, with parameter π, has mean $1/\pi$ and variance $(1-\pi)/\pi^2$.

The **probability generating function** (pgf) of a discrete random variable, which takes the values $r = 0, 1, 2, \ldots$, with probabilities $p_0, p_1, p_2, \ldots$, respectively, is defined as

$$G(t) = p_0 t^0 + p_1 t^1 + p_2 t^2 + \cdots + p_r t^r + \cdots$$

It may also be written $E[t^R]$.

The pgf of a binomial distribution, with parameters n, π, is $[1 + \pi(t-1)]^n$.

The pgf of a geometric distribution, with parameter π, is $\pi t / [1 - t(1 - \pi)]$.

REVISION EXERCISES B

1. A die is loaded in such a manner that the probability of face $k = 1, 2, 3, 4, 5$ or 6 coming up is proportional to k. Find the probability that an even number comes up.

2. (i) Construct a tree diagram to determine the number of ways in which a coin can be tossed five times in succession so that throughout this series of tosses there are never fewer heads than tails.

   (ii) Suppose that in a survey concerning the reading habits of students it is found that:

   > 60% read magazine A,
   > 50% read magazine B,
   > 50% read magazine C,
   > 30% read magazines A and B,
   > 20% read magazines B and C,
   > 30% read magazines A and C,
   > 10% read all three magazines.

   (a) What percentage read exactly two magazines?

   (b) What percentage do not read any of the magazines?

3. A probability space consists of the numbers $1, 2, \ldots, 8$ each with probability $\frac{1}{8}$. If $A = \{1, 2, 3, 4\}$, $B = \{3, 4, 5, 6\}$, $C = \{5, 6, 7, 8\}$, find $P(A \mid B)$, $P(A \mid C)$, $P(A \mid B \cap C)$, $P(A \mid B \cup C)$. Are A, B, C independent?

4. A man decides to continue tossing a fair coin until he has thrown a total of three heads. Find the probability a_n that exactly n tosses will be needed.
   Show that a_n is greatest when $n = 4$ or 5.

5. In the game of Ludo a player may not move his counter until he has thrown a six on the die. Calculate the probability that he has to throw the die more than three times before this occurs. *(SMP)*

6. A die is cast repeatedly until a six is thrown. Given that an even number of throws is required, calculate the probability that two throws are required. *(IOS)*

7. What is meant by saying that two events, A and B, are *mutually exclusive*? What is meant by saying that A and B are *independent*? Discuss whether two events can be both mutually exclusive and independent.

Explain what is wrong with the following statements, giving an amended calculation where you think it is appropriate.

(i) Since accident statistics show that the probability that a person will be involved in a road accident in a given year is 0.02, the probability that he will be involved in two accidents in that year is 0.0004.

(ii) Four persons are asked the same question by an interviewer. If each has, independently, probability $\frac{1}{6}$ of answering correctly, the probability that at least one answers correctly is $4 \times \frac{1}{6} = \frac{2}{3}$. *(JMB)*

8. (a) What is the probability that there are not more than three boys in a family of five children? (Assume that the probability of a male or a female birth is $\frac{1}{2}$.)

(b) Three persons A, B and C, in that order, successively throw an unbiased coin and the first person to throw a head wins. What are the respective probabilities that A, B and C will win?

(c) An urn contains three red balls and four blue balls. Two balls are taken at random from these seven. What is the probability that they are both of the same colour?

9. A single die is thrown six times. Find the probability that: (i) exactly one 'six' is thrown; (ii) at least two 'sixes' are thrown; (iii) each face of the die turns up once.

Find the probability that, when a die is thrown repeatedly, a 'six' appears for the first time at the rth throw. Find the least value of r such that there is more than a 50% chance of obtaining a 'six' on the rth throw or earlier.

Write down expressions, in terms of $p = \frac{1}{6}$ and $q = \frac{5}{6}$, for the expectations $E[r]$, $E[r^2]$. Calculate the numerical values of these expressions: for this purpose it may be helpful to consider the results of differentiating twice with respect to q the expansion, valid for $|q| < 1$,

$$(1 - q)^{-1} = 1 + q + q^2 + \cdots + q^r + \cdots .$$ *(MEI)*

10. A bird has flown into a room through an open window. The room has two other windows of the same size and these are closed. The open window is the bird's only way out. It flies at windows until it finds the open one and escapes. Assuming that the bird has no memory, so that each fresh attempt may be at any window, obtain the probability distribution of the number of attempts it makes in escaping.

The householder claims that his pet budgerigar would have had the intelligence not to try any window more than once. Assuming this to be true, obtain the corresponding probability distribution for the pet. Find the probabilities that: (i) the first bird would escape with fewer attempts than the pet; (ii) the pet would take fewer attempts than the first bird. *(JMB)*

11. In a premium bond competition, prize money of £10 000 is awarded each month for every 2 666 667 bonds (each costing £1) in the competition.

The prize money is divided into 295 prizes with the following values:

£1000	1
£500	1
£250	2
£100	3
£50	20
£25	268

The winning bonds are chosen at random by a computer.
(a) What is the probability of winning a prize, with one bond, at each draw?
(b) If a prize is won, what is the probability that it is one of £100 or more?
(c) What is the expected gain, each month, for each bond held?
(d) What annual rate of interest does this correspond to?
(e) A man wishes to buy enough bonds so that he will have a one in ten chance of winning at least one prize at each monthly draw. How many bonds should he buy?

12. Prove that the number of different ways in which r objects can be selected from a group of n is

$$\frac{n!}{r!(n-r)!}.$$

A hand of 5 cards is selected without replacement from a well-shuffled pack of 52 cards consisting of 4 suits each of 13 denominations. Find the probability that the 5 cards are of 5 different denominations and show that it is approximately 0.5. (*Oxford*)

13. Define the mean and variance of a discrete random variable X which has a given probability function $p(x)$. In a 'roll-a-penny' game at a fairground a penny is rolled onto a table marked in squares of equal size. The length of the sides of the squares is twice the diameter of the penny. If the penny lies fully within a square it is returned to the player. The centre of the penny may be assumed to lie at random on the marked part of the table, and the lines marking the squares are of negligible thickness. Determine the probability that the penny is returned to the player.
For some squares an additional 'prize' of threepence or sixpence is given. If one third of the squares yield a prize of threepence, and one ninth of the squares yield a prize of sixpence, calculate the mean and variance of the profit to the player which arises from rolling a single penny. (*JMB*)

14. To start a game, the four players, A, B, C, D in that order, take turns to shake an unbiased six-sided die. The first to throw a six has first turn in the game. Find for each player the probability of having first turn.
D suspects that he is at a disadvantage by this procedure and proposes that, instead, A and B shall toss a coin, the winner shall toss with C, and the winner of these two shall toss with D. The winner of the last toss is to have

first turn. Find for each player the probability of having first turn under this arrangement. (*JMB*)

15. An examination paper consists of 20 questions to each of which the candidates have to answer 'true' or 'false'. The pass mark for the examination is 75%.

    Show that the probability that a candidate passes entirely by guessing is 0.02.

16. In n independent trials with constant probability p of success at each trial, show that the ratio of the probabilities of $r+1$ and r successes is

    $$\frac{(n-r)p}{(r+1)(1-p)} \quad (0 \leqslant r \leqslant n-1).$$

    A manufacturer produces bracelets consisting of six sections linked together. After this linking-up process each section independently has a probability q of having been scratched. The bracelets are inspected and those which are unscratched are sold for £2 each. Those with a single scratched section are sold as substandard for £1 each. Any with more than one scratched section are scrapped. It is found that the ratio of the number of bracelets sold as perfect to those sold as substandard is 4:1. Calculate q and hence find to three decimal places the proportion of the bracelets that are scrapped.

    Each bracelet costs £1 to produce. Find to the nearest £1 the expected profit on a batch which consists of 1000 bracelets before inspection. (*JMB*)

17. Show that the mean and the variance of a binomial distribution of n trials are np and $np(1-p)$, where p is the probability of success in any trial.

    In an experimental trial, a biased coin is tossed 5 times. 500 such trials are carried out and the number of heads in each trial is recorded, as follows:

No. of heads per trial	0	1	2	3	4	5	Total
Observed frequency	20	75	162	149	77	17	500

    Fit a binomial distribution to these data.

18. Three people independently toss a fair coin. Each person continues tossing until he has obtained a head.

    If N and M are the length of the longest and the shortest of the three series of tosses respectively, show that $P(N \leqslant k) = (1 - (\frac{1}{2})^k)^3$.

    Obtain expressions for $P(N = k)$ and $P(M = k)$. (*IOS*)

19. A binomial distribution has probabilities of $0, 1, 2, \ldots, n$ successes in n trials given by the coefficients of powers of t in the expansion of $(q + pt)^n$. Show that the number of successes has a mean np and variance npq.

    At a pottery, when running normally, 20% of the teacups made are defective. Find the probability that a sample of five will contain:
    (i) no defective;
    (ii) exactly one defective;
    (iii) at least two defectives. (*AEB*)

20. Each of the two ends of a domino has a number of spots between 0 and 6, and in a complete set of 28 dominoes each possible combination occurs once. Find the mean and standard deviation for the total number of spots on a domino.

Two dominoes are said to *match* if they can be placed end-to-end so that the two touching ends have the same number of spots. Show that the probability that two dominoes (selected at random from a complete set) match is $\frac{7}{18}$.

Given two matching dominoes neither of which is a double, with their matching ends touching, find the probability that a third domino (drawn at random from the remainder of the set) will match one or both of their free ends. *(SMP)*

21. Each morning ten commuter trains pass a signal box in quick succession on their way to London. The trains all start their journeys from different towns, and seven of them are bound for Charing Cross Station, the rest for Victoria Station, but they pass the signal box in random order. The signal box controls a nearby set of points which needs to be changed if the destination of the next train is different from the destination of the previous one, and initially the points are always set for Charing Cross. If no other trains use the track at that time calculate the probabilities that on any morning the points will have to be changed (a) once, (b) twice for the commuter trains to reach the right destinations. *(IOS)*

22. An interviewer selects people at random from a large population. The probability that any person he selects is suitable for his survey is p.

Find the probability that the first person who is suitable for interview is the rth person selected and show that the mean number of people the interviewer must select to find the first suitable person is $1/p$.

Given that

$$1^2 + 2^2 a + 3^2 a^2 + 4^2 a^3 + \cdots = \frac{1 + a}{(1 - a)^3}$$

find the variance of the number of people to be selected before the first suitable person is found. *(IOS)*

23. A company uses machinery which fails periodically. When a failure occurs the company may contact either of two contractors to repair the fault but must wait n days for the contractor to arrive, whereupon the fault is cured immediately. The probability distribution of the number of days each contractor takes to arrive after being called is shown in the table.

n (days)	1	2	3	4	5
Contractor A	0.1	0.1	0.5	0.3	0
Contractor B	0.2	0.2	0.3	0.2	0.1

The number of days on which there is lost production is equal to the number of days that a contractor takes to arrive. Lost production costs the company £2000 per day.

(i) Discuss whether the company should prefer to contact Contractor A or Contractor B in the event of a failure if it wishes to minimize its long-term loss.

(ii) If the contractors operate independently find the expected loss if both contractors are contacted and whichever is available first effects the repair. (The contractors may arrive on the same day but they never arrive simultaneously.) (*IOS*)

24. A random sample of n items is drawn from a batch production process, and the batch is accepted if there are no faulty items in the sample. It is rejected if there are two or more faulty, and a further sample of n is taken when there is just one faulty. The same decision procedure is used as often as necessary until a decision can be made (and it may continue indefinitely).

The proportion of faulty items in a batch is π, and P(0), P(1) are the probabilities of $r = 0, 1$ respectively in a sample. Show that a batch is accepted with probability $P(0)/\{1 - P(1)\}$. Find also the average number of batches inspected before one is rejected.

Carry out the calculations for the case $n = 100, \pi = 0.02$.

25. The number of days required to complete a project is X, which is assumed to have the distribution

$$P(X = x) = \begin{cases} 0.4 & \text{if } x = 11 \\ 0.2 & \text{if } x = 10 \text{ or } 12 \\ 0.1 & \text{if } x = 13 \text{ or } 14 \\ 0 & \text{otherwise.} \end{cases}$$

The return from the project is $Y = £2000(12 - X)$.

(i) Find the probability distribution of Y, and obtain the probability of making a loss.

(ii) Find E[X], Var[X], E[Y] and Var[Y]. (*RSS*)

10 JOINT DISTRIBUTIONS

OVERVIEW

In order to investigate the properties of a random sample we consider the n observations as the values taken by n random variables $X_1, X_2, \ldots, X_n$. This leads us to define and investigate the joint distribution of many random variables. We look closely at the joint distribution of two random variables and extend the results to many random variables.

10.1 BIVARIATE DISTRIBUTIONS

There are many occasions when we want to measure two variates on each member of a population or sample, and to study the two variates together rather than look at each one separately. For example, a mathematics examination might consist of one paper in arithmetic and one paper in algebra. The mark obtained by a candidate on the arithmetic paper is one variate, and that on the algebra paper is another variate. In order to assess the overall performance of the candidates in mathematics, the examiner must find ways of considering the two variates together. If we are to be able to set up mathematical models that can be used in situations like this, we must first extend our theory of random variables to cases where we consider two (or more) random variables simultaneously.

We begin by considering a simple situation in which each one of the pair of variates observed, and the random variables corresponding to them, can take only two values. We suppose that the study population is made up of the children in a school, and that the variates observed are handedness (i.e. whether the child writes with the right hand or the left – we will ignore ambidexterity) and sex (i.e. whether the child is a boy or a girl).

Let us define a random variable for handedness, which takes values L (left-handed) and R (right-handed):

Value	L	R
Probability	0.1	0.9

and let us define another random variable for sex, which takes values M (male) and F (female):

Value	M	F
Probability	0.5	0.5

Table 10.1 *Joint probability distribution of handedness and sex*

	M	F	Total
L	0.05	0.05	0.10
R	0.45	0.45	0.90
Total	0.50	0.50	1.00

The probabilities given in these definitions are approximately equal to the relative frequencies found in the British population.

Considering these two random variables together, we can derive from them a new random variable which takes four possible values. These values are ML (corresponding to male, left-handed), FL, MR, FR. If the probability of being male is independent of the probability of being left-handed, then

$$P(ML) = P(M)P(L)$$
$$= 0.5 \times 0.1 = 0.05.$$

Similarly, assuming that handedness and sex are completely independent, we can write down the full table of probabilities of this new derived variable (Table 10.1). We call the set of probabilities the *joint distribution* of handedness and sex.

The totals in the margins of this table recover the total probabilities of the two basic variables: the row totals show the probabilities for handedness and the column totals show the probabilities of the two sexes. In the context of joint distributions, these probabilities in the margins are called the *marginal distributions*.

We were able to calculate the joint distribution because we assumed that the two random variables, sex and handedness, were independent. We may often be in the converse situation, where we are given the probabilities in the joint distribution and wish to know if the two random variables involved are independent. If each probability in the two-way table is simply the product of the two corresponding marginal probabilities, as each was in Table 10.1, then we can say that the variables *are* independent.

If we observe two variates, each of which can take only two possible values, on a large number of members of a population, we can make a table of the same sort as Table 10.1, showing the frequencies of members that fall into each of the four possible categories. Dividing these four frequencies by the total number of members observed, we obtain the *relative frequencies* in the four categories. We can obtain the marginal relative frequencies also in a similar way. In Table 10.2(a) we have the results of 160 candidates for an examination in statistics, which consists of a theory paper and an applied statistics paper. The relative frequencies shown in Table 10.2(b) are found by dividing each entry of Table 10.2(a) by 160.

Now if we set up a mathematical model of candidates' performance which says that the two variables 'result on theory' and 'result on applied statistics' are independent, we can check whether our observations are consistent with this by looking at the four relative frequencies within the table. If they differ substantially from the products of the corresponding marginal relative frequencies,

Table 10.2 *(a) Frequencies, and (b) relative frequencies, of results of 160 candidates in a statistics examination*

(a) *Frequencies*

		Applied statistics		
		Pass	Fail	Total
Theory	Pass	32	30	62
	Fail	24	74	98
	Total	56	104	160

(b) *Relative frequencies*

		Applied statistics		
		Pass	Fail	Total
Theory	Pass	0.2000	0.1875	0.3875
	Fail	0.1500	0.4625	0.6125
	Total	0.3500	0.6500	1.0000

we conclude that our model is not a good one. Assuming independence, the relative frequency (or proportion) passing both papers should be $0.3875 \times 0.35 = 0.1356$; we compare this with the 0.2000 actually observed. The relative frequency failing both is, assuming independence, $0.6125 \times 0.65 = 0.3981$, and we compare this with the 0.4625 actually observed. The other relative frequencies, assuming independence, are $0.3875 \times 0.65 = 0.2519$ passing theory only and $0.6125 \times 0.35 = 0.2144$ passing applied statistics only; we compare these with the observed relative frequencies of 0.1875 and 0.15 respectively. We may conclude that this assumption of independence is not justified, since it leads to relative frequencies that are not very close to those actually observed. (There is a proper way of making the comparison, which we shall see in Chapter 19.)

In most of this book we shall restrict ourselves to independent random variables, because they are much easier to deal with than random variables which are not independent. However, it is still of interest to give an example of two random variables which are not independent. If we wish to make a mathematical model to describe the occurrence of colour-blindness in both sexes in the population, then independence is not a reasonable assumption, since it is known that men are more likely to be colour-blind than women are. We might model a specific type of colour-blindness by a random variable that takes the two values C (colour-blind) and N (not colour-blind). To estimate probabilities for use in this model, we might observe the numbers in the four categories corresponding to the values MC, MN, FC, FN in a large population. (M and F of course stand for male and female, as before.) These give estimates of the probabilities in the joint distribution of sex and colour-blindness. Table 10.3 shows values that are reasonable for this distribution.

Table 10.3 *Joint probability distribution of colour-blindness and sex*

	M	F	Total
C	0.040	0.002	0.042
N	0.460	0.498	0.958
Total	0.500	0.500	1.000

Note, for example, that the value 0.040 in the top left corner of Table 10.3 is very far from being equal to 0.042×0.500 (which is 0.021). Another way of seeing that sex and colour-blindness are not independent is to note that the ratio C:N for males has a value very different from its value for females, i.e. the ratios in the two columns of Table 10.3 are different.

Example 10.1 *Delivery time of letters*

Suppose that the number of days taken by a letter to travel through the post is given by one of the following distributions, depending on whether the letter is sent by first or second class post:

Days taken	1	2	3
Probability for first class post	0.9	0.1	0
Probability for second class post	0.2	0.5	0.3

A letter is sent by first class post. The reply is sent on the same day as the letter is received, and is sent by second class post. On the assumption that the times taken by the two letters are independent:

(i) construct the joint distribution of X, Y, the times in days taken by the letter and the reply respectively;

(ii) derive the distribution of the total time taken by the letter and the reply.

(i) The table below is constructed by entering the marginal probabilities, which are given above, and calculating probabilities in the body of the table by taking the product of the corresponding marginal probabilities.

	Y			
X	1	2	3	Total
1	0.18	0.45	0.27	0.90
2	0.02	0.05	0.03	0.10
Total	0.20	0.50	0.30	1.00

(ii) $P(X + Y = 2) = P(X = 1, Y = 1) = 0.18$, from the table in (i).
$P(X + Y = 3) = P\{(X = 1, Y = 2) \quad \text{or} \quad (X = 2, Y = 1)\}$

$$= P(X = 1, Y = 2) + P(X = 2, Y = 1)$$

$$= 0.45 + 0.02 \quad \text{from the table}$$

$$= 0.47.$$

Proceeding in this way we construct the table

$X + Y$	Probability	
2	0.18	$= 0.18$
3	$0.45 + 0.02$	$= 0.47$
4	$0.27 + 0.05$	$= 0.32$
5	0.03	$= 0.03$
	Total	1.00

10.1.1 Exercises

1. The random variables X and Y are specified by $P(X = 1) = \frac{2}{3}$, $P(X = 0) = \frac{1}{3}$ and $P(Y = 1) = \frac{1}{4}$, $P(Y = -1) = \frac{3}{4}$. Construct the joint distribution of X and Y, assuming that X and Y are independent.

2. The random variables X and Y take the values $1, 2, 3$ and $-1, 0, 1$ respectively with the probabilities shown in the following table. Are X and Y independent?

	Y		
X	-1	0	1
1	0.20	0.05	0.15
2	0.10	0.20	0.10
3	0.05	0.10	0.05

3. Find the marginal distributions of X and Y in Question 2. Find also the distribution of: (i) $X + Y$; (ii) $X - Y$; (iii) XY.

10.2 MEAN OF THE SUM OF TWO RANDOM VARIABLES

What is the mean time taken to send a letter and receive the reply, assuming the probability distributions of Example 10.1? We can work this out from the distribution of $X + Y$:

$$E[X + Y] = 2 \times 0.18 + 3 \times 0.47 + 4 \times 0.32 + 5 \times 0.03$$
$$= 0.36 + 1.41 + 1.28 + 0.15 = 3.20.$$

It is a reasonable conjecture, and perhaps intuitively obvious to some, that the mean of the sum of the two random variables X, Y is the sum of the means of X and Y. Let us test this:

$$E[X] = 1 \times 0.90 + 2 \times 0.10 = 0.90 + 0.20 = 1.10$$
$$E[Y] = 1 \times 0.20 + 2 \times 0.50 + 3 \times 0.30 = 0.20 + 1.00 + 0.90 = 2.10.$$

Table 10.4 *Joint probability distribution of two random variables X, Y together with their marginal distributions*

	Y			
X	1	2	3	Total
1	$P(1,1)$	$P(1,2)$	$P(1,3)$	$P_X(1)$
2	$P(2,1)$	$P(2,2)$	$P(2,3)$	$P_X(2)$
Total	$P_Y(1)$	$P_Y(2)$	$P_Y(3)$	1

Thus $E[X+Y]=E[X]+E[Y]$, and we have verified the conjecture for this particular case.

We may obtain a better insight into the result by carrying through the working with symbols, rather than with numerical values. The table may be represented as in Table 10.4.

In Table 10.4, we have written $P(1,1)$ for $P(X=1, Y=1)$, and similarly for the other entries; obviously we have to be careful to keep the numbers in the brackets in the proper order because there is no reason for e.g. $P(1,2)$ to equal $P(2,1)$. We may now obtain the marginal distributions of X and Y by adding up the appropriate entries in Table 10.4. Thus if $P_X(1), P_X(2)$ are the probabilities in the marginal distribution of X, we have

$$P_X(1) = P(1,1) + P(1,2) + P(1,3)$$

and

$$P_X(2) = P(2,1) + P(2,2) + P(2,3).$$

So $P_X(1)$ is equal to $\sum_{\text{all } j} P(1,j)$, the sum of the probabilities associated with $X = 1$ over all possible Y values, and $P_X(2)$ is equal to $\sum_{\text{all } j} P(2,j)$. Similarly, $P_Y(1), P_Y(2), P_Y(3)$ are the probabilities in the marginal distribution of Y, and are equal to $\sum_{\text{all } j} P(j,1), \sum_{\text{all } j} P(j,2)$ and $\sum_{\text{all } j} P(j,3)$ respectively.

To find the expected, or mean, value of a single random variable we had to multiply each possible value of the variable by the probability of that value. When working with a joint distribution, the method is the same. The mean value of $(X+Y)$ is

$$E[X+Y] = (1+1)P(1,1) + (1+2)P(1,2) + (1+3)P(1,3)$$
$$+ (2+1)P(2,1) + (2+2)P(2,2) + (2+3)P(2,3).$$

Each of the terms on the right-hand side is of the form $(x+y)P(X=x, Y=y)$; we find the probability that X and Y take the particular pair of values x and y, and multiply this probability by $(x+y)$. When multiplying out the brackets, we can simplify the result by remembering this. We pick out first the terms going with the first number in each bracket, i.e. those going with X, and group them according to the X value. This is the same as grouping by rows in Table 10.4. We have

$$1\{P(1,1) + P(1,2) + P(1,3)\} + 2\{P(2,1) + P(2,2) + P(2,3)\}$$
$$= 1P_X(1) + 2P_X(2) = E[X].$$

The remaining terms we group by the particular Y value, which is the same as grouping by columns in Table 10.4. We obtain

$$1\{P(1,1) + P(2,1)\} + 2\{P(1,2) + P(2,2)\} + 3\{P(1,3) + P(2,3)\}$$
$$= 1P_Y(1) + 2P_Y(2) + 3P_Y(3) = E[Y].$$

Hence

$$E[X + Y] = E[X] + E[Y].$$

This result may be proved in general.

(i) We let X take values $1, 2, \ldots, m$ and Y take values $1, 2, \ldots, n$. The argument used above can then be followed through in exactly the same way for this general case.

(ii) A more concise proof is as follows. Let X take the value x with probability $P_X(x)$ and Y take the value y with probability $P_Y(y)$ and in the joint distribution the value (x, y) is taken with probability $P(x, y)$. Then

$$E[X + Y] = \sum_{\text{all } x} \sum_{\text{all } y} (x + y)P(x, y)$$

$$= \sum_{\text{all } x} \sum_{\text{all } y} xP(x, y) + \sum_{\text{all } x} \sum_{\text{all } y} yP(x, y).$$

But

$$\sum_{\text{all } y} xP(x, y) = xP_X(x),$$

and

$$\sum_{\text{all } x} yP(x, y) = yP_Y(y).$$

Hence

$$E[X + Y] = \sum_{\text{all } x} xP_X(x) + \sum_{\text{all } y} yP_Y(y) = E[X] + E[Y].$$

This result may be extended in two ways.

Theorem 10.1 Expected value of $aX + bY$

If X and Y are discrete random variables and a and b are constants, then $E[aX + bY] = aE[X] + bE[Y]$.

Proof. Follow through proof (ii) above with aX replacing X and bY replacing Y. Alternatively, define random variables $U = aX, V = bY$. Then

$$E[aX + bY] = E[U + V] = E[U] + E[V] = E[aX] + E[bY] = aE[X] + bE[Y].$$

Theorem 10.2

If $A, B, C, \ldots$ are a finite number of discrete random variables then

$$E[A + B + C + \cdots] = E[A] + E[B] + E[C] + \cdots$$

The proof is a simple extension of the argument above.

10.2.1 Exercises

1. The joint distribution of the random variables X and Y is given by the following table:

	Y		
X	3	4	5
0	$\frac{1}{8}$	$\frac{1}{8}$	0
1	$\frac{1}{3}$	0	$\frac{1}{6}$
2	0	$\frac{1}{8}$	$\frac{1}{8}$

Find the marginal distributions of X and Y. Find $E[X + Y]$: (i) directly from the table; and (ii) by first obtaining $E[X]$ and $E[Y]$.

2. R and S are two random variables, each with mean μ. Find the expected value of: (i) $R + S$; (ii) $R - S$; (iii) $\frac{1}{2}(R + S)$; (iv) $4R - 3S$.

3. The independent random variables R, S have the following distributions. Tabulate the distribution of $\frac{1}{2}(R + S)$ and find its expected value.

x	-2	-1	0	$+1$	$+2$
$P(R = x)$	$\frac{1}{15}$	$\frac{2}{15}$	$\frac{3}{15}$	$\frac{4}{15}$	$\frac{5}{15}$
$P(S = x)$	$\frac{5}{15}$	$\frac{4}{15}$	$\frac{3}{15}$	$\frac{2}{15}$	$\frac{1}{15}$

4. $\overline{R}$ is the mean of a random sample $r_1, r_2, \ldots, r_n$, drawn from a distribution with mean μ. Find $E[\overline{R}]$.

5. (a) R_1, R_2, R_3 and R_4 each have mean μ. Find E[U], where

$$U = \tfrac{1}{6}(R_1 + 2R_2 + R_3 + 2R_4).$$

(b) $R_1, \ldots, R_{2N}$ each have mean μ. Find E[V], where

$$V = \frac{1}{3N}(R_1 + 2R_2 + R_3 + 2R_4 + \cdots + R_{2N-1} + 2R_{2N}).$$

10.3 MEAN OF THE PRODUCT OF TWO INDEPENDENT RANDOM VARIABLES

A game is played in the following way. A coin is tossed to decide which of the two players wins; heads means that player A wins, and tails means that B wins. Then a die is thrown to decide the amount the winner shall receive; a '1' means that 1p is to be paid to the winner by the other player, a '2' that 2p is to be paid, etc. Provided that the coin tossed is a fair coin, so that the probability of either face turning up is $\tfrac{1}{2}$, the game is 'symmetrical' in the sense that A and B are equally likely to win or lose. From this it is clear that A's *expected, or mean, winnings* will be zero. To see this formally, define the random variable X that takes values +1, corresponding to a head and so to a win for A, and -1, corresponding to a tail and so to a win for B. Each of these values, +1 and -1, has probability $\tfrac{1}{2}$. We define a second random variable Y, which takes the values 1 to 6, each with probability $\tfrac{1}{6}$, corresponding to the faces of the die. The twelve possible outcomes of the game, expressed as the number of pence paid to A, are shown in Table 10.5. They are equal to the twelve values of the product XY.

Let us call this new random variable, shown in Table 10.5, $U = XY$. Each possible value of U has probability $\tfrac{1}{12}$, since tossing the coin and throwing the die are independent of one another (p. 145). The mean amount received by A is then

$$\mathrm{E}[U] = \left[1 \times \tfrac{1}{12} + 2 \times \tfrac{1}{12} + \cdots + 6 \times \tfrac{1}{12} + (-1) \times \tfrac{1}{12} + (-2) \times \tfrac{1}{12} + \cdots + (-6) \times \tfrac{1}{12}\right]$$

$$= 0.$$

Table 10.5 *Twelve possible outcomes of a game expressed as the number of pence paid to A (first set of rules of Section 10.3)*

X	Y					
	1	2	3	4	5	6
+1	1	2	3	4	5	6
−1	−1	−2	−3	−4	−5	−6

Now suppose the rules of the game are changed. In the new rules, heads means a win for A of *double* the die value in pence, while tails means a win for B *equal* to the die value in pence. What would A's expected winnings now be, and how much should A be willing to pay to enter the game?

This time we shall define Y as before, but X is now the random variable taking values $+2$ and -1, each with probability $\frac{1}{2}$. The table of outcomes, expressed as the number of pence paid to A, now becomes Table 10.6.

Table 10.6 *Twelve possible outcomes of a game expressed as the number of pence paid to A (second set of rules of Section 10.3)*

	Y					
X	1	2	3	4	5	6
+2	2	4	6	8	10	12
−1	−1	−2	−3	−4	−5	−6

Again these outcomes are the products XY, so again we define $U = XY$, a random variable each of whose possible values has probability $\frac{1}{12}$. The mean amount received by A in this version of the game is

$$\begin{aligned} E[U] &= 2 \times \tfrac{1}{12} + 4 \times \tfrac{1}{12} + \cdots + 12 \times \tfrac{1}{12} + (-1) \times \tfrac{1}{12} + (-2) \\ &\quad + \tfrac{1}{12} + \cdots + (-6) \times \tfrac{1}{12} \\ &= 42 \times \tfrac{1}{12} - 21 \times \tfrac{1}{12} = \tfrac{21}{12} = \tfrac{7}{4} = 1.75. \end{aligned}$$

So A may expect to win $1\frac{3}{4}$ pence per game. He therefore should be willing to pay up to this amount to play, for in the long run he should not lose by doing so.

There is a more direct way of obtaining this result for independent variables, which we state as a theorem.

Theorem 10.3 Expected value of product of independent random variables

If X and Y are *independent* discrete random variables then

$$E[XY] = E[X]E[Y].$$

Proof.

(i) We shall first write out a proof for the special case where X takes the values 1, 2 and Y takes the values 1, 2, 3. The probabilities of these values are represented as in Table 10.4. The key step in the proof is that since X and Y are independent we know that $P(x, y) = P_X(x) \times P_Y(y)$.

$$E[XY] = (1 \times 1)P(1,1) + (1 \times 2)P(1,2) + (1 \times 3)P(1,3) + (2 \times 1)P(2,1)$$
$$+ (2 \times 2)P(2,2) + (2 \times 3)P(2,3)$$
$$= (1 \times P_X(1))(1 \times P_Y(1)) + (1 \times P_X(1))(2 \times P_Y(2))$$
$$+ (1 \times P_X(1))(3 \times P_Y(3)) + (2 \times P_X(2))(1 \times P_Y(1))$$
$$+ (2 \times P_X(2))(2 \times P_Y(2)) + (2 \times P_X(2))(3 \times P_Y(3))$$
$$= [(1 \times P_X(1)) + (2 \times P_X(2))][(1 \times P_Y(1)) + (2 \times P_Y(2)) + (3 \times P_Y(3))]$$
$$= E[X]E[Y].$$

For a proof in the more general case when X takes values $1, 2, \ldots, m$ and Y takes values $1, 2, \ldots, n$ we can extend this argument.

(ii) For a more concise proof, let X take the value x with probability $P_X(x)$ and Y take the value y with probability $P_Y(y)$. Then

$$E[XY] = \sum_{\text{all } x} \sum_{\text{all } y} xy P(x,y)$$

$$= \sum_{\text{all } x} \sum_{\text{all } y} xy P_X(x) P_Y(y) \qquad \text{because of independence}$$

$$= \sum_{\text{all } x} x P_X(x) \sum_{\text{all } y} y P_Y(y) = E[X]E[Y].$$

This theorem can be applied to the two versions of our game. In the first version, X took the values $+1$ and -1, each with probability $\frac{1}{2}$, so that

$$E[X] = 1 \times \frac{1}{2} + (-1) \times \frac{1}{2} = 0.$$

The variable Y took the values $1, 2, 3, 4, 5, 6$, each with probability $\frac{1}{6}$, so that

$$E[Y] = 1 \times \frac{1}{6} + 2 \times \frac{1}{6} + 3 \times \frac{1}{6} + 4 \times \frac{1}{6} + 5 \times \frac{1}{6} + 6 \times \frac{1}{6} = \frac{21}{6} = 3.5.$$

Therefore $E[X]E[Y] = 0$, which we have already found is the value of $E[XY]$.

In the second game, Y is unchanged, so $E[Y] = 3.5$, but now X takes the values $+2$ and -1, each with probability $\frac{1}{2}$, so $E[X] = 2 \times \frac{1}{2} + (-1) \times \frac{1}{2} = 0.5$. Therefore $E[X]E[Y] = 0.5 \times 3.5 = 1.75$, which again is the value already found for $E[XY]$.

10.3.1 Exercises

1. Find the distribution of XY, and the expected value of XY, when: (i) X and Y are as defined in Exercise 10.1.1(1); and (ii) $X = R$ and $Y = S$, as defined in Exercise 10.2.1(3).

2. Two fair coins are tossed. A random variable X is defined, which takes the value $+1$ if a head appears at the first toss and 0 if a tail appears. Similarly, Y is $+1$ if a head appears at the second toss and 0 if a tail. Tabulate the joint

distribution of X and Y, find the distribution of $X + Y$ and the distribution of XY. Find $E[X + Y]$ and $E[XY]$.

3. Two dice are thrown. The random variable X takes the value of the score on the first die, and Y takes the value on the second. Find the distribution of XY and its expected value.

10.4 THE VARIANCE OF THE SUM OF TWO INDEPENDENT RANDOM VARIABLES

A major reason why the variance is so widely used by statisticians for measuring the spread in a set of observations is that a simple relation exists for the variance of a sum of independent random variables. We now state this important result in a theorem.

Theorem 10.4 Variance of sum of independent random variables

The variance of the sum of two *independent* discrete random variables equals the sum of the variances of the two random variables:

$$\text{Var}[X + Y] = \text{Var}[X] + \text{Var}[Y]$$

provided X, Y are independent.

Let us first prove the theorem for two independent random variables whose means are zero. We write the variances of the two random variables as σ_X^2, σ_Y^2 and the variance of the sum as σ_{X+Y}^2. Recalling the general definition of variance in the form $E[X^2] - (E[X])^2$, we see that the special case when the means are zero leads to simple results. We have

$E[X] = 0, E[Y] = 0$, and so $\text{Var}[X] = E[X^2] = \sigma_X^2$ and $\text{Var}[Y] = E[Y^2] = \sigma_Y^2$.

$E[X + Y]$ is zero, being equal to $E[X] + E[Y]$. Also X, Y are independent and so (Theorem 10.3)

$$E[XY] = E[X]E[Y] = 0.$$

Therefore

$$
\begin{aligned}
\sigma_{X+Y}^2 &= E[(X + Y)^2] = E[X^2 + 2XY + Y^2] \\
&= E[X^2] + E[2XY] + E[Y^2] \\
&= E[X^2] + 2E[XY] + E[Y^2] \\
&= E[X^2] + E[Y^2].
\end{aligned}
$$

That is, $\sigma_{X+Y}^2 = \sigma_X^2 + \sigma_Y^2$.

Now, for a general proof of the theorem, write

$$E[X] = \mu_X, \quad \text{Var}[X] = E[(X - E[X])^2] = \sigma_X^2,$$
$$E[Y] = \mu_Y, \quad \text{Var}[Y] = E[(Y - E[Y])^2] = \sigma_Y^2.$$

Thus, putting $U = X - \mu_X$ and $V = Y - \mu_Y$, we have

$$E[U] = E[X - \mu_X] = E[X] - E[\mu_X] = E[X] - \mu_X = 0$$

and similarly

$$E[V] = 0.$$

Also,

$$\text{Var}[U] = E[\{U - E(U)\}^2] = E[U^2] = E[(X - \mu_X)^2] = \sigma_X^2$$

and similarly

$$\text{Var}[V] = \sigma_Y^2.$$

Finally,

$$
\begin{aligned}
E[(X - \mu_X)(Y - \mu_Y) &= E[XY - \mu_X Y - X\mu_Y + \mu_X \mu_Y] \\
&= E[XY] - E[\mu_X Y] - E[X\mu_Y] + E[\mu_X \mu_Y] \\
&= E[XY] - \mu_X E[Y] - \mu_Y E[X] + \mu_X \mu_Y \\
&= E[XY] - E[X]E[Y] = 0 \qquad \text{by independence}
\end{aligned}
$$

so that $E[UV] = 0$ and U, V are also independent.

The general proof follows the same lines as for the special case above, with U and V replacing X and Y.

$$
\begin{aligned}
\sigma_{X+Y}^2 &= E[\{(X + Y) - (\mu_X + \mu_Y)\}^2] \\
&= E[\{(X - \mu_X) + (Y - \mu_Y)\}^2] \\
&= E[(U + V)^2] \\
&= E[U^2] + 2E[UV] + E[V^2] \\
&= \sigma_X^2 + 0 + \sigma_Y^2.
\end{aligned}
$$

Therefore

$$\sigma_{X+Y}^2 = \sigma_X^2 + \sigma_Y^2.$$

This result can be generalized in similar ways to the result for $E[X + Y]$. We state a very useful theorem about variances.

Theorem 10.5

If X and Y are *independent* discrete random variables and a, b are constants, then

$$\text{Var}[aX + bY] = a^2 \text{Var}[X] + b^2 \text{Var}[Y].$$

Proof. Define two random variables U, V such that $U = aX, V = bY$. Then

$$\text{Var}[U] = \text{Var}[aX] = a^2 \text{Var}[X] \quad \text{(see p. 211)}.$$

and

$$\text{Var}[V] = \text{Var}[bY] = b^2 \text{Var}[Y].$$

Since X and Y are independent, so are U and V. Thus

$$\begin{aligned}
\text{Var}[aX + bY] &= \text{Var}[U + V] \\
&= \text{Var}[U] + \text{Var}[V] \quad \text{(Theorem 10.4)} \\
&= a^2 \text{Var}[X] + b^2 \text{Var}[Y].
\end{aligned}$$

Note that we *add* the variances σ_X^2, σ_Y^2 to obtain σ_{X+Y}^2. But we also add σ_X^2, σ_Y^2 if we are finding σ_{X-Y}^2. This can easily be seen by writing $a = 1, b = -1$ in Theorem 10.5, or by carrying through the same algebra as that involved in proving Theorem 10.4. This is such an important result that we state it as a corollary to the previous theorem.

Corollary: Variance of difference of independent random variables

If X and Y are *independent* discrete random variables, then

$$\text{Var}[X - Y] = \text{Var}[X] + \text{Var}[Y].$$

Finally, the results can be generalized to more than two independent discrete random variables. We state the simplest result, and leave the proof to the reader.

Theorem 10.6

If $A, B, C, \ldots$ are a finite set of *independent* discrete random variables, then

$$\text{Var}[A \pm B \pm C \pm \cdots] = \text{Var}[A] + \text{Var}[B] + \text{Var}[C] + \cdots.$$

10.4.1 Exercises

1. R and S are two independent random variables, with means μ_R, μ_S and variances σ_R^2, σ_S^2 respectively. Find the variance and standard deviation of:
   (i) $2R + S$; (ii) $\frac{3}{4}(R - S)$; (iii) $4S - R$.

2. Two independent measurements are taken, both being from a discrete uniform distribution with parameter $\kappa = 4$. Find the variance and standard deviation of their sum.

3. The variable U follows a binomial distribution whose parameters are $n = 4$, $\pi = \frac{1}{3}$, while the random variable V follows a binomial distribution with parameters $n = 6$ and $\pi = \frac{1}{4}$. The variables U and V are independent of one another. Find the mean, variance and standard deviation of $U + V$.

4. $\overline{R}$ is the mean of a random sample $r_1, r_2, \ldots, r_n$, from a distribution with mean μ and variance σ^2. Find the variance and standard deviation of $\overline{R}$. (Note that, in a random sample, all the r_i are independent of one another.)

10.5 EXERCISES ON CHAPTER 10

1. A sequence of five random digits is read from a table. Find the expected value and the variance of their sum.

2. If x and y are independent random single digit numbers other than zero, calculate the variance of: (i) x; (ii) $x + y$; (iii) $2x$; (iv) $2x - y$.   (*SMP*)

3. A die in the form of a prism, whose cross-section is an equilateral triangle, has its rectangular faces numbered 1, 2 and 3 respectively. When the die is rolled on a horizontal surface the score obtained is the number on the rectangular face in contact with the surface when the die comes to rest. It may be assumed that the die is perfectly balanced and such that in any roll each rectangular face has probability $\frac{1}{3}$ of being the face in contact with the surface when the die comes to rest. The die is rolled three times. Let X denote the least, and Y denote the greatest, of the three scores thrown.

   (i) Show that the joint distribution of X and Y is as displayed in the following table, in which $p = \frac{1}{27}$:

	X		
Y	1	2	3
1	p	0	0
2	$6p$	p	0
3	$12p$	$6p$	p

   (ii) Determine the expected value of X.
   (iii) State, with reasons, whether or not X and Y are independent.
   (iv) Find the probability distribution of $Z = Y - X$, and hence, or otherwise, determine the mean value of Z.   (*Welsh*)

4. X, Y are discrete random variables each taking values 0, 1, 2 with joint probability distribution function

$$P(X = x \text{ and } Y = y) = \begin{cases} k & \text{if } x \neq y \\ 0 & \text{if } x = y, \end{cases}$$

where k is a constant.
   Find the value of k, show that X, Y have the same marginal distribution function, and calculate the expected values and variances.

Find the expected value of XY, and calculate the probability that XY exceeds 1. (*IOS*)

5. X_1 and X_2 are independently distributed random variables with

$$P(X_1 = \theta + 1) = P(X_1 = \theta - 1) = \tfrac{1}{2},$$
$$P(X_2 = \theta - 2) = P(X_2 = \theta + 2) = \tfrac{1}{2}.$$

Find the values of a and b which minimize the variance of $Y = aX_1 + bX_2$ subject to the condition that $E[Y] = \theta$. What is the minimum value of this variance? (*IOS*)

6. Three fair coins are tossed simultaneously and each coin showing a head is removed. The remaining coins, when there are any, are then tossed again. Let X denote the number of heads showing on the first toss and let Y denote the number of heads showing on the second toss; Y is taken to be zero if there is no second toss.

Find the values of p, a, b, c and d in the following tabular representation of the joint probability distribution of X and Y.

	X			
Y	0	1	2	3
0	p	$6p$	$12p$	ap
1	$3p$	$12p$	bp	0
2	$3p$	cp	0	0
3	dp	0	0	0

(*Welsh*)

7. The discrete random variables X, Y have the joint distribution shown in the following table. Find the expected value and the variance of X given that $Y = 0$, and also the expected value and variance of Y given that $X = 0$:

	Y		
X	-1	0	1
-1	0.12	0.12	0.04
0	0.20	0.17	0.05
1	0.14	0.09	0.07

8. The discrete random variable R has the distribution

$$P(R = r) = kr, \quad r = 1, 2, 3, 4$$

where k is a constant. Find the value of k.

If S has the same distribution and is independent of R, find the mean and the variance of: (i) $R + S$; (ii) $R - S$.

Write down the distribution of the variate $U = RS$, and find $E[U]$ and $Var[U]$.

10.6 COMPUTING EXERCISES

1. There are two schools, A and B, in a town. A group of eight senior school-children, four from each school, is to be chosen to represent the town. Each school is asked to choose four senior pupils at random. The proportion of girls among the senior children is 0.6 at school A and 0.4 at school B.

Let the numbers of girls be X from school A, Y from school B and Z in the combined sample. Simulate the situation and obtain 500 values of each of these three variables. Verify that

$$E[Z] = E[X] + E[Y]$$

and

$$Var[Z] = Var[X] + Var[Y],$$

approximately.

Suppose that the method of choice is varied. The four children from school A are chosen at random as before; but the first choice from school B is made to be a girl if two or fewer girls were chosen from school A, and a boy if there were more than two girls chosen from school A. The remaining three children from school B are chosen at random, with the probability of obtaining a girl being 0.4 as before. As previously, label the numbers of girls X, Y from schools A, B respectively and let $Z = X + Y$.

Verify that

$$E[Z] = E[X] + E[Y]$$

but that

$$Var[Z] \neq Var[X] + Var[Y].$$

2. Simulate the game described in Section 10.3, namely that a fair coin is tossed and if it is heads, A wins, while if it is tails, B wins. A die is thrown to decide the amount of winnings in pence. Player A receives from B *twice* the number on the die if he wins; and B receives from A the number on the die when he wins.

Repeat the play 500 times; find the distribution of winnings, positive and negative, that A receives, and hence decide what A's expected winnings, per game, are. Compare the result with the theoretical value given in Section 10.3.

Assume A begins with 100 pence and B's resources are unlimited. Look at the progress of A's total gain over time. Compare with the case where each loser pays out, in pence, the number on the die.

10.7 PROJECTS

1 Joint distribution of arm length and collar size

Aim. To examine the relation between the size of shirt collar a person takes and the length of the person's arm.

Equipment. Measuring tape.

Collection of data. Define carefully what is meant by arm length: a suitable measurement is from the armpit to the underside of the wrist when the arm is held out straight. As many subjects should be measured as possible, and their collar sizes recorded at the same time. (Use the standard shirtmakers' numbering system printed on the garment.)

Analysis. Arrange in a two-way (rows and columns) table, with collar size as one classification. For the other classification, group arm-length measurements into a small number of categories (about 4 or 5). Choose the upper and lower limits of each category by looking at the range of all the arm-length measurements. Find the relative frequencies within your table, and also the marginal relative frequencies for collar size and arm length. Are the two measurements independent?

2 Goals for and against in football

Aim. To examine the relation between the goals an association football team scores and those scored against it during all the league games played in a season.

Equipment. A football league annual or other record book.

Collection of data. For each team in the football leagues whose records you have, obtain the total goals scored for and against. Classify these in a two-way table, choosing fairly broad limits for the row and column headings (e.g. units of 8 or 10 goals) by looking at the range for the whole set of data. Find the relative frequencies within the table, and in the margins, and examine whether the two records appear to be independent.

Notes. Alternative pairs of records are: (a) goals scored for, at home, and goals scored for, away; (b) goals scored in the first and second halves of each match; (c) total goals scored for, and total number of points obtained during the season, though in this case it might be necessary to restrict to clubs playing in leagues of the same size.

SUMMARY

The **means** of the **sum**, and **difference**, of the random variables X and Y are

$$E[X + Y] = E[X] + E[Y],$$
$$E[X - Y] = E[X] - E[Y].$$

The **variances** of the **sum**, and **difference**, of two *independent* random variables X and Y are

$$\mathrm{Var}[X + Y] = \mathrm{Var}[X] + \mathrm{Var}[Y],$$
$$\mathrm{Var}[X - Y] = \mathrm{Var}[X] + \mathrm{Var}[Y].$$

The **mean** of the **product** of two *independent* random variables is

$$\mathrm{E}[XY] = \mathrm{E}[X]\mathrm{E}[Y].$$

For n random variables $X_1, X_2, \ldots, X_n$,

$$\mathrm{E}[X_1 + X_2 + \cdots + X_n] = \mathrm{E}[X_1] + \mathrm{E}[X_2] + \cdots + \mathrm{E}[X_n].$$

Further, if the random variables are *independent*, then

$$\mathrm{Var}[X_1 + X_2 + \cdots + X_n] = \mathrm{Var}[X_1] + \mathrm{Var}[X_2] + \cdots + \mathrm{Var}[X_n],$$
$$\mathrm{E}[X_1 X_2 \cdots X_n] = \mathrm{E}[X_1]\mathrm{E}[X_2] \cdots \mathrm{E}[X_n].$$

11 ESTIMATION

OVERVIEW

If we use a random sample to estimate a population parameter such as a mean μ or a proportion π, how good is the estimator? We investigate the sampling distribution of the sample mean $\overline{X}$ and the sample proportion P, and find unbiased estimators.

11.1 DISTRIBUTION OF ESTIMATES OF A PROPORTION

Suppose we wish to estimate the proportion π of households that have a telephone, in a large town. There will not be time to go round to every household to find the true value of π, so, as we have seen, we shall choose a random sample consisting of n households. We shall find the number r, among this sample of n, that have a telephone and calculate the sample proportion, $p = r/n$. The value of p in the sample will serve as an estimate of π in the whole population. Obviously the actual value we obtain for p will vary according to which n households happen to make up our random sample. So we ask how good the estimates obtained by this method will be. Are the values of p likely to be close to the true π? To what extent will they vary from one sample to another?

In Chapter 8 we investigated the question of how many pairs of left-handed scissors a teacher should provide. In our present problem we can use models similar to those proposed in Chapter 8. In both cases, the population we are interested in has a fixed proportion π of members with a particular attribute (people who are left-handed, or households with a telephone), while the remainder of the population does not have this attribute. As in the earlier case, we could make a physical model using beads (white, say, to stand for 'no telephone' and red to stand for 'telephone'), take a large number of samples and set up a frequency distribution for p. An alternative is to use random number tables to construct the samples, and this is what we have done. We fixed the true proportion π at 0.4. Using digits one at a time from the table, we decided to regard '1', '2', '3' and '4' as corresponding to households that had telephones, and the other six digits to correspond to those that did not. One row of the table used contained 50 digits, so each row corresponds to a random sample consisting of 50 households. In each row, the number r of digits 1 to 4 was counted, and $p = r/50$ computed. One hundred such samples (i.e. 100 rows) gave the distribution shown in Table 11.1.

Table 11.1 *Frequency distribution of r, the number of digits 1 to 4, and of p = r/50, in 100 samples each of 50 random digits*

r	0–9	10–12	13–15	16–18	19–21	22–24	25–27	28–30	31–50	Total
p (central value)		0.22	0.28	0.34	0.40	0.46	0.52	0.58		
Frequency	0	3	9	26	31	16	9	6	0	100

This gives us a good idea of the sort of distribution of values of p that might turn up, though we would of course get a better idea by taking 1000 samples rather than 100. Using the complete data – frequencies of individual values of r rather than the grouped frequencies shown in Table 11.1 – measures of the centre of this distribution have been worked out. The mean of the 100 values of p is 0.398, the median 0.38 and the mode 0.38. The centre of the distribution is therefore close to the true $\pi(=0.4)$ for the whole population. Whether the extent of the spread among these values of p is considered satisfactory would depend on how we intended to use the estimate p. Of our values of p, 91% were in a range 0.26–0.54; if the main requirement when taking one sample is to estimate π, as wide a range as this in the likely estimates is undesirable and the method does not seem sufficiently precise. This, however, could be overcome by taking a sample size larger than the 50 we chose. Since any large town would have tens or hundreds of thousands of households, one would expect to sample considerably more than 50.

Although our simulation method has given an impression of how the estimates of p behave in this particular case, it has not really increased our knowledge of how to estimate π in general. We had to fix π before we could use the method; that is we had to *know* something about the population already. We also had to fix the sample size to be taken. So we can only look at particular cases one at a time in this way. We shall obtain a much deeper insight into the problem of estimation by setting up a theoretical model. Before doing so, it is useful to define some terms.

In the example above, we calculated p *from the sample* of random digits; in earlier chapters we have calculated the mean, or the median, or the variance, or the standard deviation, *from a sample* of measurements. We call any of these quantities a *statistic*.

Definition 11.1

A **statistic** is a quantity calculated from the observations in a sample.

If we take repeated samples, with replacement, from a population, and calculate the same statistic each time we sample, we shall build up a distribution of values of this statistic. The distribution generated in this way is called the *sampling distribution* of the statistic.

So in our example, what we in fact wish to do is to find the sampling distribution of a proportion in samples of a fixed size. Our exercise above, based on a physical model using random digits, merely gave an approximate idea of the sampling distribution in one particular case ($\pi = 0.4, n = 50$).

11.1.1 The theoretical model

In our theoretical model, we define a population in which a given proportion π have the special attribute T. (Of course we shall make this correspond to households having a telephone.) If we choose random samples of size n from this population, and count the number r in the sample which have attribute T, this r would be the same as the number of 'successes' in n independent trials when there is a probability π of 'success' on each occasion. The number r therefore has a binomial distribution whose parameters are n, π. The value of p is found by dividing r by the constant n, and so the distribution of p consists of the same set of probabilities as for r:

$$P(p = r/n) = \binom{n}{r} \pi^r (1 - \pi)^{n-r}, \qquad r = 0, 1, 2, \ldots, n.$$

This gives a *complete* specification of the sampling distribution of p, from which we can calculate the distribution for any particular case. Using binomial distribution results, the mean of p is π, since that of r is $n\pi$; and the variance of p is $\pi(1 - \pi)/n$, since that of r is $n\pi(1 - \pi)$. So the mean of the sampling distribution of p is π, as we would wish. The variance of the sampling distribution is inversely proportional to n, the sample size; so, as we should expect, the larger the sample size the smaller the spread of estimates p about the true value π. The variance of p also depends on the true value π of the proportion in the population; the relation is best shown by a graph (Fig. 11.1). The variance is greatest when $\pi = 0.5$.

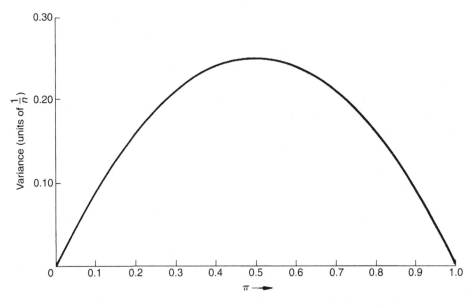

Figure 11.1 *Graph of variance of sampling distribution of p as a function of π.*

11.2 DISTRIBUTION OF ESTIMATES OF A MEAN

In the survey of journey times discussed in Chapter 3 we estimated the mean of the population of journey times by taking a random sample of journey times, and finding the mean of this sample. The reader will have used a similar procedure in many of the practical projects. We are interested in how well this procedure works, and in particular how close the sample mean is likely to be to the population mean. To appreciate this, we need to investigate how the sample mean changes from one possible random sample to another, just as we investigated the behaviour of the sample proportion. As we take each possible random sample, of fixed size n, from a population, we calculate $\bar{x}$, the mean of the sample. We wish to study the distribution of values of $\bar{x}$ generated by taking all possible random samples of n members from the population: this distribution is the *sampling distribution of the mean*.

Consider a situation in which we take samples consisting of two observations; we can make a mathematical model of this situation. We define a random variable X which has mean μ and variance σ^2. We shall assume that the two observations are chosen in sequence and that the first is labelled X_1 and the second X_2. We might, for example, be throwing a die. This corresponds to sampling from the discrete uniform distribution which takes the values 1, 2, 3, 4, 5, 6 each with probability $\frac{1}{6}$; the mean of this distribution, μ, is 3.5 and the variance, σ^2, is 2.92. We take a sample of two observations by throwing the die twice. When calculating probabilities we have seen that it is important to count separately the two throws of a die that give $(6, 1)$ and the two throws that give $(1, 6)$; we keep the first throw of the die distinct from the second throw. For a similar reason we regard the first throw of the die as generating one random variable, which we label X_1, and the second throw as generating a second random variable, which we label X_2. But, of course, the distribution of X_1 is identical with that of X_2; they both have mean μ and variance σ^2. The result of the first throw has no influence on the result of the second throw, and vice versa, so the two random variables are independent. More generally, taking *random* samples of a distribution ensures the *independence* of the sequence of random variables involved. By looking at a sample mean as if it is the mean of a set of independent random variables, we can use the results on joint distributions from Chapter 10. We assume we are sampling with replacement but if we are dealing with a large population we can treat sampling without replacement as equivalent.

The sample mean is the random variable $\overline{X} = \frac{1}{2}(X_1 + X_2)$. So, using Theorem 10.1,

$$\mathrm{E}[\overline{X}] = \mathrm{E}[\tfrac{1}{2}(X_1 + X_2)] = \tfrac{1}{2}\mathrm{E}[(X_1 + X_2)] = \tfrac{1}{2}\mathrm{E}[X_1] + \tfrac{1}{2}\mathrm{E}[X_2] = \tfrac{1}{2}(\mu + \mu) = \mu.$$

This looks a very reasonable result. For the die example it is saying that the sampling distribution of the mean score, from two throws of a die, has a mean of 3.5, the same as the mean score obtained from throwing a die only once.

The result for the variance does not look quite so obvious: using Theorem 10.5 we find

$$\text{Var}[\overline{X}] = \text{Var}[\tfrac{1}{2}(X_1 + X_2)] = \tfrac{1}{4}\text{Var}[(X_1 + X_2)]$$
$$= \tfrac{1}{4}(\text{Var}[X_1] + \text{Var}[X_2]) = \tfrac{1}{4}(\sigma^2 + \sigma^2) = \tfrac{1}{2}\sigma^2.$$

For the die example, we see that the variance of the sampling distribution of the mean score from two throws of a die is one-half of the variance of the score obtained from a single throw of a die, i.e. it is $\tfrac{1}{2}(2.92) = 1.46$. This illustrates a very useful property of the distribution of the mean, namely that the variance of the distribution of the mean of n observations is smaller than the variance of a single observation. We prove this below. So the distribution of the mean has smaller spread than the distribution of single observations, but both distributions have the same expected value μ.

11.3 THE MEAN AND VARIANCE OF $\overline{X}$

Write

$$\overline{X} = \frac{1}{n}(X_1 + X_2 + \cdots + X_n)$$

for the mean of a random sample of n observations, all drawn from the same distribution which has mean and variance μ and σ^2. Applying results for several independent random variables (pp. 236, 242), and calling the ith variable drawn X_i, we have

$$E[\overline{X}] = E\left[\frac{1}{n}(X_1 + X_2 + \cdots + X_n)\right] = \frac{1}{n}E[(X_1 + X_2 + \cdots + X_n)]$$

$$= \frac{1}{n}(E[X_1] + E[X_2] + \cdots + E[X_n])$$

$$= \frac{1}{n}(\mu + \mu + \cdots + \mu) = \mu.$$

$$\text{Var}[\overline{X}] = \text{Var}\left[\frac{1}{n}(X_1 + X_2 + \cdots + X_n)\right] = \frac{1}{n^2}\text{Var}[X_1 + X_2 + \cdots + X_n]$$

$$= \frac{1}{n^2}(\text{Var}[X_1] + \text{Var}[X_2] + \cdots + \text{Var}[X_n])$$

$$= \frac{1}{n^2}(\sigma^2 + \sigma^2 + \cdots + \sigma^2) = \frac{1}{n^2}n\sigma^2 = \frac{\sigma^2}{n}.$$

These two results are extremely important, and we shall use them over and over again. They apply not only to discrete variables but also to continuous random variables, which we shall use to model continuous variates.

Note that we have said nothing specific about the form of the random variable X, or what distribution it might follow. We have specified only its mean μ and its variance σ^2. By implication, in the calculations we have done above, μ and

σ^2 must both be finite, which does not seem to be asking much of a distribution that is going to be useful in a mathematical model of real data!

When we *do* know the precise distribution of X and $\overline{X}$, our knowledge about that population is complete. Often, however, we may not know a satisfactory distribution to suggest for X; and, as we shall see, it is also quite possible to know about X and not be able to work out the distribution for $\overline{X}$ simply. We shall also see later (Chapter 16) that we can often say what type or *family* of distributions approximates that followed by $\overline{X}$; thus knowing its mean (μ) and its variance (σ^2/n) will allow us to obtain a very good approximation to the actual distribution of $\overline{X}$.

The standard deviation of the sampling distribution of a statistic such as the mean or a proportion is often quoted as a measure of the reliability of the statistic. This standard deviation is therefore given a special name, the *standard error*.

Definition 11.2

The **standard error** of a statistic is the standard deviation of the sampling distribution of that statistic.

The standard error of the mean is of particular interest. We have shown above that $\mathrm{Var}(\overline{X}) = \sigma^2/n$, and so the standard error of the mean in a random sample of size n is $\sigma/\sqrt{n}$.

11.3.1 Exercises

1. (i) Find the mean μ and variance σ^2 of the random variable that takes the values 1 and 2, each with probability $\frac{1}{2}$.
   (ii) List all possible samples of size 3 that may be chosen, with replacement, from this population (remember that $(1, 1, 2)$ and $(1, 2, 1)$ are distinct samples) and hence construct the sampling distribution of the mean of these samples. Find the mean and variance of the sampling distribution and verify that they equal μ and $\frac{1}{3}\sigma^2$ respectively.

2. Show that if a random sample of size n is drawn from a distribution with mean μ and variance σ^2, then the sample total has a sampling distribution which has mean $n\mu$ and variance $n\sigma^2$. (Use E[.], the expectation operator, to show this.) Verify the result from the data on samples assembled in Question 1(ii).

3. (i) Find the mean μ and variance σ^2 of random digits (i.e. the random variable that takes the values $0, 1, 2, \ldots, 9$ each with probability 0.1).
   (ii) List all possible samples of size 2 that may be chosen, with replacement, from this population, and hence construct the sampling distribution of the mean of these samples. Find the mean and variance of the sampling distribution.
   (iii) Use a table of random digits to choose 100 samples of size 2 from the population (i.e. record 100 successive pairs of values from the table). Find the mean of these samples and construct the frequency table of this distribution. Compare the result with the theoretical values determined in (ii).

11.4 UNBIASED ESTIMATORS

Earlier in the chapter we examined situations in which we used a statistic (the sample proportion or the sample mean) in order to produce an estimate of a population parameter. When investigating estimation procedures, it is useful to define two terms, *estimator* and *estimate*. The first of these refers to a statistic when it is considered as a mathematical function of the sample observations, e.g. the mean of two observations is $(X_1 + X_2)/2$; this expresses the general relation between any two observations X_1, X_2 and the mean derived from them. On the other hand, an estimate is the particular value which the estimator takes in a set of data, e.g. $(6+4)/2 = 5$.

Definition 11.3

An **estimator** is a statistic used to estimate a population parameter; it is a function of the sample observations.

Definition 11.4

An **estimate** is the value an estimator takes for a particular sample.

We have seen in this chapter that we can investigate the properties of a statistic by looking at its sampling distribution. In what senses may a statistic be considered a *good* estimator, or not? What properties would we like the sampling distribution to have?

We would obviously like the sampling distribution to cluster around the true value of the parameter. A reasonable, and mathematically convenient, way of expressing this is to say that we would like the mean of the sampling distribution to equal the population parameter we are estimating. For our particular examples we would like $E[p] = \pi$, and $E[\overline{X}] = \mu$. If this property does not hold we say we have a biased estimator of the parameter; the estimates are 'out of true' because they do not centre on the true value. The statistics that we shall use in this book for estimating parameters will all be unbiased estimators.

Definition 11.5

An estimator is **unbiased** if the mean, or expected value, of its sampling distribution equals the parameter it is estimating.

We may find an estimator that is unbiased but leads to estimates that are very widely spread; such an estimator is of little use. For an unbiased estimator to be regarded as 'good', we require also that the variance of its sampling distribution – and hence its standard error – shall be small.

The first result in Section 11.3 shows that $\overline{X}$ is an unbiased estimator of μ. Also p is an unbiased estimator of π (see p. 53).

11.5 UNBIASED ESTIMATION OF VARIANCE

If we record the values x_i $(i = 1, 2, \ldots, n)$ when we take a random sample of the variable X, which has unknown mean μ and unknown variance σ^2, then an obvious way of estimating σ^2 is to use the variance of the set of observations $\sum_{i=1}^{n} (x_i - \bar{x})^2 / n$. Is this an unbiased estimator of σ^2?

The major step in answering this question is to find the expected value of the sum of squares, i.e. $\mathrm{E}[\sum_{i=1}^{n} (x_i - \bar{x})^2$. Before we evaluate this we shall express the sum of squares in a more convenient form. We have the identity

$$\sum (x_i - \mu)^2 \equiv \sum \{(x_i - \bar{x}) + (\bar{x} - \mu)\}^2$$
$$\equiv \sum (x_i - \bar{x})^2 + 2 \sum (x_i - \bar{x})(\bar{x} - \mu) + \sum (\bar{x} - \mu)^2$$
$$\equiv \sum (x_i - \bar{x})^2 + \sum (\bar{x} - \mu)^2$$

since

$$\sum (x_i - \bar{x})(\bar{x} - \mu) = (\bar{x} - \mu) \sum (x_i - \bar{x}) = 0.$$

It follows therefore that

$$\sum (x_i - \bar{x})^2 \equiv \sum (x_i - \mu)^2 - \sum (\bar{x} - \mu)^2$$
$$\equiv \sum (x_i - \mu)^2 - n(\bar{x} - \mu)^2.$$

We shall also use the facts that

$$\mathrm{E}[(x_i - \mu)^2] = \sigma^2 \quad \text{(Section 9.5, p. 208)}$$

and, because $\mathrm{E}[\bar{x}] = \mu$,

$$\mathrm{E}[(\bar{x} - \mu)^2] = \mathrm{E}[\{\bar{x} - \mathrm{E}(\bar{x})\}^2] = \mathrm{Var}[\bar{x}] = \sigma^2 / n \quad \text{(Section 11.3)}.$$

It therefore follows that

$$\mathrm{E}\left[\sum (x_i - \bar{x})^2\right] = \mathrm{E}\left[\sum (x_i - \mu)^2 - n(\bar{x} - \mu)^2\right]$$
$$= \sum \mathrm{E}[(x_i - \mu)^2] - n\mathrm{E}[(\bar{x} - \mu)^2]$$
$$= n\sigma^2 - n(\sigma^2 / n) = n\sigma^2 - \sigma^2 = (n - 1)\sigma^2.$$

We see that

$$\mathrm{E}\left[\sum (x_i - \bar{x})^2 / n\right] = \frac{(n - 1)}{n} \sigma^2$$

and that $\sum (x_i - \bar{x})^2 / n$ is a biased estimator of σ^2. We therefore define the *estimated variance* as $s^2 = \sum (x_i - \bar{x})^2 / (n - 1)$, to make s^2 an unbiased estimator of σ^2, i.e. $E[s^2] = \sigma^2$. The necessity to divide the corrected sum of squares by $(n-1)$, rather than by n, derives from the fact that μ is unknown. If we *knew* μ then we would estimate σ^2 by $\sum_{i=1}^{n}(x_i - \mu)^2/n$. This would be an unbiased estimator of σ^2 because

$$E\left[\sum(x_i - \mu)^2/n\right] = \frac{1}{n}E\left[\sum(x_i - \mu)^2\right] = \frac{n\sigma^2}{n} = \sigma^2.$$

The divisor $(n-1)$ is often called the *degrees of freedom* on which the estimated variance is based. If μ is known then the estimator $\sum(x_i - \mu)^2/n$ is based on n degrees of freedom; it is proportional to the sum of squares of n *independent* deviations $(x_i - \mu), i = 1, 2, \ldots, n$. But if we are obliged, as we usually are, to estimate μ from the sample, the estimated variance s^2 is proportional to the sum of squares of n deviations $(x_i - \bar{x}), i = 1, 2, \ldots, n$. These are *not* independent because they sum to zero as a result of the fact that $\sum(x_i - \bar{x}) = 0$: if we are told $n-1$ of the deviations we can find the remaining deviation. The number of degrees of freedom is decreased by one because of this limitation, and hence s^2 is based on $n-1$ degrees of freedom.

11.5.1 Exercises

1. Suppose I choose a random sample of three observations from a population and obtain the values $2, 5, 3$. From these values I estimate the centre of the population by ranking the observations and taking the middle one. What *estimator* am I using, and what is my *estimate*?

2. A random sample of n observations, x_i $(i = 1, 2, \ldots, n)$, is chosen from a population with unknown mean μ and unknown variance σ^2. From this sample $\bar{x} = \sum x_i/n$ and $s_X^2 = \sum(x_i - \bar{x})^2/(n-1)$ are calculated. Another random sample of m observations, y_i $(i = 1, 2, \ldots, m)$, is chosen and similar quantities $\bar{y}$ and s_Y^2 calculated. These two sets of observations are then pooled and the mean μ is estimated by

$$\frac{n\bar{x} + m\bar{y}}{n + m}$$

and the variance σ^2 by

$$\frac{(n-1)s_X^2 + (m-1)s_Y^2}{n + m - 2}.$$

Show that these two estimators are unbiased estimators of μ and σ^2 respectively.

3. A random sample of n observations is taken from a population with mean μ. The estimate of the population mean is derived from the first $n/2$ of these observations by taking their mean. This estimator is an unbiased estimator of μ, as would be the estimator based on the whole n observations. In what way is the former estimator a poorer estimator than the latter?

11.6 MINITAB: SAMPLE VARIANCE

The variance of a set of data, calculated by MINITAB, is the sum of squares of deviations about the mean divided by $(n-1)$, where n is the number of observations. More specifically, it may be called the estimated (population) variance or the sample variance. As we have shown in this chapter, it is an unbiased estimator of the population variance and therefore a good statistic to use. This is the statistic that is included in those produced by choosing **Stat**>**Basic Statistics**>**Display Descriptive Statistics** (or the session command DESCRIBE) and in all cases where MINITAB uses the term 'variance' of a set of data. 'Standard deviation' in MINITAB is the square root of this variance.

11.7 EXERCISES ON CHAPTER 11

1. An integer variable takes the value r with probability p_r defined as

$$p_r = \begin{cases} \alpha r & r = 1, 2, 3; \\ \alpha(8 - r) & r = 4, 5, 6, 7; \\ 0 & \text{all other } r. \end{cases}$$

   Determine the value of the constant α and obtain the mean and variance of r. Write down
   (i) the mean and variance of $2r - 4$;
   (ii) the mean and variance of $2r_1 + 3r_2$, where r_1 and r_2 are independent observations of r. (*Cambridge*)

2. State briefly what is meant by *mathematical expectation*. A variable x is distributed with mean μ and variance σ^2. Random samples of size n are drawn from this population. Show that the sample means $\bar{x}$ are distributed with mean μ and variance σ^2/n.

   A bag contains 4 black and 7 white balls. Random samples of 5 balls are drawn from the bag and replaced for subsequent sampling. If x denotes the number of black balls drawn per sample, find the probability distribution of x, and hence find the expected value and the variance of x. (*AEB*)

3. A cubical die has two faces numbered 1, two numbered 2 and two numbered 3, and is such that when it is rolled the probability of scoring 1 is $\frac{1}{4}$, of scoring 2 is $\frac{1}{2}$ and of scoring 3 is $\frac{1}{4}$. If the die is rolled three times show that the probability that the smallest score will be 1 and the largest will be 3 is equal to $\frac{9}{32}$.

   The average of the smallest and the largest values in a sample of n values is called the *mid-range*. Find the sampling distribution of the mid-range of the scores obtained when the above die is rolled three times. In this case show that the mid-range is an unbiased estimator of the mean score per throw, and find the ratio of its sampling variance to that of the mean of the three scores. (*Welsh*)

4. Explain what is meant by the *expectation* of a random variable.

A random sample is drawn from a large population in which a proportion p have a certain rare disease. Sampling continues until a predetermined number a of the sample are found to have the disease, and at this stage the sample size is r. Find the probability distribution of r, and show that $(a-1)/(r-1)$ is an unbiased estimate of p, i.e. that

$$E[(a-1)/(r-1)] = p. \qquad (Oxford)$$

5. If the random variables $x_1, x_2, \ldots, x_n$ are independent and have variances $\sigma_1^2, \sigma_2^2, \ldots, \sigma_n^2$, show that the mean

$$\bar{x} = (x_1 + x_2 + \cdots + x_n)/n$$

has variance

$$(\sigma_1^2 + \sigma_2^2 + \cdots + \sigma_n^2)/n^2.$$

An experimenter takes the successive readings of an instrument, but owing to fatigue the variance of his readings increases linearly with each observation, so that $\sigma_r^2 = A + Br$. Find the variance of the mean of n successive readings. Show that this can never be less than a certain value however many readings he takes.

Discuss what would happen if fatigue made the experimenter's variance increase even more rapidly with each successive observation. $\qquad (SMP)$

6. (a) An urn contains 3 black balls and 2 red balls. Two balls are chosen at random from the urn (i) simultaneously, (ii) in succession, without the first ball being replaced. Show that in each case the probability of obtaining one ball of each colour is 0.6.

(b) An urn contains N balls of which ρ are red $(0 \leqslant \rho \leqslant N)$, and the remainder are black. Show that the probability that a random sample of n balls will contain r red balls is

$$\frac{\binom{\rho}{r}\binom{N-\rho}{n-r}}{\binom{N}{n}}, \qquad r = 0, 1, \ldots, n.$$

7. The *hypergeometric distribution* has probability mass function

$$P(R = r) = \frac{\binom{\rho}{r}\binom{N-\rho}{n-r}}{\binom{N}{n}}, \qquad r = 0, 1, \ldots, n.$$

where ρ, n and N are integers such that $0 \leqslant \rho \leqslant N, 0 < n \leqslant N$ (compare Exercise 6(b)).

By writing

$$rP(R = r) = \frac{n\rho}{N} \frac{\binom{\rho - 1}{r - 1} \binom{N - 1 - (\rho - 1)}{n - 1 - (r - 1)}}{\binom{N - 1}{n - 1}}$$

show that $E[R] = n\rho/N$. Find a similar expression for $r(r - 1)P(R = r)$ and hence show that

$$\text{Var}[R] = \frac{n\rho(N - n)(N - \rho)}{N^2(N - 1)}.$$

8. A sample of n people is chosen at random from a population of size N in which the proportion of people with red hair is π. The number of people with red hair in the sample is r and the proportion of people with red hair in the population is estimated by r/n. Show that:
   (i) if sampling is with replacement then the estimated proportion r/n has probability mass function

   $$\binom{n}{r} \pi^r (1 - \pi)^{n-r};$$

   (ii) if sampling is without replacement then the estimated proportion has probability mass function

   $$\frac{\binom{N\pi}{r} \binom{N(1 - \pi)}{n - r}}{\binom{N}{n}}$$

   (iii) Using the result of Question 7, show that the variance of the random variable in (ii) tends to $\pi(1 - \pi)/n$ as $n \to \infty$ and π remains constant. Finally, show that the variance of the estimated proportion based on sampling without replacement (as in (ii)) is approximately equal to that based on sampling with replacement (as in (i)) provided the population is large.

11.8 COMPUTING EXERCISES

1. Let X have a discrete uniform distribution in which the ten digits 0–9 are equally probable. Generate a random sample of 500 values from the distribution.

   Generate random samples of size k from X, find the mean and repeat 500 times to generate the distribution of the mean. Carry this out for $k = 4, 9, 16, 25$ and call the variables generated M4, M9, M16, M25 respectively.

   Compare the five distributions using histograms, means and standard deviations.

2. Carry out Project 11.9.1 by computer simulation.

1 Sampling distribution of the estimate of a proportion

Aim. To estimate the proportion of beads of a given colour in a large population of beads, and to see how this estimate varies from one sample to another.

Equipment. A large quantity of beads of two colours, of the same size and shape.

Collection of data. Mix the beads thoroughly and, without looking at the colours, take a random sample of 20 beads. Count the number of (say) white beads in these 20. Put the 20 beads back into the population, mix again and repeat the sampling. Do this as many times as possible. The same person may repeat, and as many people as possible should take part.

Analysis. In each sample of 20, calculate p, the proportion of white beads (or other chosen colour). Combine these results to give a single estimate ($\hat{\pi}$) of the proportion based on all the samples. Work out the variance of the estimates p, and compare this with $\hat{\pi}(1 - \hat{\pi})/20$. Find how many of the values of p were less than two standard deviations away from $\hat{\pi}$.

Notes. (1) If the true value of π in the whole population is known, use it instead of $\hat{\pi}$.

(2) Any other available material may be used instead of beads, provided it contains two visibly distinguishable types that are physically the same (otherwise sampling is unlikely to be random).

Data. The data referred to in Table 8.1 can be used as an example. Here n, the sample size, was 30, and 500 such samples were taken. The numbers of blue beads in samples of 30 were counted. The samples were grouped into 10 fifties and a summary of the numbers of blue beads in each group of 50 samples is as follows:

Number blue, r	0	1	2	3	4	5	6	7	8	>8
Frequencies of r										
First 50 samples	2	6	8	14	11	4	1	3	1	0
Second 50 samples	0	11	10	14	7	4	2	2	0	0
Third 50 samples	1	5	14	12	9	5	3	1	0	0
Fourth 50 samples	5	6	10	11	12	3	3	0	0	0
Fifth 50 samples	2	8	13	12	9	3	3	0	0	0
Sixth 50 samples	0	8	13	15	9	2	2	1	0	0
Seventh 50 samples	1	3	15	8	11	8	2	2	0	0
Eighth 50 samples	2	4	9	10	9	11	2	2	1	0
Ninth 50 samples	4	4	8	13	10	6	3	2	0	0
Tenth 50 samples	4	4	8	9	10	8	3	2	2	0

Note that the true value of π in this experiment was 0.1.

2 Sampling distribution of an estimate of a mean

Aim. To estimate the mean length of pencils, and to see how much the estimate varies from one sample to another.

Equipment and collection of data. See Project 3.12.1. Use method (b) to collect the data. Measure also the lengths of all 40 pencils.

Analysis. Compute the mean, μ, and variance, σ^2, in the whole population of 40 pencils. Compute $\bar{x}$ and s^2 for each sample of five. Compare the average value of $\bar{x}$, for all samples, with μ. Work out the variance of $\bar{x}$ from all samples and compare it with $\sigma^2/5$. Compare the average value of s^2, for all samples, with σ^2.

3 Distribution of prime numbers

Aim. To examine how the proportion of prime numbers between N and $N + 99$ behaves as N increases.

Equipment. Table of primes or a computer program for their generation.

Collection of data. Find the number of prime numbers between 1 and 100, 101 and 200, 201 and 300, ...

Analysis. Arrange the data in a table showing p, the proportion of prime numbers between N and $N + 99$, for values of N increasing in steps of 100. Illustrate the results on a graph, and consider:

 (i) whether p becomes stable for large N, and whether it behaves randomly or not in succeeding hundreds;
(ii) whether p has a limit as $N \rightarrow \infty$.

SUMMARY

Sampling distribution of a proportion. Let a random sample of size n be chosen from a population in which a given proportion π have a particular attribute. Let a number r in the sample have the attribute. Then

$$P(P = r/n) = \binom{n}{r} \pi^r (1 - \pi)^{n-r}, \quad r = 0, 1, 2, \ldots, n.$$

The sampling distribution of the proportion P has mean value π and variance $\pi(1 - \pi)/n$.

    Sampling distribution of a mean. The sampling distribution of the mean of a random sample of n observations, drawn from a population with mean μ and variance σ^2, has mean μ and variance σ^2/n.

    The **standard error** of a statistic is the standard deviation of the sampling distribution of that statistic.

    An **estimator** is a statistic used to estimate a population parameter; it is a function of the sample observations. An **estimate** is the value an estimator takes for a particular sample. An estimator is **unbiased** if the mean, or expected value, of its sampling distribution equals the parameter it is estimating.

Let a random sample $X_1, X_2, \ldots, X_n$ be chosen from a population with unknown mean μ and unknown variance σ^2.

$$\overline{X} = \sum X_i/n \quad \text{is an unbiased estimator of the mean } \mu.$$

$$S^2 = \sum (X_i - \overline{X})^2/(n-1) \quad \text{is an unbiased estimator of the variance } \sigma^2.$$

12 COLLECTING DATA

OVERVIEW

Data may be collected by various sampling methods, or by experiments. In this chapter we study some of the ways of achieving data that are reliable and of good quality.

12.1 INTRODUCTION

The method used to collect data affects enormously what we can infer from data. We have seen, for example, that if we estimate the proportion of left-handed people in the population by finding the proportion of left-handed people in a random sample we do not know exactly how close our estimate is to the true proportion. We do know quite a lot about the range of possible estimates that we might obtain if we took a large number of random samples (the sampling distribution), and we know that the possible estimates centre upon the true proportion (an *unbiased* estimator). But if the sample is not chosen at random we can say virtually nothing. In this chapter we shall look at the methods by which data are collected. We shall begin by adding to what we have already said about surveys. Later in the chapter we shall look at one of the other major methods by which data are collected, namely from experiments, and we have already discussed the third major method, simulation, in Section 7.10.

12.2 SURVEYS

12.2.1 Questionnaires

In industry, business, agriculture and many other areas of official inquiry, information is collected by sending people a form or *questionnaire*, and asking them to fill it in and return it. Farmers will be asked what area of land they are using to grow wheat this year, or how many dairy cows they are keeping; how many people they employ, or how much machinery they own. The questions on a questionnaire must be asked *clearly* and *unambiguously*, with the least possible scope for misunderstanding and, therefore, for wrong answers. If a farmer is asked 'How many animals do you own?', he will no doubt include the farm dog, if he remembers! 'How many dairy cattle?', 'How many are giving milk at present?', 'How many calves?', are precise questions, and it is clear what answer is needed.

If the same farmer is asked how many gallons of milk he sold last week, he will have the records at hand, and will be able to answer reliably. But if we ask him how many gallons he sold in the last year, of course he cannot remember, nor is he going to waste his time looking it up – it is not a good question, because it covers *too long a time scale*. Even if he gives an answer, it will be a guess, and so an inaccurate piece of data; it would be better to have no data at all.

A serious problem involving questionnaires sent by post is that they are not always returned! *Non-response* is not usually randomly spread over all the people who were sent the questionnaire. Thinking of the farmer again, if his is a fairly large business he will have someone to do his office work, and the form will get filled in and sent back quickly. But if he does his own paperwork, he may leave the form on one side to deal with more urgent work, and by the time he gets a spare moment the inquiry is finished. Small farmers are different in many ways from large, and it would be quite wrong to ignore non-response, especially if it affected one section of the population more than others. Next time you look at the results of a survey, find out what the response rate was. If it was low, you ought to suspect that only one section of the target population actually responded; they might easily have been those who felt most strongly about the questions being asked, or those who held one particular view. This view is then over-represented in the sample.

In postal surveys, 70% is usually considered quite a good response rate. Instead of sending questionnaires by post, a person conducting a survey may often go and interview the people selected for his sample. When he does this, he hopes to minimize non-response. This time the problem is reduced to those who refuse to reply and those who are not at home. If an area is surveyed too often, about topics that people do not find at all interesting, they will cease to cooperate. Beware of multiplying surveys in areas that are easy to get to! And always explain, briefly and clearly, the aim of the survey before asking questions. As regards people not being at home, a second visit is needed: they may be out because either their leisure activities or their working habits differ from those of their neighbours, so that they form a distinct section of the population which it is wrong to ignore. When the survey is carried out in this way, the *interviewer* will usually fill in the form. It is necessary to take great care to ask questions in a completely neutral way, not to load or *bias* them so that the interviewer almost tells the sample member what answer is expected. Interviewing is a skill that has to be learnt. The *wording of questions* on any questionnaire, whether administered by an interviewer or filled in directly by the person to whom it is sent, is very important. Bias is so easily caused by bad wording that this is a serious criticism of many social surveys.

The method of *quota sampling* is often used to overcome difficulties of non-response and inability to get hold of people selected for a sample. In this method, an interviewer is told to find so many housewives, so many retired people, so many business men, so many manual workers, etc. This may appear to be a very good way of making sure all the groups are represented properly (see the notes below on stratified sampling), but its weakness lies in the non-random selection of individuals within the groups. Those selected will simply be the easiest to get at, and the method has no sound statistical properties. The fact that it has worked pretty well in opinion surveys on limited topics, in the hands of experienced survey workers, does *not* mean that it is wise to use quota sampling for a new problem handled by an unskilled survey worker.

12.3 STRATIFIED RANDOM SAMPLING

Suppose we want to estimate the mean number of pairs of shoes that adults have. From our earlier discussion of sampling we know that an appropriate method is to choose a random sample of the population under study (perhaps the adult inhabitants of a large town), record the number of pairs of shoes of each member of the sample, and find the mean. If the sample is of size n, and the variance of number of pairs of shoes in the population is σ^2, then the standard error of the sample mean will be $\sigma/\sqrt{n}$; this gives us an indication of how reliable an estimate we can obtain. But the population variance σ^2 is likely to be large. There may be in the population poor people with only one pair of shoes, and, at the other extreme, fashion-conscious women with thirty pairs. In fact, if we think a little about how many pairs of shoes are likely to be possessed by different types of people we realize we could probably split up the inhabitants of a town into subpopulations, or *strata*. We might expect the members of one particular stratum to have similar numbers of pairs of shoes to one another, while there would be substantial differences between members of different strata. Thus richer people are likely to have more shoes than poorer people, and it is probably true that women have more shoes than men. We therefore might divide the inhabitants of a town according to the district in which they live, which is often related to wealth and is in any case a convenient grouping for an investigator to use, and by sex. If four districts were chosen, and men and women were kept separate in each district, this would give eight groups into which the population would be divided. We might well expect the people in one of these subpopulations to be very much more like one another than people picked from the whole population. So if we estimated the mean of each subpopulation, or stratum, separately and then combined these estimates to give an overall mean, the standard error of that mean would be much less than that obtained by using a simple random sample. Let us look at a simple numerical example to make the point more forcibly.

We shall assume that we are interested in a population which splits up naturally into just two strata: men and women. The distribution of the number of pairs of shoes among men is:

r	3	4	5
p_r	$\frac{1}{3}$	$\frac{1}{3}$	$\frac{1}{3}$

This subpopulation has a mean of 4 and a variance of 2/3. The distribution of pairs of shoes among women is:

r	7	8	9
p_r	$\frac{1}{3}$	$\frac{1}{3}$	$\frac{1}{3}$

This subpopulation has a mean of 8 and a variance of 2/3. If there are equal numbers of men and women in the population then the distribution of the number of pairs of shoes in the whole population is:

r	3	4	5	7	8	9
p_r	$\frac{1}{6}$	$\frac{1}{6}$	$\frac{1}{6}$	$\frac{1}{6}$	$\frac{1}{6}$	$\frac{1}{6}$

The mean of this population is 6 and the variance is 28/6 or 4.67.

If we took a simple random sample of 50 from the whole population the standard error of the mean would be $\sigma/\sqrt{n}$ or $\sqrt{4.67/50} = 0.306$. If, instead, we took a simple random sample of 25 men and another of 25 women, we would obtain two means $\overline{X}_m, \overline{X}_w$ each with a standard error of $\sqrt{(2/3)/25} = \sqrt{2/75} = 0.163$. Our estimate of the mean number of shoes in the whole population would be $\frac{1}{2}(\overline{X}_m + \overline{X}_w)$. To obtain the standard error of this mean we first find

$$\text{Var}\left[\tfrac{1}{2}(\overline{X}_m + \overline{X}_w)\right] = \frac{1}{4}(\text{Var}[\overline{X}_m + \overline{X}_w]) = \frac{1}{4}(\text{Var}[\overline{X}_m] + \text{Var}[\overline{X}_w])$$

$$= \frac{1}{4}\left(\frac{2}{75} + \frac{2}{75}\right) = \frac{1}{75}.$$

Hence the standard error of the mean, using the stratification method, is $\sqrt{1/75} = 0.115$ which is about one third the size of the standard error of the mean obtained by simple random sampling. We have therefore improved the accuracy of the mean substantially by the use of stratification.

Another virtue of stratification is that we can ensure that we obtain sufficiently large samples from each stratum so that we can compare the strata. This may be interesting of itself, but it also might lead us to conclude that the strata differ so much from one another that it is necessary to quote a mean for each stratum; one overall mean for the whole population may give very little information.

12.4 EXPERIMENTS

The discussion, in Chapter 1, of the effect of women's smoking habits on the weight of their babies was based on survey data. We pointed out that one could not be sure that smoking was causing the difference in birth-weight. There might be some other factor associated with women who chose to smoke, and this might have been the effective factor, so that these women might have had babies with similar weights even if they had not smoked. If the data had been collected in an *experiment* and the investigator had chosen two similar groups of women and told one group to smoke and the other group not to smoke, then we might be confident that any difference in babies' birth-weight was a result of the mothers smoking. Basically an experiment differs from a survey in that the investigator has much more control; instead of simply selecting which data are to be collected in a given situation he actually creates the conditions he wants to investigate.

Experiments, if they are possible, are much better than surveys if the aim is to establish what *causes* particular effects.

Experimentation with human beings may often be difficult or impossible. One cannot simply tell people what to do. Moreover, if people are to be asked to volunteer for an experiment it must be made clear what the consequences are, and no unfair pressure must be put upon people to take part. An investigator trying out a new medical treatment, such as a new drug, may wish to include in the experiment, for scientific comparison purposes, a treatment that he would not use if he were acting solely as a doctor and not also as an investigator. But he is not justified in using such an experimental treatment, because it is not in the best interests of all the patients in the trial. The reader will see that there are substantial ethical questions in medical experimentation; there are strict safeguards over such experiments.

Experimentation is an essential part of the scientific method and the number of experiments conducted has increased with the rise of science. But it was not until the 1920s that serious study was begun, by R. A. Fisher (British, 1890–1962), of how best to do experiments. He was concerned with agricultural field trials, and the materials – plants and soil – available for these were often highly variable. The new ideas met with great success in agricultural and biological research; they have since also been applied in industrial and medical research, and to a lesser extent in other fields of study.

When planning an experiment the investigator must have a clear idea what his aim is. The next step is to decide the experimental *treatments*. This word is obviously an appropriate term if we are comparing, say, the effects of different fertilizer treatments on cabbage plants or blackcurrant bushes; it has come to be used as a technical term standing for *any* two or more things that are being compared in an experiment. We would call the different breeds of dog the 'treatments' if we were comparing the growth of terriers and dachshunds given similar diets. Next the *experimental units* must be chosen. These might be *plots* of land in a field, in an agricultural experiment, and 'plot' is often used as a convenient alternative term to 'experimental unit' in general. Thus 'plots' may be the individual patients of a doctor, or the individual bars of metal that a machine acts on, or mixtures of chemicals in individual test tubes. Next, suitable variates must be chosen, to measure the *response* of the experimental units to the treatments; the measurements might be yield of wheat in kilograms, or number of headaches a patient has per day, or the length of a metal bar. A plan must be prepared to show which treatment each plot receives; this is called a *layout*. The ideas discussed in the next section, which are often called the principles of experimental design, are concerned with the choice of layout.

12.5 BASIC PRINCIPLES OF EXPERIMENTAL DESIGN

Suppose a manufacturer of sticking plaster, used for covering minor wounds and cuts, produces a new type that he thinks is better than his current product. How does he test whether it really is better? Let us suppose that the manufacturer believes that the fabric used in the new type of plaster allows the wound to heal up more quickly. Obviously comfort when the sticking plaster is being worn, and ease of removal, would also be relevant. Thus one might test the plaster, using people suffering minor cuts or wounds as subjects. One would record how many days it takes a wound to heal, using the plaster, and ask the subjects to comment on how comfortable it is, and whether it is painful when the plaster is removed.

12.5.1 Replication

One might compare the new plaster with the standard type by finding two people with similar cuts, using the new type of plaster with one subject and the standard type with the other subject. Suppose the wound on the subject who was given the new type of plaster healed two days before that on the subject with the standard plaster. Is this an adequate test? It is not, because it is easy to argue that this result does not mean very much. It could be said that the subject who was given the new plaster is a person who heals quickly, and the speedier healing was a characteristic of the person rather than of the plaster. And there is likely to be such a huge variation in the speed of healing from person to person and from wound to wound, even using the same plaster, that this criticism would have to be accepted. Each type of plaster should be tried on a large number of people and the distributions of healing times obtained with the two plasters compared. The difference between the healing times with the two plasters must be assessed against the natural variability of healing times of similarly treated subjects. The use of two or more experimental units for each experimental treatment is called *replication*. One basic principle of experimental design is that the experimental treatments should be replicated.

12.5.2 Randomization

The manufacturer might give samples of his products to a busy first aid post. Suppose they used up the box of standard plasters first and then used up the box of new plasters. (We shall assume they are able to see the patients later and find out how many days the wounds take to heal.) The wounds of patients with the new plasters might heal more quickly, but it could be objected that perhaps all the patients with particularly bad wounds came to the post first and the comparison is not a fair one. In order to ensure that both treatments are being tested under similar conditions, and to demonstrate that this has been done, the treatments should be assigned to the patients by a *random* process. This is similar to what happens in choosing a random sample and, as in that case, random numbers are frequently used. Human beings cannot be relied on to choose things in an unbiased way. Thus the second basic principle we propose is that the treatments should be allocated to the experimental units by a random process, subject only to restrictions such as that there should be a specified number of units receiving a given treatment.

12.5.3 Completely randomized design

A design which is based on the principles of replication and randomization is the *completely randomized design*. Two (or more) treatments are assigned at random to a number of experimental units. Usually there are an equal number of experimental units for each treatment, but this is not necessary.

These designs are usually analysed by estimating the difference between the two means, and the standard error of this difference (p. 383), and by testing the difference (p. 415). For several treatments, see Section 22.2 (p. 573).

12.5.4 Paired comparisons

We would most easily pick up a difference in the response of the subjects to two types of sticking plaster if the subjects were very similar to each other, say all of the same age, same sex, similar occupation and similar habits of life. In fact, the subjects that can be used are likely to vary enormously in age, occupation and living condition, and are likely to be of both sexes. Can the experiment be done in such a way as to take account of this variability? One way to achieve this is by making all the comparisons of the two plasters be between similar experimental units. Thus, providing this is possible, we divide the subjects into pairs so that members of a pair are very similar to each other. Then we assign one member of a pair to the old treatment, and the other member of the pair to the new treatment; we toss a coin (or use random numbers) to fix which member receives which treatment. This process of random allocation is repeated independently for each pair. An analysis is based on the difference in response between the members of each pair. For example, we would find the difference in healing times of the two members of each pair and see if they are consistently in favour of one of the plasters. A particularly good pair to use is two parts of the same person. Thus we might put one plaster on a cut on the right arm and the other plaster on a cut on the left arm.

The appropriate method of finding a standard error for the difference of the treatments is that for paired data (p. 419). The difference is tested by a sign test (p. 279), a Wilcoxon signed-rank test (p. 290), or by the methods for paired samples (pp. 419, 428).

The idea of the paired-comparison design may be generalized to experiments with t ($\geqslant 2$) treatments; instead of pairs of similar units we use blocks of t similar units. The resulting design is called a randomized complete block design (Section 22.6, p. 583).

Section 22.7 includes more discussion of planning and design of experiments; see also Clarke and Kempson (1997).

12.6 EXERCISES ON CHAPTER 12

Note. No special computer exercises are provided for this chapter (but a computer could be used in Exercise 12.6(4)).

1. You are asked to survey the 2500 people in a village to find out what their leisure activities are. The village is within reach of industrial and business centres, and also contains a number of retired people.
   (a) How would you obtain a list upon which to base a sample for this purpose?
   (b) Would your sampling unit be an individual person or a whole household?
   (c) How would units be chosen for the sample?
   (d) What form of sampling would you use, and why?
   (e) What difficulties are there in defining 'leisure'?
   (f) Suggest some of the questions to be asked. Pay attention to wording them carefully.
   (g) What extra information would you try to collect about each sample member, in the expectation that it might shed light on leisure activities?
   (h) How could the results of such a survey be used to pinpoint any lack of facilities locally?

2. A psychologist claims that he has devised a method that enables people to remember very well the names of a large collection of objects; the method can be described on one sheet of paper. Explain how you would conduct an experiment to test this claim, paying particular attention to:
   (a) specification of the treatments;
   (b) choice of experimental units;
   (c) measure of the response;
   (d) allocation of treatments to experimental units;
   (e) analysis of data.

3. A survey is about to be launched to find out the public's book reading habits. Design a questionnaire for this survey which is suitable to be mailed to the chosen respondents. Your questionnaire should contain an introductory message and should seek to find out the basic demographic characteristics of the respondent, such as age and sex. Items of particular interest to the survey team include the frequency with which people read books on different kinds of subjects, how often people buy books for themselves and as gifts for others, how much people spend on books and how often people visit shops specifically to purchase books. As far as possible the answers should be coded to facilitate easy transfer to a computer. (*IOS*)

4. A survey of shops has been carried out. Data were collected by 3 enumerators who were each sent to different towns to walk the streets and choose a number of shops to include in the survey. The only control put upon their work was that they should each choose 2 grocery, 2 clothing and 2 hardware shops. The enumerators were asked to find out, from each shopkeeper, how many days each week his shop stays open and his operating profit in the last year. The following results were obtained.

Enumerator number	Town number	Type of shop*	No. of days open/week	Operating profit (£'000)
1	1	1	6	51
1	1	1	7	59
1	1	2	5	41
1	2	2	6	49
1	2	3	6	71
1	2	3	6	68
2	3	1	7	83
2	3	1	6	80
2	3	2	6	74
2	3	2	5	68
2	3	3	6	95
2	3	3	7	114
3	4	1	7	58
3	4	1	6	48
3	4	2	6	39
3	4	2	6	48
3	4	3	7	81
3	4	3	6	72

* 1 = grocery, 2 = clothing, 3 = hardware.

(a) Construct a table of number of days open per week against type of shop, including both row and column totals. Your table should show both the frequencies and the mean operating profit for each cell of the table, including totals.

(b) Write a short report to outline the main findings of the survey.

(c) Comment upon any shortcoming of this survey. (*IOS*)

12.7 PROJECTS

1 Trisection of lines

Aim. To investigate the distribution of the error made when lines are trisected by eye.

Equipment. Pencil, ruler, paper (of different colours).

Collection of data. Draw a line 15 cm long on a sheet of paper. Ask a subject to mark in a trisection point of the line by looking at it, without doing any measuring. Then measure the distance of the estimated point from the true point (recording it positive or negative according as it is to the right or left).

Repeat with as many subjects as possible, using a different sheet of paper for each. Also, if time allows, carry out the experiment under different conditions, e.g. changing the colour of paper, thickness of line, direction of line (horizontal, vertical, etc.), intensity of lighting. Lines could have arrows at their ends, pointing in the same direction or in opposite directions; other markings could also be used.

Analysis. For each set of conditions, draw up a frequency table, and illustrate the results in a histogram.

Notes. (1) If it is difficult to find many subjects, just one may be used. A 'warming-up' run of a few trials should be allowed before recording begins, in case a learning effect exists. The experimenter, not the subject, must draw all the lines.

(2) If a small number of subjects is used, each contributing several results, the data may be pooled for analysis provided the individual distributions are compared and found to be similar.

2 Survey of reading habits

Aim. To estimate the proportion of students who read each of the major national newspapers.

Collection of data. Identify your target population, design a suitable questionnaire and present it to a random sample of the population. It is wise to try out the questionnaire on a small sample first, i.e. carry out a pilot survey, before embarking on the full survey.

Analysis. Find the percentage of students who fall into each category. It may be necessary to group some of the newspapers together, to avoid small numbers in some categories.

Note. This project may be extended to become a stratified sample, collecting data from different strata of a large population, e.g. students, teaching staff, administrators, technicians; or for other types of reading matter stratified by age, sex, etc.

SUMMARY

Surveys are improved by using appropriate *methods of sampling* which allow for any known structure within the study population. An important case is **stratified random sampling** when the population can be split into parts (**strata**).

Questionnaires need careful design if the data are to be of any value.

Experiments can be conducted when an experimenter has control over allocating different **treatments** to different members of a population (experimental units or plots). *Replication* is the use of two or more experimental units for each experimental treatment. *Randomization* is the allocation of treatments to the experimental units by a random process, subject to the design chosen. A satisfactory experiment includes both replication and randomization.

In situations where the experimental units may be divided into similar pairs, a *paired-comparison* design may be used. The two treatments are allocated at random to the two units in each pair. This improves the accuracy of the comparison of treatments. The idea may be generalized to situations where *blocks* of t (>2) similar units may be formed and t treatments compared.

13 SIGNIFICANCE TESTING

OVERVIEW

Before applying statistical methods to make inferences using sample data, precise hypotheses are required, laying down what is to be studied and measured. Probability and statistical models can then be set up, and in this chapter we examine some (nonparametric) methods of testing whether or not the observed data are consistent with a hypothesis.

13.1 TESTING A HYPOTHESIS

'I am sure I can tell margarine from butter', says your friend, when he sees an advertisement saying that most people cannot tell any difference. How would you set about testing his claim?

First, you would have to make your friend express his claim more precisely. Does he mean that he can tell whether margarine or butter has been used in cooking? Not usually: it is eating bread and butter versus eating bread and margarine that is usually meant. Often, bread and butter (or margarine) is not eaten by itself but, e.g., with jam or marmalade, or as the outside of a sandwich. The type of sandwich filling would obviously affect a person's taste, and influence how sensitive he or she is to butter or margarine, whichever has been used. And if you are asked to assess an egg sandwich, and you do not happen to like egg anyway, the assessment probably does not mean much.

In order to test your friend's claim, he must taste butter and margarine in conditions which are as simple and standard as possible, and in no way confused by any outside factors such as those we have suggested. Even bread, as a base, can have too much flavour. A plain cream-cracker type biscuit is about the best base to use. One such biscuit should be spread with butter, another with margarine, and these presented in random order to the person doing the tasting. We are therefore making a paired comparison (p. 269). The taster carries out the tasting blindfold, so that he cannot use colour to help decide which is butter. The taster says which is the one with butter on; for a correct result we record a '+' and for a wrong result we record '−'. Of course one tasting gives very little information, and we should get our taster to repeat the process. We expect some variation in the results of repeated tastings, unless it is very easy indeed to tell the difference between butter and margarine. It might be very easy if there were a large difference between butter and margarine, or if there were some extra character, like colour or texture, which the taster was using to help his judgement; but if there were a large difference we should not be doing an experiment to test the claim at all, and if we allowed colours and textures to be very different we should not be controlling the experimental conditions carefully enough.

Suppose that your friend repeats the experiment ten times and obtains the results $++-+--+++++$, seven right and three wrong. We will consider the variate r, the number of correct results; so $r = 7$ in this case. How can we now use these results to test the claim that, under these tasting conditions, he can tell the difference between butter and margarine? If your friend had chosen correctly ten times, we would have no hesitation in saying that he could tell the difference; and if he had five or fewer correct results we would say he was just guessing. If he has made only one mistake, shall we ignore it? What about two or three mistakes?

This way of thinking suggests that we ought to be able to split up the possible outcomes $r = 0, 1, 2, \ldots, 10$ into two groups, with a dividing point at about 7 or 8. All the outcomes with the smaller values of r suggest that the taster *cannot* distinguish between butter and margarine, whereas the outcomes with the larger values of r support the hypothesis that the taster *can* discriminate. In order to be able to state what is the cut-off point between 'larger' and 'smaller' r-values, we must set up a mathematical model of the situation.

If we are sceptical of your friend's claim, and assume he cannot tell the difference between butter and margarine, plus and minus results will be equally likely to occur at each tasting. Using this statement, we can define a random variable which models r, the number of correct results (plus signs) in ten tastings. For an experiment in which there is no great difference between the things being compared, in which the taster cannot use other characteristics such as colour to help in judgement, and in which he is not told whether he is right or wrong each time, it seems reasonable to assume that each tasting is independent of every other tasting and that the probability of a plus result remains the same from one tasting to another. In these circumstances, a binomial random variable with $n = 10$ and $\pi = \frac{1}{2}$ is appropriate.

Underlying our argument there is an idea or *hypothesis*, namely that no difference can be detected between butter and margarine. On this hypothesis we shall base all our calculations. It is a specific hypothesis, giving definite values to all the parameters in the model (n and π in this case). This basic hypothesis is called the *null hypothesis* (NH). If it is true, then the probabilities of the possible results for r are given by the binomial formula

$$ P(r) = \binom{10}{r} \left(\frac{1}{2}\right)^r \left(\frac{1}{2}\right)^{10-r} = \binom{10}{r} \frac{1}{2^{10}} \quad (r = 0, 1, \ldots, 10). $$

Number of + signs, r	0	1	2	3	4	5	6	7	8	9	10	Total
Probability of r on NH	$\frac{1}{1024}$	$\frac{10}{1024}$	$\frac{45}{1024}$	$\frac{120}{1024}$	$\frac{210}{1024}$	$\frac{252}{1024}$	$\frac{210}{1024}$	$\frac{120}{1024}$	$\frac{45}{1024}$	$\frac{10}{1024}$	$\frac{1}{1024}$	1

On this model, based on the null hypothesis, it is very unlikely that all ten tastings will give + signs, and unlikely too that only one mistake will be made.

In fact, the probability that no more than one mistake is made is $P(9) + P(10) = \frac{11}{1024}$, about 0.01. Even allowing up to two mistakes, $P(8) + P(9) + P(10) = \frac{56}{1024}$, still only about 0.05. To go further in assessing our observed result, we must ask whether we have any other ideas to offer for explaining it, if its probability on the null hypothesis is very low. If so, we should clearly use these in the assessment.

We must set up, in more or less precise terms, an *alternative hypothesis* (AH), which states what we suppose will happen if our null hypothesis is not correct. The alternative hypothesis is an essential part of a problem, and both it and the null hypothesis must be set up at the beginning when we formulate the problem, *before we look at the results of the experiment*. An alternative hypothesis is based on what we know (or think we know) about the topic being studied. This knowledge may come, in part, from *previous* experiments or surveys, as well as from more theoretical arguments; but the alternative hypothesis is certainly *not* suggested by looking at the present data themselves. That could be an extremely misleading way to proceed if we happened to have collected an 'untypical' set of results in our particular experiment.

In our example, we chose our null hypothesis to express scepticism of your friend's claim. If his claim is true, we might expect him to be able to discriminate between butter and margarine. Then the probability of a plus result would be greater than that of a minus result, i.e. greater than $\frac{1}{2}$. In ten tastings, as we have already remarked, we should expect more pluses than minuses. Thus we take as alternative hypothesis 'there is discrimination between butter and margarine'. But when we try to put this mathematically, what value ought we to assign to π in the binomial model? How much greater than $\frac{1}{2}$ shall it be? We simply do not know. We do not need to know, because the alternative hypothesis is *not* the basis for any calculations: it is used as a guide to what calculations ought to be made, but all the calculations that are made are based on the null hypothesis.

When we test whether the null hypothesis is a reasonable explanation of the observed experimental result, we must come to one of the following decisions. *Either* we decide to reject the null hypothesis (that there is no discrimination between butter and margarine) because we consider that it does not explain the observed result satisfactorily, *or* we do not reject the null hypothesis. If we reject the null hypothesis we accept the alternative hypothesis (that there is discrimination) as a more satisfactory explanation of the data.

No value of r between 0 and 10 is *impossible* on the null hypothesis. So the decision whether or not to reject the null hypothesis on the basis of our observed value of r cannot be absolutely clear-cut. The decision has to be based on the idea that some values of r are unlikely on the null hypothesis, whereas they can better be explained by the alternative hypothesis.

In our example, the null hypothesis states that there is no discrimination; the model for this is the binomial distribution with $\pi = \frac{1}{2}$. On this model, it is very unlikely that 9 or 10 pluses will be recorded in 10 tastings. Using an alternative hypothesis in which $\pi > \frac{1}{2}$, it is more likely that there will be 9 or 10 pluses. If therefore we believed the null hypothesis, that the taster could tell no difference between butter and margarine, we would be very surprised if he picked out butter correctly 9 or 10 times (i.e. $r = 9$ or 10). If instead we believed the alternative hypothesis, that he could discriminate, we would be less surprised to observe a large value of r.

We shall thus want to reject the null hypothesis when the observed value of r is *unlikely* to occur on the null hypothesis, and is likely on the alternative hypothesis. Exactly what shall we regard as 'unlikely'? We cannot lay down a *precise* method for rejecting the null hypothesis until we have a definite idea of what we mean by 'unlikely'. To some extent this can be a subjective matter, and there are also special types of problem which need special treatment. But there is a well-established working rule or convention, used by statisticians generally, which is *to regard events having probability 0.05 or less as unlikely*. In addition, events with probability 0.01 or less may be regarded as 'very unlikely'.

In Fig. 13.1, we show the probabilities of $r = 0, 1, \ldots, 10$ on our null hypothesis; these we have already tabulated on p. 274. The random variable r is plotted along the horizontal axis, and this r-axis may be divided into two regions. The outcomes (i.e. values of r) in one region, which we call the *critical region*, are unlikely on the null hypothesis; the total probability of $r = 8$, 9 and 10 is about 0.05, as we have already seen. Furthermore, the probability of the r-values in this critical region will be greater on the alternative hypothesis, in which $\pi > \frac{1}{2}$, than it is on the null hypothesis (in which $\pi = \frac{1}{2}$). The critical region is set up before the experiment is carried out, as soon as the null and alternative hypotheses have been stated. Then the experiment is done, and a value of r, the number of correct results in 10 tastings, is observed. If this observed value falls in the critical region, the null hypothesis is rejected and the alternative hypothesis automatically accepted without further testing. If the observed r-value is not in the critical region, i.e. if it is 7 or less, we do not reject the null hypothesis. This is what happens in our example, since $r = 7$; we have not shown in this case that the null hypothesis is sufficiently unlikely to consider rejecting it.

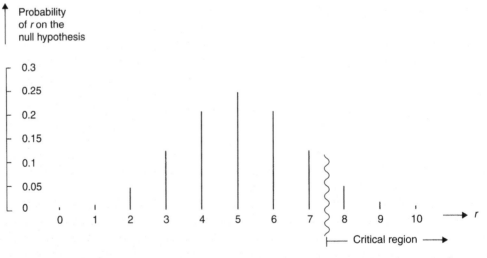

Figure 13.1 *Bar diagram of probabilities of number of correct results in tasting tests on the null hypothesis of no discrimination, showing critical region for one-tail test.*

Although your friend, by picking out butter correctly 7 times out of 10, did better than the mean number of 5 correct out of 10 that might be expected on the null hypothesis of no discrimination, he did not do well enough for us to consider his performance evidence of discrimination. Our verdict is rather like a 'not proven' result in a Scottish court. For this reason we prefer to see the dichotomy of choice as *either* 'reject NH and accept AH' *or* 'do not reject NH'. The dichotomy is often expressed as 'accept AH' or 'accept NH', but the testing situation is not a symmetrical one: if we use the words 'accept NH' they must be interpreted in a special way. If we do not reject the NH, our conclusion is certainly not that the probability with which your friend can pick out butter from margarine is 0.5; if we were asked to estimate this probability our best estimate would be $7/10 = 0.7$.

It might be asked 'Why not reject the NH for any value of r greater than 5?' The answer is that using this procedure, with a critical region containing the values 6, 7, 8, 9, 10, the probability of making the error, of saying someone can discriminate when really they cannot, is too high. This probability, which is the probability of falling in the critical region when the NH is true, is in fact $(210 + 120 + 45 + 10 + 1)/1024 = 386/1024 = 0.38$. Such a procedure would mislead us too frequently. In our tasting example, concluding that your friend is able to discriminate when he really cannot, is not likely to have serious consequences. The testing procedure is framed so that we accept the alternative hypothesis – and this is the hypothesis that really interests us – only if there is strong evidence for it; otherwise the case is 'not proven'.

Note that we used the alternative hypothesis only to decide that it was high values of r which would lead us to reject the null hypothesis, and hence to define the critical region of values of r. Having done this we based all the calculations on the null hypothesis. Note also that we did *not* look directly at our observed result and its probability on the null hypothesis; we did not look at the result *by itself* at all. The important question was whether or not it fell in the critical region.

Before giving examples of other situations where the same test can be used, let us look more critically at the alternative hypothesis we set up, which was that the taster could discriminate between butter and margarine. It led us to reject the null hypothesis when *large* values of r occurred; small values of r (0, 1, 2) do not fall in the critical region, even though they are unlikely on the null hypothesis, because the alternative hypothesis cannot give a better explanation of them. Therefore the set of r-values forming the critical region is concentrated *at one end only* of the set of all possible r-values, and we say that we have a *one-tail test* of the null hypothesis against this alternative.

13.2 ANOTHER ALTERNATIVE HYPOTHESIS

It is quite possible to imagine that our taster can tell the difference between butter and margarine but cannot say which is which. So another reasonable hypothesis is 'the taster can discriminate consistently between butter and margarine but may not name them correctly'. He may tend to pick out butter correctly most of the time, or to pick out margarine most of the time but call it butter: both of these outcomes are evidence of ability to discriminate. When describing this alternative hypothesis mathematically, we certainly need a model in which $\pi \neq \frac{1}{2}$. But this time we do not know whether to make $\pi > \frac{1}{2}$ or $\pi < \frac{1}{2}$; $\pi > \frac{1}{2}$ implies that the taster is able to discriminate and also to tell which is butter, while $\pi < \frac{1}{2}$ implies that he discriminates but names them incorrectly. As before, the null hypothesis is described by the binomial model with $\pi = \frac{1}{2}$. The critical region this time includes the very largest *and* the very smallest possible values of r. It still satisfies the rule that the r-values it contains must be unlikely on the null hypothesis and better explained on the alternative hypothesis, for the largest r need an alternative hypothesis '$\pi > \frac{1}{2}$' and the smallest need '$\pi < \frac{1}{2}$', both of which are included in our general alternative hypothesis '$\pi \neq \frac{1}{2}$'.

On the null hypothesis of no discrimination, $P(r=0) = P(r=10) = \frac{1}{1024}$, and $P(r=1) = P(r=9) = \frac{10}{1024}$, so that $P(r=0, 1, 9$ or $10) = \frac{22}{1024}$, which is about 0.02. This is *unlikely*, within our definition of the word, but if we add in the next most extreme r-values we do not any longer have an unlikely set of values. $P(r=8) = P(r=2) = \frac{45}{1024}$, giving $P(r=0, 1, 2, 8, 9$ or $10) = \frac{112}{1024}$, which is about 0.11. Our critical region, shown in Fig. 13.2, is thus $r=0, 1, 9$ and 10. As it contains values at both ends of the possible range of r, this critical region gives what we call a *two-tail test* of the null hypothesis. We reject the null hypothesis (of no discrimination) in favour of the alternative hypothesis (that there is discrimination without necessarily correct naming) if we observe $r=0, 1, 9$ or 10. As before, the actual observation $r=7$ will lead us not to reject the null hypothesis.

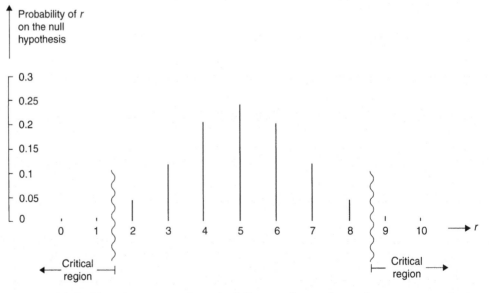

Figure 13.2 *Bar diagram of probabilities of number of correct results in tasting tests on the null hypothesis of no discrimination, showing critical region for two-tail test.*

Tests such as we have been considering *prove* neither hypothesis; we have no certain results but simply a weighing of probabilities. The result is often best summed up by saying that the weight of the evidence is for one hypothesis or the other.

The whole basis of our analysis has been the null hypothesis, even though it is the alternative hypothesis that is our real concern. What guides us in our choice of a null hypothesis? First, it must permit of precise specification so that the probabilities of all outcomes of the experiment, or sampling process, may be calculated exactly. Second, we usually frame the null hypothesis to express scepticism of the claim being made or of the result we would like to be true. The null hypothesis is therefore often a hypothesis of no difference, and this is probably the source of its name. But it is not necessary to specify that the difference is zero; a null hypothesis might claim that there is a difference of a given amount. What matters is that calculations can be based on it.

Example 13.1 Test of a median

The weekly incomes, to the nearest £, of 25 people living in a block of houses are 43, 88, 36, 65, 62, 77, 50, 38, 47, 61, 84, 30, 45, 65, 54, 73, 76, 47, 55, 68, 62, 66, 44, 52, 59. At the time these records were taken, the median weekly income in the region containing the block of houses was £50. Do these people differ from those in the whole region?

We regard the weekly incomes of the 25 people as a random sample of a parent population of weekly incomes. Our null hypothesis is really that this parent population is identical with the population of weekly incomes in the region. We usually express the null and alternative hypotheses in terms of the *parameters* of the population, or distribution; so here we express them in terms of the value of the median. We therefore state our hypotheses:

$$\text{NH} \quad \text{Median} = 50,$$
$$\text{AH} \quad \text{Median} \neq 50.$$

Each observation is equally likely to be above or below the median of the population from which it comes, by virtue of Definition 2.1 (p. 25) of the median. Since, on the null hypothesis, this median is 50 we look at each observation to see whether it is above or below 50; we neglect observations which are exactly equal to 50.

We attach a + sign to each observation greater than 50 and a − sign to each observation less than 50. We therefore allocate 24 signs: − + − + + + · − − + + − − + + + + − + + + + − + +. There are 16 + signs and we set $r = 16$. The alternative hypothesis does not indicate whether our observations are from a population with a median less than 50 or greater than 50; thus we carry out a two-tail test.

On the null hypothesis, the model for r is a binomial distribution with $n = 24$, $\pi = \frac{1}{2}$. This is tedious to compute in full, but there is no need to do so because we only need to know the two-tail critical region:

$$P(r = 0) = P(r = 24) = \left(\frac{1}{2}\right)^{24}$$

$$P(r=1) = P(r=23) = 24\left(\frac{1}{2}\right)^{24}$$

$$P(r=2) = P(r=22) = \frac{24 \cdot 23}{1 \cdot 2}\left(\frac{1}{2}\right)^{24}$$

$$P(r=3) = P(r=21) = \frac{24 \cdot 23 \cdot 22}{1 \cdot 2 \cdot 3}\left(\frac{1}{2}\right)^{24}$$

$$P(r=4) = P(r=20) = \frac{24 \cdot 23 \cdot 22 \cdot 21}{1 \cdot 2 \cdot 3 \cdot 4}\left(\frac{1}{2}\right)^{24}$$

$$P(r=5) = P(r=19) = \frac{24 \cdot 23 \cdot 22 \cdot 21 \cdot 20}{1 \cdot 2 \cdot 3 \cdot 4 \cdot 5}\left(\frac{1}{2}\right)^{24}$$

$$P(r=6) = P(r=18) = \frac{24 \cdot 23 \cdot 22 \cdot 21 \cdot 20 \cdot 19}{1 \cdot 2 \cdot 3 \cdot 4 \cdot 5 \cdot 6}\left(\frac{1}{2}\right)^{24}.$$

Thus the total probability of obtaining an r-value of 6 or less, or 18 or more, is twice the sum of the expressions above, which is

$$\frac{380\,102}{16\,777\,216} = 0.0227.$$

This set of values then forms the critical region, because if we include $r=7$ and 17 the total probability becomes 0.0639, which is greater than our conventional level of 0.05.

Our observed value $r=16$ does not therefore fall in the critical region, and so we do not reject the null hypothesis. The median income of this group of people is £59 compared with the median income of £50 for the region. But the evidence is not strong enough for us to assert that the median income of the group differs from that of people in the region.

Example 13.2 A paired comparison

Sixteen pairs of children are chosen in a primary school so that the members of each pair are as nearly alike as possible in age and background. One child of each pair, chosen at random, is taught to read by a standard method S, and the other by a new method N, which it is supposed will produce better results. The scores in a standard test taken at the end of a year's teaching are as follows. Test the hypothesis that N is producing better results.

Pair	1	2	3	4	5	6	7	8	9	10	11	12	13	14	15	16
Score on N	37	51	46	46	55	38	43	53	50	41	47	60	32	40	41	48
Score on S	40	48	50	46	53	42	47	54	49	44	51	56	34	39	41	50

We begin by stating our hypotheses:

NH: The methods produce similar scores.

AH: Method N produces higher scores than method S.

The test is made by looking at the sign of the difference of the scores for each pair. We attach a + sign to the pair if N had the higher score, and a − sign if S did; pairs in which the scores are equal are ignored. The signs are

$$- + - \cdot + - - - + - - + - + \cdot -$$

We have allocated 14 signs, of which 9 are − and 5 +.

On the null hypothesis, + and − signs will occur with equal probability, while the alternative hypothesis (that N is the better method) implies more + than − signs; hence a one-tail test is appropriate. To find the critical region, we need to compute the individual probabilities in a binomial distribution with $n = 14$, $\pi = \frac{1}{2}$, beginning at the top end, where $r = 14$, and working down. In fact this gives $r \geqslant 11$ as critical region, since $P(r \geqslant 11) = 0.0287$ and $P(r \geqslant 10)$ is well above 0.05. However, our observations showed a preponderance of − signs, not of + signs, so without doing the test in detail it is clear that we should not reject the null hypothesis in favour of an alternative hypothesis that would require more + than − signs. We conclude that there is no evidence that method N is superior to method S.

Notice that although the individual scores vary a great deal, the pairing has brought together children with similar scores and so has removed a very considerable element of child-to-child variation in the analysis of the whole collection of results.

Since many of the examples of this test depend on analysing a collection of plus and minus signs, as we have done, the test is commonly called the *sign test*. We shall refer to it again on p. 419, and give a simple, but approximate, method for computing the probability contained in the critical region on p. 377.

13.2.1 Exercises

1. What null and alternative hypotheses would you set up in the following situations? (Do not try to suggest detailed testing methods.)
   (a) It is believed a coin has been tinkered with to make it biased. It is required to test this.
   (b) A man always puts two teaspoonfuls of sugar in a cup of tea but never stirs it up. His wife believes he could not detect any difference in taste if he put in only one teaspoonful. She wishes to test this.
   (c) Of two similar groups of rats, one is fed with a standard diet, and the other with the standard diet except that one ingredient is modified. It is required to test if the modification affects the lifespan of the rats.
   (d) The time taken to travel to work by employees of a large store has been surveyed, and the mean travelling time computed for a period of several months. On a particular day there are roadworks due to a burst water main, and employees' travelling times are recorded for that day. It is desired to compare these with the records of past months.

2. The time taken by five subjects to respond when a light came on was measured when they were fully rested and also when they had been awake for 24 hours. (The order of doing these two tests was decided at random for each subject.) For all subjects the response time was longer when the subject had been awake 24 hours. Assume a null hypothesis that staying awake had no effect on response times, and an alternative hypothesis that it had an effect. What is the probability on the null hypothesis of all the differences being positive? What would you conclude using a 5% significance level? (Be careful! Is it a one- or two-tail test?)

3. I know that an attempt has been made to load a die so that 6s are more frequent. I throw it six times and 6 comes up three times. Using a null hypothesis that the probability π of a 6 coming up on a single throw is $\frac{1}{6}$, test, at the 5% significance level, whether the die has been successfully loaded.

13.3 TWO TYPES OF ERROR

Why has it become an established convention to regard a set of outcomes having probability 0.05 or less as 'unlikely'? Let us look again at what we actually do when testing a null hypothesis against an alternative. So long as the variate-value which we have observed is not part of the critical region (i.e. that set of values which have small probability on the null hypothesis), we decide without further question that there is no reason to reject the null hypothesis. Apart from using the alternative hypothesis to decide whether a one-tail or a two-tail test is more appropriate, i.e. in forming the critical region, we do not look mathematically at the alternative hypothesis at all. It is only when our observation is not consistent with the null hypothesis that we accept the alternative hypothesis, and then we do so automatically, without further examination. So we should consider the decision process (whether or not to reject the null hypothesis) from the standpoint of the null hypothesis.

Because no observations, not even those in the critical region, are *impossible* on the null hypothesis, we will sometimes make the mistake of rejecting the null hypothesis when it is true. The probability of making this error is exactly the probability of all the outcomes contained in the critical region when the null hypothesis is true. We may think of this in another way. When we actually do observe a value in the critical region, there are two possible explanations for this happening. *Either* the null hypothesis is true, and our experiment has resulted in an unusual, i.e. an unlikely, observation *or* the null hypothesis is not really true at all, so that we have based our calculation on a faulty model. We choose to believe the second explanation. We just do not believe we could have been so unlucky as to obtain an unlikely result – it could not happen to us! But it *does* happen to us, with exactly the probability of the set of outcomes forming the critical region on the null hypothesis. Therefore this probability should not be very large. On the other hand, if we choose a critical region that has a very low probability attached to it, we shall hardly ever reject the null hypothesis, even when the alternative hypothesis ought to be accepted instead. This point is discussed further on p. 293.

The probability of rejecting the null hypothesis when it is true, i.e. the total probability of the outcomes in the critical region, is called the probability of *type I error*. If we make this too small, we behave in a very conservative way, tending nearly always not to reject null hypotheses. On the other hand, if the rule for rejecting the null hypothesis is less strict, and the critical region contains a set of outcomes whose total probability is greater than 0.05, we can easily waste much time and experimental effort following up false trails. Experience in applied statistics has shown that by setting the probability of a type I error equal to 0.05, we achieve a satisfactory balance between conservatism and chasing false trails.

Another error we may make is not to reject the null hypothesis when it is false, i.e. when the alternative hypothesis is true. This is called a *type II error*. Unless we have the alternative hypothesis stated exactly, e.g. an exact value of π to put in an alternative binomial model, we cannot compute the probability of a type II error. We return to this on p. 294.

13.4 SIGNIFICANCE

This chapter contains the word *significance* in its title. We have not mentioned the word again until this section. If this seems perverse, it is because we have preferred to use a more specific idea, that of *testing a hypothesis*. However, we have made the test by seeing whether our observed variate provides evidence against the null hypothesis, in favour of the alternative hypothesis, and we can reword this testing process to contain the idea of significance. If our observation falls in the critical region, we shall say that it provides *significant evidence against the null hypothesis*, and call its numerical value simply 'significant'. When using a critical region that contains outcomes whose total probability is 0.05, we say that an observation falling in the critical region is 'significant at the 5% level'. The total probability of all the outcomes contained in the critical region is then called the *significance level*, or sometimes 'the level of the test'.

The word 'significant' can be qualified to show how strong the evidence is against the null hypothesis. If we use a critical region that contains outcomes whose total probability is 0.01, and an observed variate-value falls in this region, we consider this *strong* evidence against the null hypothesis, stronger than using a significance level of 0.05. We say that such an observation provides *highly significant* evidence against the null hypothesis, and we call it 'highly significant' or 'significant at the 1% level'. Finally, if we set up a critical region containing only those outcomes whose total probability on the null hypothesis is 0.001, we use the phrases 'very highly significant' and 'significant at the 0.1% level'. Although these purely verbal descriptions have their place, the normal practice is to quote the numerical significance level, e.g. 5%, 1% or 0.1%.

In this way, tests of hypotheses have commonly come to be known as *significance tests*.

13.4.1 Tests of simple hypotheses

In this chapter we have considered the practical use of significance tests. Each application of a significance test and the conclusion obtained from it must be considered carefully in the context in which it is used; unthinking use of a significance test can be very misleading. It is possible to construct artificial examples of hypothesis testing where the situation is much simpler. For example, imagine that there are two bags of marbles. Each contains 100 marbles but in bag A 50 are red and 50 blue while in bag B 80 are red and 20 blue. We are told that a random sample of 10 marbles has been chosen from one of the bags; 7 of the marbles in the sample are red and 3 blue. Which bag do we believe the marbles came from?

Here we have two hypotheses, which we can call H_A and H_B and which are of equal status.

$$H_A \quad \text{Sample is from bag A}$$
$$H_B \quad \text{Sample is from bag B.}$$

We do not have the asymmetry of a null hypothesis and an alternative hypothesis. In this case we would first calculate the probability of obtaining this sample on each hypothesis. On H_A, P(7 red, 3 blue) $= \binom{10}{7}(0.5)^7(0.5)^3 = 0.117$; on H_B, P(7 red, 3 blue) $= \binom{10}{7}(0.8)^7(0.2)^3 = 0.201$. We would accept H_B as being more consistent with the sample.

13.5 THE WILCOXON RANK-SUM (OR MANN–WHITNEY U) TEST

Earlier in this chapter we looked at various forms of the sign test; the main example of its use was the analysis of data from a taste experiment. The response recorded for each trial was whether the subject identified butter correctly or not. This judgement may be regarded as a primitive form of measurement, namely the assignment of a response to one of two possible categories (right or wrong, in this case). A slightly more sophisticated form of measurement is when the responses are put in rank order so that each response is recorded as a rank number. With this method of assessment we may add new significance tests to our repertoire without requiring additional probability theory. The two tests we consider, in this section and the next, were proposed by F. Wilcoxon (American, 1892–1965) in 1945. The first one allows us to compare two experimental treatments when there is no pairing of experimental units (as there was in Example 13.2). Let us consider an example.

A hospital pharmacist believes that by changing one of the ingredients in an analgesic (i.e. a pain-killer, aspirin is an example) that is commonly given to patients after an operation, he will make the analgesic much more effective and hence improve the recovery of patients. He plans a trial with a surgeon. Seven patients are to be given the same operation on a particular day; the plan is to let three of the patients, chosen at random, receive the new analgesic while the remaining four will receive the standard analgesic. After an appropriate time the surgeon will assess the recovery of the patients by listing them in order; he will assign a rank of 1 to the patient who has made the worst recovery, a rank of 2 to the next and so on to a rank of 7 to the patient who has made the best recovery.

Precautions must be taken in a trial such as this to avoid human bias. If a patient is aware he is receiving a new treatment he might gain a psychological fillip from this (advertisers believe that to call a product 'new' helps to sell it) and recover more quickly because of this belief. But, if the treatment is generally adopted, it will cease to be new and this advantage will be lost. For this and related reasons it is better if the patient does not know which treatment he is receiving. Moreover, the surgeon might believe, even if only subconsciously, that one treatment is better than the other and this might influence his assessment if he knew which treatment each patient had received. Therefore it is desirable that the surgeon does not know how the treatments are assigned. The pharmacist might decide, using a randomization procedure, which treatment each patient is to receive and put the pills, all made to look alike, in a box labelled with each patient's name for the nurse to administer them. So the surgeon would have to assess recovery without knowing which treatment each patient had received. When both patient and doctor do not know how the treatments have been assigned the trial is said to be a 'double-blind' one.

Let us label N and S respectively those patients receiving the new treatment and those receiving the standard treatment. Suppose the assessments of recovery came out in the following rank order:

$$
\begin{array}{ccccccc}
1 & 2 & 3 & 4 & 5 & 6 & 7 \\
S & S & S & N & S & N & N
\end{array}
$$

Out of the four patients who did best, three received the new treatment. On the surface this appears good evidence that the new treatment is an improvement, but as usual we need an objective method of judging. We shall devise a significance test using the same general principles that guided us in setting up the sign test.

We need a number (a statistic) which is a measure of how well the new treatment performs. A reasonable statistic is the sum of the ranks of patients receiving the new treatment, $T_N = 4 + 6 + 7 = 17$. (We could equally well work with the sum of the ranks of patients receiving the standard treatment, $T_S = 1 + 2 + 3 + 5 = 11$, but since the two rank sums always add up to 28 when there are 7 observations, the result obtained using one of the rank sums also gives us all the information we could obtain from the other.)

We choose a *null hypothesis* that the new and standard treatments have identical effects. (The *alternative hypothesis* is that the new treatment does better than the standard one.) The basis for the significance test is the distribution of the rank-sum statistic, assuming the null hypothesis is true. The new treatment was assigned at random to three of the seven patients, and if the null hypothesis is true the ranks 1 to 7 will also be assigned at random to the seven patients: it will not matter which treatment they were given. So the actual ranks scored by the three patients who were on the new treatment will be any selection of three from the numbers 1 to 7. There are $\binom{7}{3} = \dfrac{7 \cdot 6 \cdot 5}{1 \cdot 2 \cdot 3} = 35$ distinct ways in which we can choose these three from seven, and they are all equally likely: the probability of any one of them is $\frac{1}{35}$. Let us list those selections which give the lowest and highest rank sums for patients on the new treatment:

1	2	3	4	5	6	7	Rank sum
N	N	N					6
N	N		N				7
N	N			N			8
N		N	N				8
⋯	⋯	⋯	⋯	⋯	⋯	⋯	⋯
			N	N		N	16
	N				N	N	16
		N			N	N	17
				N	N	N	18

By listing all the 35 possible selections and calculating their rank sums, we can find the distribution of the rank-sum statistic, on the null hypothesis, as follows:

Rank sum, T_N	6	7	8	9	10	11	12	13	14	15	16	17	18
Probability	$\frac{1}{35}$	$\frac{1}{35}$	$\frac{2}{35}$	$\frac{3}{35}$	$\frac{4}{35}$	$\frac{4}{35}$	$\frac{5}{35}$	$\frac{4}{35}$	$\frac{4}{35}$	$\frac{3}{35}$	$\frac{2}{35}$	$\frac{1}{35}$	$\frac{1}{35}$

We use this distribution to determine a critical region (i.e. a set of values of the rank sum for which we reject the null hypothesis). We consider only the right-hand tail since it is only these values that have a high probability of occurring if the alternative hypothesis is true. On the null hypothesis the probabilities in this tail are:

$$P(T_N = 18) = \frac{1}{35} = 0.029;$$

$$P(T_N \geqslant 17) = \{P(T_N = 18) + P(T_N = 17)\} = \frac{2}{35} = 0.057.$$

The significance level is closest to 5% if we make the critical region consist of the values 17 and 18. Our data gave us a rank sum of 17. For the significance level we have chosen, we therefore reject the null hypothesis. There is evidence for us to say that the new treatment is an improvement. (But note that the improvement does have to be substantial if we are to detect it with such a small number of patients.)

13.5.1 The test in general

The validity of the test in our example rests upon the random assignment of treatments to patients. The patients need not be a random sample of a population, though this *is* necessary if we wish the results to apply generally to a population;

otherwise they apply only to the particular set of patients used. The test may also be used when we wish to compare two populations by choosing random samples from each.

The original data for the test are often what we could regard as typical measurements, e.g. heights of boys, number of faulty products in a batch. They are converted into ranks to make the test. Thus, in general, we can think of the initial data as two samples, labelled 1 and 2, made up of observations X_i $(i = 1, 2, \ldots, m)$ and Y_j $(j = 1, 2, \ldots, n)$, respectively. The observations from the two samples are combined, and a single joint ranking is formed from this combination. Care is taken to hold on to the label of each observation, so that we know whether it came from sample 1 or sample 2. In the joint ranking, $\{R_i\}$ are the ranks of the $\{X_i\}$ $(i = 1, 2, \ldots, m)$ and $\{S_j\}$ the ranks of the $\{Y_j\}$ $(j = 1, 2, \ldots, n)$. The two rank-sum statistics we can calculate are $T_1 = \sum_{i=1}^{m} R_i$ and $T_2 = \sum_{j=1}^{n} S_j$, though we use only one to make the test. Usually it will be more convenient to work with the smaller-sized sample. Note that $T_1 + T_2 = (m + n)(m + n + 1)/2$, the sum of the first $m + n$ natural numbers.

Two American statisticians, Mann and Whitney, proposed a test that is in essence identical with Wilcoxon's rank-sum test, though they used a different statistic which they called U. It is a count of the total number of times each observation of one sample comes before each observation of the other sample (a strict definition is given in Theorem 13.1). The value of U can be calculated without ranking all the observations and, since U and T are simply related as shown below, for a large number of observations T is often most easily calculated from U. Also the distribution of U on the null hypothesis can be tabulated more concisely than that of T.

Theorem 13.1

Let X_i $(i = 1, 2, \ldots, m)$ and Y_j $(j = 1, 2, \ldots, n)$ be two samples such that when all the observations are ranked, X_i takes rank R_i and Y_j takes rank S_j (no ranks are tied). The rank-sum statistics are $T_1 = \sum_{i=1}^{m} R_i$ and $T_2 = \sum_{j=1}^{n} S_j$. Define the statistics

$$U_{12} = \text{number of pairs } (X_i, Y_j) \text{ with } X_i < Y_j,$$
$$U_{21} = \text{number of pairs } (X_i, Y_j) \text{ with } Y_j < X_i.$$

Then

$$U_{12} = T_2 - \frac{n(n+1)}{2},$$
$$U_{21} = T_1 - \frac{m(m+1)}{2},$$
$$U_{12} + U_{21} = mn.$$

Proof. List the Y observations in increasing magnitude; we put the subscript in brackets to show that the subscripts may need changing so that they are also in increasing order.

Ordered observations	$Y_{(1)}$	$<$	$Y_{(2)}$	$< \cdots <$	$Y_{(j)}$	$< \cdots <$	$Y_{(n)}$
Rank	S_1		S_2		S_j		S_n

Since $Y_{(1)}$ is of rank S_1, the number of observations less than $Y_{(1)}$ is $S_1 - 1$ and, in general, the number of observations less than $Y_{(j)}$ is $S_j - j$. Hence defining U_{12} as the number of pairs (X_i, Y_j) with $X_i < Y_j$,

$$U_{12} = \sum_{i=1}^{n} (S_i - i) = \sum_{i=1}^{n} S_i - \sum_{i=1}^{n} i = T_2 - \frac{n(n+1)}{2}.$$

The result for U_{21} follows when the samples are reversed. $U_{12} + U_{21}$ equals the total number of pairs of the form (X_i, Y_j) which is mn (assuming we have no 'tied' pairs in which $X_i = Y_j$).

To make the test there is no need to derive the complete distribution on the null hypothesis; we need consider only one of its tails. (The distribution is symmetric, so for a two-tail test we simply double the probabilities obtained for one tail.) But even doing this, the calculations become prohibitive when the number of observations is large. In this case, we simplify the calculations by using the fact that the distribution of T on the null hypothesis is approximately normal (see p. 425 for the method). We note here the mean and variance of T_1 and T_2:

$$E[T_1] = \frac{1}{2}m(m+n+1); \qquad E[T_2] = \frac{1}{2}n(m+n+1);$$

$$\text{Var}[T_1] = \frac{1}{12}mn(m+n+1) = \text{Var}[T_2].$$

The normal approximation method may be used if both m and n exceed 10. For smaller values tables exist (Sprent and Smeeton, 2001).

13.5.2 Ties

Sometimes the investigator cannot distinguish between the responses of some of the subjects and wishes to give them the same rank; or some of the measurements to which ranks are to be assigned are equal. When *ties* such as these occur, the general rule is to replace the ties by the average rank of all the observations involved in the tie. For example, if five subjects are to be ranked and the middle three subjects are indistinguishable, the ranks assigned would be: $1, 3, 3, 3, 5$. In a similar way, if the observations $1.2, 1.3, 1.3, 1.5, 1.7$ were to be ranked, the ranks would be $1, 2.5, 2.5, 4, 5$. The sum of all the ranks of the set is the same as it would have been had no ties existed.

After the ranks, including tied ranks, have been assigned, the procedure for carrying out the test is identical with that when there are no tied ranks (though the presence of ties does make a slight difference to the distribution of T which we shall ignore when samples are not too small).

Example 13.3 Keeping quality of flowers

A florist tested the effect on the keeping quality of flowers of adding tablets of either type A or type B to the water in which the flowers were kept. From 11 freshly cut bunches of flowers, five were chosen at random to be put in water in separate vases to which tablets of type A had been added. The remaining six bunches were put in vases to which tablets of type B had been added. Flowers were thrown away as they became unsightly. The number of days until 50% of the flowers remained was recorded:

$$
\text{A} \quad 14,\ 17,\ 9,\ 11,\ 12
$$
$$
\text{B} \quad 13,\ 23,\ 19,\ 18,\ 16,\ 21.
$$

Test, using an approximate 5% significance level, if A and B had different effects on the keeping quality of the flowers.

NH: A and B have similar effects. AH: A and B have different effects. Two-tail test. We rank the treatments:

Rank	1	2	3	4	5	6	7	8	9	10	11
A	9	11	12		14		17				
B				13		16		18	19	21	23

The rank sum for A is $T_A = 1 + 2 + 3 + 5 + 7 = 18$.

Alternatively, we may derive T_A from U_{BA}. We count the number of times that B observations come before each A observation, and sum them. The number before each of '9', '11' and '12' is 0; the number before '14' is 1 and before '17' is 2. Hence $U_{BA} = 0 + 0 + 0 + 1 + 2 = 3$, and $T_A = U_{BA} + (5 \times 6)/2 = 3 + 15 = 18$, as before.

We look at the lower tail of the distribution of T_A on the null hypothesis. The number of ways in which 5 ranks may be chosen out of 11 is $\binom{11}{5} = 462$; hence each arrangement of ranks has a probability $\frac{1}{462}$. The arrangements which give a T value of 18 or less are:

Value of T_A	Possible arrangements	Probability	Cumulative probability
15	1, 2, 3, 4, 5	$\dfrac{1}{462} = 0.0022$	0.0022
16	1, 2, 3, 4, 6	$\dfrac{1}{462} = 0.0022$	0.0043
17	1, 2, 3, 4, 7; 1, 2, 3, 5, 6	$\dfrac{2}{462} = 0.0043$	0.0087
18	1, 2, 3, 4, 8; 1, 2, 3, 5, 7; 1, 2, 4, 5, 6	$\dfrac{3}{462} = 0.0065$	0.0152

Thus the probability that $T_A \leqslant 18$ or $T_A \geqslant 63$ (remember the test is two-tail) equals 0.030. Therefore our observed value of $T_A = 18$ will fall within the critical region for a 5% significance level. We therefore reject the null hypothesis and conclude that the two substances do have a different effect on the keeping quality of flowers.

13.5.3 Exercises

1. (a) Four values are chosen at random from the ranks $1, 2, 3, 4, 5, 6, 7, 8$. Find the probability that the sum of the four ranks chosen equals (i) 10; (ii) 11; (iii) 12; (iv) 24; (v) 25; (vi) 26.

   (b) Consider a rank-sum test in which the observations are eight distinct ranks, four derived from one experimental treatment and four from another. State the critical regions which most closely approximate to a (i) 5% two-tail test; (ii) 1% one-tail test using the lower tail; (iii) 5% one-tail test using the upper tail.

2. Four subjects were chosen at random from a group of eight and required to learn a task under normal conditions (A), while the remaining four subjects had to learn the task when there was substantial background noise (B). The number of attempts required until the task was consistently done satisfactorily is given below. Use a rank-sum test to investigate whether the experimental treatments A and B had different effects.

   A $3, 4, 2, 6$

   B $5, 10, 8, 7$.

13.6 THE WILCOXON SIGNED-RANK TEST

This is an alternative to the sign test for the analysis of data from a paired comparison experiment. If we have information on the *magnitude* of the difference between the responses of the members of each pair, as well as knowing the sign, then this test is usually to be preferred to the sign test.

Let us consider an example. A teacher writes some notes on how to use calculators more efficiently and wishes to test his advice. He constructs pairs from the members of one of his senior classes so that the members of each pair have approximately the same ability to handle a calculator; he can find 7 such pairs. He chooses one member of each pair at random and to these pupils he gives sheet A which contains exercises illustrating the new techniques he has devised. The remaining pupils are given sheet B which contains standard exercises. The pupils work at the sheets in the morning and in the afternoon they are all given the same test sheet. The teacher records the time taken for each pupil to obtain a complete set of correct results; pupils repeat a calculation if they do not obtain the right answer. The results obtained are shown in the following table, which also contains various derived quantities needed in the analysis.

Pair	I	II	III	IV	V	VI	VII
Time after sheet A (secs)	304	421	412	420	451	399	372
Time after sheet B (secs)	360	383	520	480	427	441	451
Difference, A − B	−56	38	−98	−60	24	−42	−79
Sign	−	+	−	−	+	−	−
Rank of absolute diff.	4	2	7	5	1	3	6
Signed rank	−4	+2	−7	−5	+1	−3	−6

We must begin our test procedure by stating the hypotheses:

NH: A and B lead to similar times. AH: A gives shorter times than B. One-tail test.

We might use the sign test for which the statistic is the number of positive signs; the value here is 2. But the sign test makes no use of the information we have on the magnitude of the differences. Notice that both the positive differences in the table above are relatively small differences; their ranks are 1 and 2. In other words, they are only marginally positive. Thus a simple inspection of the differences suggests reasonable evidence for the alternative hypothesis of there being a real difference. If, on the other hand, the positive differences were large (even though there were still only two of them) then the evidence for the alternative hypothesis would be much less strong. We require a statistic that takes into account both the signs of the differences and their magnitudes. The final line of the table above gives the *signed ranks* which are the ranks of the absolute differences preceded by the sign of the differences. A satisfactory statistic is the sum of the positive signed ranks $T_+ = 1 + 2 = 3$; low values lead us to reject the null hypothesis while higher values would be consistent with the null hypothesis. (We can work with the sum of the ranks with negative signs $T_- = 4 + 7 + 5 + 3 + 6 = 25$. But since, when there are N differences, $T_+ + T_- = N(N+1)/2$, results using one statistic can easily be reinterpreted in terms of the other. A good general rule is to work with the smaller sum.)

To carry out the test we use the distribution of T_+, on the null hypothesis. Since, on that hypothesis, each rank is equally likely to be associated with a positive or negative sign, we see that the sequence of signed ranks consists of all the possible combinations of:

$$\pm 1, \pm 2, \ldots, \pm N.$$

There are 2^N possible combinations, all equally probable on the null hypothesis; the distribution is obtained by calculating the value of T_+ for each combination and hence finding the probability of each value of T_+. As with the rank-sum test, we need look only at the tail to make a test. Let us derive the probability of each value of T_+ from 0 to 3:

Value of T_+	Arrangements giving value	Probability	Cumulative probability
0	No positive differences	$\left(\dfrac{1}{2}\right)^7$	0.0078
1	(+1)	$\left(\dfrac{1}{2}\right)^7$	0.0156
2	(+2)	$\left(\dfrac{1}{2}\right)^7$	0.0234
3	(+1, +2), (+3)	$2 \cdot \left(\dfrac{1}{2}\right)^7$	0.0391

The probability of a value of T_+ less than or equal to 3 is equal to 3.9% on the null hypothesis. Therefore, using a significance level of 5%, we reject the null hypothesis and conclude that the teacher's recommendations have led to a reduction in calculating time.

We can derive the mean and variance of the distribution of T_+ on the null hypothesis. We have

$$T_+ = X_1 + X_2 + \cdots + X_N$$

where X_i is a random variable which takes values i and 0 each with probability $\frac{1}{2}$. Therefore

$$\mathrm{E}[X_i] = i \cdot \frac{1}{2} + 0 \cdot \frac{1}{2} = \frac{i}{2}.$$

$$\mathrm{Var}[X_i] = \left(i - \frac{1}{2}\right)^2 \cdot \frac{1}{2} + \left(0 - \frac{i}{2}\right)^2 \cdot \frac{1}{2} = \frac{i^2}{4} \cdot \frac{1}{2} + \frac{i^2}{4} \cdot \frac{1}{2} = \frac{i^2}{4}.$$

$$\mathrm{E}[T_+] = \mathrm{E}\left[\sum X_i\right] = \sum \mathrm{E}[X_i], \quad \text{using Theorem 10.2, p. 236}$$

$$= \sum \frac{i}{2} = \frac{1}{2} \cdot \frac{N(N+1)}{2} = \frac{N(N+1)}{4}.$$

$$\mathrm{Var}[T_+] = \mathrm{Var}\left[\sum X_i\right] = \sum \mathrm{Var}[X_i], \quad \text{using Theorem 10.6, p. 242}$$

$$= \sum \frac{i^2}{4} = \frac{1}{4} \cdot \frac{N(N+1)(2N+1)}{6} = \frac{N(N+1)(2N+1)}{24}.$$

For large values of N (say >16) we may use a normal approximation to the null-hypothesis distribution of T_+ (see p. 425). For smaller values there are tables (Sprent and Smeeton, 2001).

Ties are dealt with as for the rank-sum test. They are replaced by average ranks and then the test proceeds as with untied ranks.

13.6.1 Exercises

1. Use Wilcoxon's rank-sum test on the data of Revision Exercises A, Question 10 (p. 102) to test the hypothesis that there is no difference between the weights of the populations of plants growing in the two habitats A, B.

2. Find critical regions for the signed-rank test statistic T_+, based on 10 paired comparisons, which approximate most closely to one-tailed (lower tail) tests with significance levels of (i) 2.5%; (ii) 5%.

3. Ten students were chosen at random from those taking each of two subjects A and B. The selected students of each subject were ranked according to ability, and students of the same ranking paired. The students were then asked how many textbooks they possessed. The numbers were:

Student-pair	1	2	3	4	5	6	7	8	9	10
Subject A	21	24	22	19	19	26	32	20	18	22
Subject B	14	27	17	15	29	18	20	18	19	16

Test if the number of textbooks possessed by students of the two subjects differed, using (a) the signed-rank test; (b) the sign test.

13.7 TYPE II ERRORS AND THE POWER OF A TEST

How good is a particular significance test? One method of measuring this is to determine the probability of a type II error for various values of the unknown parameter. This allows us to calculate a quantity called the *power* of the test, which offers a measure of how well a test performs.

We can compute the set of probabilities on the null hypothesis, because that contains a specific value of π. But so far we have taken as alternative hypotheses less specific values, either $\pi > \frac{1}{2}$ in a one-tail test, or $\pi \neq \frac{1}{2}$ in a two-tail test. In the one-tail test of Section 13.1, our critical region was $r = 8$, 9 or 10 correct results, and this region thus contained outcomes whose total probability on the null hypothesis was 0.05. In the two-tail test of Section 13.2, the critical region was $r = 0$, 1, 9 or 10 and its total probability was 0.02 on the null hypothesis. The probability of type I error (rejecting the null hypothesis when it is true) in our one-tail test is therefore 0.05, and in our two-tail test 0.02. Of course these probabilities depend entirely on the null hypothesis and are not affected by the actual value of π which we may choose to put in the alternative hypothesis.

Type II error is failure to reject the null hypothesis when it is false, in other words when the alternative hypothesis is true. So we need to find the total probability of all the outcomes *not* in the critical region (because these lead us not to reject the null hypothesis), but to compute this probability *on the alternative hypothesis*. This is where we need the specific value of π to put in whatever alternative hypothesis is set up.

If we have to do it by hand, the arithmetic involved in a binomial calculation when $\pi \neq \frac{1}{2}$ is tedious; tables of the binomial distribution have been published for various combinations of n and π, and there is also an approximate way of obtaining the probabilities required (see Chapter 16). However, a computer will do the work very quickly, only a short program being needed (*CCC*, Section 8.2). Calculations for $n = 10$ and $\pi = \frac{2}{3}$, also for $n = 10$ and $\pi = \frac{3}{4}$, give the following sets of probabilities of $r = 0, 1, 2 \ldots, 10$:

r	0 1 2	3	4	5	6	7	8	9	10
P(r) for $\pi = \frac{2}{3}$	0.003	0.016	0.057	0.137	0.228	0.260	0.195	0.087	0.017
P(r) for $\pi = \frac{3}{4}$	0.004		0.016	0.058	0.146	0.250	0.282	0.188	0.056

Consider how to use these calculations to find the probability of a type II error when the alternative hypothesis says '$\pi = \frac{2}{3}$' and a one-tail test is being used. Setting the probability of a type I error at 0.05, the test will be 'reject the null hypothesis if $r = 8$, 9 or 10 is observed'. Therefore we do not reject the null hypothesis on any occasion when $r \leqslant 7$ is observed. (Note that $r \leqslant 7$ is the *complement* of the critical region.) On the alternative hypothesis that $\pi = \frac{2}{3}$, $P(r \leqslant 7) = 0.701$, so the probability of making a type II error is 0.701.

So long as we do a one-tail test, and keep the probability of type I error fixed at 0.05, the type II error probability will depend only on the value of π used in the alternative hypothesis; as this π increases, the type II error probability decreases. Using the figures above, we find that for $\pi = \frac{3}{4}$ the type II error probability, $P(r \leqslant 7)$, is 0.474. We could repeat these calculations for many values of π, and draw a graph of type II error against the value of π used in the alternative hypothesis. This is shown in Fig. 13.3(a).

As the value of π on the alternative hypothesis gets further away from the value $\pi = \frac{1}{2}$ which is being used in the null hypothesis, so the probability of making a type II error drops. The further apart these two π-values are, the better we can discriminate between them. We extend this idea to define the *power* of a test, for fixed type I error probability, as $1 - P(\text{type II error})$. So in our particular example, where we fix the type I error probability at 0.05 in the one-tail test, power is a function of the actual value of π used in the alternative hypothesis, and increases steadily as π increases from $\frac{1}{2}$ towards 1 (see Fig. 13.3(b)).

If the form of the alternative hypothesis is changed, so that a two-tail test is required, then the critical region changes, as we have seen, to $r = 0, 1, 9$ or 10 when the type I error probability is set at 0.02. We shall not reject the null hypothesis so long as $2 \leqslant r \leqslant 8$. This time, in order to find type II error probabilities, we need to compute $P(2 \leqslant r \leqslant 8)$ for the π-values allowed in the alternative hypothesis, which are all values between 0 and 1 except for $\pi = \frac{1}{2}$ (which forms the null hypothesis). Figure 13.4 shows how type II error varies with π for the two-tail test in which the probability of type I error is fixed at 0.02.

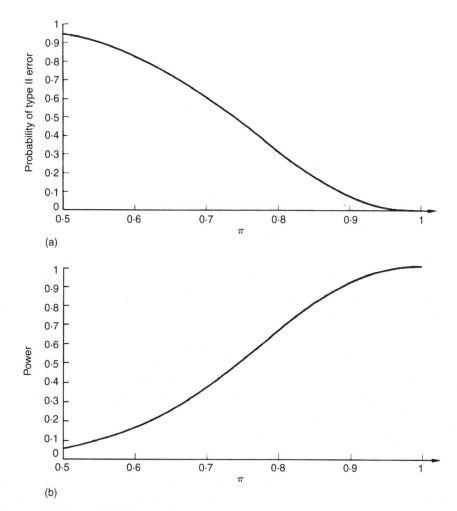

Figure 13.3 *Probability of (a) type II error, (b) power, for one-tail test.*

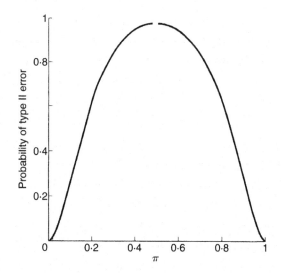

Figure 13.4 *Probability of type II error for two-tail test.*

13.7.1 Exercise

A man claims that he can, without looking, tell which face of a die is uppermost by touching it lightly. He is asked to do this on seven occasions, the die being rethrown on each occasion. The null hypothesis is that he is guessing. The number of times he is correct is recorded and the result will be declared significant if he gains four or more successes.

(a) What significance level is being used?

(b) What is the power of the test if the true probability of being correct is: (i) $\frac{1}{4}$; (ii) $\frac{1}{2}$?

13.7.2 Type I and type II errors in acceptance sampling

Acceptance sampling is a process that has many similarities to significance testing. Both can be seen as examples of decision making, although some statisticians would argue that there is a fundamental difference between gaining new knowledge, which is the concern of significance tests, and the much more mechanical process of accepting or rejecting a batch of products. Nevertheless, it is instructive to link the probabilities involved in acceptance sampling with the type I and type II errors of significance testing.

Consider a manufacturer who wants to improve his quality control. He wishes to introduce a sampling scheme which will give a small probability of offering to customers batches of material that contain too high a proportion of faulty items. But from the manufacturer's viewpoint, it is also bad business to stop the process or to scrap a batch of output when the proportion of faulty items is low and the customer would have accepted them.

If we imagine that the operating characteristic curve of the sampling scheme looks like Fig. 13.5, the rather simple ideas of *producer's risk* and *consumer's risk* can be introduced. Suppose that a customer is willing to accept batches that contain a proportion π_2 or less of faulty items; in other words the consumer wants the sampling scheme to pass batches containing π_2 faulty items with only a small probability β. Batches with a higher proportion of faulty items will, of course, have an even smaller probability of being accepted. Thus the point (π_2, β) is on the OC curve. The *consumer's risk* is β, the minimum probability that a bad batch is accepted.

Clearly the producer does not want the probability of rejecting a very good batch to be at all high; this can be measured by choosing a low value π_1 and requiring that the probability of rejecting a batch having only a proportion π_1 of faulty items shall be α, which is very small. This is then the *producer's risk*; $(\pi_1, 1 - \alpha)$ is another point on the OC curve. Once α, β, π_1 and π_2 have been specified, the next problem is to construct a sampling scheme by choosing the size n of the sample and the number of faulty items in it which will be permitted without rejecting it, say c. Wetherill and Brown (1991) give details of the mathematical methods needed to find n and c; in practice tables are often used.

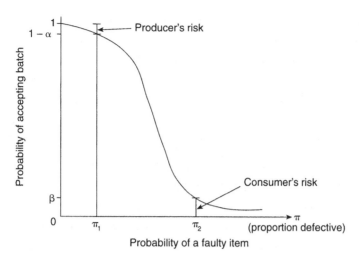

Figure 13.5 *A typical OC curve showing producer's and consumer's risks.*

In hypothesis testing, a type I error is the probability of rejecting a null hypothesis which is in fact true. Likewise, rejecting a batch with a proportion π_1 of faulty items is actually rejecting an acceptable batch, and the probability of doing this is the producer's risk α. A type II error is the probability of accepting a null hypothesis when it is false. Thus if we accept a batch with a proportion π_2 of faulty items we are saying that a batch is acceptable when in fact it is not. The probability of this occurring is the consumer's risk β. The symbols α and β as used in this section thus have meanings exactly parallel to those they carry in the theory of hypothesis testing.

13.8 SUMMARY AND DEFINITIONS

For ease of reference, we collect together the definitions of concepts introduced in this chapter, and summarize the steps required in a significance test (hypothesis test).

1. Set up a statistical model which allows the hypothesis to be expressed mathematically. State precisely the null hypothesis being tested, and the alternative hypothesis which will be accepted if the null hypothesis is rejected.

Definition 13.1

A **null hypothesis** (NH) is an assertion that a parameter in a statistical model takes a particular value.

Definition 13.2

An **alternative hypothesis** (AH) expresses the way in which the value of a parameter in a statistical model may deviate from that specified in the null hypothesis.

2. Carry out an experiment or trial to produce the observations. Calculate a test statistic (i.e. a quantity determined by the observations); in our major example it was the variate value r. Use the statistical model, assuming the null hypothesis to be true, to calculate the probabilities of all the possible values this statistic may take (i.e. determine the sampling distribution of the statistic, assuming the null hypothesis to be true).

3. Fix a significance level, e.g. 5%, and construct a critical region consisting of those values of the test statistic that are unlikely on the null hypothesis and are better explained on the alternative hypothesis. (The total probability in the critical region is 5%, or approximately 5%.)

4. Reject the null hypothesis in favour of the alternative hypothesis if and only if the value of the test statistic in the experiment falls in the critical region.

Definition 13.3

A **critical region** is a subset of the set of all possible values of the test statistic that might be observed in an experiment. The subset is chosen so that it contains values whose total probability is low on the null hypothesis, and which are better explained on the alternative hypothesis.

Definition 13.4

Type I error is to reject the null hypothesis when it is true.

Definition 13.5

The **significance level** of a test is the probability of making a type I error. It is also the total probability, on the null hypothesis, of those values of the test statistic falling in the critical region.

Definition 13.6

Type II error is to fail to reject the null hypothesis when it is false, i.e. when the alternative hypothesis is true.

Definition 13.7

The **power** of a test is the probability of accepting the alternative hypothesis when it is true; it equals $1 - P(\text{type II error})$.

13.9 MINITAB: NONPARAMETRIC TESTS

The significance tests introduced in this chapter are called nonparametric tests, as opposed to most of the other significance tests that we introduce later, which are called parametric tests (see Section 17.15 for a discussion of the difference). The particular tests are named by MINITAB **1-Sample Sign, Mann-Whitney** (which we also call Wilcoxon rank-sum) and **1-Sample Wilcoxon** (which we call Wilcoxon signed-rank). To call up the tests in MINITAB, choose initially

Stat>Nonparametrics

The test that is introduced in Sections 13.1 and 13.2, and in Examples 13.1 and 13.2, is the sign test. Suppose the 25 observations of weekly income in Example 13.1 are entered into C1. To carry out the test of whether the median differs significantly from 50, choose

Stat>Nonparametrics>1-Sample Sign/
Enter C1 in **Variables**, select **Test median** and enter 50, select **not equal** in **Alternative** (i.e. AH)/

This is a two-tail test. By selecting **less than** or **greater than**, opposite **Alternative**, one can carry out one-tail tests. In the output the NH and AH are stated. Also given are the number of values below the median, 8, the number equal to it, 1, and the number above, 16.

Under P is given what is called the *p-value*. It is the probability, on the NH, of the observed result or any more extreme result possible under the AH. In the example, the value is 0.1516; it is the probability of obtaining values 0–8 and 16–24 (remember we neglect observations that equal the median exactly). Since the probability is greater than 0.05 we do not reject the NH at the conventional level of significance.

This represents a different approach at the final stage of a significance test from the one that we have adopted. In the sign test example it is possible to calculate the *p*-value without too much effort and the reader should check MINITAB's value. But, for many significance tests, determination of the *p*-value without a computer is difficult if not impossible. Therefore tables are provided, such as the ones at the end of this book, which give critical values of statistics for conventional significance levels such as 5%, 1% and 0.1%. By comparing the test statistic with the critical value, significance is assessed. But in the *p*-value approach, which MINITAB uses throughout, the *p*-value is compared with the conventional values of 5%, etc., or any other values, to judge significance (see also Section 17.3.1).

The **1-Sample Wilcoxon** (or signed-rank test) is carried out in a similar manner. For the **Mann-Whitney** (or Wilcoxon rank-sum) test the names of the two samples must be entered. One-tail or two-tail tests may be carried out by selecting the appropriate **Alternative**.

13.10 EXERCISES ON CHAPTER 13

1. A manufacturer claims that a brand of light-bulbs that he supplies has a median burning time of 250 hours. A random sample of ten bulbs is chosen from a stock of the bulbs and the burning times (in hours) are found to

be 224, 254, 170, 206, 201, 290, 220, 150, 205, 271. Test the manufacturer's claim.

2. Suppose that in the tasting test described in this chapter, 12 tastings had been used instead of 10. State the critical regions that correspond approximately to 5%, 1% and 0.1% levels of significance for a two-tail test.

3. A method of taste testing, for comparing two substances, is to use a *triangle test*. Three samples are presented to the taster, of which two are of one substance and one is of the other substance. (It is decided at random for each triangle which substance is repeated.) The taster then has to pick out the odd sample. On the null hypothesis that the taster is choosing at random, his probability of choosing the odd sample correctly is $\frac{1}{3}$. The number of correct choices is the test statistic. If five triangles are tasted, find the critical region for a one-tail test using a significance level of: (a) 5%; (b) 1%.

4. Children in an elementary school class were paired on the basis of similarity of intelligence and home background. One of each pair was chosen at random and taught reading by method A, and the other was taught by method B. The 14 pairs of children in the experiment were then given a reading test and obtained the following scores:

Pair	1	2	3	4	5	6	7	8	9	10	11	12	13	14
A	9	8	10	3	7	5	6	8	4	7	8	9	6	9
B	7	7	8	2	4	7	5	6	5	5	7	6	5	7

Test the hypothesis that the methods are of differing effectiveness using a 5% significance level.

5. Explain what are meant by null and alternative hypotheses in significance testing. A random sample of 10 readings on an instrument gave the data: 11.7, 9.5, 10.4, 10.5, 8.9, 11.3, 12.1, 10.7, 10.8, 9.1.

It is required to test the null hypothesis that $M = 11.0$ against the alternative hypothesis that $M < 11.0$, M being the median of the distribution. Nothing is known about the form of the distribution. Use a sign test at the 10% level of significance to achieve this.

6. A man claims that he can throw a six with a fair die five times out of six on the average. Calculate the probability that he will throw four or more sixes in six throws: (i) if his claim is justified; (ii) if he can throw a six, on the average, only once in six throws.

To test the claim, he is invited to throw the die six times, his claim being accepted if he throws at least four sixes.

Find the probability that the test will: (a) accept the man's claim when hypothesis (ii) is true; or (b) reject the claim when it is justified, that is, when hypothesis (i) is true. *(AEB)*

7. A coin has probability p of falling heads when tossed. Write down, in terms of p, expressions, which need not be simplified, for:
   (i) the probability that if the coin is tossed four times all four tosses will fall alike (that is, all heads or all tails);
   (ii) the probability that if the coin is tossed seven times six or more of the tosses will fall alike.

   The following two procedures have been suggested for testing whether the coin is unbiased.

   Procedure 1 toss the coin four times and conclude that it is biased if all four tosses fall alike.

   Procedure 2 toss the coin seven times and conclude that it is biased if six or more of the tosses fall alike.

   Show that the two procedures are equally likely to lead to the conclusion that the coin is biased when in fact it is unbiased.
   Determine which of the two procedures is the less likely to lead to the conclusion that the coin is unbiased when in fact the probability of a head in any toss is equal to $\frac{2}{3}$. (*JMB*)

8. An experimenter collects 12 observations of people's reaction times to a stimulus (the flashing of a light as described on p. 63). He wishes to test a hypothesis concerning the mean value of reaction time, and does so by using a sign test. What assumptions must he make to do this, and how valid are they likely to be in this experiment?

9. Twenty-five subjects undergoing a test reported the following reaction times (in seconds) to a stimulus: 6.6, 3.6, 5.4, 7.2, 4.7, 13.1, 2.0, 7.6, 2.3, 2.8, 15.4, 4.3, 6.7, 9.5, 11.8, 1.4, 19.7, 3.0, 7.5, 6.9, 23.3, 6.4, 14.1, 6.0, 3.8. Test the hypothesis that the median reaction time is 7.8 s. Calculate the mean reaction time and comment on the result.

10. Two samples, A and B, of plants of the same species, growing on opposite slopes of a valley, were dug up and weighed. Sample A from one side gave the following 20 masses (g): 37.1, 50.3, 25.7, 13.9, 32.2, 46.4, 21.8, 26.3, 24.7, 26.2, 42.0, 25.7, 22.9, 37.5, 19.9, 24.4, 34.8, 17.2, 31.0, 28.8. Sample B from the other side gave the 12 masses (g): 21.7, 25.3, 29.1, 32.0, 16.7, 24.1, 29.1, 34.4, 25.8, 22.3, 38.7, 27.9. Test whether the distributions of masses are the same.

11. In a tasting experiment, people were presented with pairs of samples of vegetables preserved by the same method but kept for different periods of time. It was required to test the hypothesis that no difference could be detected between samples kept for a short period of time and those kept for longer. Thirty people took part. Eight said there was no difference, seven preferred the longer-kept samples, and the remaining 15 preferred the shorter-kept samples. Test the hypothesis.

12. The people taking part in the experiment described in Exercise 11 above scored each sample on a ten-point scale. They obtained these results:

Person	1	2	3	4	5	6	7	8	9	10	11	12	13	14	15
Short storage	3	6	5	7	8	5	4	2	7	6	4	5	8	6	5
Long storage	4	6	4	3	3	5	5	2	5	5	4	1	3	8	5

Person	16	17	18	19	20	21	22	23	24	25	26	27	28	29	30
Short storage	8	2	6	3	6	5	4	7	5	5	7	4	5	7	6
Long storage	2	5	5	3	7	3	7	7	6	4	4	1	5	6	2

Make a further statistical test, which uses the information contained in the scores, to examine the same hypothesis as in Exercise 11.

13. The weights of fish in two populations were compared by analysing the differences in the weights of a random sample of fish from each population matched by length. The weight differences in grams for 10 pairs of fish were: 12, −13, −125, −120, −73, 2, 3, −147, −12, −4.

Explain why a parametric test would be unsuitable for comparing the weights of fish from these two populations.

Analyse these data using a test which:
(i) ignores the magnitudes of the differences,
(ii) uses both signs and magnitudes of the differences.

Comment on the comparison of your results. (*RSS*)

13.11 COMPUTING EXERCISES

1. Carry out a sign test on the data of (i) Example 13.1, (ii) Example 13.2.

If the test is carried out using a statistical package (e.g. MINITAB), verify the probability stated by calculating individual binomial probabilities (see Example 13.1) and summing.

2. Investigate the data from the two groups in Example 13.3 by making dot plots of group A and group B and comparing them. Then carry out a Wilcoxon rank-sum (Mann–Whitney U) test on the data.

3. For the paired-comparison data sets in (i) Example 13.2 and (ii) Example 17.6, find the difference in score for each pair and make a dot plot of the differences. Compare the plotted values with zero and hence assess if the experimental methods (new and standard methods of teaching in the one example, two food-processing methods in the other) have had different effects. Follow this by making a Wilcoxon signed-rank test.

13.12 PROJECTS

1 Tasting experiment

Aim. To see whether people can tell the difference between two brands of butter (or between butter and margarine, as described in Section 13.1).

Equipment. Cream-cracker type biscuits. Two similar brands of butter, e.g. both produced in the same country, or both slightly salted, or both costing the same price. There should be only a small difference between brands, so that detection is not too easy.

Several people will be required to act as subjects. Either each will need a blindfold, or subdued lighting conditions will be required for the experiment, unless the two butter samples look exactly alike.

Collection of data. Spread the butter thinly and evenly on small pieces of biscuit. Tell the subjects the outline of the experiment, but not too much detail – not enough to simplify the comparison too greatly. Each subject is presented with two samples of one brand and one of the other. A coin is tossed to determine which brand is presented twice. The order of the three samples is randomized, with a fresh randomization for each repeat of the experiment. In this triangle of samples the subject must pick out the odd one, and may also state a preference.

Repeat as many times as possible (a subject will need a pause for the palate to clear after about three triangles, and should if possible have some water available to drink between samples if desired). Obtain as many subjects as possible. Keep the records separate for each subject.

Analysis. For each subject, record the number of triangles correctly decoded. Compare this with the number expected on a binomial distribution having $\pi = \frac{1}{3}$ (which corresponds to a null hypothesis of no discrimination). Make an appropriate significance test. Pool the data for all subjects and repeat the test.

For the correct triangles, examine the number of stated preferences. (It may be possible to do this only on the pooled data, if numbers of correct triangles and stated preferences per subject are small.) Compare this with the number expected on a hypothesis of equal preferences of each brand. Make an appropriate significance test.

Notes. (1) When differences are small, triangular tasting should give results that are more sensitive than those in a paired-comparison experiment such as that described in Section 13.1. If there is the opportunity to carry out experiments on the same materials by both methods, comparison of the results will be interesting.

(2) The experiment may use, as an alternative, different varieties of drink, e.g. brands of orange juice. Samples will need to be very small, or else the palate dulls quickly; paired comparisons rather than triangles may be possible. Great care is needed to avoid colour and texture differences being used by subjects as an aid to taste. Coloured containers, or subdued lighting, will be very helpful.

2 Comparing two strategies at darts

Aim. To see if higher scores are obtained by aiming for the bull's eye (the centre of the board) than for the treble twenty (the top of the board).

Equipment. A dart board and a set of (three) darts.

Collection of data. Throw three darts, aiming for the bull. Record the total score. Throw three darts again, this time aiming for treble twenty. Record the total score. Record a plus sign if 'bull' is the larger score, and minus if 'treble twenty'.

Repeat as many times as convenient, and have several people carrying out the experiment. Randomize the order of throwing (whether it is the first set or second set of throws that aims for the bull), deciding the order by tossing a coin before each pair of throws.

Analysis. Carry out a sign test to see whether there is any difference between the strategies: (a) for individual players; (b) overall.

SUMMARY

The statistical terms introduced in this chapter are defined in Section 13.8.

A **null hypothesis** (NH) is set up to provide a basis for calculations. Events which have small probability on the NH are regarded as 'unlikely' and lead to the NH being rejected. These events form the **critical region**, which will often contain 5% or less of the total probability of events that may be observed.

An **alternative hypothesis** (AH) is also required, showing how we propose to explain the data by changing the parameter(s) in the statistical model when the NH is rejected. This AH must be set up when the study is planned, *before* seeing the data. According to the form of the AH, either a **one-tail** or a **two-tail** test may be required.

The **sign test** examines an NH that the median of a set of data takes a given value, or that in paired samples the two measurements take similar values.

The **Wilcoxon rank-sum (or Mann–Whitney U) test** compares two independent samples of data by looking at their relative positions in a joint ranking, and the **signed-rank test** uses also the magnitudes of the ranked data.

The quality of a test for distinguishing between two hypotheses can be assessed by calculating **type I** and **type II errors**.

REVISION EXERCISES C

1. The discrete random variables X, Y are independent, and both have the same distribution:

$$P(X = 1) = P(X = -1) = \tfrac{1}{2}$$

and

$$P(Y = 1) = P(Y = -1) = \tfrac{1}{2}.$$

Another random variable, Z, is defined by $Z = XY$.
   Show that Z is independent of X and of Y.

2. In an association football match x is the number of goals scored by the home team and y is the visiting team's score. The frequency distribution for the total score, $x + y$, for first division matches in 1972–3 is as shown.

Total score	0	1	2	3	4	5	6	7	8
Frequency	35	82	138	99	61	32	7	7	1

   Calculate the mean and standard deviation of the distribution.
   Draw a frequency polygon for the data.
   It is known that the mean value of x for these games is 1.56; calculate the mean value of y.
   Assuming that x and y are independent variables, indicate how you would calculate the variance of y if that of x were known. Give a reason why the assumption of independence may not be justified in this case.   (*JMB*)

3. A discrete random variable X takes the values 0 and 1 only, with probabilities p and $1 - p$ respectively, where $0 < p < 1$. Find $E[X]$ and $Var[X]$.
   A second discrete random variable Y also takes the values 0 and 1 only. Find the joint probability distribution of X and Y in each of the cases when:
   (i) $P(Y = 0) = p$ and X and Y are independent;
   (ii) $P(Y = 1 \mid X = 1) = P(Y = 0 \mid X = 0) = p$;
        $P(Y = 1 \mid X = 0) = P(Y = 0 \mid X = 1) = 1 - p$.
   In both cases, show that $E[X + Y] = E[X] + E[Y]$, and investigate, for varying values of p strictly between 0 and 1, the equality, or otherwise, of $E[XY]$ and $E[X]\,E[Y]$.   (*Welsh*)

4. X is a discrete random variable with the following distribution:

x	-3	-1	1	3
$P(X = x)$	0.2	0.3	0.4	0.1

If $Y = X^2$, find:
(a) the distribution of Y;
(b) the joint distribution of X and Y;
(c) the expected value of $X - Y$.

5. Two random variables X and Y have the following joint distribution, in which $2\alpha + 2\beta + \gamma = 1$:

	X				
Y	-2	-1	0	1	2
0	0	0	γ	0	0
1	0	β	0	β	0
2	α	0	0	0	α

Are X and Y mutually independent? Calculate $E[X + Y]$ and the variance of $[X + Y]$.

6. A fair cubical die is tossed three times. Let X and Y denote random variables defined as follows: $X = 1$ if the score on the second toss is different from the score on the first toss, and $X = 0$ otherwise; $Y = 1$ if the score on the third toss is different from the scores on both the first and second tosses, and $Y = 0$ otherwise.
   (i) Find the joint probability distribution of X and Y, and hence show that X and Y are not independent.
   (ii) Find the probability distribution of $Z = X + Y$. Express the number of different scores obtained in the three tosses in terms of Z, and hence, or otherwise, determine the expected value and the variance of the number of different scores in the three tosses. (Welsh)

7. (a) Given a random sample of n independent observations $x_1, x_2, \ldots, x_n$ from a population distribution of unknown mean and unknown variance, prove that

$$s^2 = \frac{1}{n-1} \sum_{i=1}^{n} (x_i - \bar{x})^2,$$

where $\bar{x} = (\sum_{i=1}^{n} x_i)/n$, is an unbiased estimate of the variance.
   (b) A box contains five cards numbered 1, 2, 3, 4, 5, respectively. List the various possible pairs of numbers that may arise when two of these cards are drawn at random without replacement from the box. For each possible pair, calculate the value of s^2 given above, and hence display the sampling distribution of s^2 in this case. Show that this s^2 is not an unbiased estimate of the population variance. Explain why the result in (a) above does not apply here. (Welsh)

8. Derive the mean and the variance of the binomial distribution.

   A drawing pin was tossed n times in succession and the number of occasions, X, that it alighted with its point upright was noted. Some time later, the same drawing pin was tossed a further m times and the number of occasions, Y, that it alighted with its point upright was noted. Show that

$$Z_1 = \frac{1}{2} \left(\frac{X}{n} + \frac{Y}{m} \right) \quad \text{and} \quad Z_2 = \frac{X + Y}{n + m}$$

are both unbiased estimators of p, the probability that when tossed the drawing pin will alight with its point upright. State, with reasons, which of these two estimators you would regard to be the preferable one to use.   (*Welsh*)

9. A boy is playing a board game in which he has to throw a six with a die before he is allowed to put his counter on the board. If the probability of throwing a six at any throw is π (independently of the results of other throws), find the distribution of R, the number of throws needed until he can begin (include the one at which the six is thrown). Find the mean of the distribution of R.

   He suspects that the die is not properly balanced, and so he estimates π by calculating $1/r$. Show that this does not give an unbiased estimator of π.

10. If, in Question 9, $S = R - 1$, i.e. S is the number of throws *before* a six is obtained, find the mean and variance of S and compare with the mean and variance of R.

11. The proportion of mice with white coats, in a large population of mice bred in a laboratory, is unknown but is to be estimated as the proportion p in a sample of four observations, selected at random from this population. It is required to estimate the variance of R, the number of white mice in such samples. For this purpose, the estimator $v = 4p(1-p)$ is suggested. Show that this is biased, but that there exists an estimator kv which is unbiased, k being a constant.

12. In order to estimate the area of a large circular field, 50 people measure the radius of the circle. Their measurements will differ because of random errors of measurement. The resulting values, $X_1 \ldots, X_{50}$, may be regarded as 50 independent random variables each with mean μ (equal to the true radius) and variance σ^2.
    (a) Is the average, $\overline{X}$, of the 50 measurements an unbiased estimate of μ, the true radius?
    (b) Evaluate $E[\overline{X}^2]$. Is $A = \pi \overline{X}^2$ an unbiased estimator of the area of the circle?
    (c) Find $E[A_i]$, where $A_i = \pi X_i^2$. Is $\overline{A} = \frac{1}{50} \sum_{i=1}^{50} A_i$ an unbiased estimator of the area of the circle?

13. It is proposed to carry out a survey of an area to discover people's attitudes to education and to the educational facilities available at secondary level in that area.

   Outline how this might be done. Points to consider include how to specify the target population, whether and how to stratify, how to select individual sample members, what questions to ask and what extra information to collect.

   In what ways would you expect the population of the area to vary in respect of the information you aim to collect?

14. Choose a random sample of five observations from the geometric distribution with $\pi = \frac{1}{2}$. (Note that, since the random variable can take an infinite number of values, one sampling category will consist of all values greater than some stated value. If an observation falls in this category more random digits will be needed to decide its precise location in the category.)

15. Define the binomial distribution, stating the conditions under which it will arise.

   A biased cubical die has probability p of yielding an even score when thrown. What is the probability that no even scores are obtained in ten independent throws of the die?

   In ten independent throws of the die the probability of exactly five even scores is twice the probability of exactly four even scores. Show that $p = \frac{5}{8}$ and calculate (to 3 places of decimals) the probability of at most six even scores in eight independent throws of the die. *(JMB)*

16. A fair coin is tossed $2n$ times.
   (a) By considering the ratio of the probabilities of the occurrence of exactly k heads and of exactly $k+1$ heads, or otherwise, show that the most probable number of heads is equal to the expected number of heads.
   (b) Find the probability that a run of exactly n consecutive heads will be obtained. *(Welsh)*

17. (a) In a test taken by 12 candidates, the marks obtained by 10 of the candidates were: $13, 15, 20, 24, 25, 28, 32, 33, 38, 42$. Calculate the mean and the standard deviation of these 10 marks.

   Find the two missing marks if the mean and the standard deviation of all 12 marks are respectively equal to those of the above 10 marks.
   (b) $(x_1, y_1), (x_2, y_2), \ldots, (x_n, y_n)$ are n pairs of corresponding values of variables x and y where $y = ak^x$, a and k being positive constants. Show that when x is equal to the arithmetic mean of $x_1, x_2, \ldots, x_n$, the corresponding value of y is equal to the geometric mean of $y_1, y_2, \ldots, y_n$. *(Welsh)*

18. In a given adult human population, the heights of men have mean μ_1 and variance σ_1^2, while the heights of women have mean μ_2 and variance σ_2^2. If the proportion of men is p, calculate the mean and variance of height in the whole population.

   Assuming that a married couple is formed by choosing a man and a woman independently, at random, from the population, find the mean and variance of the difference in height between husband and wife.

19. A machine produces, on average, 20% defective bolts. A batch is accepted if a sample of five bolts taken from that batch contains no defectives, and rejected if the sample contains three or more defectives. Otherwise, a second sample is taken. What is the probability that this second sample is required? *(IOS)*

20. If blue tits lay 10 eggs, and on average three of the young survive the first year of life, what is the probability that at least three survive from a particular brood?

   Eight broods in a certain area included five that produced less than two survivors. Assuming that 10 eggs were laid, does it appear that this area has a lower survival rate?

21. A buyer wanted to know if there was any difference in quality between similar materials supplied daily by two manufacturers, A and B. Over 8 days,

A's materials were reported more satisfactory than B's 7 times. Should the buyer conclude that A's is different material from B's?

22. Define the binomial distribution, stating clearly the conditions under which it applies. Give an example of a variate you might record for which the binomial distribution is a satisfactory model.
    (i) In a fairground game three dice are thrown and a score of 16 or over is said to be a 'jack'. If four or more 'jacks' are obtained in five throws then the result is a 'jackpot' and gains a prize. What is the probability of obtaining a prize?
    (ii) In a multiple-choice examination there are five possible answers to each of the 100 questions. What is the mean and standard deviation of the number of correct answers obtained by a candidate who guesses the answer to each question?

23. The probability of success in each of a series of n_1 trials is p_1; the probability of success in each of a second series of n_2 trials is p_2. Prove that the probability of r successes in the $n_1 + n_2$ trials is the coefficient of t^r in the expansion in powers of t of

$$T = (1 - p_1 + p_1 t)^{n_1} (1 - p_2 + p_2 t)^{n_2}.$$

Five rounds are fired from a gun at a target and the probability of a hit with each round is $\frac{1}{4}$. The gun is then resighted and a further five rounds fired, the probability of a hit with each round being now $\frac{1}{2}$. Find:
    (a) the probability of more than two hits with the ten rounds;
    (b) the probability of more hits with the second five rounds than with the first five. *(O&C)*

24. An experiment consists of taking two balls without replacement from an urn containing four balls, of which θ are red and the others white. The hypothesis to be tested is H: $\theta = 3$. A proposed test is to reject H if either or both of the balls drawn are white, and not to reject H if both balls drawn are red.
    (a) Define type I and type II errors. Calculate the probability of type I error, and the probability of type II error when $\theta = 0, 1, 2$ and 4.
    (b) An alternative test is to reject H if both balls drawn are white, and not to reject H if either or both of the balls drawn is red. Calculate the probability of type I error.
    (c) Define the power of a test. Calculate the power for each of these proposed tests, for each possible value of θ.

25. An urn contains 6 balls, which are either red or white; but the number of red ones, m, is not known. Three balls are drawn, with replacement. The number, r, of red balls among these three is counted. Find the distribution of r, its mean and its variance.

    An estimator for m is to take that value of m which has maximum probability in view of the sample observed. What value of m is most likely when $r = 0, r = 1, r = 2$ and $r = 3$? In a series of 10 experiments, the observed values of r were 1, 1, 2, 1, 1, 1, 0, 0, 2, 1. Construct the probability distribution of the estimator of m.

A test of the hypothesis H: $m = 4$ is to reject H if no red balls are observed in the sample. Find the probability of type I error, and the power for each possible value of m.

26. A medical research worker is interested in the effectiveness of a drug in treating psychiatric illness. He makes an exhaustive study of journals for articles reporting trials of the drug. He finds 41 articles. In these, 29 reported indecisive results (or were badly designed). Of the remaining 12, one article reported that the drug was worse than a standard (control) treatment, and the other 11 articles reported a significant beneficial effect, at the 5% level of significance, or better, compared with controls. What conclusions should the worker draw? (*IOS*)

27. Discrete random variables X and Y have the joint probability function

$$P(X = x, Y = y) = \begin{cases} k|x - y|, & x = 0, 1, 2, 3 \\ & y = 0, 1, 2, 3, \\ 0, & \text{elsewhere}, \end{cases}$$

where k is a constant.

Determine k and obtain the marginal distributions of X and Y in tabular form. State, with reasons, whether or not X and Y are independent.

Obtain also the probability distribution of the random variable $|X - Y|$ in tabular form and find the expected value and variance of $|X - Y|$.

28. To compare a new treatment for a certain disease with the standard treatment, a group of 12 patients was divided at random into a group of 7 who received the new treatment and a group of 5 who received the standard treatment. The improvements shown by the patients after a given period are represented by the following measurements:

New treatment	3.2, 5.5, 3.4, 4.9, 3.0, 2.4, 5.1.
Standard treatment	2.3, 3.8, 2.2, 1.9, 2.6.

Use a Wilcoxon rank-sum test to decide whether or not there is evidence that the new treatment is better than the standard treatment.

29. (a) A brewery is investigating the quality of river water in two areas, R and S. Ten independent samples have been taken, five from each area. The rank order of the quality of water, according to area, is:

Quality	1 (best)	2	3	4	5	6	7	8	9	10 (worst)
Area	R	R	S	R	R	S	R	S	S	S

Analyse these results and give your conclusions.

(b) In a health club, two groups of women took part in fitness programmes, one using aerobics and one circuit training. At the outset, the fitness of each person was about the same, and allocation to groups was at random. Fitness was again assessed after four weeks, and improvement (in coded units) was as follows for the 10 women on each programme:

| Aerobics | 16, | 8, | 19, | 20, | 14, | 17, | 13, | 21, | 12, | 10; |
| Circuits | 14, | 12, | 14, | 13, | 15, | 11, | 13, | 14, | 12, | 12. |

Examine the results using a nonparametric test. State any assumptions you make and any reservations you may have about the validity of the analysis.

(*RSS*)

14 CONTINUOUS RANDOM VARIABLES

OVERVIEW

Continuous random variables (e.g. length, weight, time) are not restricted in the number of possible values they make take: any value is possible, either over an infinite range or with specified lower and upper limits. Probability density functions are required to model them; these lead also to cumulative distribution functions. Between them, these two functions allow several useful distributions to be studied, and means, variances and generating functions to be calculated.

14.1 MODELLING CONTINUOUS VARIATES

A continuous variate is one that may take all values within a given range; heights and weights of people are examples. We have constructed discrete random variables to model discrete variates; we shall now see how to construct continuous random variables as models of continuous variates.

Professor Aztec catches a train every evening to take him home from his university. He has never bothered to find out the exact times at which the trains run since there is a frequent and punctual service; a train leaves every ten minutes. When his day's work at the university is finished he walks to the station and catches the next train. One cold evening as he is waiting for a train he wonders if he would save himself much time and discomfort if he found out the train times, so that he could time his arrival at the station better. He decides to investigate his problem by setting up a simple mathematical model.

With his current practice he is equally likely to wait any time from 0 minutes to 10 minutes. He might model his waiting time by a discrete random variable that took the ten values 0.5, 1.5, 2.5, ..., 9.5, each with probability 0.1. If he were to measure the times, he would record them to the nearest minute, or half-minute, so it is not unreasonable to use a discrete random variable. But the professor thinks he will have a better model if the waiting-time variable could take *any* value within a ten-minute interval, and he rejects the use of a discrete random variable. In practice, all measurements that are recorded have only a finite number of possible values. Once we have chosen a suitable unit, such as a centimetre or a tenth of a second, we record to the nearest multiple of it. But if we are building a model to explain a variate that we know is continuous, it seems more satisfactory to think of the variable as taking *all* values in a range, while accepting that we can *measure* the variables only to a limited accuracy.

When we specify a discrete random variable we decide what values we want the variable to take and assign a probability to each one. Unfortunately, that method does not give a sensible result in the continuous case. There are an infinite number of possible waiting-times for Professor Aztec's train, each within the range 0 to 10 minutes, and each of them is equally likely. If we assigned a non-zero probability to each of the possible values then the sum of the probabilities would be infinite, and not unity as it should be. We must use a different method to specify a continuous random variable.

We shall offer two clues. The first is to say that although we cannot give the probability of any *particular* waiting-time, we *can* give the probability of the waiting-time being *less than* a stated length. For example, we would be happy to say that the probability of the professor waiting less than 5 minutes would be 0.5, and that the probability of his waiting less than 8 minutes would be 0.8. In fact, we have no difficulty in specifying the cumulative distribution function of the waiting-times in the model. (The reader will recall that the cumulative distribution function was introduced for discrete random variables on p. 185 and has values $F(b) = P(X \leqslant b)$.) Here we have $F(b) = b/10, 0 \leqslant b \leqslant 10$. The second clue comes from recalling how we investigated a frequency distribution of a continuous variate. It was pointless to consider the frequency of individual values, because most, if not all, of the values of the variates were not repeated; they had a frequency of one. We were obliged to divide the range of the variate into class-intervals, and to find the frequency with which the values of the variate fell into each of these intervals. These two clues lead us to the idea that for a continuous random variable we specify the probability with which it falls in particular *intervals*, rather than trying to specify the probability with which it takes particular values.

If we imagine the professor collecting a large number of records of the time he waits for his train, and grouping them using a class-interval of one minute, we would expect him to find approximately equal numbers in each interval. Let us idealize the situation and assume that equal numbers *do* fall in each class. We shall represent these hypothetical results by an idealized histogram (Fig. 14.1(a)). In this diagram we shall take the area above each interval as a representation of the *probability*, rather than of the frequency, of a value falling in that interval. The rectangle above each interval has therefore an area of 0.1. But the division of the whole range of values into class-intervals is an arbitrary one; our hypothetical data could as well have been grouped in intervals of ten seconds or two minutes. Therefore in the diagram we may as well remove the ordinates associated with particular intervals and represent the distribution as in Fig. 14.1(b). The rectangle has an area of unity, and a height of 0.1. From this diagram we can derive the probability of obtaining a value between a and b, assuming neither of these quantities exceeds 10, by determining the shaded area in Fig. 14.2. The probability is obviously $0.1(b - a)$. The distribution we have specified is an example of the *continuous uniform* distribution (compare it with the discrete uniform distribution, p. 171). Professor Aztec is happy that this distribution will provide a good model of his current waiting-times.

If the professor knew the train times it would influence him when he was finishing off his work, and he would expect to plan his arrival at the station so

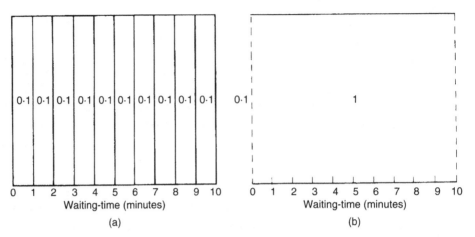

Figure 14.1 *Distribution of waiting-times represented by (a) an idealized histogram, (b) a graph.*

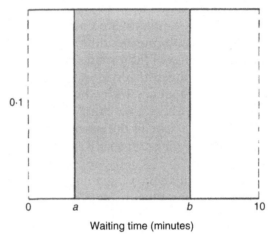

Figure 14.2 *Graph of distribution of waiting-times. Shaded area equals the probability of a waiting-time being between a and b.*

that he waited for a long time less frequently. He thinks that a reasonable model for the distribution of his waiting-times, when he does know the train times, is one that might be represented by Fig. 14.3. He wants a continuous distribution in which values are more likely to fall in intervals close to zero than in intervals further from zero, but in which all values in the range 0 to 10 are still possible. The diagram represents the simplest distribution he can think of that satisfies these conditions. Let us call it a triangle distribution, though we should warn the reader that this distribution is not one with an accepted name. The height at zero is chosen to be 0.2, so that the area of the triangle is unity. (This area represents the total probability of all possible waiting-times.)

We have begun to represent continuous random variables by diagrams that look like the graphs of continuous functions. It therefore seems natural to describe these random variables by a continuous function $f(x)$. For the

CONTINUOUS RANDOM VARIABLES

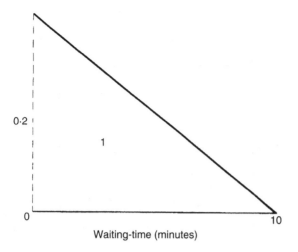

Figure 14.3 *Graph of 'triangle' distribution of waiting-times.*

continuous uniform distribution we have

$$f(x) = \begin{cases} 0.1 & 0 \leqslant x \leqslant 10 \\ 0 & \text{otherwise,} \end{cases}$$

and for the triangle distribution

$$f(x) = \begin{cases} 0.02(10 - x) & 0 \leqslant x \leqslant 10 \\ 0 & \text{otherwise.} \end{cases}$$

These are called *probability density functions* (we give a precise definition later, p. 318). With discrete random variables we defined a probability *mass* function; we pictured the probability as being made up of a number of masses localized at particular points, which are the values of the variable. With a continuous random variable we can think of the probability as being spread over the whole range of values of the variable as the mass of a rod is spread out, not necessarily uniformly, along its length. From this analogy we speak of the probability *density* function of a continuous random variable. The analogy also helps us to appreciate why any particular value of a continuous random variable has zero probability: the probability is analogous to the mass of a slice of zero length being removed from somewhere in the rod. But both the total probability and the mass of the rod take some positive, but not infinite, value.

The professor asks himself a precise question, 'What is the probability that I wait more than 5 minutes for my train?' To answer this, we must compare the probabilities of waiting more than 5 minutes for the two distributions, uniform and triangle. We wish to find the area under each of the graphs to the right of 5 (see Fig. 14.4). With the uniform distribution it follows by symmetry that this area is 0.5 (Fig. 14.4(a)). With the triangle distribution it can be shown that the area, and hence the probability, is 0.25 (Fig. 14.4(b)). If this type of comparison is to be made for more than one value of the variable x it is better done by means of the cumulative distribution $F(b)$. This, of course, gives us the cumulative probability

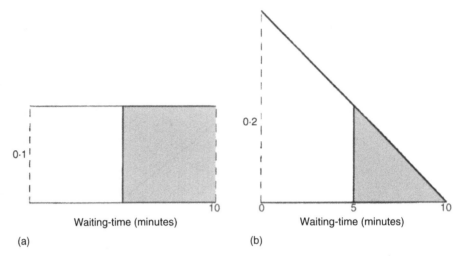

Figure 14.4 *Graphs of (a) uniform and (b) 'triangle' distributions of waiting-times. Shaded areas equal the probabilities of waiting-times being greater than 5 minutes.*

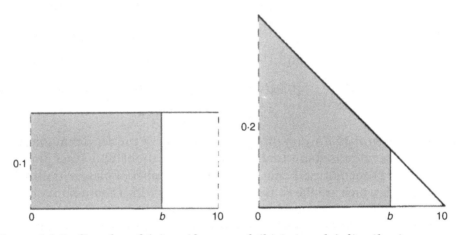

Figure 14.5 *Graphs of (a) uniform and (b) 'triangle' distributions of waiting-times. Shaded areas equal the probabilities of waiting-times being less than b minutes.*

of all values of the variable that are less than b; but from that it is very easy to obtain the probability of values greater than b. 'Less than b' and 'greater than b' are complementary events (since $P(X = b) = 0$).

For the uniform distribution, the probability of obtaining a value less than b is $0.1b$ (Fig. 14.5(a)). To find the probability for the triangle distribution we need to determine the area of a trapezium (Fig. 14.5(b)) whose parallel sides are of length 0.2 and $0.02(10 - b)$. The area of the trapezium is therefore $\frac{1}{2}b[0.2 + 0.02(10 - b)] = 0.01b(20 - b)$. Alternatively, the area beneath the line $y = 0.02(10 - x)$, $0 \leqslant x \leqslant b$, may be evaluated by integration. The two functions, $0.1b$ and $0.01b(20 - b)$, are graphed in Fig. 14.6(a). In fact, the quantity that is

CONTINUOUS RANDOM VARIABLES

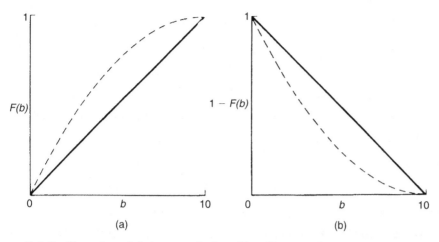

Figure 14.6 *Graphs of (a) cumulative distribution function F(b) and (b) 1 − F(b) for the uniform distribution (−) and the 'triangle' distribution (---).*

of interest to the professor is the probability that the waiting-time exceeds b, which is $1 - F(b)$. This quantity is graphed in Fig. 14.6(b). There is a substantial reduction in the probability of waiting more than 5 minutes by changing from the uniform to the triangle distribution.

The median is perhaps the best single measure to use when comparing these two distributions. The median waiting-time is given by the value M such that the probability of obtaining a waiting-time value less than or equal to M is 0.5. To derive M from the graph of the density function, we find the value of x which corresponds to the ordinate that divides the area under the graph in half. Alternatively, we find the value of b for which the cumulative distribution function $F(b)$ has the value 0.5. The medians are 5 minutes for the uniform distribution and 2.9 minutes for the triangle distribution. Professor Aztec concludes that he can reduce his median waiting-time by almost 50% if he takes account of the train times but, as he says, 'If I stopped waiting for trains, when would I find the time to think up new problems?'.

14.1.1 Exercises

1. Suppose Professor Aztec's triangle distribution has a probability density function proportional to $5 - x, 0 \leqslant x \leqslant 5$, and is zero otherwise. Find the median of this distribution.

2. Suppose that a random variable follows a continuous uniform distribution, and that its range of values is from 0 to k. What is: (i) the probability density function; and (ii) the cumulative distribution function? Find the median of this distribution.

14.2 THE SPECIFICATION OF CONTINUOUS RANDOM VARIABLES

We have seen that a continuous random variable may be represented by the graph of a continuous function $f(x)$ (Fig. 14.7). We choose the function so that the probability of the variable taking any value within a particular interval (a, b) equals the area under the curve between a and b. This area is equal to the integral $\int_a^b f(x)\mathrm{d}x$.

Definition 14.1

The **probability density function** $f(x)$ of the continuous random variable X is a function whose integral from $x = a$ to $x = b$ ($b \geqslant a$) gives the probability that X takes a value in the interval (a, b).

We can write

$$P(a \leqslant X \leqslant b) = \int_a^b f(x)\,\mathrm{d}x.$$

Note that

$$P(X = a) = P(a \leqslant X \leqslant b) = \int_a^b f(x)\,\mathrm{d}x = 0.$$

What is the probability of obtaining a value near to a particular value c? Consider a small interval of length δx around the value c. The probability of obtaining

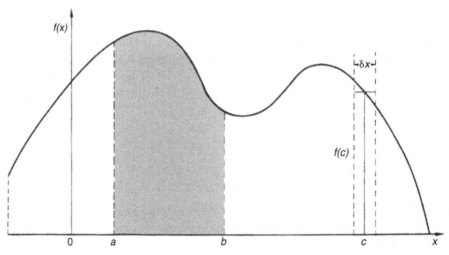

Figure 14.7 *Graph of a probability density function f(x). Shaded area equals probability of obtaining a value between a and b. Rectangle above the value c has area approximately equal to the probability of obtaining a value close to c.*

a value in this interval equals the area above the interval, and under the curve, and is therefore approximately $f(c)\,\delta x$. We see therefore that $f(c)$ is a measure of how likely it is that the random variable will be close to c (Fig. 14.7).

What conditions must a continuous function $f(x)$ satisfy if it is to be a probability density function? There are two, namely that:

(i)
$$f(x) \geqslant 0, \quad \text{for all values of } x,$$

(ii)
$$\int_{-\infty}^{\infty} f(x)\,dx = 1.$$

Condition (i) is necessary since $f(x)\,\delta x$ is the probability that the variable takes a value close to x, and a probability cannot be negative. An alternative way of stating condition (ii) is that the area under the curve over the whole range of the variable is unity. The range of integration is written as $-\infty$ to $+\infty$ so that all cases are covered. If X takes positive values only in the interval (a, b), and is zero otherwise, the condition reduces to $\int_a^b f(x)\,dx = 1$. We may verify for the triangle distribution above that

$$\int_{-\infty}^{\infty} f(x)\,dx = \int_0^{10} 0.02(10 - x)\,dx = 0.02\left[10x - \tfrac{1}{2}x^2\right]_0^{10}$$

$$= 0.02[100 - 50] = 0.02 \times 50 = 1.$$

The phrase 'probability density function' is used so often that we have a shorthand for it, 'pdf'.

We conclude this section by giving definitions of other concepts that have been introduced.

Definition 14.2

The **cumulative distribution function** $F(b)$, of the continuous random variable X with probability density function $f(x)$, is such that

$$F(b) = P(X \leqslant b) = \int_{-\infty}^{b} f(x)\,dx.$$

Since $F(x)$ is an integral of $f(x)$, it follows that $f(x)$ is the derivative of $F(x)$, i.e. $F'(x) = f(x)$.

Definition 14.3

The **median** of the continuous random variable X with probability density function $f(x)$ is that value M such that

$$\int_{-\infty}^{M} f(x)\,dx = 0.5.$$

Definition 14.4

The **mode** of the continuous random variable X with probability density function $f(x)$ is the value of x for which $f(x)$ has a relative (local) maximum.

It is wise to make a sketch of the probability density function before determining the mode. Differentiating the pdf will often yield the mode.

Example 14.1

A continuous random variable X takes values between 0 and 3 only, and its probability density function is $kx^2(3 - x)$. Find: (i) the value of the constant k; (ii) the cumulative distribution function; (iii) the mode.

(i) The integral of the pdf over the range $0 \leqslant x \leqslant 3$ must be 1. Now

$$\int_0^3 kx^2(3 - x)\,dx = 3k \int_0^3 x^2 dx - k \int_0^3 x^3 dx$$

$$= k[x^3 - \tfrac{1}{4}x^4]_0^3 = (3^3 - \tfrac{1}{4} \cdot 3^4) = \frac{27}{4}k.$$

Therefore

$$k = \frac{4}{27}.$$

(ii) The cdf

$$F(b) = \int_0^b \frac{4}{27}x^2(3 - x)\,dx = \left[\frac{4}{27}(x^3 - \tfrac{1}{4}x^4)\right]_0^b$$

$$= \frac{4}{27}b^3(1 - \tfrac{1}{4}b).$$

(iii) When $x = 0$, or when $x = 3$, $f(x) = 0$. Between $x = 0$ and 3, there is a maximum at $x = 2$, and the values of $f(x)$ increase steadily from $x = 0$ to $x = 2$, after which they decrease steadily. The mode is thus $x = 2$.

Example 14.2 Median and cdf of continuous random variable

A continuous random variable X has the pdf

$$f(x) = \begin{cases} 2x^c & 0 \leqslant x \leqslant 1, \\ 0 & \text{otherwise.} \end{cases}$$

Find: (i) the value of the constant c; (ii) the cdf; (iii) the median.

(i) We must have

$$\int_0^1 f(x)\,dx = 1, \quad \text{so} \quad 1 = \int_0^1 2x^c\,dx.$$

Thus

$$1 = \left[\frac{2}{c+1}x^{c+1}\right]_0^1 = \frac{2}{c+1},$$

i.e. $c + 1 = 2$, i.e. $c = 1$.

(ii) The pdf is $f(x) = 2x$, so the cdf is

$$F(b) = \int_0^b 2x \, dx \quad \text{for } 0 \leqslant b \leqslant 1$$
$$= \left[x^2 \right]_0^b = b^2.$$

(iii) For the median M, we require $F(M) = \frac{1}{2}$. This leads to $M^2 = \frac{1}{2}$, so $M = 0.707$. (The negative square root is not possible, being outside the range of x, which is from 0 to 1.)

14.2.1 Exercises

1. The random variable X has the pdf

$$f(x) = \begin{cases} cx & 0 \leqslant x \leqslant 1, \\ c(2-x) & 1 \leqslant x \leqslant 2, \\ 0 & \text{elsewhere.} \end{cases}$$

Determine the value of the constant c. Find the cdf, the median and the mode for this distribution.

2. By drawing a graph (or otherwise), show that $kx(x^2 - 4)$ is not a possible form for the pdf of a random variable that takes values in the range $-2 \leqslant x \leqslant 2$.

3. For each of the following functions $f(x)$, decide whether or not there is a constant k such that $kf(x)$ is a probability density function over the range of x-values indicated. Find the cumulative distribution functions for those cases where a pdf does exist.

   (i) $f(x) = x$, $0 < x < 2$.
   (ii) $f(x) = \sin x$, $0 \leqslant x \leqslant 3\pi$.
   (iii) $f(x) = x^2$, $x > 0$.
   (iv) $f(x) = \sqrt{x}$, $1 < x < 4$.
   (v) $f(x) = \cos x$, $-\frac{1}{2}\pi \leqslant x \leqslant \frac{1}{2}\pi$.
   (vi) $f(x) = e^{-2x}$, $x > 0$.

14.3 THE MEAN AND VARIANCE OF CONTINUOUS RANDOM VARIABLES

Professor Aztec might be interested in comparing the means of the uniform and triangle distributions he has specified. The mean of the uniform distribution is 5 minutes, by symmetry. What is the mean of the triangle distribution?

We defined the mean of a grouped set of observations of a continuous variate as

$$\bar{x} = \sum_{i=1}^{k} f_i x_i \Big/ \sum_{i=1}^{k} f_i \quad \text{(p. 33)}.$$

Suppose we divide the range of the continuous random variable X into small intervals, each of length δx. A typical one of these intervals, with centre x, has a probability of approximately $f(x)\,\delta x$ associated with it, the approximation improving as δx becomes smaller. We now put the probability $f(x)\,\delta x$ in place of the frequency in the variate definition. The mean, or expected value, of X is therefore approximately $\sum_{\text{all }x} x f(x)\,\delta x$. The limiting value of this, as $\delta x \to 0$, is the integral $\int_{\text{all }x} x f(x)\,\mathrm{d}x$. This leads us to the following definition.

Definition 14.5

The **mean**, or **expected value**, μ of the continuous random variable X which has probability density function $f(x)$ is

$$\int_{-\infty}^{\infty} x f(x)\,\mathrm{d}x.$$

The mean of the professor's triangle distribution is

$$\int_{0}^{10} 0.02\,x(10 - x)\,\mathrm{d}x = 0.02 \left[5x^2 - \frac{x^3}{3} \right]_{0}^{10}$$

$$= 0.02 \left[500 - \frac{1000}{3} \right] = \frac{2 \times 5}{3} = \frac{10}{3} = 3\tfrac{1}{3}.$$

A similar argument to that for the mean leads us to a definition of the variance.

Definition 14.6

The **variance** σ^2 of the continuous random variable X which has probability density function $f(x)$ and mean μ is

$$\int_{-\infty}^{\infty} (x - \mu)^2 f(x)\,\mathrm{d}x.$$

The variance of the uniform distribution is

$$\int_{0}^{10} 0.1(x - 5)^2\,\mathrm{d}x = 0.1 \int_{0}^{10} (x^2 - 10x + 25)\,\mathrm{d}x = 0.1 \left[\frac{x^3}{3} - 5x^2 + 25x \right]_{0}^{10}$$

$$= 0.1 \left[\frac{1000}{3} - 500 + 250 \right] = \frac{25}{3} = 8.33.$$

The standard deviation $\sqrt{8.33} = 2.89$.

Variances may often be worked out more easily if we use the expectation operator. This is the method we shall use to find the variance of the professor's triangle distribution.

14.4 USE OF THE EXPECTATION OPERATOR WITH CONTINUOUS RANDOM VARIABLES

We have seen that the basic definition of the expected value of a continuous random variable X is

$$E[x] = \int_{-\infty}^{\infty} xf(x)\,dx.$$

There is a theorem for the expected value of a function of a continuous random variable that corresponds to a similar theorem (9.2, p. 205) for discrete random variables. We state it without proof.

Theorem 14.1 Expected value of function of random variable

If the continuous random variable X has pdf $f(x)$ then the random variable which is a continuous function of X and is denoted by $g(X)$ has expected value

$$E[g(X)] = \int_{-\infty}^{\infty} g(x)f(x)\,dx.$$

It follows from this theorem that the expected value of the sum of any finite number of functions is equal to the sum of the expected values of the functions. We shall show this for the two functions $g(X)$ and $h(X)$.

Now

$$E[g(X) + h(X)] = \int_{-\infty}^{\infty} [g(x) + h(x)]\,f(x)\,dx$$

$$= \int_{-\infty}^{\infty} g(x)f(x)\,dx + \int_{-\infty}^{\infty} h(x)f(x)\,dx = E[g(X)] + E[h(X)].$$

We use this to obtain an expression for the variance of a continuous random variable, which is identical with the one we derived for discrete random variables (p. 208):

$$\mathrm{Var}[X] = \int_{-\infty}^{\infty} (x - \mu)^2 f(x)\,dx, \quad \text{where } \mu = E[X],$$

$$= E[(X - \mu)^2] = E[X^2 - 2\mu X + \mu^2]$$

$$= E[X^2] - E[2\mu X] + E[\mu^2]$$

$$= E[X^2] - 2\mu E[X] + E[\mu^2]$$

$$= E[X^2] - 2\mu^2 + \mu^2$$

$$= E[X^2] - \mu^2 = E[X^2] - (E[X])^2.$$

Thus, for the professor's triangle distribution,

$$E[X^2] = \int_0^{10} x^2(0.02)(10-x)\,dx = 0.02 \int_0^{10} (10x^2 - x^3)\,dx$$

$$= 0.02 \left[\frac{10x^3}{3} - \frac{x^4}{4} \right]_0^{10} = 200 \left[\frac{1}{3} - \frac{1}{4} \right] = \frac{200}{12} = \frac{50}{3}.$$

$$\text{Var}[X] = E[X^2] - \mu^2 = \frac{50}{2} - \left(\frac{10}{3} \right)^2 = \frac{150 - 100}{9} = \frac{50}{9} = 5.56.$$

Example 14.3 Mean, mode and variance of continuous random variable

A random variable X has probability density function

$$f(x) = \begin{cases} cx(6-x)^2 & 0 \leqslant x \leqslant 6, \\ 0 & \text{elsewhere.} \end{cases}$$

Calculate the arithmetic mean, mode, variance and standard deviation of X. We require $\int_0^6 cx(6-x)^2\,dx = 1$. This implies

$$c \int_0^6 x(36 - 12x + x^2)\,dx = 1.$$

Thus

$$c \left[18x^2 - 4x^3 + \frac{x^4}{4} \right]_0^6 = 1;$$

therefore

$$36c(18 - 24 + 9) = 1,$$

so that

$$c = \frac{1}{108}.$$

$$\mu = E[X] = c \int_0^6 x^2(6-x)^2\,dx = c \int_0^6 (36x^2 - 12x^3 + x^4)\,dx$$

$$= c \left[12x^3 - 3x^4 + \frac{x^5}{5} \right]_0^6 = \frac{216}{108} \left[12 - 18 + \frac{36}{5} \right] = \frac{2 \times 6}{5} = \frac{12}{5} = 2.4.$$

$$E[X^2] = c \int_0^6 x^3(6-x)^2\,dx = c \int_0^6 (36x^3 - 12x^4 + x^5)\,dx$$

$$= c \left[9x^4 - \frac{12x^5}{5} + \frac{x^6}{6} \right]_0^6 = \frac{36^2}{108} \left[9 - \frac{72}{5} + 6 \right] = \frac{36}{5}.$$

$$\sigma^2 = \text{Var}[X] = \text{E}[X^2] - \mu^2 = \frac{36}{5} - \frac{144}{25} = \frac{180 - 144}{25} = \frac{36}{25} = 1.44.$$

$$\sigma = \frac{6}{5} = 1.2.$$

The density function has a graph which is part of the cubic curve $y = cx(6-x)^2$. It passes through the origin, increases to a maximum as x increases, and then decreases to touch the x-axis at $x = 6$.

Since

$$\frac{dy}{dx} = c(36 - 24x + 3x^2) = 3c(12 - 8x + x^2) = 3c(6-x)(2-x),$$

there is a turning point at $x = 2$. This is the maximum since

$$\frac{d^2y}{dx^2} = 3c(-8 + 2x),$$

which is negative when $x = 2$. Hence the mode is 2.

We shall now look at two theoretical distributions which often are of use as models.

14.5 THE CONTINUOUS UNIFORM DISTRIBUTION

Definition 14.7

The continuous random variable X has a **uniform distribution** over the interval (α, β) if the pdf is given by

$$f(x) = \begin{cases} \dfrac{1}{\beta - \alpha} & \alpha < x < \beta, \\ 0 & \text{otherwise.} \end{cases}$$

We see this is a valid density function since $f(x) \geqslant 0$ and

$$\int_\alpha^\beta \frac{1}{\beta - \alpha}\, dx = \left[\frac{x}{\beta - \alpha}\right]_\alpha^\beta = \frac{\beta - \alpha}{\beta - \alpha} = 1.$$

We see by symmetry that the mean and median are at $(\alpha + \beta)/2$.

$$\text{E}[X^2] = \int_\alpha^\beta \frac{x^2}{\beta - \alpha}\, dx = \frac{1}{\beta - \alpha}\left[\frac{x^3}{3}\right]_\alpha^\beta = \frac{\beta^3 - \alpha^3}{3(\beta - \alpha)} = \frac{\beta^2 + \alpha\beta + \alpha^2}{3}.$$

$$\begin{aligned} \text{Var}[X] = \text{E}[X^2] - \mu^2 &= \tfrac{1}{3}(\beta^2 + \alpha\beta + \alpha^2) - \tfrac{1}{4}(\alpha + \beta)^2 \\ &= \tfrac{1}{12}[4(\beta^2 + \alpha\beta + \alpha^2) - 3(\alpha^2 + 2\alpha\beta + \beta^2)] = \tfrac{1}{12}(\beta^2 - 2\alpha\beta + \alpha^2) \\ &= \tfrac{1}{12}(\beta - \alpha)^2. \end{aligned}$$

The distribution function is

$$F(b) = \int_\alpha^b \frac{dx}{\beta - \alpha} = \frac{b - \alpha}{\beta - \alpha}, \quad \alpha < b < \beta.$$

and equals 0 for $b < \alpha$ and 1 for $b > \beta$.

The uniform distribution is commonly used as a model for the error made by rounding up or down when recording measurements. If we measure rods to the nearest centimetre, then a rod whose length we record as 72 cm has in fact a length that is somewhere between 71.5 and 72.5, and is equally likely to be anywhere in that range. The difference between the true length and the recorded length of 72 cm is a quantity that is equally likely to take any value between -0.5 cm and $+0.5$ cm. A reasonable model for rounding errors is therefore that they are uniformly distributed on the interval $(-0.5, 0.5)$.

The values of the parameters α, β of the distribution are usually obvious from the context and rarely need to be estimated.

14.5.1 Exercises

1. The probability density function of x is

$$f(x) = \begin{cases} kx(1 - ax^2) & 0 \leqslant x \leqslant 1, \\ 0 & \text{otherwise}, \end{cases}$$

where k and a are positive constants. Show that $a \leqslant 1$ and that $k = 4/(2 - a)$.

Sketch the probability density function for the case when $a = 1$ and in this case find the mean, variance and 75th percentile of the distribution. (*JMB*)

2. A continuous random variable X has the pdf

$$f(x) = \begin{cases} 3x^a & 0 < x < 1, \\ 0 & \text{otherwise}. \end{cases}$$

Find:
   (i) the value of the constant a;
   (ii) the expectation of X;
   (iii) the cdf of X;
   (iv) the probability that X lies between $\frac{1}{4}$ and $\frac{1}{2}$. (*IOS*)

3. A probability density $p(x)$ is given by

$$p(x) = \begin{cases} Cx(4 - x) & 0 \leqslant x \leqslant 4, \\ 0 & \text{otherwise}, \end{cases}$$

where C is a positive constant. Find the value of C and sketch the distribution.

Explain why the standard deviation of this distribution is the same as the standard deviation of the distribution with probability density

$$p(x) = \begin{cases} C(4 - x^2) & |x| \leqslant 2, \\ 0 & |x| > 2. \end{cases}$$

For the first distribution, calculate the mean, the standard deviation and the mode. Calculate also the probability that a value of the variable x taken at random will lie further than one standard deviation away from the mean.   (*MEI*)

14.6 THE EXPONENTIAL DISTRIBUTION

Definition 14.8

A continuous random variable X has an **exponential distribution** if it has, for some $\lambda > 0$, the pdf

$$f(x) = \begin{cases} \lambda e^{-\lambda x} & x \geqslant 0 \\ 0 & x < 0. \end{cases}$$

The pdf has the graph shown in Fig. 14.8. Notice that x may tend to infinity.
    Since $f(x) \geqslant 0$, and

$$\int_0^\infty \lambda e^{-\lambda x}\, \mathrm{d}x = \left[-e^{-\lambda x} \right]_0^\infty = 1,$$

$f(x)$ is a valid pdf. Its mean is

$$\mu = \mathrm{E}[X] = \int_0^\infty x \lambda e^{-\lambda x}\, \mathrm{d}x.$$

Figure 14.8 *Graph of e^{-x} (exponential distribution with $\lambda = 1$).*

This may be evaluated using integration by parts:

$$\int_0^\infty x(\lambda e^{-\lambda x})\,\mathrm{d}x = \left[-xe^{-\lambda x}\right]_0^\infty - \int_0^\infty (-e^{-\lambda x})\,\mathrm{d}x$$

$$= \int_0^\infty e^{-\lambda x}\,\mathrm{d}x, \quad \text{since } \lim_{x\to\infty}(xe^{-\lambda x}) = 0,$$

$$= \left[-\frac{e^{-\lambda x}}{\lambda}\right]_0^\infty = \frac{1}{\lambda}.$$

The variance is then evaluated as follows:

$$\mathrm{E}[X^2] = \int_0^\infty x^2 \lambda e^{-\lambda x}\,\mathrm{d}x = \left[-x^2 e^{-\lambda x}\right]_0^\infty - \int_0^\infty 2x(-e^{-\lambda x})\,\mathrm{d}x$$

$$= \frac{0 + 2\mathrm{E}[X]}{\lambda} = \frac{2}{\lambda^2}.$$

$$\mathrm{Var}[X] = \mathrm{E}[X^2] - \mu^2 = \frac{2}{\lambda^2} - \frac{1}{\lambda^2} = \frac{1}{\lambda^2}.$$

The cumulative distribution function of the exponential distribution is

$$F(b) = \int_0^b \lambda e^{-\lambda x}\,\mathrm{d}x = \left[-e^{-\lambda x}\right]_0^b = 1 - e^{-\lambda b}.$$

The median is given by the equation $F(M) = 0.5$, or

$$1 - e^{-\lambda M} = 0.5$$

i.e.

$$e^{-\lambda M} = 0.5;$$

thus

$$e^{\lambda M} = 2$$

so that

$$\lambda M = \ln 2 \quad (\text{ln stands for } \log_e)$$

or

$$M = \frac{\ln 2}{\lambda}.$$

The mode is at 0.

14.7 USES OF THE EXPONENTIAL DISTRIBUTION

A number of real-life situations can be modelled using an exponential distribution. If a batch of electric light-bulbs has been made in the same factory to the same specification, we may record the length of time each bulb lasts before it fails. Some of the bulbs will last a very short time, and a few will last a very long time. The exponential random variable X often provides a good model of the relation between lifetime x and its frequency, $f(x)$. Similar models often fit the lifetime distributions of other electrical and electronic components.

As a second example, consider measuring the length of time, in seconds, between vehicles as they pass an observation point. We suppose that traffic is flowing freely, so that vehicles can travel independently of each other, not restricted by the one in front. Occasionally, therefore, there will be quite long time-intervals between successive vehicles, while more often there will be smaller intervals. Again the distribution of time-intervals often follows an exponential distribution.

Another variate that can be modelled using the exponential distribution is the length of time between successive reports of an infectious (notifiable) disease, when it is spreading in a random manner through the population. There is, in fact, a general relation between the exponential distribution and the Poisson distribution (Chapter 20) which serves to explain all these examples.

Example 14.4 *Fitting the exponential distribution to data*

The times to failure of 500 electric batteries have been recorded as follows:

Time (hours), x	Frequency
0–50	208
50–100	112
100–150	75
150–200	40
200–250	30
250–300	18
300–350	11
350–400	6

Estimate the mean of these times. Draw a histogram to illustrate the data. Suggest a distribution which could be used to model these data, and find the frequencies in each time-interval on this model.

To estimate the mean, we shall take the mid-point of the interval 0–50 as 25, and so on; we have not been told how accurately the records were made, e.g. nearest hour, nearest minute, so this is as good a mid-point as we can take. Proceeding directly, the mean is

$$[(25 \times 208) + (75 \times 112) + (125 \times 75) + \cdots + (375 \times 6)]/500 = 47\,500/500 = 95.$$

An alternative calculation is to code the mid-points of the time-intervals, x, into a new variable $u = \frac{1}{50}(x - 25)$; this leads to $\bar{u} = 1.4$, so $\bar{x} = 50\bar{u} + 25 = 95$.

The histogram is shown in Fig. 14.9, and suggests an 'exponential decay' type of pattern, that is to say the exponential distribution. If an exponential distribution is to explain the data, it must have the same mean as the data, namely 95. The

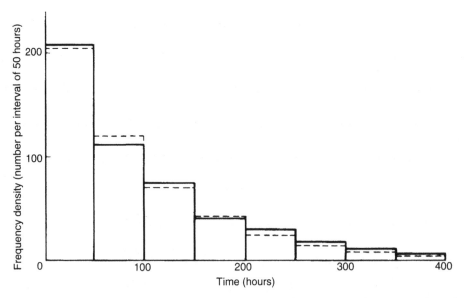

Figure 14.9 *Histogram showing lifetimes of 500 electric batteries (——). Broken lines (- - -) show frequencies based on exponential distribution model.*

exponential distribution has only one parameter λ; its mean is $1/\lambda$. Hence we need to fit an exponential distribution with $\lambda = \frac{1}{95}$. We will do this by computing $F(50)$, the cumulative probability up to $x = 50$; this gives the probability in the first interval. We saw above that $F(b) = 1 - e^{-\lambda b}$, which in this case is $1 - e^{-b/95}$. We compute $F(100)$, and then $F(100) - F(50)$; this gives the probability in the second interval. We continue in this way. Table 14.1 contains the results of these computations, in its third column.

We see at once one of the difficulties in using continuous random variables to model real data: the exponential distribution has no upper limit. There is in fact a probability of 0.0148 of obtaining values in the exponential distribution that

Table 14.1 *Frequencies in exponential distribution with $\lambda = 1/95$*

Time (hours), x	Upper limit of interval, b	$F(b)$	Probability in interval	Frequency in interval
0–50	50	0.4092	0.4092	204.6
50–100	100	0.6510	0.2418	120.9
100–150	150	0.7938	0.1428	71.4
150–200	200	0.8782	0.0844	42.2
200–250	250	0.9280	0.0498	24.9
250–300	300	0.9575	0.0295	14.7
300–350	350	0.9749	0.0174	8.7
350–400	400	0.9852	0.0103	5.1
Above 400	∞	1.0000	0.0148	7.4

are higher than anything we observed in the data we are trying to model. This is inevitable whenever we use, in the model, a random variable having one (or both) of its limits infinite. But if only such a small probability as 0.0148 (i.e. $1\frac{1}{2}\%$ of the total) falls outside the observed data, we can be satisfied that we have a very reasonable model. We shall show in Chapter 19 how to test whether the frequencies in the observed data and those in the last column of Table 14.1 are 'close together', statistically speaking. Meanwhile, we have shown both sets of frequencies in Fig. 14.9, and agreement looks good.

Observed frequencies must be whole numbers; but in the last column of Table 14.1 we have quoted the model frequencies to one place of decimals. This serves to emphasize that they *are* model frequencies, calculated on a theoretical model that really is a continuous random variable. As we have already pointed out, even continuous data can, in real life, only be recorded to the nearest unit. But this does *not* imply that model frequencies also have to be rounded to whole numbers; to make them so would impose some inaccuracy in comparing the data with the model.

14.7.1 Exercises

1. A random variable X follows the exponential distribution with mean 1. Find its (cumulative) distribution function, its median, and the probability that an observation drawn from the distribution will lie between the median and the mean.

2. The time t seconds between the arrival of successive vehicles at a point on a country road has a probability density function $ke^{-\lambda t}, t \geqslant 0$, where k and λ are positive constants. Find k in terms of λ and sketch the graph of the probability density function.

    Given that $\lambda = 0.01$, find the mean and variance of t.

    $$\left(\text{You may use the result} \int_0^\infty t^r e^{-\lambda t}\, dt = \frac{r!}{\lambda^{r+1}}, \quad r = 0, 1, 2, \ldots.\right)$$

    An elderly pedestrian takes 50 seconds to cross the road at this point. With $\lambda = 0.01$, calculate the probability that, if he sets off as one vehicle passes, he will complete the crossing before the next vehicle arrives. Calculate also the probability that, if he adopts the same procedure on the return journey, he completes each crossing without a vehicle arriving while he is doing so.

    (JMB)

3. An exponentially distributed random variable is such that its mean is equal to twice its variance. What is the value of its parameter λ?

4. Three random variables X, Y, Z are independently distributed, in exponential distributions, with parameters λ, μ, ν respectively. Find: (i) $E[X + Y + Z]$; (ii) $E[XY]$; (iii) $\text{Var}[X + Y + Z]$; (iv) $\text{Var}[Y - 3Z]$.

5. The probability that a light-bulb lasts longer than t hours is $e^{-t/\mu}$. Find the probability density function for the lifetime of a bulb.

Show that the mean lifetime is μ.

If the mean lifetime is 1500 hours, how unlikely is it that a bulb will last more than 3000 hours?

If the manufacturer wants to ensure that less than 1 in 1000 bulbs fail before 5 hours, what is the lowest mean lifetime he can allow his bulbs to have?

(SMP)

14.8 THE MOMENT GENERATING FUNCTION

We defined the probability generating function for a discrete random variable as $\sum_{\text{all } r} p_r t^r$ (p. 216). A similar function is the *moment generating function*, which can be used for discrete and continuous variables. For a discrete variable, write $M(t) = \sum_{\text{all } r} p_r e^{rt}$, replacing t in the pgf by e^t. This function $M(t)$ is called the moment generating function because from it we can obtain the *moments* of a distribution. Two classes of moments, similar in nature to the physical moments defined in mechanics, exist and can be expressed in terms of the expectation operator.

Definition 14.9

The *s*th **uncorrected moment** of a continuous random variable with pdf $f(x)$ is

$$E[X^s] = \int_{-\infty}^{\infty} x^s f(x)\,\mathrm{d}x.$$

Definition 14.10

The *s*th **corrected moment** of a continuous random variable with pdf $f(x)$ and mean μ is

$$E[(X - \mu)^s] = \int_{-\infty}^{\infty} (x - \mu)^s f(x)\,\mathrm{d}x.$$

The usual notation for moments is to write $E[X^s] = \mu'_s$ and $E[(X - \mu)^s] = \mu_s$. For discrete random variables, equivalent definitions are

$$\mu'_s = \sum_{\text{all } r} r^s p_r \quad \text{and} \quad \mu_s = \sum_{\text{all } r} (r - \mu)^s p_r.$$

Thus the first uncorrected moment, $E[X]$, is simply the mean, μ. As μ corresponds to the centre of mass in mechanics, corrected moments correspond to moments about this mass centre. The second corrected moment, $E[(X - \mu)^2]$, is the variance, σ^2. This corresponds to a moment of inertia.

The moment generating function (mgf) can likewise be defined using the E operator.

Definition 14.11

The **moment generating function** of a random variable X is $E[e^{Xt}]$.

This definition holds for both discrete and continuous variables.

14.8.1 Generating the moments

Let us assume that e^{xt} can be expanded as a series:

$$e^{xt} = 1 + xt + \frac{(xt)^2}{2!} + \frac{(xt)^3}{3!} + \cdots$$

(see Appendix II). Then we can write

$$E[e^{xt}] = \int_{-\infty}^{\infty} e^{xt} f(x)\,dx = \int_{-\infty}^{\infty} \left(1 + xt + \frac{(xt)^2}{2!} + \frac{(xt)^3}{3!} + \cdots \right) f(x)\,dx.$$

Further, assuming that we can integrate each term separately (which is true when $f(x)$ is a pdf),

$$E[e^{xt}] = \int_{-\infty}^{\infty} f(x)\,dx + t \int_{-\infty}^{\infty} xf(x)\,dx + \frac{t^2}{2!} \int_{-\infty}^{\infty} x^2 f(x)\,dx + \frac{t^3}{3!} \int_{-\infty}^{\infty} x^3 f(x)\,dx + \cdots,$$

i.e.

$$E[e^{xt}] = M(t) = 1 + t\mu + \frac{t^2}{2!}\mu_2' + \frac{t^3}{3!}\mu_3' + \cdots + \frac{t^s}{s!}\mu_s' + \cdots,$$

In this series, μ_s' is the coefficient of $t^s/s!$.

Example 14.5

Find the mgf of the exponential distribution. Express this as a power series in the dummy variable t, and use it to find the mean and variance.

The pdf is

$$f(x) = \begin{cases} \lambda e^{-\lambda x} & 0 < x < \infty. \\ 0 & \text{elsewhere,} \end{cases}$$

and so the mgf is

$$\int_0^{\infty} e^{xt} \lambda e^{-\lambda x}\,dx = \lambda \int_0^{\infty} e^{-(\lambda - t)x}\,dx.$$

Setting $u = x(\lambda - t)$, this is

$$\lambda \int_0^{\infty} e^{-u} \frac{du}{\lambda - t} = \frac{\lambda}{\lambda - t}[-e^{-u}]_0^{\infty} = \frac{\lambda}{\lambda - t}.$$

To expand, write this mgf as

$$\left(1 - \frac{t}{\lambda} \right)^{-1} = 1 + \frac{t}{\lambda} + \frac{t^2}{\lambda^2} + \frac{t^3}{\lambda^3} + \cdots.$$

Hence the mean μ is equal to $1/\lambda$, the coefficient of t. To find the variance, use the relation that $E[(x - \mu)^2] = E[x^2] - \mu^2$, as shown on p. 323. Hence the variance,

σ^2, is equal to $\mu_2' - \mu^2$. The coefficient of $t^2/2!$ in the mgf's series expansion is $2!/\lambda^2 = 2/\lambda^2$; this is μ_2'. So

$$\sigma^2 = \frac{2}{\lambda^2} - \left(\frac{1}{\lambda}\right)^2 = \frac{1}{\lambda^2},$$

as already shown.

By considering the expansion

$$M(t) = 1 + t\mu + \frac{t^2}{2!}\mu_2' + \cdots + \frac{t^s}{s!}\mu_s' + \cdots,$$

we see that

$$\left(\frac{dM(t)}{dt}\right)_{t=0} = \mu; \quad \left(\frac{d^2M(t)}{dt^2}\right)_{t=0} = \mu_2'; \ldots; \left(\frac{d^sM(t)}{dt^s}\right)_{t=0} = \mu_s'; \ldots.$$

Applying this to the exponential distribution, $M(t) = \lambda/(\lambda - t)$, so that

$$\frac{dM(t)}{dt} = \frac{\lambda}{(\lambda - t)^2};$$

when $t = 0$,

$$\frac{dM(t)}{dt} = \frac{1}{\lambda},$$

which is μ as already found. Differentiating again,

$$\frac{d^2M(t)}{dt^2} = \frac{2\lambda}{(\lambda - t)^3};$$

when $t = 0$, this is $2/\lambda^2$, which is μ_2' as already found. Thus in either of these ways, by a series expansion or by differentiation, we can generate the set of uncorrected moments from the mgf $M(t)$.

Example 14.6 Mgf of the binomial distribution

Find the mgf of the binomial distribution, and use it to show that the mean is $n\pi$ and the variance is $n\pi(1 - \pi)$.

$$P(R = r) = \binom{n}{r} \pi^r (1 - \pi)^{n-r}, \quad r = 0, 1, \ldots, n.$$

So

$$M(t) = \sum_{r=0}^{n} e^{rt} P(R = r) = \sum_{r=0}^{n} \binom{n}{r} (\pi e^t)^r (1 - \pi)^{n-r}$$

$$= [(1 - \pi) + \pi e^t]^n$$

using the binomial series (Appendix I).

$$\frac{\mathrm{d}M(t)}{\mathrm{d}t} = n[(1-\pi)+\pi e^t]^{n-1}\pi e^t = n(1-\pi+\pi)^{n-1}\pi, \quad \text{when } t=0, = n\pi.$$

$$\frac{\mathrm{d}^2 M(t)}{\mathrm{d}t^2} = n(n-1)[(1-\pi)+\pi e^t]^{n-2}\pi^2 e^{2t} + n[(1-\pi)+\pi e^t]^{n-1}\pi e^t$$

$$= n(n-1)\pi^2 + n\pi \quad \text{when } t=0.$$

This is μ_2'. We require

$$\sigma^2 = \mu_2' - \mu^2$$

$$= n(n-1)\pi^2 + n\pi - n^2\pi^2$$

$$= n\pi(1-\pi).$$

14.8.2 Exercises

1. The discrete random variable X takes the values 0 and 1, each with probability $\frac{1}{2}$. Find its probability generating function, its moment generating function and its standard deviation.

2. Find the moment generating function of the random variable X which has the continuous uniform distribution on the interval (a,b). Hence find the mean and variance of X.

14.9 SUMS OF INDEPENDENT RANDOM VARIABLES

The probability generating function (Section 9.9) and the moment generating function (Section 14.8) can both be used to give summaries of the set of probabilities in the distribution of $X+Y$, where X and Y are two independent random variables.

If X takes the values $0,1,2,\ldots$ with probabilities $p_0, p_1, p_2, \ldots$ and Y also takes the values $0,1,2,\ldots$ but with probabilities $\pi_0, \pi_1, \pi_2, \ldots$ then the sum $X+Y$ can take the value r in a variety of ways:

> if $X=0$, Y must equal r;
> if $X=1$, Y must equal $r-1$;
> if $X=2$, Y must equal $r-2$; and so on, until
> if $X=r$, Y must equal 0.

The probabilities of these ways are, respectively, $p_0\pi_r, p_1\pi_{r-1}, p_2\pi_{r-2}, \ldots$ up to $p_r\pi_0$, provided that X and Y are independent. Then the total probability that $X+Y=r$, given that X and Y are independent, is

$$p_0\pi_r + p_1\pi_{r-1} + p_2\pi_{r-2} + \cdots + p_{r-1}\pi_1 + p_r\pi_0.$$

Let the probability generating function of X be

$$G_X(t) = E[t^X] = p_0 t^0 + p_1 t^1 + \cdots + p_r t^r + \cdots$$

and of Y be

$$G_Y(t) = E[t^Y] = \pi_0 t^0 + \pi_1 t^1 + \cdots + \pi_r t^r + \cdots.$$

Thus in the product $G_X(t)G_Y(t)$, the coefficient of t^r is

$$p_0 \pi_r + p_1 \pi_{r-1} + \cdots + p_{r-1} \pi_1 + p_r \pi_0.$$

Therefore, by the definition on p. 216, $G_X(t)G_Y(t)$ is the probability generating function of $X + Y$. Because X and Y are independent, we may also write

$$G_{X+Y}(t) = E[t^{X+Y}] = E[t^X]E[t^Y] = G_X(t)G_Y(t).$$

These results may be stated as a theorem.

Theorem 14.2

Let the probability generating functions of the independent random variables X and Y be $G_X(t)$ and $G_Y(t)$ respectively. Then the **probability generating function of the sum $X + Y$** is $G_{X+Y}(t) = G_X(t)G_Y(t)$.

Example 14.7

Two fair coins are each tossed once. Let X take the value 1 if the first coin shows a head, and 0 otherwise; likewise $Y = 1$ if the second coin shows a head and $Y = 0$ otherwise. Find the distribution of the sum $X + Y$, the total number of heads when both are tossed together.

For both variables, their pgf is $\frac{1}{2}t^0 + \frac{1}{2}t^1 = \frac{1}{2}(1 + t)$; so by multiplying the two pgfs together we find $G_X(t)G_Y(t) = \frac{1}{4}(1 + t)^2$ for the pgf of the sum. This is $\frac{1}{4}t^0 + \frac{1}{2}t^1 + \frac{1}{4}t^2$, and if we write this as $P_0 t^0 + P_1 t^1 + P_2 t^2$ we see that $P_0 = \frac{1}{4}$ is the probability of no heads, $P_1 = \frac{1}{2}$ is the probability of one head, and $P_2 = \frac{1}{4}$ is the probability of two heads when both coins are tossed once.

These results confirm what we know already, but they also show a quick way of combining the results of many independent tosses.

14.9.1 The binomial distribution

We may regard the random variables defined in Example 14.7 as Bernoulli variables with $\pi = \frac{1}{2}$. From Example 9.12 (p. 217) we know that for any value π in $(0 < \pi < 1)$ the probability generating function of a Bernoulli random variable is $G(t) = 1 - \pi + \pi t$. But we also saw in Example 9.13 that the pgf of a binomial (n, π) is $\{1 - \pi + \pi t\}^n$. Now if we add together n independent Bernoulli random variables, independent of one another but all having the same parameter π, the pgf of this sum will be $\{G(t)\}^n$, because each time we add a new random variable to the sum we must include an extra factor $G(t)$ in the product.

The sum of n independent Bernoulli random variables with parameter π has the same pgf as a binomial (n, π) random variable, and therefore this sum of Bernoullis is binomial (n, π) (assuming an equivalence theorem which we will not prove here). Ignoring the mathematics, we can see that we may think of the number of successes in n independent trials *either* as being a binomial random variable *or* as the sum of n independent Bernoulli random variables.

14.9.2 The geometric and negative binomial distributions

The geometric distribution has probability mass function

$$P(r) = \pi(1 - \pi)^{r-1} \quad r = 1, 2, 3, \ldots$$

(p. 177), where $0 < \pi < 1$; π is the probability of an item having a particular characteristic and r is the number of items (the sample size) needed to obtain one with this characteristic. Thus

$$
\begin{aligned}
G(t) &= \pi t^1 + \pi(1 - \pi)t^2 + \pi(1 - \pi)^2 t^3 + \cdots \\
&= \pi t(1 + (1 - \pi)t + (1 - \pi)^2 t^2 + \cdots) \\
&= \pi t / (1 - (1 - \pi)t).
\end{aligned}
$$

Suppose now that we carry on sampling until we have two items with the special characteristic. The process can be modelled by considering the sample size R_1 needed to find the first special item, and then beginning counting again to obtain R_2, the sample size to collect the second item. Each of these, R_1 and R_2, will have a geometric distribution as specified above. The total sample size needed to obtain two items with the special characteristic is $R_1 + R_2$, and its generating function will be

$$
\begin{aligned}
\{G(t)\}^2 &= \pi^2 t^2 / (1 - (1 - \pi)t)^2 \\
&= \pi^2 t^2 (1 - (1 - \pi)t)^{-2} \\
&= \pi^2 t^2 (1 + 2(1 - \pi)t + 3(1 - \pi)^2 t^2 + 4(1 - \pi)^3 t^3 + \cdots) \\
&= \pi^2 t^2 + 2\pi^2 (1 - \pi)t^3 + 3\pi^2 (1 - \pi)^2 t^4 + 4\pi^2 (1 - \pi)^3 t^5 + \cdots.
\end{aligned}
$$

Writing $R = R_1 + R_2$, we have $P(R) = (r - 1)\pi^2 (1 - \pi)^{r-2}, r = 2, 3, 4 \ldots$.

Comparing this with the formulae given in Exercise 9.11(11), we have a *negative binomial* distribution with $k = 2$. Thus the sum of two independent geometric distributions, with parameter π, is a negative binomial distribution, with parameters π and 2.

Let us take this development a stage further. If we carry on sampling until we have k items of the special type, the above argument shows that the *total* sample size R will have probability generating function

$$\{G(t)\}^k = \pi^k t^k / (1 - (1 - \pi)t)^k$$
$$= \pi^k t^k (1 - (1 - \pi)t)^{-k}$$
$$= \pi^k t^k \{1 + kt(1 - \pi) + k(k + 1)t^2 (1 - \pi)^2 / 2!$$
$$+ k(k + 1)(k + 2)t^3 (1 - \pi)^3 / 3! + \cdots\}.$$

Hence

$$P(R = k) = \pi^k; P(R = k + 1) = k(1 - \pi)\pi^k;$$
$$P(R = k + 2) = k(k + 1)(1 - \pi)^2 \pi^k / 2!; \ldots$$

and the term in t^{k+s} will be

$$k(k + 1)(k + 2)\ldots(k + s - 1)(1 - \pi)^s \pi^k / s!.$$

This general term can be written

$$\frac{\pi^k (1 - \pi)^s (k + s - 1)!}{s!(k - 1)!};$$

this is $P(R = k + s)$, and corresponds to the negative binomial distribution which was developed in Exercise 9.11(11). So in general, the sum of any number k of geometric random variables with parameter π generates a negative binomial with parameters π and k, and gives the distribution of the sample size needed to obtain exactly k items with a special characteristic.

14.9.3 Relation with moment generating function

We have seen (p. 332) that the moment generating function $M_X(t)$ of the random variable X equals $E[e^{Xt}]$. It may be obtained from the probability generating function $G_X(t)$, which is $E[t^X]$, by replacing t by e^t. Hence formally we can write

$$M_{X+Y}(t) = G_{X+Y}(e^t) = G_X(e^t)G_Y(e^t) = M_X(t)M_Y(t).$$

This leads to the following theorem.

Theorem 14.3

Let the moment generating functions of the independent random variables X and Y be $M_X(t)$ and $M_Y(t)$ respectively. Then the **moment generating function of the sum $X + Y$** is $M_{X+Y}(t) = M_X(t)M_Y(t)$.

14.10 DISTRIBUTION OF A FUNCTION OF A RANDOM VARIABLE

We shall consider only the case where X is distributed uniformly over the range (α, β), and the random variable Y is defined by a relation such as $Y = \log X$, $Y = X^3$, $Y = +\sqrt{X}$, in which only one value of Y corresponds to each value of X (i.e. the transformation from X to Y is one-to-one).

Example 14.8 Distribution of function of random variable

The random variable Y is defined by $Y = X^3$, and X is a random variable that is uniformly distributed over the range (α, β). Find the distribution function and the density function of Y. If $\alpha = 0$ and $\beta = 1$, sketch the graph of this density function, find the mean of the distribution of Y and the probability that Y is above its mean value.

The cdf of X is given by

$$F(b) = P(X \leqslant b) = \frac{b - \alpha}{\beta - \alpha}.$$

For Y, the cdf is $P(Y \leqslant c)$ (using c rather than b to avoid any confusion between the two variables). But

$$P(Y \leqslant c) = P(X^3 \leqslant c) = P(X \leqslant c^{1/3})$$
$$= (c^{1/3} - \alpha)/(\beta - \alpha).$$

Figure 14.10 *Graph of the probability density function $f(y) = \frac{1}{3}y^{-2/3}$.*

The range of X is (α, β), and so this expression for the cdf of Y holds in the range $\alpha \leqslant c^{1/3} \leqslant \beta$.

To find the pdf, we differentiate the cdf:

$$f(y) = \frac{d}{dy}\left(\frac{y^{1/3} - \alpha}{\beta - \alpha}\right) = \frac{1}{3(\beta - \alpha)y^{2/3}}, \quad \alpha^3 \leqslant y \leqslant \beta^3.$$

In the case $\alpha = 0, \beta = 1$, the density function of Y is $f(y) = 1/(3y^{2/3}), 0 \leqslant y \leqslant 1$. This is sketched in Fig. 14.10. The mean of Y is

$$\int_0^1 y\frac{1}{3y^{2/3}}\,dy = \frac{1}{3}\int_0^1 y^{1/3}\,dy = \left[\frac{1}{3}\cdot\frac{3}{4}y^{4/3}\right]_0^1 = \frac{1}{4}.$$

$$P(Y > \tfrac{1}{4}) = \int_{1/4}^1 \frac{1}{3y^{2/3}}\,dy = \left[y^{1/3}\right]_{1/4}^1 = 1 - \sqrt[3]{\tfrac{1}{4}} = 0.37.$$

14.10.1 Exercises

1. The area, x, of a square of side y, is uniformly distributed on $(a, 4a)$. Find the probability density function of y and also the mean and variance of y. Sketch the probability density function of y.

   If a random sample of four values of y is taken, find the probability that at least three of the values are greater than $\frac{3}{2}\sqrt{a}$. *(JMB)*

2. A measuring flask is so shaped that the volume V of the liquid in it is related to the height h of the surface of the liquid above the base of the flask by $V = \pi h^3/9$. The flask is used to measure volumes V of liquid which are uniformly distributed between 0 and 1 units. Find the probability density function of the distribution of the corresponding values of h and sketch the graph of this function. Calculate:
   (i)   the probability that h is less than 1 unit;
   (ii) the mean of h;
   (iii) the variance of h;
   (iv) the probability that h is less than the mean of h.
   (Take $\sqrt[3]{9/\pi}$ as 1.42.) *(JMB)*

14.11 MINITAB: CONTINUOUS RANDOM VARIABLES

MINITAB will provide values of the probability functions for the continuous uniform, exponential and triangle distributions that we have introduced in this chapter. The values are obtained by choosing

Calc>Probability Distributions>

followed by the name of the distribution. One may select **Probability density**, for the probability density function, **Cumulative probability**, for the cumulative distribution function, or **Inverse cumulative probability**, for the inverse cumulative distribution function.

Uniform is the name MINITAB uses for the continuous uniform distribution. The parameters that must be entered in the dialogue box are the **upper** and **lower endpoints**.

MINITAB uses a slight generalization of our definition of the **exponential** distribution by allowing a minimum value, or **threshold**, to be set with a value different from zero. In most applications zero will be the appropriate value. **Scale**, when the threshold is zero, is equal to the mean. The default values of threshold and scale, which appear in the dialogue box, are 0.0 and 1.0 respectively. To obtain the values of the exponential distribution function $F(b)$, given in Table 14.1, set the upper limits (of the intervals) $50, 100, \ldots, 400$ in C1 and choose

Calc>Probability Distributions>Exponential/
Select **Cumulative probability**, enter **Scale** (i.e. mean) 95 and C1 in **Input column**/

The **triangular** distribution of MINITAB is a generalization of our triangle distribution. In visual terms, the graph of the density function is two right-angled triangles back to back. Its parameters are the **upper** and **lower endpoints**, and the **mode**.

Random samples from the three distributions may be generated by choosing

Calc>Random Data.

14.12 EXERCISES ON CHAPTER 14

1. Find the mean, μ, and the variance, σ^2, of the random variable X which has pdf

$$f(x) = \begin{cases} x & 0 \leqslant x \leqslant 1, \\ 2 - x & 1 \leqslant x \leqslant 2, \\ 0 & \text{elsewhere.} \end{cases}$$

Two independent observations of X are made. Find the probability that they both lie in the interval $(\mu - \sigma, \mu + \sigma)$.

2. A random variable Y has a uniform distribution with probability density function

$$f(y) = \begin{cases} (b - a)^{-1} & a \leqslant y \leqslant b, \\ 0 & \text{otherwise.} \end{cases}$$

Find the standard deviation, σ, and the mean deviation, β, of Y and show that $2\beta = \sigma\sqrt{3}$.

3. A random variable Y has density function

$$f(y) = \begin{cases} k(7 - 2y) & 0 \leqslant y \leqslant 3, \\ 0 & \text{elsewhere.} \end{cases}$$

Find the value of k and calculate the mean and median of the distribution.
   Given three independent observations from this distribution, calculate the probability that all of them are less than the mean. What is the probability that at least one of the three observations lies between the median and the mean? *(IOS)*

4. The random variable X has the probability density function

$$f(x) = \begin{cases} \frac{1}{2} & -1 \leqslant x \leqslant 1, \\ 0 & \text{elsewhere.} \end{cases}$$

Find the mean, μ, and the variance, σ^2, of X.
   Find the probability that, given two independent observations on X, at least one lies outside the interval $(\mu - \sigma, \mu + \sigma)$. *(IOS)*

5. The probability density function $f(x)$ of a variable x is given by

$$f(x) = \begin{cases} kx \sin \pi x & 0 \leqslant x \leqslant 1, \\ 0 & \text{for all other values of } x. \end{cases}$$

Show that $k = \pi$ and deduce that the mean and the variance of the distribution are

$$1 - \frac{4}{\pi^2} \quad \text{and} \quad \frac{2}{\pi^2}\left(1 - \frac{8}{\pi^2}\right)$$

respectively. *(AEB)*

6. A mathematical model for the fraction x of the sky covered with cloud $(0 < x < 1)$ assigns to this a probability density function

$$\phi(x) = k/\sqrt{[x(1 - x)]}.$$

Calculate:
   (i) the value of k;
   (ii) the expected fraction covered by cloud;
   (iii) the probability that not more than $\frac{1}{4}$ of the sky is covered.
   (*Hint.* Your integrations may be made easier by using the substitution $x = \sin^2\phi$. You may assume that this substitution is valid, even though the functions to be integrated may be discontinuous at the ends of the interval of integration.) *(SMP)*

7. A quality characteristic X of a manufactured item is a continuous random variable having probability density function

$$f(x) = \begin{cases} 2\lambda^{-2}x & 0 < x < \lambda, \\ 0 & \text{otherwise,} \end{cases}$$

where λ is a positive constant whose value may be controlled by the manufacturer.

(i) Find the mean and the variance of X in terms of λ.

(ii) Every manufactured item is inspected before being dispatched for sale. Any item for which X is 8 or more is passed for selling and any item for which X is less than 8 is scrapped. The manufacturer makes a profit of £$(27-\lambda)$ on every item passed for selling, and suffers a loss of £$(\lambda+5)$ on every item that is scrapped. Find the value of λ which the manufacturer should aim for in order to maximize his expected profit per item, and calculate his maximum expected profit per item. (*Welsh*)

8. Petrol is delivered to a garage every Monday morning. The weekly demand for petrol at this garage, in thousands of gallons, is a continuous random variable X distributed with a probability density function of the form

$$f(x) = \begin{cases} kx(c - 2x) & \frac{1}{2} < x < \frac{3}{2}, \\ 0 & \text{otherwise.} \end{cases}$$

(i) Given that the mean weekly demand is 900 gallons, determine the values of k and c.

(ii) Calculate the mean number of gallons *sold* per week at this garage if its supply tanks are filled to their total capacity of 1000 gallons every Monday morning. (*Welsh*)

9. A discrete random variable is such that $P(x=r)=p_r$, where $r=0,1,2,\ldots$. The probability generating function $P(t)$ is defined by $P(t) = \sum_{r=0}^{\infty} p_r t^r$. Show that the moment generating function $M(t)$ of x about the origin is given by $M(t) = P(e^t)$.

The discrete rectangular distribution is defined by

$$P(x = r) = \frac{1}{n} \quad (r = 0, 1, 2, \ldots, n - 1).$$

Show that the moment generating function $M(t)$ of x about the origin is given by

$$M(t) = (e^{nt} - 1)/[n(e^t - 1)]. \quad (\textit{Oxford})$$

10. A random variable X has probability density function $f(x)$ given by

$$f(x) = \begin{cases} ce^{-2x} & 0 < x < \infty, \\ 0 & \text{otherwise.} \end{cases}$$

Find the moment generating function of X and hence, or otherwise, show that the mean is $\frac{1}{2}$ and the variance is $\frac{1}{4}$.

Show also that the median of the distribution is $\frac{1}{2}\log_e 2$ and the inter-quartile range is $\frac{1}{2}\log_e 3$. (*Oxford*)

11. Given that $y = \log_e x$ and that x is uniformly distributed in the interval $(1, 10)$, find the probability density function of y and sketch this function. Determine the probability that a randomly chosen value of x has a natural logarithm whose characteristic is 1.

A random sample of three values of x is obtained. Find (to two decimal places) the probability that their natural logarithms all have the same characteristic. (The characteristic of a logarithm is its whole-number part.) *(JMB)*

12. A child rides on a roundabout and his father waits for him at the point where he started. His journey may be regarded as a circular route of radius 6 metres and the father's position as a fixed point on the circle. When the roundabout stops, the shorter distance of the child from the father, measured along the circular path, is S metres. All points on the circle are equally likely stopping points, so that S is uniformly distributed between 0 and 6π. Find the mean and variance of S.

The direct linear distance of the child's stopping point from the father is D metres. Show that the probability density function of D is

$$\frac{2}{\pi\sqrt{(144 - D^2)}}$$

for D between 0 and 12, and zero outside this range.

The father's voice can be heard at a distance of up to 10 metres. Find to two decimal places the probability that the child can hear his father shout to him when the roundabout stops. *(JMB)*

14.13 COMPUTING EXERCISES

1. Cars arrive at traffic lights at random. Simulate this by assuming that the *inter-arrival time* has an exponential distribution, with mean 0.5 minutes. Assume that the traffic lights turn red and halt the traffic at intervals of 5 minutes and stay red for 1 minute. Investigate the distribution of the size of queue that builds up when the lights are red.

2. Is the sum of two continuous uniform distributions a continuous uniform distribution? To examine this, generate random values: (a) X from the continuous uniform distribution on $(0, 1)$; (b) Y from the continuous uniform distribution on $(0, 2)$. Calculate $Z = X + Y$. Repeat 500 times. Compare the three distributions X, Y and Z by drawing histograms and calculating descriptive statistics.

14.14 PROJECTS

1 Time between successive hatchback cars in traffic

Aim. To find the distribution of the time-interval between one hatchback car and the next in flowing traffic.
Equipment. A watch with a second hand, or a stop-watch.

Collection of data. Choose a road with a reasonable rate of traffic flow, not obstructed by traffic lights, steep hills, bus stops, etc. Record to the nearest second the time at which each hatchback car passes. Hence find the time-intervals between successive hatchback cars. Collect as large a sample of times as possible, at least 50 and preferably nearer 100.

Analysis. Group the data on time-intervals into a frequency table, with suitably chosen intervals (which may not all be of equal width). Present the results in a histogram. Compute the mean time-interval. Find the frequencies which would occur in each interval in an exponential distribution with this mean.

Note. Depending on the nature of the traffic, it may be better to choose some other type of (easily distinguishable) vehicle, e.g. lorry or delivery van.

2 Distribution of arm lengths

Aim. To find the distribution of the length of people's arms.
Equipment and collection of data as in Project 10.7.1.
Analysis. Take the arm lengths, group them into a frequency table, draw a histogram and calculate the mean and variance of arm length (either directly from the records or from the grouped frequency table).

SUMMARY

A **continuous random variable** X is specified by an equation for its probability density function $f(x)$, and the range over which it is defined; e.g. for the exponential distribution,

$$f(x) = \begin{cases} \lambda e^{-\lambda x} & 0 \leqslant x < \infty, \\ 0 & x < 0, \end{cases}$$

where λ is a positive constant.

A **probability density function** must satisfy these conditions: $f(x) \geqslant 0$ for all values of x; and

$$\int_{-\infty}^{\infty} f(x)\,dx = 1.$$

The probability that X lies in (a, b) is $\int_a^b f(x)\,dx$, and the **cumulative distribution function** $F(b)$ is $F(b) = \int_{-\infty}^b f(x)\,dx$, which becomes $\int_0^b f(x)\,dx$ when x is not defined for $x < 0$; here b is any possible value of x.

The **mean** of X is $\mu = \int_{-\infty}^{\infty} xf(x)\,dx$ and the **variance** is $\int_{-\infty}^{\infty} (x - \mu)^2 f(x)\,dx$. These are $E[X]$ and $E[(X - \mu)^2]$.

Expectations of any function $g(x)$ can be defined: $E[g(x)] = \int_{-\infty}^{\infty} g(x)f(x)\,dx$ and in particular the (uncorrected) sth moment is $\mu'_s = E[X^s]$.

$M(t) = E[\exp(Xt)]$ is the **moment generating function**, and when expanded as a series in powers of t it gives

$$M(t) = 1 + t\mu + \frac{t^2}{2!}\mu'_2 + \cdots + \frac{t^s}{s!}\mu'_s + \cdots.$$

Sums of independent random variables can be studied using moment generating functions.

15 THE NORMAL DISTRIBUTION

OVERVIEW

The normal (Gaussian) distribution is of great importance in statistics, because it can be studied theoretically in depth and because it is often a suitable model, exact or approximate, for data.

15.1 INTRODUCTION

The most important theoretical distribution we have met so far is the binomial distribution; it was proposed by Jacob Bernoulli (Swiss, 1654–1705) about 1700. The other two important theoretical distributions in statistics were both first derived from the binomial distribution by considering particular limits of it. The normal distribution is due to de Moivre (French, 1667–1754) but it is more commonly associated with the later mathematicians Gauss (German, 1777–1855) and Laplace (French, 1749–1827). Physicists and engineers often call it the Gaussian distribution or, in France, the Laplacian distribution. K. Pearson (British, 1857–1936) appears to have coined the name 'normal'. The normal distribution is without doubt the most important distribution in theoretical statistics. We study it in this chapter and shall find it recurring frequently throughout the remainder of the book. The third important distribution is the Poisson distribution, which was published by Poisson (French, 1781–1840) in 1837; it will be discussed in Chapter 20.

We first illustrate how the normal distribution arises as a limiting form of the binomial. The distribution of a binomial variable r involves two parameters, the number of trials n and the proportion of successes π. To simplify our illustration we shall set $\pi = \frac{1}{2}$. In Fig. 15.1 we show the frequency polygons of the binomial distributions (with $\pi = \frac{1}{2}$) obtained by taking $n = 2, 4, 8$ and 16 in turn. The mean and variance of a binomial random variable are $n\pi$ and $n\pi(1 - \pi)$; we have standardized this variable (see p. 211), and plotted along the horizontal axis $(r - n\pi)/\sqrt{n\pi(1 - \pi)}$, which has mean 0 and variance 1. The probabilities have been multiplied by $\sqrt{n\pi(1 - \pi)}$, so that each frequency polygon has an area of 1 underneath it. As n increases, the polygons tend towards a smoothly changing continuous curve which has the shape of a bell or a cocked hat. We will now imagine that n increases without limit. It is possible to show mathematically that as $n \to \infty$ the equation of the limiting curve is

$$y = \frac{1}{\sqrt{2\pi}} \exp(-x^2/2).$$

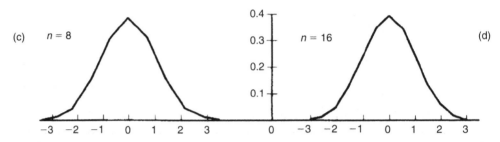

Figure 15.1 *Frequency polygons of the binomial distribution with $\pi = \frac{1}{2}$, for (a) $n=2$, (b) $n=4$, (c) $n=8$, (d) $n=16$. Horizontal scale represents the standardized variable $(r - n\pi)/\sqrt{n\pi(1-\pi)}$. Vertical scale shows probabilities adjusted to make area under polygon 1.*

We shall treat this function of x as the pdf (probability density function) of a new distribution.

15.2 THE STANDARD NORMAL DISTRIBUTION

What are the properties of a distribution that has the pdf

$$f(x) = \frac{1}{\sqrt{2\pi}} \exp\left(-\tfrac{1}{2}x^2\right), \quad -\infty < x < \infty?$$

Let us list first the mathematical properties of the function. There is a graph of $f(x)$ in Fig. 15.2 for values of x between -3 and $+3$.

(i) The graph is symmetrical about the y-axis, as can be seen either from Fig. 15.2 or from the fact that $f(-x)=f(x)$.
(ii) The range of x is from $-\infty$ to $+\infty$.
(iii) The x-axis is an asymptote to the curve at both extremes.
(iv) The curve has a maximum at the origin, where its value is $1/\sqrt{2\pi}=0.3989$, and it has points of inflexion at ± 1. (The latter may be verified by setting $f''(x)=0$.)

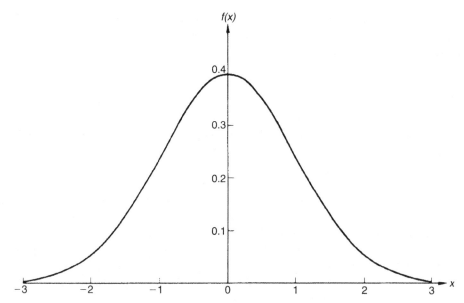

Figure 15.2 *Graph of $f(x) = \frac{1}{\sqrt{2\pi}} exp(-\frac{1}{2}x^2)$ for values of x between -3 and 3.*

We shall now verify that the function is a valid pdf by checking the properties listed on p. 319. We can see that $f(x) \geqslant 0$ for all x. The total area under the curve is

$$\int_{-\infty}^{\infty} \frac{1}{\sqrt{2\pi}} \exp\left(-\tfrac{1}{2}x^2\right) dx.$$

This integral cannot be evaluated by elementary techniques, so we must content ourselves with saying that its value can be shown to be unity. We carry out a simple numerical integration in Appendix IV(a) which shows this approximately.

The mean, median and mode are zero, because of the symmetry of the function. The standardized binomial distributions from which the distribution was derived in the previous section all had unit variance, so we would expect it to have unit variance too; this is so.

The distribution function (cdf) is

$$\Phi(b) = \int_{-\infty}^{b} \frac{1}{\sqrt{2\pi}} \exp\left(-\tfrac{1}{2}x^2\right) dx.$$

(We use the special symbol $\Phi(b)$, instead of the common $F(b)$, for the normal distribution function.) This integral cannot be evaluated explicitly. It has been exhaustively tabulated and there is a table of it on p. 716 (Table A2). Later in this chapter we shall consider in detail how to use this table. We shall content ourselves at the moment with giving the percentages of the distribution that come within certain bounds, to give some idea of how the distribution clusters about zero.

The area under the curve between the values $x = -1$ and $x = +1$ is 68% of the total area under the curve; i.e. 68% of the total probability is concentrated between $x = -1$ and $x = +1$. Between $x = -2$ and $x = +2$ it is 95%; and between

$x = -3$ and $x = +3$ it is 99.7%. So, although x is allowed to take all values from $-\infty$ to $+\infty$, there is in fact only 0.3% of the distribution which is distant more than 3 units from the mean.

15.2.1 Exercise

What percentage of the standard normal distribution, described above, lies between

(i) $x = 0$ and $x = +1$,   (ii) $x = -2$ and $x = -1$,
(iii) $x = +1$ and $x = +3$,   (iv) $x = -3$ and $x = -1$,

and what percentage lies
(v) above $x = +3$,   (vi) below $x = 0$?

15.3 THE NORMAL DISTRIBUTION AS A MODEL FOR DATA

Let us consider how we can extend the definition of the standard normal distribution so that it might serve as a model for a wide range of data. For example, look at the frequency polygon of the weights of babies born to mothers who never smoked (Fig. 1.1 on p. 5). This has a shape that is similar to the graph of the standard normal distribution, but the birth-weight data have a mean of 3353.8 and a variance of 327 866.

We can change the mean of the standard distribution by a simple shift along the x-axis (Fig. 15.3). The density function of the distribution with mean μ is

$$\frac{1}{\sqrt{2\pi}} \exp\left[-\tfrac{1}{2}(x - \mu)^2\right] ;$$

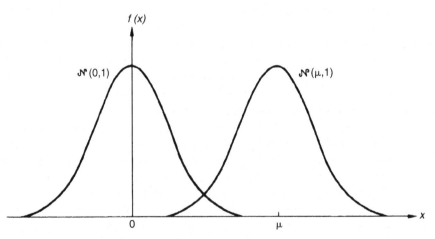

Figure 15.3 *Graph of normal distribution with mean* μ, *variance 1, compared with the standard normal distribution.*

this density function has the same value at $x = t + \mu$ as the standard normal density function has at $x = t$.

Next we must allow for a distribution with a different spread. We can do this by expanding, or contracting, the x-scale. An example of this is shown in Fig. 15.4. Since the total area under the curve must still be unity, the curve will not rise so high in the centre when the x-scale is stretched. (If the scale were contracted, the curve would of course rise higher in the centre.) In the density function we replace x by x/c, where c is a constant; the constant in front of the exponential function must also be modified so that the total area under the curve is still unity. The modified density function is

$$\frac{1}{\sqrt{2\pi c^2}} \exp\left[-\frac{1}{2}\left(\frac{x}{c}\right)^2\right].$$

We shall show in Section 15.4 that a distribution with a density function of this form has a variance c^2, so it is convenient to replace c^2 by the common symbol for variance, σ^2. Thus a change of spread, or variance, without changing the mean from 0, leads us to the density function

$$\frac{1}{\sqrt{2\pi\sigma^2}} \exp\left[-\frac{1}{2}\left(\frac{x}{\sigma}\right)^2\right].$$

The combination of the change in mean and change in variance leads us to the general form of the normal distribution.

Definition 15.1

A continuous random variable X has the **normal distribution** if it has the probability density function

$$f(x) = \frac{1}{\sqrt{2\pi\sigma^2}} \exp\left(-\frac{(x-\mu)^2}{2\sigma^2}\right), \quad -\infty < x < \infty.$$

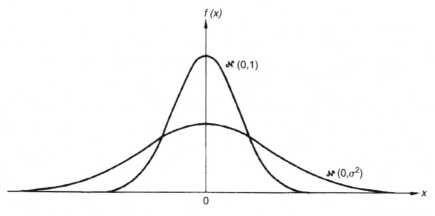

Figure 15.4 *Graph of normal distribution with mean 0, variance $\sigma^2 > 1$, compared with the standard normal distribution.*

We see that the normal distribution is a two-parameter distribution and that its two parameters are the mean μ and the variance σ^2; it is convenient to use $\mathcal{N}(\mu, \sigma^2)$ as a label for the normal distribution with mean μ and variance σ^2. In this notation, the mean and variance are specified, *in that order*; and it is *variance*, not standard deviation, that is given.

If we wish to model the birth-weight distribution by a normal distribution, we replace μ and σ^2 by their estimates, which we derive from the birth-weight data. The density function is therefore

$$f(x) = \frac{1}{\sqrt{(2\pi)(327\,866)}} \exp\left(-\frac{(x - 3353.8)^2}{2(327\,866)}\right).$$

A table which compares the frequency of the observed values and the model values is given later in this chapter (Project 15.12.2). The normal distribution fits the data reasonably well, though by no means perfectly.

Gauss used the normal distribution as a model for the errors in astronomical observations. This led to a name for the distribution which is now obsolete, 'the law of errors'. Quetelet (Belgian, 1796–1874) was an astronomer, but he was also the first to use the normal distribution as a model for biological, and in particular human, variability. He used it to describe the distribution of the chest girth of soldiers. Galton (English, 1822–1911) made a great deal of use of the normal distribution in his study of heredity in man. Among other things, he considered the relation of the heights of parents and children, and found the normal distribution could be used to describe heights in the same age-group and sex.

It has been found that data from many fields of study may often be adequately modelled by the normal distribution, although it is rarely found to hold exactly when exhaustive tests are made. The conditions which should lead us to consider the normal random variable as a model for a variate are when:

1. there is a strong tendency for the variate to take a central value;

2. positive and negative deviations from this central value are equally likely;

3. the frequency of deviations falls off rapidly as the deviations become larger.

15.4 DERIVATION OF MEAN AND VARIANCE

We need some basic integrals; the results stated here are shown in Appendix IV.

$$\int_{-\infty}^{\infty} \exp\left(-\tfrac{1}{2}x^2\right) dx = \sqrt{2\pi} \tag{15.1}$$

$$\int_{-\infty}^{\infty} x \exp\left(-\tfrac{1}{2}x^2\right) dx = 0 \tag{15.2}$$

$$\int_{-\infty}^{\infty} x^2 \exp\left(-\tfrac{1}{2}x^2\right) dx = \sqrt{2\pi}. \tag{15.3}$$

From Equation (15.1), we have at once

$$\int_{-\infty}^{\infty} \frac{1}{\sqrt{2\pi}} \exp\left(-\tfrac{1}{2}x^2\right) dx = 1,$$

confirming that the area under the standard normal distribution curve is 1 (as it must be for a properly defined pdf).

The *mean* of the standard normal distribution is

$$E[X] = \int_{-\infty}^{\infty} \frac{1}{\sqrt{2\pi}} x \exp\left(-\tfrac{1}{2}x^2\right) dx = 0,$$

using Equation (15.2).

For the general normal distribution which has the probability density function

$$f(x) = \frac{1}{\sigma\sqrt{2\pi}} \exp\left[-\tfrac{1}{2}(x - \mu)^2/\sigma^2\right]$$

we have

$$E[X] = \int_{-\infty}^{\infty} \frac{1}{\sigma\sqrt{2\pi}} x \exp\left[-\tfrac{1}{2}(x - \mu)^2/\sigma^2\right] dx.$$

On substituting $t = (x - \mu)/\sigma$, or $x = \mu + \sigma t$, we have $dx = \sigma\, dt$. Thus

$$\begin{aligned}
E[X] &= \frac{1}{\sigma\sqrt{2\pi}} \int_{-\infty}^{\infty} \sigma(\mu + \sigma t) \exp\left(-\tfrac{1}{2}t^2\right) dt \\
&= \frac{\mu}{\sqrt{2\pi}} \int_{-\infty}^{\infty} \exp\left(-\tfrac{1}{2}t^2\right) dt + \frac{\sigma}{\sqrt{2\pi}} \int_{-\infty}^{\infty} t \exp\left(-\tfrac{1}{2}t^2\right) dt \\
&= \frac{\mu}{\sqrt{2\pi}} \sqrt{2\pi} + \frac{\sigma}{\sqrt{2\pi}} 0 \quad \text{using Equations (15.1), (15.2)} \\
&= \mu.
\end{aligned}$$

The *variance* is

$$\text{Var}[X] = E[(X - \mu)^2] = \frac{1}{\sigma\sqrt{2\pi}} \int_{-\infty}^{\infty} (x - \mu)^2 \exp\left[-\tfrac{1}{2}(x - \mu)^2/\sigma^2\right] dx.$$

Using, once again, the substitution $t = (x - \mu)/\sigma$,

$$\begin{aligned}
\text{Var}[X] &= \frac{1}{\sigma\sqrt{2\pi}} \int_{-\infty}^{\infty} \sigma^3 t^2 \exp\left(-\tfrac{1}{2}t^2\right) dt \\
&= \frac{\sigma^2}{\sqrt{2\pi}} \int_{-\infty}^{\infty} t^2 \exp\left(-\tfrac{1}{2}t^2\right) dt \\
&= \frac{\sigma^2}{\sqrt{2\pi}} \sqrt{2\pi} \\
&= \sigma^2
\end{aligned}$$

using Equation (15.3).

15.4.1 Exercises

1. Write down the pdfs of the normal distributions with the following means and variances:
   (i) mean $= 3$, variance $= 4$; (ii) mean $= 0$, variance $= 5$; (iii) mean $= -2$, variance $= 1$; (iv) mean $= -6$, variance $= 10$.

2. A random variable X has pdf

$$f(x) = \frac{1}{3\sqrt{2\pi}} \exp\left(-\tfrac{1}{18}x^2\right), \quad -\infty < x < \infty.$$

   What are its mean and standard deviation?

3. A random variable X has pdf

$$f(x) = c \exp\left[-(x-4)^2/6\right], \quad -\infty < x < \infty.$$

   What is c?

15.5 THE CALCULATION AND USE OF STANDARD VARIATES WHEN DATA ARE NORMALLY DISTRIBUTED

A psychologist who advises on careers gives a series of tests to a young man. One is a test of verbal skills on which the young man scores 55. This score is of itself almost meaningless but the psychologist knows the distribution these verbal skills scores have for men in the general population. The tests have been used widely with a large and representative sample of men and women and, for men, this particular test of verbal skills has approximately an $\mathcal{N}(50, 10^2)$ distribution. (Many psychological tests are deliberately constructed so that the test scores will be approximately normally distributed when used with an appropriate population.) The distribution of the scores and the young man's score are shown in Fig. 15.5. What is of most interest to the psychologist is how the man compares with other men. This comparison could most easily be summed up by finding what percentage of men might be expected to obtain a worse score than 55 on the test. That is, we want to find the area under the normal curve $\mathcal{N}(50, 10^2)$ which lies to the left of 55. Since we have a normal curve whose parameters are specified this is possible. In fact we require the value of the cumulative distribution function $F(b)$ for this distribution when b takes the value 55. This could be obtained if we had a table of the particular distribution $\mathcal{N}(50, 10^2)$; but it would be a quite impossible task to prepare tables for all the whole family of normal distributions, since μ may take any value in $(-\infty, \infty)$ and σ^2 any value in $(0, \infty)$.

In Section 9.7 we described the process of *standardization*. If X is a random variable with mean μ and variance σ^2, then $(X - \mu)/\sigma$ has mean 0 and variance 1. In the present case, if we subtract the mean (50) from the test score, and then

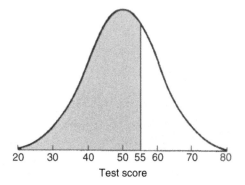

Figure 15.5 *Distribution of test scores* $N(50, 100)$. *Shaded area shows probability of obtaining a score less than 55.*

divide the result by the standard deviation (10) we obtain the standardized score. This follows the standard normal distribution $N(0, 1)$, and we can therefore use the table (Table A2) of the standard normal distribution. A standardized normal score is usually denoted by z, and is often called a z-score. We therefore write

$$z = \frac{55 - 50}{10} = \frac{5}{10} = 0.5.$$

The area below $z = 0.5$ in the standard normal distribution is shown in Fig. 15.6, and we find from Table A2 that the value of $\Phi(z)$ corresponding to $z = 0.5$ is 0.6915. We conclude that 69% of the population would be expected to obtain a worse score, and 31% a better score, than the young man. (There is no point in quoting the result to greater accuracy than the nearest percentage, and even this suggests a greater accuracy than is justified. If the young man took a similar test on another occasion it is doubtful if he would score *exactly* 55 again.)

The young man's score on numerical skills was 40. This test was also constructed to be $N(50, 10^2)$. How does the young man compare with his fellows in this test? His score when standardized is

$$z = \frac{40 - 50}{10} = \frac{-10}{10} = -1.$$

Figure 15.6 *Distribution of standardized test scores. Shaded area shows probability of obtaining a standardized score less than 0.5.*

Figure 15.7 *Distribution of standardized scores for numerical skills. Shaded area shows probability of obtaining a standardized score less than −1.*

A corresponding diagram is shown in Fig. 15.7. There is no entry in the tables of $\Phi(z)$ for $z = -1$, or for any negative value of z. Since the curve is symmetric about $z = 0$, it would be wasteful to tabulate it for negative z; from Fig. 15.7 it is easily seen that the area below $z = -1$ is equal to that above $z = +1$. This in turn is equal to $1 - (\text{area below } z = +1)$. Hence $\Phi(-1) = 1 - \Phi(1) = 1 - 0.8413 = 0.1587$. Thus about 16% of the population would be expected to do worse than this young man on this test of numerical skill, and 84% would be expected to do better.

Transforming an $\mathcal{N}(\mu, \sigma^2)$ variate into a standardized variate z is also called *expressing the variate in standard measure.*

15.5.1 Exercises

(Draw sketches similar to those in Figs 15.6 and 15.7 to help you with these exercises.)

1. In a standard normal distribution, what is the area under the curve: (i) below 1; (ii) below 2; (iii) between 1 and 2; (iv) above $-\frac{1}{2}$; (v) below -1; (vi) between -1 and $+1$?

2. Find constants a, b, c, d, e, f such that, if Z follows the standard normal distribution:
   (i) $P(Z \leqslant 1.645) = a$; (ii) $P(-1.96 \leqslant Z \leqslant 1.96) = b$; (iii) $P(0 \leqslant Z \leqslant c) = 0.3413$; (iv) $P(Z \geqslant -d) = 0.6554$; (v) $P(Z \leqslant e) = 0.2266$; (vi) $P(|Z| \leqslant 2.576) = f$.

3. If IQ scores are normally distributed with a mean of 100 and a standard deviation of 15, what proportion of people have IQs: (a) above 110; (b) above 125; (c) below 80; (d) above 75; (e) between 100 and 115; (f) between 75 and 125; (g) between 135 and 145; (h) between 60 and 90?

4. If the random variable X is distributed as $\mathcal{N}(5, 4)$ find: (i) $P(4 \leqslant X \leqslant 6)$; (ii) $P(X \geqslant 9)$; (iii) $P(|X - 5| > 3)$; (iv) $P(X > 0)$; (v) $P(|X| > 1)$.

15.5.2 Derivation of z-score from normal probability

It is often required to use the normal tables in the reverse way. For example, an employer when filling a post might require a candidate's performance in numerical skills to be in the top 25% of the population before he would consider employing him for that particular job. He could translate this requirement into a requirement that the candidate shall exceed a certain score on the test of numerical skills. We shall again assume that test scores are $N(50, 10^2)$.

The corresponding standard score z is such that $\Phi(z) = 0.75$. The nearest two entries to 0.75 in Table A2 are $\Phi(0.67) = 0.7486$ and $\Phi(0.68) = 0.7517$. Using the method of mathematical *interpolation*, we assume that $\Phi(z)$ increases linearly as z goes from 0.67 to 0.68. The total increase is $0.7517 - 0.7486 = 0.0031$. Starting from 0.7486, we require an increase 0.0014 to reach 0.75, and we assume that this is achieved when $z = 0.67 + \left(\frac{14}{31}\right)(0.01)$, i.e. $\frac{14}{31}$ of the way through the interval $(0.67, 0.68)$. Thus $z = 0.6745$.

If y is the score in the scale of the test, then $z = (y - \mu)/\sigma$, or $\sigma z = y - \mu$; so $y = \mu + \sigma z$. Hence $y = 50 + 10(0.6745) = 57$, to the nearest whole number. Thus, any candidate not scoring as much as 57 in the test would not be acceptable to the employer.

15.5.3 Exercises

1. In relation to the distribution of IQs in Exercise 15.5.1(3), assume that it is the practice in a country to provide special education for the lowest 5% of the population, and provide university education for the top 7%. Find the z-scores (i.e. standardized normal scores) corresponding to these percentages and hence state what you would expect would be the IQ cut-off points for those requiring special education, and those entering university, in that country.

2. Hens' eggs have mean mass 60 grams with standard deviation 15 grams, and the distribution may be taken as normal. Eggs of mass less than 45 grams are classified as small. The remainder are divided into standard and large, and it is desired that these should occur with equal frequency. Suggest the mass at which the division should be made (correct to the nearest gram).   (*SMP*)

3. The weights of bars of soap made in a factory are normally distributed. Last week $6\frac{2}{3}$% of bars weighed less than 90.50 grams and 4% weighed more than 100.25 grams.

   Find the mean and variance of the distribution of weights, and the percentage of bars produced which would be expected to weigh less than 88 grams.

   If the variance of the weight distribution was reduced by one-third, what percentage of the next week's production would you expect to weigh less than 88 grams, assuming the mean is not changed?   (*IOS*)

4. It is estimated that 1400 commuters regularly aim to catch the 5.30 p.m. train at a certain London terminus, that 50 will have arrived before the platform

gate is opened at 5.20 p.m., and that when the train leaves on time 70 arrive too late. Assuming the distribution of arrival times to be normal, use tables to obtain the mean and standard deviation. Hence estimate:
  (i) at what time the platform gate should be opened if not more than 20 passengers are to be kept waiting at the gate;
  (ii) how many of the commuters will miss the train on a day when (unexpectedly) it leaves two minutes late! (*SMP*)

Example 15.1 Fitting a normal distribution to data

The masses measured on a population of 100 animals were grouped in the following table, after being recorded to the nearest gram:

Mass (g)	$\leqslant 89$	90–109	110–129	130–149	150–169	170–189	$\geqslant 190$
Frequency	3	7	34	43	10	2	1

Show that the mean is 131.5 and the standard deviation is 20. (Assume the lower and upper classes have the same range as the other classes when calculating the moments.)

Find the expected frequencies in each class for a normal distribution with the same mean and standard deviation. (*Oxford*)

The mean and standard deviation are calculated using the coding

$$u = (x - 139.5)/20.$$

Centre x	u	f	uf	u^2f	$x = 139.5 + 20u$
79.5	−3	3	−9	27	$\mu_x = 139.5 + 20\mu_u$
99.5	−2	7	−14	28	$\sigma_x = 20\sigma_u$
119.5	−1	34	−34	34	$\mu_u = -\frac{40}{100} = -0.40$
139.5	0	43	−57	0	$\mu_x = 139.5 + 20(-0.40)$
159.5	1	10	10	10	$= 139.5 - 8.0 = 131.5$
179.5	2	2	4	8	$\sigma_u^2 = \frac{1}{100}\left(116 - \frac{(-40)^2}{100}\right)$
199.5	3	1	3	9	$= \frac{1}{100}(116 - 16) = \frac{100}{100} = 1$
			17		
Total		100	−40	116	$\sigma_u = 1, \sigma_x = 20 \times 1 = 20$

The upper end-point l of each interval is expressed in standard measure, $z = (l - \mu_x)\sigma_x$. The probability of obtaining a value less than z (i.e. $F(z)$) is determined from standard normal tables (p. 359); taking differences of the probabilities gives the probability of obtaining an observation in each interval. Multiplication of these probabilities by the total frequency gives the required frequencies.

Upper end-point l_i	$l_i - \mu_x$	$z_i = \frac{(l_i - \mu_x)}{\sigma_x}$	$F(z_i)$	$F(z_i) - F(z_{i-1})$	Frequency
89.5	−42	−2.1	0.018	0.018	2
109.5	−22	−1.1	0.136	0.118	12
129.5	−2	−0.1	0.460	0.324	32
149.5	18	0.9	0.816	0.356	36
169.5	38	1.9	0.971	0.155	15
189.5	58	2.9	0.998	0.027	3
∞	∞	∞	1.000	0.002	0

15.5.4 Approximating discrete variates

As we said at the beginning of this chapter, the normal distribution was first discovered as a mathematical limit of a discrete distribution, the binomial. This result makes it reasonable to use the normal distribution as an approximation to the binomial distribution (see Section 16.3). Figure 15.1(d) showed a situation which is typical of many sets of real-life data that, strictly speaking, are discrete rather than continuous: the pattern of their frequencies seems to satisfy all the conditions suggested at the end of Section 15.3 for modelling them by a normal distribution. In the remaining chapters of this book we shall meet other discrete distributions, beside the binomial, which can also be approximated by a normal distribution; and in particular in Chapter 16 many of the results we shall develop can be applied equally to continuous or discrete variates. Some of the mathematical methods needed to show these approximations are beyond the scope of this book, but we shall anticipate the results by using the normal distribution as a model for *any* set of data – preferably reasonably large – in which Equations (15.1)–(15.3) seem to hold, even when we realize that we are handling a discrete measurement.

15.6 EXPLORING DATA

The practical statistician does not spend very much time constructing new theory, but instead is using all the methods in this book and many more in attempts to extract as much information as possible from sets of data. We hope that these data have been collected as carefully as possible, according to principles discussed in Chapters 12, 22 and elsewhere, so that they are capable of leading to valid inferences. We may have some ideas about the distribution that the data could follow, but it is always useful, and wise, to make a visual, graphical display of the data to check these ideas and to note any strange or unexpected features. For example, one item of data may be a long way away from all the others on the scale of measurement: it could be an *outlier*, and its value could seriously affect estimates of the mean and variance of the data (though not of the median and quartiles). A *dot plot*, or *dot diagram*, will alert us to the possible existence of outliers, just as it does to the possibility of there being two subpopulations of data (Chapter 2).

Another piece of information needed for any useful statistical analysis is whether a set of data appears to come from a *symmetrical* population or not; the *box-and-whisker* plot, or *box plot*, helps to show this. The normal distribution is very often used for fitting to data that are apparently symmetrical, because its good theoretical properties make statistical inference relatively easy. But it is wise to look more closely at the data first. For example, a normal distribution should have the bulk of its observations near the mean, and steadily fewer as we go further away from the mean. One outlier at each end of the data set does not necessarily make it unsymmetrical, but it does affect the estimate of variance seriously, and any inferences which depend on this. Other distributions with symmetrical density functions can easily be constructed, but inferences based on the theory of the normal distribution will be of little use for small samples from them.

15.7 RESIDUALS

Many of the hypotheses that require examining in applied statistics involve the means, or medians, of populations or sets of data. Inferences are made on the basis of these *central measures* (Chapter 2), and although the measures of scatter for the data may feature in the mathematics used in a test (see Chapter 16 and later) the central measure is the summary of the data which is chiefly used in making decisions.

In Example 13.1 (p. 279), the null hypothesis being examined is that the median $M = 50$. Likewise in Exercise 16.1.1(2) attention is focused on the mean of the results of 10 chemical tests. We may think of each individual measurement in the set as being made up of two parts:

(i) the 25 incomes in Example 13.1 are '*median plus or minus deviation from median*', so that the first two, 43 and 88, are '$50 - 7$' and '$50 + 38$';

(ii) the 10 records of chemical content are '*mean plus or minus deviation from mean*', and the mean of the data is 1.57, so the observations 1.60 and 1.56 are '$1.57 + 0.03$' and '$1.57 - 0.01$'.

The first part, the median or the mean, is *fixed*, either given by the hypothesis or determined from the whole set of data, and the second part is the departure of each individual observation from this fixed point. These departures are *random*, though they will not always follow a recognizable distribution (and this is one reason why nonparametric methods are sometimes needed).

Later in this book (Chapters 22–24) we shall deal with *linear models*, as used in regression and designed experiments, and there also we want to explain data as the sum of two components. One of these will be a non-random part, though less simple than just the value of a mean or median, and the other component will contain all the random element; it is this component that follows a normal distribution or whatever model is being suggested for the data.

Anticipating the approach we shall need later, we may express each item of data as

'fitted value ± residual'.

In (i) and (ii) the median and the mean, respectively, provide the *fitted values*, while the departures from these are the *residuals*.

The residuals for the income data of Example 13.1 thus are:

$$-7, +38, -14, +15, +12, +27, 0, -12, -3, +11, +34, -20, -5,$$
$$+15, +4, +23, +26, -3, +5, +18, +12, +16, -6, +2, +9.$$

Similarly, residuals for the chemical data of Exercise 16.1.1(2) are:

$$+0.03, -0.01, +0.02, -0.03, -0.04, +0.01, -0.05, +0.02, +0.07, -0.02.$$

The second set add up to zero, since they are departures from an observed mean; but the first set do not, since they are departures from a hypothetical median.

In many applied statistical problems, where analysis is carried out on a computer using a statistical package, graphics facilities will be available and residuals can be studied. Dot plots can be very informative for small or moderate amounts of data. Figure 15.8 shows a dot plot for the residuals of the income data and Fig. 15.9 for the chemical data.

From Fig. 15.8 there is little suggestion of normally distributed data because there is no tendency to cluster round a central value, and there are large values at either end which are some distance away from their neighbours. A nonparametric approach was clearly sensible for these data.

Figure 15.9 illustrates a problem which is all too common with small samples of data. The mean of this set is 0, there is one quite large positive residual and a fairly uniform spread of the negative ones; yet several of the measurements are reasonably near their mean. Whether it is correct to regard these as a sample from a normal distribution we really cannot say, although with only 10 data items an analysis using normal theory may give as much information as any

Figure 15.8 *Residuals for income data of Example 13.1.*

Figure 15.9 *Residuals for chemical data of Exercise 16.1.1(2).*

other. However, this approach would *not* be sensible if either:

(i)   there had been obvious outliers a long way from the others, or

(ii) there had been a very skew pattern, or

(iii) the dot plot had suggested that more than one population of data may be present.

15.8 GRAPHICAL STUDY OF SYMMETRY AND NORMALITY

A very quick test of the symmetry of a set of data has been suggested by du Toit *et al.* (1986), and although symmetry by itself is not the most important property the lack of it implies that it is not worth while doing the less straightforward test for normality.

The first step in the symmetry test is to rank the data in increasing order of size, $x_{(1)}, x_{(2)}, \ldots, x_{(n)}$. A graph is then plotted containing the points $(x_{(1)}, x_{(n)})$, $(x_{(2)}, x_{(n-1)}), (x_{(3)}, x_{(n-2)}), \ldots$ as far as necessary; the smaller values are plotted in the X-direction and the larger ones in the Y-direction. If the data set has exact symmetry, the graph will show a straight line with slope -1. Obvious departures from this, of general trend or particularly of individual points, will indicate lack of symmetry, and also suggest whether outliers may be present.

The chemical data Exercise 16.1.1(2) arranged in rank order are

$$1.52, 1.53, 1.54, 1.55, 1.56, 1.58, 1.59, 1.59, 1.60, 1.64$$

and so the points to be plotted are $(1.52, 1.64), (1.53, 1.60)$ etc. Figure 15.10 leaves us just as uncertain as Fig. 15.9 about the nature of these data!

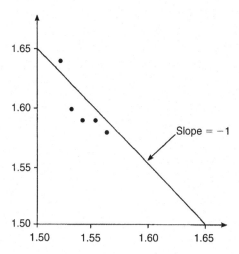

Figure 15.10 *Graph of data from Exercise 16.1.1(2) to examine symmetry.*

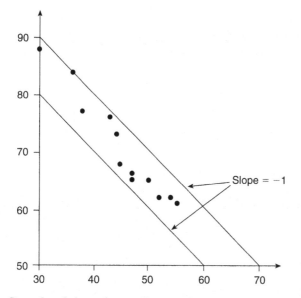

Figure 15.11 *Graph of data from Example 13.1 to examine symmetry.*

Ranking the income data in Example 13.1, we have

$$30, 36, 38, 43, 44, 45, 47, 47, 50, 52, 54, 55, 59,$$
$$61, 62, 62, 65, 65, 66, 68, 73, 76, 77, 84, 88.$$

Therefore we plot (30, 88), (36, 84), (38, 77) etc. as shown in Fig. 15.11. The slope of this line is not very near to -1, and perhaps the main reason for this is the gap to be found below the top group of five data.

15.8.1 Normal probability plots

A very useful procedure, now widely available in statistical packages, is the *normal probability plot*. The ranked observations $x_{(i)}$ are plotted in the horizontal direction on special graph paper; the horizontal scale is the usual equally spaced one but the vertical scale Z indicates the cumulative probabilities in a standard normal distribution transformed to a linear scale (an 'inverse transformation', see *CCC*, Section 9.3). (Of course, special paper is not needed when using a statistical graphics program because the scales are adjusted as part of the routine.)

If we have n observations in a sample, then $Z_i = (i - 0.5)/n$ is plotted against $x_{(i)}$, and a normally distributed set of data should yield a straight line, approximately. (Some packages use slightly different values for Z_i, such as $i/(n+1)$, but we can ignore this refinement.)

The method can equally well be used for large sets of data, which have been grouped into a number of intervals in a frequency table, remembering that cumulative frequencies must always be plotted against the *upper end-points* of the intervals as $x_{(i)}$.

After the data have been plotted, the best-fitting line could be put through them by eye; then the x-value corresponding to 0.5 (50%) on the probability scale would give an estimate of μ. The x-values corresponding to 0.025 ($2\frac{1}{2}$%) and 0.975 ($97\frac{1}{2}$%) on the probability scale estimate $\mu \pm 1.96\sigma$ in a normal distribution, so that the difference between these two x-values estimates 3.92σ.

Example 15.2 Lognormal data

A study of the times to failure of a mechanical component under a standard amount of stress is made using two machines, A and B. Eleven components, randomly chosen from a large batch, are tested on machine A and a further 12 on B. The failure times (in hours) are as follows:

$$A : 158.7, 6.7, 15.0, 2.2, 111.4, 55.4, 9.1, 10.2, 22.1, 8.5, 144.0;$$

$$B : 61.2, 4.6, 7.9, 101.0, 14.6, 4.0, 12.5, 18.7, 45.9, 4.0, 6.1, 14.0.$$

(i) Draw two dot plots to illustrate the two sets of data; place one dot plot underneath the other and use the same scale of measurement so that a visual comparison of the two sets of data may be made. Comment on the characteristics of the data (location, skewness, etc.).

(ii) Repeat (i) using the logarithms of the failure times. Comment on whether or not the use of logarithms makes the data more symmetrical.

(iii) Using the more symmetrical form of the data, calculate a 95% confidence interval for a suitable measure of location for each of the machines.

(iv) The scientist conducting the study wishes to know whether there is any difference between the average failure times on the two machines. Investigate this question statistically. *(RSS)*

(i) Figure 15.12 shows that both sets of data contain a majority of low failure times, but they are very skew; B appears to give the lower mean and also less dispersed values than A.

Figure 15.12 *Dot plots of failure time data (Example 15.2).*

(ii) Logarithms to base 10 are:

$$A : 2.20, 0.83, 1.18, 0.34, 2.05, 1.74, 0.96, 1.01, 1.34, 0.93, 2.16;$$

$$B : 1.79, 0.66, 0.90, 2.00, 1.16, 0.60, 1.10, 1.27, 1.66, 0.60, 0.79, 1.15.$$

Figure 15.13 shows distributions that are now much less skew than they were on the original scale of measurement. They still do not show a very strong tendency to cluster around a mean value, but an analysis based on the assumption that they are samples from normal distributions will give much more acceptable results than one using the original data.

Figure 15.13 *Dot plots of log₁₀ (failure time data) from Example 15.2.*

We can therefore apply methods based on normal theory, and will be able to complete this example in Section 18.5. Without this *transformation* to log X, we could only use nonparametric methods for studying the data in small samples.

If a set of observations $\{x_i\}$ is skew, but the set $\{\log x_i\}$ is normally distributed, then the variate X is said to have a *lognormal* distribution. We also say that the log transformation 'normalizes' the data. There is nothing 'unnatural' about using log units for analysis; if we are studying a biological measurement which has arisen in a *multiplicative* process, where for example cells increase by progressively doubling rather than adding constant increments over time, their size will show an *exponential* type of increase over time. A graph like Fig. 23.9(b) would then illustrate the relationship between size and time and the log transformation would linearize it. The important feature in comparing sizes would be ratios, not differences. Similar situations can arise in economic examples.

15.9 MINITAB: THE NORMAL DISTRIBUTION

Finding values of the probability functions and generating random samples for the normal distribution is done as for other theoretical probability distributions. Choose

Calc>Probability Distributions>Normal/
or
Calc>Random Data>Normal/

The values for the mean and standard deviation must be given if the default values of 0 and 1, i.e. those for the standard normal distribution, are not appropriate.

In Section 15.5 we discussed how to assess the performance of a young man on a test of verbal skills that was assumed to have a normal distribution with mean 50 and standard deviation 10. To obtain the proportion of subjects who might do worse on the test than the young man, we standardized the test score of 55 to give a z-score of 0.5. Using tables of the standard normal distribution, we found that the probability of obtaining a z-score less than 0.5 was 0.6915. With MINITAB we can adopt a more direct approach. Choose

Calc>Probability Distribution>Normal/
Select **Cumulative probability**, enter 50 for **Mean** and 10 for **Standard Deviation**, select **Input constant** and enter 55/

We obtain a value 0.691482. The extra decimal places are of no value in this situation but we escape the necessity to standardize.

To carry out the inverse process of finding the value corresponding to a given cumulative probability, **Inverse cumulative probability** replaces **Cumulative probability** in the dialogue box.

When plotting a histogram MINITAB will allow the fitted normal distribution to be displayed. More precisely, the graph of the probability density function of a normal distribution, with mean and sample standard deviation identical with those of the data, is superimposed on the histogram. To do this choose

> **Graph>Histogram/**
> Select **With Fit/**
> Enter **Graph variable/**

Normal probability plots, as described in Section 15.8.1, may be obtained by choosing

> **Stat>Basic Statistics>Normality Test/**
> Name **Variable/**

To test symmetry, choose

> **Stat>Quality Tools>Symmetry Plot/**

The graph is equivalent to that (Fig. 15.10) produced by the method of Section 15.8 but the reference line is of slope +1 and passes through the origin. The graphs can be seen as mirror images of each other. MINITAB also inserts a small histogram plot beside the symmetry plot. The symmetry plot may not be available in some versions of MINITAB.

15.10 EXERCISES ON CHAPTER 15

1. A man leaves home at 8 a.m. every morning in order to arrive at work at 9 a.m. He finds that over a long period he is late once in forty times. He then tries leaving home at 7.55 a.m. and finds that over a similar period he is late once in one hundred times. Assuming that the time of his journey has a normal distribution, before what time should he leave home in order not to be late more than once in two hundred times? *(SMP)*

2. In a certain book the frequency function for the number of words per page may be taken as approximately normal with mean 800 and standard deviation 50. If I choose three pages at random, what is the probability that none of them has between 830 and 845 words each? *(SMP)*

3. In a certain country the heights of adult males have mean 170 cm and standard deviation 10 cm, and the heights of adult females have mean 160 cm and standard deviation 8 cm; for each sex the distribution of heights approximates closely to a normal probability model. On the hypothesis that height is not a factor in selecting a mate, calculate the probability that a husband and wife selected at random are both taller than 164 cm. *(SMP)*

4. An athlete finds that in the long jump his distances form a normal distribution with mean 6.1 m and standard deviation 0.03 m.

   Calculate the probability that he will jump more than 6.17 m on a given occasion.

   Find the probability that three independent jumps will all be less than 6.17 m.

   What distance can he expect to exceed once in 500 jumps?   (*AEB*)

5. Eggs may be classified as standard if they weigh less than 46 g, medium between 46 and 56 g, or large over 56 g. Suppose that eggs laid by a particular breed of hen have mean weight 50 g, and that weights are normally distributed with standard deviation 5 g. Find the proportions of each class of egg that are laid by these hens.

   If the selling prices of standard, medium, large are 4p, 5p, 6p respectively, and the cost of production is 4p each, find the expected profit per egg sold.

   Another breed of hen lays eggs whose mean weight is 52 g, and whose weights are normally distributed with standard deviation 5 g. However, they eat more food and the cost of production is now $4\frac{1}{2}$p per egg. What is the expected profit per egg for this breed?

6. Find the mean and the standard deviation of the following set of measurements. Find also the expected frequencies in each class for a normal distribution having the same mean and standard deviation. Draw frequency polygons for this and the original distribution.

mm	29.6	29.7	29.8	29.9	30.0	30.1	30.2	30.3	30.4
Frequency	7	18	54	106	115	96	73	22	9

7. Suppose that adult male shoe size is an integer, k, which is related to foot length, y, measured in inches, by the fact that size k will accommodate feet of length between $5.5 + 0.5k$ and $6 + 0.5k$ inches; $k = 5, 6, \ldots, 14$. It may be assumed that adult male foot length is normally distributed with mean 10.2 inches and standard deviation 1.1 inches.
   (i) What proportion of the population require a shoe size greater than 14?
   (ii) What is the most common size, and what proportion take that size?
   (iii) What is the probability that two men, chosen at random, will have shoe sizes differing by 8 or more? (Assume that all men with feet smaller than $8\frac{1}{2}$ inches take size 5, and those with feet larger than $12\frac{1}{2}$ inches take size 14.)

8. A manufacturer packs screws in boxes which should contain 144 screws. One day's production of 10 000 boxes includes 668 boxes containing less than 132 screws and 62 boxes containing more than 180 screws. On suitable assumptions, which should be stated, calculate the mean and variance of the number of screws per box on that day.

9. The tolerance on a dimension of an article is set at 10.00 ± 0.05 mm, and every article produced is inspected to see whether it satisfies these limits.

The dimension is normally distributed, but there is a bias in the machine settings so that the mean dimension of articles produced is 10.01 mm. The standard deviation is 0.04 mm, and it costs 10p to produce each article.

All articles that do not meet the tolerance limits have to be scrapped. In order to reduce the loss, it is possible:
  (i) to remove the bias in the mean dimension, at an extra cost of 4p per article, or
  (ii) to reduce the standard deviation to 0.03 mm, at an extra cost of 2p per article, or
  (iii) to do both (i) and (ii), at an extra cost of 6p per article.

Which of these procedures would you recommend? *(IOS)*

10. The product specification for packets of sweets states that each packet must weigh between 140 and 160 g. The packet weights are normally distributed with variance $4\,g^2$. Where would you set the target packing weight, and why?

15.11 COMPUTING EXERCISES

1. Work through Example 15.1, *Fitting a normal distribution to data*, using a computer. Assume that the mean is 131.5 and the standard deviation 20.

2. Carry out Project 15.12.1, *Normal approximation to sum of random digits*, using a computer.

3. Display three histograms based on a random sample of n observations from each of the following three distributions:
   (i) normal distribution with mean 5 and variance 1;
   (ii) continuous uniform distribution on the interval $(2, 8)$;
   (iii) exponential distribution with mean 5.

   Present the histograms side by side for comparison. Questions to be asked are: 'Is the distribution symmetric or skew?' and 'Is the distribution peaked or flat?'

   Begin with $n = 160$, when the distributions should be easily distinguishable. Reduce n through the values 80, 40, 20, 10.

15.12 PROJECTS

1 Normal approximation to sum of random digits

Aim. To fit a normal distribution to the distribution of the sums of random digits.
Equipment. Either a table of random digits, or a die and shaker for generating random digits (see Project 8.15.2), or a computer program for generating random digits.

Collection of data. Find the total of 10 random digits. (It is recommended that the totals of the first five digits and the second five digits are found and added together to give the required total. The totals of five digits will be useful for Project 16.10.1.) Repeat at least 250 times.

Analysis. (a) The discrete uniform (or integer) distribution which takes the values 0, 1, 2, 3, 4, 5, 6, 7, 8, 9 (i.e. the distribution of random digits) has mean 4.5 and variance 8.25. The totals of ten random digits will therefore have mean 45 and variance 82.5. Choose class boundaries at $10.5, 20.5, \ldots, 90.5$ and find the expected frequency that would be obtained in each of the intervals formed for a normal distribution with mean 45 and variance 82.5.

(b) Group the observed totals using the same class boundaries. Compare the observed and expected frequencies, and illustrate by a diagram.

2 Normal approximation to data

Aim. To fit a normal distribution to data.

Collection of data. With many real-life data the normal distribution provides a reasonable approximation but it rarely gives a very good fit, though it is often difficult to collect enough data to make an adequate assessment of the fit. We suggest a number of possible sources of data:

(a) Birth-weight of babies born to mothers who smoke (p. 4). We illustrate below the analysis of the corresponding data from mothers who do not smoke.
(b) Body measurements (heights, lengths of individual limbs or fingers) in large samples of people of the same age-group, sex and social group.
(c) Diameters of glass marbles made by the same process, measured with a micrometer.

Analysis. Draw up a frequency table, compute the mean and variance of the data, and display the results in a histogram. Fit a normal distribution with the same mean and variance, and draw it on the same diagram.

We give below the table for the birth-weights of babies whose mothers did not smoke, as described in Chapter 1:

Interval	1–	501–	1001–	1501–	2001–	2501–	3001–	3501–	4001–	4501–	5001–	Total
Observed frequency	4	20	25	73	236	1089	2530	1927	551	101	7	6563
Normal frequency	0	0	4	55	388	1316	2182	1769	701	135	13	6563

SUMMARY

The **standard normal distribution** has probability density function

$$f(x) = \frac{1}{\sqrt{2\pi}} \exp\left(-\tfrac{1}{2}x^2\right) \quad (-\infty < x < \infty)$$

and is symmetrical about $x = 0$, which is its mean, with a rapid fall in $f(x)$ as x departs further from 0. The variance is $\sigma^2 = 1$.

The general **normal distribution** $\mathcal{N}(\mu, \sigma^2)$, with mean μ and variance σ^2, has

$$f(x) = \frac{1}{\sigma\sqrt{2\pi}} \exp\left(-\frac{1}{2}\left\{\frac{x-\mu}{\sigma}\right\}^2\right) \quad (-\infty < x < \infty).$$

It is symmetrical about μ.

When X has the distribution $\mathcal{N}(\mu, \sigma^2)$ then the corresponding **standardized** variable $Z = (X - \mu)/\sigma$ is $\mathcal{N}(0, 1)$.

Normality of a data set may be studied using normal probability plots and various other methods.

The normal distribution is a good *approximation* to the binomial distribution having the same mean and variance, when a sample of a large number, n, of observations is available and the parameter π is not too near 0 or 1.

Data may be *transformed* to make them (approximately) normal for analysis, e.g. skew data X sometimes become (approximately) normal when $\ln X$ (or $\log_{10} X$) is examined.

16 SAMPLING DISTRIBUTIONS OF MEANS AND RELATED QUANTITIES

OVERVIEW

Because of the central limit theorem, the standard normal distribution can be used to obtain general results concerning means or totals of measurements (including discrete measurements) in large samples of data. Control charts, used in statistical process control, are an important example. We study sampling distributions in this chapter and some of their uses in the next chapter.

16.1 THE CENTRAL LIMIT THEOREM

We discussed earlier in the book how to estimate the mean journey time of a population of schoolchildren. The method was to take a random sample of the children and use the mean journey time of the sample as an estimate of the population mean. We looked at the reliability of this estimate in Chapter 11 by considering the sampling distribution of the sample mean, and showed that this sampling distribution had the same mean μ as the population. We showed also that the variance of the sampling distribution of the mean was equal to the population variance σ^2 divided by the sample size n. These facts give us a reasonable idea of the reliability of the estimate; but it would be much better if we knew the *exact form* of the sampling distribution, and not merely its mean and variance. We can find the form of this sampling distribution, at least with a fair approximation, provided the sample is a large one, by using a remarkable theorem, the *central limit theorem*.

Theorem 16.1 The central limit theorem

Let X be a random variable with mean μ and variance σ^2. If $\overline{X}_n$ is the mean of a random sample of size n chosen from the distribution of X, then the distribution of

$$\frac{\overline{X}_n - \mu}{\sigma/\sqrt{n}}$$

tends to the standard normal distribution as $n \to \infty$.

When we draw a *large* random sample from a population whose mean is μ and whose variance is σ^2, this theorem gives us valuable information about the mean $\overline{X}_n$. It tells us that, provided n is large, the distribution of $(\overline{X}_n - \mu)/(\sigma/\sqrt{n})$ is approximately normal with mean zero and variance one, or, more directly, that the sample mean $\overline{X}_n$ has a distribution that is approximately normal with mean μ and variance σ^2/n. What is remarkable about the theorem is that there is no need to specify the nature of the distribution of X. All we need to know about that distribution is its mean and variance. This theorem is of enormous practical value, as we shall see in this chapter, and is a major reason for the importance of the normal distribution in statistics.

A proof of the theorem requires more advanced mathematics than we can assume here. But it is possible for the reader to verify the theorem, for particular cases, by investigation. When discussing random sampling, we considered (p. 50) the means of samples of size five drawn from the distribution of random digits, that is, from the discrete uniform distribution which has values $0, 1, 2, \ldots, 9$. We saw that the sample mean had a unimodal distribution, and we can now see, further, that this unimodal distribution is similar in shape to the normal. The reader has very likely carried out investigations and collected other data which help to verify this (see Project 16.10(1) at the end of this chapter).

We should mention here that *if* the distribution of X is *normal* then the distribution of the mean $\overline{X}_n$ of random samples of size n is exactly $\mathcal{N}(\mu, \sigma^2/n)$ for any value of n. We shall look more closely at this result later in the chapter.

Example 16.1 Use of central limit theorem

A harassed father, wishing to keep his son quiet, offers to pay him 10p if the mean score of 100 throws of a die exceeds 4. What is the probability that the father will have to pay out?

We model the outcome of a single throw by the random variable X which has values $1, 2, 3, 4, 5, 6$ each with probability $\frac{1}{6}$.

$$\mu = E[X] = 3.5.$$

$$\sigma^2 = \frac{1}{6}[(-2.5)^2 + (-1.5)^2 + (-0.5)^2 + (0.5)^2 + (1.5)^2 + (2.5)^2]$$

$$= \frac{17.50}{6} = 2.917.$$

From the central limit theorem, the mean $\overline{X}$ of 100 throws has approximately the distribution $\mathcal{N}(3.5, 2.917/100)$. The standard error of $\overline{X}$ is $\sqrt{2.917}/10 = 0.171$. Therefore $z = (\overline{X} - 3.5)/0.171$ has a standard normal distribution.

For $\overline{X} = 4$, the corresponding standard score is

$$z = \frac{4 - 3.5}{0.171} = \frac{0.5}{0.171} = 2.92.$$

The probability of a z-score greater than 2.92 is 0.0018. It is therefore very unlikely that the father will have to pay out.

Note that the probability of a score greater than 4 on a *single* throw is $\frac{1}{3}$. We see that a sample mean tends to stay closer to the population mean than does a single value.

Example 16.2 Use of central limit theorem

The weekly wages paid to the weekly paid staff of a manufacturing company are approximately normally distributed with a mean of £133.40 and a standard deviation of £3.00.

(a) Calculate the probability that the mean weekly wage of a random sample of 9 of the weekly paid staff is within £2 of the population mean of £133.40.
(b) How large a random sample should be taken if the sample mean is to have probability 0.9 of being within £1 of the population mean?

(a) A sample of 9 is not large, so we do not use the central limit theorem. We use the fact that the mean of a sample from a normal distribution is also normal. In this case, we are told that X is 'approximately normal', so we can say that $\overline{X}$ is approximately $\mathcal{N}(133.40, 3.00^2/9)$ which is $\mathcal{N}(133.40, 1.00)$. Write z_1 for the standardized value for the upper limit of the mean wage. Then $z_1 = 2.00/1.00 = 2.00$. $\Phi(2.00) = 0.9772$, from Table A2. The lower limit z_2 is -2.00. We have $\Phi(-2.00) = 1 - \Phi(2.00)$. Hence the *difference*

$$\Phi(2.00) - \Phi(-2.00) = 2\Phi(2.00) - 1 = 2 \times 0.9772 - 1 = 0.9544.$$

This is the required probability.

(b) Consider a sample of size n. Then $\overline{X}_n$ is approximately $\mathcal{N}(133.40, 3.00^2/n)$. The upper limit is

$$z_1 = \frac{1.00}{3.00/\sqrt{n}} = \frac{\sqrt{n}}{3};$$

the lower limit is $-z_1 = -\sqrt{n}/3$. We require $\Phi(\sqrt{n}/3) - \Phi(-\sqrt{n}/3) = 0.9$. Since $\Phi(-z) = 1 - \Phi(z)$, we have

$$2\Phi(\sqrt{n}/3) - 1 = 0.9$$
$$2\Phi(\sqrt{n}/3) = 1.9 \quad \text{or} \quad \Phi(\sqrt{n}/3) = 0.95.$$

Using linear interpolation in Table A2, we find

$$\sqrt{n}/3 = 1.645, \quad \sqrt{n} = 4.935.$$

Hence $n = 24.35$.

Taking n to the next whole number, we obtain a sample size of 25.

16.1.1 Exercises

1. Marks in a mathematics examination are normally distributed with mean 50 and standard deviation 10. What is the probability that: (i) the mean mark of

a group of 5 students will be above 60; (ii) the mean mark of a group of 20 students will be between 42 and 48?

2. A chemical test on a compound gave the following results (grams of a particular element in 1 kg of compound): 1.60, 1.56, 1.59, 1.54, 1.53, 1.58, 1.52, 1.59, 1.64, 1.55. From these figures make an estimate of the size of the sample required so that the standard error of the mean will be about 0.005. Explain what assumptions have to be made, and what sources of error there could be in the estimate.

3. In measuring the time which people take to react to a flashing light, it is estimated from a large number of previous experiments that the standard deviation is 0.5 seconds. How many observations must be taken in a new experiment in order to be 99% confident that the error in the estimate of mean reaction time will not exceed 0.1 seconds?

4. The weights of a large group of animals have mean 8.2 kg and variance 4.84 kg^2. What is the probability that a random selection of 80 from the group will have mean weight between 8.3 and 8.4 kg?

16.2 DISTRIBUTION OF SAMPLE TOTAL

The total T of the values of X measured in a sample of size n equals $n\overline{X}$, a constant times the sample mean. It follows that T has a distribution of the same form as $\overline{X}$, that is, a normal distribution. The mean and variance of T are easily found:

$$E[T] = E[n\overline{X}] = nE[\overline{X}] = n\mu,$$

$$\mathrm{Var}[T] = \mathrm{Var}[n\overline{X}] = n^2 \, \mathrm{Var}[\overline{X}] = n^2 \cdot \sigma^2/n = n\sigma^2.$$

Example 16.3

A charter aeroplane company is asked to carry regular loads of 100 sheep. The plane available for this work has a carrying capacity of 5000 kg. Records of the weights of about a thousand sheep, which are typical of those that might be carried, show that the distribution of sheep weight has a mean of 45 kg and a variance of 9 kg^2; the range of weights is from 37 to 56 kg. Can the company take the order?

A naïve approach to the problem would be to try to estimate the maximum possible load that might occur. The argument might be that the maximum recorded weight of a sheep was 56 kg and hence the maximum possible load might be estimated at $56 \times 100 = 5600$ kg; a more cautious assessor might increase 56 by a safety factor, say about 10%, and conclude that the likely maximum weight of a sheep is 62 kg and the maximum possible load might therefore be 6200 kg. Either way, the company could not take the order.

These assessors are being much too cautious. The chance of obtaining such extreme samples is very small indeed. Let us look rather at the expected distribution of the total weight T of 100 sheep.

The distribution of individual sheep weights has mean $\mu = 45$ and variance $\sigma^2 = 9$. Then

$$E[T] = n\mu = 100 \times 45 = 4500,$$

$$\text{Var}[T] = n\sigma^2 = 100 \times 9 = 900.$$

The distribution of T will be approximately normal since the samples are large. We can find the probability of a total weight of more than 5000 kg occurring by first expressing 5000 in standard measure. To do this we must calculate $z = (T - n\mu)/\sqrt{n\sigma^2}$. Therefore

$$z = \frac{5000 - 4500}{\sqrt{900}} = \frac{500}{30} = 16.7.$$

The probability of obtaining a value of z greater than this is so small that standard tables do not give it. (Note that the probability of a value of z greater than 5 is about 3×10^{-7}.) Thus the probability of the allowed load being exceeded is so small as to be negligible. The company could safely take the order.

Example 16.4

An employer has to interview 20 candidates for a job. His experience has been that he may treat the length of an interview as normally distributed with mean 10 minutes and standard deviation 3 minutes. He begins to interview at 9 a.m. At what time should he ask for coffee to be brought to him if he is to be 99% certain that he will have seen 50% of the candidates by then?

What is the probability that he will finish the interviews by 1 p.m. if he takes 15 minutes over coffee?

(Use linear interpolation in the table of the normal distribution provided.)

(*Oxford*)

The length of an individual interview is distributed as $\mathcal{N}(\mu, \sigma^2)$, i.e. $\mathcal{N}(10, 9)$. The total length of 10 interviews is $\mathcal{N}(10\mu, 10\sigma^2)$, i.e. $\mathcal{N}(100, 90)$. To determine the length of time t at which coffee should be brought (after 10 interviews), we require

$$\Phi\left(\frac{t - 100}{\sqrt{90}}\right) = 0.99.$$

From Table A2,

$$\frac{t - 100}{\sqrt{90}} = 2.328.$$

Therefore

$$t = 100 + 2.328\sqrt{90} = 122.$$

Hence coffee should be brought at 09.00 hours + 122 minutes, i.e. at 11.02 a.m.

The distribution of the total length of 20 interviews is $\mathcal{N}(200, 180)$. The time available for interviews is $240 - 15 = 225$ minutes. The probability of finishing in time is

$$\Phi\left(\frac{225 - 220}{\sqrt{180}}\right) = \Phi\left(\frac{25}{13.42}\right) = \Phi(1.86) = 0.97.$$

16.2.1 Exercises

1. The weight of luggage that aircraft passengers take with them is distributed with mean 20 kg and standard deviation 5 kg. A certain type of aircraft carries 100 passengers. What is the probability that the total weight of the passengers' luggage exceeds 2150 kg?

    To what extent does your answer depend on the distribution of individual weights being normal? What other assumptions have you made?   (*Oxford*)

2. A doctor working in a clinic finds that the consulting times of his patients are independently normally distributed with mean 5 minutes and standard deviation 1.5 minutes. He sees his patients consecutively with no gaps between them, starting at 10 a.m. At what time should the tenth patient arrange to meet a taxi so as to be 99% certain that he will not keep it waiting? If the doctor sees 22 patients in all, what is the probability that he will finish before noon?   (*Oxford*)

3. The lengths of individual pieces of wooden fencing are nominally 2 m, but in fact the lengths are normally distributed with mean 2.05 m and variance 0.01 m^2.

    In order to fence a 21 m length of his garden, a man buys 10 pieces of fencing and 11 wooden posts, each 9 cm thick, which he places between the pieces of fence.

    What is the probability that he will find that his material is (i) too short, (ii) too long, (iii) exactly the right length for the job?

16.3 NORMAL APPROXIMATION TO THE BINOMIAL DISTRIBUTION

When the parameter n of the binomial distribution is large a great deal of calculation is needed to evaluate the distribution function; many terms have to be summed, and each one of them is complicated to work out. We can simplify the calculations substantially for large n by making use of the central limit theorem.

A binomial variable may be regarded as the sum of n Bernoulli variables (compare the remarks on estimating a proportion in Section 3.3). Thus if T has the binomial distribution with parameters n and π, we may regard it as the sum of n independent random variables, each of which takes the value 1 with probability π and the value 0 with probability $1 - \pi$. Therefore when n is large we can use the central limit theorem and say that T will have a distribution that is approximately normal. The mean of the Bernoulli distribution is π and its variance is $\pi(1-\pi)$. Hence T has mean $n\pi$ and variance $n\pi(1-\pi)$.

Theorem 16.2 Normal approximation to binomial distribution

If R has a binomial distribution with parameters n and π then, for large n, the distribution of R can be approximated by a normal distribution with mean $n\pi$ and variance $n\pi(1-\pi)$.

A working rule for this approximation to be satisfactory is to require $n\pi$ and $n(1-\pi)$ both to exceed 5.

We shall use this normal approximation to evaluate the probability that the binomial variable R is less than or greater than a particular value r. This r is, of course, an integer, so we must take account of the fact that we are approximating a discrete random variable R by a continuous random variable X. We therefore think of the probability mass corresponding to the value r as being spread over the interval $(r - \frac{1}{2}, r + \frac{1}{2})$. Hence, using this *continuity correction*,

$$P(R \leqslant r) \simeq P(X \leqslant r + \tfrac{1}{2}) = P\left(z \leqslant \frac{(r + \frac{1}{2}) - n\pi}{\sqrt{n\pi(1 - \pi)}} \right)$$

and

$$P(R \geqslant r) \simeq P(X \geqslant r - \tfrac{1}{2}) = P\left(z \geqslant \frac{(r - \frac{1}{2}) - n\pi}{\sqrt{n\pi(1 - \pi)}} \right).$$

When X is continuous, as we noted on p. 313, $P(X = x) = 0$ for any x; we can specify probabilities *in intervals* only, not *at points*. However, when using a normal approximation we *can* specify $P(R = r)$; it is equal to $P(r - \frac{1}{2} \leqslant X \leqslant r + \frac{1}{2})$. Therefore $P(R \geqslant r)$ and $P(R > r)$ are *not* the same; they differ by an amount equal to $P(R = r)$. Hence

$$P(R < r) \simeq P(X \leqslant r - \tfrac{1}{2}) = P\left(z \leqslant \frac{(r - \frac{1}{2}) - n\pi}{\sqrt{n\pi(1 - \pi)}} \right)$$

and

$$P(R > r) \simeq P(X \geqslant r + \tfrac{1}{2}) = P\left(z \geqslant \frac{(r + \frac{1}{2}) - n\pi}{\sqrt{n\pi(1 - \pi)}} \right).$$

Example 16.5

A factory has 6 machines which intermittently use electric power for 20 minutes in each hour on the average. If the machines are operated independently, show that the probability that 4 or more will be using power at the same time is about 0.1.

If the factory had 60 machines, find an approximate number r, such that the probability that more than r will be using power at the same time is 0.1.

(*Oxford*)

At any time, the probability that a machine will be using power is $\frac{1}{3}$. Since the 6 machines operate independently, the probability that r of them will be using power at the same time is the binomial probability

$$\binom{6}{r}\left(\frac{1}{3}\right)^r\left(\frac{2}{3}\right)^{6-r}$$

Hence the probability that 4 or more will be using power is

$$\binom{6}{4}\left(\frac{1}{3}\right)^4\left(\frac{2}{3}\right)^2 + \binom{6}{5}\left(\frac{1}{3}\right)^5\left(\frac{2}{3}\right) + \left(\frac{1}{3}\right)^6$$

$$= 15\left(\frac{1}{3}\right)^4\left(\frac{2}{3}\right)^2 + 6\left(\frac{1}{3}\right)^5\left(\frac{2}{3}\right) + \left(\frac{1}{3}\right)^6$$

$$= \frac{1}{3^6}(60 + 12 + 1) = \frac{73}{729} \simeq 0.1.$$

When the total number of machines is 60, the distribution of R, the number of machines using power at the same time, is binomial with $n = 60$, $\pi = \frac{1}{3}$.

We may approximate this by X, which is $\mathcal{N}(n\pi, n\pi(1-\pi))$, i.e. $\mathcal{N}(20, \frac{40}{3})$ or $\mathcal{N}(20, 3.65^2)$. $P(R > r)$ is approximated by $P(X \geqslant r + \frac{1}{2})$ and we are looking for the value of r to make this probability 0.1.

We therefore require that the probability of exceeding the standardized value

$$z = \frac{(r + \frac{1}{2}) - 20}{3.65}$$

shall be 0.1, i.e. $\Phi(z) = 0.9$. The value of z such that $1 - \Phi(z) = 0.1$ or $\Phi(z) = 0.9$, is 1.28, from Table A2. Hence

$$\frac{(r + \frac{1}{2}) - 20}{3.65} = 1.28$$

i.e.

$$r = 19.5 + 1.28 \times 3.65 = 19.5 + 4.67 = 24.17.$$

Hence $r = 24$ satisfies the required condition.

Example 16.6

When appointing computer operators a firm requires that the candidates pass a written examination. The examination paper contains 100 multiple-choice questions, each with a choice of three answers of which only one is correct. A pass is obtained by answering 40 or more questions correctly. Use the normal distribution as an approximation to the binomial distribution to estimate the probability that a candidate who chooses the answer to each question randomly will pass the examination.

Estimate the smallest number of questions that the paper should contain if the pass mark is to remain at 40% but the probability of a pass by random choice is not to exceed 1%.

We take as model for the number of correct answers obtained by random choice the binomial distribution with $n = 100$, $\pi = \frac{1}{3}$. We approximate this by $\mathcal{N}(n\pi, n\pi(1 - \pi))$ or $\mathcal{N}(\frac{100}{3}, \frac{200}{9})$.

We wish to find the probability of a score of 40 or more. Since we are approximating a discrete distribution by a continuous one, we find the probability of exceeding 39.5:

$$z = \frac{39.50 - 33.33}{\sqrt{200/9}} = \frac{6.17}{4.71} = 1.31.$$

The probability is 0.095.

If the number of questions on the paper is n, and $0.4n$ must be correct, and the probability of a random choice being correct is $\frac{1}{3}$, then

$$z = \frac{(0.4n - \frac{1}{2}) - n/3}{\sqrt{2n/9}}.$$

If the probability of a pass is not to exceed 1% we require $z > 2.326$ (Table A3). We solve

$$\frac{(0.4n - 0.5) - n/3}{\sqrt{2n/9}} = 2.326.$$

i.e.

$$\frac{3(0.4n - 0.5) - n}{\sqrt{2n}} = 2.326.$$

or

$$0.2n - 1.5 = 2.326\sqrt{2n}.$$

Squaring,

$$0.04n^2 - 0.6n + 2.25 = 10.82n$$

i.e.

$$0.04n^2 - 11.42n + 2.25 = 0$$

$$n = \left(11.42 \pm \sqrt{11.42^2 - 0.36}\right)/0.08$$

$$= (11.42 \pm 11.4042)/0.08$$

$$= 285.3 \quad \text{or} \quad 0.1972.$$

Clearly the second of these answers is impossible (and in any case we can see from the first part of the question that the answer is going to be a large number). We round up n to the next whole number and take $n = 286$.

Checking this result,

$$z = \frac{(0.4 \times 286 - 0.5) - 286/3}{\sqrt{2 \times 286/9}} = \frac{18.5667}{7.9722}$$

$$= 2.329.$$

Taking $n = 285$ would give $z = 2.325$, and so our result is the smallest n to satisfy requirements.

This is a very large battery of questions, and readers will see the effect of increasing the number of possible answers to each question when they have worked Exercise 16.3.1(5) below. Of course the required n can be reduced by relaxing the probability of a pass by random choice from 1% to, say, 5%, when the critical value of z becomes 1.645 instead of 2.326.

16.3.1 Exercises

1. Apples in a store contain 5% that are damaged. From this store, boxes of 72 are packed, and a greengrocer buys three of these boxes. What is the probability that, among these three boxes, there are: (i) less than 10 damaged apples; (ii) between 10 and 20 damaged apples; (iii) more than 20 damaged apples?

2. (a) Twenty-five per cent of pupils in a school have no days' absence through illness in a particular year. In this school there are 120 pupils in the sixth form. Find a number r such that the probability that fewer than r sixth-formers had no days' absence is 0.1. What assumptions have you made? How reasonable are they?
   (b) If, in the same school, 5% suffered from influenza, and the first form contained 30 pupils, find r' such that the probability that more than r' first-formers had influenza was 0.1.

3. A canvasser for a political party claims that 42% of the voters in a given area are going to vote for that party. The whole area contains 3500 voters. Assuming that all will vote, and that the canvasser's claim is based on a proper (unbiased) sampling method, what is the probability that his party will receive less than 1500 votes? Also find $r_1, r_2 (0 < r_1 < r_2 < 3500)$ such that, if r is the number of votes received, $P(r < r_1) = 0.025 = P(r > r_2)$.

4. The probability of successfully completing a sequence of operations in an experiment is 0.44. If 65 of these sequences are begun, each independently of the others, what is the probability that fewer than 25 are successfully completed? Show that if the success probability is only 0.04, the probability of at least 4 successes is about $\frac{1}{4}$.

5. Repeat Example 16.6 above when each question has a choice between (a) 4 answers, (b) 5 answers, of which only one is correct.

16.4 PROPORTIONS

Let R have the binomial distribution with parameters n, π. We think of R as the number of successes in n independent trials. Sometimes we are interested in the *proportion* of successes, which equals the number of successes divided by the number of trials, and may be represented by the random variable $P = R/n$.

If n is large, R may be approximated by $\mathcal{N}(n\pi, n\pi(1-\pi))$. When we multiply or divide a random variable by a constant number, we do not alter the form of the distribution, but only change its scale of measurement. So $P = R/n$ is also approximately normally distributed. Also

$$E[P] = E[R/n] = \frac{1}{n}E[R] = \frac{n\pi}{n} = \pi,$$

$$\text{Var}[P] = \text{Var}[R/n] = \frac{1}{n^2}\text{Var}[R] = \frac{1}{n^2} \cdot n\pi(1-\pi) = \frac{\pi(1-\pi)}{n}.$$

Theorem 16.3

If P is the proportion of successes in a fixed number n of binomial trials, then for large n the distribution of P is approximately normal with mean π and variance $\pi(1-\pi)/n$.

16.4.1 Exercises

1. Thirty-six out of a random sample of 100 pupils in a school travelled to school by bicycle. Estimate the probability that, in a second random sample of 100 from this school, $\frac{3}{4}$ or more of the pupils will travel by bicycle.

2. Ten per cent of shirts supplied by a particular maker have faulty stitching somewhere in the garment. A buyer for a large store will only accept a batch if they contain less than 5% faulty. What is the probability that the buyer will accept a batch containing: (a) 200; (b) 500 shirts?

3. A random sample of 20 children in a large school were asked a question, and 12 answered it correctly. Estimate the proportion of children in the whole school who would answer correctly, and the standard error of this estimate.

16.5 DISTRIBUTION OF LINEAR COMBINATIONS OF NORMAL RANDOM VARIABLES

One of the many useful properties of normally distributed random variables is that the sum of two (or more) independent normal variables is also normally distributed. That is, if X is a normal random variable, and Y is another normal random variable, independent of X, then $X + Y$ is also a normally distributed random variable. Unfortunately, we cannot prove this result with the mathematics available to us at this stage. However, we shall give several instances of its use.

A manufacturer mass-produces a standard pattern of cups and saucers. The weights of the cups can be modelled by the random variable X, which is $\mathcal{N}(\mu_x, \sigma_x^2)$, and the weights of the saucers can be modelled by the random variable

Y, which is $\mathcal{N}(\mu_y, \sigma_y^2)$. The total weight of a cup and a saucer, each selected at random from the mass-produced items, can then be modelled by the random variable $W = X + Y$. From what we have just said, W will be normally distributed. Further, using the results in Chapter 10 on expectations in joint distributions, we have

$$E[W] = E[X + Y] = E[X] + E[Y] = \mu_x + \mu_y.$$

Also, since the cup and saucer were selected independently (at random from all those available), we have

$$\text{Var}[W] = \text{Var}[X + Y] = \text{Var}[X] + \text{Var}[Y] = \sigma_x^2 + \sigma_y^2.$$

The distribution of total weight W of a cup and a saucer is now completely specified: it is $\mathcal{N}(\mu_x + \mu_y, \sigma_x^2 + \sigma_y^2)$.

What can we say about the distribution of the *difference* of two normal variables? Suppose that we take the normally distributed random variable X as a model of the milk yield of a cow which is milked while music is being played, and the normal random variable Y as a model of the milk yield of a cow of the same breed which is milked without music. We might regard the difference $X - Y$ as a measure of the effect of music on milk yield. We are therefore interested in the distribution of the difference $X - Y$. Now if Y is normally distributed, a moment's thought about the graph of the normal probability density function (Fig. 15.2) will show that $Z = -Y$ is also a normal random variable. But we know that a sum of two normal variables is normally distributed, so $X + Z$ is normally distributed and hence $X - Y$ is normally distributed. If we denote the means and variances of X, Y by μ_x, μ_y and σ_x^2, σ_y^2 respectively, then

$$E[X - Y] = E[X] - E[Y] = \mu_x - \mu_y.$$

In order to calculate the variance, note first that $\text{Var}[-Y] = \text{Var}[Y]$. This follows from the result $\text{Var}[aY] = a^2 \, \text{Var}[Y]$ by setting $a = -1$. Thus assuming that the cows have been chosen independently we have

$$\text{Var}[X - Y] = \text{Var}[X] + \text{Var}[-Y] = \text{Var}[X] + \text{Var}[Y] = \sigma_x^2 + \sigma_y^2.$$

It is important to remember that we have to *add* variances in this case, not subtract one from the other. So the rule is that *the variance of the sum or difference of two independent random variables is the sum of the two variances.*

If X is a normal random variable then the random variable $Y = cX$, where c is a constant such as 3 or -6, is also normally distributed. Clearly Y arises from X simply by altering the scale of measurement, without changing the basic shape of the distribution. Generalizing these results, we can consider a linear combination of many normally distributed random variables. This leads to the following theorem.

Theorem 16.4 Distribution of linear combination of normal variables

If $X_1, X_2, \ldots, X_n$ are independent normal random variables and $c_1, c_2, \ldots, c_n$ are constants then $Y = c_1 X_1 + c_2 X_2 + \cdots + c_n X_n$ is also a normal random variable,

and

$$E[Y] = c_1 \, E[X_1] + c_2 \, E[X_2] + \cdots + c_n \, E[X_n]$$

and

$$\mathrm{Var}[Y] = c_1^2 \, \mathrm{Var}[X_1] + c_2^2 \, \mathrm{Var}[X_2] + \cdots + c_n^2 \, \mathrm{Var}[X_n].$$

Example 16.7 Distribution of difference of means

A consumer organization found the mean lifetimes $\overline{X}, \overline{Y}$ of two types of electric light-bulbs taken from large random samples of both types. The population means and variances of the lifetimes for the two types were μ_x, μ_y and σ_x^2, σ_y^2 respectively, and the sample sizes were n_x, n_y respectively. The difference in mean lifetime of the two types was estimated by $\overline{X} - \overline{Y}$. What is the distribution of this estimate?

If we sample on many occasions, taking large numbers n_x and n_y of the two types on each occasion, at random from the bulbs available we can generate the sampling distribution of $\overline{X} - \overline{Y}$. Lifetimes of individual bulbs are not very likely to be normally distributed, but because the samples are large it follows from the central limit theorem that both $\overline{X}$ and $\overline{Y}$ are approximately normally distributed. Hence $\overline{X} - \overline{Y}$ is approximately normally distributed:

$$E[\overline{X} - \overline{Y}] = E[\overline{X}] - E[\overline{Y}] = \mu_x - \mu_y,$$
$$\mathrm{Var}[\overline{X} - \overline{Y}] = \mathrm{Var}[\overline{X}] + \mathrm{Var}[\overline{Y}] = \sigma_x^2/n_x + \sigma_y^2/n_y.$$

Thus the sampling distribution of $\overline{X} - \overline{Y}$ is completely specified in terms of the population values: it is $\mathcal{N}(\mu_x - \mu_y, \sigma_x^2/n_x + \sigma_y^2/n_y)$. Of course, in a situation like this the population values are unknown; we have to replace them by estimates. The mean $\mu_x - \mu_y$ would have to be estimated by $\overline{X} - \overline{Y}$, and the variances σ_x^2 and σ_y^2 would be replaced by the unbiased estimates s_x^2 and s_y^2, respectively.

16.5.1 Exercises

1. An office manager takes, on average, 25 minutes to get to work; his travelling time is normally distributed with standard deviation 3 minutes. His secretary takes, on average, 24 minutes to get to work; her travelling time is normally distributed with standard deviation 2 minutes.
   (a) If both set out from home at the same time, what is the probability that the manager arrives first?
   (b) If the manager leaves home 2 minutes before the secretary, what is the probability that he arrives first?
   (c) If the secretary knows that the manager leaves home promptly at 8.30 a.m. every day, what time should she leave if she wants to be 99% confident that she will arrive first?

2. The means and variances of independent normal variables x and y are known. State the means and variances of $x \pm y$ in terms of those of x and y.

The values of two types of resistors are normally distributed as follows:

Type	Mean (ohm)	Standard deviation (ohm)
A	100	2
B	50	1.3

(a) What tolerances would be permitted for type A if only 0.5% were rejected?
(b) 300-ohm resistors are made by connecting together three of the type A resistors, drawn from the total production. What percentage of the 300-ohm resistors may be expected to have resistances greater than 295 ohms?
(c) Pairs of resistors, one of 100 ohms and one of 50 ohms, drawn from the total production for types A and B respectively, are connected together to make 150-ohm resistors. What percentage of the resulting resistors may be expected to have resistances in the range 150 to 151.4 ohms?

3. The mass of a biscuit is a normal variable with mean 50 g and standard deviation 4 g. A packet contains 20 biscuits and the mass of the packing material is a normal variable with mean 100 g and standard deviation 3 g. Find the probability that the total mass of the packet:
   (i) exceeds 1074 g;
   (ii) is less than 1120 g;
   (iii) lies between 1074 g and 1120 g. *(Cambridge)*

16.6 CONTROL CHARTS

In any process that is supposed to produce large numbers of 'identical' items, there is almost certain to be some variability. The modern approaches to *statistical process control (SPC)* combine several ideas and methods for deciding whether a production process is 'in control' or not; an introduction to some of these is given in Clarke and Kempson (1997) and much more detail may be found in Bissell (1994). Two of the earliest methods to be used were *acceptance sampling*, which we studied in Sections 8.8 and 9.8, and *control charts*, introduced by W. A. Shewhart in 1931.

Variability may fall into two categories. There is *random* variation which is always present in the process however well it is running, and there may also be *non-random* variation due to *assignable causes*, for example a continuously running machine goes out of its proper adjustment. Shewhart, and other pioneers in this field, developed the *control chart*, which is a very valuable graphical method for detecting possible non-random variation. Although most of its applications have been in industry, it has more recently been applied to medical, administrative and many other situations, in the same way as acceptance sampling (p. 178).

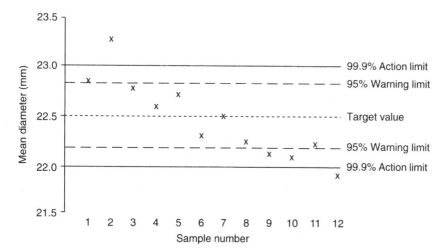

Figure 16.1 *Control chart for mean of three measurements in machine drilling process:* ———— *action limits;* — — — — *warning limits;* - - - - - - *target value.*

The idea of a control chart is shown in Fig. 16.1. A process may be running continuously, producing output that ought to have a target value *x* for some important measurement. Random samples taken at regular intervals will show the variation in the successive values of *x*. We can either use past data to establish the value of the variance of *x* when the process is operating satisfactorily or insist that *x* lies within specified tolerance limits. Some assumption about the distribution of *x* is necessary, as in the following sections. Given the target value, the variance and the distribution, we can assign limits within which sample values of *x* ought to lie, at suitable probability levels. This information can be shown on a graph and so long as *x* lies within these limits we remain satisfied that the process is working properly. Otherwise some action may be necessary.

16.6.1 Mean value of a continuous measurement

Suppose that *x* is a measurement made on a continuous scale, such as the length or diameter of a mass-produced component, or the weight of a full packet of sugar coming off a packing line. There will be a *target* value of *x* which, in the absence of variability, we intend every item to have. Because of the natural random variation which is present in every industrial process, however well adjusted or controlled it may be, *x* will not actually be the same for every item. So long as there is only *random variation* in the system, and this is as small as it can technically be made, the producer will keep the process running continuously. The Shewhart control chart is one way of monitoring the value of *x* during continuous running, with the aim of finding out as quickly as possible if there is apparently some *non-random variation* also. It assumes that the values of *x* will follow a normal distribution, which for the examples mentioned above is not unreasonable as a model.

Non-random variation may be due to sudden changes in the system, such as a new supply of the raw material (the metal alloy from which the components are made, or the source of the sugar which is being packed). It may also be due to a break in production, such as a change of shift or of operator. If the producer is aware, from previous experience, that non-random variation is likely to arise at these times, the system will be monitored carefully for a short period after every change or break.

We also need to detect systematic changes in x when there is no obvious assignable cause. The philosophy here is slightly different, because the producer will want to be fairly sure that the process does need adjustment before stopping it, and losing production time and potential profit. To design a rule for deciding when to stop the process, we must try to separate random variation and non-random variation. We therefore require (i) methods for constructing a control chart and (ii) statistical decision rules to interpret and apply what the chart shows us.

Inferences about the value of a measurement x will not usually be made on the basis of a single observation, but on the mean of a random sample $x_1, x_2, \ldots, x_n$. For a continuously running process, n items will be selected from the production line, at random, and this sampling will be repeated at fixed time-intervals. The number n in the sample at each time will usually be quite small, often no more than four or five. The value of $\bar{x}$ for each sample is plotted in the vertical direction on the control chart, against the time on the horizontal axis.

Table 16.1 shows a typical set of data. It gives diameters (in millimetres) of holes in steel plates from a continuous production process. A machine is set to drill three holes in the plate with the same drill bit, and samples from the process are taken at regular intervals. The target mean is 22.5 mm, and when in control the process standard deviation (computed using a large quantity of historical data) is 0.3 mm.

Table 16.1 *Diameters (in millimetres) of holes drilled in steel plates*

Sample	Diameters			Mean	Range
1	22.7	22.8	23.1	22.87	0.4
2	23.4	23.6	23.0	23.33	0.6
3	23.1	22.7	22.6	22.80	0.5
4	22.4	22.5	22.9	22.60	0.5
5	22.4	22.8	23.0	22.73	0.6
6	22.1	22.5	22.3	22.30	0.4
7	22.6	22.8	22.1	22.50	0.7
8	22.0	22.2	22.5	22.23	0.5
9	22.2	22.0	22.1	22.10	0.2
10	22.0	21.9	22.3	22.07	0.4
11	22.3	22.4	21.9	22.20	0.5
12	21.6	21.8	22.2	21.87	0.6

Assuming a normal distribution for the observations, we can now summarize all the necessary statistical information.

When the system is in control, x is normal with mean 22.5 and variance 0.3^2. Hence $\bar{x}$ is $\mathcal{N}(22.5, 0.3^2/3)$, i.e. $\mathcal{N}(22.5, 0.03)$. Thus with probability 0.95, $\bar{x}$ should be between the limits $\mu \pm 1.96\sigma/\sqrt{n}$, i.e. $22.5 \pm 1.96\sqrt{0.03}$, or 22.5 ± 0.34. These limits, 22.16 to 22.84, are marked on the chart shown in Fig. 16.1, and are called the *warning limits* (the upper and lower 95% warning limits).

Further, with probability 0.999, and when the system is in control, $\bar{x}$ should lie between the limits $\mu \pm 3.29\sigma/\sqrt{n}$, i.e. 22.5 ± 0.57, or 21.93 to 23.07; these are shown on the chart as *action limits* (the upper and lower 99.9% action limits). Thus when a value of the mean falls in the extreme 0.1% tails of the normal distribution it is *outside* these *action limits*; if it is in the extreme 5% tails but *not* in the extreme 0.1%, then it is *outside* the *warning limits*.

We now need a rule for deciding when to stop the process. The value of $\bar{x}$ observed for this set of data at the first time-point is just outside the warning limits, but still inside the action limits. In these circumstances, the process would usually be allowed to go on until the next sampling time, unless the set of three x-values showed some other cause for concern such as being very variable (see Section 16.6.3).

The second time-point gives a value of $\bar{x}$ clearly outside the action limits; action would indeed be taken, the process would be stopped and examined. The next six points (time 3 to time 8 inclusive) are all within the warning limits and no action would be taken, but those at times 9 and 10 both fall below the lower warning limit and the process should be stopped. At time 11 the mean is inside the warning limits but at time 12 it has again fallen outside the action limits and the process needs to be examined.

Rules commonly used for deciding that an assignable cause of variation may be present, in addition to normal random variation, are to stop the process if:

(i) an observed value of $\bar{x}$ falls outside the 99.9% action limits, calculated using the target value for μ and the given σ; or
(ii) two or more successive values of $\bar{x}$ fall between the 95% warning limits and the 99.9% action limits.

These rules should be accompanied by an intelligent look at the control chart as it is being plotted over time. For this set of data it seems very likely that even when $\bar{x}$ is within the warning limits there is a downward trend, and someone who is familiar with the working of this process may regard this as evidence of poor performance – and perhaps even know what to do about it.

16.6.2 Other probabilities and conventions for limits

The three conventional levels used in statistical inference are 95%, 99% and 99.9% (p. 283). The corresponding critical values in the normal distribution are 1.960, 2.576, 3.291. A common alternative to the method described above is to take *warning limits* at $\mu \pm 2\sigma/\sqrt{n}$ and *action limits* at $\mu \pm 3\sigma/\sqrt{n}$. These 'two-sigma' and 'three-sigma' limits are not exactly at the 95% and 99.9% limits, but reasonably close, and for all practical purposes inferences based on 2 and 3 as multipliers will be just as satisfactory as those using 1.96 and 3.29. In the context of particular applications, different probabilities may be chosen, and quite often the action limits are calculated at the 99% level, using 2.576 as the critical value rather than 3.29. The conventions used by American and British industrialists have often differed from one another. Because of these variations, we recommend always quoting the probabilities that are being used for warning and action limits, as in the previous section.

16.6.3 Measuring variability

We usually measure variability by calculating s^2, the estimate of σ^2 for the sample of data that is available. If data are assumed to follow a normal distribution, the variance is the 'natural' estimate of variability, in the sense that the parameter σ^2 appears directly in the density function. But when the sample size n is small, say less than 10 and especially when it is as small as 3 or 4, an estimate of σ based on the range in the sample is likely to be just as good as that based on the usual formula for s^2. It is also much easier to calculate.

If the range of a sample of n observations from a normal distribution is R, then $s = R/d_n$ is an estimate of the standard deviation of an observation; d_n is a number which depends on n, and the values of d_n for $n = 2$ to 10 are shown in Table 16.2.

Example 16.8

For the 12 samples of data in Table 16.1, compute the average range, $\overline{R}$, and use it to estimate s.

The range in each sample is given in Table 16.1; the average range

$$\overline{R} = \sum_{i=1}^{12} R_i/12 = 5.9/12 = 0.49.$$

To convert $\overline{R}$ to an estimate of s we divide by 1.693, which is the value corresponding to sample size $n = 3$ in Table 16.2. This gives $s = 0.49/1.693 = 0.29$.

When several samples of data are available, as here, $\overline{R}$ gives the best possible estimate of s that can be found by this method. (The alternative, of finding an estimate s_i from R_i in each sample and then combining those estimates, gives less satisfactory results.) In this case, we were actually told that $\sigma = 0.3$, calculated from a large quantity of historic data. The estimate found using $\overline{R}$ agrees very well with this. If we had not been told σ, we could have used this estimate s in calculating warning and action limits.

Table 16.2 *Values of d_n, the ratio of range to standard deviation, in a sample from a normal distribution*

Sample size, n	2	3	4	5	6	7	8	9	10
d_n		1.128	1.693	2.059	2.326	2.534	2.704	2.847	2.970	3.078

Control chart for range

From the result of Example 16.8 we might be tempted to say that the level of variability among these 12 samples agrees very well with what had been found from the historic data. However, we can do more than this: a control chart can be constructed to monitor the range of the observations in each sample. We assume that n observations have been sampled on each occasion, that the data are normally distributed, and that variability is measured by the average range R in past data samples when the process was judged to be running satisfactorily.

Table A10, giving the factors required to find limits for the control chart, will be found on p. 728. Factors have to be provided for all four warning and action limits since the underlying distribution of $\bar{R}$ is not symmetrical.

Figure 16.2 shows the control chart. Successive values of the range are given in Table 16.1 and have been plotted in the vertical direction in Fig. 16.2 against time (sample number) on the horizontal axis. The calculation of the limits in Fig. 16.2 has been based on the estimate of variability given by the average range 0.49, for illustration, though in practice we would require previous data for an independent estimate of R. (An estimate, in this case, could be found

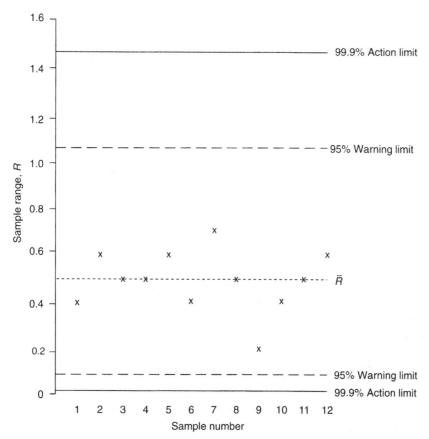

Figure 16.2 Control chart for range (Example 16.8).

from the value $\sigma = 0.3$ which was given. Since $\sigma = R/d_n = R/1.693$, we have $R = 0.3 \times 1.693 = 0.508$.)

With $n = 3$, the four factors for the 99.9% and 95% limits are 0.04, 0.18, 2.17, 2.99, and when multiplied by the average range 0.49 they lead to the limits 0.02, 0.09, 1.06, 1.47 for R; from these it is clear that there is no suggestion of variability being out of control.

16.6.4 Exercises

1. At regular time intervals, samples of four components are taken from a production line. Each component is weighed, and the mean weights are plotted on a Shewhart control chart. The process has a target mean of 50 g with a known standard deviation of 5 g.
   (a) What are the warning and action limits for the chart?
   (b) Comment on the implications of the following, on the chart for means:
       (i) a value of 53.25 g;
       (ii) a value of 43.25 g;
       (iii) two consecutive values 55.75 and 54.00 g;
       (iv) two consecutive values 55.25 and 44.75 g.

2. Samples of metal castings were drawn five at a time from a continuous production process. The diameter of each casting was measured and the mean diameters of the samples are given in the following table, in units of 0.1 mm.

Sample	Mean diameter	Sample	Mean diameter
1	118	9	126
2	122	10	121
3	117	11	116
4	126	12	122
5	119	13	125
6	124	14	119
7	121	15	124
8	117		

The target mean and standard deviation of the process are respectively 120 and 0.8 mm, and the standard deviation is believed to be under control.

Draw a control chart for mean diameter, and comment on the information that it gives.

3. Test resistances ($\mu\Omega$) for 20 random samples, each of three resistors, from batches produced on a continuous production process are shown in the table below. The resistors are intended to operate at $40\,\mu\Omega$. The standard deviation of a measurement of resistance is $3.2\,\mu\Omega$. Construct two control charts and use them to discuss whether you might have concluded that the process was out of control as the batch test results came in.

Batch no.	Resistances			Total	Batch no.	Resistances			Total
1	38	41	38	117	11	38	45	42	125
2	39	42	40	121	12	43	46	46	135
3	36	39	35	110	13	40	43	47	130
4	35	42	43	120	14	41	46	39	126
5	39	37	38	114	15	46	38	47	131
6	37	43	40	120	16	49	44	43	136
7	42	45	40	127	17	45	48	42	135
8	37	39	44	120	18	48	46	49	143
9	42	45	39	126	19	49	50	44	143
10	41	40	44	125	20	51	45	52	148

16.6.5 Proportion faulty ('fraction defective')

Individual items may be inspected visually, or given a pass/fail test, and as a result each item is classified 'sound' or 'faulty'. A binary classification like this gives very little information per item; much larger samples are therefore needed than when numerical measurements x can be taken.

The aim generally is to check the proportion of faulty items being produced. If a process is in control, the probability of an item being faulty should be constant, and the conditions for a binomial distribution should apply to random samples of n items from the production line. Therefore if n is reasonably large and the probability of an item being faulty is π, the distribution of $\hat{p} = r/n$, that is, the proportion faulty in a random sample of n, may be approximated as $N(\pi, \pi(1-\pi)/n)$. The sample size n taken at random at each time point may be 50, or 100, or more.

However, the recommendation often made (see p. 377) that the normal approximation to the binomial can be applied when the product $n\pi$ is 5 or more is not always good for the very small values of π expected when a process is in proper control. This can be improved by requiring that $n\pi \geqslant 10$, instead of 5. For example, when $\pi = 0.1$ we would like to have n at least 100, and if $\pi = 0.05$, which can often be the sort of value desired, $n \geqslant 200$. Obviously inspection of large batches like this, or even of the whole production, needs to be done by a rapid method, visual or mechanical, otherwise it will be too costly.

The observed proportion of faulty items in each sample, $\hat{p} = r/n$, the *fraction defective*, is plotted in the vertical direction on a control chart against time on the horizontal axis. The same warning and action limits can be used as for means:

99.9% upper action limit is $\pi + 3.29 \sqrt{\pi(1-\pi)/n}$,
95% upper warning limit is $\pi + 1.96 \sqrt{\pi(1-\pi)/n}$,
95% lower warning limit is $\pi - 1.96 \sqrt{\pi(1-\pi)/n}$,
99.9% lower action limit is $\pi - 3.29 \sqrt{\pi(1-\pi)/n}$.

These are the limits we shall use, although the same remarks as in Section 16.6.2 apply about alternative conventions for limits.

With very small values of π and particularly if n is not really quite large enough for a satisfactory normal approximation, it is possible for the lower action or warning limits when calculated to appear negative. In such a situation the lower limit must clearly be taken as 0; but it is a reminder that the approximation we are using may be barely adequate in that particular problem.

Example 16.9

Products are inspected in batches of 100. Past experience indicates that the fraction defective is 0.1 in normal operation. A sequence of 30 batches gives the following numbers of defectives:

$$10, 8, 9, 6, 17, 12, 18, 9, 14, 8, 7, 13, 16, 10, 11,$$
$$10, 7, 8, 13, 15, 11, 10, 5, 11, 4, 3, 8, 12, 9, 11.$$

Construct a control chart to monitor the working of the process.

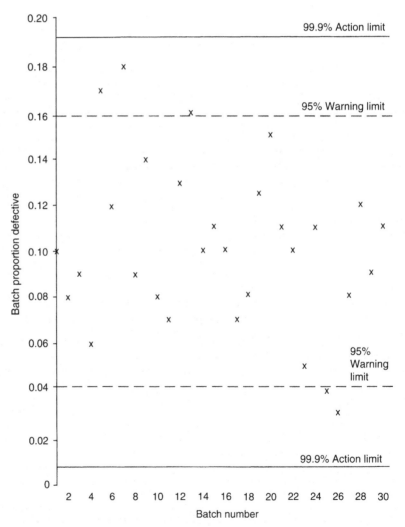

Figure 16.3 *Control chart for fraction defective (Example 16.9).*

The value of $\sqrt{\pi(1-\pi)/n}$ when the process is in control is $\sqrt{0.1 \times 0.9/100} = 0.03$. Warning limits are therefore set at $0.1 \pm 1.96 \times 0.03$, i.e. 0.1 ± 0.0588, giving 0.0412 and 0.1588. For the action limits, we need $0.1 \pm 3.29 \times 0.03$, i.e. 0.1 ± 0.0987, giving 0.0013 and 0.1987.

Figure 16.3 shows the successive values of $\hat{p} = r/100$, and the control limits. At the fifth observation, $\hat{p}$ is 0.17, which is above the 95% upper warning limit but not above the 99.9% action limit. Neither of the observations next to it suggests any problem, so the process is allowed to continue. Likewise the seventh observation gives a value of $\hat{p} = 0.18$, above the 95% upper warning limit but not above the 99.9% upper action limit; again the process may be allowed to continue. There is one further observation, the thirteenth, which is just above the 95% upper warning limit (it is $\hat{p} = 0.16$) but as with the fifth there is no other evidence to warrant stopping the process.

In batches 25 and 26, $\hat{p} = 0.04$ and 0.03. Both of these are below the 95% lower warning limit, although not below the 99.9% lower action limit. It appears that, by analogy with the usual conventions for $\bar{x}$ charts, we should regard the process as *out of control* here also, and stop it. It could be argued that having $\hat{p}$ very small in a sampled batch is an excellent thing, which raises the question of why the lower limits are calculated at all. But experience has shown that virtually no continuous production process can operate without some variability in quality and without producing some defectives. Further, no inspection system is likely to be perfect. Therefore very small numbers of defectives do also require to be followed up. Rather than stop the process immediately, which could well be a drastic and costly step, the first thing to do in these circumstances is to check the inspection method.

16.6.6 Exercises

1. Twenty-five samples, each of size $n = 250$, were taken at regular intervals from a production line and the number r showing any fault or defect was recorded: the results were as follows.

    10, 8, 11, 7, 13, 9, 7, 12, 16, 10, 5, 4, 4, 14, 9, 7, 14, 9, 12, 10, 17, 17, 12, 10, 6.

    If the expected value of π is 0.04, construct a control chart and use it to assess whether the process remains in control.

2. Small, cheap glass components were made at one factory and were then transported to a second factory where 100% inspection was carried out and the non-defective components were then used in a manufactured product. The transport between the two factories could handle batches of 500 at a time. Twenty such batches, on successive working days, gave the following numbers of defective items at the inspection stage:

    18, 24, 27, 17, 36, 34, 15, 24, 21, 18, 30, 21, 19, 21, 20, 26, 28, 31, 33, 28.

    If there are only 5% defective items the manufacturer who is going to use the components will not complain. Investigate the situation using a control chart.

16.7 MINITAB: CONTROL CHARTS

16.7.1 Mean value of a continuous measurement

To construct a control chart similar to that in Section 16.6.1, the data should be entered into the worksheet with each sample on a row, the n members $x_1, \ldots, x_n$ being in C1 ... Cn. Choose

> **Stat>Control Charts>Variables Charts for Subgroups>X-bar/**
> Select **Observations for a subgroup are in one row of columns** and enter C1–Cn, choose **Xbar Options/**
> Enter known **Mean** and **Standard deviation**, choose **S limits/**
> Opposite **control limits at these multiples of the standard deviation** enter 2 and 3//

A chart similar to that in Fig. 16.1 will be produced.

If no values are entered for the **mean** and **standard deviation** they will be estimated from the data. The **S limits** specify the warning and action limits as in Section 16.6.2. Note that *only* whole numbers are permitted.

S (standard deviation) and R (range) charts are obtained in exactly the same way; replace **Xbar** by **R** or **S**. The lower limit may be set at zero if appropriate.

16.7.2 Proportions or numbers defective

A closely related choice of menus will allow the production of a chart similar to that produced in Example 16.9. Assume all batches sampled are the same size n. Enter numbers faulty all in C1, in the order the batches were sampled. Choose

> **Stat>Control Charts>Attributes Charts>NP/**
> Enter **Variable** C1 and **Subgroup size** n, select **NP Chart Options/**
> Enter known **Proportion**, select **S Limits/**
> Opposite **control limits at these multiples of the standard deviation** enter 2 and 3//

A chart is produced in which the numbers themselves are plotted. If proportions are required then choose **P** instead of **NP** in the menu choice. If no target value is entered for **Proportion** it will be estimated from the data.

16.8 EXERCISES ON CHAPTER 16

1. Packets of a certain detergent have a nominal net weight of 10 ounces. They are packed in batches of 100 packets. A machine fills the packet with detergent in such a way that the weights are normally distributed with mean net

weight 10.06 ounces and standard deviation 0.2 ounces. Find the probabilities: (i) that a batch does not contain any underweight packets; (ii) that the mean net weight of the packets in a batch does not fall below the nominal net weight. (*SMP*)

2. A coin is tossed 1000 times, the probability of it falling heads being π. Give an approximate value for the probability that the total number of heads exceeds 550, in the two cases $\pi = 0.5$ and $\pi = 0.6$.

3. The independent variables x_1 and x_2 have means μ_1, μ_2 and variances σ_1^2, σ_2^2 respectively. State the mean and variance of the variable $m_1 x_1 + m_2 x_2$ where m_1 and m_2 are constants.

   The scores of boys of sixteen years of age in a certain aptitude test form a normal distribution with mean 70 and standard deviation 3. The scores of boys of fifteen years of age in the same test form a normal distribution with mean 64 and standard deviation 4. What percentage of the boys of each age will score 62 or less?

   Find the mean and standard deviation of the excess score of a sixteen-year-old boy over that of a fifteen-year-old boy, both taken at random.

   What is the probability that the younger boy will obtain a higher score in the test? (*AEB*)

4. A chemical test involves measuring two independent quantities, A and B, and then expressing the result of the test as C, where $C = A - \frac{1}{4}B$.

   If A is normally distributed with mean 10 and standard deviation 1, and B is also normally distributed with mean 34 and standard deviation 2, what proportion of C values would you expect to be negative? (*IOS*)

5. Which of the following measurements will allow of normal approximations? For those that will, give the mean and variance of the approximating normal distribution.
   (i) The mean of 500 records of time taken by subjects to react to a visual stimulus. The mean and variance in a very large population of such records are 8 and 125 respectively.
   (ii) The number of heads in 200 tosses of a fair coin.
   (iii) The average weight of a random sample of 6 men from a population in which mean weight is 160 lb and variance 600 lb^2.
   (iv) The number of white mice in a sample of 20 from a population in which $\frac{1}{16}$ of the mice are white.
   (v) The number of words per page of a book, which contains on average $8\frac{1}{2}$ words per line, with standard deviation 3 words per line. There are 40 lines per page. (Neglect incomplete pages.)
   (vi) The total mark of a candidate in an examination consisting of 8 papers, each of which is marked out of 50. The mean marks for the papers are 22, 19, 36, 28, 24, 27, 25, 25 respectively, and the standard deviation on each paper is 6.

6. The life (in hours) of a valve is normally distributed with mean 200 hours. If a purchaser buys 10 valves, and requires that, with probability 0.95, the

mean life of the 10 shall be at least 190 hours, what is the largest value that the standard deviation of the lifetime can have?

7. A torch uses two batteries. The distribution of lifetime of batteries of this type is $N(24, 4)$ hours. What is the probability (assuming that the batteries fail independently of one another) that the torch will operate for: (i) less than 24 hours; (ii) between 25 and 26 hours; (iii) more than 28 hours?

8. A mechanism contains three springs, end to end. The natural length of one is normally distributed with mean 2.5 cm and standard deviation 0.02 cm, and, of the other two, normal with mean 3 cm and standard deviation 0.01 cm. Find the probability that the sum of the natural lengths of a set of three springs in the mechanism is greater than 8.58 cm.

9. The random variables X, Y are independent, X being distributed as $N(0,4)$ and Y as $N(1,9)$. A sample of 10 observations is taken from each. Find $P(\overline{X} < \overline{Y})$.

10. A factory is illuminated by 2000 bulbs. The lives of these bulbs are normally distributed with a mean of 550 hours and a standard deviation of 50 hours. It is decided to replace all the bulbs at such intervals of time that only about 20 bulbs are likely to fail during each interval. How frequently should the bulbs be changed?

    When the manufacturing process is improved so that the mean life of bulbs is increased to 600 hours and the standard deviation is reduced to 40 hours, the replacement interval is changed to 500 hours. Show that it will now be necessary to tolerate the failure of only about 12 bulbs per interval. (*AEB*)

11. If the probability of a male birth is 0.514, what is the probability that there will be fewer boys than girls in 1000 births? (You may assume that $0.514 \times 0.486 \approx 0.25$.)

    How large a sample, to the nearest hundred, should be taken to reduce the probability of fewer boys than girls to less than 5%? (You may assume that the sample size in this part of the question is sufficiently large for a continuity correction to be unnecessary.) (*SMP*)

12. X_1 and X_2 are two independent and independently distributed normal variables with means μ_1, μ_2 and variances σ_1^2 and σ_2^2 respectively. State the means and variances of Y and Z where

$$Y = X_1 + X_2 \quad \text{and} \quad Z = X_1 - X_2.$$

Alf, Bill and Charlie are three bricklayers working on the three walls of a garage. The bricks used have a thickness which is normally distributed with mean 65 mm and standard deviation 0.9 mm; Alf spreads mortar with a mean thickness of 16 mm and standard deviation of 0.5 mm while the corresponding figures for Bill are 14.8 mm and 0.6 mm.
   (i) Calculate the mean height and the standard deviation of the height of Alf's wall when he has laid three courses (i.e. three rows of bricks and three layers of mortar alternately).

(ii) What is the probability that, after three courses, the heights of Alf's and Bill's walls will be within 2 mm of each other?

(iii) Charlie is new to the job. When he has laid three courses it is found that his wall is 241 mm high. Calculate the mean thickness of mortar in the courses he has laid. *(Southern)*

13. The following table gives the cumulative frequency distribution of the masses x in kilograms of a group of 200 eighteen-year-old boys:

x	30	35	40	45	50	55	60	65	70	75	80	85	90	95
Number with mass less than x	0	1	4	11	25	47	79	114	146	171	187	195	198	200

Draw a cumulative frequency graph and from this estimate the median. Compile a frequency distribution from the data and hence estimate the mean and standard deviation of the sample. State a well-known probability distribution which you would expect to fit such data. *(JMB)*

14. A machine is producing components whose lengths are normally distributed about a mean of 6.50 cm. An upper tolerance limit of 6.54 cm has been adopted and, when the machine is correctly set, 1 in 20 components is rejected as exceeding this limit. On a certain day, it is found that 1 in 15 components is rejected for exceeding this limit.

(a) Assuming that the mean has not changed but that the production has become more variable, estimate the new standard deviation.

(b) Assuming that the standard deviation has not changed but that the mean has moved, estimate the new mean.

(c) If 1000 components are produced in a shift, how many of them may be expected to have lengths in the range 6.48 to 6.53 cm if the machine is set as in (a)? *(AEB)*

16.9 COMPUTING EXERCISES

1. Let X have a continuous uniform distribution on the interval $(0, 12)$. Generate a random sample of size k from X, find the mean of the sample and standardize it by subtracting the distribution mean 6 and dividing the result by the distribution standard deviation $\sqrt{12/k}$. Repeat this 500 times for each of the values $k = 1, 12, 48$.

Compare each of the distributions generated with the normal distribution whose mean is zero and variance one, by:

(i) plotting histograms;

(ii) calculating means, medians, variances, standard deviations;

(iii) calculating the frequencies that fall in the intervals formed by dividing $(-3, 3)$ into equal intervals of length 0.5.

2. Choose a random sample of 500 observations from the binomial distribution with parameters n and π. Standardize each value x by subtracting the mean $n\pi$ and dividing the deviation $x - n\pi$ by the standard deviation $\sqrt{n\pi(1-\pi)}$. Compare the resulting distribution with a normal distribution having mean 0 and variance 1.

   Repeat this for the three sets of values of the parameters (n, π): (i) $(9, 0.1)$; (ii) $(49, 0.1)$; (iii) $(81, 0.1)$. Note the improvement of the normal approximation as n increases.

16.10 PROJECTS

1 Sampling distribution of an estimate of a mean

Aim. To study the sampling distribution of the mean of discrete uniform observations.

Collection of data. As for Project 15.12(1).

Analysis. Use the frequency table of totals of 10 random digits that was made in Project 15.12(1), but divide the variate values by 10 so that the table may be interpreted as a frequency distribution of means of 10 random digits. Find the mean, variance and standard deviation of the data.

Find the frequency table of the means of 5 random digits, using the data collected in Project 15.12(1). Use the same total number of means of 5 as you did for means of 10 digits. Find the mean, variance and standard deviation of these data.

Draw histograms for the two sets of data. Standardize the values of the variates, by subtracting the mean and dividing by the standard deviation, so that the two histograms may be compared. Fit normal distributions to each set of data and draw on the same graph as the histogram.

2 Sampling distribution of the estimate of a proportion

Aim. To study the sampling distribution of an estimated proportion.

Equipment. Beads of two colours, or other equipment as described for Project 11.9(1).

Collection and analysis of data. Obtain a large number of samples of beads, as explained in Project 11.9(1), each sample to consist of 20 beads. Find the estimate of the proportion of blue beads in each sample. Arrange these estimates of proportions in a frequency table, compute the mean and variance of this collection of estimates, and display the results in a histogram. Fit a normal distribution with the same mean and variance, and draw it on the same diagram.

Repeat the calculations for samples of 40 (i.e. combine the samples of 20 in pairs to give samples of 40). Draw all diagrams to the same scale so that the results for samples of 20 can be compared with those for samples of 40.

Note. The data quoted in Project 11.9(1) may also be useful.

SUMMARY

If X is a random variable with finite mean μ and variance σ^2, and $\overline{X}$ is the **mean** of a **random sample** of n observations, then $(\overline{X} - \mu)/(\sigma/\sqrt{n})$ approaches the standard normal distribution $\mathcal{N}(0,1)$ as n increases.

If T is the **total** of n observations from X, then $E[T] = n\mu$ and $\mathrm{Var}[T] = n\sigma^2$; therefore $(T - n\mu)/(\sigma\sqrt{n})$ approaches $\mathcal{N}(0,1)$.

For π not too near 0 or 1 and n large, a **binomial variable** R is *approximately* $\mathcal{N}(n\pi, n\pi(1 - \pi))$ and the proportion $P = R/n$ is approximately $\mathcal{N}(\pi, \pi(1 - \pi)/n)$.

The distribution of any **linear combination** of *independent* normally distributed variables is normal. If

$$Y = \sum_{i=1}^{n} c_i X_i$$

then

$$E[Y] = \sum c_i E[X_i] \quad \text{and} \quad \mathrm{Var}[Y] = \sum c_i^2 \mathrm{Var}[X_i].$$

The **difference** $\overline{X} - \overline{Y}$ has mean $\mu_x - \mu_y$ and variance $\sigma_x^2/n_x + \sigma_y^2/n_y$. If X and Y both have *independent* normal distributions then $\overline{X} - \overline{Y}$ is exactly normally distributed, and otherwise this is approximately so in large samples.

In **statistical process control**, measurements at successive time intervals, of $\bar{x}$ or of $\hat{p}$ as characteristics of a process, may be plotted on a **control chart** which has *warning* and *action* limits marked on it to help locate occasions when the process may be 'out of control'. Range may be studied as a measure of process variability.

REVISION EXERCISES D

1. A random variable x, whose probability density function is $f(x)$, has a mean μ, and μ_2' is the expected value of x^2. Show that the standard deviation σ is given by

$$\sigma^2 = \mu_2' - \mu^2.$$

Determine c in order that

$$f(x) = \begin{cases} c \exp(-x/\sigma) & x \geqslant 0 \\ 0 & x < 0 \end{cases}$$

may represent the probability density function of a random variable x.

Show that the mean and the standard deviation of the variable x are both equal to σ and that the probability that x exceeds σ is e^{-1}. *(AEB)*

2. Independent observations are taken on a random variable X with density function

$$f(x) = \frac{1}{8}x \quad 0 \leqslant x \leqslant 4.$$

Find the expected number of observations that will have to be taken before a value of x greater than 3 is observed. (IOS)

3. A chain is made up of eight links. Each link is drawn independently from a batch of links whose tensile strength is uniformly distributed over the range 100 to 120 lb. What is the probability that the chain breaks when it is subjected to a load of 102 lb?

4. A random variable has an exponential distribution with parameter λ. Compare the values of its standard deviation and its semi-inter-quartile range.

5. The distribution of annual incomes (in pounds sterling) of people employed in a certain occupation may be represented by the continuous probability density function,

$$f(x) = \begin{cases} kx^{-3.5} & x \geqslant x_0 \\ 0 & x < x_0 \end{cases}$$

where k and x_0 are constants. Show that $k = 2.5x_0^{2.5}$ and sketch the graph of $f(x)$. What is the interpretation of the constant x_0?

Given that $x_0 = 1900$, find the mean and variance of the annual incomes. Determine the proportion of people employed in this occupation who earn between £2600 and £ 3500 per year.

$$\left(\int_a^\infty x^{-p} \, dx = \frac{1}{(p-1)a^{p-1}} \quad \text{for} \quad a > 0 \quad \text{and} \quad p > 1. \right) \quad (JMB)$$

6. The diameters of a number of perfectly spherical balls, all of the same uniform material, follow the rectangular distribution

$$f(x) = \begin{cases} \dfrac{1}{b-a} & a \leqslant x \leqslant b \\ 0 & \text{otherwise.} \end{cases}$$

For the case $a = 1, b = 2$, find the distribution of the weights of the balls (assuming they have unit density), and the mean and standard deviation of this distribution. (IOS)

7. By considering the series expansion, term by term, of

$$\int_a^\infty (x - \mu)^3 f(x) \, dx,$$

show that $\mu_3 = \mu_3' - 3\mu\mu_2' + 2\mu^3$. How does this extend to give an expression for μ_r in terms of $\mu_r', \mu_{r-1}', \ldots, \mu$?

8. X, Y are independent, continuous random variables with common cdf F. If Z is the maximum of X and Y, show that the probability that Z is less than or equal to z is $(F(z))^2$.

Hence find the pdf of Z when X and Y both have the rectangular distribution on the interval $(0, 1)$.

What is the probability that the minimum of X and Y exceeds $\frac{1}{2}$?

(IOS)

9. The random variables $x_1, x_2, \ldots, x_{300}$ are independent and identically distributed, each having pdf

$$f(x) = \begin{cases} 1 & 0 \leqslant x \leqslant 1 \\ 0 & \text{elsewhere.} \end{cases}$$

Let $y = x_1 + x_2 + \cdots + x_{300}$. Find the expected value and the variance of y. Use these results to give an approximation to $P(y \leqslant 155)$ in terms of the normal integral.

(IOS)

10. A continuous random variable X is uniformly distributed over the interval $(a - c, a + c)$. Determine the expected values of X and X^2.

An instrument for measuring the length of a line is such that the recorded value for a line of length a cm is equally likely to be any value in the interval $(a - c, a + c)$, where $c(<a)$ is a known constant. The instrument is used to obtain two independent determinations, X_1 and X_2, of the length of a side of a square. Consider the following two methods for estimating the area A of the square:

$$\text{Method 1:} \quad A = X_1 X_2$$
$$\text{Method 2:} \quad A = \tfrac{1}{4}(X_1 + X_2)^2.$$

Show that method 1 is the only one of these methods which gives an unbiased estimator of the area of the square, and determine the variance of this unbiased estimator.

(Welsh)

11. A shopkeeper's fortnightly sales of a commodity can be regarded as having a normal distribution with mean 200 kg and variance 225 kg^2. Find the probability that the sales of the commodity are less than 185 kg. When he reorders, delivery takes two weeks. Determine to the nearest kilogram the stocks which he should have in hand when he reorders so that the probability that he will not run out before the new delivery arrives is 95%.

Five shops combine for bulk buying purposes and their fortnightly sales of the product are independent and have normal distributions with means 200, 240, 180, 260 and 320 kg and variances 225, 240, 225, 265 and 270 kg^2 respectively. Write down the mean and variance of their total fortnightly demand and determine to three significant figures the total level of stock at which they must reorder so that the probability of not running out before the new delivery is 99%.

Find the probability that the total amount sold by the shops in ten weeks is more than 6200 kg.

(JMB)

12. The length (in centimetres) of a piece of material produced by process A follows the density function $f(x) = 1/x^2$ $(x \geqslant 1)$.

The length of a piece produced by process B follows $f(y) = 2/y^3 (y \geqslant 1)$.

A piece of material can be used for a particular job if its length is between 1.2 and 2.4 cm. When two pieces of material are produced, one by each process, what is the probability that exactly one of them will be usable?

13. A jam manufacturer produces a pack consisting of 8 assorted pots of differing flavours, each nominally weighing 50 g. In practice, the actual weight of each pot is independently normally distributed with mean 52 g and standard deviation 1.10 g.
   (i) What proportion of pots weigh less than 50 g?
   (ii) What proportion of packs weigh less than 400 g?
   (iii) What is the probability of one or more pots in the pack weighing less than 50 g?
   (iv) What should the standard deviation of the weight of a pot be if 99% of packs are to weigh above 400 g? (*IOS*)

14. The random variable X follows a normal distribution with mean 0 and variance σ^2. Determine σ if $P(|X| < 6 = 0.77)$.

15. A job takes T minutes to complete, T being an $\mathcal{N}(28, 4)$ random variable. Another job, independent of the first, takes S minutes to complete, and is begun 5 minutes after the first. S is an $\mathcal{N}(25, 1)$ random variable. Show that the probability that the job which was begun last is finished first is approximately 0.185.

16. Two independent random variables, X and Y, follow normal distributions with means $\mu, 2\mu$ respectively and a common variance equal to 1.
   If $P(4X + 3Y \leqslant 5.8) = 0.2$, find μ.
   Using this value of μ, what is the probability that X is less than Y?

17. A random variable Y has the pdf

$$f(y) = ay + b \quad -1 \leqslant y \leqslant 2.$$

If the probability that Y lies between 0 and 1, given that Y is not negative, is $\frac{1}{3}$, determine a and b.
   Calculate the median of the distribution.

18. The weights (in grams) of the males of a species of bird are normally distributed with mean 15 and variance 4. The weights of the females are $\mathcal{N}(10, 1)$. Sketch the distribution of:
   (a) the weights of a population of half male and half female birds; and
   (b) the total weight of a pair of breeding birds.

   Find the probability that a bird drawn at random from population (a) will have a weight less than 10 g, and also the probability that the total weight of a pair drawn from population (b) will be less than 22 g. What assumptions have you made? (*IOS*)

19. A manufacturer produces packets of paper which are supposed to contain, on average, 100 sheets each. On checking a week's output, he finds that 4%

of packets contain less than 99 sheets of paper and $22\frac{1}{2}\%$ contain more than 104. What are the mean and variance of the number of sheets of paper in a packet? Do you think the manufacturer is meeting his specification? What assumptions have you made?

20. The diameters of 100 rods, measured in millimetres to the nearest millimetre, are classified in the following frequency table:

Diameters (mm)	62–64	65–67	68–70	71–73	74–76	77–79	Total
Class frequency	6	14	31	25	17	7	100

Taking the class marks to be $63, 66, \ldots$ show that the mean and the standard deviation of this population are respectively 70.62 and 3.85 approximately.

Using the same class marks allocate sampling numbers from 00 to 99 to these data. By reading consecutive pairs of digits from left to right along the following four rows of random numbers, select 10 samples each of 4 items.

$$
\begin{array}{cccc}
50532 & 25496 & 95652 & 42457 \\
73547 & 07136 & 40876 & 79971 \\
54195 & 25708 & 27989 & 64728 \\
10744 & 08396 & 56242 & 85184
\end{array}
$$

Calculate the mean and the standard deviation of the distribution of sample means so determined and compare them with the theoretical values expected for a large number of samples. Comment on the result.   (*AEB*)
(*Note*: the *class mark* is the centre of a class.)

21. Weights of persons using a certain lift are normally distributed with mean 150 lb and standard deviation 20 lb. The lift has a maximum permissible load of 650 lb.
   (i) Four persons arising at random from this population are in the lift. Determine the probability that the maximum load is exceeded.
   (ii) One person arising at random from this population is in the lift. He is accompanied by luggage weighing three times his own weight. Determine the probability that the maximum load is exceeded.
   Explain any difference in your answers to (i) and (ii).   (*JMB*)

22. Show that the variance of the rectangular distribution of a variable which lies at random between $-a$ and $+a$ is $a^2/3$.
   The n numbers of a set are to be added together. Before they are added, each number is rounded off to the nearest integer, thus introducing an error whose distribution is rectangular. Find the mean and variance of the total error of the sum.
   Assuming that for large enough values of n the total error is distributed normally, find the least value of n for which the total rounding-off error is more likely than not to lie outside the limits ± 10.   (*AEB*)
   (*Note:* the *rectangular* distribution is another name for the *continuous uniform* distribution.)

23. Describe what is meant by saying that an estimator of a population parameter is unbiased. State unbiased estimators of the mean and of the variance of a normal distribution, based on a random sample of observations $x_1, x_2, \ldots, x_n$.

Independent random samples of sizes n_1 and n_2 are drawn from normal distributions with means μ and 2μ, and variances σ^2 and $2\sigma^2$ respectively. To estimate μ we consider estimators of the form $a\bar{x}_1 + b\bar{x}_2$, where a and b are constants and $\bar{x}_1$ and $\bar{x}_2$ are the sample means.

Show that we need $a + 2b = 1$ for such an estimator to be unbiased. Find the value of a which minimizes the variance of this unbiased estimator.

(*JMB*)

24. The operating lifetime, X hours, of a certain brand of battery is distributed with probability density function

$$f(x) = \begin{cases} \frac{1}{9000}(x - 30)(70 - x) & 40 \leqslant x \leqslant 70, \\ 0 & \text{otherwise.} \end{cases}$$

A battery whose operating lifetime is less than 50 hours is classified as being of poor quality, one whose lifetime is from 50 hours to 60 hours as being of average quality and one whose lifetime exceeds 60 hours as being of good quality. Show that the proportions of poor, average and good quality batteries are in the ratio $11:11:5$.

A simple electronics test has been devised in an attempt to predict the quality of a new battery. A battery subjected to this test may give either a positive or a negative response and it has been observed from applications of the test to new batteries that the probabilities of a positive response from a poor, an average and a good quality battery are respectively $\frac{1}{4}, \frac{1}{2}$ and $\frac{3}{4}$. On the basis of these results, find the probability that a new battery which gives a positive response when tested is of good quality. (*Welsh*)

25. A certain ingredient may be extracted from raw material by either one of two methods, A and B. For a fixed volume of raw material, the amount X cm^3 of ingredient extracted using method A is normally distributed with mean 13 and standard deviation 2, and the amount Y cm^3 extracted using method B is distributed with probability density function

$$g(y) = \begin{cases} \frac{2}{25}(y - 10) & 10 \leqslant y \leqslant 15, \\ 0 & \text{otherwise.} \end{cases}$$

(a) Determine which of the two methods: (i) has the greater probability of extracting more than 14 cm^3 of the ingredient; (ii) extracts, on average, the greater amount of ingredient.

(b) Suppose that the cost of applying method A is 3p per cm^3 extracted and that the cost of extracting Y cm^3 using method B is $(15 + 2Y)$p. If the extracted ingredient is sold at 5p per cm^3, determine which of the two methods gives the higher expected profit per fixed volume of raw material. (*Welsh*)

26. The random variables X_1 and X_2 are independent, and Z is defined to be the greater of X_1 and X_2. Explain why

$$P(Z \leqslant z) = P(X_1 \leqslant z)P(X_2 \leqslant z).$$

Hence, or otherwise, find the cumulative distribution function $F(z)$ of Z, given that X_1 and X_2 both have a rectangular distribution taking values between 0 and a.

Find the probability density function of Z and find also the expected value of Z. *(Cambridge)*

27. In a certain electrical circuit a nominal resistance of 250 ohm is obtained by connecting in series two resistors each nominally 100 ohm and one resistor nominally 50 ohm. Resistors nominally 100 ohm are known to vary about a mean of 100 ohm with standard deviation 0.6 ohm and resistors nominally 50 ohm are known to vary about a mean of 50 ohm with standard deviation 0.4 ohm. Assuming normal variation, calculate the percentage of combined resistances in the range 250 ± 1.5 ohm.

Calculate also the percentage in the same range if:
(i) the mean resistance of the resistors nominally 50 ohm changes to 50.5 ohm, all other values remaining unchanged;
(ii) the mean resistances do not alter but both standard deviations increase by 50%. *(MEI)*

28. (a) Define the *expected value* of a function $g(X)$ of a random variable X.
Define the *variance* of X, and show that

$$\text{Var}[X] = E[X(X-1)] + E(X) - [E(X)]^2.$$

(b) A discrete random variable X assumes all positive integral values with probabilities

$$P(X = x) = k\theta^x/x!, \qquad x = 1, 2, \ldots,$$

where k and θ are positive constants.
Determine k in terms of θ and show that

$$E[X] = \theta(1 - e^{-\theta})^{-1},$$
$$\text{Var}[X] = \theta(\theta + 1)(1 - e^{-\theta})^{-1} - \theta^2(1 - e^{-\theta})^{-2}. \qquad (IOS)$$

29. A manufacturer of television sets offers a one-year warranty which covers free replacement if the picture tube fails. He estimates the time, T years, until the tube fails to be a random variable with probability density function $f(t) = \frac{1}{5}\exp(-t/5)$ for $t \geqslant 0$. Find the probability that a randomly chosen set will require at least one replacement tube under the warranty.

If the gross profit per sale is £100 and a tube replacement costs £70, find the expected profit per set sold, assuming that no set requires more than one replacement under the warranty. *(RSS)*

30. If a farmer grows cereals, his profit X_1 is normally distributed with mean $1750 and standard deviation $300. If he grows beans, his profit X_2 is

normally distributed with mean \$2000 and standard deviation \$400. If he grows a proportion p of cereals and a proportion $1 - p$ of beans, where $0 \leqslant p \leqslant 1$, the profit Y is $pX_1 + (1 - p)X_2$.

  (i) Assuming that X_1 and X_2 are independent, what is the distribution of Y?

 (ii) What is the value of p which maximizes the expected profit $E[Y]$?

(iii) Find the value of p which minimizes the variance of Y.

(iv) Calculate the expected profit for each of the values of p in (ii) and (iii).

 (v) The farmer says that he will face ruin if this profit is less than \$1480.

Calculate the probabilities of ruin for each of the values of p in (ii) and (iii). Which course of action is better for the farmer, and why?   (*RSS*)

17 SIGNIFICANCE TESTS USING THE NORMAL DISTRIBUTION

OVERVIEW

We use the results from Chapters 15 and 16 to make hypothesis tests about single observations and means from normal distributions, and extend these to situations where the central limit theorem allows normal approximations to be applied. These include proportions and binomial variables.

When small samples of data are available from normal distributions, but the variance has to be estimated from the data (rather than being known), another distribution is required: the t-distribution.

Paired comparisons, as well as samples of independent data, are considered.

17.1 INTRODUCTION

We considered in Chapter 13 how to test the hypothesis that someone could tell the difference between butter and margarine. The procedure that we used, the sign test, could be applied in some other situations too, as examples showed; but it has serious limitations as a test for continuous variates. It does not use the exact values measured and therefore wastes some information. Many significance tests (i.e. tests of hypotheses) are based on the normal distribution, and we are now in a position to investigate these.

Let us consider an example. A biologist, investigating primate remains in Africa, has found many skulls of one species in a particular valley. He moves his area of investigation to the next valley, and on his first day he finds only one skull. He would like to know whether this skull is from an animal similar to those in the other valley (let us call it ape-man A), or whether the new skull differs from the others sufficiently to indicate that the creature in this new valley is from a different population. We wish to use statistical methods to assess the hypotheses that the biologist has in mind.

The biologist makes measurements of several different characteristics on each skull. He has found, however, that the width of the skull is the most useful single variate for characterizing a particular species; for ape-man A, the measurements of skull widths in centimetres have mean 12 and variance 0.64. A histogram of the frequency distribution of skull widths for ape-man A suggests that the normal distribution will provide a good model for the data. If the new skull, which is of width 13.3 cm, is from the same population (i.e. the same species) as the earlier ones, then its width measurement should take a value that we might expect if we were to pick, at random, one observation from the normal distribution with mean 12 and variance 0.64. What values are likely, and what unlikely, to occur when we pick at random from this distribution?

To answer this question we need to develop an appropriate method for making a significance test. The method is most clearly developed by first considering a random variable that, on the null hypothesis, has a standard normal distribution $N(0, 1)$. Real-life data rarely, if ever, follow the standard normal distribution but many measurements, such as the ape-man data, can easily be transformed to $N(0, 1)$, as we showed in Chapters 15 and 16. We return to the ape-man problem later (p. 411). The reader will find it useful to look again at Section 13.8, on the general procedure for setting up a significance test, before beginning the next section.

17.2 THE STANDARD NORMAL DISTRIBUTION

We used tables of the standard normal distribution in Chapter 15; they showed us, for example, that when z is $N(0, 1)$,

$$P(-1.96 \leqslant z \leqslant +1.96) = 0.95.$$

That is to say, 95% of the probability under the $N(0, 1)$ curve is contained between the values -1.96 and $+1.96$ of z. Another way of stating this is to say that if one single observation is drawn at random from $N(0, 1)$, it has probability 0.95 of lying within the interval $(-1.96; +1.96)$ and probability 0.05 of lying outside it. The event that such an observation *does* lie outside $(-1.96; +1.96)$ is therefore an *unlikely* event in the sense defined in Chapter 13 (p. 276).

Suppose that we have taken a single measurement, x, of a variate, and have set up the null hypothesis that x should be distributed as the random variable $N(0, 1)$. In other words, our single observation has been picked at random from the standard normal distribution. We note the value of x; if it is between -1.96 and $+1.96$, and we are carrying out a two-tail test at the 5% significance level, we shall not reject the null hypothesis. If x is less than -1.96 or greater than $+1.96$, we do reject the null hypothesis.

There are of course two parameters in the normal distribution, namely its mean and its variance. It might therefore appear that we could use an alternative hypothesis in which both mean and variance have been changed from the values laid down in the null hypothesis. Testing in such a situation would not be easy, and we shall not consider it. We will confine attention to comparing $N(0, 1)$ with $N(\mu, 1)$; that is, we will look only at the null hypothesis 'mean $= 0$' against an alternative hypothesis of the form 'mean $= \mu$'; μ is usually not given a precise numerical value but is specified by statements such as $\mu \neq 0$ or $\mu < 0$. We assume all the time that we are dealing with a normal distribution having unit variance.

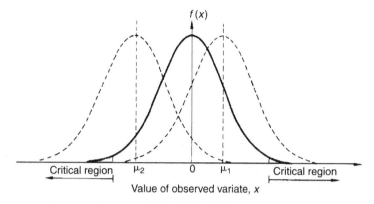

Figure 17.1 *Graph showing the curve (——) for a null hypothesis that an observation is drawn from $\mathcal{N}(0, 1)$, and two possible curves (- - -) for alternative hypotheses $\mathcal{N}(\mu, 1)$, $\mu \neq 0$, viz. $\mu = \mu_1 > 0$ and $\mu = \mu_2 < 0$.*

Figure 17.1 illustrates the situation when the null hypothesis says 'x is taken from $\mathcal{N}(0, 1)$' and the alternative hypothesis says 'x taken from $\mathcal{N}(\mu, 1)$ where $\mu \neq 0$': the two broken curves are examples of distributions that are possible on the alternative hypothesis while the continuous curve shows the null hypothesis $\mathcal{N}(0, 1)$.

The form of the alternative hypothesis tells us that we need a two-tail test (see Section 13.2). As we have already stated, this two-tail test, when carried out at the 5% significance level, requires us to reject the null hypothesis if our observed variate-value x lies below -1.96 or above $+1.96$, and not to reject the null hypothesis if x lies between -1.96 and $+1.96$. The critical region for this test is thus all values of observations x below -1.96 or above $+1.96$.

If we preferred to use a 1% significance level, the form of test would be exactly the same, save that ± 1.96 would have to be replaced by ± 2.576; and for a 0.1% level we must use ± 3.291.

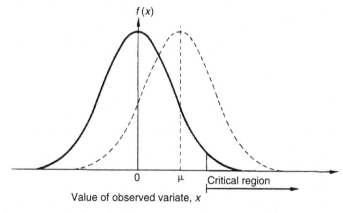

Figure 17.2 *Graph showing the curves for a null hypothesis that an observation is drawn from $\mathcal{N}(0, 1)$ (——) and an alternative hypothesis $\mathcal{N}(\mu, 1)$, $\mu > 0$ (- - -).*

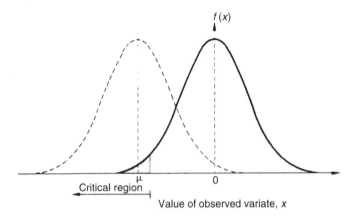

Figure 17.3 *Graph showing the curves for a null hypothesis that an observation is drawn from $\mathcal{N}(0, 1)$ (——) and an alternative hypothesis $\mathcal{N}(\mu, 1)$, $\mu < 0$ (- - -).*

When the null hypothesis says 'x is taken from $\mathcal{N}(0, 1)$' and the alternative hypothesis says 'x is taken from $\mathcal{N}(\mu, 1)$, where $\mu > 0$', the situation is as shown in Fig. 17.2. We need a one-tail test that rejects large positive x-values only, since these are the only values which are better explained on the alternative hypothesis. From tables, we find $P(z \geqslant +1.645) = 0.05$ when z is $\mathcal{N}(0, 1)$. Hence the appropriate test, when a 5% significance level is being used, is to reject the null hypothesis whenever the observed value x is greater than $+1.645$, and not to reject the null hypothesis otherwise. The critical region is then all values of x from $+1.645$ upwards.

Finally, if the null hypothesis says 'x is taken from $\mathcal{N}(0, 1)$' and the alternative hypothesis says 'x is taken from $\mathcal{N}(\mu, 1)$, where $\mu < 0$', Fig. 17.3 shows the situation. Now we need a one-tail test that rejects large negative values of x only, since this time it is only these values that can better be explained on the alternative hypothesis. At the 5% level of significance, the null hypothesis is rejected when $x < -1.645$ and is not rejected otherwise. The critical region is then all values of x from -1.645 downwards.

In the two one-tail tests, we must replace 1.645 by 2.326 if it is required to test at the 1% significance level, and by 3.09 if a 0.1% significance level is to be used.

17.3 A SINGLE OBSERVATION FROM $\mathcal{N}(\mu, \sigma^2)$

We saw (p. 355) that if X is distributed normally with mean μ and variance σ^2, then $Z = (X - \mu)/\sigma$ is a standard normal variable. That is, when X is $\mathcal{N}(\mu, \sigma^2)$ then $Z = (X - \mu)/\sigma$ is $\mathcal{N}(0, 1)$. So a null hypothesis that a single observed value x has been drawn from $\mathcal{N}(\mu, \sigma^2)$ is equivalent to a null hypothesis that $z = (x - \mu)/\sigma$

has been drawn from $\mathcal{N}(0, 1)$. Therefore tests of this hypothesis about x can be reduced to the tests described in the previous section.

We now return to our ape-man problem. We set up a statistical model that skull widths are normally distributed with mean μ and variance 0.64. Our null hypothesis (NH) is that the width of the newly found skull comes from a population as specified in the model with $\mu = 12$, which is the appropriate value for ape-man A. Our alternative hypothesis (AH) is that the width of the new skull is from a population of the same type except that $\mu \neq 12$: it may be greater than or less than 12. A two-tail test is therefore appropriate. Since the width of the new skull x is 13.3, we calculate

$$z = \frac{x - \mu}{\sigma} = \frac{13.3 - 12}{0.8} = \frac{1.3}{0.8} = 1.63.$$

On the null hypothesis, the random variable Z, of which z is a particular value, is $\mathcal{N}(0, 1)$. The critical region, using a 5% significance level, contains those values such that $z > 1.96$ and $z < -1.96$, which we can summarize in the statement $|z| > 1.96$. Our observed value of Z is not in the critical region so we are not given reason to reject the null hypothesis. On the basis of this test we could agree that the skull was that of ape-man A.

Example 17.1 Two-tail test on single observation

Over a long period, the weights of pots of jam made by a standard process have been normally distributed with mean 345 g and standard deviation 2.8 g. A pot produced just before the process closed for the day weighs 338.5 g. Is the process working correctly? (Assume the standard deviation has not changed.)

We test whether the mean of the process has changed. The change may be up or down so a two-tail test is appropriate. We summarize the specification for the test.

Model: pot weights are $\mathcal{N}(\mu, 2.8^2)$.
NH: $\mu = 345$. AH: $\mu \neq 345$. Two-tail test.
Significance level: 5%

$$z = \frac{338.5 - 345.0}{2.8} = -\frac{6.5}{2.8} = -2.321.$$

On the NH, Z is $\mathcal{N}(0, 1)$. The critical region is $|z| > 1.96$. Hence we reject the NH. We conclude the process is not working correctly.

Example 17.2 One-tail test on single observation

The height to which a particular variety of plant grows in a nurseryman's plantation during its first season of growth has for a long time been normally distributed with mean 53 cm and variance 12 cm^2. In one year, owing to an error, one such plant received three fertilizer applications instead of two, and grew to a height of 60 cm. Has the fertilizer had a beneficial effect?

This is a typical example of the questions and statements statisticians have to examine. Careful reading of the background information, and a look at the claim that the fertilizer might have a beneficial effect (not just 'an effect') shows that an alternative hypothesis should specify a greater height (not just a different one from what has been observed before). A one-tail test rejecting the null hypothesis when large heights are observed is therefore the right one.

> Model: heights are $\mathcal{N}(\mu, 12)$.
> NH: $\mu = 53$.   AH: $\mu > 53$.   One-tail test.
> Significance level: 5%.
>
> $$z = \frac{60 - 53}{\sqrt{12}} = \frac{7}{3.46} = 2.02.$$

On the NH, Z is $\mathcal{N}(0, 1)$. The critical region is $z > 1.64$. The value 2.02 is in the critical region. We therefore reject the NH in favour of an AH which says that the fertilizer induced more growth.

17.3.1 p-values

The final stage of a test such as this is sometimes carried out differently. Instead of specifying a critical region and deciding whether or not z comes within it, the probability of obtaining a value more extreme than z is calculated; this probability is then compared with the significance level probability. Thus, in the ape-man problem discussed above, the probability of the random variable Z being greater than 1.63 or less than -1.63 is found; it is 0.103. This value, 0.103, is often called the *p-value* of the data. Since it is greater than 0.05, the null hypothesis is not rejected. Calculating *p*-values is difficult, in general, unless a computer is available.

17.3.2 Exercises

1. The weight of packs of unwrapped chocolates is normally distributed with mean 508 g and variance $16\,\text{g}^2$. I am told that a pack of chocolates weighs 516 g, but it is not clear whether the chocolates were wrapped or unwrapped. Make a significance test to provide evidence of which is likely to be the case.

2. Jim produces components whose length is normally distributed with mean 2.500 cm and standard deviation 0.005 cm. One day he is not very well and a component chosen at random from those he has made has length 2.493 cm. Test whether his health has affected the length of the components.

17.4 THE MEAN OF n OBSERVATIONS FROM $\mathcal{N}(\mu, \sigma^2)$

In practice a decision about rejecting, or not rejecting, a null hypothesis will hardly be made by looking at one single observation. A random sample of

n observations will be collected, and a decision made on the basis of looking at the whole sample. We noted in Chapter 16 that if this random sample was drawn from an $\mathcal{N}(\mu, \sigma^2)$ distribution, then the mean of the sample, $\overline{X}$, followed an $\mathcal{N}(\mu, \sigma^2/n)$ distribution.

So, given a null hypothesis which states that a random sample $\{X_1, X_2, \ldots, X_n\}$ has been drawn from $\mathcal{N}(\mu, \sigma^2)$, we calculate

$$\overline{X} = \frac{1}{n} \sum_{i=1}^{n} X_i,$$

and test whether the value of $\overline{X}$ could reasonably have come from $\mathcal{N}(\mu, \sigma^2/n)$. To do this, $Z = (\overline{X} - \mu)/(\sigma/\sqrt{n})$ is computed, and tested as $\mathcal{N}(0, 1)$.

Example 17.3 One-tail test on mean

Scores on a standard IQ test are normally distributed with standard deviation 15. In the whole population the mean of this distribution is 100. A company has a selection test of its own, which it claims will pick out people whose IQ is better than average. Twenty of its employees, selected using this test, are given the standard IQ test also, and on this their mean score is 104.2. Examine the company's claim.

Model: IQ scores are $\mathcal{N}(\mu, 15^2)$. We assume the variance remains unchanged.

NH: $\mu = 100$. AH: $\mu > 100$. One-tail test.

Significance level: 5%.

On the NH, the mean $\overline{X}$ is $\mathcal{N}(100, 15^2/20)$, and

$$Z = \frac{\overline{X} - 100}{15/\sqrt{20}}$$

is $\mathcal{N}(0, 1)$. But, using the observed values,

$$z = \frac{104.2 - 100}{15/\sqrt{20}} = \frac{4.2 \times \sqrt{20}}{15} = \frac{4.2 \times 4.47}{15} = 1.25.$$

The critical region is $z > 1.64$. We therefore do not reject the NH. The evidence does not support the company's claim.

Example 17.4 Two-tail test on mean

From a survey of a large quantity of writing by a certain author, a research worker has concluded that this author writes sentences of mean length 31.5 words. The standard deviation of sentence length is 6.8 words, and the records of sentence length follow a normal distribution. The research worker now reads another piece of writing which he thinks may be by the same author. In this, the mean length of 80 sentences is 34.0 words. Is this mean sentence length consistent with the known work of the author?

> Model: sentence length is $\mathcal{N}(\mu, 6.8^2)$.
> NH: $\mu = 31.5$.   AH: $\mu \neq 31.5$.   Two-tail test.

The AH is that length follows a distribution with a different mean, which may be larger or smaller – we cannot say which.

On the NH, the mean $\overline{X}$ is $\mathcal{N}(31.5, 6.8^2/80)$.

$$z = \frac{\bar{x} - \mu}{\sigma/\sqrt{n}} = \frac{34.0 - 31.5}{6.8/\sqrt{80}} = \frac{2.5 \times 8.94}{6.8} = 3.29$$

and is the observed value of the random variable Z which is $\mathcal{N}(0, 1)$. The value 3.29 is just on the 0.1% significance point; it is very unlikely to have arisen by chance if the NH is true. Therefore we reject the NH. On measurements of sentence length, the evidence is against the new piece of writing being by the same author.

17.4.1 Exercises

1. A random sample of 9 observations has mean 53. The variance of the normal population from which they were chosen is 16. Test the hypothesis that the mean of the population is 50, using a two-tail test with a significance level of: (a) 5%; (b) 1%.

2. A normal population has unknown mean and variance 9. Four observations chosen at random from the population have values 9, 16, 11, 12. Test whether the population mean is 10.

3. The marks of all the candidates in an English examination may be taken to be normal with mean 64 and standard deviation 8. From a certain school, 54 candidates obtained a mean mark of 68. Does this suggest they differ in English performance from the norm?

4. The length of time for which a wheel will spin is normal with mean 150 s and variance $100 \, \text{s}^2$. Some oil is put on the bearings and five spins take times 160, 140, 160, 180, 170 s. Use a one-tail significance test to see if oiling the bearings has increased the spinning time.

17.5 DIFFERENCE BETWEEN TWO MEANS FROM NORMAL DISTRIBUTIONS

If two samples are taken at random from normal distributions, the first of size n_1 from $\mathcal{N}(\mu_1, \sigma_1^2)$ and the second of size n_2 from $\mathcal{N}(\mu_2, \sigma_2^2)$, the means of the samples may be calculated and compared. If the means are $\overline{X}_1$ and $\overline{X}_2$ respectively, then the difference $\overline{X}_1 - \overline{X}_2$ has the normal distribution with mean $\mu_1 - \mu_2$ and variance $\sigma_1^2/n_1 + \sigma_2^2/n_2$ (Example 16.7, p. 383). We may use this result to test the difference between two observed means, provided the means come from normally distributed data.

Example 17.5 Test on difference of means

A manufacturer is comparing the settings of two machines, A and B, which should produce rods of the same length. Both have, over a long period, given rods whose lengths were normally distributed with variance $0.20\,\mathrm{cm}^2$. Although the two machines are supposed to give the same length of rod, he suspects that this is not so. Examine this suspicion, if the total length of 20 rods from machine A is 824.0 cm, and the total length of 10 rods from machine B is 395.0 cm.

Model: the lengths of rods produced by machines A and B are denoted by the random variables X_A and X_B, respectively. X_A is $\mathcal{N}(\mu_A, 0.20)$ and X_B is $\mathcal{N}(\mu_B, 0.20)$.

$$\text{NH: } \mu_A = \mu_B. \quad \text{AH: } \mu_A \neq \mu_B. \quad \text{Two-tail test.}$$

We assess the difference by looking at the difference in the means $\overline{X}_1 - \overline{X}_2$, which on the NH should have a mean of zero. In order to make the test we need to know the variance of this difference:

$$\mathrm{Var}[\overline{X}_1 - \overline{X}_2] = \mathrm{Var}[\overline{X}_1] + \mathrm{Var}[\overline{X}_2] = \frac{\sigma_A^2}{n_A} + \frac{\sigma_B^2}{n_B} = \frac{0.20}{20} + \frac{0.20}{10} = 0.03.$$

Hence, on the NH, $\overline{X}_1 - \overline{X}_2$ is $\mathcal{N}(0, 0.03)$.

The observed means are $\bar{x}_A = 824/20 = 41.2\,\mathrm{cm}$, $\bar{x}_B = 395/10 = 39.5\,\mathrm{cm}$, and $\bar{x}_A - \bar{x}_B = 41.2 - 39.5 = 1.7\,\mathrm{cm}$.

$$z = \frac{1.7 - 0}{\sqrt{0.03}} = \frac{1.7}{0.173} = 9.83.$$

On the NH this is drawn from an $\mathcal{N}(0, 1)$ population. Using a 5% significance level the critical region is $|z| > 1.96$. The value falls in the critical region and we reject the NH. The evidence is in favour of the AH and helps to confirm the manager's suspicions.

Example 17.6 Test on difference of means

A fruit grower has, for several seasons, found that the yield per plant of one variety has on average been 12 g greater than that of a second variety when grown under identical conditions. He now proposes to use a new field, and is not sure whether the soil conditions in it will affect the performance of both varieties equally; but he is prepared to assume that the crop figures will be normally distributed, as in the past, with variance $66\,\mathrm{g}^2$ for each variety. He grows 25 plants of each variety under the usual standard conditions to test their performance in this field. The first variety yields on average 116 g per plant and the second variety yields on average 109 g per plant.

Model: the yield per plant of varieties 1 and 2 is $\mathcal{N}(\mu_1, 66)$ and $\mathcal{N}(\mu_2, 66)$ respectively.

$$\text{NH}: \mu_1 - \mu_2 = 12. \quad \text{AH}: \mu_1 - \mu_2 \neq 12. \quad \text{Two-tail test.}$$

On the NH, $\overline{X}_1 - \overline{X}_2$ is $\mathcal{N}(12, \frac{66}{25} + \frac{66}{25})$ i.e. $\mathcal{N}(12, 5.28)$.
The difference in observed mean yields was $\bar{x}_1 - \bar{x}_2 = 116 - 109 = 7\,\text{g}$.

$$z = \frac{7 - 12}{\sqrt{5.28}} = \frac{-5}{2.30} = -2.17.$$

On the NH, this is a value from an $\mathcal{N}(0, 1)$ distribution. The critical region is $|z| > 1.96$, using a 5% significance level. We therefore reject the NH in favour of the AH that the two varieties are not affected equally by moving to the new field.

17.6 LARGE-SAMPLE TESTS

The central limit theorem (Theorem 16.1, p. 371) tells us that whatever the form of the distribution we are sampling from, the mean of a large enough sample will be approximately normally distributed. (There is a minor theoretical qualification which we shall ignore.) We need to have an idea what 'large enough' means in practice before we apply this theorem to examples. Broadly speaking, if the distribution of a random variable X is symmetrical (even though not normal), sample size may be quite modest (say 30, even smaller in some cases) but the normal approximation is good none the less. If, however, X has a very skew distribution, we need a sample of considerable size (say, at least 200) before the distribution of $\overline{X}$ approaches reasonably closely to a normal distribution. When a normal approximation can be used, its mean and variance will be equal to μ and σ^2/n, where μ is the mean of X, σ^2 the variance of X and n the sample size. In this way we can test hypotheses about the means of distributions which are not themselves normal, provided a large sample of observations is available. We shall use exactly the same methods as in the earlier part of the chapter, except that we must remember that our test statistics will be only approximately $\mathcal{N}(0, 1)$. Usually this causes no problems at all, but if a result is very near to the borderline of significance, we may prefer not to make a judgement without more information.

Example 17.7 Large-sample test on mean
The flying-times of a particular type of aircraft on a given route have averaged $16\frac{1}{4}$ hours ever since that type of aircraft began to be used on the route. The distribution of flying-times has had standard deviation $1\frac{1}{2}$ hours, though it has not been normal. The most recent year's figures, covering 120 flights, showed an average flying-time of 15 hours 56 minutes. Is this different from the overall average?

Since we have 120 observations, we shall use a large-sample test. We assume that the flying-time distribution is not *very* skew, and this is probably quite safe to do. (If we wished to, we could make a histogram of the actual distribution first.) We therefore feel justified in making our initial assumption that the mean flying-time $\overline{X}$ is approximately normal.

$$\text{Model: } \overline{X} \text{ is approximately } \mathcal{N}\left(\mu, \frac{(1.5)^2}{120}\right).$$

NH: $\mu = 16.25$.   AH: $\mu \neq 16.25$.   Two-tail test.

$$z = \frac{15.933 - 16.250}{1.5/\sqrt{120}} = -\frac{0.317 \times 10.95}{1.5} = -2.31$$

On the NH, z is chosen from a distribution that is approximately $\mathcal{N}(0, 1)$. The critical region is $|z| > 1.96$, using a 5% significance level. We reject the NH. The value of $\bar{x}$ is significant. Although z comes from a distribution that is only approximately $\mathcal{N}(0, 1)$, the value -2.31 leaves us in no doubt that there *is* evidence to reject the NH in favour of an AH that some change has occurred in average flying-time.

Example 17.8 Large-sample test on difference of means

The mean number of days' absence through sickness, over a two-month period, is 2.03 days per worker in one part of a factory and 1.28 days per worker in another part of the same factory; there are 65 workers in each of these parts of the factory. The estimates of the variances of the number of days' absence are 3.0125 and 2.1482 respectively. Examine whether the means in the two parts of the factory differ.

Note here, first, that the original measurement is not a continuous variate at all, but is discrete, being a number of days; and, second, that we do not have a value of a variance given by any theory or previous knowledge, but must use the sample variances. Neither of these points is a very serious drawback in large samples. In the first place, the probability that means of large samples are approximately normally distributed is not restricted to samples from continuous distributions: we have already set up a normal approximation to a binomial as an example of this. Then secondly, when a sample is large the estimate of variance calculated from it is a very good estimate of the true σ^2 in the whole population, so that an approximate significance test of means is not made noticeably worse by using the sample estimate of variance.

Model: let the number of days' absence per worker in the two parts of the factory, which we label 1 and 2, be X_1 and X_2 respectively. Then X_1 is approximately $\mathcal{N}(\mu_1, \sigma_1^2)$ and X_2 approximately $\mathcal{N}(\mu_2, \sigma_2^2)$.

NH: $\mu_1 = \mu_2$.   AH: $\mu_1 \neq \mu_2$.   Two-tail test.

On the NH, $\overline{X}_1 - \overline{X}_2$ is approximately

$$\mathcal{N}\left(0, \frac{\sigma_1^2}{n_1} + \frac{\sigma_2^2}{n_2}\right).$$

We replace σ_1^2 by the estimate $s_1^2 = 3.0125$ and σ_2^2 by $s_2^2 = 2.1482$. The variance of $\overline{X}_1 - \overline{X}_2$ is therefore estimated to be

$$\frac{3.0125}{65} + \frac{2.1482}{65} = 0.0794.$$

Hence $\overline{X}_1 - \overline{X}_2$ is approximately $\mathcal{N}(0, 0.0794)$ on the NH;

$$\bar{x}_1 - \bar{x}_2 = 2.03 - 1.28 = 0.75.$$

$$z = \frac{0.75 - 0}{\sqrt{0.0794}} = \frac{0.75}{0.28} = 2.68,$$

and is significant at the 1% level. We therefore have evidence to reject the NH, and to suspect that the mean numbers of days' absence are *not* the same in the two parts of the factory.

17.7 ONE-TAIL AND TWO-TAIL TESTS

We have given examples of both one-tail and two-tail tests. Which of these is appropriate in any given context is not always clear and the reader should appreciate that there are two main approaches used when deciding the matter.

One approach is that a one-tail test is appropriate only when it is known that deviations from the null hypothesis will be in a particular direction. For example, an educational research worker might wish to test the effect of giving pupils additional mathematics instruction with a computer teaching program. The test is to be between pupils who receive normal lessons and those who receive normal lessons plus computer instruction. The research worker expects the computer instruction to improve performance in an examination; but whether he uses a one-tail or a two-tail test depends, in this approach, on what his attitude is to a result in which pupils receiving the extra instruction do worse than those receiving normal lessons only. If he is quite certain that extra instruction cannot lead to a worse performance and that any such result must be a chance one, then he can use a one-tail test, rejecting the null hypothesis only when the pupils receiving computer instruction do appreciably better than those receiving normal lessons only. But if he allows the possibility that the computer instruction might lead to a worse performance, then a two-tail test is the appropriate one. This approach is linked to the idea that a test of significance is a technique to improve the understanding of a particular phenomenon; it is the approach that is usually appropriate in scientific as opposed to technological research. Using this approach, one-tail tests are relatively rare since one cannot often be sure that a result may be in one direction only.

The other approach is linked to the idea that a statistical test is a matter of choosing between two hypotheses. Suppose a seed merchant offers a new variety of tomato, saying that it yields better than a standard variety, and we wish to test his claim. We might say that our aim, after collecting appropriate data, is either to accept the claim as a reasonable one, or to reject the claim because there is no evidence of an increase in yield or because the yield is lower. Suppose the mean yield of the new variety is μ and that of the standard variety is μ_0. We would accept the claim if we have evidence that $\mu > \mu_0$ but if $\mu \leqslant \mu_0$ we would wish to reject the claim; in particular, any indication that μ is less than μ_0 is strong evidence against the claim. Thus the hypotheses that we wish to test are:

$$\text{NH}: \mu \leqslant \mu_0. \quad \text{AH}: \mu > \mu_0.$$

This conflicts with our requirement that the distribution on the null hypothesis must be completely specified. But we act as if the null hypothesis is $\mu = \mu_0$ and determine a one-tail critical region (at the 5% level, say) on that basis. If we consider the same critical region for use with any specific null hypothesis in which $\mu < \mu_0$, then our significance level with that test would be less than 5%. Therefore the procedure ensures that the type I error, or probability of rejecting the null hypothesis when it is true, will never exceed 5%. The usual terminology is still to say that the significance level of the test is 5%: i.e. the maximum type I error over the possible values of μ on the null hypothesis is used.

17.8 PAIRED-COMPARISON DATA

We analysed paired-comparison data in Example 13.2 (p. 280) by the sign test. In that example, two methods of teaching reading were being compared; pairs of similar children were chosen and one was taught by one method and the other child by the other method. Although test scores, say X and Y, existed for each pair the test used only the information of which method did the better in each pair. The information on *how much better* one method was doing than the other was not used. How can we use this information? The Wilcoxon signed-rank test (Section 13.6) did make use of the numerical scores, but not of the distribution theory which holds in large samples.

We would find the value of the difference $D = Y - X$ for each pair. We test the null hypothesis that the two methods of teaching reading are producing similar scores by testing whether the mean $\overline{D}$ is from a distribution with mean zero. If the variance of D is not known then we estimate it from the separate values of D found for each pair. (There are only 16 pairs in Example 13.2, which is not a large enough sample for estimating σ^2 in the significance test; a small-sample test, as described in Example 17.16 below, would have to be used.) We illustrate the test for a *large* sample in Example 17.9.

Note that $\text{Var}[\overline{D}] = \text{Var}[\overline{Y} - \overline{X}]$ but it is not likely to equal $\text{Var}[\overline{Y}] + \text{Var}[\overline{X}]$, for X and Y are not independent random variables since the values come in pairs. Any estimation of $\text{Var}[D]$ should therefore be based on the individual values of D rather than by first calculating $\text{Var}[X]$ and $\text{Var}[Y]$.

Example 17.9 Test of paired comparison

Two types of razor, X and Y, were compared using 100 men as subjects. Each subject shaved one side, chosen at random, of his face using one razor and the other side using the other razor. The times taken to shave, x and y minutes corresponding to the razors X and Y respectively, were recorded. From the data for X a mean of 2.84 and an estimated variance of 0.48 were calculated. The corresponding values for Y were 3.02 and 0.42 respectively.

Assuming x and y to be independent and normally distributed, use the data to test at the 5% level of significance whether the razors led to different shaving times.

When the value of $y - x$ was calculated for each subject the variance of the values was found to be 0.60. Use this further information to apply an improved test to the data.

The first part of the question asks us to treat x and y as if they were values of independent random variables, which we shall call X and Y respectively.

Model: X is $\mathcal{N}(\mu_X, 0.48)$, Y is $\mathcal{N}(\mu_Y, 0.42)$, and X and Y are independent.

NH: $\mu_X = \mu_Y$.   AH: $\mu_X \neq \mu_Y$.   Two-tail test.

On the NH, $\overline{Y} - \overline{X}$ is distributed as

$$\mathcal{N}\left(0, \frac{\sigma_X^2}{100} + \frac{\sigma_Y^2}{100}\right)$$

or $\mathcal{N}(0, 0.0090)$. Hence

$$z = \frac{(3.02 - 2.84) - 0}{\sqrt{0.0090}} = \frac{0.18}{0.0949} = 1.90.$$

Using a 5% significance level, the critical region is $|z| > 1.96$. We do not reject the NH, and so we may conclude that there is no evidence of a difference in shaving times.

The second part of the question asks for an improved method of analysing the data, when the variance of the *differences* is known.

Model : $Y - X$ is $\mathcal{N}(\mu, 0.60)$.

NH: $\mu = 0$.   AH: $\mu \neq 0$.   Two-tail test.

On the NH, $\overline{Y} - \overline{X}$ is $\mathcal{N}(0, 0.60/100)$ or $\mathcal{N}(0, 0.0060)$.

$$z = \frac{0.18}{\sqrt{0.0060}} = \frac{0.18}{0.0775} = 2.32.$$

The critical region is $|z| > 1.96$ as above. We therefore reject the NH and conclude there is a difference in the shaving times with the two razors. Note that the variance of $\overline{Y} - \overline{X}$ was smaller by the second method. One would usually expect this to be the case, because taking differences removes the systematic component of variation due to the differences between subjects.

17.8.1 Exercises

1. The times it takes two soldiers to dismantle a given type of gun are measured for each soldier on 49 occasions. For both soldiers the times are normally distributed with variance 100 seconds^2, the mean for soldier A is 140 and for soldier B 138 seconds. Test if there is a significant difference in the mean performance of the soldiers.

2. A manufacturer claims that his light-bulbs have a life of 1800 hours. 250 bulbs are tested and found to have a mean life of 1790 hours; the estimated variance from the tested bulbs is 2500 hours^2. Test the manufacturer's claim. State the assumptions you make.

3. Suppose that an appropriate model for the lifetimes of the light-bulbs in Question 2 is an exponential distribution, and that the manufacturer claimed that the *median* of this distribution was 1500 hours. Suppose further that a sample of 9 bulbs were tested and found to have a median lifetime of 1400 hours. State what null and alternative hypothesis you would use in this situation, explain what sampling distribution you would require to know, and describe how to make the test assuming the sampling distribution was known. (Do not attempt to work this question numerically.)

4. A test of reasoning ability has been devised for eight-year-old children and standardized so that for British children the test scores are normally distributed with a mean of 60 and a standard deviation of 10. A group of 12 eight-year-old children are selected at random and given individual instruction in solving mathematical problems and use of English. After the instruction the mean score of the children was 65. Use a one-tail test to see if the instruction improved the children's performance on the test.

   Would your conclusion be valid if the test scores were not normally distributed? Can you suggest any improvements in the design of the investigation?

5. The wages of two groups of workers are to be compared. Group A contains 300 workers; their mean wage is £100 per week and the standard deviation is £10. Group B contains 400 workers; their mean wage is £110 and the standard deviation is £12. Test if there is a significant difference between the two means.

6. Five hundred people are asked to do an arithmetic test at 9 a.m. (represent the score by the variable X) and a test of similar difficulty at 6 p.m. (represent the score by Y). From the data the means of X and Y are calculated to be 50.6 and 49.2 respectively; Var[X], Var[Y] and Var[$X - Y$] are estimated to be 200, 200 and 300 respectively. Test whether the time of day affects performance on the arithmetic test.

One of the most useful normal approximations is that for a binomially distributed variable. We introduced this on p. 376, and earlier (p. 279) we had seen that it can be very tedious to compute the tail probabilities in binomial distributions. Provided we have quite a large sample of observations (larger if π is near to 0 or 1 than if π is near $\frac{1}{2}$), a binomial random variable with parameters n, π follows the normal distribution $\mathcal{N}(n\pi, n\pi(1-\pi))$, i.e. the normal distribution with the same mean and variance. This is an approximation, but a very adequate one when n is at all large; we recommend the use of a continuity correction (see Section 16.3).

We remarked on p. 381 that if a binomial variable r can be approximated by a normal distribution, then so can r/n, for this is simply r divided by a constant (the sample size) and will have the same distribution except for a scale factor. With large samples, then, we have the further result that a sample proportion $p(=r/n)$ follows approximately $\mathcal{N}(\pi, \pi(1-\pi)/n)$.

Example 17.10 Test of proportion

A supplier of components to the motor industry makes a particular product which sometimes fails immediately it is used. He controls his manufacturing process so that the proportion of faulty products is supposed to be only 4%. Out of 500 supplied in one batch, 28 prove to be faulty. Has the process gone out of control to produce too many faulty products?

Model: the number of faulty products is binomial with parameters π and $n = 500$.

$$\text{NH: } \pi = 0.04. \quad \text{AH: } \pi > 0.04. \quad \text{One-tail test.}$$

On the NH, the number faulty R has mean $n\pi = 500 \times 0.04 = 20$, and variance $n\pi(1-\pi) = 500 \times 0.04 \times 0.96 = 19.2$. The direct approach is to find $P(R \geqslant 28)$.

Using a computer (CCC, Section 8.2.1) we find $P(R \leqslant 27) = 0.9511$, and hence $P(R \geqslant 28) = 1 - 0.9511 = 0.0489$. Since this is less than 0.05 we reject the null hypothesis. Alternatively, since the sample is large the distribution of R is approximately $\mathcal{N}(20, 19.2)$:

$$z = \frac{27.5 - 20}{\sqrt{19.2}} = \frac{7.5}{4.38} = 1.71,$$

which corresponds to a probability of 0.0435. The critical region is $z > 1.64$. We therefore reject the NH and conclude that the process is out of control.

Example 17.11 Test of difference of proportions

Two types of pain-relieving drug are given to patients in hospital after an operation. Of 100 patients given drug A, 38 claim that it reduces pain, while of 120 patients given B, 56 claim that it reduces pain. Is there evidence of a difference in effectiveness between A and B?

Model: the number of patients for which pain is reduced is, for drug A, binomial with parameters $n = 100$ and π_A, and, for drug B, binomial with parameters $n = 120$ and π_B.

$$\text{NH: } \pi_A = \pi_B = \pi, \text{say.} \quad \text{AH: } \pi_A \neq \pi_B. \quad \text{Two-tail test.}$$

On the null hypothesis, there is no difference between the two sets of patients, and as regards the effectiveness of the drug they received they can all be treated as one set. Instead of estimating two proportions π_A and π_B, their response to treatment can therefore be summarized by finding an estimate of π, which is the overall proportion of patients who claimed a reduction in pain.

The best estimate of π is thus found by using all the data, and is

$$\frac{\text{total with pain reduced}}{\text{total treated}} = \frac{94}{220} = 0.427.$$

To decide if there is a difference we look at the difference between the observed proportions $P_A - P_B$:

$$\text{Var}[P_A - P_B] = \text{Var}[P_A] + \text{Var}[P_B] = \frac{\pi_A(1 - \pi_A)}{n_A} + \frac{\pi_B(1 - \pi_B)}{n_B}.$$

On the NH, $\pi_A = \pi_B = \pi$ so instead of π_A and π_B we use the common estimated value of π which is 0.427. Using an estimated value is reasonable since the samples are large. Hence, on the NH, $\text{Var}[P_A - P_B]$ is estimated to be

$$0.427 \times 0.573 \left(\frac{1}{100} + \frac{1}{120} \right) = 0.004\,477.$$

Also, on the NH, $P_A - P_B$ is approximately $\mathcal{N}(0, 0.004\,477)$. The observed difference in proportions

$$(p_A - p_B) = \frac{38}{100} - \frac{56}{120} = 0.380 - 0.467$$

$$= -0.087.$$

Thus

$$z = \frac{-0.087 - 0}{\sqrt{0.004\,477}} = \frac{-0.087}{0.067} = -1.30.$$

Using a 5% significance level, the critical region is $|z| > 1.96$. There are therefore no grounds for rejecting the NH. We have no evidence that the drugs differ in effectiveness.

17.9.1 Exercises

1. I am told that the proportion of left-handed people in the population is 10%. I choose a random sample of 400 people and find that 47 are left-handed. Does this contradict what I have been told?

2. A margarine firm has invited 200 men and 200 women to see if they can distinguish margarine from butter. It is found that 120 of the women, but only 108 of the men can. Investigate whether there is any evidence of a sex difference in taste discrimination. (*SMP*)

3. A die is thrown 240 times and a '6' turns up 30 times. Test if the die is biased.

4. In a game of cards on a particular occasion the probability that I receive at least one ace in a hand may be calculated to be 0.2. In fact in 100 games this happens 12 times. Is there reason to think that the cards are not being dealt fairly?

5. An experimenter offers four sweets of differing sizes to subjects and asks them to take one. When children were used as subjects 60 out of 100 chose the biggest sweet. When adults were used as subjects 60 out of 120 chose the biggest sweet. Is there evidence of a difference in the proportions choosing the biggest sweet in the two groups?

17.10 NORMAL APPROXIMATIONS TO NONPARAMETRIC TEST STATISTICS

If sample sizes are large, the distributions of the nonparametric test statistics (or even just the tails of the distributions) are tedious to calculate. Moreover, even tables cover only relatively small sample sizes. To overcome this difficulty we may use the fact that for large samples the test statistics of the common nonparametric tests are approximately normally distributed. For each test we give minimum sample sizes for the normal approximation and illustrate its use. We give specific sample sizes, simply to be definite, but would claim no authority for the values; in fact, what is appropriate would depend on the precise circumstances of a test. We recommend that readers use a continuity correction when calculating the z statistic (see Section 16.3).

Example 17.12 Sign test

In Example 13.2 (p. 280), data were given from a paired comparison of two methods (standard S and new N) of teaching children to read. Of the 14 pairs of children in which one child did better than the other, the child taught by the new method did better in 5 cases (labelled + in Example 13.2) and did worse in 9 cases (labelled −). Test if there was a *difference* in effectiveness of the two methods.

On the null hypothesis of no difference in effectiveness of the two methods, the number of + signs R has a binomial distribution with n equal to the number of pairs and $\pi = \frac{1}{2}$. We may use a normal approximation if both $n\pi$ and $n(1 - \pi)$ exceed 5. The approximating distribution has the same mean $n\pi$ and variance $n\pi(1 - \pi)$ as the binomial distribution.

In our example, $n = 14$ and a two-tail test is appropriate. The critical region for a 5% significance test is $|z| > 1.96$. The normal statistic derived from $r = 5$, using a continuity correction, is

$$z = \frac{5.5 - 7}{\sqrt{14 \times \frac{1}{2} \times \frac{1}{2}}} = \frac{-1.5}{1.871} = -0.802.$$

The result is not significant and we do not reject the null hypothesis. There is no evidence of a difference between the two methods.

Example 17.13 Wilcoxon rank-sum (or Mann–Whitney U) test

A psychology teacher when interviewing 24 of his students gave 11 of them, chosen at random, a glass of sherry. At the end of each interview he measured the reaction time of the student. The times were ranked and the rank sum of the students who received the sherry was 80. Test if the sherry had an effect on the reaction time of the students.

Let the samples be labelled 1 and 2 and be of sizes m and n respectively; the rank sum of sample 1 is T_1. If m and n both exceed 10 then the distribution of T_1, on the null hypothesis of no difference between the two groups, is approximately normal with mean $m(m+n+1)/2$ and variance $mn(m+n+1)/12$ (p. 288); the U statistic has mean $mn/2$ and the same variance as T.

For the example $m = 11, n = 13$ and a two-tail test is appropriate. The normal statistic, using a continuity correction, is

$$z = \frac{79.5 - 137.5}{\sqrt{3575/12}} = \frac{-58}{17.260} = -3.360.$$

The result exceeds even the critical value 3.291 of a 0.1% significance test. We conclude that the sherry had an effect on reaction time.

Example 17.14 Wilcoxon signed-rank test

The concentration of a certain chemical in the blood of 20 subjects was measured before and after the subjects took exercise. For each subject the difference in concentration of the chemical before and after exercise was found; the sum of the positive signed ranks was calculated from the differences and was 185. Test if there was a change in the concentration of the chemical after exercise.

We may use a normal approximation to the distribution of T_+ if the number of pairs N exceeds 16. The mean of T_+, on the null hypothesis of no difference between samples, is $N(N+1)/4$ and the variance is $N(N+1)(2N+1)/24$.

The particular null hypothesis is that the concentration of the chemical is the same before and after exercise. The test is two-tailed. Putting $N = 20$, the normal statistic is

$$z = \frac{184.5 - 105}{\sqrt{717.5}} = \frac{79.5}{26.786} = 2.968.$$

The critical value for a 5% significance test is 1.96, and for a 1% test 2.576, so we reject the null hypothesis. There is quite strong evidence that the concentration of chemical changes after exercise.

17.11 TEST ON SAMPLE MEAN WHEN VARIANCE ESTIMATED FROM SAMPLE

When testing the significance of a mean of normally distributed variables, we use the statistic

$$Z = \frac{\overline{X} - \mu}{\sigma/\sqrt{n}}.$$

This contains only one variable element in it, $\overline{X}$ which varies from sample to sample: n is fixed and σ and μ are known. But if (as is common in practice) σ is not known, it has to be estimated from the sample using

$$s^2 = \frac{1}{n-1} \sum_{i=1}^{n} (x_i - \bar{x})^2$$

as the unbiased estimator of σ^2. If the sample is large s^2 is likely to be very close to σ^2 so we treat s^2 as if it is σ^2 (as we did in Section 17.6, for example). But we cannot do this in small samples: s^2 has a high chance of being very different from σ^2. How do we cope with this difficulty?

We use the quantity

$$\frac{\overline{X} - \mu}{s/\sqrt{n}},$$

to test significance but we cannot assume that it has an $\mathcal{N}(0,1)$ distribution. It has two variable elements in it, $\overline{X}$ and s, both of which alter each time a new sample of size n is taken. Its sampling distribution cannot then be exactly the same as Z which has only the one variable element, $\overline{X}$, in it. In fact the expression

$$\frac{\overline{X} - \mu}{s/\sqrt{n}},$$

which is usually denoted by t in this situation, represents a whole family of distributions because n (or, more strictly, $n-1$) is the parameter in its probability density function. Thus a different member of the family is needed for each different value of n. These distributions were studied first by W. S. Gosset (British, 1876–1937), who worked as a statistician for Guinness, writing under the pen-name 'Student', and are known as Student's t-distributions. They are very similar in shape to the standard normal $\mathcal{N}(0,1)$, for they have mean 0 and are bell-shaped. In fact as n becomes fairly large the distribution of t becomes very close to that of $\mathcal{N}(0,1)$, while for very small n the relation of t to $\mathcal{N}(0,1)$ is as shown in Fig. 17.4.

The expression

$$t = \frac{\overline{X} - \mu}{s/\sqrt{n}}$$

is distributed as t with $n-1$ *degrees of freedom*, as the parameter is always called. The notation $t_{(n-1)}$ is useful to denote this random variable. Because there is a

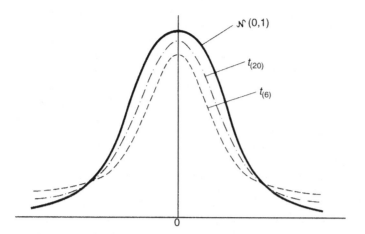

Figure 17.4 *Comparative shapes of the distributions $t_{(6)}$, $t_{(20)}$, $\mathcal{N}(0, 1)$.*

different curve for each value of $(n-1)$, it is not possible to give the same sort of tables as Table A2 for $\mathcal{N}(0, 1)$. Instead the common form of t-table (Table A3) gives only the *percentage points* (i.e. percentiles).

To summarize, if σ^2 is unknown and we replace it by the estimate s^2, then the appropriate distribution to use in the test is the t-distribution, with $n-1$ degrees of freedom, rather than the normal distribution. In large samples the two distributions are very similar so we tend to use the normal distribution for convenience. Note that the derivation of the t-distribution assumes that the basic variable measured, X, is normally distributed.

Example 17.15 Small sample test of mean

The widths of a sample of ten beetles, chosen from a particular locality, were measured and found to be 28, 21, 26, 16, 18, 13, 15, 22, 19, 22 mm. Previous extensive measurements of beetles of the same species had shown the width to be normally distributed with mean 23 mm. Test whether the beetles from the chosen locality have a different mean width from the main population.

> Model: the width X is $\mathcal{N}(\mu, \sigma^2)$.
> NH: $\mu = 23$.   AH: $\mu \neq 23$.   Two-tail test.

From the data we find $\bar{x} = 200/10 = 20$, and $s^2 = \frac{1}{9}\sum(x_i - 20)^2$
$$= 204/9 = 68/3 = 22.67.$$

The test statistic

$$t = \frac{\bar{x} - \mu}{s/\sqrt{n}} = \frac{(20 - 23)\sqrt{10}}{\sqrt{22.67}} = \frac{-3}{\sqrt{2.267}} = \frac{-3}{1.51} = -1.99.$$

On the NH, t is distributed as Student's t with $n-1 = 9$ degrees of freedom. Using a 5% significance level, the critical region is $|t| > 2.26$. We therefore do not reject the null hypothesis. We do not have evidence that the mean width of the beetles in the chosen locality differs from that of the main population.

17.11.1 Exercises

1. The lives of six candles are found to be 8.1, 8.7, 9.2, 7.8, 8.4, 9.4 hours. Estimate the population mean and show that the estimate of the population variance is 0.388.

   The manufacturer claims that the average life is $9\frac{1}{2}$ hours. Making a suitable assumption concerning the nature of the distribution of the life of a candle, carry out a statistical test of the manufacturer's claim. Give full details of your test. (*Cambridge*)

2. A random sample of 16 values $x_1, x_2, \ldots, x_{16}$, was drawn from a normal population and the following results were obtained:

$$\sum_{i=1}^{16} x_i = 51.2, \quad \sum_{i=1}^{16} x_i^2 = 243.19.$$

Test the hypothesis that the population mean is 4.0 against the alternative hypothesis that the population mean is less than 4.0, (i) using a known population standard deviation of 1.9; (ii) when the population standard deviation is unknown. Comment briefly on why it would be better to use the procedure in (i) if the population standard deviation were known. (*IOS*)

17.12 PAIRED COMPARISONS USING ESTIMATED VARIANCE

We showed, in Section 17.8, that paired-comparison data may be analysed by considering the differences of the pairs of observations. Thus, if the observations made on the distinct members of the n pairs are represented by the random variables X_1 and X_2, we consider the derived random variable $D = X_1 - X_2$. The mean of the distribution of D is μ_D which equals $\mu_1 - \mu_2$. A null hypothesis that μ_1 and μ_2 differ by a stated amount may be written $\mu_D = \mu_0$ (in most applications $\mu_0 = 0$). If we assume that D is normally distributed then, on the null hypothesis, D is $N(\mu_0, \sigma_D^2)$. If σ_D^2 has to be estimated from the observed differences the situation is equivalent to that in Section 17.11, except that we are dealing with D instead of X. Hence the significance test is made with the statistic

$$t = \frac{\overline{D} - \mu_0}{s_D/\sqrt{n}}$$

where $s_D^2 = \sum (d_i - \bar{d})^2/(n-1)$. The statistic is distributed as t with $n-1$ degrees of freedom.

Example 17.16 Paired comparison (small sample)

Ten joints of meat are each cut in half; one half is frozen and wrapped by process A and the other half is frozen and wrapped by a new process B. The halves are placed in ten freezers, halves of the same joint being put in the same freezer. The number of days to spoilage, which can be detected from a change in the colour of a pack, are found for each pack:

Joint number	1	2	3	4	5	6	7	8	9	10
Process A	63	109	82	156	161	155	47	141	92	149
Process B	129	105	76	207	253	146	62	160	90	177

Assuming the differences between these number-pairs are random samples from a normal distribution, test the hypothesis that the processes A and B are equally effective.

If the above data were misinterpreted as ten random selections from meat treated by process A, and ten from meat treated by process B, taken in random order, would the conclusion be affected by this error?

This is a paired comparison, so the first step is to calculate the difference of the days to spoilage between A and B. We take $d = $ (days for B – days for A).

Pair number	1	2	3	4	5	6	7	8	9	10	Total	Mean
d	66	−4	−6	51	92	−9	15	19	−2	28	250	25
$d - \bar{d}$	41	−29	−31	26	67	−34	−10	−6	−27	3	0	
$(d - \bar{d})^2$	1681	841	961	676	4489	1156	100	36	729	9	10678	

$$s^2 = \frac{1}{9} \sum (d - \bar{d})^2 = \frac{10\,678}{9} = 1186.44.$$

Model: the difference D is $\mathcal{N}(\mu, \sigma^2)$.

NH: $\mu = 0$.   AH: $\mu \neq 0$.   Two-tail test.

On the NH, $\bar{D}$ is $\mathcal{N}(0, \sigma^2/n)$. Hence,

$$t = \frac{25 - 0}{\sqrt{1186.44/10}} = \frac{25}{10.89} = 2.30.$$

On the NH, t is drawn from Student's t-distribution with $n - 1 = 9$ degrees of freedom. Using a 5% significance level, the critical region is $|t| > 2.26$. We therefore reject the NH and conclude that there is a difference in the effectiveness of the two processes.

If the experimental method was misunderstood, as is stated in the question, then a different method of analysis is appropriate (the test is fully discussed in Section 17.13). Let us write X_A for the results from process A and X_B for the results from process B. We would test $\bar{X}_B - \bar{X}_A$, which equals $\bar{D}$, but use $\mathrm{Var}[\bar{X}_B - \bar{X}_A] = \sigma_A^2/10 + \sigma_B^2/10$ instead of $\sigma^2/10$. Therefore for known variances, σ^2 in the test above would be replaced by $\sigma_A^2 + \sigma_B^2$. From the data it is obvious that $\sigma_A^2 + \sigma_B^2$ is much greater than the variance of the differences, σ^2. In fact, to carry the test through, we need to assume $\sigma_A^2 = \sigma_B^2$ and we would derive a pooled estimate of the common variance with 18 degrees of freedom. But this pooled estimate of variance would be much bigger than s^2 above and t would not be significant. The conclusion would therefore be changed.

17.12.1 Exercises

1. To compare two diets A and B for pigs, eight pairs of pigs were used, where the pigs in each pair were littermates. One pig in each pair was chosen at random and given diet A and the other pig in each pair was given diet B. The gains in weight (in kilograms) of the pigs over the period of the experiment were as follows:

Pair	1	2	3	4	5	6	7	8
Diet A	25.6	20.7	14.0	21.5	21.6	25.6	26.6	22.2
Diet B	24.1	17.7	14.3	19.7	22.2	23.5	25.4	21.3

Is there evidence of a difference between the effects of diets A and B on the weight-gain of pigs? State any assumptions that you make.

Explain briefly why the above experimental design would be superior to one in which the pigs given diet A were chosen independently of those given diet B. *(IOS)*

2. Eight machines, in an experiment under standard conditions, produced quantities $x_1, x_2, \ldots, x_8$ of their product. When the experiment was repeated, under the same conditions except for the use of a different lubricant, the machines produced $y_1, y_2, \ldots, y_8$.

If

$$\sum_{r=1}^{8} x_r = 8.73, \quad \sum_{r=1}^{8} y_r = 9.21 \quad \text{and} \quad \sum_{r=1}^{8} (y_r - x_r)^2 = 0.051,$$

is there evidence that the lubricant has improved productivity? Explain the assumptions made in the analysis. *(IOS)*

17.13 TESTS ON MEANS OF TWO UNPAIRED SAMPLES USING ESTIMATED VARIANCES

The situation is similar to that considered in Section 17.5 except that the variances of the two populations involved are both assumed equal to σ^2 and this variance has to be estimated from the samples. Thus we assume we have two samples chosen at random from normal distributions, the first of size n_1 from $\mathcal{N}(\mu_1, \sigma^2)$ and the second of size n_2 from $\mathcal{N}(\mu_2, \sigma^2)$. Take particular note of the assumption that both populations have the *same* variance. If it is not valid the test becomes more complicated and we do not discuss it in this book (it is not based simply on the *t*-distribution).

The obvious quantity to consider is the difference of sample means $\overline{X}_1 - \overline{X}_2$. On a null hypothesis that $\mu_1 - \mu_2 = \mu_0$, the variable $\overline{X}_1 - \overline{X}_2$ is

$$\mathcal{N}\left(\mu_0, \sigma^2\left(\frac{1}{n_1} + \frac{1}{n_2}\right)\right)$$

(cf. Section 17.5). If σ^2 were known the appropriate statistic for an NH of '$\mu_1 - \mu_2 = \mu_0$' would be

$$Z = \frac{(\overline{X}_1 - \overline{X}_2) - \mu_0}{\sigma\sqrt{\left(\frac{1}{n_1} + \frac{1}{n_2}\right)}},$$

which would have a normal distribution. In fact σ^2 has to be replaced by s^2, calculated from the samples. We may obtain separate estimates of σ^2 from the two samples:

$$s_1^2 = \frac{1}{n_1 - 1}\sum(x_{1i} - \bar{x}_1)^2, \quad s_2^2 = \frac{1}{n_2 - 1}\sum(x_{2i} - \bar{x}_2)^2.$$

How should these be combined? If $n_1 = n_2$, we simply take the mean, $s^2 = (s_1^2 + s_2^2)/2$. But if the sample sizes are not equal we must give greater weight to the larger sample; the appropriate weights are the degrees of freedom corresponding to each estimate of variance. Thus, in general, s^2 is estimated by

$$\frac{(n_1 - 1)s_1^2 + (n_2 - 1)s_2^2}{(n_1 - 1) + (n_2 - 1)}.$$

This is commonly called a *pooled* estimate of the variance σ^2. Referring to the definitions of s_1^2 and s_2^2 above, we see that a more convenient formula for calculation is that

$$s^2 = \frac{\sum(x_{1i} - \bar{x}_1)^2 + \sum(x_{2i} - \bar{x}_2)^2}{n_1 + n_2 - 2}.$$

We calculate the sums of squares of deviations of the observations, measuring each deviation from the appropriate mean, and divide by the sum of the degrees of freedom corresponding to the two samples. The estimate s^2 may be said to be based on $n_1 + n_2 - 2$ degrees of freedom. The test statistic is

$$t = \frac{(\bar{x}_1 - \bar{x}_2) - \mu_0}{s\sqrt{\left(\frac{1}{n_1} + \frac{1}{n_2}\right)}},$$

which, on the NH, has a *t*-distribution with $n_1 + n_2 - 2$ degrees of freedom.

Example 17.17 Test on difference of means (small sample)

The mean reaction times, in hundredths of a second, of two groups of subjects to a flashing-light stimulus are given below. The first group consisted of subjects who were new to the experimental investigations while the subjects in the second group had taken part in previous experiments. Test if experience has had an effect on the mean response time.

New subjects	2.7	3.0	3.3	2.9	3.5	2.7	3.0	3.1	2.8	3.0
Experienced subjects	2.7	2.5	3.0	2.7	2.6	2.5	2.9	2.7		

Model: The response time of the new subjects X_1 is $\mathcal{N}(\mu_1, \sigma^2)$; the response time of the experienced subjects X_2 is $\mathcal{N}(\mu_2, \sigma^2)$.

$$\text{NH: } \mu_1 = \mu_2. \quad \text{AH: } \mu_1 \neq \mu_2. \quad \text{Two-tail test.}$$

$$n_1 = 10, \quad \bar{x}_1 = 30.0/10 = 3.0. \qquad \sum (x_{1i} - \bar{x}_1)^2 = 0.58.$$

$$n_2 = 8, \quad \bar{x}_2 = 21.6/8 = 2.7. \qquad \sum (x_{2i} - \bar{x}_2)^2 = 0.22.$$

The pooled estimate of variance is

$$s^2 = \frac{\sum (x_{1i} - \bar{x}_1)^2 + \sum (x_{2i} - \bar{x}_2)^2}{n_1 + n_2 - 2} = \frac{0.58 + 0.22}{16} = \frac{0.80}{16} = 0.05.$$

$$t = \frac{(3.0 - 2.7) - 0}{\sqrt{0.05(\frac{1}{10} + \frac{1}{8})}} = \frac{0.3}{\sqrt{0.01125}} = \frac{0.3}{0.1061} = 2.828.$$

On the NH t has a t-distribution with 16 degrees of freedom. The critical region for a 5% significance test is $|t| > 2.120$. (For a 1% significance test the critical region is $|t| > 2.921$.) We therefore reject the null hypothesis and conclude there is a difference in response time of the two groups.

17.13.1 Exercises

1. A company manufactures steel cable at two factories X and Y. A random sample of 9 lengths, each of 10 metres of cable, was taken from factory X and a random sample of 16 lengths, again each of 10 metres, from factory Y. The breaking loads $(x_i, i = 1, 2, \ldots, 9; y_j, j = 1, 2, \ldots, 16)$ were determined in kilonewtons. The results were:

$$\bar{x} = \sum_{i=1}^{9} x_i/9 = 30.11, \qquad \sum_{i=1}^{9} (x_i - \bar{x})^2 = 0.8013,$$

$$\bar{y} = \sum_{j=1}^{16} y_j/16 = 29.63, \qquad \sum_{j=1}^{16} (y_j - \bar{y})^2 = 3.0206.$$

Do these data give evidence that the mean breaking loads of steel cable from the two factories differ?

State any assumptions that you make. *(IOS)*

2. Twenty-four individually potted seedlings were used in a greenhouse experiment to compare the effects of two different plant hormone preparations on growth. Twelve seedlings were allocated at random to each of the two hormone preparations. Two of those given hormone A were accidentally damaged. The following results were obtained concerning the dry matter contents (in grams) of the seedlings that remained:

	n	$\bar{x}$	S.E. of the mean
Hormone A	10	16.54	0.475
Hormone B	12	21.13	0.573

The experimenter concluded that there was no difference between the effects of the two hormones on the growth of the seedlings. Do you agree?

17.14 INFERENCE WHEN THE NULL HYPOTHESIS IS THE HYPOTHESIS OF INTEREST

The reader has, by this point in the book, seen many examples of significance tests. We stress again that significance tests must be used with care or they can be very misleading. We wish to point out a common misuse of significance tests.

We made the point in Section 13.1 that a significance test is framed so that we accept the alternative hypothesis only if there is strong evidence for it; otherwise the case against the null hypothesis is 'not proven'. The procedure assumes that it is the alternative hypothesis that interests us; we set up the null hypothesis in the hope that we can knock it down. But what if the hypothesis that interests us is the natural null hypothesis in a particular situation?

Let us return to the example of Chapter 1: the effect of a mother smoking on the birth-weight of her baby. Suppose a cigarette manufacturer hears that some research workers are asserting that smoking by mothers has the effect of reducing the birth-weight of their babies; the manufacturer would like to disprove this. Assuming also that the manufacturer has data (such as that in Chapter 1) on the birth-weight of babies born to mothers who (i) smoked, and (ii) have never smoked, an obvious procedure is to carry out a suitable significance test. The natural null hypothesis is one of no difference, namely that the mean birth-weights of babies born to mothers who smoke, and to those who have never smoked, are identical.

If the manufacturer went ahead and made the test, and the null hypothesis was rejected, at an appropriate significance level, then the conclusion would be clear. The manufacturer would have to accept that there was evidence of an effect of smoking on birth-weight. But if the null hypothesis is not rejected can the manufacturer really say he has good evidence that there is no effect whatever of smoking on birth-weight? We must emphasize firmly that he cannot. A significance test does not provide evidence *in support of* the null hypothesis. We might have chosen as null hypothesis the assertion that 'smoking has the effect of reducing mean birth-weight by 10 grams'; let us suppose again that the significance test does not reject this null hypothesis. In fact, we would find that there is a whole range of values we could use in place of the '10' and still obtain a non-significant result. We are led to a ridiculous situation; the difference cannot *equal* all these values! The fault is that we are misusing the mechanism of significance tests.

Another point is that with variable material we can more or less guarantee obtaining a non-significant result (i.e. not rejecting the null hypothesis) by taking very few observations. Such a procedure cannot be valid. But all too often the point is overlooked when data are being collected, especially when an experimenter fails to take statistical advice to begin with.

The way out of this dilemma is to use a confidence interval. This is the subject of the next chapter. The particular example of this section is taken up again in Section 18.6.

17.15 PARAMETRIC VERSUS NONPARAMETRIC TESTS

Three tests that we introduced in Chapter 13 (the sign test, the rank-sum test and the signed-rank test) are commonly called *nonparametric* tests. By contrast, the tests in this chapter are called *parametric* tests; they are based on the normal distribution which is a parametric distribution.

We remarked in Section 8.3 that the binomial probability expression represented a whole family of distributions; the same is true of the normal (Definition 15.1, p. 351), and in the same way we need to be given the numerical values of its parameters before we know which particular normal distribution we are using. The parameters for the normal distribution are the mean μ and the variance σ^2, and we have been making tests on one of them, namely μ. (Parametric tests of σ^2 are also available.)

If a normal model cannot be assumed for the data then the tests on means in this chapter are not applicable. Nonparametric tests were created to overcome this difficulty. In a nonparametric test, very few assumptions are made about the distribution underlying the data and, in particular, it is not assumed to be a normal distribution. Some statisticians prefer to use the term *distribution-free* rather than nonparametric to describe these tests.

Nonparametric tests are often (but not always) based on the use of ranks: this is so for those in Sections 13.5 and 13.6, and also for rank correlation in Section 21.9.

If the original observations are in the form of ranks then tests devised for ranks *must* be used. But if the data are quantitative variates there may be the option to use either a parametric or a nonparametric test. We shall consider cases where both types of test might be used.

For unpaired random samples from two populations we have introduced the *t*-test (or normal *z*-test if the variance is known) and the rank-sum test. If the data are normal or near-normal the *t*-test is the better: by using the assumption of normality we allow that test of the mean to be made which gives the most precise results for the available number of observations. But for non-normal data we have two choices: if we know the appropriate transformation we may transform the data (e.g. take their logarithms) so that they become normal and the *t*-test applies; or we may express the data in rank form and use the rank-sum test. It has been shown that if the rank-sum test is used with normal data, the power of the test (see Section 13.7) is little less than that of the *t*-test. Thus if we have any doubt that a set of data really are normal, we may as a precaution use the rank-sum test, and can lose little by doing so.

For paired-comparison data we have introduced the *t*-test (or normal *z*-test if the variance is known), the signed-rank test and the sign test. The comparison of *t*-test and signed-rank test closely parallels that of *t*-test and rank-sum test: for normal data the *t*-test is better but the power of the signed-rank test is only a little less than that of the *t*-test, whereas for non-normal data the signed-rank test should be used. The sign test is substantially less powerful than either the *t*-test (with normal data) or the signed-rank test (with normal or non-normal data). Its only practical merit is that it is a quick test and may be useful when making an initial scan of data.

The authors prefer to use parametric tests (perhaps after a transformation of the data) when they are valid, because the quantities estimated in order to carry out the test are descriptive of the data and may be used to construct confidence intervals (see Chapter 18). By contrast, a nonparametric method often gives information only in the form of a significance test. However, this situation is being studied and nonparametric estimation methods are being developed; but so far most of them still seem rather awkward, and some are quite imprecise. If data are definitely non-normal and no suitable transformation can be found, then a nonparametric test must be used.

17.16 MINITAB: SIGNIFICANCE TESTS ON MEANS AND PROPORTIONS

17.16.1 Tests on means

MINITAB calculates a confidence interval at the same time as it produces the test statistic and *p*-value that is used in a significance test. We explain confidence intervals in the next chapter but ignore them here.

An assumption of many significance tests, and certainly of the tests in this chapter, is that the underlying distribution of the data is normal. The first step, therefore, in an analysis is to check whether this is a reasonable assumption for the data in hand. A starting point, for continuous data, is to plot a histogram and consider if the data look as if they are normally distributed. MINITAB helps here (see Section 1.6 on histograms). Other relevant diagrams are a normal probability plot (Sections 15.8.1 and 15.9) and a symmetry plot (Sections 15.8 and 15.9).

If data are normally distributed then means derived from the data will be normally distributed. But remember that the *critical* assumption for many significance tests is that the *mean* be normally distributed. The underlying data may not be normally distributed but, because of the central limit theorem, the mean may well be approximately normally distributed. This is discussed at the beginning of Section 17.6 on large-sample tests. There are also many examples through the book (e.g. Section 16.3 on the binomial and Section 20.10 on the Poisson) of distributions that may be approximated by the normal distribution. Remember as well that a proportion is the mean of a variate taking values 0 and 1.

In some cases, data that are distinctly non-normal may be transformed to produce a new variate that is approximately normal. Example 15.2 shows how taking the logarithm of a variate does this.

Having decided that the normality assumption is valid, the next important step is to decide whether the variance is known or whether it is estimated from the sample data. MINITAB offers tests for the following cases. (The section references are to this book.)

Data	Variance	Section	Menu name	Session command
One sample	known	17.4	**1-sample Z**	ONEZ C
One sample	estimated	17.11	**1-sample t**	ONET C
Two samples	estimated	17.13	**2-sample t**	TWOSAMPLE C C
Paired samples	estimated	17.12	**Paired t**	PAIR C C

Note that MINITAB does not offer tests when the variance is known and there are two samples (Section 17.5) or paired samples (Section 17.8). These situations are relatively uncommon in practice. The large-sample test on difference of means (Example 17.8) would be carried out with **2-sample t**.

Data are usually presented to MINITAB in columns but some versions accept summarized data. The latter demands giving the number of observations and the mean and standard deviation calculated from the data, except for **1-sample Z** where the standard deviation is known independently and must always be inserted. The **2-sample t** menu also allows all the data to be entered in one column with subscripts, or labels, for the two samples entered in another column. The procedures for using the tests are summarized in the table that follows.

Menu	Dialogue box	Options
1-sample Z	Enter **Std dev** and **Test mean**	**Alternative** (AH)
1-sample t	Enter **Test mean**	**Alternative** (AH)
2-sample t	Select **equal variances** or not	**Alternative** (AH)
		Enter **Test mean**
Paired t		**Alternative** (AH)
		Enter **Test mean**

The one-sample tests request that the mean on the NH be stated whereas the default assumption for the two-sample tests is that the mean on the NH (**Test mean**) is zero. For all tests, the default assumption is that a two-tail test is appropriate. Choice of the **Options** box, in the menu dialogue box, allows one-tail tests to be used and the mean on the NH to be non-zero in the two-sample cases. The usual assumption, though it should be considered, with the **2-sample t**, is that the samples have equal variances.

17.16.2 Tests on proportions

MINITAB can carry out the tests on proportions that are discussed in Section 17.9. The tests are usually carried out with summarized data, i.e. the observed proportions, but a list of the outcomes of individual trials can also be used. To carry out the test on one proportion in Example 17.10, choose

> **Stat>Basic Statistics>1 Proportion/**
> Select **Summarized data**, enter 500 for **Number of trials** and 28 for **Number of Events**, select **Options/**
> Enter 0.04 in **Test proportion** and choose the **Alternative** 'greater than'/

This carries out a one-tail test with NH $p = 0.04$, and AH $p > 0.04$. The default case, which is carried out if **Options** is not chosen, is a two-tail test with NH $p = 0.5$. The exact probability using the binomial distribution is calculated unless **Use test and interval based on normal distribution** is selected under **Options**.

To carry out a test on the difference of two proportions, as in Example 17.11, choose

> **Stat>Basic Statistics>2 Proportions/**

The default case is a two-tail test with NH that the proportions are not equal, using a normal approximation, as in the example. But, to obtain an estimate of the test variance, we used an estimate of the proportion based on all the data. To do this select **Options** and, in the dialogue box, **Use pooled estimate of p for test**.

17.16.3 The *t*-distribution

Use of MINITAB to evaluate the probability functions of the *t*-distribution is discussed in Section 18.9.

1. Bill has used a particular type of razor for shaving for a long time. The length of time (in seconds) he takes to shave has a normal distribution with mean 240 and standard deviation 20. He changes to a new type of razor and finds his shaving times on 9 days are 210, 230, 220, 220, 250, 230, 260, 210, 240. Assuming the standard deviation of his shaving time has not changed, test if his mean shaving time has changed with the new razor.

2. Explain briefly what is meant by a significance level of $\alpha\%$.

   A student was asked to mark, by eye, the centre of a line drawn on a sheet of paper. He repeated this 50 times, using a new test sheet each time. The signed deviations from the centre, d, were measured in millimetres and the quantities $\sum d = -60.0, \sum d^2 = 197.44$, calculated. Is there evidence that the student does not locate the mean position of his bisection marks on the centre of the line? *(Oxford)*

3. A new method of treating a skin complaint is to be tested against a standard treatment. The success rate of the standard is known to be 60%. A random sample of 100 subjects is given the new treatment and 72 respond successfully. Is the new treatment significantly better than the standard treatment? Estimate the probability that at least 80% of 100 subjects receiving the new treatment respond successfully. *(IOS)*

4. A random sample of n values is taken from a normal distribution with mean μ and variance σ^2. State the distribution of the sample mean.

   Two random samples, each of size n, from this same distribution have means $\bar{x}_1$ and $\bar{x}_2$. State the distributions of: (i) $\bar{x}_1 + \bar{x}_2$; (ii) $\bar{x}_1 - \bar{x}_2$.

   Slabs of toffee are specified by a manufacturer to have a mean mass of 200 grams and the standard deviation is known to be 5 grams. Assuming that the mean mass has the specified value, find the greatest whole number of grams which the manufacturer could state to be the mass of the bar so that no more than 1% of the bars have less than the stated mass.

   It is found that two samples, each of 100 bars, have mean 199.2 grams and 199.3 grams. Test each mean separately and also their sum to see whether or not they are significantly less at the 5% level than their expected values according to the manufacturer's specification. State which of the three results you consider most important, giving a reason for your choice. *(JMB)*

5. Explain what are meant by the *quartiles* of a frequency distribution, and verify that, for a normal distribution with zero mean and unit variance, the quartiles are ± 0.67.

   An attainment test is standardized to provide a normally distributed score with mean 100 and standard deviation 16. In a sample of 1000 scores, 540 are found to lie within the range 89 to 111. Do you regard this as evidence that the population from which the sample is drawn may be unusual?

6. Two independent random samples, of n_1 and n_2 observations, are drawn from normal distributions with common variance σ^2. If s_1^2 and s_2^2 are unbiased

estimates of σ^2 based on the first and second sample, respectively, show that $(n_1 s_1^2 + n_2 s_2^2)/(n_1 + n_2)$ is an unbiased estimate of σ^2.

Two makes of car safety belt, A and B, have breaking strengths which are normally distributed with the same variance. A sample of 140 belts of make A and a sample of 220 belts of make B were tested, and the sample means, and sums of squares about the means, of the breaking strengths (in lbf units) were $(2685, 1.9 \times 10^4)$ for make A and $(2680, 3.4 \times 10^4)$ for make B. Examine whether these results provide significant evidence that belts of make A are stronger, on average, than belts of make B.   (*JMB*)

7. You are engaged as an expert witness for the prosecution in a court case in which a gaming club is accused of running an unfair roulette wheel. The evidence is that out of 3700 trial spins, zero (on which the club wins) turned up 140 times. There are 37 possible scores on a trial spin, labelled 0 to 36, and these should have equal probability.

   Test whether there is evidence that the wheel is biased. Explain briefly what this test of significance means, bearing in mind you have to convince a non-mathematical jury.   (*Oxford*)

8. Describe briefly what is meant by the *sampling distribution of a statistic*, illustrating your remarks by reference to the sampling distribution of the mean of a random sample of n observations from a normal distribution with mean μ and variance σ^2.

   Weights of a certain type of biscuit are normally distributed with mean 0.5 oz and standard deviation 0.05 oz. Random samples of 16 biscuits are chosen to produce for sale packets of nominal net weight 8 oz. Calculate the probability of such a packet weighing less than 7.5 oz.

   Similar biscuits made in a different mould are also sold in packets of 16. Again the weights of individual biscuits are normally distributed with standard deviation 0.05 oz. A random sample of 30 packets had a sample average net weight of 7.9 oz. Examine whether there is significant evidence that packets of biscuits produced in this new mould have mean net weight different from the others.   (*JMB*)

9. A librarian, looking over his stock of books, found that of the 500 fiction books, 320 had been borrowed in the previous year, whereas out of the 700 non-fiction books, 400 had been borrowed. Is there a significant difference in the proportions borrowed in the two categories?

10. The proportion of blue-eyed persons in a certain large population is 0.2. A group of ten persons is selected at random. Calculate the probability that the number of blue-eyed persons in the sample is: (i) exactly three; (ii) at least three.

    A second sample of ten persons is examined; calculate the probability that the total number of blue-eyed persons in the two samples combined is exactly two.

    A random sample of 200 persons from a second large population is examined for eye colour. In this sample 50 persons are found to have blue eyes. Is this result consistent with the hypothesis that the proportion of blue-eyed persons in this second population is 0.2? What would have been your conclusion if there had been 200 blue-eyed persons in a random sample of 800 persons from this second population? (*MEI*)

11. Explain how Student's *t*-distribution may be used to test the hypothesis that a random sample of *n* observations is derived from a population whose mean is μ, assuming the population to be normal.

Ten athletes ran a 400 metres race at sea level and at a later meeting ran another 400 metres race at high altitude.

Their times in seconds were as follows:

Runner	A	B	C	D	E	F	G	H	I	J
Time at sea level	48.3	47.6	49.2	50.3	48.8	51.1	49.0	48.1	50.7	47.9
Time at high altitude	50.4	47.3	50.8	52.3	47.7	54.5	48.9	49.9	54.8	48.5

Test the hypothesis that the athletes' performance is not affected by the altitude.

State the assumptions made in applying the test. (*AEB*)

12. Two random samples of sizes n_1 and n_2 have means $\bar{x}_1$ and $\bar{x}_2$ and standard deviations s_1 and s_2. Show how the hypothesis, that the two samples are drawn from the same normal population, may be tested by the statistic

$$t = \frac{\bar{x}_1 - \bar{x}_2}{\hat{\sigma}} \sqrt{\frac{n_1 n_2}{n_1 + n_2}}$$

where $\hat{\sigma}^2$ is an unbiased estimate of the population variance based on pooled data.

Two groups of students take the same mathematics examination. The first group, comprising 10 students, scores an average mark of 66 with a standard deviation of 8 marks, whilst the second group of 14 students scores an average mark of 60 with a standard deviation of 10 marks. Test the hypothesis that the first group is not better at mathematics than the second group. (*AEB*)

13. Measurements are made of the tensile strengths of 12 steel rods of which 6 rods have received special treatment and the remaining 6 have not received any treatment.

The coded results of these measurements are tabulated below:

Tensile strengths						
No treatment: x	3.5	2.6	3.4	3.9	3.4	3.8
Special treatment: y	4.5	3.9	3.2	4.8	4.3	4.1

On the evidence of these measurements, does the special treatment increase the tensile strength?

Comment upon a procedure for improving the design of this test. (*AEB*)

14. Two alternative hypotheses concerning the probability density function of a random variable are

$$H_0: f(x) = \begin{cases} 2x & 0 < x < 1, \\ 0 & \text{otherwise}, \end{cases}$$

$$H_1: f(x) = \begin{cases} 2(1-x) & 0 < x < 1, \\ 0 & \text{otherwise}. \end{cases}$$

Give a sketch of the probability density function for each case.

The following test procedure is decided upon. A single observation of X is made and if X exceeds a particular value a, where $0 < a < 1$, then H_0 is accepted, otherwise H_1 is accepted. Find the value of a if the probability of accepting H_1 given that H_0 is true is $\frac{1}{9}$. With this value of a, find the probability of accepting H_0 given that H_1 is true.   (*Cambridge*)

15. A machine is intended to produce piston heads of mean diameter 72.5 mm. A sample of 10 items is drawn at random from those produced by the machine on one day, and has the following diameters:

$$72.8, 72.4, 72.9, 72.5, 72.8, 72.7, 72.5, 72.8, 72.7, 72.5 \text{ mm}.$$

The distribution of the diameters of piston heads produced on this machine is normal. Is there evidence that the machine is wrongly set on this day?

16. The following data give the weights (in kilograms) of the contents of 10 bags of chemical from a certain company:

$$0.98, 0.94, 0.95, 1.01, 0.94, 0.99, 1.01, 0.96, 0.97, 0.95.$$

(For these data $\bar{x} = 0.97$, $\sum (x - \bar{x})^2 = 0.0064$, where $\bar{x} = \sum x/10$.) Use (i) a sign test and (ii) a t-test to decide whether or not there is evidence that the median weight of the bags of chemical supplied by the company is less than 1 kg. What conditions must be satisfied for the t-test to be strictly valid? If these conditions were satisfied, comment briefly on why the t-test would be superior to the sign test.   (*IOS*)

17. Six sets of twin monkeys of a particular type are given a dexterity test. One monkey in each twin pair has the right eye blindfolded during the test, whilst the other monkey has the left eye blindfolded during the test. The following times(s) were taken by the monkeys to perform the dexterity test.

Twin pair	A	B	C	D	E	F
Right eye blindfolded	196	208	172	193	181	202
Left eye blindfolded	192	210	186	199	179	210

Test whether, for this type of monkey, those with the right eye blindfolded show less dexterity than those with the left eye blindfolded. Discuss any assumptions that you have made.

18. (a) Explain carefully, as you would if writing for a non-statistician, the differences between the paired t-test, the sign test and the Wilcoxon signed-rank test. Include in your account

  (i) the various assumptions underlying the tests,
  (ii) the procedures for carrying out the tests,
  (iii) the uses for the tests.

(b) Two different methods for assessing candidates' knowledge of certain test material were compared by applying both methods to twelve randomly chosen candidates, with the following results. (For each candidate the methods were applied in random order.)

	A	B	C	D	E	F	G	H	I	J	K	L
Method I score	46	48	55	46	89	45	44	42	50	63	41	49
Method II score	42	44	40	36	59	42	39	45	41	47	43	41

  (i) The investigator suspects that such scores are not normally distributed; however, he considers that the distributions of the scores for the two different methods differ only in location. State which test you would advise him to use and carry it out.

  (ii) The investigator considers that the score of 89 for candidate E is an outlier, and hence wishes to ignore both scores for candidate E. What do you advise? *(IOS)*

19. Packets of a certain variety of sweets are supposed to have a mean mass of 30 g. Nine such packets had masses of 32, 26, 23, 33, 28, 27, 30, 29, 24 g. Assuming that the masses are normally distributed, test whether the mean mass of these packets differs from the stated mean.

17.18 COMPUTING EXERCISES

1. *Pebble sampling data.* Test whether the mean of the judgement samples and the mean of the random samples (Exercises 3.11(1)) differ from the true mean weight of the pebbles.

   Also test whether the mean of the judgement samples differs from the mean of the random samples.

2. *Ape-man problem.* Simulate the test described in Section 17.3.
   (i) Assume the skulls under test come from a normal population with $\mu = 12$ and $\sigma = 0.8$. Repeat the test 1000 times, by calculating $z = (x - 12)/0.8$, rejecting the null hypothesis when $z < -1.96$ or $z > 1.96$. Verify that approximately 5% of trials lead to rejection of the null hypothesis.
   (ii) Repeat as in (i) but set $\mu = 13$. Find the proportion of trials that lead to rejection of the null hypothesis.
   (iii) Repeat as in (ii) but set $\mu = 14$.

1 Word length comparisons (Data from earlier project)

Aim. To examine whether there is a significant difference in the mean length of words used by different authors.

Collection of data. As for Project 4.11(2).

Analysis. (a) Call the authors A and B and their works 1 and 2. Test whether the means of A1 and A2 differ and whether the means of B1 and B2 differ.

(b) Combine samples 1 and 2 for each author and test the mean for A against the mean for B.

Note. If sentence length instead of word length is considered for this project, the analysis is best done using log (sentence length) since this is usually approximately normally distributed.

2 Proportion of hatchback cars (Data from earlier project)

Aim. To investigate whether there has been a change in the proportion of hatchback cars.

Collection of data. As for Project 7.14(3).

Analysis. Split the data into two or more age-groups. Find the proportion of hatchback cars in each group. Each group should contain at least 100 cars. Consider the oldest and the youngest of the age-groups: test whether the proportions in these two groups differ.

Note. Cars of different colours could be studied, to see if some colours are more popular in one year than in another.

SUMMARY

Test of single observation. If an observation X is drawn from a normal distribution with variance 1, and whose mean is thought to be 0, then with probability 0.95 the value of X lies between -1.96 and $+1.96$. Otherwise it is 'unlikely' and the null hypothesis 'mean $= 0$' would be rejected.

Test of mean. When X follows $\mathcal{N}(\mu, \sigma^2)$, σ^2 being known though μ is not, then $Z = (\overline{X} - \mu)/(\sigma/\sqrt{n})$ follows $\mathcal{N}(0, 1)$; $\overline{X}$ is the mean of a random sample of n observations on X. If Z is outside $(-1.96, +1.96)$ the null hypothesis 'mean $= \mu$' would be rejected.

Test of difference of means. If X_1 follows $\mathcal{N}(\mu_1, \sigma_1^2)$ and X_2 independently follows $\mathcal{N}(\mu_2, \sigma_2^2)$ then $\overline{X}_1 - \overline{X}_2$ is

$$\mathcal{N}((\mu_1 - \mu_2), (\sigma_1^2/n_1 + \sigma_2^2/n_2)),$$

which allows us to test a null hypothesis '$\mu_1 = \mu_2$' with the help of random samples of n_1 and n_2 observations respectively on X_1 and X_2. Supposing $\bar{x}_1, \bar{x}_2$ to be the values of the sample means, the test statistic is

$$Z = \frac{(\bar{x}_1 - \bar{x}_2) - 0}{\sqrt{\sigma_1^2/n_1 + \sigma_2^2/n_2}},$$

which is $\mathcal{N}(0, 1)$.

Large-sample tests use the normal approximations developed in Chapter 16. A particularly useful case is for proportions.

Normal approximations can be found for several of the test statistics calculated in nonparametric testing.

If data come in pairs, such as measurements on the same units before and after some activity, the two measurements will not be independent. However, the differences of the two measurements in each pair may often be treated as a single sample of normally distributed data.

If the *variance* σ^2 is *not known*, but has to be estimated by s^2 from the data, we cannot use $Z = (\overline{X} - \mu)/(\sigma/\sqrt{n})$ in tests but must replace it by $t = (\overline{X} - \mu)/(s/\sqrt{n})$ which follows Student's t-distribution with $n - 1$ degrees of freedom. The values of critical points from a t-table must be used instead of the unit normal values 1.96, etc.

18 ESTIMATION OF INTERVALS AND PARAMETERS

OVERVIEW

Samples of data have been used to give point estimates of parameters, such as the mean and the variance, in earlier chapters. We now study interval estimation; confidence intervals can provide much additional information. Normal distribution theory is the basis for several types of exact and approximate confidence interval. We also consider general methods of estimating parameters: the methods of moments and of maximum likelihood.

18.1 INTRODUCTION

In Chapter 11, we considered how to estimate μ, the true but unknown value of the mean in a population. A random sample of n observations from the population was available, the values of the observations being $x_1, x_2, \ldots, x_n$. Our method was to calculate the sample mean,

$$\bar{x} = \frac{1}{n} \sum_{i=1}^{n} x_i,$$

and use this as an estimate of μ. We call the value of $\bar{x}$, from a sample, a *point estimate* of μ. Now of course the actual numerical value we obtain for $\bar{x}$ depends on which particular members of the population fall in the sample; we know that we shall obtain a different value of $\bar{x}$ from a different sample (from the same population). That is to say, we know that our point estimate will vary from sample to sample, and so we cannot say how near to μ a point estimate from a single sample will be.

What we can say is that $\overline{X}$ (using, as usual, a capital letter for the random variable) provides an *unbiased estimator* of μ, $E[\overline{X}] = \mu$, so that if many samples are taken, each of size n, from the same population, the mean of the sampling distribution of $\overline{X}$ will be μ. In practice, we usually have a single sample of n observations, from which we have calculated the mean $\bar{x}$, and we want to extract as much information as possible from it. For example, a manufacturer of a food product which is packed automatically has to set his machinery to give a correct mean weight μ. After servicing or resetting the machine, he wishes immediately to take a sample of packets from his production line and check that the mean weight is satisfactory. He does not want to wait for many samples, and many values of $\bar{x}$; he wants to find out about μ from this one sample.

Or suppose that a doctor is giving a drug to a patient in order to control the level of the patient's blood sugar. The doctor knows that in order to avoid danger to the patient the true blood sugar level has to be within certain limits. He obviously cannot take a very large number of blood samples: he must be satisfied with one sample of n observations, n being perhaps quite a small number, and on the basis of this he must decide whether the true mean blood sugar level in the whole bloodstream is within the acceptable limits.

The methods we shall set up in this chapter will make it possible to give an *interval estimate* for μ: using our random sample we find upper and lower limits for μ, and hence an interval (i.e. a range of values) which is very likely to contain μ. Because we can say just how likely our interval is to contain the true value of μ, this type of estimate gives more information than the previous point estimate.

When estimating proportions, we have a similar problem. So far, a true proportion π in a population has been estimated by calculating the sample proportion p; this gives us a point estimate of π. Suppose, for example, that we are carrying out a survey to see what proportion of residents in a village would use a local bus service if it ran at a certain time, say at 8.30 a.m. We can either ask every resident or, much more practically, we can base an estimate on the proportion found in a sample. Obviously here too it will be very useful to be able to give limits that are very likely to contain π. These would enable us to know whether the bus is likely to be nearly empty or very full.

18.2 CONFIDENCE INTERVAL FOR MEAN OF A NORMAL DISTRIBUTION

If we take a random sample of n observations from a normal distribution whose mean is μ and whose variance is σ^2, the sample mean $\overline{X}$ will follow the $\mathcal{N}(\mu, \sigma^2/n)$ distribution. We have already carried out a significance test (p. 413) of the null hypothesis that the true mean really is μ, a particular specific value, assuming that σ^2 is known.

Now we shall use the same random sample in a different way. We calculate $\bar{x}$, we know σ^2 and, of course, n; but this time we do *not* have a specific value of μ to test. Instead we shall ask what values μ could take that would make our sample a likely one.

In sampling from a normal distribution whose mean takes some unknown value μ, it is still true that $\overline{X}$ will follow $\mathcal{N}(\mu, \sigma^2/n)$ and therefore that $Z = \sqrt{n}(\overline{X} - \mu)/\sigma$ will be $\mathcal{N}(0, 1)$. Using the table of the standard normal distribution, we can then say that

$$P\left(-1.96 \leqslant \frac{\overline{X} - \mu}{\sigma/\sqrt{n}} \leqslant +1.96\right) = 0.95. \qquad (18.1)$$

If we are given the values of σ^2 and n, we can find the set of values of μ that satisfy the inequality within the brackets. Now

$$-1.96 \leqslant \frac{\overline{X} - \mu}{\sigma/\sqrt{n}} \leqslant +1.96$$

if and only if

$$-1.96\sigma/\sqrt{n} \leqslant \overline{X} - \mu \leqslant 1.96\sigma/\sqrt{n},$$

since we have multiplied through the inequality by a positive number. Rearranging, $\overline{X} - 1.96\sigma/\sqrt{n} \leqslant \mu \leqslant \overline{X} + 1.96\sigma/\sqrt{n}$. The reader may understand the derivation better if he works first with the left-hand inequality, then with the right-hand one, and finally combines them. So statement (18.1) is equivalent to

$$P(\overline{X} - 1.96\sigma/\sqrt{n} \leqslant \mu \leqslant \overline{X} + 1.96\sigma/\sqrt{n}) = 0.95. \qquad (18.2)$$

The inequality in brackets in (18.1) is true with probability 0.95, and hence the inequality for μ inside the brackets in (18.2) is also true with probability 0.95. Any μ within the limits given in (18.2) is acceptable, at this level of probability, as the true mean in the normal distribution from which the sample of n has been drawn.

We call the interval in the brackets in (18.2) a 95% *confidence interval* for μ, and the two ends of the interval the 95% *confidence limits*. Note that μ is fixed (though unknown), and it is the interval which is the random variable: for if we take a new sample of n from the same distribution μ does not change, but $\bar{x}$ does and therefore so does the interval. From any one sample, with mean $\bar{x}$, we claim that the interval given by (18.2) contains the true (but unknown) μ. From many samples we would obtain many intervals; if we claimed in each case that the interval contained μ then in the long run 95% of such claims would be true.

Example 18.1 Confidence interval for mean

A machine which packs sugar has for a long time given a normal distribution of weights of filled packets, and the standard deviation of weight has been 2.5 grams. It is adjusted to give a new metric size of pack, and 20 of the new packets are weighed. Their mean is 1002 grams. Set up 95% confidence limits for the true mean weight after adjustment.

The best point estimate of μ is of course $\bar{x} = 1002$ grams. We can calculate 95% confidence limits to μ in the form $\bar{x} \pm 1.96\sigma/\sqrt{n}$, namely

$$1002 \pm 1.96 \times 2.5/\sqrt{20} = 1002 \pm 4.90/4.472 = 1002 \pm 1.1.$$

So with probability 0.95 of being correct, we can claim that

$$1000.9 \leqslant \mu \leqslant 1003.1.$$

Sometimes we wish it to be more than 95% probable that our interval really does contain μ. When setting up (18.1), we took the 95% central area of $\mathcal{N}(0,1)$; we could equally well have taken 99% or 99.9%. If we wish to set up 99%

confidence limits for the true μ, we replace 1.96 by 2.576 in (18.1), and hence in (18.2) also.

For the previous example, 99% limits are

$$1002 \pm 2.576 \times 2.5/\sqrt{20} = 1002 \pm 6.44/4.472 = 1002 \pm 1.44.$$

So with probability 0.99 of being correct, we assert that $1000.56 \leqslant \mu \leqslant 1003.44$. The more nearly certain we want to be that our interval really does contain μ, the wider we have to make the interval – as we should expect. If we require as strong a confidence level as 99.9%, we take 3.291 instead of 1.96 in (18.2), and this leads to the interval $1000.16 \leqslant \mu \leqslant 1003.84$.

Example 18.2 Sample size

It is required to estimate μ, the mean length of a population of mass-produced screws, to within limits of $\pm \frac{1}{2}$ mm. What is the minimum sample size needed to achieve this, if the standard deviation of length is known to be 1.2 mm, and the probability attached to the limits is to be 95%?

For 95% confidence limits, μ is in the interval $\bar{x} \pm 1.96\sigma/\sqrt{n}$, and the last term in this expression is to be not more than $\frac{1}{2}$. If $1.96\sigma/\sqrt{n} = \frac{1}{2}$, and $\sigma = 1.2$, then $2 \times 1.96 \times 1.2 = \sqrt{n} = 4.704$, so that $n = 22.1$. If we take $n = 22$, the limits will be slightly wider than $\pm\frac{1}{2}$: we must round up to the next integer for n to keep within the specified size of limits. Therefore n must be at least 23.

18.3 CONFIDENCE INTERVAL FOR POPULATION MEAN BASED ON A LARGE SAMPLE

In Chapters 16 and 17, we have used the central limit theorem. This says that if we take a large sample $(X_1, X_2, \ldots, X_n)$ from (almost) *any* distribution the mean $\overline{X}$ will follow a distribution which is approximately $\mathcal{N}(\mu, \sigma^2/n)$, where μ and σ^2 are the mean and variance in the population from which the random sample of size n has been drawn. The quality of the approximation improves as n increases and is better when the original X-distribution is symmetrical rather than skew. We have carried out approximate significance tests using this theorem, and we can also use it to give *approximate* confidence intervals.

Let us suppose that a large sample is available, of n observations from a non-normal distribution of X. The statement in brackets in (18.2) can still be made, but is now true with probability *approximately* 0.95, rather than exactly. For this reason, it is often said that approximate 95% confidence limits to μ are $\bar{x} \pm 2\sigma/\sqrt{n}$, 1.96 being rounded to 2 to ease arithmetic since we know the result is going to be approximate anyway.

Example 18.3 Confidence interval for mean (large sample)

A particular component in a transistor circuit has a lifetime which is known to follow a skew distribution. A random sample of 250 components from a week's production gives an average lifetime of 840 hours, and the variance of lifetimes is 483 hours^2. Find approximate 95% confidence limits to the true mean lifetime in the whole population of the product.

We are not actually told σ^2 for the whole population, but in a large sample such as this little is lost in the approximation by using the sample variance instead. So for approximate 95% limits we shall use $\bar{x} \pm 2s/\sqrt{n}$, i.e. $840 \pm 2\sqrt{483/250}$ or $840 \pm 2\sqrt{1.932} = 840 \pm 2 \times 1.39$ or 837.2 to 842.8 hours. A sample as large as this can give very good precision, perhaps more than is really needed; but the large sample size is necessary so that the distribution of $\overline{X}$ can be approximated using the central limit theorem.

18.3.1 Exercises

1. A random sample of one observation x is chosen from a normal distribution with unknown mean μ and variance 4. Give a 95% confidence interval for μ.

2. A random sample of 16 observations, with mean $\bar{x}$, is chosen from a normal distribution with unknown mean μ and variance 25. Give a (i) 95%, (ii) 99%, (iii) 99.9% confidence interval for μ.

3. A random sample of 100 packs of apples with a professed weight of 1 kg were found to have mean 1020 g; the estimated variance from the sample was 144 g^2. Give an approximate 95% confidence interval for the mean of the apple packs.

18.4 CONFIDENCE INTERVAL FOR A POPULATION PROPORTION

We have estimated π, the true proportion of some special type contained in a population, by calculating $P = R/n$, the proportion of that type in a random sample of size n drawn from the population. The value of P gives an unbiased estimate of π. In Section 17.9 we saw that there was a normal approximation to the distribution of P which allowed us to test hypotheses about the true π. The same normal approximation allows approximate confidence limits to be set to π on the basis of a sample estimate p.

Since the distribution of P is approximately $\mathcal{N}(\pi, \pi(1-\pi)/n)$, we may develop an argument of the same type as that which led to (18.2) to show that an approximate 95% confidence interval for π is

$$p - 2\sqrt{\frac{p(1-p)}{n}} \leqslant \pi \leqslant p + 2\sqrt{\frac{p(1-p)}{n}}. \tag{18.3}$$

We have approximated 1.96 by 2, as in the previous section; and we do not know π so have had to replace $\sqrt{\pi(1-\pi)/n}$ by $\sqrt{p(1-p)/n}$. As the approximation (18.3), like that of the previous section, is only satisfactory in large samples, neither of these processes makes it any worse. (In fact the unbiased estimate of $\pi(1-\pi)/n$ is $p(1-p)/(n-1)$, but in large samples the use of n rather than $n-1$ obviously makes very little difference.)

Example 18.4 Confidence interval for proportion

Out of 300 households in a large town, 123 have microwave ovens. Set approximate 95% limits to the true value of the proportion of households with microwave ovens in the whole town.

Assuming our sample of $n = 300$ households to be a random selection from the whole town, the point estimate of the proportion is $p = 123/300 = 0.41$. The value of $\sqrt{p(1-p)/n}$ is

$$\sqrt{\frac{0.41 \times 0.59}{300}} = \sqrt{0.000\,806\,33} = 0.0284,$$

so that approximate 95% limits to π are $0.41 \pm 2 \times 0.0284$, i.e. 0.41 ± 0.057, or 0.353 to 0.467.

Example 18.5 Sample size

What sample size should be taken in order to estimate to within ± 0.05 the proportion of supporters for a political party among the whole population? It is thought that the proportion is about 0.4, and the confidence level attached to the estimate is to be 95%.

Note that since the width of limits depends on p as well as n, we need to make a guess at p before we can estimate n. The guess need not be highly accurate, but the value of $\sqrt{p(1-p)}$ does have a maximum at $p = \frac{1}{2}$, and we will need larger samples for these middle values of p than for those nearer 0 or 1.

Using $p = 0.4$,

$$\sqrt{\frac{p(1-p)}{n}} = \sqrt{\frac{0.24}{n}}.$$

We require that $2\sqrt{0.24/n}$ shall not be greater than 0.05. Setting

$$2\sqrt{\frac{0.24}{n}} = 0.05,$$

we have

$$\sqrt{\frac{0.24}{n}} = 0.025 \quad \text{or} \quad \frac{0.24}{n} = 0.000\,625,$$

which gives

$$n = \frac{0.24}{0.000\,625} = 384.$$

This is the smallest sample size that fulfils the requirement.

18.4.1 Exercises

1. A random sample of 250 of the 5000 students in a university contained 30 left-handed students. Give an approximate 95% confidence interval for the proportion of left-handed students in the university.

2. A random sample of 200 voters in a constituency included 110 who said they would vote for Mr A. Assuming all the 15 000 voters in the constituency would vote, give an approximate 95% confidence interval for the number who would vote for Mr A.

18.5 CONFIDENCE INTERVALS FOR MEANS OF NORMAL DISTRIBUTIONS WHEN THE VARIANCE σ^2 IS UNKNOWN AND THE SAMPLE SIZE IS SMALL

When following through the argument at the beginning of this chapter without knowing σ, the estimated standard deviation must be written for σ everywhere. When the sample size is small the ratio

$$\frac{\overline{X} - \mu}{s/\sqrt{n}}$$

has a t-distribution, with $n-1$ degrees of freedom, rather than a normal distribution. The statement equivalent to (18.1) is

$$P\left(-t_{(n-1,0.05)} \leqslant \frac{\overline{X} - \mu}{s/\sqrt{n}} \leqslant t_{(n-1,0.05)}\right) = 0.95$$

where $t_{(n-1,0.05)}$ is found from Table A3 by looking along the row corresponding to degrees of freedom $(n-1)$ until we reach the column headed '$P = 0.05$'. The appropriate confidence interval is

$$\bar{x} - \frac{s}{\sqrt{n}}t_{(n-1,0.05)} \leqslant \mu \leqslant \bar{x} + \frac{s}{\sqrt{n}}t_{(n-1,0.05)}. \tag{18.4}$$

Example 18.6 Confidence interval for mean

Fifteen tomato plants have their heights measured after a period of growth in a greenhouse. The mean height in the sample is 83 cm and the standard deviation of height in the sample is 5.8 cm. Set 95% confidence limits to the mean height in the population from which the random sample of 15 was drawn, assuming height to be normally distributed in the population.

Since $n = 15$, we need the 5% point of $t_{(14)}$. From the table this is 2.145. We must use this value instead of 1.96 in (18.2), and we must also write s (which is 5.8) instead of σ. The 95% limits for μ now become $83 \pm 2.145 \times 5.8/\sqrt{15}$, i.e. $83 \pm 12.441/3.873$ or 83 ± 3.21, which are 79.8 to 86.2.

Small samples *from a normal distribution* can be used in this way; but if a small sample is all that is available and we do *not* know that it comes from a normal distribution, it cannot (except in one or two very special cases) be used at all to provide limits. Then the only information that could be extracted would be the point estimate, $\bar{x}$, of μ.

As n becomes large, s becomes a steadily better estimate of σ, and we can use the large-sample theory of Section 18.3. The density function for $t_{(n-1)}$ approaches that for $\mathcal{N}(0, 1)$ and, mathematically speaking,

$$\lim_{n \to \infty} t_{(n-1)} = \mathcal{N}(0, 1).$$

Hence the t-table contains '∞' as its last row, the percentage points being those of $\mathcal{N}(0, 1)$. This is often a neat way to avoid printing a separate table. Comparing this last row with percentage points for other degrees of freedom indicates how close any particular $t_{(n-1)}$ curve is to $\mathcal{N}(0, 1)$.

Example 18.7 Confidence interval for mean of paired comparisons

A paired-comparison experiment was made to compare the effect of two processes A and B on freezing and wrapping meat (see Example 17.16, p. 428). The mean difference $\bar{d}$, between A and B, in number of days to spoilage, was 25. The estimated variance s_D^2, derived from the 10 differences, was 1186.44. Find a 95% confidence interval for the unknown mean difference of the processes A and B assuming that the observed differences were normally distributed.

The 95% confidence interval has the form $\bar{d} \pm (s_D/\sqrt{n})t_{(n-1, 0.05)}$. Hence the interval is

$$25 \pm (\sqrt{1186.44/10}) \times 2.262 = 25 \pm 24.64 = (0.36, 49.64).$$

Example 18.8 Confidence interval for difference of means of unpaired samples

In a comparison of 10 new and 8 experienced subjects in an experiment, the mean reaction times, in hundredths of a second, of the two groups were 3.0 ($\bar{x}_1$) and 2.7 ($\bar{x}_2$) respectively (see Example 17.17, p. 432). The pooled estimate of variance s^2 was found to be 0.05. Find a 95% confidence interval for the difference between the unknown mean reaction times of the two groups.

The 95% confidence interval has the form

$$(\bar{x}_1 - \bar{x}_2) \pm t_{(n_1+n_2-2, 0.05)} \times \sqrt{s^2(n_1 + n_2)/n_1 n_2}.$$

Hence the interval is

$$(3.0 - 2.7) \pm (2.120)\sqrt{(0.05)(18)/80} = 0.3 \pm 0.225 = (0.075, 0.525).$$

Example 18.9 Lognormal data

We have seen already, from the dot plots in (i) and (ii) on p. 364, that the logarithms (to base 10 – but any base would have served) of the failure-time data may be analysed as samples from normal distributions.

(iii) In logarithmic units,

$$n_A = 11, \quad \bar{x}_A = 1.34, \quad s_A^2 = 0.37832,$$
$$n_B = 12, \quad \bar{x}_B = 1.14, \quad s_B^2 = 0.22229.$$

Hence $\sqrt{s_A^2/n_A} = 0.1855$ and $\sqrt{s_B^2/n_B} = 0.1361$. The critical values at the 5% level for $t_{(10)}$ and $t_{(11)}$ respectively are 2.228 and 2.201 (Table A3). The 95% confidence limits for A are $1.34 \pm 2.228 \times 0.1855$, i.e. 1.34 ± 0.41 or $(0.93, 1.75)$. Those for B are $1.14 \pm 2.201 \times 0.1361$, i.e. 1.14 ± 0.30 or $(0.84, 1.44)$.

In order to understand the results better, these limits should be transformed back to the original failure-time units (hours) by taking antilogs. For A we find 8.5 to 56.2 hours, and for B we have 6.9 to 27.5 hours. These are still 95% limits, but of course are not symmetrical about the antilogs of the mean values on this scale.

(iv) An approximate method for looking at the difference between 'average' failure times is to use a two-sample t-test on their logarithms. First we need to check the ratio of the variances (p. 580): $s_A^2/s_B^2 = 1.70$, which is not significant as $F_{(10,11)}$, so it is valid to calculate a pooled $s^2 = (10s_A^2 + 11s_B^2)/21 = 0.29659$. The null hypothesis is that the means (of the logarithms of times) are equal and the alternative hypothesis is that they are not. On the null hypothesis,

$$\frac{1.34 - 1.14}{\sqrt{s^2(1/11 + 1/12)}}$$

follows $t_{(21)}$; its value is

$$\frac{0.20}{\sqrt{0.05168}} = \frac{0.20}{0.227} = 0.88,$$

giving no suggestion of any significant difference.

18.5.1 Exercises

1. Ten observations $x_1, x_2, \ldots, x_{10}$ were chosen at random from a normal population; $\sum_1^{10} x_i = 10.7$ and $\sum_1^{10} (x_i - \bar{x})^2 = 3.24$. Find a 95% confidence interval for the population mean.

2. Information was collected from a random sample of 20 of the 2000 employees in a factory about the amount of money they were carrying. The mean amount was found to be £5.25 and the estimated standard deviation was £2. On the assumption that the amount of money per person is normally distributed, find a 95% confidence interval for the total amount of money carried by the 2000 employees of the factory.

3. An experiment was carried out to compare the corrosion of two grades, A and B, of zinc plate. Corrosion was measured by determining the loss in mass (grams per square foot) over a period of 12 months. The recorded losses for pairs of plates put in a random choice of locations were:

Pair	1	2	3	4	5	6
A	10.8	4.3	5.4	10.5	2.6	2.6
B	13.0	3.4	4.5	8.2	3.4	2.3

Find a 99% confidence interval for the mean difference in corrosion of the two grades of plate.

4. Standard amounts (S) of fertilizer were applied to eight plots of oilseed, while decreased amounts (D) were applied to four plots. The resulting yields (tonnes per hectare) were measured:

S	1.62	1.80	1.80	1.90	1.88	2.02	1.92	1.76
D	1.52	1.80	1.90	1.66				

Find a 95% confidence interval for the difference between the mean yields of the plots receiving a standard amount of fertilizer and those receiving a decreased amount.

18.6 INVESTIGATION OF A NULL HYPOTHESIS

We have stressed that in a significance test one cannot 'accept' the null hypothesis; the outcomes of a significance test are *either* to reject the null hypothesis and accept the alternative hypothesis *or* not to reject the null hypothesis. But, as we showed in Section 17.14, in some circumstances it is the natural null hypothesis that is the hypothesis of interest. Let us take up again the example given there: a cigarette company would like to counter the claim that if a woman smokes during pregnancy the weight at birth of her baby is reduced.

Let us assume that the cigarette company has data which show that the mean weight at birth of babies born to 100 mothers who smoked during pregnancy was 3200 g, while the mean weight of babies born to 100 mothers who did not smoke was 3300 g. We shall also assume that the standard deviations of both sets of weights were identical and equalled 600 g. If we make a significance test of the difference between the two mean weights, using a null hypothesis that they are identical, we obtain a z-value of 1.18 and conclude that we cannot reject the null hypothesis. But this does not allow the company to assert that they have shown that a mother's cigarette smoking has no effect on the birth-weight of her baby.

What is informative is a confidence interval for the difference in mean birth-weight of babies born to mothers who smoked and those who did not smoke.

A 95% confidence interval for this difference equals

$$(3200 - 3300) \pm 1.96\sqrt{(2 \times 600^2)/100} = -100 \pm 166 = (-266, +66).$$

We interpret this to mean that there is strong evidence that the consequence of smoking by the mother is a change in mean birth-weight of the baby which is between a reduction of 266 g and an increase of 66 g. If the company could say that a reduction in mean birth-weight of 266 g is negligible, then it could reasonably assert that smoking has had no detrimental effect; but if, as is likely to be the case, such a reduction is considered to be a serious decrease, then though the case is 'not proven' the weight of the evidence is against the cigarette company. What the company needs to do is to collect more data and obtain a narrower confidence interval. For interpretation one needs to know what is the maximum size of decrease that might be regarded as negligible; it might be 10 g in our example. Increases might not be regarded as detrimental. If the company collected more data and obtained a 95% confidence interval equal to $(-8, 90)$ then they could regard their case as shown, but if the interval were $(-110, -20)$ they would have to accept that the evidence was against them.

18.7 CONFIDENCE INTERVAL FOR VARIANCE OF A NORMAL DISTRIBUTION

In Section 18.2, we assumed that we knew the value of σ^2, the variance in a normal distribution, and used it to help in working out a confidence interval for the mean μ. In practice, if we do not know μ we are very unlikely to know σ^2 either, so that the method of Section 18.5 for μ is much more often needed in real life.

It is also possible to calculate a confidence interval for the true value of σ^2 from a sample of n observations. Given a random sample $(x_1, x_2, \ldots, x_n)$ from $\mathcal{N}(\mu, \sigma^2)$, we first find the sample mean and variance,

$$\bar{x} = \frac{1}{n}\sum_{i=1}^{n} x_i \quad \text{and} \quad s^2 = \frac{1}{(n-1)}\sum_{i=1}^{n}(x_i - \bar{x})^2.$$

A result from more advanced distribution theory is needed: it uses the chi-squared distribution (Section 19.2), which arises from taking the squares of normal deviates $(x_i - \bar{x})$. It can be shown that $(n-1)s^2/\sigma^2$, which equals $\sum_{i=1}^{n}(x_i - \bar{x})^2/\sigma^2$, follows the chi-squared distribution with $n-1$ degrees of freedom.

Denoting by χ_L^2 and χ_U^2 the lower and upper $2\frac{1}{2}\%$ points of χ^2 with $n-1$ degrees of freedom, which are given in Table A4, we can say that

$$P\left(\chi_L^2 \leqslant \frac{(n-1)s^2}{\sigma^2} \leqslant \chi_U^2\right) = 0.95. \tag{18.5}$$

This arises by the same sort of argument as we have already seen earlier in this chapter, and in the same way we shall wish to rearrange inequality (18.5) to make σ^2 the subject. From the first part of the inequality,

$$\sigma^2 \chi_L^2 \leqslant (n-1)s^2$$

and so

$$\sigma^2 \leqslant (n-1)s^2/\chi_L^2.$$

From the second part,

$$\sigma^2 \chi_U^2 \geqslant (n-1)s^2$$

and so

$$\sigma^2 \geqslant (n-1)s^2/\chi_U^2.$$

The chi-squared random variable exists only in the range $(0, \infty)$ since it is a sum of squares; therefore χ_L^2 and χ_U^2 must both be positive and there is no need to reverse the signs in the inequality.

Now we may combine the two statements, obtaining an inequality which is true with the same probability as (18.5), namely

$$P\left(\frac{(n-1)s^2}{\chi_U^2} \leqslant \sigma^2 \leqslant \frac{(n-1)s^2}{\chi_L^2}\right) = 0.95 \qquad (18.6)$$

giving a 95% confidence interval for the true value of σ^2 based on the given random sample.

The expression $(n-1)s^2$ is the corrected sum of squares (p. 32), and this is what we require for calculation; an estimate of s^2 itself is not needed.

Example 18.10 *Confidence interval for variance*

In Example 18.6, the mean height of a random sample of 15 plants was found, and also the standard deviation in the sample. We have already used this information to give a 95% confidence interval for the true value of μ. In order to do the same for σ^2, we require $(n-1)s^2 = 14 \times (5.8)^2 = 470.96$, and also the lower and upper $2\frac{1}{2}$% points of χ^2 with 14 degrees of freedom; these are 5.63 and 26.12 (Table A4).

Using (18.6), we can say

$$P\left(\frac{470.96}{26.12} \leqslant \sigma^2 \leqslant \frac{470.96}{5.63}\right) = 0.95,$$

i.e. $P(18.0 \leqslant \sigma^2 \leqslant 83.7) = 0.95$. A 95% confidence interval for σ^2 is thus 18.0 to 83.7. It is a wide interval; this is usually the case, since estimates of variances are themselves very variable.

This interval is for the true value of the variance, not the standard deviation. It is therefore in squared units of the measurement. This is what we normally give, because the necessary distribution theory involves σ^2. But we may, if we wish, take square roots and say that the interval 4.2 to 9.1 contains the true standard deviation σ with probability 0.95 *approximately*.

Note (1) If an interval with a different probability is required, the appropriate lower and upper χ^2 points must be looked up; for 99% probability in this example they would be 4.07 and 31.32, giving an interval from 15.0 to 115.7.

(2) In the event that μ is known, the unbiased estimate of variance is

$$\frac{1}{n}\sum_{i=1}^{n}(x_i-\mu)^2.$$

Therefore $\sum_{i=1}^{n}(x_i-\mu)^2$ is used as the sum of squares, instead of $\sum_{i=1}^{n}(x_i-\bar{x})^2$, and when this is done the χ^2 distribution must be looked up with n degrees of freedom, not $n-1$.

(3) Unless we sample from a *normal* distribution, we cannot make statements about variances; there are no reliable 'approximate' results.

18.7.1 Exercises

1. Find a 95% confidence interval for the variance in the population from which the data of Exercise 18.5.1(1) were drawn.

2. A random sample of 12 observations from a normal distribution gave the values 11.4, 10.8, 12.1, 10.6, 11.3, 10.5, 12.2, 11.0, 12.5, 10.7, 11.9, 11.6. Find (i) 95%, (ii) 99% confidence intervals for the mean and for the variance.

18.8 GENERAL METHODS OF ESTIMATING PARAMETERS

We have already given many examples of estimating the parameters of a statistical distribution using sample data. In Example 18.4 we estimated the proportion of households with microwave ovens in a town by equating it to the proportion 123/300 found in a sample. In Example 18.6, we modelled the heights of tomato plants by a normal distribution. We equated the unknown mean, μ, and variance, σ^2, to the mean, $\bar{x}$, and variance, s^2, respectively of the sample data. The general principle we have been using is to identify some property (i.e. a proportion or a mean or a variance) of the population or statistical distribution and estimate it by the value of the same property in the sample. There is an element of arbitrariness about this, depending on which property we choose to study. In fact, all the cases we mention above may be regarded as examples of the **method of moments**, which will be described in Section 18.8.1.

On p. 255, we stated a further principle in methods of estimation, the desirability of using *unbiased* estimators; this must be kept in mind when applying general methods.

18.8.1 The method of moments

The *uncorrected* (or 'raw') moments of a distribution were defined in Section 14.8; the sth uncorrected moment $\mu'_s = \mathrm{E}[X^s]$. Thus the mean is the first moment. Often the mean of a probability distribution is a simple function of the parameters in the mass or density function, especially so when there is only one parameter involved.

The **method of moments** was introduced by K. Pearson in the late nineteenth century. When there is only one parameter to be estimated, the procedure is to equate the sample mean $\bar{x} = \sum_{i=1}^{n} x_i/n$ to the mean in the population, μ. This leads to the *method-of-moments estimator* of the single parameter in the mass or density function, which will be denoted by the symbol $\sim$, so that $\tilde{\mu}$ represents the method-of-moments estimator of μ.

Example 18.11 *Estimating the parameter of a geometric distribution*

A random sample of n observations of the value of r (see p. 177) is available, from the geometric distribution

$$P(r) = \pi(1 - \pi)^{r-1} \qquad (0 < \pi < 1).$$

The parameter π is not known, and is to be estimated from the data.

The theoretical value of the mean in this distribution is $\mu = 1/\pi$. Writing

$$\tilde{\mu} = \sum_{i=1}^{n} r_i/n = \bar{r},$$

we obtain

$$1/\tilde{\pi} = \tilde{\mu} = \bar{r} \quad \text{or} \quad \tilde{\pi} = 1/\bar{r}.$$

Example 18.12 *Estimating the parameter of an exponential distribution*

There is available a random sample of n observations of the variable x whose probability density function is

$$f(x) = \lambda e^{-\lambda x} \qquad (\lambda > 0; \quad 0 < x < \infty).$$

Use it to find an estimate of the parameter λ.

In order to estimate λ, use the fact that $\mu = 1/\lambda$. This leads to $\tilde{\mu} = 1/\tilde{\lambda}$; and $\tilde{\mu} = \bar{x}$, the sample mean, so that finally $\bar{x} = \tilde{\mu} = 1/\tilde{\lambda}$ and $\tilde{\lambda} = 1/\bar{x}$.

Example 18.13 Estimating the mean of a normal distribution with unit variance

A random sample of n observations has been taken from the normal distribution $N(\mu, 1)$, with known variance equal to 1 but with unknown mean. The probability density function is thus

$$f(x) = \frac{1}{\sqrt{2\pi}} \exp\left(-\frac{1}{2}\{x - \mu\}\right)^2$$

and the only parameter to be estimated is μ. Therefore immediately $\tilde{\mu} = \bar{x}$.

18.8.1.1 Estimating more than one parameter

The method of moments uses the *raw* moments; in the sample these are $\sum_{i=1}^{n} x_i^s/n$, and the corresponding theoretical values in the population are $\mu_s' = E[X^s]$. The method of moments can be extended very simply to more than one parameter by constructing as many equations as there are parameters to be estimated:

$$\mu_1' \equiv \mu = \bar{x}$$

$$\mu_2' = \sum_{i=1}^{n} x_i^2/n$$

$$\mu_3' = \sum_{i=1}^{n} x_i^3/n \ldots \text{and so on.}$$

Example 18.14 Estimating the two parameters of a normal distribution

Use the method of moments to find estimators for μ and σ^2 based on a random sample $x_1, x_2, \ldots, x_n$ drawn from the normal distribution $N(\mu, \sigma^2)$.

In the general normal distribution $N(\mu, \sigma^2)$, whose probability density function is

$$f(x) = \frac{1}{\sigma\sqrt{2\pi}} \exp\left(-\frac{1}{2}\left\{\frac{x - \mu}{\sigma}\right\}^2\right),$$

the first of the estimating equations gives

$$\tilde{\mu} = \bar{x}.$$

For the second, we have $\mu_2' = \sum_{i=1}^{n} x_i^2/n$. But we know (p. 323) that $\mu_2 = \mu_2' - \mu^2$, i.e. $\sigma^2 = \mu_2' - \mu^2$. Using $\tilde{\mu}$ which has already been found,

$$\tilde{\sigma}^2 = \sum_{i=1}^{n} x_i^2/n - \bar{x}^2.$$

This estimator is equal to $\sum_{i=1}^{n}(x_i - \bar{x})^2/n$. So although $\tilde{\mu}$ is unbiased, $\tilde{\sigma}^2$ is not (see Section 11.5), and in general there is no theoretical reason for method-of-moments estimators to be unbiased.

But in practice we would normally choose to use the unbiased estimator

$$s^2 = \sum_{i=1}^{n}(x_i - \bar{x})^2/(n - 1) \quad \text{(p. 255).}$$

18.8.2 The method of maximum likelihood

We now introduce another method, which is more generally applicable than the method of moments. It was introduced by R. A. Fisher in a paper to the Royal Society in 1922 and is called the **method of maximum likelihood**.

Let us consider a guessing game. An urn contains ten balls; an unknown number of these are black and the remainder are white. You are allowed to see three balls in turn. Each is selected at random from the ten balls in the urn; balls are returned after inspection and are thoroughly shuffled before each choice. Knowing the number of black balls in the sample of three, you are asked to guess how many of the ten balls are black.

Given these conditions, the number of black balls in the sample will follow a binomial distribution with $n = 3$ and π, the proportion of black balls in the urn, unknown. But because there are only ten balls altogether, π must take one of the eleven values $0, 0.1, 0.2, \ldots, 0.9, 1$. Suppose that, in fact, you have seen in the sample two black balls and one white ball. In order to derive an estimate of π, we may work out the probability $3\pi^2(1 - \pi)$ for this sample using each of the eleven possible values of π in turn. We find:

π	0	0.1	0.2	0.3	0.4	0.5	0.6	0.7	0.8	0.9	1
$3\pi^2(1-\pi)$	0	0.027	0.096	0.189	0.288	0.375	0.432	0.441	0.384	0.243	0

It is reasonable to choose as our estimate of π that value of π which, if it had been the true population value, would have been most likely to give rise to the actual sample that was obtained. From the table above, we see that that value is 0.7, giving a probability 0.441 of finding two black balls and one white ball, and we therefore guess that the number of black balls in the urn is 7.

If we had applied the method of moments, which we could still do even with only one sample from the binomial distribution, we would have set $\tilde{\mu} = n\tilde{\pi}$ equal to the observed value $r = 2$ of black balls, giving $3\tilde{\pi} = 2$ so that $\tilde{\pi} = \frac{2}{3} = 0.67$. Again we would have guessed 7, as the nearest possible value of π to the estimate $\tilde{\pi}$ is 0.7.

The probabilities which we have used in this problem, $3\pi^2(1-\pi)$, were worked out from the probability mass function of the binomial distribution, which we have previously thought of as a function of the variate value r (p. 173). But when we use the probabilities in finding an estimate of π, our view changes and we think of them as functions of π, the actual function in this case being $3\pi^2(1-\pi)$.

To emphasize this change of viewpoint, the values of the probability for the eleven possible values of π are called the **likelihoods**, and the estimate of π is found by looking for the **maximum likelihood**.

We will look more closely at the method of maximum likelihood in the following sections. In Chapters 22 and 23 we introduce another general method for estimating parameters which is based on the **principle of least squares**. The various methods often give the same, or closely related, values as estimates. Theoretical relations exist between the methods, but we do not have space to go into these.

Example 18.15 The exponential distribution

Two observations, x_1 and x_2, have been chosen at random from the exponential distribution with probability density function

$$f(x) = \lambda e^{-\lambda x} \quad (x > 0; \ \lambda > 0).$$

Find (i) the joint pdf of the sample;
(ii) the likelihood function, $L(\lambda)$;
(iii) the value of λ that maximizes the likelihood function.

(i) Since x_1 and x_2 are independent, the joint pdf is

$$f(x_1)f(x_2) = \lambda \exp(-\lambda x_1)\,\lambda \exp(-\lambda x_2)$$
$$= \lambda^2 \exp(-\lambda\{x_1 + x_2\}).$$

(ii) The likelihood function is the joint pdf interpreted as a function of λ:

$$L(\lambda) = \lambda^2 \exp(-\lambda\{x_1 + x_2\}).$$

In this function the *numerical* values of x_1 and x_2 are known.

(iii) In order to maximize L, we may calculate its derivative $dL/d\lambda$. However, the task is often simplified (especially for functions used in statistics) by noting that since $\ln L$ is a monotonic function of L the maxima of L and $\ln L$ will be at the same value of λ. If

$$L(\lambda) = \lambda^2 \exp(-\lambda\{x_1 + x_2\})$$

then

$$\ln L(\lambda) = 2\ln \lambda - \lambda(x_1 + x_2)$$

and

$$\frac{d}{d\lambda}(\ln L) = \frac{2}{\lambda} - (x_1 + x_2)$$

which is zero when

$$\frac{1}{\lambda} = \frac{x_1 + x_2}{2} = \bar{x}.$$

Hence $\hat{\lambda} = 1/\bar{x}$ is a turning point of $L(\lambda)$. Since

$$\frac{d^2}{d\lambda^2}(\ln L) = -\frac{2}{\lambda^2},$$

and $\lambda > 0$ by definition, the second derivative is certainly negative and we have a maximum of $\ln L$ and hence of L. Thus $\hat{\lambda} = 1/\bar{x}$ is the **maximum likelihood estimator** of λ. Maximum likelihood estimators are usually denoted by using the symbol $\wedge$, which helps to distinguish them from moment estimators.

18.8.2.1 Notation

In general we shall be concerned with a random sample of n observations, $x_1, x_2, \ldots, x_n$ drawn from a distribution with probability density (or mass) function $f(x)$. The random sample may be denoted by $\{x_i\}$, or as a vector $\mathbf{x}$. Since the individual sample values are independent, the joint pdf

$$f(\mathbf{x}) = f(x_1)f(x_2)\ldots f(x_n).$$

Corresponding to $\sum$ (p. 29) for 'sum' we can use $\prod$ (Greek capital pi) for 'product' and write this as

$$f(\mathbf{x}) = \prod_{i=1}^{n} f(x_i).$$

When the pdf involves one unknown parameter θ, we may write it as $f(x \mid \theta)$, and the joint pdf as

$$f(\mathbf{x} \mid \theta) = \prod_{i=1}^{n} f(x_i \mid \theta).$$

The **likelihood function** is obtained by regarding this as a function of θ:

$$L(\theta \mid \mathbf{x}) = f(\mathbf{x} \mid \theta),$$

where the value of the sample $\mathbf{x}$ is known.

18.8.2.2 Discrete distributions

The same method may be used with discrete mass functions. Consider estimating the parameter π of a geometric distribution. The probability mass function is

$$P(r) = \pi(1 - \pi)^{r-1} \qquad r = 0, 1, 2, \ldots; \quad 0 < \pi < 1.$$

Suppose that we have a sample of n values $r_1, r_2, \ldots, r_n$. The likelihood function for this is

$$L(\pi) = P(r_1)P(r_2)\ldots P(r_n)$$

$$= \prod_{i=1}^{n} (\pi\{1 - \pi\}^{r_i-1}) = \pi^n \prod_{i=1}^{n} (1 - \pi)^{r_i-1}$$

and so

$$\ln L(\pi) = n \ln \pi + \sum_{i=1}^{n} (r_i - 1) \ln(1 - \pi)$$

$$= n \ln \pi + \left(\sum_i r_i - n \right) \ln(1 - \pi).$$

Because π is a continuous quantity, we may differentiate with respect to it:

$$\frac{\mathrm{d}}{\mathrm{d}\pi}(\ln L) = \frac{n}{\pi} - \frac{\sum r_i - n}{1 - \pi}$$

which is zero when $n(1 - \pi) = (\sum r_i - n)\pi$, or $n = \pi \sum r_i$. Hence the maximum likelihood estimator is $\hat{\pi} = n / \sum_{i=1}^{n} r_i$. Again we do have a maximum, because the second derivative is

$$-\frac{n}{\pi^2} - \frac{\sum r_i - n}{(1 - \pi)^2}$$

which must be negative.

18.8.2.3 The normal distribution

The normal distribution in general depends on two parameters, μ and σ^2, in the probability density function

$$f(x \mid \mu, \sigma^2) = \frac{1}{\sigma\sqrt{2\pi}} \exp\left\{ -\frac{1}{2} \left(\frac{x - \mu}{\sigma} \right)^2 \right\}.$$

If one of these is known, the other may be estimated by the procedure we have been discussing above.

Suppose we know that $\mu = 0$. We wish to estimate σ^2; since it is σ^2, rather than σ, that has to be estimated, the mathematics is greatly simplified by writing $\sigma^2 = \omega$. The density function for $\mathcal{N}(0, \omega)$ then becomes

$$f(x \mid \omega) = (2\pi\omega)^{-1/2} \exp(-x^2/2\omega)$$

and when a sample $\mathbf{x} = (x_1, x_2, \ldots, x_n)$ is available the likelihood function becomes

$$L(\omega) = f(\mathbf{x} \mid \omega) = (2\pi\omega)^{-n/2} \exp\left(-\sum_{i=1}^{n} x_i^2/2\omega \right).$$

Hence

$$\ln L = -\frac{1}{2} n \ln(2\pi) - \frac{1}{2} n \ln \omega - \sum_{i=1}^{n} x_i^2/2\omega$$

and

$$\frac{\mathrm{d}}{\mathrm{d}\omega}(\ln L) = 0 - \frac{n}{2\omega} + \frac{\sum x_i^2}{2\omega^2}$$

which is zero when $n = \sum x_i^2 / \omega$ or $\hat{\omega} = \sum x_i^2 / n$. The second derivative is

$$\frac{n}{2\omega^2} - \frac{\sum x_i^2}{\omega^3} = -\frac{n}{2\omega^2}$$

when the value of $\hat{\omega}$ is inserted; so we have a maximum.

Note that, once again, this method (like the method of moments) does not always give unbiased estimators; but for this particular problem *when μ is known* we do have an unbiased estimator $\hat{\omega}$.

18.8.2.4 *Estimating two parameters. The normal distribution*

When there is more than one parameter, the method of maximum likelihood requires *partial differentiation* of the likelihood function with respect to each of the parameters being estimated. A brief discussion of partial differentiation is given in Appendix VII (p. 679).

Suppose that a random sample of observations $x_1, x_2, \ldots, x_n$ is available from $N(\mu, \sigma^2)$, and is to be used to find estimates of both μ and σ^2. Again it is simpler to write $\sigma^2 = \omega$, giving the probability density function as

$$f(x \mid \mu, \omega) = \frac{1}{\sqrt{2\pi\omega}} \exp\left\{ -\frac{(x - \mu)^2}{2\omega} \right\}.$$

$$L(\mu, \omega) = f(\mathbf{x} \mid \mu, \omega) = \prod_{i=1}^{n} (2\pi\omega)^{-1/2} \exp\{-(x_i - \mu)^2 / 2\omega\}$$

$$= (2\pi\omega)^{-n/2} \exp\left(-\sum_{i=1}^{n} (x_i - \mu)^2 / 2\omega \right)$$

and

$$\ln L = -\frac{1}{2} n \ln(2\pi\omega) - \sum_{i=1}^{n} (x_i - \mu)^2 / 2\omega$$

$$= -\frac{1}{2} n \ln 2\pi - \frac{1}{2} n \ln \omega - \sum_{i} (x_i - \mu)^2 / 2\omega.$$

Taking partial derivatives with respect to μ and to ω,

$$\frac{\partial}{\partial \mu} (\ln L) = +\sum_{i=1}^{n} (x_i - \mu)/\omega$$

and

$$\frac{\partial}{\partial \omega} (\ln L) = -\frac{n}{2\omega} + \sum_{i} (x_i - \mu)^2 / 2\omega^2.$$

The first of these is zero when $\sum_i (x_i - \mu)$ is zero, and gives $\hat{\mu} = \bar{x}$. (Note that ω must be positive, and, using second derivatives, it can be shown that this gives a maximum.)

When we solve the second equation, resulting from $(\partial/\partial\omega)(\ln L) = 0$, we have to use $\hat{\mu}$ instead of μ as necessary. We obtain

$$n\hat{\omega} = \sum_i (x_i - \hat{\mu})^2 = \sum_i (x_i - \bar{x})^2$$

so that

$$\hat{\omega} = \frac{1}{n} \sum_{i=1}^{n} (x_i - \bar{x})^2.$$

(Again it can be shown to give a maximum.)

The functions $\hat{\mu}, \hat{\omega}$ are the maximum likelihood estimators whose numerical values can be found from the sample values $\{x_i\}$. In this case, one $(\hat{\mu})$ has turned out to be an unbiased estimator while the other $(\hat{\omega} \equiv \hat{\sigma}^2)$ is biased. In this problem we would generally prefer the unbiased estimator whose divisor is $n - 1$ rather than n.

18.8.3 Exercises

1. A random sample of five observations from the continuous uniform distribution on the interval (α, β) gave the values: 4.3, 3.1, 4.0, 3.7, 4.7. Find the method of moments estimates of α and β.

2. A random sample of n items is taken from a binomial distribution in which the probability of 'success', π, is not known. The number of 'successes' found in the sample is r. By using the expression for P(r) as a likelihood function, find the maximum likelihood estimator for π.

3. A normal distribution with variance 1 has unknown mean μ. Find the maximum likelihood estimator of μ when a random sample of observations $(x_1, x_2, \ldots, x_n)$ is available.

18.9 MINITAB: CONFIDENCE INTERVALS FOR MEANS AND PROPORTIONS; t, χ^2 AND F

18.9.1 Confidence intervals

MINITAB usually calculates a confidence interval simultaneously with the calculation of the associated test statistic. Thus the procedures for making a test on a mean, a difference of means, a mean of paired differences, a proportion and a difference of proportions (Section 17.16) are those used to obtain confidence intervals for these quantities. That section should be referred to when confidence intervals are to be calculated.

For the **1-sample** cases, if the entry in the relevant dialogue box opposite **Test mean (required for test)** is deliberately left empty, then only a confidence interval is presented in the Session window and no test statistic. This option is not available with the other procedures.

A 95% two-sided confidence interval is calculated by default in all cases. Choice of the **Options** box offers a one-sided interval: **less than** or **greater than** is selected opposite **Alternative**. The confidence level may also be changed.

18.9.2 t, χ^2 and F

These three probability distributions are derived from the normal distribution. The t-distribution was introduced in Chapter 17 (p. 426); χ^2 and F will be introduced in later chapters (pp. 475 and 579, respectively). For each distribution MINITAB will calculate, with appropriate selection, (i) the value of the probability density function $f(x)$, (ii) the value of the cumulative distribution function $F(b) = P(X \leqslant b)$, and (iii) the value of the inverse cumulative distribution function, i.e. the value of b such that $F(b)$ equals a given probability.

There are tables of a limited number of percentage points for each of the three distributions at the end of the book (Tables A3, A4 and A7). With MINITAB it is possible to find values for *any* chosen percentage point. As an example of the method, consider the value in Table A3 (p. 717) corresponding to the 5% value of t with 4 degrees of freedom. We seek the value of the inverse cumulative distribution function corresponding to $1 - 0.05/2 = 0.975$. (Note that values are for a two-tail test; look at the graph below the table.) Hence choose

> **Calc>Probability Distributions>t/**
> Select **Inverse cumulative probability**, enter **Degrees of freedom** 4,
> select **Input constant** and enter 0.975/

MINITAB outputs a value of 2.776 45 which is consistent with the value of 2.776 in the table. Note that the **noncentrality parameter**, listed in the dialogue box, has a default value of 0. That is appropriate for all the t-distributions in this book.

For a percentage point P, as defined in our tables, the input constants to enter in the dialogue box of the inverse cumulative distribution functions are $1 - P/2$ for t, and $1 - P$ for χ^2 and F. For the χ^2 distribution the parameter to be entered is called the **degrees of freedom**, as with t. For the F-distribution the parameters are the **numerator degrees of freedom** and the **denominator degrees of freedom**. For all three distributions the **noncentrality parameter** should be set at 0.

18.10 EXERCISES ON CHAPTER 18

1. A random sample from a normal distribution with standard deviation 3 produced the observations 7, 1, 4, 5, 9, 10, 13, 4. Find an 80% confidence interval for the mean of the distribution.

2. The span in centimetres of 150 adult males is given in the following grouped frequency distribution:

Centre of interval	156	160	164	168	172	176	180	184	188	192
Frequencies	2	9	12	20	25	30	25	14	10	3

Calculate the mean and the standard deviation, making an allowance for the grouping of the data.

Give 95% confidence limits for the mean of the population from which the sample is taken.   (O&C)

(*Note*. When the variance, and standard deviation, are calculated from grouped data, a correction is sometimes made in order to compensate for the effect of the grouping. It is called *Sheppard's correction* and consists in reducing the computed variance by $h^2/12$, where h is the width of the class-interval.)

3. Describe how the normal distribution may be used as an approximation to the binomial distribution.

A random sample of 20 children in a school were asked a 'general knowledge' question and 12 answered it correctly. Estimate the proportion of children in the whole school who would answer correctly, and find the standard error of this estimate. Calculate a 95% confidence interval for this proportion.   (IOS)

4. (a) The error made when using a certain length-measuring instrument is known to be normally distributed with mean zero and standard deviation 1 mm.

(i) Calculate the probability that the error made in one use of the instrument will be numerically less than 0.5 mm.

(ii) If the instrument is used independently 9 times to measure a particular length, calculate the probability that the mean of the 9 readings will be within 0.5 mm of the true length.

(b) Another instrument was used independently 10 times to measure a particular length and the observed errors (in mm) were

$$-0.2, \quad +0.1, \quad -0.3, \quad +0.1, \quad +0.2,$$
$$-0.3, \quad -0.1, \quad +0.1, \quad +0.2, \quad +0.3.$$

Assuming that these errors are normally distributed, calculate a 95% confidence interval for the mean error using this instrument.   (*Welsh*)

5. In observations of a particular type of event, the probability of a positive result of any one observation is independent of the results of other observations and has the value θ, the same for all observations. In n observations the proportion giving positive results is p. State the mean and standard deviation of the probability distribution of p. Say also how and in what circumstances this probability distribution can be approximated by a normal distribution.

Show that, according to this approximation, the probability that p satisfies the inequality

$$|p - \theta| < 1.96\sqrt{\frac{\theta(1 - \theta)}{n}}$$

is 95%.

In a set of 100 observations of this type, 90 gave a positive result. Obtain an inequality of the above form, and by squaring both sides of the inequality calculate from the roots of a quadratic equation an approximate 95% symmetric confidence interval for the value of θ for the type of event observed. *(JMB)*

6. A company has two factories which produce a certain material. Their daily outputs are independent and are normally distributed with means 12 and 15 tonnes and standard deviations 0.5 and 0.6 tonnes respectively.

   Some changes are made in the staffing of the plants and it is thought that the means may be affected but the standard deviation will not. A check over 100 days reveals mean production figures of 11.92 and 15.10 tonnes respectively. Test at the 5% level to see if either of these is significantly different from its established mean daily value.

   Test also whether or not the difference of the means is significantly different from the value which would have been expected.

   Calculate a symmetrical two-sided 95% confidence interval for the difference of the means based on these 100-day samples of output figures. *(JMB)*

7. A medical investigator states that half the elderly people given anaesthetics for operations suffer from complications. On examining his records, it is found that 36 such people out of a random sample of 100 in fact have complications. Is this evidence that his statement is justified? Explain very carefully each step of your argument and point out why you have used the particular theoretical distributions you have in your answer.

   What is a 95% confidence interval for the proportion of elderly people suffering from such complications? *(Oxford)*

8. Explain carefully what is meant by a *confidence interval*.

   Twelve pairs of identical twins of similar age are observed as to their gains in height over a certain period. Of the rth pair $(r = 1, 2, \ldots, 12)$, one is given a certain drug, and has gain x_r; the other is not given the drug, and has gain y_r. It is found that (in suitable units)

$$\sum_{r=1}^{12} x_r = 113.7, \qquad \sum_{r=1}^{12} y_r = 86.3, \qquad \sum_{r=1}^{12} x_r^2 = 1513.2, \qquad \sum_{r=1}^{12} y_r^2 = 911.3,$$

$$\sum_{r=1}^{12} (x_r - y_r)^2 = 123.7.$$

Construct a 99% confidence interval for the difference in average height gain between a person given the drug and one not given it. What assumptions underlie your analysis? *(IOS)*

9. Eleven rabbits were examined for the total fatty acid content of their plasma, with the following results (in units of milligrams per 100 ml):

$$160, 168, 154, 156, 172, 163, 169, 175, 150, 167, 166.$$

Find a 90% confidence interval for the mean of the population from which this sample has been taken, giving the conditions and assumptions necessary for this calculation to be valid.

An experimenter wishes to estimate the true fatty acid content of the plasma to within ±1% of the true value, with 90% confidence. How large a sample would you recommend?

10. An experiment was conducted to compare the performance of two varieties of wheat. Seven farms were chosen for the experiment, and the yields (in tonnes per hectare) for each variety on each farm were as in the table.

Variety	Farm						
	A	B	C	D	E	F	G
1	4.63	4.81	3.21	4.69	4.25	3.70	4.08
2	4.14	4.02	3.49	4.11	4.45	3.17	3.79

(a) Analyse the data to investigate whether the population mean yield for variety 1 is superior to that for variety 2.
(b) Describe the conditions necessary for your analysis to be valid, and discuss any assumptions you make.
(c) Obtain a 99% confidence interval for the overall difference in population mean yield between the two varieties. *(IOS)*

11. An egg packing company grades eggs for size and then packs them into boxes of 12. The weights in two such boxes are (in grams):

Box 1: 64.1, 60.2, 53.8, 67.2, 56.9, 58.6, 60.0, 66.3, 50.7, 56.0, 73.3, 48.2;
Box 2: 58.7, 62.8, 67.2, 68.0, 68.4, 64.4, 60.9, 73.2, 51.6, 63.5, 74.3, 71.2.

Assuming both sets of data follow normal distributions, find 95% confidence intervals for the true variance in each set.

12. (a) The survival time after diagnosis, T, of an individual suffering a certain illness is exponentially distributed with $f(t) = \lambda e^{-\lambda t} (t \geqslant 0)$. Show that the probability that an individual survives for at least t_0 is $P(T \geqslant t_0) = e^{-\lambda t_0}$.
(b) The survival times (days) of a group of 12 patients in a study of this illness are

$$1327, 1464, 241, 1027, 20, 332, 308, 20, 100, 71, 889, 229.$$

Find the maximum likelihood estimate of λ. *(RSS)*

13. The Pareto distribution is often used to model household income in a country. It has the probability density function

$$f(x) = \frac{\alpha c^{\alpha}}{x^{\alpha+1}}, \quad \text{for } x \geqslant c,$$

with α and c positive constants. Sketch the curve.

Find moment estimators of α and c when a random sample of n observations is available. (RSS)

14. A random sample $15, 12, 7, 28, 5, 11$ is available from the continuous uniform distribution on $[0, a]$, a being unknown. Explain why the likelihood is zero for $a < 28$.

Find maximum likelihood and method of moments estimates for a. Comment on the results. (RSS)

18.11 COMPUTING EXERCISES

1. *Left-handedness.* One hundred university students were given a questionnaire to assess their left-handedness. Based on which hand was used for common activities (e.g. writing, throwing, striking a match) a score was recorded for each student on an integer scale from 0 to 20, a high score indicating a very left-handed person. The scores were:

```
0, 20,  3, 0,  8, 7, 3, 6,  4, 3, 6,  3, 1, 2, 6, 5, 8, 3,  1,  3, 2, 17, 7,  4, 3,
3,  4, 11, 5, 18, 4, 6, 6, 15, 4, 3, 18, 7, 1, 5, 6, 5, 3, 14, 18, 4,  4, 4,  8, 6,
1,  0, 10, 8,  7, 0, 0, 6, 15, 3, 6,  3, 4, 5, 3, 0, 7, 4,  0,  8, 5,  9, 9, 19, 8,
0,  4, 15, 2,  3, 5, 7, 3,  6, 5, 3,  7, 7, 1, 8, 9, 8, 5,  2,  6, 2,  4, 1,  4, 4.
```

(i) Construct a diagram to show the distribution of the scores.
(ii) Defining a left-handed person as one with a score of 11 or greater, find a 95% confidence interval for the proportion of left-handed people in the population from which the sample was drawn. Base the interval on the normal approximation to the binomial.

2. A machine which packs sugar gives a normal distribution of weights with mean 1000 g and standard deviation 2.5 g. Choose a random sample of ten weights from the distribution, and from it calculate a 95% confidence interval for the mean, assuming the standard deviation, 2.5 g, is known (i.e. use z-values as multipliers).

Repeat 1000 times and verify that the proportion of cases in which the confidence interval does not include the true mean (1000 g) is approximately 5%.

3. *A comparison of three unbiased estimators of the mean of a continuous uniform distribution, based on a sample of $(2n+1)$ observations.* The three estimators are (i) the mean, (ii) the median, (iii) $(n+1)h/(2n+1)$, where h is the maximum value of the observations in the sample.

Generate a random sample of size 15 from the uniform distribution on $(0, 2)$. Find the values of the three estimators described above. Repeat 500 times and compare the distributions of the three estimators by (i) plotting histograms, (ii) calculating means, variances and standard deviations.

18.12 PROJECTS

1 Pulse rate

Aim. To find a 95% confidence interval for mean pulse rate.
Equipment. Watch with second hand.
Collection of data. Each subject should have been sitting for about 5 minutes. (Pulse rate changes when people stand or do any activity.) Take the pulse at the wrist and count the number of pulses in one minute. Data from at least 50 subjects should be collected. The subjects will be treated as a random sample from a well-defined population, so they should be chosen with this in mind, though practical difficulties may make it hard to achieve.
Analysis. Make a frequency table of the data. Calculate the mean and estimated variance. Find a 95% confidence interval for the unknown mean pulse rate of the population sampled.

2 Proportion of left-handed people

Aim. To find a 95% confidence interval for the proportion of left-handed people in a population.
Collection of data. There are many degrees and forms of left-handedness so it is necessary to be specific. It is probably best to limit the left-handedness to writing. Ask at least 50 subjects. As with the previous project, it is desirable that they are a random sample from a well-defined population.
Analysis. Find the proportion of left-handed people in the sample and, by use of the normal approximation, find a 95% confidence interval for the unknown proportion of left-handed people in the population.

SUMMARY

A **95% confidence interval** for μ in a normal distribution, based on a random sample of n observations and knowing the value of σ^2, is

$$\bar{x} - 1.96\sigma/\sqrt{n} \leqslant \mu \leqslant \bar{x} + 1.96\sigma/\sqrt{n}$$

where $\bar{x}$ is the sample mean. For a 99% or 99.9% confidence interval, replace 1.96 by 2.576 or 3.291 respectively.

In a **large sample** from any distribution to which the central limit theorem applies, the same inequality gives an approximate interval for μ; the value of σ^2 may either be known or estimated from the large sample of data.

Using a large sample, n, an approximate 95% confidence interval for the **parameter π in a binomial distribution** is

$$p - 2\sqrt{p(1-p)/n} \leqslant \pi \leqslant p + 2\sqrt{p(1-p)/n}$$

where π is not too near 0 or 1 and p is the point estimate of π from the sample.

If σ^2 is not known, the 95% confidence interval for **μ in a normal distribution** becomes

$$\bar{x} - ts/\sqrt{n} \leqslant \mu \leqslant \bar{x} + ts/\sqrt{n}$$

where t is the critical value at the 5% level in a t-distribution with $n-1$ degrees of freedom and s^2 is the unbiased estimator of σ^2 from the sample. This inequality *must* be used for samples of size less than about 30 observations.

The χ^2 distribution provides confidence intervals for a **variance in a normal distribution**. If χ^2_L and χ^2_U are the lower and upper $2\frac{1}{2}\%$ points of a $\chi^2_{(n-1)}$ distribution then a 95% confidence interval for σ^2 is

$$(n-1)s^2/\chi^2_U \leqslant \sigma^2 \leqslant (n-1)s^2/\chi^2_L.$$

The **method of moments** for estimating a single parameter in a density function is to equate the sample mean $\bar{x}$ to the population mean μ. For two parameters, $\bar{x}$ is equated to μ and the raw second moment $\sum x_i^2/n$ is equated to $E[X^2] = \mu'_2$. This can be extended for more parameters.

The **likelihood function** is the joint probability density function of the sample values $(x_1, x_2, \ldots, x_n)$. If there is only one parameter, say λ, in the pdf for X, the likelihood function is considered as a function of λ, and the value of λ which maximizes this likelihood is found in terms of $(x_1, x_2, \ldots, x_n)$. This is the **maximum likelihood** estimator of λ.

With two or more parameters to estimate, partial differentiation will be required to maximize the likelihood with respect to all of them.

19 HYPOTHESIS TESTS USING THE χ^2 DISTRIBUTION

OVERVIEW

Hypotheses about the goodness-of-fit of data to a discrete or continuous distribution can be examined using the χ^2 statistic. The test of independence of row and column classifications in a two-dimensional table also uses χ^2. We study a nonparametric goodness-of-fit test, due to Kolmogorov and Smirnov, in this chapter also.

19.1 RANDOM DIGITS

We have used random numbers quite often in projects; usually these have either been taken from a table, such as Table A1, or generated by a computer program. What is the underlying property of random number sequences, and how can we test whether a particular set of numbers is effectively random?

Definition 19.1

A set of digits is a sequence of **random digits** if every position in the sequence is equally likely to be occupied by any one of the digits being used, and positions are filled independently.

Thus if we use all the digits $0, 1, \ldots, 9$ every position has a probability $\frac{1}{10}$ of containing a 0, or $\frac{1}{10}$ of containing a 1, and so on. Further, every position has to be filled independently of every other, to satisfy the definition; so a 0 is equally likely to be followed by a 0 or a 1 or a 2, and so on. In fact we are defining random digits by saying that they must conform to a particular statistical model: the digits are independent observations drawn from the discrete uniform distribution whose probability mass function is

$$P(r) = \frac{1}{10} \qquad r = 0, 1, 2, \ldots, 9.$$

It is very difficult to produce a random sequence by writing down a run of numbers 'haphazardly' out of the head, because we tend not to follow any digit with the same digit again. In a list like this, there is often a lack of 00s, 11s and so on, and 000s, 111s, ..., 999s are usually *much* too rare. We remarked in Chapter 3 how hard it was not to show some sort of bias, however unconsciously, when not using a purely objective method of random selection. This is a good illustration of that remark.

Tests based on the properties of random numbers were proposed when the originally published sets of random digits were being prepared (see Chapter 3). The simplest of these is the *frequency test*. This takes long runs (one or more pages) of digits and examines whether each digit $0, 1, \ldots, 9$ has equal frequency, within acceptable statistical limits.

Example 19.1 Test of randomness of digits

A run of 1000 digits had the following frequencies of $0, 1, 2, \ldots, 9$. Can they be regarded as 'random'?

Digit, r	0	1	2	3	4	5	6	7	8	9	Total
Frequency, f_r	106	88	97	101	92	103	96	112	114	91	$N = 1000$

The frequencies f_r are not *exactly* equal, but no statistician expects them to be. In repeated samples, each sample consisting of 1000 digits, the sets of frequencies will (like this one) show variations from exact equality, and also they will show variations from one another. What we wish to do is to compare these *observed* frequencies with those *expected* if our statistical model is correct.

If these digits are a sequence of random digits, the discrete uniform distribution models them. The null hypothesis therefore is that $P(r) = \frac{1}{10}$ for each r from 0 to 9, and so in a sample of N the frequencies of $r = 0, 1, \ldots, 9$ are all equal to $NP(r) = N/10$, which is 100 in this example. The frequencies predicted by the null hypothesis are the *expected frequencies*, which we will call $\{E_r\}$. The *observed frequencies* are almost always labelled $\{O_r\}$, in the situations considered in the rest of this chapter, and we shall follow this notation from now on.

We add a row to the table already given; this row will contain the expected frequencies. This helps us to construct a test which compares each O_r with the corresponding E_r.

Digit, r	0	1	2	3	4	5	6	7	8	9	Total
Observed frequency, O_r	106	88	97	101	92	103	96	112	114	91	1000
Expected frequency, E_r	100	100	100	100	100	100	100	100	100	100	1000

The statistic

$$X^2 = \sum_{r=0}^{9} \frac{(O_r - E_r)^2}{E_r}$$

is the basis of this test, and of several others like it. For each cell of the table (each r-value in this case) we calculate (observed *minus* expected)^2/expected.

All these contributions are summed, for all cells of the table. Thus

$$X^2 = \frac{(106-100)^2}{100} + \frac{(88-100)^2}{100} + \frac{(97-100)^2}{100} + \frac{(101-100)^2}{100} + \frac{(92-100)^2}{100}$$
$$+ \frac{(103-100)^2}{100} + \frac{(96-100)^2}{100} + \frac{(112-100)^2}{100} + \frac{(114-100)^2}{100} + \frac{(91-100)^2}{100}$$
$$= (36 + 144 + 9 + 1 + 64 + 9 + 16 + 144 + 196 + 81)/100$$
$$= 7.00.$$

Obviously we shall obtain a different value of X^2 each time we take a fresh sample of 1000 digits: X^2 has a *sampling distribution* in the sense defined in Chapter 16. On the assumption that the null hypothesis is true, we can find the sampling distribution of X^2, at least approximately. It does not follow any of the distributions we have already studied. We need to introduce a new family of distributions. (We shall complete this example on p. 478.)

19.2 THE χ^2 FAMILY OF DISTRIBUTIONS

We shall not quote the exact density function for this family, because it is too complicated to throw any light on the properties of the distributions. However, we do need to know that it contains only a single parameter, which we shall call v. The usual name for v is *degrees of freedom*, and the family of distributions is called χ^2 (the Greek letter *chi* squared: chi, or khi, represents a hard k sound followed by a breathing, and should *never* be pronounced as in 'church'!). We indicate a particular member of the χ^2 family, that with v degrees of freedom, by $\chi^2_{(v)}$. Figure 19.1 shows the general shape of χ^2 distributions: they begin at 0

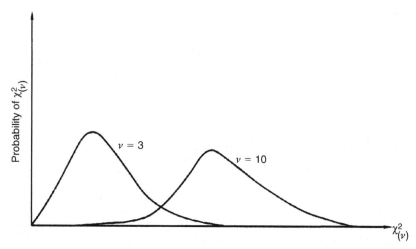

Figure 19.1 *Comparative shapes of the distributions $\chi^2_{(3)}$ and $\chi^2_{(10)}$.*

and have long tails on the right. The mean of $\chi^2_{(\nu)}$ is ν, and the variance is 2ν. Thus for large values of ν the distribution is very slow to rise to its peak, and is also very widely scattered. Distributions in the χ^2 family are skew to the right, so that the mean is to the right of the mode. These distributions only become approximately symmetrical when ν is very large, and a normal approximation to χ^2 has to be found indirectly when needed (see p. 532). We have not shown $\chi^2_{(1)}$ or $\chi^2_{(2)}$ in Fig. 19.1; although they are not quite the same shape as other members of the family they are used in the same way. (They do not rise to a peak; instead they take their highest value at the beginning and fall off from it in an exponential-decay pattern.)

The χ^2 family of distributions is usually tabulated as in Table A4. One row of the table corresponds to one member of the family, that with degrees of freedom ν. On this row, the entry in a particular column is the value of $\chi^2_{(\nu)}$ which is exceeded with the probability shown in that column heading. For example, $\chi^2_{(3)}$ is greater than 7.81 with probability 0.05, $\chi^2_{(10)}$ is greater than 23.21 with probability 0.01, $\chi^2_{(5)}$ is greater than 12.83 with probability 0.025, and $\chi^2_{(5)}$ is greater than 0.83 with probability 0.975. As the diagram beneath Table A4 shows, we tabulate the values of $\chi^2_{(\nu)}$ which have to the *right* of them the areas stated in the column headings. The table is one-tail, since we are given the probability of being *above* a certain critical value.

It is, however, easy to use the table in a two-tail fashion when needed; the first two columns are there for this purpose. Suppose that we wish to find lower and upper values of $\chi^2_{(5)}$, which we will call χ^2_L, χ^2_U, such that $P(\chi^2_L < \chi^2_{(5)} < \chi^2_U) = 0.95$. Let us make these limits *symmetrical*, in the sense that the probability of being below χ^2_L is the same as the probability of being above χ^2_U, i.e. 0.025. (Although limits do not have to be symmetrical, it seems reasonable always to make them so unless there is some special reason not to.) If $P(\chi^2_{(5)} < \chi^2_L) = 0.025$, then

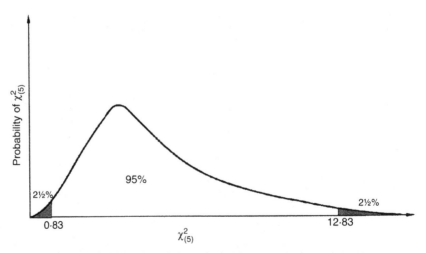

Figure 19.2 *The distribution $\chi^2_{(5)}$, showing the points above and below which the proportion of the total area under the curve is $2\frac{1}{2}$%.*

$P(\chi^2_{(5)} > \chi^2_L) = 0.975$. As we noted above, $P(\chi^2_{(5)} > 0.83) = 0.975$, so $\chi^2_L = 0.83$. We find this lower limit in the second column of the table, on the row for $\chi^2_{(5)}$. On the same row, in the column headed 0.025 the entry is 12.83. So $P(\chi^2_{(5)} > 12.83) = 0.025$. Thus $\chi^2_U = 12.83$. Figure 19.2 illustrates what we have found, which is that $P(0.83 < \chi^2_{(5)} < 12.83) = 0.95$.

19.2.1 Exercises

1. Find constants a, b, c, d, e, f, g, h such that: (i) $P(\chi^2_{(1)} > a) = 0.05$; (ii) $P(\chi^2_{(2)} > 5.99) = b$; (iii) $P(\chi^2_{(c)} > 59.70) = 0.001$; (iv) $P(\chi^2_{(4)} < d) = 0.025$; (v) $P(23.68 < \chi^2_{(14)} < 29.14) = e$; (vi) $P(f < \chi^2_{(6)} < g) = 0.025$; (vii) $P(16.79 < \chi^2_{(h)} < 46.98) = 0.95$. (*Note:* f, g not unique.)

2. Find a symmetric (two-sided) confidence interval for: (a) $\chi^2_{(8)}$ at the 95% confidence level; (b) $\chi^2_{(11)}$ at the 99% confidence level.

3. What is $P(0 < \chi^2_{(7)} < 16.01)$?

4. What is $P(58.30 > \chi^2_{(29)} > 49.59)$?

The sampling distribution of X^2 is approximately a χ^2 distribution (see Section 19.4). Most uses of the χ^2 distribution in significance testing are with test statistics which, like our X^2, are only approximately like the theoretical χ^2 distribution. (This is similar to the way in which the normal distribution was used approximately in the tests on pp. 416–425.) Our X^2 is approximated well by χ^2 so long as none of the expected values E_r falls below 5 (and in fact the approximation is still adequate if only one or two E_r are not too far below 5).

The null hypothesis which gave us $\{E_r\}$ in Example 19.1 was that all digits occur equally frequently. A suitable alternative is that all digits do not occur equally frequently. The nearer each O_r is to the corresponding E_r the more consistent this is with the null hypothesis; therefore the smaller the value of X^2 the less we shall want to reject this null hypothesis. It is only when X^2 becomes large that the evidence of the observations points against the null hypothesis and the alternative hypothesis provides a better explanation of the observations. The rule for carrying out the significance test is thus to reject the null hypothesis when X^2 is large.

19.3 THE RULE FOR FINDING DEGREES OF FREEDOM IN χ^2 TESTS

In order to find v, the degrees of freedom of the χ^2 distribution which approximates X^2, we first count the number of cells in the table, that is the number of pairs (O_r, E_r) we have available for comparison. This is 10 in Example 19.1.

Then we ask what restrictions or *constraints* we have placed on the set $\{E_r\}$. In the present case we require that the total of the members of the set $\{E_r\}$ shall equal the observed total N: clearly we cannot compare $\{E_r\}$ with $\{O_r\}$ unless they both represent the same total collection of items. Thus $\sum_{r=0}^{9} E_r = N$. This is a *linear constraint* on the members of the set $\{E_r\}$, in other words a linear equation which they must satisfy. In any χ^2 test, this constraint applies. In our present example, it is the only one: given N and the null hypothesis, the $\{E_r\}$ can be calculated:

Degrees of freedom = number of cells in table − number of linear constraints on the expected frequencies.

For this example, d.f. $= v = 10 - 1 = 9$. (A common abbreviation for degrees of freedom is d.f.)

Example 19.1 *(concluded)*

X^2 will be approximately distributed as $\chi^2_{(9)}$ if the null hypothesis is true. The computed value of X^2 was 7.00. By reference to Table A4 we see that such a value is by no means an unlikely one for a $\chi^2_{(9)}$ random variable, and we cannot reject the null hypothesis. On the evidence of this test, it is reasonable to regard the observed set of 1000 digits as random.

19.4 WHY THE X^2 STATISTIC HAS AN APPROXIMATE χ^2 DISTRIBUTION

To justify using χ^2 as an approximation to X^2 for all the tests we shall describe in this chapter demands more mathematics than we may assume. However, we will show the derivation in a simple case.

A slot machine is supposed to be constructed so as to give a prize in a proportion π of 'plays' of the machine, and so as to make each play independent of every other. A particular player makes n plays and wins in w of them. We may use the statistic X^2 to test if the machine has fulfilled its specification. If it has, then the expected number of wins is $n\pi$; the n plays should divide into $n\pi$ wins and $n(1 - \pi)$ losses. So we construct the table:

	Wins	Losses	Total
Observed	w	$n - w$	n
Expected	$n\pi$	$n(1 - \pi)$	n

Notice that we must classify and account for *all* the n plays, not just consider the wins. In any χ^2 test, it is the *complete* subdivision of n that has to be examined.

Thus

$$X^2 = \sum \frac{(O_r - E_r)^2}{E_r} = \frac{(w - n\pi)^2}{n\pi} + \frac{[n - w - n(1 - \pi)]^2}{n(1 - \pi)}$$

$$= \frac{(w - n\pi)^2}{n\pi} + \frac{(n\pi - w)^2}{n(1 - \pi)} = \frac{(1 - \pi)(w - n\pi)^2 + \pi(n\pi - w)^2}{n\pi(1 - \pi)}$$

$$= \frac{(w - n\pi)^2}{n\pi(1 - \pi)}.$$

So

$$X = \frac{w - n\pi}{\sqrt{n\pi(1 - \pi)}}.$$

Now on the null hypothesis that the machine fulfils its specification, w is binomially distributed, with mean $n\pi$ and variance $n\pi(1 - \pi)$; and thus (p. 377), provided n is large, X has approximately a standard normal distribution. Thus X^2 is approximately the square of a standard normal variable. The square of a standard normal variable is $\chi^2_{(1)}$, this being the basic mathematical definition of the χ^2 family; also, if we add together the squares of v independent standard normal variables, this sum of squares is distributed as $\chi^2_{(v)}$. So

$$X^2 = \frac{(w - n\pi)^2}{n\pi(1 - \pi)}$$

is approximately $\chi^2_{(1)}$. This agrees with our rule for calculating degrees of freedom: there are just two cells in the table, and one constraint on the expected frequencies.

The derivation we have given for this simple case makes it clear that the theoretical χ^2 distribution is only an *approximation* to the sampling distribution of the X^2 statistic, and that n must be large for this approximation to be satisfactory. In less simple cases, where tables contain more than two cells, a distribution more complicated than the binomial is required to model the expected frequencies. However, if we were able to carry out the necessary mathematics we would find that X^2 proved to be the sum of v independent expressions, each of which was approximately $\chi^2_{(1)}$. This sum would then follow the distribution $\chi^2_{(v)}$, approximately, and the value of v would be one less than the number of cells in the table. We give some more general rules for finding degrees of freedom in Sections 19.6 and 19.7.

19.4.1 Exercises

1. Fifty tosses of a coin produced 30 heads; is this evidence that the coin is biased?

2. A botanist expects $\frac{3}{4}$ of the plants of a particular species to have red flowers, and the rest white. He grows 100 such plants from seed, and obtains 15 with white flowers. Should he change his ideas about these plants?

3. In 240 throws of a die, there are 35 scores of one, 43 twos, 39 threes, 45 fours, 44 fives and 34 sixes. Is there evidence that the die is not properly balanced?

4. A pair of guinea pigs has produced a total of 64 offspring of which 30 have been red, 9 black and 25 white. A geneticist has predicted that, in a cross of this type, such progeny should occur in the ratios 9:3:4. Use a χ^2 test to determine whether or not the observed numbers are in agreement with his theory.

19.5 TESTING THE FIT OF A THEORETICAL DISTRIBUTION TO OBSERVED DATA

Another test applied to random digits is called the 'gap' test. Take each zero that appears in the sequence, and count the number of non-zero digits between it and the next zero in the sequence. For example, if we have 0 2 3 7 0 5 2 0 6 6 3 8 3 4 9 6 0 0 1 3 2 0..., there are 3 other digits (237) between the first 0 and the second, then 2 others before the third 0, then 8 before the fourth 0, none between the fourth and fifth 0's, 3 before the sixth.... If we call R the number of non-zeros between successive zeros in the sequence we build up the set of observations $3, 2, 8, 0, 3, \ldots$ on R. What random variable will model R, if every position in the sequence is equally likely to be occupied by any one of the digits 0 to 9 and all positions are filled independently?

Beginning at a 0, there is a probability $\frac{1}{10}$ that a 0 will follow it immediately, i.e. that $R = 0$. If R is not 0, then we may obtain $X0$, where X stands for any of the digits 1 to 9, or $XX0$, or $XXX0, \ldots$. Since 0 occurs with probability $\frac{1}{10}$, X with probability $\frac{9}{10}$, and every position is filled independently of every other, the sequence $X0$ has probability $\frac{9}{10} \times \frac{1}{10}$, $XX0$ has probability $\frac{9}{10} \times \frac{9}{10} \times \frac{1}{10}, \ldots$. So we see that $P(R = 1) = (\frac{9}{10})(\frac{1}{10})$, $P(R = 2) = (\frac{9}{10})^2(\frac{1}{10})$, and in general $P(R = r) = (\frac{9}{10})^r(\frac{1}{10})$ for $r = 0, 1, 2, \ldots$ without upper limit. This is a geometric distribution, though not exactly the one defined in Chapter 8 because R begins at 0 instead of 1.

There were 106 zeros in the sequence of digits of Example 19.1, and so there are 105 gaps to look at. The observed values of R for these were:

R	0	1	2	3	4	5	6	7	8	9	10	11	12	13
Frequency	14	10	10	6	5	9	7	2	2	3	5	3	2	2

R	14	15	16	17	18	19	20	21	22	23	24	28	38	Total
Frequency	1	4	3	2	1	1	3	2	1	3	2	1	1	105

The null hypothesis, that digits are random, leads by the argument above to the set of expected frequencies $NP(R = r) = 105(\frac{9}{10})^r(\frac{1}{10})$ for $r = 0, 1, 2, \ldots$.

r	0	1	2	3	4	5	6	7	8	9	10	etc.
$NP(r)$	10.50	9.45	8.51	7.65	6.89	6.20	5.58	5.02	4.52	4.07	3.66	

The values of $NP(r) = E_r$ reduce only slowly as r increases, but we must remember the rule that in a χ^2 test the values of E_r should not in general be allowed to fall below 5. Usually we can ensure this by grouping together the top tail of the distribution and treating, say, $R \geqslant 9$ as one class. However that would contain an expected frequency of 40.68, which would give a rather badly balanced set $\{E_r\}$ to use in testing. In this case it seems better to group $R = 8$ with $R = 9$; 10 with 11; 12 with 13 and 14, and so on to achieve a better balance in the individual values E_r. A possible grouping is shown below, together with the set $\{O_r\}$ that corresponds to the set $\{E_r\}$ after grouping.

r	0	1	2	3	4	5	6	7
Observed frequency, O_r	14	10	10	6	5	9	7	2
Expected frequency, E_r	10.50	9.45	8.51	7.65	6.89	6.20	5.58	5.02

r	8–9	10–11	12–14	15–17	18–21	22–28	$\geqslant 29$
Observed frequency, O_r	5	8	5	9	7	7	1
Expected frequency, E_r	8.59	6.96	8.04	5.86	5.42	5.40	4.93

This has given us a table containing 15 cells, i.e. 15 pairs (O_r, E_r). These expected frequencies were computed subject to the same linear constraint as in the previous example, namely that they must add to the same total as the observed, which is 105. So X^2 will this time be distributed as $\chi^2_{(14)}$ approximately. We obtain

$$X^2 = \sum \frac{(O_r - E_r)^2}{E_r} = \frac{3.50^2}{10.50} + \frac{0.55^2}{9.45} + \frac{1.49^2}{8.51} + \frac{1.65^2}{7.65} + \frac{1.89^2}{6.89} + \frac{2.80^2}{6.20}$$
$$+ \frac{1.42^2}{5.58} + \frac{3.02^2}{5.02} + \frac{3.59^2}{8.59} + \frac{1.04^2}{6.96} + \frac{3.04^2}{8.04} + \frac{3.14^2}{5.86}$$
$$+ \frac{1.58^2}{5.42} + \frac{1.60^2}{5.40} + \frac{3.93^2}{4.93} = 14.33.$$

The value of X^2 is again not significant at the 5% level, when tested as a $\chi^2_{(14)}$ variable. So on this test also the null hypothesis of randomness is not rejected.

19.6 TESTING THE FIT OF DATA TO A BINOMIAL DISTRIBUTION

In Chapter 8, we stated the conditions under which an observed variate could be modelled by a binomial random variable R. If we have collected a set of observations and wish to test whether they do conform to a binomial distribution, a goodness-of-fit test is appropriate.

Example 19.2 Goodness-of-fit test

A survey was made of the numbers of boys among families having five children altogether. In 320 families, the number of boys R occurred with the following frequencies:

Number of boys	0	1	2	3	4	5	Total
Observed number of families, O_r	8	40	88	110	56	18	$320 = N$
Expected frequencies, E_r	10	50	100	100	50	10	320

If births are all independent of one another, and the probability π of a male birth is the same from one family to another, R should be binomially distributed with parameters $n = 5$ and π. First let us suppose that $\pi = \frac{1}{2}$. The null hypothesis is now fully specified: 'R is binomial with parameters $n = 5$ and $\pi = \frac{1}{2}$'. This gives the set of expected frequencies

$$E_r = NP(r) = N \frac{n!}{r!(n-r)!} \left(\frac{1}{2}\right)^5, \quad r = 0, 1, \ldots, 5.$$

The set of binomial coefficients $n!/r!(n-r)!$ is 1, 5, 10, 10, 5, 1 and $(\frac{1}{2})^5 = \frac{1}{32}$. The values of E_r are thus 10, 50, 100, 100, 50, 10, and $\sum_{r=0}^{5} E_r = 320$. We have put one linear constraint on the E_r, the usual one that their total must equal N, the total number of observations. Now

$$X^2 = \sum_{r=0}^{5} \frac{(O_r - E_r)^2}{E_r} = \frac{(8-10)^2}{10} + \frac{(40-50)^2}{50} + \frac{(88-100)^2}{100}$$

$$+ \frac{(110-100)^2}{100} + \frac{(56-50)^2}{50} + \frac{(18-10)^2}{10}$$

$$= \frac{68}{10} + \frac{136}{50} + \frac{244}{100} = 11.96.$$

This statistic is based on six pairs (O_r, E_r), and the E_r are subject to one linear constraint, so X^2 is approximately χ^2 with $6 - 1 = 5$ degrees of freedom. It is significant at the 5% level (the 5% point for $\chi^2_{(5)}$ is 11.07), so at this level we reject the null hypothesis.

We did not give a specific alternative hypothesis, but simply assumed that if the null hypothesis were not true there would be some other set of probabilities, not given by the binomial distribution with $n = 5$ and $\pi = \frac{1}{2}$, that would be more appropriate. Let us consider more carefully what might happen. It is quite possible that births in a family may not be independent events, but that if the first child is a girl the later children are more likely to be girls. In that case, a basic condition for the binomial is violated and we cannot assume either that π is constant for all births or that all observations are independent of one another. No simple model can be set up in such a case. However, another alternative is that the binomial conditions do still hold, with $\pi \neq \frac{1}{2}$. This is easy to deal with, and does also appear to explain many sets of data.

If $\pi \neq \frac{1}{2}$, and there is no theoretical reason which gives the exact value of π, we must estimate π from the data. If the data do follow a binomial distribution, the mean, $\bar{r}$, of the observed data will estimate the mean, $n\pi$, of the distribution. Thus $\bar{r}/n$ will estimate π. We find $\bar{r} = \frac{860}{320} = \frac{43}{16}$. The estimate of π is then $\frac{1}{5} \times \frac{43}{16} = 0.5375$; call this p. The set of expected frequencies on the null hypothesis that the observations follow a binomial distribution (with π not specified in the hypothesis) is therefore

$$NP(r) = N \binom{5}{r} p^r (1-p)^{5-r}$$

$$= N \frac{5!}{r!(5-r)!} (0.5375)^r (0.4625)^{5-r}, \quad r = 0, 1, \ldots, 5.$$

Hence the values of E_r now are 6.8, 39.3, 91.5, 106.3, 61.8, 14.4. The statistic X^2 is calculated for the following table of O_r and E_r.

r	0	1	2	3	4	5	Total
O_r	8	40	88	110	56	18	$320 = N$
E_r	6.8	39.3	91.5	106.3	61.8	14.4	(320.1)

$$X^2 = \frac{(8 - 6.8)^2}{6.8} + \frac{(40 - 39.3)^2}{39.3} + \frac{(88 - 91.5)^2}{91.5} + \frac{(110 - 106.3)^2}{106.3}$$

$$+ \frac{(56 - 61.8)^2}{61.8} + \frac{(18 - 14.4)^2}{14.4} = 1.93.$$

The expected values were calculated subject to *two* constraints this time: as usual, $\sum_r E_r = N$, but also this time the mean value of r calculated using the expected frequencies had to equal the mean using the observed frequencies, because this was the equation that we used to estimate π. This additional constraint is also linear: $\sum r E_r = \sum r O_r$. Thus X^2 will be distributed approximately as χ^2 with $6 - 2 = 4$ degrees of freedom.

Since we need one equation for each parameter to be estimated, and each of these equations imposes a constraint on the E_r, another way of counting degrees of freedom is as 'number of cells in table *minus* one for total *minus* one for each parameter estimated'. The value 1.93 is certainly not significant as $\chi^2_{(4)}$ and we shall not reject the null hypothesis that the data were binomially distributed.

This result suggests that π is greater than $\frac{1}{2}$, but that otherwise the binomial conditions are reasonable.

19.6.1 Exercises

1. Four coins are tossed 160 times, and the distribution of the number of heads is observed to be

x (number of heads)	0	1	2	3	4
f (frequency)	5	35	67	41	12

   Find the expected frequencies if the coins are unbiased. Compare the observed and expected frequencies and apply the χ^2 test. Is there any evidence that the coins are biased? (*AEB*)

2. If P(x) denotes the probability of x successes in a binomial distribution for n trials for which the probability of a success in each trial is p, show that P$(x+1)/$P$(x)=p(n-x)/(1-p)(x+1)$.

   A bag contains a very large number of black marbles and white marbles. 8192 random samples of 6 marbles are drawn from the bag. The frequencies of the number of black marbles in these samples are tabulated below:

Number of black marbles per sample	0	1	2	3	4	5	6	Total
Frequencies	3	42	255	1115	2505	2863	1409	8192

   Test the hypothesis that the ratio of the numbers of black to white marbles in the bag is 3:1. (*AEB*)

3. Four players meet weekly and play eight hands of cards. In a year one of the players finds that he has won x of the eight hands with a frequency given in the following table:

x	0	1	2	3	4	5	6	7	8
f	4	13	12	12	6	3	2	0	0

   Find the frequencies of the number of hands he would expect to win if the probability of winning any hand were 1/4.
   Use the χ^2 distribution to test this hypothesis. (*AEB*)

19.7 CONTINGENCY TABLES

Suppose that two characteristics are observed on each of N members of a sample, and that each characteristic is classified into types rather than having an actual

Table 19.1 *Contingency table for people classified by hair colour and eye colour (observed frequencies)*

Colour of hair	Colour of eyes			
	Brown	Green/grey	Blue	Total
Black	50	54	41	145
Brown	38	46	48	132
Fair	22	30	31	83
Ginger	10	10	20	40
Total	120	140	140	$400 = N$

measurement recorded. For example, in a human population, we might record colour of hair and colour of eyes for each of N persons. Eyes would be classified 'brown, green/grey, blue' and hair would be classified 'black, brown, fair, ginger'. A summary table would be drawn up, the column headings giving the categories for eye colour and the row headings those for hair colour. Each cell of the table would give the number of people, among the population of N, who had a particular eye colour/hair colour combination. Table 19.1 shows a set of results classified in this way.

Do the two characteristics, eye colour and hair colour, tend to go together, or is the colour of a person's hair quite independent of the colour of eyes? It seems quite possible, on genetic grounds, that these two characteristics might *not* be independent. We shall now set up a null hypothesis that hair colour and eye colour are independent, and an alternative hypothesis that they are not. We wish to calculate a table of the expected frequencies on the null hypothesis. On this hypothesis, the ratio of the three eye colours, brown:green/grey:blue, should be the same for each one of the hair colours. That is, the ratio should be the same on each row of the table. If the ratio *is* the same on each row, then the best estimate of it is from the totals of the eye colours, namely 120:140:140. This ratio 120:140:140 should then apply to each individual row of the table, so that on each row there should be a proportion $\frac{120}{400}$ of the row total who have brown eyes, a proportion $\frac{140}{400}$ with green/grey eyes and a proportion $\frac{140}{400}$ with blue eyes. There were 145 people altogether who had black hair, so on the null hypothesis $\frac{120}{400} \times 145 = 43.50$ of these should have brown eyes, $\frac{140}{400} \times 145 = 50.75$ of these should have green/grey eyes, and 50.75 also should have blue eyes (Table 19.2). Similarly, in the second row of the table, there should be $\frac{120}{400} \times 132$ in the first column and $\frac{140}{400} \times 132$ in the second and also in the third column, to account for all the 132 people having brown hair. The third row of the table is dealt with in the same way.

There is a general rule for finding the expected frequencies from the table of observed frequencies, as follows. Let us call the hair-colour totals (145, 132, 83, 40) in the right-hand margin and the eye-colour totals (120, 140, 140) at the foot of the table the *marginal totals*. The total number of observations is N (=400 in the present example). The expected frequency in the cell in row i and column j

Table 19.2 *Expected frequency table for people classified by hair colour and eye colour, assuming these two characteristics are independent*

	Colour of eyes			
Colour of hair	Brown	Green/grey	Blue	Total
Black	43.50	50.75	50.75	145.00
Brown	39.60	46.20	46.20	132.00
Fair	24.90	29.05	29.05	83.00
Ginger	12.00	14.00	14.00	40.00
Total	120.00	140.00	140.00	400.00

of the table is equal to $(1/N) \times$ the marginal total of row $i \times$ the marginal total of column j. This gives an alternative derivation of Table 19.2.

We now compare Table 19.2 cell by cell with Table 19.1, using

$$X^2 = \sum_{\substack{\text{all} \\ \text{cells}}} \frac{(O_{ij} - E_{ij})^2}{E_{ij}}.$$

(We write i, j as suffixes to indicate that summing goes over all rows and all columns of the table – but not, of course, the marginal totals!) The value of X^2 is therefore

$$\frac{(50 - 43.50)^2}{43.50} + \frac{(54 - 50.75)^2}{50.75} + \frac{(41 - 50.75)^2}{50.75} + \frac{(38 - 39.60)^2}{39.60} + \frac{(46 - 46.20)^2}{46.20}$$

$$+ \frac{(48 - 46.20)^2}{46.20} + \frac{(22 - 24.90)^2}{24.90} + \frac{(30 - 29.05)^2}{29.05} + \frac{(31 - 29.05)^2}{29.05}$$

$$+ \frac{(10 - 12.00)^2}{12.00} + \frac{(10 - 14.00)^2}{14.00} + \frac{(20 - 14.00)^2}{14.00} = 7.74.$$

In Table 19.2, there are several linear constraints on the calculated frequencies. On the first row, the three expected frequencies must add to 145, the total observed with black hair; this is one constraint, and there is a similar one in the second row and in the third row. But in the fourth row, there is no freedom at all to the expected frequencies; all are constrained by the need for the expected column frequencies to add up to the observed column totals. The full number of constraints applied in the table of expected frequencies is then 1 (first row) + 1 (second row) + 1 (third row) + 3 (last row) = 6. There are 12 cells in the table; hence the degrees of freedom are $12 - 6 = 6$. Thus X^2 is approximately $\chi^2_{(6)}$, so its value in this example is not significant. In spite of one or two noticeable discrepancies between an O_{ij} and its corresponding E_{ij}, the whole set of observations gives no ground for rejecting the null hypothesis.

The same process can be applied to a table with any number r of rows and c of columns. *The degrees of freedom of the χ^2 variable which will approximate X^2 in this general case are $(r-1)(c-1)$.* There are rc cells, there is one constraint for each of the first $(r-1)$ rows, and there are c constraints for the last row; this leaves $rc - (r-1) - c = (r-1)(c-1)$ degrees of freedom.

19.7.1 Exercises

1. Television sets are manufactured at three different factories to the same specification. Records are kept on batches of sets produced at the three factories, to find out how many need service during their first six months' working. From the following results, is there any evidence that the factories differ in the quality of their product?

	Service required	Service not required
Factory 1	120	60
Factory 2	90	30
Factory 3	60	40

2. People from two different racial groups were classified according to their blood type, with the following results:

	Blood type			
	O	A	B	AB
Race I	176	148	96	72
Race II	78	50	45	12

Test the hypothesis that there is no relationship between racial group and blood type.

3. A sample of 250 seedling plants, growing in a nursery, is classified for vigour and for leaf colour, the results being summarized in the following table. Test whether these two characteristics appear to be related.

	Vigour		
	Good	Average	Weak
Leaf colour:			
Green	55	79	4
Yellow-green	11	60	15
Yellow	1	6	19

19.8 THE 2 × 2 TABLE

In the previous section, we set up the hypothesis that the characteristic classified in rows was independent of that classified in columns. If we reduce the number of rows and the number of columns each to two, we have a special case. We are in fact now comparing two proportions.

Example 19.3 2 × 2 table

On a certain day, 74 trains were on time arriving at one London terminus during the rush hour and 83 were not; at a second terminus there were 65 on time and 107 not. Is there any difference in the proportions arriving on time at the two termini?

We could of course treat this, as in Chapter 17 (p. 422), as the difference between two proportions, and use the normal approximation to the binomial distribution. The χ^2 test which we now propose is equivalent to the test described in Example 17.11.

Classify the observations by rows as 'terminus 1' and 'terminus 2', and by columns as 'on time' and 'late'.

	Observed frequencies		
	On time	Late	Total
Terminus 1	74	83	157
Terminus 2	65	107	172
	139	190	329

Proceeding as in the previous section, the expected frequencies are (139 × 157)/329 = 66.33 in row 1, column 1;...; (190 × 172)/329 = 99.33 in row 2, column 2. The expected frequencies, on the null hypothesis that the proportions on time at both termini are the same, are shown in the table below. This null hypothesis is equivalent to the one previously used in the $r \times c$ table, namely that the classification by rows (termini) is independent of that by columns (lateness):

	Expected frequencies		
	On time	Late	Total
Terminus 1	66.33	90.67	157.00
Terminus 2	72.67	99.33	172.00
	139.00	190.00	329.00

$$X^2 = \frac{(74 - 66.33)^2}{66.33} + \frac{(83 - 90.67)^2}{90.67} + \frac{(65 - 72.67)^2}{72.67} + \frac{(107 - 99.33)^2}{99.33}$$

$$= 2.94$$

and this is distributed approximately as $\chi^2_{(1)}$ by the usual rule for degrees of freedom in an $r \times c$ table. This X^2 is less than 3.84, and so (from Table A4) is not significant at the 5% level: there is no evidence to reject the null hypothesis that the proportions of trains arriving on time at the two termini are the same.

It is particularly important when analysing a 2×2 table not to allow expected values to fall below 5. The possible values that can be taken in the sampling distribution of the X^2 statistic are finite in number, in any particular problem, because the observations from which X^2 is calculated must be whole numbers. But we approximate this distribution, which is discrete, by χ^2 which is continuous. The approximation can be improved for 2×2 tables by using a *continuity correction* due to Yates. This is done by reducing each difference (observed *minus* expected) by $\frac{1}{2}$ in *absolute* value before squaring it.

Example 19.4 2×2 *table*

Two areas of heathland are examined; in the larger area 66 sampling units are examined and 58 of them contain a particular species of heather, while in the smaller area 22 units are examined and 12 of these contain that species. Is the species occurring at the same density over both areas?

The null hypothesis here is that the proportion of units containing the species is the same for both areas. We may arrange the observations in a frequency table, and calculate also an expected table in the usual way, on the null hypothesis.

Observed frequencies	Number of units in which		Total
	Species present	Species absent	
Area A	58	8	66
Area B	12	10	22
	70	18	88

Expected frequencies	Number of units in which		Total
	Species present	Species absent	
Area A	52.5	13.5	66
Area B	17.5	4.5	22
	70.0	18.0	88

$$X^2 = \frac{(|58 - 52.5| - 0.5)^2}{52.5} + \frac{(|8 - 13.5| - 0.5)^2}{13.5} + \frac{(|12 - 17.5| - 0.5)^2}{17.5}$$

$$+ \frac{(|10 - 4.5| - 0.5)^2}{4.5} = (5.0)^2 \left(\frac{1}{52.5} + \frac{1}{13.5} + \frac{1}{17.5} + \frac{1}{4.5} \right)$$

$$= 25 \times 0.3725 = 9.312.$$

The statistic X^2 is distributed approximately as $\chi^2_{(1)}$; the 5%, 1% and 0.1% points of $\chi^2_{(1)}$ are 3.84, 6.63 and 10.83 respectively. The calculated value of X^2 is therefore significant at 1%, and provides strong evidence for rejecting the null hypothesis that the densities are equal.

In all these χ^2 tests a *one-tail test* has been used; the better the data agree with the null hypothesis, the smaller the value of X^2 should be, and it is only when agreement becomes bad, with observed and expected frequencies differing considerably, that we shall want to reject the null hypothesis.

Note that the symbol χ^2 is often used both for the statistic that we have called X^2 and for its theoretical distribution.

19.8.1 Exercises

1. Two groups of people are asked what their opinion was of a television programme about farming. One group, of 120 people, lived in a town, and of these 42 liked the programme. The other group, of 80 people, lived in the country, and of these 18 liked it. Is this evidence that the programme was equally popular in town and country?

2. Carry out the analysis of the data in Exercise 1 by the method of Example 17.11 (p. 422). Also set 95% confidence limits to the true proportions of people who liked the programme: (i) in the town; and (ii) in the country, assuming the proportions to be different.

3. Guinea pigs drawn from two populations, A and B, are classified as having or not having a particular genetic character. The observed results are arranged in a table as shown below: a, b, c and d represent the numbers observed in each cell of the table and $a + b + c + d = N$. Show that the X^2 statistic, for testing whether both populations contain the same proportion with the character, is equal to

$$\frac{N(ad - bc)^2}{(a + b)(a + c)(b + d)(c + d)}.$$

	With characteristic	Without characteristic
Sample A	a	b
Sample B	c	d

4. Extend the formula for X^2 in Exercise 3 to the case where Yates' correction is applied, as described above.

5. Two strains of the same plant were grown from seed in a nursery. Two hundred seeds of an old strain showed 80% germination, while 100 of a new strain showed 90% germination. Is this evidence that the germination rates differ?

Repeat the calculation if 2000 and 1000 seeds, respectively, had been available.

19.9 TESTING THE FIT OF A CONTINUOUS DISTRIBUTION

If a set of observations is being modelled by a continuous random variable, we shall obviously want to find out whether this model fits the data well. In this section we look at two ways of making such a test.

19.9.1 The chi-squared test

This test was developed by Karl Pearson (British, 1857–1936). It compares the frequencies observed, in suitable intervals, with those predicted by the model. Example 14.4 (p. 329) contained data on the times to failure of 500 electric batteries, and later in Section 14.7 we computed the frequencies expected when the exponential distribution was fitted to the data. In order to calculate these expected frequencies, we first had to estimate one parameter, λ. Table 19.3 gives observed and expected frequencies in the intervals 0–$50, 50$–$100, \ldots$ hours.

Since no expected frequencies in Table 19.3 are less than 5, we can use the intervals given there as the cells of the table, and compare the observed and expected frequencies for these nine cells. As usual, we compute

$$X^2 = \sum_{\text{all } r} \frac{(O_r - E_r)^2}{E_r} = \frac{(3.4)^2}{204.6} + \frac{(-8.9)^2}{120.9} + \frac{(3.6)^2}{71.4} + \frac{(-2.2)^2}{42.2} + \frac{(5.1)^2}{24.9}$$

$$+ \frac{(3.3)^2}{14.7} + \frac{(2.3)^2}{8.7} + \frac{(0.9)^2}{5.1} + \frac{(-7.4)^2}{7.4} = 10.96.$$

This statistic X^2 is distributed approximately as $\chi^2_{(7)}$; there are nine cells in the table (nine intervals used in classifying the observations), and the expected frequencies have *two* constraints imposed on them. One is the usual constraint that the totals of the expected and observed frequencies must be the same, and the other is that the mean of the expected frequencies must equal the mean of the observed. The second constraint arises because the parameter, λ, that had to be estimated before the expected frequencies could be found, is equal to 1/(mean of observed data). As $\chi^2_{(7)}$, the calculated value $X^2 = 10.96$ is not significant, so there is no reason to reject the exponential distribution as a model for these data.

If there had been some reason to fix what the value of the parameter λ ought to be, there would be no need to estimate it from the data, and no need to lose one

Table 19.3 *Observed and expected frequencies of lifetimes of 500 electric batteries*

Time (hours)	0–	50–	100–	150–	200–	250–	300–	350–	$\geqslant$400
Observed frequencies	208	112	75	40	30	18	11	6	0
Expected frequencies	204.6	120.9	71.4	42.2	24.9	14.7	8.7	5.1	7.4

degree of freedom by doing so. Thus if we had been told, from long experience of observing this type of battery in the past, that $\lambda = 0.01$, we would calculate expected frequencies using this given value for λ, without making an estimate from the data.

We took as one interval in Table 19.3 all times above 400 hours. No such times were actually observed, but the expected frequency of 400 upwards was 7.4. In continuous distributions, where frequencies tail off, it is always necessary to group the frequencies at the end (or ends) of the distribution to avoid having any intervals that contain expected frequencies of less than five.

In fitting a normal distribution, we need to know μ and σ^2 before the expected frequencies can be calculated. So when μ and σ^2 are not given, we need to estimate these by $\bar{x}$ and σ^2, and to lose two degrees of freedom in doing so.

19.9.2 Exercises

1. Compare the data of Example 14.4 (p. 329) with an exponential distribution having $\lambda = 0.01$ (given). Does this exponential model seem adequate?

2. A supply of rope is thought to have a breaking stress that is uniformly distributed in the range 30 to 40 kg. Two hundred samples gave the following distribution of breaking stress when tested:

Breaking stress (kg)	30–32	32–34	34–36	36–38	38–40
Number of samples	32	47	50	36	35

Test the hypothesis that these data follow the uniform distribution.

3. Test how well the data of Example 15.1 (p. 358) are fitted by a normal distribution.

19.9.3 Kolmogorov–Smirnov test

In Section 19.9.1 we examined the set of data from Example 14.4 to see whether it could be modelled by an exponential distribution. With 500 observations, a χ^2 test could be used satisfactorily. However, if the sample size is not large, a continuous distribution cannot be tested in this way with any accuracy. A method devised by A. N. Kolmogorov (Russian, 1903–1987) and N. V. Smirnov (Russian, 1900–1966) is available. They construct what is called an *empirical* (cumulative) distribution function $S(x)$ from the sample data $x_1, x_2, \ldots x_n$ and compare it with the cumulative distribution function $F(x)$ of the specified model.

Example 19.5 Survival time
Twenty observations were made of the survival time, under standard testing conditions, for an industrial component, using items produced in the same batch at the same factory. The survival times, in hours, were:

17, 28, 33, 37, 44, 46, 51, 60, 72, 80, 95, 105, 119, 137, 148, 176, 186, 242, 300, 338.

These have already been arranged in ascending order, so that the empirical cdf can be calculated immediately.

The null hypothesis is that these survival times may be modelled by an exponential distribution

$$f(x) = \lambda e^{-\lambda x} \quad (0 < x < \infty)$$

in which $\lambda = 0.01$.

The cumulative distribution function for this model is (Section 14.6)

$$F(b) = 1 - e^{-0.01b}$$

when $x = b$. This is plotted as a continuous function in Fig. 19.3.

The empirical distribution function is evaluated by ranking the sample observations in increasing magnitude (which has already been done for these data), and then calculating the cumulative proportional frequencies $S(x)$ at each of the n sample values of x:

Ranked sample value	x_1	x_2	x_3	$\ldots$	x_{n-1}	x_n
$S(x)$	$1/n$	$2/n$	$3/n$		$(n-1)/n$	1

If the differences $F(b) - S(b)$, for the n sample values b of x, are never 'too large', the model is acceptable.

The information required is summarized in Table 19.4. We calculate $F(x)$ from the theoretical distribution and so it is a continuous function. In particular its values $F(b)$ at each of the 20 sample observations $b = 17, 28, \ldots, 338$ are required. Figure 19.3 shows $F(x)$ and also $S(x)$, which is the empirical cumulative distribution calculated from the sample values; $S(x)$ is a *step function*, because it only changes in value at each observation b, and remains constant in between the successive values of b. Since there are $n = 20$ observations, $S(x)$ is 0 up to the lowest sample value $b = 17$, where it becomes $1/n$, that is $1/20$; it retains this value until $b = 28$ where it becomes $2/20$; at $b = 33$ it jumps to $3/20$, and so on. The last

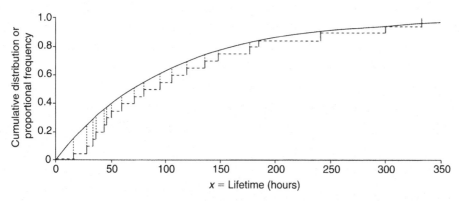

Figure 19.3 *Cumulative distribution $F(x)$ and cumulative proportional frequencies $S(x)$ for Kolmogorov–Smirnov test of Example 19.5:* —— $F(x)$; – – – – $S(x)$. *Differences* $|F(x) - S(x)|$|-----|.

Table 19.4 *Nonparametric Kolmogorov–Smirnov test of goodness-of-fit to an exponential distribution with $\lambda = 0.01$*

Sample observation, b	17	28	33	37	44	46	51	60	72	80
Cumulative distribution in population, $F(b)$	0.156	0.244	0.281	0.309	0.356	0.369	0.400	0.451	0.513	0.551
Sample proportional cumulative frequency, $S(b)$	0.050	0.100	0.150	0.200	0.250	0.300	0.350	0.400	0.450	0.500
Absolute difference $|F(b) - S(b)|$	0.106	0.144	0.131	0.109	0.106	0.069	0.050	0.051	0.063	0.051
Sample observation, b	95	105	119	137	148	176	186	242	300	338
Cumulative distribution in population, $F(b)$	0.613	0.650	0.696	0.746	0.772	0.828	0.844	0.911	0.950	0.966
Sample proportional cumulative frequency, $S(b)$	0.550	0.600	0.650	0.700	0.750	0.800	0.850	0.900	0.950	1.000
Absolute difference $|F(b) - S(b)|$	0.063	0.050	0.046	0.046	0.022	0.028	0.006	0.011	0	0.034

row of Table 19.4 contains the values of $D = |F(x) - S(x)|$ at each of the sample observations $b = 17, 28, \ldots, 338$. These *absolute differences* are the basis of a Kolmogorov–Smirnov test; they are included in Fig. 19.3.

A special table is needed for this test, which is presented as Table A9 (p. 724); it gives the critical values of the absolute difference $|F(x) - S(x)|$ for various significance levels and for sample sizes $n = 1, 2, 3, \ldots, 40$ together with an approximation that can be used for $n > 40$. The *maximum* value of $D = |F(x) - S(x)|$ for the n sample values b is used in the test, and if this is less than the critical value given in Table A9 then the null hypothesis is not rejected at the chosen significance level.

For our example, the maximum value of D is 0.144 (at the second smallest observation) and the sample size is $n = 20$; there is no suggestion at all that we should reject the null hypothesis because the critical value at the 5% level of significance is 0.294. Our result is not even 'significant' at the 10% level, and so we may regard the hypothesis of an exponential distribution with parameter $\lambda = 0.01$ as a very reasonable explanation of the data.

19.10 MINITAB: CHI-SQUARED TESTS

Values of the probability distributions for chi-squared, and the generation of random samples from the distribution, may be obtained by choosing

Calc>Probability Distributions>Chi-square/

and

Calc>Random Data>Chi-Square/

The number of degrees of freedom is entered in the dialogue box. There is also a box labelled **Noncentrality parameter** set at 0.0. This should be left alone. The chi-squared distributions we meet in this book all have value zero for the parameter.

The analysis of contingency tables using a chi-squared test is very straightforward. The numbers in the body of the table are entered into the worksheet. Thus the data in Table 19.1, on hair and eye colour, would be entered in three columns, say C1, C2, C3, with four rows in each column. The analysis is carried out by choosing

Stat>Tables>Chi-Square Test (Tables in Worksheet)/
Enter names under **Columns containing the table**/

The observed and expected values, and the corresponding component of the test statistic, are presented in a table together with the value of X^2 (labelled Chi-Sq), the degrees of freedom and the probability of obtaining a value of chi-squared greater than X^2.

A slightly different choice in the menu allows much more flexibility in the output. The table produced by the choices of the previous paragraph may be seen as rather cluttered. To obtain a table containing just the observed and expected values proceed as follows. Put all the frequencies in C1; put row and column numbers in C2 and C3 opposite the appropriate frequencies. Then choose

Stat>Tables>Cross-Tabulation and Chi-Square/
Under **Categorical Variables** enter names **for rows** and **columns**,
enter column of **frequencies**, select **Counts** for **Display**, select **Chi-Square**/
Select for **Display, Chi-Square analysis** and **Expected cell counts**//

The procedure is not so automatic with a goodness-of-fit test. The steps of the calculation have to be worked through. Consider Example 19.2, in which it was tested whether the observed distribution of number of boys, in families of five children, followed a binomial distribution. The possible number of boys, 0 to 5, might be put in C1 and the corresponding observed numbers put in C2 and named O. The expected frequencies can be calculated using **Calc>Probability Distributions>Binomial**, and the probabilities, multiplied by 320, put into C3 and named E. To calculate X^2, or CHISQ, choose

Calc>Calculator/
Enter CHISQ in **Store result in variable**, enter SUM((O-E)**2/E) in
Expression/

The value of CHISQ appears in the worksheet. The value of the cumulative distribution function with argument CHISQ, for the appropriate degrees of freedom, can be found using **Calc>Probability Distribution>Chi-Square**. Taking the value from one gives the p-value.

MINITAB offers the Kolmogorov–Smirnov test for the particular case of testing the normality of a set of data. Choose

Stat>Basic Statistics>Normality Test/

and it is one of the options.

1. According to a simple genetic hypothesis, *morning glory* plants should fall into four classes I, II, III, IV in the ratio 9:3:3:1. In one family of plants, there are 75, 14, 14, 11 respectively of the four classes. Examine whether this family appears to follow the hypothesis.

2. A new method of treating a foot complaint is to be tested against a standard treatment. The success rate of the standard treatment is known to have been 64% over a long period. A random sample of 100 subjects is given the new treatment, and 75 respond successfully. Is the new treatment significantly better than the standard treatment?

3. A music teacher believes that half of his students will prefer works by Mozart and the other half will prefer works by Schubert. He further believes that, of those who prefer Schubert, half will prefer songs to piano music. He plays three works to his students, one (M) by Mozart, one (SS) a Schubert song, and one (SP) a Schubert piano piece. Eighty-four students state preferences: 48 for M, 17 for SS, 19 for SP. Does this support his idea?

4. Windfall apples (those falling from the tree before harvest) were examined to see whether they had been attacked by insects. Of 60 windfalls in an orchard, 32 had been attacked, while of 105 apples harvested from the trees in the same orchard 41 had been attacked. Does this indicate that the proportion of attacked apples is different in windfalls and harvested apples?

5. In 120 throws of a die, the scores obtained were as follows. Is there any evidence that the die is 'unfair'?

Score	1	2	3	4	5	6
Frequency	15	25	18	15	23	24

6. Three coins are tossed, and the number of heads is recorded. The process is repeated 80 times, with the following results. Is it reasonable to claim that the coins were fair?

Number of heads	0	1	2	3
Frequency	10	25	34	11

7. In a vegetation survey, 46 units of area of fixed size were marked on the ground, and the presence or absence of several species of plant was recorded. Two species, A and B, occurred together in 20 of these unit areas, A alone in 2, B alone in 13, and neither A nor B in the remainder. Is there evidence that these two species are associated?

8. Two psychologists, P and S, interviewed the same 64 patients and recorded whether or not the patient showed symptoms of self-reproach. Their results were as follows. Are the two psychologists agreeing in their assessment?

| | P's diagnosis of symptoms | |
S's diagnosis of symptoms	Present	Absent
Present	21	8
Absent	20	15

9. A chain store offers rain hats for sale in three colours: red, blue and green.
A sales manager wonders if the colour preference of customers differs in
London and a county town. In the London store 48 hats were sold in a week
and of these 19 were red, 14 blue and 15 green. In the county town branch 32
hats were sold and of these 6 were red, 16 blue and 10 green. Is there evidence
of differential colour preference between London and the county town?
   Are any colours significantly preferred to the others?   (*Oxford*)

10. A doctor's records for a three-month period show the social classes of the
500 people who attended his surgery. He compares these with the social
class distribution in the whole area covered by his practice. Do the following
results suggest that those who attended surgery are a representative sample
of the whole area in respect of social class?

Social class	I	II	III	IV	V	Not known
Patients	10	85	275	78	6	46
Population in whole area	300	690	1500	470	190	710

11. Sixty-four observations on a random variable X gave the following grouped
frequency distribution:

Interval	0–	1–	2–	3–	4–	5–	6–	7–	8–
Frequency of observations	0	2	7	7	8	11	16	13	0

Use the χ^2 goodness-of-fit test to test if X could have the pdf

$$f(x) = \begin{cases} x/32 & 0 \leqslant x \leqslant 8 \\ 0 & \text{otherwise.} \end{cases} \qquad (IOS)$$

12. The following frequency table gives the heights in inches of 100 male students,
classified in intervals of 2 in, with the corresponding class frequencies:

Height in inches	61–63	64–66	67–69	70–72	73–75	Total
Observed class frequencies	5	20	39	28	8	100

   By coding the above data, show that the mean and the standard deviation
of the distribution are 68.42 in and 2.97 in respectively. Fit a normal curve to
the above data. Compare the theoretical class frequencies with the observed
values. Use a χ^2 test to check that the fit is very good.   (*AEB*)

13. Traffic is passing freely along a road, and the time interval between successive vehicles is measured (in seconds). Fit an exponential distribution to the following observations, and test whether it is a satisfactory explanation of them:

Time interval	Up to 15	15–30	30–45	45–60	60–75	75–90	90–105
Frequency	63	25	14	7	6	3	2

14. The data below relate to the daily coke yield (per cent) of a coke oven plant over a period of 260 days:

Daily coke yield (per cent to nearest unit)	67	68	69	70	71	72	73
Number of days	7	23	61	87	56	19	7

Calculate the mean daily coke yield (per cent) for the period and also the standard deviation of the yield.

Calculate the expected frequency in the interval 68.5–70.5 (per cent) for a normal distribution having the same mean and standard deviation.

Estimate the frequency of observations in the above table within one standard deviation of the mean value and find also the expected frequency in the same interval for the normal distribution. (*MEI*)

15. Test whether the normal distribution is a good explanation of the data in Exercise 14.

16. A plant breeder grows three strains of corn which are thought to be the same genetically. A number of seedlings of each strain are grown, and classified as type a, b, c or d in respect of a certain characteristic. The results are shown below. Test whether the strains are the same in respect of this characteristic, and test also whether the ratio of a to $(b + c + d)$ in strain II could be 9:7.

	Type				
	a	b	c	d	Total
Strain I	75	15	25	5	120
Strain II	85	37	26	12	160
Strain III	60	28	19	13	120

17. A machine shop produces specialized tools. It is suspected that more faulty tools are produced towards the end of a working day. Over the last month, the numbers of faulty items per hour in the day totalled:

Working hour	1	2	3	4	5	6	7	8
Number faulty	2	4	3	3	3	7	8	10

Test the hypothesis that these data come from a population having a uniform distribution, using (i) a χ^2 test, (ii) a Kolmogorov–Smirnov test. Compare and comment on the results. (*RSS*)

19.12 COMPUTING EXERCISES

1. By using a chi-squared goodness-of-fit test, decide whether the data on blood pressure of workers (Revision Exercises E, Question 16) may be fitted by a normal distribution. Use a mean of 129.25 and standard deviation of 13.804, as may be calculated from the data, for fitting the normal distribution.

2. Generate 500 observations from the standard normal distribution. Form a frequency table with lowest class centre at -3.25 and a class-interval of 0.5. Carry out a chi-squared goodness-of-fit test, based on a comparison of the data with a normal distribution with mean 0 and standard deviation 1.

3. Use a computer to test whether the two classifications in the contingency table in Section 19.7 are independent.

19.13 PROJECTS

1 Testing fit of theoretical distributions to observed data (Data from earlier projects)

Aim. To use a χ^2 goodness-of-fit test.
Collection of data. Use one or more sets of data from the projects:
(a) Sex distribution of children in families (8.15(1)). Binomial distribution.
(b) Throwing sixes with a die (2.13(2)). Geometric distribution.
(c) Time between successive hatchback cars in traffic (14.14(1)). Exponential distribution.
(d) Total (or mean) of ten random digits (15.12(1)). Normal distribution.
Analysis. Fit the theoretical distribution named for each set of data. Test the fit using a χ^2 test.

2 Testing independence of two variates (Data from earlier projects)

Aim. To test independence of two variates using a contingency table and a χ^2 test.
Collection of data. Use data from the projects:
(a) Joint distribution of arm length and collar size (10.7(1)).
(b) Goals for and against in football (10.7(2)).

Analysis. Arrange the data in contingency tables ensuring that none of the expected values in cells are small. Use a χ^2 test.

SUMMARY

Suppose that an observation or measurement is classified into k different categories, and the frequencies observed in the categories are O_r ($r = 1, 2, \ldots, k$) such that $\sum O_r = N$. A null hypothesis states that the corresponding expected values are $\{E_r\}$, and $\sum E_r = N$. The statistic for testing this hypothesis is $\sum (O_r - E_r)^2 / E_r$, which is distributed approximately as χ^2, with degrees of freedom equal to the number of categories *minus* the number of constraints placed upon $\{E_r\}$ (one of which is $\sum E_r = N$). This is a **goodness-of-fit test**.

Because the χ^2 test is approximate, no expected value should fall very far below 5, and categories may have to be combined to achieve this.

Two-way contingency tables, with a null hypothesis that the row classification is independent of the column classification, may also be examined in a χ^2 test with $\{E_r\}$ calculated from the same marginal row and column totals as in the observed table. A 2×2 table is a simple way of comparing two estimated proportions against an NH that they are equal.

A **nonparametric test** that a set of observed data follow a specified distribution finds an empirical cumulative distribution function from the observed data and compares this with the cdf at the corresponding points in the specified distribution. If the maximum difference between the two cdfs is less than a tabulated value, the specified distribution is not rejected as a model for the data.

REVISION EXERCISES E

1. The degree of cloudiness of the sky may be measured on an eleven-point scale, with the value 0 corresponding to a clear sky and the value 10 corresponding to a completely overcast sky. Observations of the degree of cloudiness were recorded at a particular meteorological station at noon every day during the month of June over ten consecutive years. The frequency distribution of the 300 observations was as follows:

Degree of cloudiness	0	1	2	3	4	5	6	7	8	9	10
Number of days	64	24	13	9	6	8	7	9	14	27	119

Plot a frequency polygon for these data, and calculate the sample mean and sample variance.

Comment briefly on the shape of the distribution.

Discuss the usefulness (or otherwise) of the sample mean as a measure of location of such a distribution. *(JMB)*

2. In a mechanism a plunger moves inside a cylinder and the difference between the diameters of the plunger and the cylinder must be at least 0.02 cm. Cylinders and plungers are manufactured separately; the diameters of the cylinders are normally distributed about a mean of 4.05 cm with standard deviation 0.004 cm, while the diameters of the plungers are normally distributed about a mean of 4.02 cm with standard deviation 0.006 cm.
   (i) Find the proportion of cylinders that will have inadequate clearance for plungers of diameter 4.02 cm.
   (ii) Find the proportion of plungers that will have inadequate clearance in cylinders of diameter 4.05 cm.
   (iii) If plungers and cylinders are assembled at random, find the proportion of pairs in which the clearance is insufficient.
   (iv) If a set of ten plungers and ten cylinders are taken at random and assembled, find the probability of the clearance being sufficient in all cases. *(O&C)*

3. When an experienced kingfisher tries to catch a fish the probability that he is successful is 1/3. Find the probability that:
   (i) he catches exactly 2 fish in 5 attempts;
   (ii) he catches at least 2 fish in 5 attempts.
   A less experienced kingfisher is only successful, on the average, in one attempt out of 10. Using a suitable approximation, calculate the probability that he catches more fish in 200 attempts than the experienced kingfisher catches in 20 attempts. *(Cambridge)*

4. A random experiment has the three possible outcomes 0, 1, 2 occurring with probabilities $p, 1 - 2p$, and p, respectively, where $0 < p < \frac{1}{2}$. In three independent trials of the experiment show that the probability of at least two zeros occurring is $p^2(3 - 2p)$.
   Let M denote the median of the outcomes of three independent trials of the experiment. Derive the sampling distribution of M and hence show that the variance of M is less than the variance of the mean of the three outcomes only if $6p^2 - 9p + 1 > 0$. *(Welsh)*

5. A random sample of observations is taken from a population which is normally distributed, with unknown mean μ and known standard deviation equal to 1.5 units.
   Calculate the size of sample required to ensure, with probability 0.95, that the sample mean lies within 0.5 units of μ.

6. A survey of 500 boxes arriving at a factory showed 15% to be damaged. How accurate is this result? What sample size would be needed so that, with a probability of 99.9%, the proportion of defective boxes found is accurate to within 5%?

7. An experiment was conducted to test whether or not treatment with a new rust preventative reduces oxidation. The experimenter took ten test pieces of iron of equal weight and divided each piece in two. One half was given the new treatment, and the other half a standard treatment, the choice being

made at random. After exposure to test conditions, the pairs were weighed to discover the loss in weight, giving the results shown below. Analyse these results and state what conclusion you reach. Give a 95% confidence interval for the mean difference in weight loss.

| | Weight loss (g) | |
Test piece	Standard	New
1	6.3	3.7
2	5.2	2.1
3	4.0	4.2
4	4.5	2.8
5	6.2	5.6
6	6.1	4.9
7	5.8	3.6
8	5.3	4.2
9	4.9	5.1
10	5.3	4.7

(*IOS*)

8. A survey of a filing system indicated that, out of 500 invoices selected at random from a very large number in an office, 80 were incorrectly completed.

What are 95% confidence limits on the proportion of incorrect invoices in this office?

9. State the conditions in which the normal approximation to the binomial distribution may be used.

In trying to establish differences in taste between products A and B, the following test was used. Members of the public were each given three coded but unidentified samples, which were in fact two samples of A and one of B. They were asked to pick out the odd sample by taste alone, but to make a definite choice even if they were uncertain. In 1000 trials, 360 identified B correctly. Test whether the result is indicative of a real difference in taste between the two products.

Using the sample result, obtain an approximate 99% confidence interval for the probability of picking out B.

Estimate the number of trials necessary for the sample estimate of the probability of picking out B to lie within 0.01 of the true value with 99% confidence. (*JMB*)

10. A method of chemical analysis is used to determine the iron content in a compound. Over a long period of use it is known to have a standard deviation of 0.2 units of iron, and to give normally distributed measurements.

Nine samples give 34.2, 34.6, 34.6, 34.7, 34.4, 35.2, 34.6, 34.5, 34.6, units. Set 95% confidence limits to the true content in the population from which these samples were drawn. (*IOS*)

11. Derive the expressions for the mean and the variance of the sampling distribution of the mean of a random sample of n independent observations from a population distribution whose mean is μ and whose variance is σ^2.

   In the case when $\sigma = 1$, show that for the standard error of the sample mean to be less than 0.3, it is necessary that n should be at least 12. For $n = 12$, and assuming that the population distribution is normal, calculate the probability that the observed value of the sample mean will be within 0.3 units of the population mean. Interpret your result as a confidence interval for the population mean. *(Welsh)*

12. Soap cartons are manufactured to a target width of 240 mm. If the manufacturing process results in widths having a standard deviation of 5 mm, what size samples are required to check that the mean carton width falls between 237.5 and 242.5 mm with 95% probability?

13. Two types of seed, X and Y, difficult to distinguish, are such that in the long run under given conditions 80% of type X but only 60% of type Y will germinate. Samples of 100 seeds of type X are selected at random. Find the mean and standard deviation of the number germinating.

   A package arrives at a nursery without its label. It contains just one type of seed done up in packets each containing 100 seeds. The nurseryman knows that the seed is either of type X or of type Y, and thinks that it is more likely to be the former. He decides to plant a packet and to accept the seed as type X provided that at least 68 germinate, but otherwise to regard it as type Y. Using the normal approximation, find the probability that the nurseryman will wrongly label type X seed as type Y.

   What is the probability that he will wrongly label type Y seed as type X? *(SMP)*

14. Random samples of size n are chosen from a normal distribution with mean μ and known variance σ^2. In each case the sample is used to test, at the 5% level, the hypothesis that $\mu = \mu_0$. If μ_0 happens to be the true value of the mean of the normal distribution, in what proportion of the tests will the hypothesis be accepted?

   In an industrial experiment to study the speed of performing a specific task, 120 workmen were chosen at random to perform this task. The sample mean was 17 min 18 sec and the sum of squares about the mean was 368 $(min)^2$. Is there significant evidence that the mean time taken to perform the task is more than 17 minutes? Determine a 95% symmetric two-sided confidence interval for this mean time. *(JMB)*

15. Fifteen pupils experimented to find the value of g, the acceleration due to gravity. Their results were as follows:

9.806	9.807	9.810	9.802	9.805	9.806	9.804	9.811	9.801
9.804	9.805	9.808	9.803	9.809	9.807			

Calculate the mean and standard deviation of these results. Give 95% confidence limits for the value of g based upon them.

Estimate the number of experimenters needed to give a confidence interval of less than 0.001. *(Southern)*

16. The following table gives the distribution of systolic blood pressure, in millimetres of mercury, for 254 male workers aged 30 to 39 years.

Blood pressure (mid-interval value)	85	95	105	115	125	135	145	155	165	175
Number of men		1	2	12	47	77	65	32	13	4	1

Calculate the mean and the standard deviation of these pressures. Determine the standard error of the mean and calculate 99% confidence limits for the mean blood pressure of the population of which the above group is a random sample. *(MEI)*

17. Five metal strips taken at random from a large consignment are tested for hardness. The hardness figures obtained in the five tests, in VDH units, were

$$720 \quad 724 \quad 727 \quad 723 \quad 729.$$

Make an estimate of the mean hardness of the strips and give 95% confidence limits for the mean.

Subsequently a further eight strips from the same consignment were tested and the hardness figures, in the same units, were

$$722 \quad 716 \quad 720 \quad 721 \quad 721 \quad 715 \quad 722 \quad 724.$$

Is there a significant difference between the means of the two samples?
(O&C)

18. Let $\overline{X}$ denote the mean of a random sample of 16 observations from a normal distribution having mean μ and standard deviation σ. Show that 95% confidence limits for μ are given by $\overline{X} \pm 0.49\sigma$.

Let $\overline{Y}$ denote the mean of an independent random sample of 25 observations from another normal distribution having mean λ but the same standard deviation σ. Write down expressions for the mean and the variance of $W = \overline{X} - \overline{Y}$.

The observed values of $\overline{X}$ and $\overline{Y}$ are to be used to test the hypothesis that $\lambda = \mu$, and it is decided to reject this hypothesis only if the 95% confidence intervals for μ and λ do not overlap. If, in fact, $\lambda = \mu$, calculate the probability that the hypothesis will be rejected. (You may assume that W is normally distributed.) *(Welsh)*

19. The manufacturers of a product A wish to compare the mass of the contents of a packet of their product with that of a competitor's product B which is

nominally of equal mass. For 100 packets of each product the mass x grams of each packet is measured. The following data are collected:

	Product A	Product B
$\sum x$	45 551	45 404
$\sum (x - \bar{x})^2$	3 300	2 475

Find a symmetrical two-sided 95% confidence interval for the amount by which the mean mass of a packet of A exceeds that of a packet of B. State whether the observed difference of the sample means is significant at the 5% level.

The nominal net mass of either packet is 454 grams. Estimate the probability that a packet of product A contains less than this stated quantity, and find the value (to 3 significant figures) of the mean mass which would be necessary for less than 1% of the packets of product A to fall below this nominal net mass, assuming that the variance remains unaltered.   (*JMB*)

20. In an experiment to test the effectiveness of a slimming diet, 100 persons followed the diet for one month and the losses in their weights, in kilograms, were recorded. The results were summarized by stating that 95% confidence limits for the mean loss in weight were 2.27 kg and 3.63 kg. Which, if any, of the following interpretations of this statement is correct? If you disagree with both interpretations, give your own interpretation.

 (i) Exactly 95 of the persons in the experiment lost between 2.27 kg and 3.63 kg.

 (ii) If the diet was followed for one month by several groups of 100 persons per group, then 95% of these groups would have a mean weight loss between 2.27 kg and 3.63 kg.

In an independent experiment, 10 persons followed this diet for two months and the weight losses (in kg) were respectively

$$0.6, \quad 1.1, \quad 1.5, \quad 1.9, \quad 2.6, \quad 2.8, \quad 3.0, \quad 3.5, \quad 3.7, \quad 4.3.$$

Use these results to determine 95% confidence limits for the mean loss in weight when the diet is followed for two months. State clearly any assumptions you make in order to determine these limits.   (*Welsh*)

21. The breaking loads (in kilograms) of 50 steel rods are given in the following table:

6.70	7.04	7.21	7.29	7.35	7.45	7.59	7.70	7.72	7.74
7.84	7.88	7.94	7.99	7.99	8.04	8.10	8.12	8.15	8.17
8.20	8.21	8.24	8.24	8.26	8.28	8.31	8.37	8.38	8.42
8.49	8.51	8.55	8.56	8.58	8.59	8.60	8.66	8.66	8.69
8.70	8.74	8.81	8.84	8.86	8.90	9.00	9.15	9.55	9.80

Taking equal intervals of 0.5 kg, of which the first is from 6.25 to 6.75, group the results as a frequency distribution and draw a histogram to illustrate the data.

On a separate diagram draw a cumulative frequency curve to fit the measurements and deduce the values of the median and quartiles.

Assuming the results agree with a normal distribution of loads, compare the semi-inter-quartile range with that of a normal distribution and make a rough estimate (to two significant figures) of the standard deviation. (*O & C*)

22. (i) If $\bar{x}$ is the mean of the n values $x_1, x_2, \ldots, x_n$, show that for any constant a,

$$\sum_1^n (x_i - a)^2 = \sum_1^n (x_i - \bar{x})^2 + n(\bar{x} - a)^2.$$

(ii) Let $x_1, x_2, \ldots, x_{10}$ denote 10 random values from a normal distribution with unknown mean μ and unknown variance σ^2. Given that the mean of these values is $\bar{x} = 3.2$ and that $\sum_1^{10} (x_i - \bar{x})^2 = 1.80$, calculate 99% symmetric confidence limits for μ.

Suppose that y_1, y_2, y_3, y_4, y_5 are a further 5 random values from the above normal distribution; the mean of these values is $\bar{y} = 2.9$ and $\sum_1^5 (y_i - \bar{y})^2 = 1.44$. Regarding all 15 values as comprising one random sample from the above normal distribution, recalculate 99% symmetric confidence limits for μ. (*Welsh*)

23. Two schools, A and B, enter candidates for an examination in which the result is honours, pass or fail. Of the 70 candidates from A, 2 obtain honours, 17 pass and 51 fail. Of the 30 candidates from B, 1 obtains honours, 5 pass and 24 fail. Comment on these results from a statistical point of view.

24. The numbers of two types of moth caught in light traps during 1973 at three sites were as follows:

	Site A	Site B	Site C
Turnip moth	30	56	78
Tomato moth	214	132	193

Analyse these results and explain carefully how they help to indicate any differences between moths or between sites.

25. The table below gives details of accidents occurring in different areas of a factory last year. Is there a significant difference between the chances of a worker being involved in an accident in the different areas?

Area	Number of people having accidents	Number of workers employed
1 Production	169	700
2 Packaging	32	100
3 Quality control	3	20
4 Engineering	15	40
5 Laboratory	16	60
6 Warehouse	15	80

26. The table shows the month of onset of a particular disease for all cases reported in a region during 1946 to 1960.

Jan.	Feb.	March	April	May	June	July	Aug.	Sept.	Oct.	Nov.	Dec.
40	34	30	44	39	58	51	55	36	48	33	38

(a) Test whether the disease appears to arise uniformly through the year.
(b) Test whether the months May–October show a different rate of onset from the months November–April, but the disease arises uniformly (at one rate) in May–October and uniformly (at another rate) in November–April.

27. Describe the χ^2 test as applied to a 2×2 table and explain how the degrees of freedom and expected values are obtained. Why is an allowance made for discontinuity in calculating the value of χ^2 in such problems?
   The following table gives the results of alternative medical treatments A and B on 110 patients in all:

	Successful	Unsuccessful
Treatment A	55	5
Treatment B	39	11

Is the difference between the results of the two treatments significant?
   A second similar trial of the two treatments also showed A slightly more successful than B and gave $\chi^2 = 3.21$. What conclusion may be drawn from the two trials together?   (O & C)
(*Note.* When X, Y are independent random variables following χ^2 distributions with m, n degrees of freedom, then $X + Y$ is a χ^2 variable with $m + n$ degrees of freedom.)

28. In a comparison of the wearing quality of two types of stocking, A and B, ten wearers each tested one pair of A and one pair of B. From the results below, decide whether there is any difference in wearing quality, using a parametric test and a nonparametric test. Which test is the better one?

Wearer		1	2	3	4	5	6	7	8	9	10
Life in days	type A	20	18	21	14	16	29	16	14	11	12
	type B	25	19	24	17	19	28	18	15	13	14

29. An experimenter takes 20 pairs of identical male twins. In each pair he chooses one at random, and includes a particular additive in his diet. After a time he weighs the 40 individuals: the weight gain over the experimental period of the twin in the rth pair with the additive is x_r, and that of his brother is y_r.

The experimenter then calculates

$$\bar{x} = \frac{1}{20} \sum_{r=1}^{20} x_r, \quad \bar{y} = \frac{1}{20} \sum_{r=1}^{20} y_r,$$

and using

$$\sum_{r=1}^{20} (x_r - \bar{x})^2 \quad \text{and} \quad \sum_{r=1}^{20} (y_r - \bar{y})^2$$

he estimates the standard error of $\bar{x}$ and $\bar{y}$. Pooling these he derives the standard error of $\bar{x} - \bar{y}$ and is thus able to use Student's t-distribution (with 38 degrees of freedom) to test whether the additive had any significant effect on weight gain.

Comment on this analysis. If you think the right statistical procedure was used, describe the assumptions implicit in it. Otherwise, describe the procedure that should have been adopted.

30. Suppose a random sample of n_1 observations is obtained from one population and an independent random sample of n_2 observations from a second population, where the means and variances of both populations are unknown. Explain in detail how you would test the hypothesis that the population means are equal, against a two-sided alternative hypothesis, for each of the following cases:
    (i) if n_1 and n_2 are small and the populations may be taken to be normally distributed with the same variance;
    (ii) if the populations are not assumed normally distributed or to have the same variance, but n_1 and n_2 are large.
    In each case state whether the test that you give is exact or approximate.

31. Eleven sites, randomly chosen within city A, gave smoke concentration readings with a mean of $72 \, \mu\text{g/m}^3$ and a standard deviation of $6.9 \, \mu\text{g/m}^3$. In city B, fifteen randomly chosen sites gave smoke concentration readings with a mean of $51 \, \mu\text{g/m}^3$ and a standard deviation of $5.8 \, \mu\text{g/m}^3$.
    (i) Give a 95% confidence interval for the difference between the mean smoke concentrations for the two cities.
    (ii) It is now claimed that the true concentration in A is at least $10 \, \mu\text{g/m}^3$ greater than that in B. Test this claim.
    (iii) Discuss critically any assumptions that you have made.

32. What is meant by a *nonparametric test*?
    A random sample of ten observations gave the values, in centimetres, 96.7, 84.3, 101.8, 78.3, 110.6, 93.4, 87.8, 91.3, 98.2, 88.7. Use (i) a parametric; (ii) a nonparametric; test of the hypothesis that the sample came from a distribution whose median value was 100 cm.

33. In a paired-comparison experiment the variate recorded was the percentage of seeds that germinated. Discuss which of the three tests (i) t-test, (ii) sign

test, (iii) Wilcoxon signed-rank test, is most appropriate for the analysis of these data.

In a comparison of two composts, trays of the different composts were put together in pairs. One hundred seeds were sown on each tray, the percentages that had germinated after two weeks being:

Pair	1	2	3	4	5	6	7	8	9	10
Compost I	25	26	38	27	26	28	32	23	30	28
Compost II	21	25	26	20	24	22	24	35	33	23

Test whether there was a difference in germination rate in the two composts.

34. Twenty seedling plants are raised from the same source of material. After a fixed period of growth their heights are measured (in centimetres). The measurements are 22, 16, 17, 19, 24, 20, 21, 20, 28, 25, 13, 15, 19, 22, 17, 18, 23, 27, 20, 15. Find (i) 95%, (ii) 99% confidence intervals for the true values of mean and variance in the population (assumed normally distributed) from which these were drawn.

35. Patients in five different disease groups were classified according to whether they smoked or not, and if so how heavily (measured by number of cigarettes per day on average). The following data were obtained. Test whether extent of smoking and disease classification are related.

Disease group	Extent of smoking			
	Non-smoker	Light	Medium	Heavy
Cancer (other than lung)	236	78	237	167
Respiratory (not cancer)	42	33	128	132
Heart	22	19	64	61
Gastric	39	31	143	115
Others	38	24	91	62

36. The lengths X of offcuts of timber in a carpenter's workshop follow the continuous uniform distribution with probability density function (pdf)

$$f(x) = \begin{cases} 1/\theta & (0 \leqslant x \leqslant \theta; \ \theta > 0); \\ 0 & \text{(elsewhere).} \end{cases}$$

(i) Obtain the cumulative distribution function, $F(x)$, and sketch the graphs of $f(x)$ and $F(x)$. Also find $E[X]$ and $Var[X]$.

(ii) A random sample of offcuts have lengths $x_1, x_2, \ldots, x_n$. Explain why

$$P(X_{(n)} \leqslant x) = (x/\theta)^n, \quad 0 \leqslant x \leqslant \theta,$$

where $X_{(n)}$ is the longest offcut in the sample. Show that

$$E[X_{(n)}] = n\theta/(n+1) \quad \text{and} \quad \text{Var}[X_{(n)}] = n\theta^2/\{(n+1)^2(n+2)\}.$$

(iii) Write down a multiple of $X_{(n)}$ which is an unbiased estimator of θ and obtain its variance. Is $X_{(n)}$ the maximum likelihood estimate?

(iv) Show that $2\sum_{i=1}^{n} X_i/n$ is the method-of-moments estimator of θ and obtain its variance. How would you advise the carpenter to estimate θ?

20 THE POISSON DISTRIBUTION

OVERVIEW

The Poisson distribution models the number of random events occurring in unit intervals of time or space. In this chapter we study its properties and uses, and methods of testing hypotheses about it.

20.1 INTRODUCTION

A discrete distribution which has many applications in statistics was discovered by the French mathematician S. D. Poisson in 1837. We will derive the distribution first of all, in the way that he did, as a limit of the binomial distribution.

20.2 TWO MATHEMATICAL RESULTS

We shall need to use two mathematical results involving the exponential function, exp x or e^x. Their proofs appear in Appendices II and V, and readers unfamiliar with the results should consult these before proceeding.

(1) The exponential function may be defined as a series:

$$e^x = 1 + x + \frac{x^2}{2!} + \frac{x^3}{3!} + \cdots + \frac{x^r}{r!} + \cdots$$

and this is valid for all values of x.

(2)
$$\lim_{n \to \infty} \left(1 + \frac{x}{n}\right)^n = e^x.$$

20.3 THE BINOMIAL DISTRIBUTION AS $n \to \infty, \pi \to 0$

Six hundred children attend a school. What is the probability that on a particular day, say 1 May, r of the children have a birthday? It is not difficult to set up a reasonable model for this situation. We assume that the probability that one child has a birthday on 1 May is 1/365; this is not likely to be exactly true since there are seasonal differences in births, but it will be close to the true value. If we make the further assumption that the birthdays of separate children are independent, we have the conditions for a binomial distribution, and hence

$$P(r \text{ births on 1 May}) = \binom{600}{r} \left(\frac{1}{365}\right)^r \left(\frac{364}{365}\right)^{600-r}.$$

The probabilities of $0, 1, 2, \ldots$ birthdays are:

$$\left(\frac{364}{365}\right)^{600}, \binom{600}{1}\left(\frac{1}{365}\right)\left(\frac{364}{365}\right)^{599}, \binom{600}{2}\left(\frac{1}{365}\right)^2\left(\frac{364}{365}\right)^{598}, \ldots$$

It would be extremely tedious to compute this set of probabilities. We shall show how the computations may be simplified by using Poisson's result.

Let us write π for the probability of a birthday on 1 May and n for the total number of children. The probability of no birthdays in the school on 1 May is $p_0 = (1 - \pi)^n$. Poisson considered the limit of this as $n \to \infty$ and $\pi \to 0$, in such a way that their product $n\pi$ remained constant. We call this product λ. We require the limit of

$$p_0 = (1 - \pi)^n = \left(1 - \frac{\lambda}{n}\right)^n.$$

Now from the second mathematical result quoted above, we see that

$$\lim_{n \to \infty} p_0 = \lim_{n \to \infty} \left(1 - \frac{\lambda}{n}\right)^n = e^{-\lambda}.$$

This therefore gives us an approximation to p_0 provided that n is large and π is small. For our example, $\lambda = n\pi = 600 \times (1/365) = 1.6438$. Hence an approximate value of p_0 is $e^{-\lambda} = e^{-1.6438} = 0.193\,237$. The exact value $(364/365)^{600} = 0.192\,802$. Thus the approximation is a very good one, the two values being the same when expressed correct to three decimal places.

We may derive an approximation for p_1 in a similar way. We have

$$p_1 = n\pi(1 - \pi)^{n-1} = \frac{n\pi}{(1 - \pi)}p_0 = \frac{\lambda}{1 - (\lambda/n)}p_0.$$

Hence

$$\lim_{n \to \infty} p_1 = \lim_{n \to \infty} \frac{\lambda}{1 - (\lambda/n)}p_0 = \lim_{n \to \infty} \frac{\lambda}{1 - (\lambda/n)} \lim_{n \to \infty} p_0 = \lambda e^{-\lambda}.$$

The approximate value for p_1, in our example, is $\lambda e^{-\lambda} = 1.6438\,e^{-1.6438} = 0.317\,651$. The exact value is $0.317\,805$. Again the approximation gives a value that is very close indeed to the exact one, and differs from it only in the fourth decimal place.

We next examine how the approximate value of p_r is related to the approximate value of p_{r-1}. Since for a binomial distribution

$$p_r = \frac{n - r + 1}{r} \frac{\pi}{1 - \pi} p_{r-1} \quad (\text{see p. 193}),$$

we have, writing $n\pi = \lambda$, that

$$p_r = \frac{1 + [(1-r)/n]}{r} \frac{n\pi}{1-\pi} p_{r-1} = \frac{1 + [(1-r)/n]}{r} \frac{\lambda}{1-(\lambda/n)} p_{r-1}$$

and hence

$$\lim_{n\to\infty} p_r = \lim_{n\to\infty} \frac{1 + [(1-r)/n]}{r} \frac{\lambda}{1-(\lambda/n)} \lim_{n\to\infty} p_{r-1} = \frac{\lambda}{r} \lim_{n\to\infty} p_{r-1}.$$

This allows us to generate the following set of approximations to the binomial probabilities $\{p_0, p_1, p_2, p_3, \ldots, p_r, \ldots\}$:

$$\left\{ e^{-\lambda}, \left(\frac{\lambda}{1} \cdot e^{-\lambda} \right), \left(\frac{\lambda}{2} \cdot \frac{\lambda}{1} \cdot e^{-\lambda} \right), \left(\frac{\lambda}{3} \cdot \frac{\lambda}{2} \cdot \frac{\lambda}{1} \cdot e^{-\lambda} \right), \ldots, \right.$$

$$\left. \left(\frac{\lambda}{r} \cdot \frac{\lambda}{r-1} \cdots \frac{\lambda}{2} \cdot \frac{\lambda}{1} \cdot e^{-\lambda} \right), \ldots \right\} = \left\{ e^{-\lambda}, \lambda e^{-\lambda}, \frac{\lambda^2}{2!} e^{-\lambda}, \frac{\lambda^3}{3!} e^{-\lambda}, \ldots, \frac{\lambda^r}{r!} e^{-\lambda}, \ldots \right\}.$$

We give some of the values of these approximations to 5 decimal places, and the exact values, for our example in which $n = 600$ and $\pi = 1/365$:

	p_0	p_1	p_2	p_3	$P(r \geqslant 4)$
Poisson approximation	0.193 24	0.317 65	0.261 08	0.143 06	0.084 97
Exact binomial value	0.192 80	0.317 81	0.261 49	0.143 20	0.084 70

The approximations form the probabilities of a distribution, since each is positive and their sum is

$$e^{-\lambda} \left(1 + \lambda + \frac{\lambda^2}{2!} + \frac{\lambda^3}{3!} + \cdots + \frac{\lambda^r}{r!} + \cdots \right) = e^{-\lambda} \cdot e^{\lambda} = 1,$$

using the first mathematical result above. We use these probabilities to define a new distribution.

Definition 20.1

A discrete random variable R is said to have a **Poisson distribution** if

$$P(R = r) = \frac{\lambda^r}{r!} e^{-\lambda} \quad r = 0, 1, 2, \ldots; \lambda > 0.$$

Note that the Poisson distribution is a discrete distribution that has an infinite number of possible values. It has a *single* parameter λ. When we use it as an approximation to the binomial distribution, we must not think of n, π individually, but only of their product, which is λ.

Example 20.1 Poisson approximation to binomial

A sweet manufacturer wishes to ensure that the probability, P, of there being at least 2 sweets with marzipan filling in packs of 30 sweets should be high. The sweets to go in the packet are chosen at random from a large collection of sweets. Use the Poisson approximation to the binomial distribution in order to calculate P when the proportion of marzipan-filled sweets in the whole collection is: (a) 0.1; (b) 0.2; (c) 0.3.

The number of marzipan-filled sweets in packs of 30 will be binomially distributed with $n = 30$ and $\pi = 0.1, 0.2$ or 0.3.

$$P(\text{at least two of the particular type}) = 1 - P(\text{none or one of the particular type})$$
$$\simeq 1 - (e^{-\lambda} + \lambda e^{-\lambda}),$$

using the Poisson approximation.
We set out the results in a table:

π	$\lambda = n\pi$	P(0 or 1)	P(at least 2)
0.1	3	$e^{-3}(1+3) = \ 4e^{-3} = 0.199\,15$	0.800\,85
0.2	6	$e^{-6}(1+6) = \ 7e^{-6} = 0.017\,35$	0.982\,65
0.3	9	$e^{-9}(1+9) = 10e^{-9} = 0.001\,23$	0.998\,77

The corresponding binomial probabilities are respectively $0.816\,31$, $0.989\,48$, $0.999\,69$.

Example 20.2 *Probability of misprints*

The mean number of misprints per page in a book is 1.2. What is the probability of finding on a particular page: (a) no misprints; (b) three or more misprints?

What is a reasonable model? We might assume that there is a constant probability π of making a mistake with each letter, and that the probabilities of making mistakes with different letters are independent. This latter assumption is more reasonable if we count the common mistake of interchanging two letters as one misprint. The number of misprints on a page is then binomially distributed with n equal to the number of letters on a page (say, about 2000) and probability π (say, about 0.0006). The value of n is large and that of π is small so we may use the Poisson approximation. We are told that the mean number of misprints per page is 1.2; on the binomial model this is equal to $n\pi$. This is sufficient information to apply the Poisson approximation; we do not need the individual values of n and π, because we only need to replace the product $n\pi$ by λ. We thus have $\lambda = 1.2$.

$$P(\text{no misprints}) \simeq e^{-\lambda} = e^{-1.2} = 0.301.$$

$$P(3 \text{ or more misprints}) = 1 - P(0 \text{ or } 1 \text{ or } 2 \text{ misprints})$$

$$\approx 1 - e^{-\lambda}\left(1 + \lambda + \frac{\lambda^2}{2}\right) = 1 - e^{-1.2}(1 + 1.2 + 0.72)$$

$$= 1 - 2.92e^{-1.2} = 1 - 0.879 = 0.121.$$

The Poisson approximation to the binomial distribution works quite well for moderate sample sizes, as Example 20.1 shows, though when the individual terms (rather than sums of terms) are needed it is best to have $n = 50$ or more and π not too large (say 0.1 or less). We may of course relax these conditions if we do not want a highly accurate approximation.

20.3.1 Exercises

1. Six dice are thrown and the number of sixes is observed. Use the Poisson approximation to the binomial distribution to estimate the probability of: (a) no sixes; (b) one six; (c) at least two sixes. Compare these values with the exact binomial probabilities.

2. A rare disease affects, at random, 0.5% of babies born. One hundred babies are born in a hospital in one week. What is the probability that exactly three of them have the disease?

3. Estimate the mean number of misprints per page that makes the probability of at least one misprint on a page equal to 0.05. (Use the Poisson approximation and equate the probability of no misprints to 0.95.)

20.4 THE POISSON DISTRIBUTION AS THE OUTCOME OF A RANDOM PROCESS

Suppose that cars are being counted as they pass along a road, and that the road is adequate to carry smoothly the volume of traffic that uses it: it has a dual carriageway, perhaps, and does not have hills, blind bends or other obstructions which will cause traffic to bunch together rather than allowing it to flow smoothly. In these conditions, we may expect to count each car as a unit in its own right, and not as one of a batch or group. We decide to count the number of cars that pass each minute. We choose the time of day for our study so that the average rate of flow is not likely to alter during the time that we continue counting.

If all these conditions are met it is reasonable to assume that each car passes the observation point independently of every other car, and at random in time. What probability model might we choose to model the variate we are observing, which is the number of cars that pass in a one-minute interval?

As a first, very rough, approximation we might try to simulate the passage of these cars by imagining a time axis divided into one-minute intervals (Fig. 20.1). We decide on a value for λ, the probability that a car will pass in a one-minute interval; let us take $\lambda = 0.4$. We will first consider the simplest case, where not more than one vehicle can pass each minute. This is simulated using random digits from a table; take a run of digits and associate each digit with an interval. So that λ shall be 0.4, we will assume that a car passes in any interval that has the digits 1, 2, 3 or 4 associated with it, and not otherwise. The run of digits

Figure 20.1 *Time axis divided into one-minute intervals, showing (•) those intervals in which a vehicle passes.*

Figure 20.2 *Time axis divided into half-minute intervals, showing (•) those intervals in which a vehicle passes.*

6 2 0 3 5 4 2 2 1 9 3 7 . . . gives the pattern shown in Fig. 20.1, dots indicating the intervals in which a car passes. Since $\lambda = 0.4$, the *mean rate* at which cars pass is 0.4 per minute.

This is not a satisfactory model since in each one-minute interval the only possible numbers of cars that may pass are 0 and 1, whereas in practice we might observe many cars in a minute. Suppose we reduce the intervals to be half a minute long. We repeat the simulation, but now make the probability of a car passing in any half-minute interval equal to 0.2, so that the mean rate, λ, is still 0.4 cars per minute. Again we use a run of random digits, this time associating 1 and 2 with a car passing. The run of digits

0 7 2 1 4 3 1 1 8 5 2 0 9 2 7 2 9 8 1 6 1 6 6 5 . . .

leads to the pattern shown in Fig. 20.2, in which we have marked intervals in which cars pass by a dot. This is an improvement, since now we can obtain 0, 1 or 2 cars in any one-minute interval, and all these possibilities have occurred in Fig. 20.2. But again the maximum possible number of cars per one-minute interval is not high enough. Suppose we divide each minute into n intervals. If the mean rate per minute at which cars pass is to be a constant λ, which is 0.4 in our example, and π is the probability that a car passes in an interval of $1/n$ minute, then we require that $n\pi = \lambda$ or $\pi = \lambda/n$. For a practical simulation we might be satisfied to choose $n = 10$.

If, in our first simulation model (where $n = 1$, so that $\pi = \lambda$), we count the number of cars in a one-minute interval we obtain a random variable that takes the value 1 with probability π and the value 0 with probability $(1 - \pi)$, i.e. it is a Bernoulli variable (see p. 172). When we use half-minute intervals, the number of cars in a *one-minute* interval is a binomial variable with $n = 2$ and $\pi = \frac{1}{2}\lambda = 0.2$. (It is the number of successes, i.e. the number of half-minute intervals in which a car passes, out of a total of two half-minute intervals, when the probability of success in any interval is 0.2.) Similarly, if we use $(1/n)$-minute intervals, the number of cars in a one-minute interval is a binomial variable with parameters n and π, subject to the condition that $n\pi = \lambda$. The probability of r cars in a minute interval is

$$\binom{n}{r} \pi^r (1 - \pi)^{n-r}.$$

For a practical simulation we would fix n at a suitable value. This has the theoretical disadvantage that two cars must be separated by a distinct time interval of at least $1/n$ minutes. We would prefer, since time is continuous, that one car could follow immediately after another. (We exclude the possibility of two cars passing simultaneously.) We can obtain what we want if we let the number of intervals n, into which we divide each one-minute interval, tend to infinity while π tends to zero and $n\pi$ remains at a constant value λ. We already know the result of taking this limit. We obtain a Poisson distribution and find that the probability of r cars in a one-minute interval is $\lambda^r e^{-\lambda}/r!$, where λ is the mean number of cars passing per minute. This has proved to be a good model for traffic flow provided that the conditions given in the first paragraph of this section are met.

20.5 THE CONDITIONS FOR A POISSON DISTRIBUTION

Let us list, in a general form, the assumptions we have used in the derivation above. They give us the conditions under which a Poisson distribution is obtained.

1. Events occur at *random* in continuous space or time.

2. Events occur *singly*, and the probability of two events occurring simultaneously is zero.

3. Events occur *uniformly*, i.e. the expected number of events in a given interval is proportional to the size of the interval.

4. Events occur *independently*, i.e. the probability of an event occurring in any small interval is independent of the probability of the event occurring in any other small interval.

5. The variable is the *number of events* that occur in an interval of a given size.

From these assumptions it is possible to derive the Poisson distribution by a strict mathematical argument. The mathematics required is more advanced than we may use so we have contented ourselves with the illustration above. There are many situations in which these assumptions are reasonable and the Poisson distribution is a widely used model. Some examples are the number of currants in buns, the number of plants of a given species found in areas one metre square in a meadow, the number of people who arrive at a doctor's surgery in a five-minute interval. It is important to realize that although we derived the Poisson distribution initially from the binomial distribution, it is, like the normal distribution, a distribution in its own right. When we wish to decide whether the Poisson distribution is likely to provide a reasonable model we will usually do best to check the conditions above, of which the key ones are randomness, independence and constant rate of occurrence, rather than try to link the conditions with a special case of a binomial distribution. Let us look more closely at further examples.

20.5.1 Radioactive emission

When a radioactive material is placed in a counting chamber, and the number of particles r emitted per minute by the material is counted for many minutes, say 100, we shall be able to set up a frequency table for the variate r. An appropriate random variable to model this will often be the Poisson distribution because the appropriate conditions are usually satisfied: (1) each particle is emitted at a random instant in time; (2) for practical purposes we may take it that particles are emitted singly; (3) unless we have a radioactive material with a very short lifetime, or go on observing for a very long time, the rate of emission will not change noticeably during the study; (4) the fact that a particle comes off at a certain instant neither prevents another from coming off immediately afterwards nor induces one to do so, i.e. the particles are emitted independently. Hence the number of particles emitted per minute should follow the Poisson distribution.

20.5.2 Blood cell counts

Biologists may take a drop of a well-stirred sample of blood, put it on a slide, and look at the slide under a microscope to count the number of a particular type of blood cell. The unit here is the *volume* of blood on the slide, i.e. the distribution is in space rather than in time. But we can again see the conditions of randomness, independence and constant rate at work. If the sample is well stirred the particular cells of interest are distributed through the sample of blood at random and independently. Since the total number of cells per unit volume in the blood sample is constant, there is a constant probability of finding a particular type of cell in drops of a given volume.

20.6 CONDITIONS IN WHICH A POISSON MODEL BREAKS DOWN

If we are counting vehicles on a road leading away from a business or industrial centre at the end of the working day, they are certainly not arising at random in time – they all left work as nearly as they physically could at the same time! – nor are they independent. Even if a Poisson model appears to fit traffic counts on this road during the rest of the day, it will not do so at the beginning and end of the working day. Vehicles will not be coming along independently if we count just past traffic lights either, or if we have a single-carriageway road which suffers occasional hold-ups due to buses or heavy lorries.

Blood cells, or yeast cells, or any other organism held in suspension in a liquid, will not remain uniformly spaced through the liquid unless the liquid is shaken up in its container before each drop is taken out to place on a microscope slide. This lack of uniformity is a change of mean, in the same way as the mean rate of traffic flow can change in the rush-hour. Physical arguments like these are always very valuable in deciding whether a Poisson model is appropriate. There are also ways of testing if a given set of data does fit the Poisson distribution, and we look at these in Sections 20.9 and 20.14.

20.6.1 Exercise

1. State, with reasons, whether you think that each of the following variates is best modelled by: (a) a Poisson distribution; (b) a binomial distribution; (c) some other distribution. If the answer is (a) or (b), estimate the parameters of the distribution from the information given, or by making a reasonable guess.

   (i) The number of calls passing through a telephone exchange every 20 seconds, in a uniformly busy period when calls average 45 per minute.
   (ii) The number of raindrops that fall into an open dustbin in a period of 5 seconds during a shower.
   (iii) The number of sixes in ten successive throws of a die during a game of snakes and ladders.
   (iv) The number of sixes thrown in five minutes' playing time of a game of snakes and ladders.
   (v) The number of occupants in cars that pass me as I wait at the bus stop in the morning.
   (vi) The number of deaths by suicide in a large town in a year.

20.7 THE MEAN AND VARIANCE OF THE POISSON DISTRIBUTION

The probability mass function of the Poisson random variable is given by

$$P(R = r) = e^{-\lambda}\lambda^r/r! \quad r = 0, 1, 2, \ldots.$$

It contains only one parameter λ, so we may expect that λ determines both the mean and variance. In fact, the Poisson distribution has the striking property that both its mean and variance equal λ, as we shall show. The mean is

$$E[R] = \sum_{r=0}^{\infty} rP(r) = \sum_{r=0}^{\infty} re^{-\lambda}\lambda^r/r!$$

$$= e^{-\lambda} \sum_{r=1}^{\infty} \lambda^r/(r-1)! \quad \text{(the term for } r = 0 \text{ is zero)}$$

$$= \lambda e^{-\lambda} \sum_{r=1}^{\infty} \lambda^{r-1}/(r-1)! = \lambda e^{-\lambda} \cdot e^{\lambda} = \lambda.$$

$\text{Var}[R] = E[R^2] - (E[R])^2$. We also use the identity $E[R^2] \equiv E[R(R-1)] + E[R]$, which helps in the evaluation. We already know that $E[R] = \lambda$, so we need only $E[R(R-1)]$.

$$E[R(R-1)] = \sum_{r=0}^{\infty} r(r-1)P(r) = \sum_{r=0}^{\infty} r(r-1)e^{-\lambda}\lambda^r/r!$$

$$= e^{-\lambda} \sum_{r=2}^{\infty} \lambda^r/(r-2)! \quad \text{since the terms for } r=0 \text{ and } r=1 \text{ are both zero,}$$

$$= \lambda^2 e^{-\lambda} \sum_{r=2}^{\infty} \lambda^{r-2}/(r-2)!$$

$$= \lambda^2 e^{-\lambda} \cdot e^{\lambda} = \lambda^2.$$

Thus $E[R^2] = \lambda^2 + \lambda$. So

$$\text{Var}[R] = (\lambda^2 + \lambda) - (\lambda)^2 = \lambda.$$

It would, of course, have been perfectly possible to write

$$\text{Var}[R] = \sum_{r=0}^{\infty} (r-\lambda)^2 P(r)$$

and attempt to expand this formula. But in doing so we should find it helpful to write $r^2 = r(r-1) + r$ so as to sum the series involved, and this is equivalent to the work we have set out above.

Note that, because λ is the mean of the Poisson distribution, many writers prefer to use μ rather than λ for the parameter in the probability mass function.

The fact that mean and variance are equal is a very important characteristic of the Poisson distribution, and can be used as a basis for testing whether a set of observations appear to follow the Poisson distribution (p. 531).

Example 20.3 Poisson probabilities

If a large grass lawn contains on average 1 weed per 600 cm^2, what will be the distribution of r, the number of weeds in an area of 400 cm^2?

A suitable model for this situation is a Poisson distribution whose mean is $\frac{400}{600} = \frac{2}{3}$. So $P(r) = e^{-2/3}\left(\frac{2}{3}\right)^r/r!$: each term may either be found directly or calculated from the preceding one.

$$P(r=0) = e^{-2/3} = 0.5134,$$

$$P(r=1) = \frac{2}{3}e^{-2/3} = 0.3423 = \frac{2}{3}P(r=0),$$

$$P(r=2) = \left(\frac{2}{3}\right)^2 e^{-2/3}/2! = 0.1141 = \frac{1}{2} \cdot \frac{2}{3}P(r=1),$$

$$P(r=3) = \left(\frac{2}{3}\right)^3 e^{-2/3}/3! = 0.0254 = \frac{1}{3} \cdot \frac{2}{3}P(r=2),$$

$$P(r \geqslant 4) = 0.0048.$$

20.7.1 Exercises

1. On the average it is found that 1% of the items of a particular manufactured product are defective. Calculate the probability that: (i) not more than one defective item is found in a sample of 100; and (ii) not more than two defective items are found in a sample of 200 of the product.

2. Describe the Poisson distribution, stating clearly the meanings of the symbols used. Show that the variance of the distribution is equal to its mean.

   A vehicle has an essential component. In a journey of K km the number of times this component needs replacing has a Poisson distribution with a mean $K/1200$. If the driver is setting out on an expedition which involves driving 3600 km, how many *spare* components should he take with him so as to have a 95% assurance of being able to complete his journey? (*O&C*)

3. The numbers of telephone calls received at an exchange in 90 successive half-minute intervals are shown below:

```
3  5  2  4  6  5  6  3  4  7
5  6  5  3  1  6  6  1  5  3
2  1  3  1  5  1  3  0  4  5
1  3  1  5  3  2  3  6  4  7
2  0  6  3  4  1  1  2  7  0
4  4  3  4  5  5  2  3  4  4
7  6  6  4  2  2  4  4  3  4
5  3  5  4  4  5  3  2  4  3
3  6  4  4  6  2  5  7  0  6
```

Construct a frequency distribution from these data and draw a frequency polygon. Calculate the sample mean.

There is reason to believe that the data arise from a Poisson distribution with mean m. Using the sample mean as an estimate of m, estimate the probability that at least two calls will be received at the exchange in a comparable half-minute interval. (*JMB*)

20.8 THE SUM OF TWO POISSON RANDOM VARIABLES

Suppose that two different types of blood cell are being counted, and that both occur rarely, one with mean λ per unit area when viewed under the microscope on a slide, and the other with mean κ per unit area. If we let r be the sum of the number of blood cells of both types, per unit area, how will r be distributed?

We will assume that the numbers of the two types existing in unit area are independent of one another; if we are dealing with two types only, out of a large number of possible types, and if they are both fairly rare, this is a reasonable assumption.

Thus the probability of finding s of the first type and $r - s$ of the second is

$$\frac{e^{-\lambda}\lambda^s}{s!} \frac{e^{-\kappa}\kappa^{(r-s)}}{(r-s)!}.$$

In order that the total of both types shall be r, we may have 0 of the first type and r of the second, or 1 and $r - 1$, or 2 and $r - 2$, and so on. In fact s may take any value from 0 to r. So the total probability of obtaining r is the sum of the probabilities of all these ways of splitting up the r between the two types:

$$P(r) = \sum_{s=0}^{r} \frac{e^{-\lambda}\lambda^s}{s!} \frac{e^{-\kappa}\kappa^{(r-s)}}{(r-s)!}$$

$$= \frac{e^{-(\lambda+\kappa)}}{r!} \sum_{s=0}^{r} \frac{r!}{s!(r-s)!} \lambda^s \kappa^{(r-s)}$$

$$= e^{-(\lambda+\kappa)}(\lambda + \kappa)^r / r!$$

since the sum is simply the binomial expansion $(\lambda + \kappa)^r$. This shows that the sum of two independent Poisson random variables, one with mean λ and one with mean κ, is another Poisson random variable, this time with mean $\lambda + \kappa$.

If we look back to the conditions for a Poisson distribution, stated at the beginning of Section 20.5, we see that these are perfectly reasonable for the sum, because (1) if the two cell types occur at random, considered separately, they cannot be other than random when considered together; (2) if they still occur singly, there cannot be two in exactly the same place; (3) uniformity of each cell type implies uniformity of the sum; (4) if individual cells of each type are independent they will certainly be independent when we look at the two together; (5) the variable being counted is still the number of events in an interval of given size.

Example 20.4 Sum of Poisson random variables

A random sample of leaves is taken from tomato plants which had radioactive phosphorus incorporated into their applied fertilizer. When placed in a counting chamber, the sample emits radioactive particles at a mean rate of 6 per second. Supposing that it is removed to another laboratory in which there is already 'background radiation' present at the rate of 1 particle per second, what should the distribution of radioactive counts be in this laboratory?

We assume that both sources of radioactivity can be modelled by Poisson distributions, with means 6 and 1 particle(s) per second respectively, and that the sources are independent. Using the result above, the total radioactive count is then distributed as a Poisson random variable with mean $6+1=7$ particles per second.

20.8.1 Generating functions and sums of Poisson random variables

The probability generating function of a random variable which follows the Poisson distribution with mean λ is

$$
\begin{aligned}
G(t) &= p_0 t^0 + p_1 t^1 + p_2 t^2 + p_3 t^3 + \cdots \\
&= e^{-\lambda} + \lambda e^{-\lambda} t + \lambda^2 e^{-\lambda} t^2/2! + \lambda^3 e^{-\lambda} t^3/3! + \cdots \\
&= e^{-\lambda}\{1 + \lambda t + (\lambda t)^2/2! + (\lambda t)^3/3! + \cdots\} = e^{-\lambda} e^{\lambda t} \\
&= \exp\{-\lambda(1 - t)\}.
\end{aligned}
$$

Thus if R and S are independent Poisson variables with means λ and κ, the generating function for their sum $R + S$ will be (p. 336) the product

$$
\exp\{-\lambda(1 - t)\} \cdot \exp\{-\kappa(1 - t)\} = \exp\{-(\lambda + \kappa)(1 - t)\}.
$$

This is of exactly the same form as the generating function for a Poisson random variable whose mean is $\lambda + \kappa$, which provides another way of showing that sums of (any number of) Poisson random variables are themselves Poisson random variables – all we have to do is add together the parameters (means).

Related results may be derived for the moment generating function. By replacing t by e^t in the probability generating function $G(t)$ (see p. 338) we may show that the moment generating function $M(t)$ of a Poisson random variable with mean λ is

$$
M(t) = \exp\{-\lambda(1 - e^t)\}
$$

and that the sum of independent Poisson random variables with means λ and κ has the moment generating function

$$
\exp\{-(\lambda + \kappa)(1 - e^t)\}.
$$

20.8.2 Exercises

1. (a) Find the moment generating function of a Poisson distribution from first principles, and use it to find the values of the mean and variance.
   (b) Use the probability generating function of a Poisson distribution to derive its mean and variance by the method of Examples 9.12 and 9.13, pp. 217–218.

2. The number of plants of a particular species which are counted in sample quadrats of a standard size follows a Poisson distribution whose mean is λ. Some of these have a special marking, which occurs independently with probability π on plants of this species. Show that the number of *marked* plants per quadrat follows a Poisson distribution whose mean is $\lambda\pi$.

20.9 TESTS OF GOODNESS-OF-FIT TO A POISSON DISTRIBUTION

The χ^2 test may be used to test whether data fit a Poisson model. We follow the process described in Section 19.6, with only two special points which need mention. First, in order to calculate the frequencies of values of r given by a Poisson hypothesis, we need to know the parameter λ. Since we do not know it, the best we can do is to estimate it from the data; and since λ is the mean of the Poisson distribution we estimate it by $\bar{r}$. Second, the χ^2 test will have $k - 2$ degrees of freedom if we have k classes: one degree of freedom has, as usual, been lost for ensuring that observed and expected frequencies have the same total, and a second is lost because we have had to estimate one parameter, λ, before expected frequencies can be computed.

Example 20.5 Goodness-of-fit test

The number, r, of letters received each day for 100 days, as presented in Example 1.1, p. 6, was observed as follows. Do these observations fit a Poisson distribution?

No. of letters, r	0	1	2	3	4	5	Total
Frequency of r	48	32	17	2	0	1	100

The first step is to calculate the mean, $\bar{r}$; this is 0.77. Now a Poisson distribution with mean 0.77 and total frequency 100 is found by using $P(r) = e^{-0.77}(0.77)^r/r!$ and multiplying by 100 the probabilities so calculated:

r	0	1	2	$\geqslant 3$
Frequency of r on Poisson	46.30	35.65	13.73	4.32

When applying a χ^2 test, the expected values in the classes should not as a rule fall far below 5. In this case we accept 4.32, which is the expected value for all r-values of 3 or more; if we used only the three classes $r = 0$, $r = 1$ and $r \geqslant 2$ we should lose another degree of freedom and therefore some sensitivity in the test. With four classes, there will be two degrees of freedom, one lost for making the sum of expected frequencies 100 and one for estimating the mean to use in the Poisson formula.

r	0	1	2	$\geqslant 3$	Total
Observed frequency	48	32	17	3	100
Expected frequency	46.30	35.65	13.73	4.32	100

Note that we do *not* round the expected frequencies to whole numbers, even though the observed must be. The expected frequencies are theoretical ones assuming the Poisson model is obeyed exactly; we are aiming to test whether it is obeyed within reasonable statistical limits.

$$\chi^2 = \sum (\text{observed} - \text{expected})^2/\text{expected}$$

$$= \frac{(1.70)^2}{46.30} + \frac{(3.65)^2}{35.65} + \frac{(3.27)^2}{13.73} + \frac{(1.32)^2}{4.32} = 1.618.$$

(We have denoted the statistic X^2 of Chapter 19 by χ^2, as is common in practice.) This is certainly not significant as $\chi^2_{(2)}$, so we can accept the hypothesis that the observations follow a Poisson distribution.

If we find, in this way, that a set of data do fit a Poisson model, we often take this as evidence that the data arose under the conditions of randomness, independence and constant rate that we listed in Section 20.5. For if these conditions *are* satisfied, then a Poisson distribution should result. (Logically, however, it is not valid to claim that the existence of a Poisson distribution proves the truth of the conditions. Statement A (the conditions) implies B (the Poisson); but B does not *necessarily* imply A.) In the example above, the data were collected at a time when the postal service operated at weekends; since then, the 'constant rate' condition may have changed, so a fresh set of similar data may not follow a Poisson distribution.

Sometimes it *is* possible to say what λ should be in a Poisson distribution that is used as a model for data. In that case, this theoretical λ is used to calculate expected frequencies, and we do not therefore have to subtract a degree of freedom for estimating λ.

20.9.1 Exercises

1. Given that X and Y are independent and have Poisson distributions with means a and b respectively, show that the sum of these quantities, $X + Y$, has probabilities of values $0, 1, 2$ which are in accordance with those of a Poisson distribution with mean $a + b$.

   A shopkeeper has two shops which are supplied from a central store. For a particular product each shop asks the store for a complete box when required. The numbers of boxes per week requested by the two shops are independent and have Poisson distributions with means 4/5 and 1/5 respectively. Find the probabilities that two or more boxes are requested from the store in a week: (i) by the first shop; (ii) by the second shop; (iii) altogether.

   Calculate the lowest level of stock at the central store for which there is a probability greater than 90% that all demands from the shops in the next week can be met. (*JMB*)

2. Test whether the following data could have come from a Poisson distribution with mean 3:

Value of variate	0	1	2	3	4	5	6	7	Total
Observed frequency	2	7	12	14	7	5	2	1	50

If you had not been given the mean, indicate how you would have proceeded.

3. It is proposed to install a set of traffic lights on a dual carriageway. A preliminary traffic count yields the following data:

Vehicles passing in 10 seconds	0	1	2	3	Total
No. of occasions observed	63	24	10	3	100

Test whether these data support the highway engineer's hypothesis that the number of vehicles in 10-second intervals follows a Poisson distribution with mean $\frac{1}{2}$. If the lights are to be set to remain at red for one minute, what is the probability that not more than three vehicles will be held up at the lights?

(*IOS*)

20.10 THE NORMAL APPROXIMATION TO THE POISSON DISTRIBUTION

In Chapter 15, we showed that under the right conditions it was possible to approximate very closely to the binomial distribution by using a normal distribution with the same mean and variance as the binomial. A similar approximation is possible for the Poisson distribution. If the random variable R has a Poisson distribution whose mean is λ, then for large λ the random variable $(R - \lambda)/\sqrt{\lambda}$ is approximately $\mathcal{N}(0,1)$. The approximation is in fact very good for $\lambda > 10$, and perfectly usable for $\lambda > 5$: some improvement may be made by applying a *continuity correction* and writing

$$P(R \leqslant r) = P(Z \leqslant (r + \frac{1}{2} - \lambda)/\sqrt{\lambda}),$$

in which Z has the $\mathcal{N}(0,1)$ distribution and the $\frac{1}{2}$ represents the continuity correction. (Compare p. 376 for the binomial distribution.)

Example 20.6 Normal approximation to Poisson

A large computer installation suffers breakdowns at the rate of 9 per week. Assuming these to occur at constant rate, at random in time and independently of one another, find the probability that there will not be more than 12 breakdowns in any given week.

The conditions as stated indicate a Poisson distribution for the number of weekly breakdowns, r. Its mean $\lambda = 9$, so

$$P(r \leqslant 12) = \sum_{r=0}^{12} e^{-9} \frac{9^r}{r!} = 0.8758,$$

by direct calculation of the 13 terms involved, which is quite tedious. Using the normal approximation, $z = (r - 9)/3$ has the standard normal distribution, and so

$$P(r \leqslant 12) = P(z \leqslant (12 - 9)/3) = P(z \leqslant 1) = 0.8413.$$

If we include the continuity correction, we calculate

$$P(z \leqslant (12 + \frac{1}{2} - 9)/3) = P(z \leqslant 1.167) = 0.8783;$$

clearly this gives a much better approximation.

It is also possible to approximate individual probabilities. $P(r)$ in the Poisson distribution is approximated by finding the probability between $r - \frac{1}{2}$ and $r + \frac{1}{2}$ in the corresponding normal distribution. As an example, in a Poisson distribution with mean 15, $P(r = 22) = 0.020\,36$ by direct calculation. Applying the normal approximation including continuity correction, we find

$$P \left(\frac{21\frac{1}{2} - 15}{\sqrt{15}} < z < \frac{22\frac{1}{2} - 15}{\sqrt{15}} \right),$$

z denoting an $\mathcal{N}(0, 1)$ random variable. This is

$$P \left(\frac{6.500}{3.873} < z < \frac{7.500}{3.873} \right) = P(1.678 < z < 1.936) = \Phi(1.936) - \Phi(1.678)$$

$$= 0.9735 - 0.9533 = 0.0202.$$

20.11 TESTING A HYPOTHESIS ABOUT THE MEAN IN A POISSON DISTRIBUTION

Suppose that in the computer installation mentioned above, the mean rate of breakdowns has been 9 per week over a long period. A new piece of equipment is installed, and in the first week of operation afterwards there are 13 breakdowns. It is claimed that the new equipment led to more breakdowns. We shall test this claim.

The number of breakdowns per week, R, is Poisson with mean 9. We must carry out a one-tail test of the null hypothesis '$\lambda = 9$' against an alternative hypothesis which states '$\lambda > 9$'. We require to know the probability of values as extreme as our observation or more so; if this probability is small we reject the null hypothesis. The probability is $P(R \geqslant 13)$; we would do best to calculate $P(R \leqslant 12)$ and find the complement. The calculation is not difficult if a hand-held calculator or computer is used (see *CCC*, Chapter 8); it is most conveniently done using a recursive method (see Example 20.3). We found, by computer, that $P(R \leqslant 12) = 0.876$, so that the required probability $P(R \geqslant 13) = 0.124$. For significance this needs to be less than 5%; therefore we do not reject the null hypothesis.

Alternatively, when a Poisson distribution with reasonably large mean λ is being studied, we may approximate the distribution by $\mathcal{N}(\lambda, \lambda)$, here $\mathcal{N}(9, 9)$. The tail of the normal distribution from $12\frac{1}{2}$ upwards corresponds to $P(R \geqslant 13)$ in the Poisson distribution. The standard normal variable corresponding to $12\frac{1}{2}$ is

$$z = \frac{12.5 - 9}{\sqrt{9}} = \frac{3.5}{3} = 1.167.$$

From normal tables, the probability of a value exceeding this is 0.122, which is in close agreement with the exact value of 0.124. We therefore do not reject the null hypothesis.

As usual, we will hardly be happy to base an inference on one week's operation only, and would prefer to collect data for several weeks. If there are n weeks' records, and the mean number of breakdowns is $\overline{R}$, then this $\overline{R}$ will have an approximately normal distribution, the mean and variance in this approximation being λ and λ/n. Suppose we observe the same computer installation for 4 weeks, during which $\overline{r} = 12$. Once again we will make a one-tail test, using the same null and alternative hypotheses. The mean $\overline{R}$ is approximated by $\mathcal{N}(9, \frac{9}{4})$. In order to incorporate a continuity correction in the calculation, it is best to convert this to an integer-valued variable, and deal with $4\overline{R}$ which will be $\mathcal{N}(36, 36)$ and whose observed value is 48. We wish to find $P(4\overline{R} > 47\frac{1}{2})$, including the continuity correction, and the standard normal variable corresponding to $47\frac{1}{2}$ is thus

$$z = \frac{47.5 - 36}{\sqrt{36}} = \frac{11.5}{6} = 1.917.$$

As we are carrying out a one-tail test, this result is significant at the 5% level and so leads us to reject the null hypothesis: there is some evidence in support of the claim that was made.

20.12 COMPARING TWO MEANS

There is a similar method for comparing the means of two Poisson distributions, which applies when both the means are greater than 5.

Example 20.7 Normal approximation to Poisson
A large area of grassland is examined for weed content. Parts of it have had chemical weedkiller applied to the soil, and 20 separate square metres are selected at random from these parts. The mean number of weeds per square metre is 12. The remainder of the area has received no weedkiller, and in 20 randomly chosen square metres from this section the mean number of weeds per square metre is 16. Test whether the weedkiller had any effect.

The null hypothesis here will be that the two means are equal, and the alternative that the weedkiller reduces mean weed number. If weeds are scattered at random through the whole area, as in quite a number of situations they do seem to be, a suitable model for the number per unit area is given by the Poisson distribution. In the present case, the mean is large enough for a normal approximation to be made to the Poisson distribution. Thus we have approximately the same test of difference between two means as in Section 17.5; although we do not, strictly speaking, know σ^2, we do know that if our Poisson model is the right one σ^2 will equal the mean. So we take the two sets of data, and estimate from them the common mean which the null hypothesis claims they have. This is clearly 14, since both samples were the same size and their means were 12 and 16. Hence each observed mean, based on 20 observations, is approximately normal with variance $14/20 = 0.70$. The difference between two means is $\mathcal{N}(0, 1.40)$. We actually observed a difference of 4 between the means, and the direction of this difference is consistent with the alternative hypothesis that has been stated. A one-tail test then gives

$$z = \frac{4}{\sqrt{1.40}} = \frac{4}{1.183} = 3.38.$$

We reject the null hypothesis at the 0.1% significance level, in favour of the alternative hypothesis that some reduction has been achieved by the weedkiller.

20.12.1 Exercises

1. A storekeeper in the spare-parts department of a garage finds that the weekly demand for a particular component follows a Poisson distribution, the mean of which is 25. He stocks up at the beginning of each week. About how many items of this component should he hold each Monday morning so that he can be 95% certain of meeting the demand that week?

2. A newspaper report states that last year there were only 9 fatal accidents on a certain stretch of road. Over the previous several years, the average had been 15. Drivers must therefore be becoming more careful. What justification does the newspaper have for its statement?

3. The same newspaper states that on another stretch of road in the same county, where the average had also been 15 over a period of several years, the last three years' total number of accidents was 30. So drivers are more careful here too, the improvement being about the same on both stretches of road. Comment.

4. Samples of plant material from the leaves of bean plants and from their roots are being examined. The plants have been treated with fertilizer that contained a radioactive isotope. There were five leaf samples, and the average radioactive emission count of these was 15 particles per minute. There were four root samples, and their average count was 12 particles per minute. Is there evidence that the two parts of the plant are different in radioactivity?

20.13 GIVING LIMITS TO AN ESTIMATED MEAN

Using the normal approximation that we have described, it is possible to give confidence limits to an estimated mean so long as this mean is not too small. For $\lambda \geqslant 10$, the method is close enough provided we work at the 95% confidence level; if we wish to set up 99% limits in this way we should ensure that $\lambda \geqslant 20$.

When $\bar{r}$ has been calculated on the basis of n observations, drawn at random from a Poisson distribution whose mean λ is to be estimated, $(\bar{r} - \lambda)/\sqrt{\lambda/n}$ is approximately $\mathcal{N}(0, 1)$. To obtain rough 95% limits therefore, we could write the variance of $\bar{r}$ as $\bar{r}/n$ and thus $|\bar{r} - \lambda| \leqslant 1.96\sqrt{\bar{r}/n}$ with probability approximately 0.95 (compare p. 446). Approximate 95% confidence limits for the true value of λ are then $\bar{r} \pm 1.96\sqrt{\bar{r}/n}$ (and we would usually replace 1.96 by 2, as in computing limits for the binomial).

However, we can do better than this. The 95% limits are found by requiring that $|\bar{r} - \lambda|/\sqrt{\lambda/n} \leqslant 1.96$, the 99% limits by writing 2.576 instead of 1.96; and so in general when we have laid down what limits we wish to compute we use the table of the standard normal distribution to find the number to put on the right of the inequality. Let us therefore solve the general inequality $|\bar{r} - \lambda|/\sqrt{\lambda/n} \leqslant c$ first. When using this, c may be 1.96, 2.576 etc., according to the confidence level chosen. To obtain the limits for λ, replace $\leqslant$ by $=$; then square to give $n(\bar{r} - \lambda)^2 = c^2\lambda$, which is a quadratic in λ:

$$n\lambda^2 - (2n\bar{r} + c^2)\lambda + n\bar{r}^2 = 0.$$

The two roots of this quadratic will give the upper and lower limits for the true value of λ, at a confidence level determined by the value of c used. The limits are

$$\lambda = \frac{1}{2n}[2n\bar{r} + c^2 \pm \sqrt{(2n\bar{r} + c^2)^2 - 4n^2\bar{r}^2}]$$

$$= \bar{r} + \frac{c^2}{2n} \pm \frac{c}{2n}\sqrt{c^2 + 4n\bar{r}}.$$

Example 20.8 Confidence limits for Poisson mean

In a traffic survey on a motorway, the mean number of vehicles passing per minute for 1 hour was 18. Set 95% confidence limits to the true mean rate of vehicles passing per minute.

Since 95% limits are required, $c = 1.96$. The survey gave $\bar{r} = 18$ over $n = 60$ minutes. Limits to λ are

$$18 + \frac{3.84}{120} \pm \frac{1.96}{120}\sqrt{3.84 + (240 \times 18)} = 18.032 \pm 0.0163\sqrt{4323.84}$$

$$= 18.032 \pm 0.0163 \times 65.76$$

$$= 18.032 \pm 1.072,$$

$$\text{i.e. } 16.96 \text{ to } 19.10.$$

The rough limits $\bar{r} \pm 2\sqrt{\bar{r}/n}$ are $18 \pm 2\sqrt{0.30}$, i.e. 16.90 to 19.10, which are very close to the more accurate ones since n is large as well as $\bar{r}$.

20.13.1 Exercises

1. (a) Set 99% confidence limits to the true λ in the above example; and (b) find the value $\bar{\lambda}$ such that, on the data of the example, $P(\lambda > \bar{\lambda}) = 0.1$.

2. A computer manager has a large job on hand. If he does not experience more than 12 machine breakdowns in the next week he can finish the job on schedule. The rate of machine breakdown at present is 8 per week. What is the probability that he will finish the job on schedule?

20.14 THE 'INDEX OF DISPERSION'

There is one other method of studying a set of data to see if they could have come from a Poisson distribution. We have so far used a χ^2 goodness-of-fit test (p. 524). But the special property of the Poisson distribution, namely *mean = variance*, can also be used. If we divide the sample variance by the sample mean, we should obtain a number which is roughly 1 when observations really do come from a Poisson distribution. It is the sum of squares, rather than sample variance, that is actually used in the *index of dispersion*:

$$I = \sum_{i=1}^{n} (x_i - \bar{x})^2 / \bar{x}.$$

For reasonably large values of λ, I follows a $\chi^2_{(n-1)}$ distribution approximately. Since we cannot hope to test the fit to a particular distribution well using only a few observations, n will generally be quite large, and the degrees of freedom for χ^2 therefore much larger than in other uses of the χ^2 test statistic.

Example 20.9 Index of dispersion test

The number of printing errors per page was counted in a sample of 100 pages from several months' issues of a magazine, as shown in the following frequency table:

Number of errors per page, r	0	1	2	3	4	5	6	7	Total
Frequency of r	8	18	30	25	12	4	1	2	100

Test the hypothesis that errors occur randomly, and independently of one another.

If the hypothesis is true, r should follow a Poisson distribution. We find $\bar{r} = 2.41$. We could then apply the method of Section 20.9. Instead, we shall compute

$$\sum_{i=1}^{100} (x_i - \bar{x})^2 = \sum x_i^2 - \frac{(\sum x_i)^2}{100} = 789 - 580.81 = 208.19.$$

Note that we need the sum of squares, not the sample variance. Now the index of dispersion $I = 208.19/2.41 = 86.386$; it is distributed as $\chi^2_{(99)}$ if our hypothesis is true. Although we do not have tables of $\chi^2_{(99)}$ it is clear from looking at the values for 90 and 100 degrees of freedom that 86.386 is in the central part of the distribution of $\chi^2_{(99)}$. It is therefore 'not significant', and so there is no good reason to reject the hypothesis of randomness and independence.

20.14.1 Comparison with the goodness-of-fit test

Does this test give us the same result as that in Section 20.9 if we apply it to the same data? Consider the set of data

r	0	1	2	3	4	5	6	7	8	9	Total
Frequency	9	15	15	4	2	1	1	1	1	1	50

The mean of these data is 2.00 and the variance 4.041, with 49 degrees of freedom. $I = 198.0/2.0 = 99.0$, and as a $\chi^2_{(49)}$ is significant even at the 0.1% level: we have very strong evidence to reject a null hypothesis that says these data follow a Poisson distribution.

However, the expected values of a Poisson variate with mean 2.0, in a sample of 50 observations, are 6.767, 13.534, 13.534, 9.022, 4.511, 1.804, 0.828 for $R = 0, 1, 2, 3, 4, 5, \geqslant 6$ respectively. The last two classes, at least, have to be amalgamated before carrying out a χ^2 test, so that $P(R \geqslant 5) = 2.632$. Some people would recommend using $R \geqslant 4$ as the final class, but (see p. 524) the χ^2 approximation is still adequate without doing so. Using the six classes $R = 0, 1, 2, 3, 4, \geqslant 5$ the χ^2 statistic will have 4 degrees of freedom since the mean is estimated from the data; the goodness-of-fit test of Section 20.9 gives $\chi^2_{(4)} = 7.38$ which is not significant. We need *not* this time reject the null hypothesis that the data follow a Poisson distribution.

Although, as a bar diagram displaying both observed and expected frequencies would show, the data are not too far from the expected Poisson pattern for $\lambda = 2$, the few large observations together with the excess of observed zeros spread the data out far enough to make the variance considerably greater than the mean. So the two tests are examining different characteristics of the data: I compares mean and variance while the earlier test looks at the general pattern of frequencies.

20.15 SIGNIFICANCE TESTS WHEN χ^2 HAS A VERY LARGE NUMBER OF DEGREES OF FREEDOM

In the index of dispersion test, degrees of freedom may exceed 100, and so be outside the range of available tables. Unfortunately, χ^2 itself only approaches normality very slowly indeed. But there is a function of it which leads to a better normal approximation. When the degrees of freedom, f, for a χ^2 distribution are more than 100, $Y = \sqrt{2\chi^2_{(f)}}$ is approximately $\mathcal{N}(\sqrt{2f-1}, 1)$.

When using tables we must remember that this form of the χ^2 test is always a *one-tail* test.

Example 20.10 Critical value for χ^2 with large number of d.f.

A statistic whose value was 252 has been calculated. On a null hypothesis that is being tested this should come from a χ^2 distribution with 120 degrees of freedom. Examine this.

The normal approximation requires us first to calculate Y; this is $\sqrt{2 \times 252} = \sqrt{504} = 22.45$. The value of $\sqrt{2f - 1}$ is $\sqrt{239} = 15.46$. Therefore Y should be $\mathcal{N}(15.46, 1)$, and so the corresponding standard normal value is $z = (22.45 - 15.46)/1 = 6.99$, which is very highly significant. There is very strong evidence to reject the hypothesis being tested.

20.15.1 Exercises

1. In a forest, trees are occasionally struck by lightning. The owner's family have kept records over the past 120 years, of the yearly number of trees struck:

Number of trees	0	1	2	3	4	5	Total
Number of years	36	42	23	11	7	1	120

Calculate the mean and the variance of the annual number of trees struck, and test the hypothesis that this number follows a Poisson distribution.

2. The numbers of snails found in each of 100 sampling units in a garden were as follows:

Number of snails	0	1	2	3	4	5	8	15
Number of sampling units	69	18	7	2	1	1	1	1

Examine whether these data follow a Poisson distribution.

20.16 MINITAB: THE POISSON DISTRIBUTION

To obtain values of the probability functions of the Poisson distribution and to generate random samples from it, choose

  Calc>Probability Distributions>Poisson/

or

  Calc>Random Data>Poisson/

The mean of the distribution must be stated.

20.17 EXERCISES ON CHAPTER 20

1. Draw a rough sketch of the frequency polygon of the Poisson distribution for each of the cases where the mean is 2, 1 and $\frac{1}{2}$.

   In an inspection scheme for items produced by a factory, a sample of 50 is selected at random from a large batch and examined. If the number of faulty articles found in the sample is less than 3, the batch is accepted. If the number found is exactly 3, a second sample, this time of 30, is examined and the batch is accepted if not more than 1 defective article is found in this sample. In all other cases the batch is rejected. If the proportion of defective articles in the batch is 2%, find, using the Poisson distribution, the probability of the batch being accepted and the average number of articles examined in such a batch. *(O&C)*

2. State the conditions under which the Poisson probability distribution may be used as an approximation to the binomial distribution.

   Nuts and bolts are manufactured in a factory. It has been found over a long period that on average 1 defective bolt is produced in every 100. If a random sample of 100 bolts contains 3 defectives, determine, using a 5% level of significance, whether this indicates that the process has broken down.

   The bolts are packed in cartons of 50. When the process is working normally, determine, to 3 decimal places, the probability that if a man buys 5 packets he will not get any defectives.

   The nuts are also packed in cartons of 50 items and here again on average one in 100 is defective. If a man buys a packet each of bolts and nuts, find, to 3 decimal places, the probability that he can get at least 49 matching pairs of non-defective items from the two packets. *(JMB)*

3. Lemons are packed in boxes, each box containing 200. It is found that, on average, 0.45% of the lemons are bad when the boxes are opened. Use the Poisson distribution to find the probabilities of 0, 1, 2, and more than 2 bad lemons in a box.

   A buyer who is considering buying a consignment of several hundred boxes checks the quality of the consignment by having a box opened. If the box opened contains no bad lemons he buys the consignment. If it contains more than 2 bad lemons he refuses to buy, and if it contains 1 or 2 bad lemons he has another box opened and buys the consignment if the second box contains fewer than 2 bad lemons. What is the probability that he buys the consignment?

   Another buyer checks consignments on a different basis. He has one box opened; if that box contains more than 1 bad lemon he asks for another to be opened and does not buy if the second also contains more than 1 bad lemon. What is the probability that he refuses to buy the consignment? *(Southern)*

4. Write down the probability that a random variable which is Poisson distributed with parameter λ takes the value r ($r = 0, 1, 2, \ldots$).

   The number of defects, r, in a piece of equipment is Poisson distributed with parameter λ. The values of r for each such piece of equipment passing through a service station are recorded, but, naturally, items for which $r = 0$ are not observed, and the resulting observed distribution is a Poisson distribution for which all values $r = 0$ are missing. Find the theoretical probabilities that $r = 1, 2, 3 \ldots$ for such a distribution, and show that its mean is $\lambda/(1 - e^{-\lambda})$. *(Oxford)*

5. If $p(x)$ is the probability of x successes in a Poisson distribution with mean μ, show that

$$p(x + 1)/p(x) = \mu/(x + 1).$$

A car hire firm has 3 cars to be hired out by the day. Assuming the number of cars ordered has a Poisson distribution with mean 2, calculate, for a period of 100 days, the expected number of days when:
(a) no car will be ordered;
(b) some orders cannot be met;
(c) one particular car is not used, assuming each car is used equally frequently. *(AEB)*

6. Define the Poisson distribution and derive its mean and variance.
   The number of telephone calls received at a switchboard in any time interval of length T minutes has a Poisson distribution with mean $\frac{1}{2}T$. The operator leaves the switchboard unattended for five minutes. Calculate to three decimal places the probabilities that there are: (i) no calls; (ii) four or more calls in her absence.
   Find to three significant figures the maximum length of time in seconds for which the operator could be absent with a 95% probability of not missing a call. *(JMB)*

7. State the conditions under which the binomial distribution approximates to the Poisson distribution. Hence derive the Poisson distribution of mean m and show that its variance is also m.
   Tests for defects are carried out in a textile factory on a lot comprising 400 pieces of cloth. The results of the tests are tabulated below:

Number of faults per piece	0	1	2	3	4	5	6	Total
Number of pieces	92	142	96	46	18	6	0	400

Show that this is approximately a Poisson distribution and calculate the frequencies on this assumption.
   How many pieces from a sample of 1000 pieces may be expected to have 4 or more faults? *(AEB)*

8. In a test of car components in which the failure rate is thought to be constant, the numbers of failures in a series of 100-mile intervals are recorded in the following table:

Number of failures	0	1	2	3	4	5	6	7	Total
Frequency	25	30	26	18	9	5	6	1	120

Calculate the mean number of failures per interval and the expected distribution of numbers of failures for the 120 intervals that would be given by a Poisson distribution with this mean.

A good fit between the sets of observed and expected frequencies is taken as evidence of the constancy of the failure rate. Use the χ^2 test to test the goodness-of-fit.

(In using the χ^2 test, cells with expected frequency less than seven should be amalgamated with neighbouring cells.)   (O&C)

9. A company has two machines, A and B. On average there are 0.8 breakdowns per week on machine A and 1.2 breakdowns per week on machine B. What is the probability of there being no breakdowns on machine A during the next two weeks? What is the probability of a total of two breakdowns on these two machines next week? You may assume that machines which break down can be repaired immediately. What further assumption is necessary to answer this question?

10. Give an account of the main characteristics of the Poisson distribution, and give two examples of variables that might be distributed in this way.

A sample of 50 observations from a distribution gives a mean of 6.25 and a variance of 5.78. Without performing a significance test, state whether you would consider this evidence in support of the hypothesis that the sample was drawn from a Poisson distribution, and give a 95% confidence interval for the population mean.   (Oxford)

11. Prove that the moment generating function about the origin of a random variable x which is Poisson distributed with parameter μ is $\exp[\mu(e^t - 1)]$. Hence prove that the mean of x is μ and find the moment generating function about the mean. Use this to show that the variance of a Poisson distribution is also μ.   (Oxford)

(Note: mgf about the mean is $E[\exp\{t(x - \mu)\}]$.)

12. One hundred squares were chosen at random in a large area of heathland, and the number of plants of a certain species falling in each square recorded. The records were as follows. Calculate an 'index of dispersion' for them. Do they suggest that the plants of this species were growing at random in the heathland?

0, 4, 1, 1, 1, 3, 0, 1, 3, 4, 2, 1, 2, 0, 3, 5, 2, 0, 5, 2,
1, 3, 5, 1, 4, 0, 3, 1, 0, 5, 3, 1, 2, 1, 2, 4, 2, 0, 3, 1,
0, 3, 3, 3, 0, 1, 2, 8, 1, 5, 3, 2, 0, 3, 2, 0, 2, 1, 2, 5,
2, 3, 2, 5, 1, 5, 3, 0, 0, 3, 1, 1, 2, 3, 2, 1, 2, 4, 2, 1,
3, 2, 2, 0, 2, 2, 0, 2, 2, 4, 2, 0, 4, 0, 2, 2, 3, 4, 1, 2.

13. State the conditions under which the binomial distribution may be approximated by the Poisson distribution and the conditions under which it may be approximated by the normal distribution.

For a certain teleprinter the probability that an incorrect character will be printed is 0.003 each time a character is printed. For printing errors that occur independently, find an approximate value for the probability that a message of 150 characters will contain

(i) no errors;

(ii) four or more errors.

If 100 messages each of 150 characters are printed find an approximate value for the probability that 60 or more of the messages will contain no errors. *(IOS)*

14. The random variable X has the Poisson probability mass function with mean λ. What is the variance of X?

   A random sample $x_1, x_2, \ldots, x_n$ is available from this distribution. Obtain the maximum likelihood estimate $\hat{\lambda}$, of λ, and say whether it is unbiased. Find an unbiased estimator based on $\text{Var}[X]$, say λ^*. Which of $\hat{\lambda}, \lambda^*$ would you prefer?

   For the sample $2, 2, 3, 3, 3, 4, 5, 6, 35$ find the mean, variance, $\hat{\lambda}$ and λ^*, and comment on the result.

   It was subsequently found that '35' was an error; it should have been 8. Repeat the calculations and comment on the results. *(RSS)*

20.18 COMPUTING EXERCISES

1. Generate 500 random values from a Poisson distribution with parameter λ, and construct a bar diagram or dot plot for the data. Use, successively, values of $\lambda = 1, 2, 4, 8, 16$. Note that the distributions for small λ are skew, and for large λ are approximately normal. Verify that the mean and the variance are approximately equal in each case.

2. By using a chi-squared goodness-of-fit test, decide whether the data in Exercise 20.7.1(3) could come from a Poisson distribution.

3. *Poisson approximation to binomial.* Construct a table of the probabilities that a variable takes values $0, 1, 2, \ldots$ assuming that the distribution is (a) binomial, and (b) the approximating Poisson, with the following parameters:

(i)	$n = 10$,	$\pi = 0.1$,	$\lambda = 1$;	(iv) $n = 200$,	$\pi = 0.05$, $\lambda = 10$;

   (i) $n = 10$, $\pi = 0.1$, $\lambda = 1$;   (iv) $n = 200$, $\pi = 0.05$, $\lambda = 10$;

   (ii) $n = 100$, $\pi = 0.01$, $\lambda = 1$;   (v) $n = 5$, $\pi = 0.02$, $\lambda = 0.1$;

   (iii) $n = 20$, $\pi = 0.5$, $\lambda = 10$;   (vi) $n = 50$, $\pi = 0.002$, $\lambda = 0.1$.

   Quote probabilities to 3 decimal places.

20.19 PROJECTS

1 Distribution of numbers of vehicles passing an observation point

Aim. To examine whether the number of vehicles passing in unit time follows a Poisson distribution.

Equipment. Watch with second hand; pencil and paper.

Collection of data. As in Project 1.9(2).

Record R, the number of vehicles passing in each unit time interval.

Analysis. Make a frequency table and illustrate with a bar diagram. Calculate the mean and variance of R. Calculate the frequencies to be expected in a Poisson distribution having the same mean. Compare these with the observed frequencies, using a χ^2 test. Also examine the fit to a Poisson distribution using an index of dispersion test.

Note. If the variance, s^2, is greater than the mean $\bar{r}$ of the observed data, and a Poisson distribution does not fit, a possible distribution to explain the data is the negative binomial. In this,

$$P(R = r) = \frac{k^k m^r}{(k+m)^{k+r}} \frac{k(k+1)\dots(k+r-1)}{r!} \quad \text{for } r = 0, 1, 2, \dots,$$

where m is estimated by $\bar{r}$ and k is estimated by $\bar{r}^2/(s^2 - \bar{r})$. This allows for some bunching or grouping in the traffic, so that vehicles arrive in batches rather than independently. The observed frequencies could be compared with this new set of frequencies using a χ^2 test.

2 Birthday dates

Aim. To examine the distribution of birthday dates in a large population.

Collection of data. Ask as many people as possible their date of birth (or obtain the dates from school or college records etc.). Record the dates for each person in a table. Omit any whose birthday falls on 29 February (why?).

Analysis. (1) Make a frequency table showing the numbers of people having a birthday on each day of the year. Count the numbers of days with $r = 0, 1, 2, \dots$ people and make a frequency table of r. Compare the distribution of r with that expected in a Poisson distribution whose mean is $\lambda = n\pi$, where $\pi = 1/365$ and n is the number of people studied.

(2) Group the data for individual days into weeks (or longer periods if there are not very many data). Calculate the frequency of people represented in each cell of the grouped table. Compare, using a chi-squared test, the frequencies in this table with those expected if the birth rate is constant throughout the year.

SUMMARY

The probability mass function of a **Poisson distribution** is $P(r) = \exp(-\lambda)\lambda^r/r!$, for $r = 0, 1, 2, \dots$. Its mean and variance are both equal to λ.

It arises when these conditions are satisfied: events occur at random in space or time, singly, independently of one another, uniformly and at constant rate; r is the count of the number of events in unit space or time.

A **binomial** distribution with n large and π small can be *approximated* by a Poisson with $\lambda = n\pi$.

The **sum** of *independent* random variables distributed as Poissons with means λ and κ is also a Poisson, with mean $\lambda + \kappa$.

Goodness-of-fit to a Poisson distribution may be tested by the usual χ^2 test where the final category is 'all observations $\geqslant$ some specified r' since there is no theoretical upper limit to r.

A Poisson random variable with mean > 5 is adequately *approximated* by a normal distribution $N(\lambda, \lambda)$. Significance tests and confidence intervals for means can use this approximation.

The Poisson **index of dispersion** compares mean and variance of a sample of data:

$$I = \sum_{i=1}^{n} (x_i - \bar{x})^2 / \bar{x}$$

is approximately distributed as $\chi^2_{(n-1)}$ when λ is not too small.

A χ^2 with a very large number of degrees of freedom, f, has a normal approximation: $\sqrt{2\chi^2_f}$ is approximately $N(\sqrt{2f - 1}, 1)$.

21 CORRELATION

OVERVIEW

When two variables X and Y are measured on the same units they may not be independent but instead correlated. We study graphical and numerical methods of measuring the relation between X and Y, and some pitfalls in interpreting the meaning of correlation.

21.1 INTRODUCTION

Most of the methods we have developed so far have been for dealing with one variate only. However, often several different characteristics are measured on each member of a sample, and it may be of great interest to ask whether the variates are interrelated. In this chapter, we shall look at methods which investigate whether two quantitative variates are related. (The χ^2 test of Section 19.7 allowed us to examine qualitative or categorized variates.)

21.2 SCATTER DIAGRAMS

Let us imagine a form master reviewing the examination results of his class of 11 senior science pupils. Two of the sets of marks he has in front of him are those for mathematics and physics. He wonders whether the mathematics mark X and the physics mark Y are related: did a good performance in the one subject go with a good performance in the other? He decides that he can most easily discover this by plotting the marks for all the pupils on a sheet of graph paper. The ith pupil $(i = 1, 2, \ldots, 11)$ scored x_i in mathematics and y_i in physics; the performance of this pupil is simply shown by marking on the graph the point with coordinates (x_i, y_i). The two sets of marks are given in Table 21.1, and are shown plotted on a graph in Fig. 21.1. This graph is called a *scatter diagram*.

The reader is no doubt familiar with graphs in which one variable is plotted against another, e.g. temperature may be plotted against time during a scientific experiment or from a series of weather records. Graphs often have straight lines, or curves, or continuous jagged lines in them. But Fig. 21.1 does not come into any of these categories. Each variable X, Y in that diagram is subject to considerable variability, so we obtain a cloud, or scatter, of points rather than a collection of points falling on a well-defined line. Sometimes there may in fact be an underlying

Table 21.1

Pupil	A	B	C	D	E	F	G	H	J	K	L	Total	Mean
Mathematics mark, x	41	37	38	39	49	47	42	34	36	48	29	440	40
Physics mark, y	36	20	31	24	37	35	42	26	27	29	23	330	30

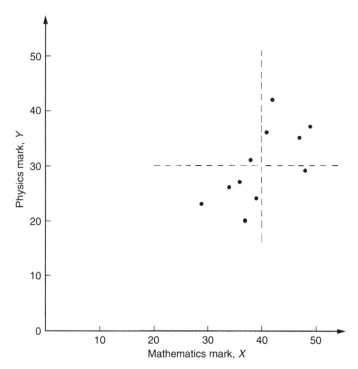

Figure 21.1 *Scatter diagram of mathematics and physics marks for 11 pupils.*

linear relation between the variables, but the points in the graph are scattered all round it rather than falling on it.

We cannot hope to detect subtle or precisely specified relations in diagrams where there is so much scatter. We shall look for the simplest, most basic relation: we ask if above-average marks in mathematics usually go with above-average marks in physics, and below-average mathematics marks usually with below-average physics marks. The dotted horizontal and vertical lines in Fig. 21.1 show the mean marks (40 for mathematics, 30 for physics), and this helps us to see that above-average mathematics marks (those to the right of 40) do usually go with above-average (above 30) physics marks. Below-average marks in the two subjects also go together; we see that almost all the points fall in either the first (top-right) or third (bottom-left) quadrant in Fig. 21.1. Only two of the eleven points fall outside these two quadrants. This indicates that there is a relation between the mathematics mark X and the physics mark Y of the same student.

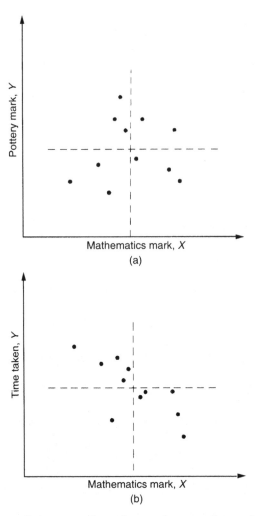

Figure 21.2 *Scatter diagram of mathematics marks and (a) pottery marks, (b) time taken to answer the mathematics paper.*

Two other sets of records interested the form master, and he wondered whether either of them would be related to the mathematics marks. One was the marks given for products made during a pottery course, and the other was the time each pupil took to answer the mathematics paper (the examination had been open-ended as regards time). He plotted two other scatter diagrams, those of Figs 21.2(a) and 21.2(b). The first is of mathematics marks X against pottery marks Y. We have again put in dotted lines showing the means of X and Y, as these help us to assess whether a relation exists or not. (Scatter diagrams do *not* usually have these dotted lines: they are certainly not a standard part of a scatter diagram.) In Fig. 21.2(a), showing mathematics and pottery, the points appear spread at random all over the graph; there is no indication of a relation and we may infer that ability in mathematics is not associated with ability in pottery.

In Fig. 21.2(b), however, there does appear to be a relation between the mathematics mark and the time taken in answering the examination. But this time above-average mathematics marks go with below-average times, so that the majority of points appear in either the second (top-left) or fourth (bottom-right) quadrant.

The types of relation we have detected are called *correlations*. That between mathematics and physics marks is said to be a *positive* correlation (since a line that followed the general trend of the points would have a positive slope), while the relation between mathematics marks and time taken is called a *negative* correlation (a line showing the general trend would have a negative slope). In the case of the mathematics and pottery marks we would say there is no correlation.

21.2.1 Exercises

For each of the sets of data below, draw a scatter diagram and assess whether there appears to be correlation between the two measurements labelled X and Y.

1. Vehicles and road deaths – latest available figures for each country.

Country	Vehicles per 100 population, X	Road deaths per 100 000 population, Y
Great Britain	31	14
Belgium	32	29
Denmark	30	22
France	47	32
West Germany	30	25
Irish Republic	19	20
Italy	36	21
Netherlands	40	22
Canada	47	30
USA	58	35

2. An owner of greyhounds notes the dogs' weights when they enter a race and their finishing positions in the race.

Dogs' weights (lb), X	60	63	70	65	60	64	67	73	56	58	60	60	66	55	61	60	64	53	68	65	55	60	60	65
Finishing position, Y	2	6	2	4	6	5	4	2	3	2	1	3	3	3	1	2	3	1	3	2	2	3	7	2

3. The weights and average daily food consumption were measured for 12 obese adolescent girls.

Weight (kg), X	84	93	81	61	95	86	90	78	85	72	65	75
Food consumption, Y (hundred calories per day)	32	33	33	24	39	32	34	28	33	27	26	29

4. The marks gained in tests by the same students in chemistry and in history.

Chemistry mark, X	4	22	4	45	22	34	5	27	17	26	27	36	24	25	27	43	19	39	12	5	19	21	35	7	33
History mark, Y	54	17	10	34	26	25	27	38	7	25	28	34	12	57	21	31	50	14	43	16	13	56	35	25	51

5. Data for 33 sections of major roads in Minnesota, USA, on the accident rate Y (per million vehicle miles) and the frequency of access points or junctions X (number per mile of road).

X	Y	X	Y
4.6	4.58	4.3	2.34
4.4	2.86	11.1	2.83
4.7	3.02	6.8	1.81
3.8	2.29	53.0	9.23
2.2	1.61	18.0	2.93
24.8	6.87	30.2	7.48
11.0	3.85	10.3	2.57
7.5	3.29	18.2	5.77
8.2	5.88	12.3	2.90
5.4	4.20	7.1	2.97
11.2	4.61	14.0	1.84
5.4	3.85	11.3	3.78
7.9	2.69	16.3	2.76
11.0	2.01	9.6	4.27
8.9	4.22	9.0	3.05
7.8	2.55	10.4	4.12
9.6	1.89		

21.3 THE CORRELATION COEFFICIENT

As we have seen before in this book, diagrams help us to obtain a general idea of the information contained in a set of data, but for precision we need a numerical measure. We now use the ideas suggested by the diagrams to define a *correlation coefficient* which measures the degree of correlation. First we shall redraw the scatter diagram, Fig. 21.1, showing mathematics and physics marks, and we shall make the dotted lines in that figure become the axes in the new one (Fig. 21.3(a)). The new diagram will therefore show the deviation $d_x = x - \bar{x}$ of a mathematics mark from the mean of all mathematics marks and the deviation $d_y = y - \bar{y}$ of a physics mark from the mean of all physics marks. For pupil A, $d_x = 41 - 40 = +1$ and $d_y = 36 - 30 = +6$; for pupil B, $d_x = 37 - 40 = -3$ and

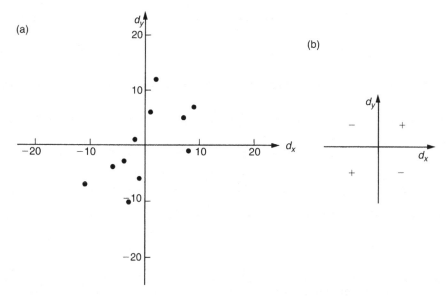

(a)

(b)

Figure 21.3 (a) Scatter diagram showing the deviation d_x from the mean of the X observations plotted against the corresponding Y deviation d_y. (b) Diagram showing the signs of the product $d_x d_y$ in each quadrant.

Table 21.2 Deviations from mean marks in mathematics and physics for eleven pupils

Pupil	A	B	C	D	E	F	G	H	J	K	L	Total (or mean)
Deviation of mathematics marks from the mean, d_x	1	−3	−2	−1	9	7	2	−6	−4	8	−11	0
Deviation of physics marks from the mean, d_y	6	−10	1	−6	7	5	12	−4	−3	−1	−7	0
Product $d_x d_y$	6	30	−2	6	63	35	24	24	12	−8	77	$\sum d_x d_y = 267$

$d_y = 20 - 30 = -10$, and so on: the complete set of $\{d_x, d_y\}$ is shown in Table 21.2. As a check $\sum d_x = 0 = \sum d_y$.

We said on p. 541–543 that there was a correlation if most of the points fell either in the first and third quadrants or in the second and fourth quadrants. When there is no correlation the points fall more or less uniformly in all four quadrants. Figure 21.3(b) shows that in quadrants 1 and 3 the product $d_x d_y$ is positive, since d_x and d_y are either both positive (quadrant 1) or both negative (quadrant 3). In quadrants 2 and 4, one of d_x, d_y is positive and the other is negative, so that the product $d_x d_y$ is negative. Thus in the case of positive correlation, when most points lie in the first or the third quadrant, most of the products $d_x d_y$ will be positive; while in the case of negative correlation, when most points lie in the second or the fourth quadrant, most of the products $d_x d_y$ will be negative. When there is no correlation, roughly half of the points will have $d_x d_y$ positive and the other half will have $d_x d_y$ negative. So the sum of products of the deviations, $\sum d_x d_y$, will be large in magnitude (either positive or negative in sign) when there is correlation, and small in magnitude where there is no correlation since in this case positive and negative products mostly cancel one another out.

Therefore this sum of products of deviations from means, $\sum d_x d_y$, is made the basis of our correlation coefficient. It has, however, one serious disadvantage, namely that its size depends on the units in which X and Y are measured. These mathematics and physics marks were actually out of 50. If the master had decided to express them as percentages, X and Y would both double, and so would d_x and d_y. The products $d_x d_y$ would each be multiplied by 4, and so then would $\sum d_x d_y$. It is possible to define a coefficient which is independent of both scales of measurement by dividing $\sum d_x d_y$ by $\sqrt{(\sum d_x^2)(\sum d_y^2)}$.

Definition 21.1

The **correlation coefficient** between n pairs of observations, whose values are (x_i, y_i) is

$$r = \frac{\sum_{i=1}^{n}(x_i - \bar{x})(y_i - \bar{y})}{\sqrt{[\sum_{i=1}^{n}(x_i - \bar{x})^2][\sum_{i=1}^{n}(y_i - \bar{y})^2]}} = \frac{\sum d_x d_y}{\sqrt{(\sum d_x^2)(\sum d_y^2)}}.$$

From Table 21.2, we have $\sum d_x d_y = 267$. We find also that $\sum d_x^2 = 386$ and $\sum d_y^2 = 466$. So, for the example of mathematics and physics marks, $r = 267/\sqrt{386 \times 466} = 0.630$.

This correlation coefficient r is often referred to as the *product-moment* correlation coefficient, to distinguish it from some other measures of correlation which we shall mention later.

It can be shown that r takes values only in the interval -1 to $+1$. It takes the values ± 1 when there is an exact straight-line relation $Y = mX + c$ connecting the two variates; $r = +1$ when m is positive and $r = -1$ when m is negative. When there is no correlation r will take a value close to 0; when there is strong correlation r will take a value near to ± 1, being positive for situations like Fig. 21.1, as we have found, and negative for those like Fig. 21.2(b). We need to remember an important property of r, namely that it only measures the extent of *linear* relation between X and Y. It is not helpful in detecting more general curved relationships, and we shall return to discuss this point in Section 21.6.

21.3.1 Exercises

1. Ten people were asked a set of test questions designed to measure their attitude to television as a news medium, and a further set to measure their attitude to newspapers. A higher overall score shows greater satisfaction. The scores are shown in the following table. Calculate the correlation coefficient between the two scores. Draw a scatter diagram to illustrate them.

Person	1	2	3	4	5	6	7	8	9	10
TV score, X	5	0	3	1	2	2	5	3	5	4
Newspaper score, Y	1	2	1	3	3	4	3	1	0	2

2. (a) Two variables X, Y are measured. The correlation coefficient between them is found to be r_0. If, instead of X and Y, the variables $U = aX$ and $V = bY$ are measured, find the correlation coefficient between U and V, in terms of r_0. (a and b are constants.)

(b) X, Y are now replaced by $W = aX + c$ and $Z = bY + d$ (where a, b, c, d are all constants). Find the correlation coefficient between W and Z in terms of r_0.

3. Show that $r = \pm 1$ when $Y = \pm mX + c$ (m being a positive number).

4. Calculate correlation coefficients between X and Y for the data of Exercises 21.2.1(1) and (2).

In real life, of course $\bar{x}$ and $\bar{y}$ are most unlikely to be convenient numbers, so that $(x_i - \bar{x})$ and $(y_i - \bar{y})$ must either carry several decimal places for an exact result or must be rounded off. We met the same problem in computing variances, and rewrote the formula for $\sum (x_i - \bar{x})^2$ to avoid this (see p. 32). The formula defining r can be dealt with in the same way. We leave it as an exercise for the reader to show that

$$\sum_{i=1}^{n} (x_i - \bar{x})(y_i - \bar{y}) = \sum_{i=1}^{n} x_i y_i - \frac{1}{n} \left(\sum x_i \right) \left(\sum y_i \right)$$

$$= \sum_{i=1}^{n} x_i y_i - n\bar{x}\bar{y}.$$

Therefore an alternative formula for r, useful for calculation purposes, is

$$r = \frac{\sum_{i=1}^{n} x_i y_i - (\sum_{i=1}^{n} x_i)(\sum_{i=1}^{n} y_i)/n}{\sqrt{[\sum_{i=1}^{n} x_i^2 - (\sum_{i=1}^{n} x_i)^2/n][\sum_{i=1}^{n} y_i^2 - (\sum_{i=1}^{n} y_i)^2/n]}}.$$

Readers will find a discussion of efficient methods for calculating r, using a computer, in *CCC*, Chapter 6.

As an illustration, we may carry out the calculation of the correlation coefficient between mathematics and physics marks using the data as originally given in Table 21.1, without first working out $(x_i - \bar{x})$ and $(y_i - \bar{y})$. From that table

$$\sum_{i=1}^{11} x_i^2 = 17\,986, \quad \sum_{i=1}^{11} x_i y_i = 13\,467 \quad \text{and} \quad \sum_{i=1}^{11} y_i^2 = 10\,366;$$

also

$$n = 11, \quad \sum_{i=1}^{11} x_i = 440, \quad \sum_{i=1}^{11} y_i = 330.$$

Thus

$$r = \frac{13\,467 - 440 \times 330/11}{\sqrt{(17\,986 - 440^2/11)(10\,366 - 330^2/11)}} = \frac{267}{\sqrt{386 \times 466}}$$

$$= 0.630, \quad \text{as before.}$$

21.3.2 Exercises

Calculate the correlation coefficient between the measurements X and Y for the three sets of data in Exercises 21.2.1(3), (4) and (5).

21.4 CORRELATION BETWEEN TWO RANDOM VARIABLES

So far we have considered how to make calculations using observed values (x_i, y_i) of variates. As usual, if we are to make inferences about the correlation between measurements in a whole population we need to set up models, and this requires us to define what we mean by the correlation of two random variables.

We recall the formal definition of the mean and variance of a random variable as expectations: the mean of X is $E[X]$ (p. 198) and the variance of X is $\text{Var}[X] = E[(X - E[X])^2]$ (p. 208). Now if we look at the definition of the correlation coefficient r on p. 546, and divide the numerator and the denominator in that definition by n, the various sums involved become averages. Finally, we replace these averages by expected values, to arrive at a definition as follows.

Definition 21.2

The **correlation coefficient** between two random variables X and Y is

$$\rho = \frac{E[(X - E[X])(Y - E[Y])]}{\sqrt{E[(X - E[X])^2]E[(Y - E[Y])^2]}}.$$

The denominator is equal to $\sqrt{\text{Var}[X]\,\text{Var}[Y]}$. The numerator takes a form very similar to that of a variance except that it contains a product of deviations (of X and Y) rather than a square of a deviation (of one of the variables). It is called a *covariance*.

Definition 21.3

The **covariance** between two random variables X and Y is

$$\text{Cov}[X,\ Y] = E[(X - E[X])(Y - E[Y])].$$

We may therefore write $\rho = \text{Cov}[X,Y]/\sqrt{\text{Var}[X]\,\text{Var}[Y]}$. Writing $\text{Var}[X] = \sigma_x^2$, $\text{Var}[Y] = \sigma_y^2$, another form is $\rho = \text{Cov}[X,\ Y]/\sigma_x\sigma_y$.

The covariance may be written in an alternative way:

$$\text{Cov}[X,\,Y] = \text{E}[XY - Y\text{E}[X] - X\text{E}[Y] + \text{E}[X]\text{E}[Y]]$$
$$= \text{E}[XY] - \text{E}[Y\text{E}[X]] - \text{E}[X\text{E}[Y]] + \text{E}[\text{E}[X]\text{E}[Y]].$$

Now $\text{E}[X]$, $\text{E}[Y]$ are simply constant numbers, and it will help us to remember this if we write them as $\text{E}[X] = \mu_X$, $\text{E}[Y] = \mu_Y$. We also remember that for any constant number c, it is true that $\text{E}[c] = c$ (p. 199). The expression for covariance now becomes

$$\text{Cov}[X, Y] = \text{E}[XY] - \text{E}[Y\mu_X] - \text{E}[X\mu_Y] + \text{E}[\mu_X\mu_Y]$$
$$= \text{E}[XY] - \mu_X\text{E}[Y] - \mu_Y\text{E}[X] + \mu_X\mu_Y$$
$$= \text{E}[XY] - \mu_X\mu_Y - \mu_Y\mu_X + \mu_X\mu_Y = \text{E}[XY] - \mu_X\mu_Y$$
$$= \text{E}[XY] - \text{E}[X]\text{E}[Y].$$

We saw in Theorem 10.3 (p. 238) that when X and Y are independent random variables, $\text{E}[XY] = \text{E}[X]\text{E}[Y]$. Thus, independent random variables have zero covariance and so also a zero correlation coefficient. However, as we have also seen, it is possible to find $\text{E}[XY] = \text{E}[X]\text{E}[Y]$, in particular cases, without X and Y being statistically independent. So the converse is not true: zero covariance (or correlation) does *not* imply independence in general.

Referring back to the definitions of covariance and variance, it is easy to see that

$$\text{Cov}[X, X] = \text{E}[(X - \text{E}[X])(X - \text{E}[X])] = \text{Var}[X].$$

So, for example,

$$\text{Cov}[X, X + Y] = \text{E}[(X - \text{E}[X])(X + Y - \text{E}[X + Y])]$$
$$= \text{E}[(X - \text{E}[X])(X - \text{E}[X] + Y - \text{E}[Y])]$$
$$= \text{E}[(X - \text{E}[X])^2] + \text{E}[(X - \text{E}[X])(Y - \text{E}[Y])]$$
$$= \text{Var}[X] + \text{Cov}[X, Y],$$
$$= \sigma_X^2 + \rho\sigma_X\sigma_Y.$$

Note that the algebra in calculations like this simplifies because we may write

$$\text{Cov}[X, X + Y] = \text{Cov}[X, X] + \text{Cov}[X, Y].$$

21.4.1 Exercises

1. Find $\text{Cov}[X + Y, X - Y]$.

2. Show that if $Y = c$ (which is constant) then $\text{Cov}[X, Y] = 0$ for any random variable X.

3. If X, Y are random variables such that $\sigma_X^2 = 1, \sigma_Y^2 = 4$, and their correlation coefficient $\rho = \frac{1}{4}$, find the correlation coefficient between X and $X + Y$. (*Note.* $\mathrm{Var}[X + Y] = \mathrm{Var}[X] + \mathrm{Var}[Y] + 2\,\mathrm{Cov}[X, Y]$.)

4. The variances of the random variables X, Y are known, and so is the correlation coefficient ρ between them. New random variables U, V are defined by $U = X + Y, V = X - Y$. Find the variances of U, V and their correlation coefficient, in terms of ρ and of the known variances of X, Y.

21.5 TESTING VALUES OF r FOR SIGNIFICANCE

If a calculated value of r is very near to 0, we shall be satisfied that no relation exists between X and Y, while if r is very near to ± 1 we shall be happy to claim that there is correlation. As usual, intermediate values leave us in doubt and we need an objective way of assessing and testing the results of calculations. A model is needed, in which the observed variates are modelled by random variables. We assume that X and Y are jointly normally distributed with correlation coefficient ρ. Data must therefore have been collected so as to be a random sample from the whole population of values of X and Y; if so, we can validly treat the data as a random sample from the joint theoretical distribution, in which ρ represents the true correlation coefficient. The sample correlation coefficient r then provides an unbiased estimate of ρ. (Provided X and Y seem to consist of observations that are symmetrically distributed, we usually accept the assumption of a joint normal distribution: but see also p. 560.)

The sampling distribution of r depends on the actual value of ρ. When $\rho = 0$, r is relatively easy to deal with: it has a distribution which is symmetrical about 0. The distribution of r also depends, as for example does χ^2 (Chapter 19), on the amount of data available for calculating r. The density function for r contains a parameter $(n - 2)$, where n is the number of pairs of observations. As in χ^2, this parameter is called the *degrees of freedom*. A table of r, such as the one on p. 719, therefore needs a separate row for each value of the degrees of freedom.

Table A5 on p. 719 shows the values of r which will be exceeded numerically (positively or negatively), with the probabilities shown at the head of the columns, if ρ really is 0. For example, with $n = 18$ pairs of observations, r has $n - 2 = 16$ degrees of freedom, and in random samples from a distribution in which $\rho = 0$, r has only 5% probability of lying outside the range -0.468 to $+0.468$. This information is used in the usual way to test the null hypothesis '$\rho = 0$' against the alternative '$\rho \neq 0$' in a two-tail test that rejects numerically large values of r, i.e. those furthest from 0 in either direction.

In our example, there were 11 pairs of observations, and r was $+0.630$. This has 9 degrees of freedom; from Table A5, the probability of r exceeding 0.602 numerically is 5%, and of exceeding 0.735 is 1%. Our value is thus significant at the 5% level (but not at 1%), sufficiently unlikely for us to reject the null hypothesis in favour of an alternative hypothesis that there *is* correlation.

21.5.1 Use of the *t*-distribution

When X and Y are jointly normally distributed, the null hypothesis that $\rho = 0$ may be tested using an expression which follows the *t*-distribution (Chapter 17). It can be shown that

$$r\sqrt{\frac{n-2}{1-r^2}}$$

is distributed as t with $n-2$ degrees of freedom, where n is the number of pairs of observations available. For the usual alternative hypothesis that $\rho \neq 0$ a two-tail *t*-test will be needed. If there is a good reason to claim that ρ must be positive (or that it must be negative), an alternative hypothesis in one direction only, then a one-tail test is appropriate; but this is an unusual situation. The tables of the correlation coefficient, such as Table A5, are based on this *t*-distribution.

21.5.2 Exercises

Test for significance the values of r calculated in Exercises 21.3.2.

21.6 FALLACIES IN INTERPRETING CALCULATED CORRELATION COEFFICIENTS

21.6.1 Correlation does not imply cause

It is very tempting to make claims that are much too sweeping when a significant value of r is found. We are certainly at liberty to discard the null hypothesis that $\rho = 0$, in favour of an alternative which says that there is *some* relation between X and Y, the two variates observed. But this does *not* imply that X caused Y, or that Y caused X. It is not a good explanation of the relation between mathematics and physics marks in our example to claim that physics marks were high *because* mathematics marks were high, or that mathematics marks were low *because* physics marks were low. Maybe some of the same skills are useful in obtaining good marks in both these subjects, but that is a very different situation from claiming a direct cause.

Nonsense correlation is the name often given to a relation found between two variates that happen to be changing together, for possibly related reasons but certainly not as mutual cause and effect. A widely quoted example is to take, say for the 20 years 1945–1964, before TV was so widely available in the UK, x_i = the number of TV licences taken out in year i, and y_i = the number of convictions of 'juvenile delinquents' in year i. The calculated value of r for these 20 pairs of observations turns out to be significant and positive, so we are tempted to argue that TV causes increased delinquency. However, as a warning against doing this without further thought or research, consider another example. In the 15 years 1925–1939 let x_i = the number of radio licences taken out in year i and y_i = the number of certified lunatics in year i. The value of r for these data is also significant and positive. Do we argue that radio caused an increase in lunacy? We have as much, or as little, right to do this as we had to argue from the 1945–1964 data about TV and delinquency. The most plausible argument in the second example, and perhaps also in the first, seems to be that we are measuring two variates which are changing together in time, perhaps for somewhat similar reasons such as the increasing complexity of life or simply because population figures were increasing steadily too. Medical skill and scientific skill were obviously also increasing together in time, during the period of time in question. Whenever (x_i, y_i) are a pair of measurements taken at a point in time, and time changes from one pair to another, great care should be taken to look for underlying forces influencing X and Y together.

21.6.2 Correlation is a measure of linear relation only

The value of r measures only the strength of *linear* relation between X and Y. If there happened to be a perfect quadratic relation we should in theory have $\rho = 0$. Figure 21.4(a) shows a typical set of data that might be collected in these conditions. For this type of data, the computed value of r would not be very far from 0: it is possible to find larger than average values of X that go with larger than average Y, and also larger than average X going with smaller than average Y. Therefore positive $(x_i - \bar{x})$ go sometimes with positive $(y_i - \bar{y})$ and sometimes with negative $(y_i - \bar{y})$, and the sum $\sum (x_i - \bar{x})(y_i - \bar{y})$ of the products contains both positive and negative terms which more or less cancel one another out. Each of the four quadrants in Fig. 21.4(a) contains about the same number of points.

Figure 21.4(b) illustrates the point that when the relation between X and Y is not purely linear, it is critical to know whether we have a sample whose values do properly represent the whole population. Is the set of Fig. 21.4(b) taken from only part of the population illustrated by Fig. 21.4(a), or is it from a population complete in its own right? If the latter, the underlying relation does seem to be curved, yet most of the points fall in the first and third quadrants and this will lead to a positive, fairly large, correlation coefficient. This indicates that the relation between X and Y has a *linear component* to it, not that it is *purely* linear. We cannot emphasize too strongly that a scatter diagram should be drawn, as well as a correlation coefficient calculated, when trying to interpret the relation between two variates.

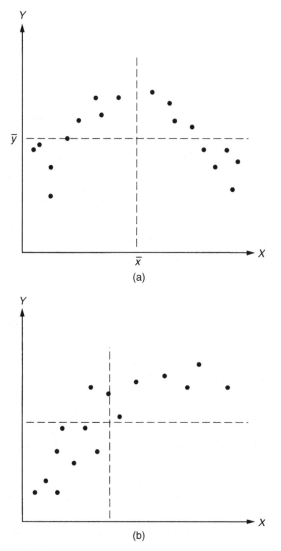

Figure 21.4 *Scatter diagrams obtained when the relation between variables X and Y is not linear.*

21.6.3 The misleading influence of a third variable

An underlying third variable, to which both X and Y may be related, can often be detected. Time is such a variable, as has been remarked above, but in scientific studies a more specific measurable variable Z can sometimes be detected. So X and Y appear related to one another because both are related, more or less directly, to Z. For example, X may be the measurement of blood pressure of a patient and Y may be the patient's heart rate. We may very well find that X and Y are related numerically to one another, but this could be because both are related to Z, the patient's weight.

It is clearly very important to try and forecast what 'third variables' ought to be examined, so that they can be measured at the same time as X and Y; and it is also clear that there may be plenty of scope for arguing about which variables really are directly related to which, and which indirectly. Correlation studies in these circumstances need to be supported wherever possible by planned scientific experiments, in each of which one variable is altered in a known way and the effect of this alteration on another variable is observed.

A good example of this last point is the argument over smoking and lung cancer. Do smokers run a higher risk of dying from lung cancer than non-smokers? Correlation studies first made this seem quite likely, but many other possible causes of lung cancer had to be examined, such as air pollution and exposure to certain types of substance at a person's place of work. Small groups of people, consisting of smokers and non-smokers but otherwise as alike as possible in living and working conditions, had to be examined to see whether the correlation persisted when other suggested possible causes had been removed. It is very difficult, from correlation studies only, to refute the suggestion that, for some unknown physiological reason, those more prone to lung cancer have an urge to smoke.

21.6.4 Spurious correlation of a part with the whole

If we try to correlate two variates, one of which forms part of the other, we shall find an arithmetic relation which may easily lead to a significant value of r. For example, we may obtain a set of monthly figures of X, the amount of personal savings held in Trustee Savings Bank and similar accounts, and Y, the total personal savings. Clearly, $Y = X + W$, where W represents personal savings held in accounts other than Trustee and similar savings. Exercise 21.4.1(3) indicates that even when X and W bear no relation to one another, X and Y will still show a non-zero (and quite possibly significant) correlation. A moment's thought before doing a correlation analysis would show us that it is the relation between X and W which can provide interesting information, not that between X and Y.

Spurious relationships also appear if we measure two variates X and Y, and then try to correlate X with the ratio X/Y, or Y with this ratio. Another important (and not immediately obvious) case is when we measure X and Y, and relate these to the same base Z. For instance, X and Y may be the cost of UK imports of food and of oil respectively in the same month, and Z the total cost of imports in that month. We may well compute the proportions X/Z and Y/Z because they give interesting information. But a spurious relationship appears if we correlate X/Z with Y/Z, and this will happen even if X and Y are independent. Ratios, and indices, need careful scrutiny before applying correlation analyses to them.

21.6.5 Combination of unlike populations

A rather subtle fallacy is to combine data from two or more separate groups as though they were a single homogeneous population. A scatter diagram, such as Fig. 21.5, helps to show what will occur if we do this. The data in group I show a good positive correlation, measured about the means $\bar{x}_1, \bar{y}_1$, whereas the data in group II show hardly any relationship, measured about their means $\bar{x}_2, \bar{y}_2$. If we

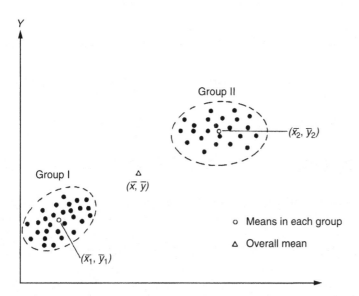

Figure 21.5 *Scatter diagram including two distinct populations in which the correlations are different.*

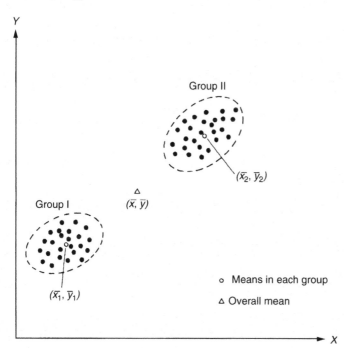

Figure 21.6 *Scatter diagram including two distinct populations in which the correlations are similar.*

try to compute a single correlation coefficient for the whole of the data, we are of course making the calculation relative to the overall means $\bar{x}, \bar{y}$, which in Fig. 21.5 happens to fall between the two groups of observations; this in itself should be warning enough. The r value will be very large because all the observations fall in the first and third quadrants relative to $(\bar{x}, \bar{y})$. The same is true in Fig. 21.6, which will give a similarly high r value although the nature of the correlations shown in Figs 21.5 and 21.6 is very different. Even if the correlation in group I in Fig. 21.5 is negative (the 'ellipse' being turned through a right angle), or both correlations in groups I and II in Fig. 21.6 are negative, the combined correlations will be strongly positive.

21.6.6 Inference to an unsampled population

Suppose that an agricultural research worker has records of the average yield Y of a particular variety of wheat in several different areas where it is grown. In some of these areas, he has set up experiments to measure Y and also (in the same units) X, which is the crop yield of a new variety of wheat. He finds that on these experimental sites the yields Y and X are significantly positively correlated, and so he concludes that this correlation will hold over all the areas in which the old variety is grown. Therefore he does not need to extend his experiments to cover all of these areas.

It is possible that he would be proved right if experiments were set up in all areas. It is also possible that he would find the relation between X and Y broke down in certain conditions of soil and climate which he did not happen to have covered in his experiments. He is making a statistically dangerous leap in inferring something about a wider population than that covered by his experiments, unless he knows a very great deal indeed about the factors that influence the yield of the new variety in varying conditions.

21.6.7 Exercises

Discuss what information, if any, can be gained from the following correlation analyses. Point out any fallacies.

1. Over a period of years, the correlation coefficient between the annual number of human births X and the estimated population of storks Y in Sweden is positive and significant.

2. The correlation coefficient between the length of cane L produced by a raspberry plant in its first growing season and the crop C of the plant in its second season is positive and significant for a large number of plants of the same variety. (A question for gardeners.)

3. The correlation coefficient between the intelligence quotient Q of a child measured at a standard age and the age A of that child's mother when the child was born is positive and significant for a large group of children.

4. At certain stages in school, girls tend to be superior to boys in verbal skills. In a group of girls, the correlation coefficient between X, which is the score in a spelling test, and Y, their IQ, is +0.50. In a group of boys, the correlation

coefficient between X and Y is also $+0.50$. The data are combined by calculating the correlation coefficient between X and Y for the whole set of children (boys and girls). Will this coefficient be 0.5? (Draw a diagram.)

5. There is a significant positive correlation between Y, the total annual cost of UK imports, and Z, the cost of UK food imports; and also between Y and X, the cost of UK oil imports.

6. What would you expect the correlation to be between the variables X and Z of Exercise 5?

7. A sample survey was carried out in three large towns in south-east England. Two of the variates recorded were the time X spent travelling between home and work, and the number Y of hours spent watching TV. The correlation coefficient between X and Y has a value similar to that found in a survey in one large Midland town.

21.7 COMPARING TWO CORRELATION COEFFICIENTS

We remarked that the distribution of r was only easy to deal with in the case $\rho = 0$. However, there is a normal approximation to the distribution of r in general, which can be used whatever the value of ρ.

The expression $Z = \frac{1}{2}\ln[(1+r)/(1-r)]$ is approximately normally distributed with mean $\frac{1}{2}\ln[(1+\rho)/(1-\rho)]$ and variance $1/(n-3)$, n being as usual the number of pairs of observations from which r was calculated. This approximation is not in fact all that good for small samples, say for $n < 50$, although it does quite often seem to be used. As in the earlier part of this chapter, we must assume that the two measurements being correlated come from a bivariate normal distribution.

Occasionally $\frac{1}{2}\ln[(1+r)/(1-r)]$ is written $\tanh^{-1} r$, where $\tanh^{-1}$ is the inverse hyperbolic tangent.

Example 21.1 Test of correlation coefficient

A correlation coefficient is calculated, based on 84 pairs of observations, between X and Y; a null hypothesis states that $\rho = \frac{1}{2}$. Test this hypothesis if the calculated $r = +0.34$.

The alternative hypothesis here is clearly '$\rho \neq \frac{1}{2}$', and so we shall need to make a two-tail test. $Z = \frac{1}{2}\ln(1.34/0.66) = \frac{1}{2}\ln 2.0303 = 0.3541$. Also $\frac{1}{2}\ln(1.5/0.5) = \frac{1}{2}\ln 3 = 0.5493$ and $\sqrt{1/(84-3)} = \frac{1}{9} = 0.1111$. If the null hypothesis is true, the value Z has been drawn at random from a normal distribution whose mean is 0.5493 and standard deviation 0.1111. Therefore $(Z - 0.5493)/0.1111$ is an $\mathcal{N}(0,1)$ random variable; we have

$(0.3541 - 0.5493)/0.1111 = -1.76$, which is not significant in a two-tail test. This leads us not to reject the null hypothesis.

Example 21.2 *Comparison of correlation coefficients*

The correlation coefficient between X and Y, the mathematics mark and general science mark in classes of fifteen-year-olds in a large school, is $+0.67$ for a group of 75 boys and $+0.42$ for a group of 63 girls. Test the hypothesis that the true correlation coefficient in the whole population of fifteen-year-old girls is the same as that in the population of fifteen-year-old boys.

Note that we can only generalize from our observations to the populations from which we have taken random samples. These might be all fifteen-year-olds studying courses based on the same syllabuses. Also, we do not know the exact values of ρ in either population; all we have is a null hypothesis that says ρ is the same (not necessarily zero) value in both populations. This implies that Z has the same distribution in both populations. So we test the null hypothesis that Z_1 and Z_2 have the same mean, i.e. $Z_1 - Z_2$ has mean 0. Provided that the two samples have been drawn independently of one another,

$$\text{Var}[Z_1 - Z_2] = \text{Var}[Z_1] + \text{Var}[Z_2]$$

$$= \frac{1}{75-3} + \frac{1}{63-3} = \frac{1}{72} + \frac{1}{60} = 0.0306.$$

From the observed data

$$z_1 = \tfrac{1}{2} \ln\left(\frac{1.67}{0.33}\right) = \tfrac{1}{2} \ln 5.0606 = 0.8107,$$

and

$$z_2 = \tfrac{1}{2} \ln\left(\frac{1.42}{0.58}\right) = \tfrac{1}{2} \ln 2.4483 = 0.4497.$$

Using the normal approximation,

$$\frac{(Z_1 - Z_2) - 0}{\sqrt{\text{Var}[Z_1 - Z_2]}}$$

should be $\mathcal{N}(0,1)$. The value is

$$\frac{0.8107 - 0.4497}{\sqrt{0.0306}} = \frac{0.3610}{0.1749} = 2.064$$

which is greater than 1.96, thus significant at the 5% level, leading us to reject the null hypothesis of equal correlations.

21.8 CONFIDENCE INTERVAL FOR ρ

The normal approximation can be used to provide a confidence interval for the true ρ in a population, given that a random sample of size n has yielded a calculated value r. Again we must assume that the population of measurements being correlated follows a bivariate normal distribution.

We know that $\frac{1}{2} \ln [(1+r)/(1-r)]$ is approximately

$$\mathcal{N} \left(\frac{1}{2} \ln \left(\frac{1+\rho}{1-\rho} \right), \frac{1}{n-3} \right),$$

so that

$$\frac{1}{2} \left[\ln \left(\frac{1+r}{1-r} \right) - \ln \left(\frac{1+\rho}{1-\rho} \right) \right] / \sqrt{\frac{1}{n-3}}$$

is approximately $\mathcal{N}(0,1)$. Hence, proceeding along the same lines of argument as in Chapter 18, we may say

$$P \left(\frac{1}{2} \ln \left(\frac{1+r}{1-r} \right) - \frac{1.96}{\sqrt{n-3}} \leqslant \frac{1}{2} \ln \left(\frac{1+\rho}{1-\rho} \right) \leqslant \frac{1}{2} \ln \left(\frac{1+r}{1-r} \right) + \frac{1.96}{\sqrt{n-3}} \right)$$

$$= 0.95 \text{ approximately.}$$

Very often 1.96 will be replaced by 2 in this expression. We obtain limits to $\frac{1}{2} \ln[(1+\rho)/(1-\rho)]$, and hence limits to ρ: these give an approximate 95% confidence interval for the true ρ. We may of course choose to make our confidence level 99% or 99.9%, by using 2.576 or 3.291 instead of 1.96.

Example 21.3 Confidence interval for correlation coefficient

The correlation coefficient between X, the number of daily train movements at a railway terminus, and Y, the number of trains arriving on time, is -0.21 for a random sample of 50 termini in a large country. Set approximate 95% limits to the value of the true correlation coefficient through the whole country.

We find

$$Z = \frac{1}{2} \ln \left(\frac{1+r}{1-r} \right) = \frac{1}{2} \ln \left(\frac{0.79}{1.21} \right) = \frac{1}{2} \ln 0.6529 = -0.2132.$$

Also

$$\sqrt{\frac{1}{n-3}} = \sqrt{\frac{1}{47}} = 0.1459,$$

and so (for use in finding approximate limits) $2/\sqrt{47} = 0.2917$. Hence $\frac{1}{2} \ln [(1+\rho)/(1-\rho)]$ lies between $(-0.2132 - 0.2917)$ and $(-0.2132 + 0.2917)$ with probability approximately 0.95. These limits are -0.5049 to $+0.0785$. From these we find limits to $(1+\rho)/(1-\rho)$, by doubling and taking exponentials. So

$$0.3643 \leqslant \frac{1+\rho}{1-\rho} \leqslant 1.1700,$$

finally giving

$$-0.466 \leqslant \rho \leqslant +0.078,$$

which is an approximate 95% interval for the true value of ρ.

21.8.1 Exercises

1. Using the data of Exercise 21.2.1(3), examine whether there appears to be any difference between this group of girls and a large population previously studied, in which the correlation coefficient was found to be 0.9.

2. Set 95% confidence limits to the true value of the correlation coefficient between weight of dogs and finishing position using the data of Exercise 21.2.1(2).

3. Two different varieties of wheat are grown on the same set of 33 experimental sites. For one variety, the correlation coefficient between mean soil temperature and time to germination is -0.47, and for the other variety it is -0.64. Examine whether the correlation coefficients are different.

21.9 RANK CORRELATION

We have so far assumed that X and Y can both be measured on a continuous scale, and that they are jointly normally distributed. On occasions, however, neither of these assumptions may appear safe to make. Suppose that a food manufacturer is experimenting with different methods of making strawberry jam, and is using different varieties of strawberry in the experiments. He will set out a sample of each jam, and will ask several judges to place these samples in an order of preference. The judges will hardly ever agree exactly with one another, so the manufacturer will end up with several different orders of preference, i.e. *rankings* of the samples. He will want to see how similar these rankings are, that is to say he will want to see whether those samples ranked highly by judge A are also ranked highly by judge B. Table 21.3 shows two rankings.

Rank 1 is the highest; judge A gave this to sample (iii) while judge B preferred sample (vii).

Clearly X, the rank assigned by judge A, and Y, the rank assigned by judge B, are discrete variates, certainly not jointly normally distributed. So while we may still calculate a correlation coefficient in the same way, its distribution will not have the same form as a product-moment correlation. Spearman (British, 1863–1945) studied this distribution, and as a result the rank correlation coefficient is usually called after him. It is denoted by r_s.

Table 21.3 *Ranking of ten samples of strawberry jam by two judges*

Sample	i	ii	iii	iv	v	vi	vii	viii	ix	x
Judge A's ranking	7	8	1	6	3	9	2	4	5	10
Judge B's ranking	10	9	4	3	6	8	1	2	5	7

When n objects are placed in rank order, the sum of ranks is $1 + 2 + \cdots + n = \frac{1}{2}n(n+1)$, so that if X denotes rank then $\bar{x} = \frac{1}{2}(n+1)$. Also

$$\sum_{i=1}^{n} x_i^2 = 1^2 + 2^2 + \cdots + n^2 = \frac{1}{6}n(n+1)(2n+1).$$

Thus

$$\sum_{i=1}^{n}(x_i - \bar{x})^2 = \sum_{i=1}^{n} x_i^2 - n\bar{x}^2 = \frac{1}{6}n(n+1)(2n+1) - \frac{1}{4}n(n+1)^2 = \frac{n(n^2-1)}{12}.$$

If (x_i, y_i) stands for the ranks of sample i by judges A and B respectively, then also

$$\bar{y} = \frac{1}{2}(n+1) \quad \text{and} \quad \sum_{i=1}^{n}(y_i - \bar{y})^2 = \frac{n(n^2-1)}{12},$$

since y_i is a rank just as x_i is. Thus

$$n\bar{x}\bar{y} = \frac{1}{4}n(n+1)^2 \quad \text{and} \quad \sqrt{\left[\sum(x_i - \bar{x})^2\right]\left[\sum(y_i - \bar{y})^2\right]} = \frac{1}{12}n(n^2-1).$$

Using now the same formula as for the product-moment correlation coefficient,

$$r_s = \frac{\sum(x_i - \bar{x})(y_i - \bar{y})}{\sqrt{[\sum(x_i - \bar{x})^2][\sum(y_i - \bar{y})^2]}},$$

we obtain

$$r_s = \frac{\sum x_i y_i - n\bar{x}\bar{y}}{\frac{1}{12}n(n^2-1)}$$

$$= \frac{\sum x_i y_i - \frac{1}{4}n(n+1)^2}{\frac{1}{12}n(n^2-1)}.$$

Although this is a perfectly good way of calculating r_s, the form in which r_s is usually written involves d_i, the difference in ranks $(x_i - y_i)$ allocated to the same sample by the two judges:

$$\sum_{i=1}^{n} d_i^2 = \sum_{i=1}^{n}(x_i - y_i)^2 = \sum_{i=1}^{n} x_i^2 - 2\sum_{i=1}^{n} x_i y_i + \sum_{i=1}^{n} y_i^2$$

$$= \frac{1}{3}n(n+1)(2n+1) - 2\sum_{i=1}^{n} x_i y_i.$$

Thus

$$\sum_{i=1}^{n} x_i y_i = \frac{1}{6}n(n+1)(2n+1) - \frac{1}{2}\sum_{i=1}^{n} d_i^2$$

and so

$$r_s = \frac{\frac{1}{6}n(n+1)(2n+1) - \frac{1}{4}n(n+1)^2 - \frac{1}{2}\sum_{i=1}^{n} d_i^2}{\frac{1}{12}n(n^2-1)}$$

$$= \frac{\frac{1}{12}n(n^2-1) - \frac{1}{2}\sum_{i=1}^{n} d_i^2}{\frac{1}{12}n(n^2-1)} = 1 - \frac{6\sum_{i=1}^{n} d_i^2}{n(n^2-1)}.$$

Table 21.3 may be completed to show the set of values $\{d_i\}$: check that $\sum_{i=1}^{10} d_i = 0$.

Sample	i	ii	iii	iv	v	vi	vii	viii	ix	x
Judge A's rank, X	7	8	1	6	3	9	2	4	5	10
Judge B's rank, Y	10	9	4	3	6	8	1	2	5	7
$d = (x - y)$	−3	−1	−3	3	−3	1	1	2	0	3: $\sum d_i = 0$

Here $n = 10$, so $n(n^2 - 1) = 990$; we find $\sum_{i=1}^{10} d_i^2 = 52$, and so

$$r_s = 1 - \frac{312}{990} = 0.685.$$

21.9.1 Exercises

1. Show that when both rankings agree exactly, $r_s = +1$, while when one ranking is the exact reverse of the other (X is 1 when Y is 10, X is 2 when Y is 9, and so on), then $r_s = -1$.

2. For the example of mathematics and physics marks (p. 541), convert the actual marks to rankings (i.e. order in class). From these calculate the value of r_s.

21.10 TESTING r_s FOR SIGNIFICANCE

As we remarked, r_s does not have the same distribution as r. We cannot therefore test hypotheses about it in the same way, at least for small samples. A suitable null hypothesis is that the judges are not in any predictable agreement with one another, and a mathematical model for this is that the two rankings are such that one is a random permutation of the other. When the number of objects n which are being ranked is less than 10, the distribution of r_s on this hypothesis is not a standard one, even approximately. Special tables are needed, such as Table A6 (p. 719). However, when n is 10 or more, we can find a function of r_s which is distributed approximately as a t-statistic (Chapter 17): it can be shown that

$$r_s \sqrt{\frac{n-2}{1 - r_s^2}}$$

is approximately t with $n - 2$ degrees of freedom.

Thus in the example of Section 21.9, $0.685\sqrt{8/[1 - (0.685)^2]}$ is approximately a $t_{(8)}$ variable. Its value is $0.685\sqrt{15.0723} = 0.685 \times 3.882 = 2.66$. We have not yet stated our alternative hypothesis: in general it will be that $r_s \neq 0$, i.e. there is *some* relation, positive or negative, between the rankings. In such a case the test must be a two-tail one. The two-tail, 5% point for $t_{(8)}$ is 2.306, and our value of 2.66 is therefore significant evidence against the null hypothesis, so we reject the suggestion that there is no relation between the rankings given by the two judges.

21.10.1 Exercises

1. Test for significance the value found for r_s in Exercise 21.9.1(2), and compare the inference which you would make from r_s with that already made for the same data using r.

2. In a certain paper in a professional examination, the examiners reported that candidates' median marks were highest for Question 4, next highest for Question 3, then for Questions 1, 2, 10, 6, 12, 5, 8, 9, 7, 11 in descending order. However, the highest number of answers was submitted to Question 3, next highest for Question 1, and the order of popularity of questions continued 5, 9, 4, 6, 11, 8, 2, 7, 10, 12. Test the suggestion that candidates are good at picking out questions which will yield high marks.

3. Twelve typists entered a proficiency test, and their times for typing a standard piece of work were noted. Each piece of work was also assessed for quality (accuracy, general layout, etc.), and the best piece of work was ranked 1, the next best 2, and so on. Does the following information indicate a significant association between the quality of the work and the time taken to type it?

Typist	A	B	C	D	E	F	G	H	I	J	K	L
Time taken (seconds)	106	92	98	102	99	110	130	113	104	86	100	91
Ranking given for quality	6	9	3	10	12	1	2	5	4	8	7	11

State briefly when you would measure association using Spearman's coefficient of rank correlation rather than the product-moment correlation coefficient.

4. The mean temperature and the amount of milk sold to manufacturers during the months of a certain year were:

Month	Mean temperature (°C)	Milk sold (10 million gal)
April	9	11
May	13	15
June	14	13
July	17	10
August	16	9
September	15	8

(i) Calculate a product-moment correlation coefficient, and test its significance.

(ii) Calculate a rank correlation coefficient. In order to test whether it is significantly different from 0, use Table A6, which gives critical values of r_s that are exceeded in absolute value with the probabilities stated. (It is used exactly as Table A5 for r.)

21.11 MINITAB: CORRELATION COEFFICIENTS

Any investigation of correlation should begin with plotting a scatter diagram. Using MINITAB to do this has been discussed in Section 5.5. We have introduced two correlation coefficients: the Pearson product-moment correlation coefficient (Section 21.3) and Spearman's rank correlation coefficient (Section 21.9). The Pearson coefficient is usually called simply the correlation coefficient, and the Spearman, the *rank* correlation coefficient.

Suppose the marks in Table 21.1 are entered into a MINITAB worksheet in columns labelled X and Y. To obtain the correlation coefficient, choose

Stat>Basic Statistics>Correlation/
Enter X and Y in **Variables/**

The output gives the correlation coefficient 0.630 and the *p*-value. The test is a two-tailed test with NH: $\pi = 0$ and AH: $\pi \neq 0$.

To calculate the rank correlation coefficient, the observations in each column are first ranked, using

Data>Rank/

When ranking, the data may be stored in the same column or in a different one. Then proceed as with the correlation coefficient except that **Display p-values** is deselected since the values are not correct for the rank correlation coefficient.

21.12 EXERCISES ON CHAPTER 21

1. A textile firm manufactures a particular fabric in a variety of different qualities at correspondingly different prices. In 1970 various amounts of the different qualities were sold; the amounts (in thousands of metres) and prices (in pence per metre) are shown below.

Amount sold	75	62	71	61	70	59	65	69
Price	37	55	25	20	47	62	24	52

Evaluate the product-moment correlation coefficient.

Draw a scatter diagram of the data, and contrast the information this provides about the relationship between prices and sales with that given by the product-moment correlation coefficient. *(JMB)*

2. The random variables x and y are distributed with means μ_x, μ_y and variances σ_x^2, σ_y^2 respectively. The correlation coefficient of x and y is ρ. If $z = x + y$, write down:
   (i) the expected value of z;
   (ii) the variance of z.

   The running times in minutes of a cross-city bus service, routed through the city centre, were observed over a long period. The following data were obtained:

Stage of journey	Mean running time	Standard deviation of running time
Terminus X to city centre	12	1
City centre to terminus Y	20	2

   Given that the correlation coefficient of the running times on the two stages of the journey is $\frac{1}{2}$, find the mean and variance of the total time taken to complete a journey. If the scheduled running time is 33 minutes, estimate, using normal probability tables, the proportion of buses which take longer than the scheduled time. *(MEI)*

3. Obtain an expression for the covariance of the random variables $X - Y$ and $X + Y$ in terms of the variances of X and Y. If the value of this covariance is 2 and the covariance of $2X - Y$ and $2X + Y$ is 11, calculate the variances of X and Y. *(IOS)*

4. Define the product-moment correlation coefficient and explain how you would interpret values of 0 and 1.

   Guess the value of the correlation coefficient between the variables in the following situations and interpret your estimate.
   (i) The height of water and the volume of road traffic at London Bridge, if high tide is at 7 a.m. The interval between successive high tides is about 12 hours.
   (ii) The marks in paper I and total marks in a two-paper examination. *(Oxford)*

5. Explain how you would calculate the product-moment correlation coefficient from a sample of n pairs of values x_i, y_i $(i = 1, 2, \ldots, n)$.

   The number of eggs laid, x, by a certain species of bird is either 1 or 2, and this is related to the age of the bird, y, which is also either 1 or 2. The frequencies of the four possible combinations of values x, y are recorded for a total of n birds, as shown below. Show that the product-moment correlation coefficient between x and y is

$$(ad - bc)/[(a+b)(c+d)(a+c)(b+d)]^{\frac{1}{2}}.$$

		Number of eggs laid, x		
		1	2	Total
Age y	1	a	b	$a+b$
(years)	2	c	d	$c+d$
Total		$a+c$	$b+d$	n

(Oxford)

6. The correlation coefficient between tensile strength and hardness for two different synthetic materials was estimated as follows:

Material	Number of pairs of observations	r
A	20	0.75
B	30	0.85

Do these estimates differ significantly: (a) from zero; (b) from each other? Justify the test used in (a).

7. Each of the variables x and y takes the values $1, 2, \ldots, n$ but not in the same order as each other.

Prove that the covariance of x and y is

$$\sigma_{xy} = \frac{n^2 - 1}{12} - \frac{1}{2n} \sum_{i=1}^{n} (x_i - y_i)^2,$$

and further that

$$0 \leqslant \sum_{i=1}^{n} (x_i - y_i)^2 \leqslant \frac{1}{3}n(n^2 - 1).$$

The orders of merit of 10 individuals in two separate trials were:

Individual	A	B	C	D	E	F	G	H	I	J
Order in first trial	1	2	3	4	5	6	7	8	9	10
Order in second trial	9	3	4	7	10	1	2	5	6	8

Find Spearman's coefficient of rank correlation between the two orders and discuss the correlation obtained.
(O&C)

8. It is believed that a patient who absorbs a drug well on one occasion will do so on another occasion. Twelve patients gave the following results for percentage absorbed on two days:

Patient	1	2	3	4	5	6	7	8	9	10	11	12
Day 1	35.5	16.6	13.6	42.5	39.0	29.5	28.5	36.0	19.7	42.0	30.3	24.5
Day 2	27.6	15.1	12.9	30.5	23.1	14.5	35.5	27.5	16.1	18.9	32.5	24.5

Calculate a rank correlation coefficient, and use it to decide whether the belief appears justified.

9. Records of the annual crop of sugar beet and summer rainfall (June to August) for 1970–1975 are as follows:

Year	1970	1971	1972	1973	1974	1975
Crop (in 100 000 tons)	63	77	61	73	45	62
Summer rainfall (cm)	20	26	17	22	24	14

Calculate (i) the product-moment correlation coefficient, (ii) a rank correlation coefficient, between rainfall and crop.

Comment on the results.

When would you use each of the two coefficients as a general measure of the association between observed variates?

10. In an experiment on perception, two subjects are told that each will be presented with eight objects, all of different weights, and that the task is to sort them into order of weight. In fact the objects, while of different composition and shape, all weigh the same amount. The subjects are, separately, presented with the objects all together on a table, and required to place them in a row in descending order of weight; ties are not allowed. The rankings of the two subjects are as follows.

Object	1	2	3	4	5	6	7	8
Subject A	7	3	6	1	2	5	4	8
Subject B	8	1	2	3	5	4	7	6

Analyse the data to determine whether the assessments of the subjects are essentially unrelated, or whether the subjects tend to rank the objects similarly.

11. For the following data, use suitable correlation coefficients to measure the degrees of association between
    (i) latitude ranking and mean high temperature;
    (ii) mean high temperature and mean low temperature.

City	Latitude ranking	Mean high temp. (°F)	Mean low temp. (°F)
Bombay	4	87	74
Calcutta	6	88	69
Hong Kong	5	77	68
Karachi	7	86	70
Manila	2	89	72
Rangoon	3	90	73
Saigon	1	91	75
Shanghai	8	69	53

In each case give reasons for your choice of correlation coefficient.

21.13 COMPUTING EXERCISES

1. A physical fitness specialist weighed 19 participants in a fitness programme, and also used skin fold measurements to estimate their total body fat. He then calculated their percentages of body fat, with the following results. Participants 1 to 10 are men; 11 to 19 are women.

Participant	Body weight (kg)	Total body fat (kg)	% body fat	Participant	Body weight (kg)	Total body fat (kg)	% body fat
1	89	25	28.1	11	57	17	29.8
2	88	24	27.3	12	68	22	32.4
3	66	16	24.2	13	69	24	34.8
4	59	14	23.7	14	59	18	30.5
5	93	27	29.0	15	62	18	29.0
6	73	18	24.7	16	59	15	25.4
7	82	24	29.3	17	56	16	28.6
8	77	19	24.7	18	66	22	33.3
9	100	30	30.0	19	72	24	33.3
10	67	15	22.4				

Verify that the sample correlation coefficients between body weight and percentage fat for men, women and all participants are $+0.90$, $+0.79$ and $+0.07$ respectively. Carry out further graphical and numerical analysis of the data in order to provide an explanation of this apparently anomalous result. *(IOS)*

2. The following four sets of data I–IV were devised by F. J. Anscombe (*American Statistician*, **27**, pp. 17–21). For each set, plot a scatter diagram and find the

correlation coefficient. Comment on the results.

I		II		III		IV	
x	y	x	y	x	y	x	y
10	8.04	10	9.14	10	7.46	8	6.58
8	6.95	8	8.14	8	6.77	8	5.76
13	7.58	13	8.74	13	12.74	8	7.71
9	8.81	9	8.77	9	7.11	8	8.84
11	8.33	11	9.26	11	7.81	8	8.47
14	9.96	14	8.10	14	8.84	8	7.04
6	7.24	6	6.13	6	6.08	8	5.25
4	4.26	4	3.10	4	5.39	19	12.50
12	10.84	12	9.13	12	8.15	8	5.56
7	4.82	7	7.26	7	6.42	8	7.91
5	5.68	5	4.74	5	5.73	8	6.89

21.14 PROJECTS

1 Relation between goals scored for and against in football

Aim. To examine whether the goals scored by football clubs are correlated with those scored against them.

Equipment and collection of data. As for Project 10.7(2).

Analysis. Calculate a product-moment correlation coefficient between goals for (X) and goals against (Y) of individual clubs, and test it for significance (taking $\rho = 0$ as null hypothesis).

Note. Although X and Y are not jointly normally distributed, which is strictly necessary for the calculation of r to be valid, the results of the calculation for a large number of clubs can, as a good approximation, be treated in the usual way.

Extensions. (1) A rank coefficient between league position and goals (ranked in order of scores) could also be calculated.

(2) Differences between product-moment coefficients in different leagues could be tested for significance.

2 Ranking of 'stars' by different judges

Aim. To see whether different people's rankings of sports stars, pop stars, politicians etc. agree.

Equipment. Pictures of different sports stars, pop stars, etc.; or a list of names of well-known players, politicians or other people 'in the news'.

Collection of data. Present two people with a gallery of pictures or a list of names, preferably containing at least 12 items. Ask each person, independently of the other, to rank the items in order of preference. Record the two rankings.

Analysis. Calculate a rank correlation coefficient between the rankings assigned by the two people.

Extension. This can be extended to include more than two people acting as judges, and correlation coefficients between all pairs of judges can be calculated. Also, the same judges may try ranking more than one set of items, to see whether they tend to have the same degree of agreement with one another whatever items they are judging.

3 Other data

Correlation studies may also be made of the measurements collected in Project 10.7(1), and in the projects of Chapter 23.

SUMMARY

The **covariance** between X and Y is defined as

$$\text{Cov}[X, Y] = E[\{X - E(X)\}\{Y - E(Y)\}]$$

and the **correlation coefficient** $\rho = \text{Cov}[X, Y]/\sigma_X \sigma_Y$.

The **product-moment correlation coefficient** calculated from n pairs of observations (x_i, y_i) is

$$r = \frac{\sum (x_i - \bar{x})(y_i - \bar{y})}{\sqrt{[\sum (x_i - \bar{x})^2][\sum (y_i - \bar{y})^2]}}.$$

To test a null hypothesis that the true value of ρ is 0, use

$$t_{(n-2)} = r\sqrt{\frac{n-2}{1-r^2}}.$$

For comparing coefficients or finding confidence intervals the following normal approximation is useful:

$$Z = \frac{1}{2}\ln\left(\frac{1+r}{1-r}\right) \quad \text{is approximately} \quad \mathcal{N}\left(\frac{1}{2}\ln\left(\frac{1+\rho}{1-\rho}\right), \frac{1}{n-3}\right).$$

Correlation is a measure of *linear relationship* only and does not imply a direct cause and effect.

Spearman's coefficient is the correlation between two rankings of the same objects:

$$r_s = 1 - \frac{6\sum d_i^2}{n(n^2 - 1)}$$

where d_i is the difference between the two rankings for the ith unit.

22 THE ANALYSIS OF VARIANCE

OVERVIEW

The analysis of variance is a general method for studying linear models by subdividing the total sum of squares $\sum(y_{ij} - \bar{y})^2$. In this chapter we consider one-way and two-way analyses for removing systematic variation, and the practical applications in controlled experimentation. The planning and design of experiments are also discussed.

22.1 INTRODUCTION

Consider this set of data on the number of hours six people slept on a certain night:

Person	A	B	C	D	E	F
Number of hours' sleep	8	10	10	8	8	10

We might summarize these observations by noting that the mean is $54/6 = 9$ hours and that the estimated variance, assuming they come from a single population, is

$$\{(-1)^2 + 1^2 + 1^2 + (-1)^2 + (-1)^2 + 1^2\}/5 = 6/5.$$

But if we had the extra information that the data were all from one family and that A, D and E were adults while B, C and F were children, we might set out the data in the table:

	Number of hours' sleep		
Adults	8(A)	8(D)	8(E)
Children	10(B)	10(C)	10(F)

We see that the variability in the set of data is 'explained' by the factor *age*: the difference between adults and children. Another way of looking at the data is to say that we have data not from one population but from two. Our measure of variability is picking up the fact that the data come from two populations, of different location (i.e. different mean), rather than that the data are intrinsically variable.

With real sets of data the variability of the observations can rarely be completely explained by the factors of which we are aware; there is usually some residual unexplained variability. Let us suppose, in our example, that if length of sleep is measured to the nearest tenth of an hour the values are:

	Number of hours' sleep		
Adults	8.4	7.7	7.9
Children	9.8	9.9	10.3

There is an individual variation about the mean of each age-group. This becomes clear if we write the observations:

Adults	$8+0.4$	$8-0.3$	$8-0.1$
Children	$10-0.2$	$10-0.1$	$10+0.3$

We can now see that we have samples from two statistical populations with different means, 8 and 10; the differences from the means, $+0.4, -0.3, \ldots, +0.3$, are the *residuals* (p. 360), because they measure the extent of the natural variation after the means are accounted for. We will assume this variation is the same for children as for adults, so that all six observations come from populations with variance σ^2.

If we further analyse each population mean, into a general mean for the whole of the data and an age deviation of each population from this, we obtain:

Adults	$9-1+0.4$	$9-1-0.3$	$9-1-0.1$
Children	$9+1-0.2$	$9+1-0.1$	$9+1+0.3$

Thus each observation consists of three components:

$$y_{ij} = \bar{y} + (\bar{y}_i - \bar{y}) + (y_{ij} - \bar{y}_i) \qquad i = 1,2; \; j = 1,2,3.$$

In this linear model, the first component represents the general mean of the whole set of observations; the second is a deviation from the general mean which represents an age difference; and the third is the residual variation for each individual person. The first two terms in the model are *systematic*, not random, components; only the third is random.

We will obviously be interested in assessing the relative importance of the age and residual variability components: is there really a systematic variation due to age, or is this small compared with the residual variability in the observations? We need some overall measure of variation which can be split up into useful parts. A suitable measure is the sum of squares of deviations $\sum_i \sum_j (y_{ij} - \bar{y})^2$ which is a measure of the variability of the complete set of deviations. Since

$$(y_{ij} - \bar{y}) = (\bar{y}_i - \bar{y}) + (y_{ij} - \bar{y}_i), \quad \text{from above,}$$

$$(y_{ij} - \bar{y})^2 = (\bar{y}_i - \bar{y})^2 + (y_{ij} - \bar{y}_i)^2 + 2(\bar{y}_i - \bar{y})(y_{ij} - \bar{y}_i), \quad \text{and}$$

$$\sum_i \sum_j (y_{ij} - \bar{y})^2 = \sum_i \sum_j (\bar{y}_i - \bar{y})^2 + \sum_i \sum_j (y_{ij} - \bar{y}_i)^2$$

$$+ 2 \sum_i \sum_j (\bar{y}_i - \bar{y})(y_{ij} - \bar{y}_i).$$

The cross-product term on the right equals

$$2\sum_i (\bar{y}_i - \bar{y}) \sum_j (y_{ij} - \bar{y}_i) = 2\sum_i (\bar{y}_i - \bar{y}) \times 0 = 0.$$

Hence

$$\sum_i \sum_j (y_{ij} - \bar{y})^2 = 3\sum_i (\bar{y}_i - \bar{y})^2 + \sum_i \sum_j (y_{ij} - \bar{y}_i)^2.$$

We may set this down in table form:

Source of variation	Sum of squares
Age	$3\sum_i(\bar{y}_i - \bar{y})^2 = 6.00$
Residual	$\sum_i \sum_j(y_{ij} - \bar{y}_i)^2 = 0.40$
Total	$\sum_i \sum_j(y_{ij} - \bar{y})^2 = 6.40$

This type of decomposition is the fundamental idea used in an analysis of variance. It allows us to estimate the residual variability and, using this as a yardstick, assess whether there are differences in the means of the populations involved.

In this chapter we shall show some of the information that may be obtained from an analysis of variance. It is a technique that is widely used in analysing data from experiments or surveys in biology, engineering, the chemical industry, medicine, market research and many other fields.

22.2 COMPARISON OF THE MEANS OF POPULATIONS: ONE-WAY ANALYSIS OF VARIANCE

A consumer research organization decided to investigate the price of unleaded petrol in a particular county. They chose to look at the city that was in the county, a country area and a small town. They chose, at random, five filling stations from each area and recorded the following prices (in pence per litre):

						Total
City	65	64	63	66	67	325
Country area	71	68	70	70	71	350
Town	65	67	64	69	65	330
						1005

We have random samples from three populations. We shall assume that the populations have means μ_1, μ_2, μ_3 that may, or may not, be different; and also that the variance of each of the three populations is equal to σ^2. The research organization wishes to compare the means of the three populations, but to make this possible we must first estimate the common variance σ^2 of the populations. We shall do this by using the idea, introduced in Section 22.1, of splitting up the variability of the 15 observations into two components: a component that depends on the differences between the means of the three populations, and a component that measures the residual variability (or variability within the populations).

22.2.1 Estimation of residual variability by the analysis of variance

We may represent the data by the following table of symbols:

						Total	Mean
City	$y_{11},$	$y_{12},$	$y_{13},$	$y_{14},$	y_{15}	T_1	$\bar{y}_1$
Country area	$y_{21},$	$y_{22},$	$y_{23},$	$y_{24},$	y_{25}	T_2	$\bar{y}_2$
Town	$y_{31},$	$y_{32},$	$y_{33},$	$y_{34},$	y_{35}	T_3	$\bar{y}_3$
						G	$\bar{y}$

One method of estimating the residual variability would be to estimate the variance of each of the three populations, and combine these estimates. Thus for the city data, which have a mean $\bar{y}_1$, we would calculate the five deviations:

$$y_{11} - \bar{y}_1, \quad y_{12} - \bar{y}_1, \quad y_{13} - \bar{y}_1, \quad y_{14} - \bar{y}_1, \quad y_{15} - \bar{y}_1,$$

square them, sum and divide by 4. This would be an estimate of σ^2. The same might be done for the other areas and the three estimates combined. But the decomposition of the sum of squares of deviations of the observations from the overall mean, introduced in the previous section, produces the same result and the calculation is easier. Recall that the deviation of an observation from the overall mean was written:

$$y_{ij} - \bar{y} = (\bar{y}_i - \bar{y}) + (y_{ij} - \bar{y}_i).$$

On squaring, and summing over the petrol data, we obtain:

$$\sum_{i=1}^{3}\sum_{j=1}^{5}(y_{ij} - \bar{y})^2 = 5\sum_{i=1}^{3}(\bar{y}_i - \bar{y})^2 + \sum_{i=1}^{3}\sum_{j=1}^{5}(y_{ij} - \bar{y}_i)^2.$$

Our main interest, at the moment, is in the final term on the right, since this is the sum of squares of deviations *within* each population and this may be used in the estimation of σ^2. In fact the other two terms in the expression are easier to calculate and we obtain the term we want from them.

The term on the left-hand side of the equation is called the *total sum of squares*. Let us represent the *grand total* of the observations, $\sum_{i=1}^{3}\sum_{j=1}^{5} y_{ij}$ by the symbol G. Then, using the identity which expresses a sum of squares of deviations as the difference of a raw sum of squares and a correction term (p. 32), we write

$$\sum_{i=1}^{3}\sum_{j=1}^{5}(y_{ij}-\bar{y})^2 = \sum_{i=1}^{3}\sum_{j=1}^{5}y_{ij}^2 - \frac{G^2}{15}.$$

The numerical calculation is

$$65^2 + 64^2 + 63^2 + \cdots + 64^2 + 69^2 + 65^2 - \frac{(1005)^2}{15} = 67\,437 - 67\,335 = 102.$$

The first term on the right-hand side of the equation is called the *between-groups sum of squares*; it measures how much the means of the individual groups differ from the overall mean, and would be zero if all the means were the same. Using again the identity for the sum of squares of deviations, we write

$$5\sum_{i=1}^{3}(\bar{y}_i - \bar{y})^2 = 5\left\{\sum_{i=1}^{3}\bar{y}_i^2 - \frac{(3\bar{y})^2}{3}\right\}$$

$$= \frac{\sum_{i=1}^{3}T_i^2}{5} - \frac{G^2}{15},$$

using the facts that $T_i = 5\bar{y}_i$ and $G = 15\bar{y}$. This final form is the most convenient for hand calculation. For the petrol data the between-groups sum of squares is:

$$\frac{325^2 + 350^2 + 330^2}{5} - \frac{1005^2}{15} = \frac{337\,025}{5} - 67\,335 = 67\,405 - 67\,335 = 70.$$

The numerical values are inserted into Table 22.1. This table is called the *analysis of variance table*, or often simply the analysis of variance. The *residual sum of squares*, which is the remaining component of the sums of squares of deviations, is calculated as the difference of the total sum of squares and the between-groups sum of squares. Another column is needed: 'degrees of freedom'. The reader will recall (Section 11.5) that when dealing with a single sample, an unbiased estimate of variance is found by dividing the *sum of squares* $\sum_{i=1}^{n}(x_i - \bar{x})^2$ by the *degrees of freedom* $(n-1)$; we can think of $n-1$ as the appropriate degrees of freedom that correspond to the sum of squares of a sample of n observations. Thus, with the petrol data, the degrees of freedom corresponding to the total sum of squares are $15 - 1 = 14$. The between-groups sum of squares is based on a sample of three group totals and the corresponding degrees of freedom are $3 - 1 = 2$. The degrees of freedom of the residual sum of squares is a more awkward quantity to calculate directly and is obtained as the difference between

Table 22.1 *Analysis of variance of petrol data in Section 22.2*

Source of variation	Sum of squares	Degrees of freedom	Mean square
Between groups	70	2	35.000
Residual	32	12	2.667
Total	102	14	7.286

the total and the between groups degrees of freedom, i.e. $14 - 2 = 12$. An alternative method of derivation is to say that a residual sum of squares for variability within *each* population would be based on $5 - 1 = 4$ degrees of freedom and that the residual sum of squares in the analysis of variance table equals the sum of these; hence it corresponds to $4 + 4 + 4 = 12$ degrees of freedom.

Finally, we obtain the quantity we want, the estimate of the residual variance σ^2; it is the residual sum of squares divided by the residual degrees of freedom, $32/12 = 8/3 = 2.667$. This quantity is called the residual mean square.

Definition 22.1

A **mean square** is a sum of squares divided by its degrees of freedom.

Although both the examples we have considered in this section had the same number of observations in each group, this is not necessary in a one-way analysis, and the summary 22.2.2 below is written for the general case of k groups, with n_i observations from group i ($i = 1, 2, \ldots, k$).

22.2.2 Calculation of the one-way analysis of variance

We now summarize, for reference, the calculation of a one-way analysis of variance. The meaning of the term 'one-way' will become clearer by comparison with the two-way analysis of variance which we develop in Section 22.5. It is one-way because there is only one source of systematic variation (e.g. the difference between city, country or town areas).

Data: observations $y_{ij}(i = 1, \ldots, k; \ j = 1, \ldots, n_i)$ from k groups, sample size of n_i from group i; $\sum n_i = N$.

We list the steps in the calculation beginning with the raw data (e.g. as at the beginning of Section 22.2).

1. Calculate group totals $T_i = \sum_j y_{ij}(i = 1, \ldots, k)$; grand total $G = \sum_i T_i$.
2. Calculate correction term G^2/N.
3. Calculate total sum of squares S.
4. Calculate between-groups sum of squares S_G.
5. Calculate residual sum of squares $S_R = S - S_G$.
6. Insert degrees of freedom.
7. Calculate mean squares.

Source of variation	Sum of squares	Degrees of freedom	Mean square
Between groups	$S_G = \sum_i T_i^2/n_i - \frac{G^2}{N}$	$k - 1$	$S_G/(k-1)$
Residual	$S_R = S - S_G$	$N - k$	$S_R/(N-k)$
Total	$S = \sum_i \sum_j y_{ij}^2 - \frac{G^2}{N}$	$N - 1$	$S/(N-1)$

Calculation of a one-way analysis of variance using a computer is discussed in *CCC*, Chapter 12.

22.2.3 Exercises

1. The prices of three types of biscuit, in pence per pound, were recorded at randomly chosen shops in a city. Construct an analysis of variance table to split up the total variation into a component between types and a residual variation component.

      Biscuit A: 33.9, 34.1, 33.7, 33.9, 34.0, 34.1, 33.8, 33.6, 34.2, 34.0.
      Biscuit B: 34.2, 34.8, 34.6, 34.8, 35.0, 35.0, 34.7, 34.6.
      Biscuit C: 34.3, 34.1, 34.7, 34.5, 34.6, 34.4, 34.6, 34.8, 34.3.

2. Four groups of children in a primary school are chosen at random from children of similar age and intelligence. Each group is taught the same topic, but by four different methods. A test score y is recorded at the end of the period of teaching; although the groups originally contained 10 children each, some were absent for part of the time and their scores are not to be used in any analysis. The remaining scores are as follows:

Teaching method	Scores
A	4.3, 6.4, 5.1, 3.2, 4.1, 5.5, 7.0, 4.4
B	5.4, 7.5, 6.9, 3.2, 7.2, 3.8, 6.3, 5.7, 8.1, 5.9
C	7.8, 8.0, 6.3, 8.7, 3.9, 8.8, 5.5
D	5.8, 6.4, 8.3, 9.2, 8.9, 7.8, 8.3, 8.4, 8.9

Construct a one-way analysis of variance table for these results.

22.3 CONFIDENCE INTERVALS AND TESTS ON MEANS

The research organization carrying out the investigation described in Section 22.2 is particularly interested in two comparisons: city versus country and city versus town. To obtain confidence intervals and carry out significance tests we often make the further assumption that the data are normally distributed. The numbers of observations in our example are typical of many practical investigations, far fewer than would be needed for any formal tests of goodness-of-fit to a normal distribution. There are some helpful graphical methods of checking whether normality is a reasonable assumption, by studying *residuals* (see Clarke and Kempson (1997), Chapter 4); and we can also consider whether the conditions listed on p. 352 are reasonable ones to impose on data of this type. Often, in scientific work particularly, an experimenter will have studied quite large collections of data from which ideas on appropriate distributions have already emerged.

For this example, the assumption of normality seems not unreasonable, and so we can calculate a 95% confidence interval for the mean difference in petrol prices in city and country by the method of Example 18.8 (p. 452). The estimate of variance is $s^2 = 2.667$ (Table 22.1), and it has 12 degrees of freedom; thus we use $t_{(12)}$ in constructing a confidence interval. The difference observed in the means (country *minus* city) was $350/5 - 325/5 = 5.0$, and the standard error of this observed difference is $\sqrt{2s^2/5}$ since each mean was based on five observations. Limits for the true difference in means are thus

$$5.0 \pm t_{(12;0.05)}\sqrt{2 \times 2.667/5}$$

or

$$5.0 \pm 2.179 \times 1.033$$

i.e.

$$5.0 \pm 2.25 \quad \text{or} \quad 2.75 \text{ to } 7.25.$$

With 95% probability of being correct we can claim that the true difference is contained in the interval 2.75 to 7.25 pence per litre, country being in excess of city by these amounts.

When we consider (town *minus* city), the observed mean difference is 1.0, and by the same method we find 95% limits to the true mean difference as

$$1.0 \pm 2.25 \quad \text{or} \quad -1.25 \text{ to } +3.25.$$

In this case town may be greater than city by up to 3.25, or city may be greater than town by up to 1.25 pence per litre.

Carrying out significance tests of the NH 'city mean = country mean' and the NH 'city mean = town mean', against corresponding AHs with '$\neq$' in them (i.e. tests are two-tail), we find by the method of Section 17.13 that the first gives a result significant at the 1% level while the second gives a result that is not significant. We cannot claim any true difference between mean city and town prices, but country does appear to be different from city. This is consistent with our confidence interval calculations: the interval for town versus city difference contains the value zero while that for country versus city does not.

If the two means being compared are based on different numbers, n_1 and n_2, of observations, the standard error of their difference is

$$s\sqrt{\frac{1}{n_1} + \frac{1}{n_2}};$$

otherwise it is just as straightforward to calculate confidence intervals or carry out significance tests.

22.3.1 Exercises

1. Set 95% confidence limits to the true price per pound, in the whole population of shops, of biscuit A (Exercise 22.2.3(1)); and also to the true difference between the prices of biscuits B and C.

2. Set 95% confidence limits to the true mean score under each of teaching methods A and C (Exercise 22.2.3(2)); and set 99% confidence limits to the true difference between the scores under methods B and D.

22.4 THE F-DISTRIBUTION AND THE F-TEST

One of the most useful functions of an analysis of variance such as that shown in Table 22.1 is to provide an estimate of the variance of an individual observation, so that the operations described in Section 22.3 can be carried out. Occasionally, however, interest centres not on individual means but on testing a null hypothesis that the means of the groups are all equal, and for this purpose we need a new distribution.

22.4.1 Fisher's F-distribution

Our new distribution was originally discovered by R. A. Fisher (British, 1890–1962) – though in a slightly different form: it was the American statistician G. W. Snedecor who developed what he called F in honour of Fisher. It is concerned with the ratio of two estimates of variance in random samples from normal distributions: if both samples are drawn from normal distributions with the *same* variance then the ratio of the two estimates follows the F-distribution.

As with χ^2 and t, F is in fact a whole family of distributions; it is more complicated than χ^2 or t because it depends on two parameters for its shape, namely the degrees of freedom of each of the two estimates of variance. This shape is, in general, skew to the right like $\chi^2_{(3)}$ shown in Fig. 19.1; but the position of any given percentile depends on both degrees of freedom, and this creates a problem when we wish to tabulate F (see Table A7, pp. 720–721).

Let us suppose that the two samples consist of n_1 and n_2 observations respectively, the estimates of variance calculated from these being s_1^2 (with $n_1 - 1$ degrees of freedom) and s_2^2 (with $n_2 - 1$ degrees of freedom). Let us suppose that $s_1^2 \geqslant s_2^2$. If both samples have in fact been taken from normal distributions with the *same* variance then the ratio s_1^2/s_2^2 will follow the F-distribution with $n_1 - 1$ and $n_2 - 1$ degrees of freedom, written $F_{(n_1-1, n_2-1)}$. It is important to put the degrees of freedom in the right order, those for the numerator in the ratio coming first. When using Table A7, the degrees of freedom for the numerator are the 'upper d.f.' denoted by v_1 and are shown along the top of the table. The 'lower d.f.' v_2, for the denominator, are shown in the left-hand column. The two pages of Table A7 give respectively the upper 5% and 1% points of F, that is, those values which would by chance be exceeded with probability 5% or 1% if both samples were in truth drawn from normal distributions with the same variance. Standard statistical tables, such as those of Fisher and Yates, give additional pages of tables including 2.5% and 0.1% levels; these levels may also be found in Clarke and

Kempson (1997). Statistical computer packages such as MINITAB allow the calculation of values for any chosen percentage points.

Example 22.1 Test on variances

Measurements were made of the length (in micrometres) of components produced by a machine. On a particular day the machine was operated by many people; 11 components chosen at random from the production were found to have an estimated variance of 136.68. On the next day the machine was operated by one man only; 9 components selected at random had an estimated variance of 94.72. Assuming that the lengths of the components are normally distributed, test whether more operators led to an increase in variability of the product.

We have $s_1^2 = 138.68$, with $v_1 = n_1 - 1 = 10$ degrees of freedom; and $s_2^2 = 94.72$, with $v_2 = n_2 - 1 = 8$ degrees of freedom. On the NH, 138.68/94.72 is $F_{(10,8)}$; its value is 1.46. From the first section of Table A7, we find that the upper 5% point of $F_{(10,8)}$ is 3.35, so our calculated F-value is well below this and certainly not significant.

The test above is a one-tail test. F-tables are constructed for one-tail tests since that is the form of the test used in the analysis of variance, the most common application of the F-test. But if we wish to investigate whether two normal distributions *differ* in variance, a two-tail test is required. Tables give only upper percentage values, so to carry out the test we always put the larger estimated variance in the numerator when calculating F. Let the degrees of freedom of the variance in the numerator be v and those of the variance in the denominator be v^*. We compare the calculated F with the tabulated $F_{(v,v^*)}$ and reject the null hypothesis of no difference in variance if the calculated F exceeds the tabulated F; the significance level is *twice* that stated in Table A7. Thus our tables allow only 10% or 2% two-tail tests to be made.

22.4.2 *F*-tests in analysis of variance

Let us now return to test the null hypothesis which we mentioned at the beginning of this section, namely that all group means are equal, i.e. $\mu_i = \mu$ for all i. In spite of the fact that this test is a common one, it is really rather limited in usefulness, particularly in scientific work; often the experimenter knows there are differences between the groups as a whole, and is more interested in making particular comparisons between means. A full discussion of the logic of comparing many means is beyond the scope of this book, but will be found in Clarke and Kempson (1997).

In Table 22.1, we included a between-groups mean square as well as a residual mean square. It can be shown (for mathematical justification see Clarke and Kempson (1997)) that on the NH 'all $\mu_i = \mu$' *both* of these mean squares provide estimates of σ^2, the variance of an individual observation, and that these estimates have the degrees of freedom shown on the corresponding rows in Table 22.1. The AH 'not all μ_i are equal' leads to a situation in which the residual mean square still gives an estimate of σ^2 (this is why it could be used without question in Section 22.3); however, the expected value of the between-groups mean square is now greater than σ^2, so that a one-tail test is appropriate.

From Table 22.1, therefore, we can test the NH 'the mean prices are the same in all three areas' against the AH 'there are some differences between mean prices' by comparing the between groups mean square of 35.000 with the residual mean square of 2.667. On the NH the ratio of these two estimates of σ^2 will follow $F_{(2,12)}$; the ratio is 13.13 which, from Table A7, is well above the critical 1% point of $F_{(2,12)}$ (which is 6.93). The NH is rejected in favour of the AH that there are some differences (not specified in detail), for that is consistent with a large F-value.

22.4.3 Exercise

Carry out F-tests to test whether
 (i) the three mean prices of biscuits A, B, C in Exercise 22.2.3(1),
 (ii) the four mean scores in children taught by methods A–D in Exercise 22.2.3(2),
differ in the populations from which these samples were drawn.

22.5 A MATHEMATICAL MODEL FOR COMPARING POPULATION MEANS

We have so far worked from the data, with the aim of calculating quantities to estimate the interesting properties of the data. But we may also follow the same approach as we have already done many times, namely set up a mathematical model of the underlying situation and seek ways of estimating the parameters in this model.

In the example of petrol prices (Section 22.2) we chose random samples from three populations. As a more general example, consider taking random samples of sizes n_i $(i = 1, 2, \ldots, k)$ from k populations whose true means are μ_i respectively. Observations in all the populations are normally distributed with the same variance σ^2. Thus any observation Y_{ij} in the random sample from the ith population is explained as

$$Y_{ij} = \mu_i + U_{ij} \qquad (j = 1, 2, \ldots, n_i),$$

where μ_i is the mean of that population and U_{ij} is a deviation from that mean. Y_{ij} and U_{ij} are random variables but μ_i is a fixed quantity. We have specified the statistical properties of Y_{ij} (above) by saying Y_{ij} follows $\mathcal{N}(\mu_i, \sigma^2)$; an equivalent statement is that U_{ij} follows $\mathcal{N}(0, \sigma^2)$.

Looking now at the $N(= \sum_i n_i)$ observations actually obtained, these will be explained as

$$y_{ij} = \mu_i + u_{ij} \qquad (i = 1, 2, \ldots, k; \ j = 1, 2, \ldots, n_i).$$

Applying the principle of least squares (see p. 603), we estimate the means $\{\mu_1, \mu_2, \ldots, \mu_k\}$ by those functions of the observations which minimize

$$Q = \sum_i \sum_j (y_{ij} - \mu_i)^2.$$

The minimization is best carried out by partial differentiation of Q with respect to each of the $\{\mu_i\}$ in turn, setting the partial derivatives equal to 0 and solving the resulting k equations. (See Appendix VII for some notes on partial differentiation.)

We obtain

$$\frac{\partial Q}{\partial \mu_i} = -2 \sum_j (y_{ij} - \mu_i) = 0 \qquad (i = 1, 2, \ldots, k)$$

which is equivalent to

$$\sum_j (y_{ij} - \mu_i) = 0 \qquad (i = 1, 2, \ldots, k).$$

Denoting the solution of this for μ_i by the symbol m_i, we thus have

$$\sum_j y_{ij} = \sum_j m_i = n_i m_i;$$

hence m_i, which is the *estimate* of the mean of the ith population, is given by

$$m_i = T_i/n_i = \bar{y}_i,$$

T_i and $\bar{y}_i$ being (as in Section 22.2) respectively the total and mean of the observations in the sample from the ith population.

Intuitively this is the obvious estimator; we see that it is unbiased since

$$E\left[\sum_j y_{ij}/n_i\right] = \frac{1}{n_i} E\left[\sum_j y_{ij}\right] = \frac{1}{n_i} \sum_j E\left[y_{ij}\right]$$

$$= \frac{1}{n_i} \sum_j \mu_i = n_i \mu_i / n_i = \mu_i.$$

The estimate of variance is based on the sum of squares of deviations of observations from their means, $\sum_i \sum_j (y_{ij} - \bar{y}_i)^2$. As a step towards finding an unbiased estimator of the population variance σ^2, consider $E[\sum_i \sum_j (y_{ij} - \bar{y}_i)^2]$. Using the result on p. 255,

$$E\left[\sum_j (y_{ij} - \bar{y}_i)^2\right] = (n_i - 1)\sigma^2.$$

Hence

$$E\left[\sum_i \sum_j (y_{ij} - \bar{y}_i)^2\right] = \sum_i E\left[\sum_j (y_{ij} - \bar{y}_i)^2\right]$$

$$= \sum_i (n_i - 1)\sigma^2 = (N - k)\sigma^2.$$

So $\sum_i \sum_j (y_{ij} - \bar{y}_i)^2/(N - k)$ is an unbiased estimator of σ^2; this is the residual mean square that appears in the analysis of variance table.

22.6 TWO-WAY ANALYSIS OF VARIANCE

Like many other methods of applied statistics, the analysis of variance arose first in experimental work in biological and agricultural science. Let us therefore take an example of its use in a modern experimental situation.

Many of the crops grown in Britain are grown in other parts of the world too; but often a different variety of the same crop is favoured by the growers in another country. For example, the very well-known English variety of apple, Cox's Orange Pippin, is not well known in other European countries or other parts of the world. Other varieties are preferred there, and it is a natural question to ask if those other varieties would grow well if we tried them in Britain. Work on fruit trees takes a long time, because trees grow and come to maturity relatively slowly; but recently the same question has been asked concerning many other crops, and experiments on some of these can be done in a single growing season (i.e. a single year).

Experiments to compare different varieties of wheat are carried out by dividing a field into a number of small units ('plots' is the standard name for them), and planting the different varieties in randomly chosen units in the field. Thus if we had six varieties of wheat to compare, and a field big enough to split into 24 plots, we might assign at random each variety to four of the plots. All the time they were growing, all the plots would receive exactly the same fertilizers, and be cultivated in exactly the same way, so that the only known reason for systematic variation between plots would be differences between varieties. This is an example of *design of experiments*, to which we return in Section 22.7. This particular experimental scheme, with varieties assigned to plots at random, is known as a *completely randomized* design; it is based on a model exactly like those used earlier in this chapter, with varieties corresponding to groups, and its analysis is a one-way analysis of variance.

But we would be very fortunate to find a field large enough to do an experiment like this which was not itself varying in fertility in a systematic way, so that even if the whole field were planted with the *same* variety we would still observe systematic trends in the way the crop grew. In order to use such a field for *comparing* varieties, we clearly have to take into account the effect of its systematic fertility trends if our variety comparison is to mean anything.

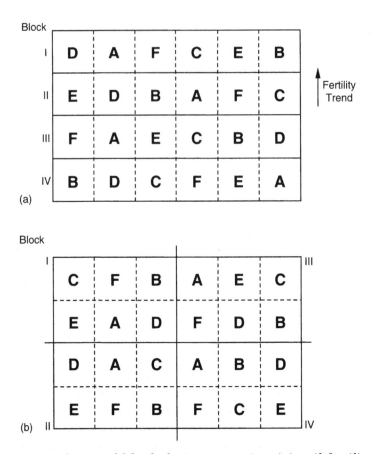

Figure 22.1 *Randomized block design removing (a) soil fertility trend running in the direction shown, (b) gradual fertility changes through a field.*

Two types of fertility trends are commonly found. One is a systematic trend across the experimental area; it may be due to the prevailing wind, to sloping land, to varying depth of soil or trend in soil drainage – good at one end but poor at the other. Another type is where plots near to one another tend to be more alike than those further apart. Our solution in each case is to divide the field into *blocks* (see Fig. 22.1). Each block consists of sufficient plots to contain each variety once; plots within a block are as alike as possible, so that the variation between the blocks effectively removes the systematic fertility variation present in the field. Figure 22.1 shows how we would allocate six varieties A–F: each block would contain each variety once and the order of varieties would be randomized within each block, using a different randomization for each. This type of scheme is called a *randomized block* design. The analysis for it, which we now develop, is called a two-way analysis since it must recognize variation between blocks as well as between varieties.

Agricultural experiments were the first to be designed and analysed according to statistical principles, and so several words which are appropriate to field trials have been given technical meanings that are now applied to experiments in any other area of study. *Blocks* is one of these. A block of units is simply a collection of units of experimental material that are as alike one another as possible, so that the only known or suspected systematic variation is *between* blocks, only random variation remaining within. *Plots*, likewise, are simply the experimental units: the people whose length of sleep is recorded, the petrol filling stations whose prices are noted. Another such word is *treatments*. In agriculture, the aim of an experiment is often to compare different ways of treating the same crop; these may be applying different amounts of fertilizer, or of water, or different methods of ground control of weeds, or times of sowing seeds, or depths of ploughing soil before planting. These are the *treatments* that will lead to systematic differences between the performance of the crop. In market research, different ways of packaging a product will be the 'treatments'. In Section 22.2, we called the three sets of filling stations *groups*, and we have just been considering *varieties* (of the same crops); in future we will use the word *treatments* for all of these types of systematic variation.

As before, we take a measurement on each plot (i.e. each experimental unit), and call it y_{ij}. This time the label i denotes different treatments and the label j indicates which block the plot lies in: in Fig. 22.1, i runs from 1 to 6 corresponding to varieties A–F (which we are now calling 'treatments'), while j runs from 1 to 4 corresponding to the four blocks I–IV. We look at the deviations from the overall mean $\bar{y}$, as in Section 22.1, but this time we write

$$y_{ij} - \bar{y} = (\bar{y}_i - \bar{y}) + (\bar{y}_j - \bar{y}) + (y_{ij} - \bar{y}_i - \bar{y}_j + \bar{y}),$$

in which $\bar{y}_i$ is the mean value of all y_{ij} for a given value of i (a treatment – that is, variety – mean) and $\bar{y}_j$ is that for a given j (a block mean). Recall that the total sum of squares $\sum_i \sum_j (y_{ij} - \bar{y})^2$ is the measure of variation among all members of $\{y_{ij}\}$. Thus we consider

$$
\begin{aligned}
(y_{ij} - \bar{y})^2 = {} & (\bar{y}_i - \bar{y})^2 + (\bar{y}_j - \bar{y})^2 + 2(\bar{y}_i - \bar{y})(\bar{y}_j - \bar{y}) \\
& + (y_{ij} - \bar{y}_i - \bar{y}_j + \bar{y})^2 + 2(y_i - \bar{y})(y_{ij} - \bar{y}_i - \bar{y}_j + \bar{y}) \\
& + 2(\bar{y}_j - \bar{y})(y_{ij} - \bar{y}_i - \bar{y}_j + \bar{y}).
\end{aligned}
$$

Summing this over both suffixes i, j gives the total sum of squares. The last term is zero, being

$$2\sum_j (\bar{y}_j - \bar{y}) \sum_i (y_{ij} - \bar{y}_i - \bar{y}_j + \bar{y}) = 2\sum_j (\bar{y}_j - \bar{y}) \sum_i \{(y_{ij} - \bar{y}_j) - (\bar{y}_i - \bar{y})\},$$

so that, carrying out summation over i first, we see that each of the terms within the curly brackets sums to zero. The last but one term is also zero by a similar argument, and so is the term $2\sum_i \sum_j (\bar{y}_i - \bar{y})(\bar{y}_j - \bar{y})$. This leaves three terms:

(i) $\sum_i (\bar{y}_i - \bar{y})^2$, which we will call S_T because it measures treatment differences; it is exactly the same as S_G in the one-way analysis in Section 22.2.2;
(ii) $\sum_j (\bar{y}_j - \bar{y})^2$, another term of the same type which measures variation between blocks: call it S_B;

(iii) $\sum_i \sum_j (y_{ij} - \bar{y}_i - \bar{y}_j + \bar{y})^2$, the residual sum of squares in this analysis, S_R, measuring the variation on individual plots after between-treatment and between-block variation is removed.

Once again it is the residual sum of squares that we really want, but this is most easily calculated as $S_R = S - S_T - S_B$. The formulae for calculating S and S_T are

$$S = \sum_i \sum_j y_{ij}^2 - \frac{G^2}{N} \qquad \text{(as in Section 22.2.2)}$$

and

$$S_T = \sum_i T_i^2 / n_i - \frac{G^2}{N}$$

(the same as S_G in Section 22.2.2). In a simple two-way analysis, all $\{n_i\}$ will be the same, as in the analysis in Section 22.6.1.

In order to find S_B, we define *block totals* B_j as the sum of all observations in block j, i.e. $\sum_i y_{ij}$ for a given value of j. This experimental scheme requires every variety to appear just once in every block, so we use a slightly different notation for the numbers of observations. We suppose in general that there are v varieties $(i = 1, 2, \ldots, v)$, and each of these appears once in each of the b blocks $(j = 1, 2, \ldots, b)$. The total number of plots and observations, N, therefore equals the product bv.

22.6.1 Calculation of the two-way analysis of variance

We summarize, in a form similar to Section 22.2.2, the steps in a two-way analysis. The 'treatments' (like the 'varieties' in Section 22.5 above) and the 'blocks' simply represent two systematic sources of variation and the same form of analysis can be used whenever data arise in this way.

Data: observations y_{ij} $(i = 1, 2, \ldots, v; \ j = 1, 2, \ldots, b)$ from v treatments, each treatment represented once in each of b blocks; $bv = N$.

1. Calculate   Treatment totals $T_i = \sum_j y_{ij} \ (i = 1, 2, \ldots, v)$;

           Block totals $B_j = \sum_i y_{ij} \ (j = 1, 2, \ldots, b)$;

           Grand total $G = \sum_{i=1}^{v} \sum_{j=1}^{b} y_{ij} = \sum_i T_i = \sum_j B_j$.

2. Calculate correction term G^2/N.

3. Calculate total sum of squares

$$S = \sum_i \sum_j y_{ij}^2 - G^2/N.$$

4. Calculate between-treatments sum of squares

$$S_T = \sum_i T_i^2/b - G^2/N.$$

5. Calculate between-blocks sum of squares

$$S_B = \sum_j B_j^2/v - G^2/N.$$

6. Calculate residual sum of squares

$$S_R = S - S_T - S_B.$$

7. Insert degrees of freedom.

8. Calculate mean squares.

Source of variation	Sum of squares	Degrees of freedom	Mean square
Between treatments	S_T	$v - 1$	$S_T/(v-1)$
Between blocks	S_B	$b - 1$	$S_B/(b-1)$
Residual	S_R	$(v-1)(b-1)$	$S_R/(v-1)(b-1)$
Total	S	$N-1$ $(=bv-1)$	

When considering degrees of freedom, the argument for treatments is as in Section 22.2.1, and by an exactly parallel argument the degrees of freedom for blocks will be $b - 1$. This leaves $(b-1)(v-1)$ for residual since

$$(b-1) + (v-1) + (b-1)(v-1) = b + v - 2 + bv - b - v + 1$$
$$= bv - 1 = N - 1.$$

Mean squares can now be calculated, and the residual mean square will, as before, provide the estimate of the variance of an individual observation, enabling confidence intervals to be calculated for treatment means and their differences just as in Section 22.3. Because every treatment is represented in every block, the comparisons between treatments can be made independently of any block differences since these affect every treatment in the same way. If we did *not* ensure that every treatment was represented once in every block, the analysis would be a very great deal more complicated and not to be attempted by the amateur on his own! Calculation of two-way analysis of variance using a computer is discussed in *CCC*, Chapter 12.

Example 22.2 *Analysis of randomized block design*

The recorded yields of grain (in kilograms), from experimental plots of a standard size, of wheat of six varieties A–F in an experiment laid out in five randomized blocks, were as shown in the following table. (The table also shows the block and variety totals which are a necessary first step in the computation.) Construct an analysis of variance table and test the hypothesis that there is no difference in the mean yields of the six varieties.

Block	Variety						Block totals
	A	B	C	D	E	F	
I	6.35	6.72	7.51	6.48	5.85	6.03	38.94
II	5.90	6.14	6.95	6.06	5.75	5.94	36.74
III	6.24	7.13	7.25	6.59	5.86	6.00	39.07
IV	6.81	7.02	7.60	6.83	6.13	6.22	40.61
V	6.63	6.95	7.64	6.75	6.03	6.15	40.15
	31.93	33.96	36.95	32.71	29.62	30.34	$G = 195.51$

When presented with the basic table of data, the first operation is to find the block totals B_j (the rows in the table above) and the variety or 'treatment' totals T_i (the columns); to save space we have included these in the table. The total of all 30 observations is $G = 195.51$, and we check that $G = \sum_j B_j = \sum_i T_i$, in the notation we have used previously.

$$S_B = \frac{1}{6}(38.94^2 + 36.74^2 + 39.07^2 + 40.61^2 + 40.15^2) - 195.51^2/30$$

$$= 1275.6351 - 1274.1387 = 1.4964.$$

$$S_T = \frac{1}{5}(31.93^2 + 33.96^2 + 36.95^2 + 32.71^2 + 29.62^2 + 30.34^2) - 195.51^2/30$$

$$= 1281.1826 - 1274.1387 = 7.0439.$$

It is necessary to sum the squares of all individual observations: this is 1283.1469. Hence

$$S = 1283.1469 - 1274.1387 = 9.0082.$$

We are now in a position to complete an analysis of variance table:

Source of variation	Sum of squares	Degrees of freedom	Mean square
Blocks	$S_B = 1.4964$	4	0.374
Treatments	$S_T = 7.0439$	5	1.409
Residual	$S_R = 0.4679$	20	0.0234
Total	$S = 9.0082$	29	

The null hypothesis that all variety (treatment) means are effectively equal requires that the treatments mean square be compared with the residual mean square in an F-test; on this NH $1.409/0.0234 = 60.21$ follows $F_{(5,20)}$. From Table A7 the upper 1% point of $F_{(5,20)}$ is 4.10, so our result is far into the top 1% tail of the distribution and we reject the NH.

Though we were not asked to do so, we might look at another null hypothesis, namely that the block means are all the same. For this, we test in an exactly similar way $0.374/0.0234 = 15.98$ as $F_{(4,20)}$, which also leads to rejection of the NH since the upper 1% point of $F_{(4,20)}$ is 4.43. Therefore we did well to lay this experiment out in randomized block form, for the blocks have been success-ful in removing some systematic variation that was present in the land being used.

We may complete our study of this example in various ways. The methods of Section 22.3 are often employed, and so a list of the variety means is needed; the standard error of a single mean, or of the difference between two means, is useful additional information when all the means have been based on the same number of observations.

The variety means are

Variety	A	B	C	D	E	F
Mean	6.39	6.79	7.39	6.54	5.92	6.07

Each of these is based on five observations, and the standard error for any mean is thus $\sigma/\sqrt{5}$; our estimate of σ^2 is the residual mean square and so our estimate of standard error is $\sqrt{0.0234/5} = 0.068$. The standard error of a difference between two means is $\sqrt{2\sigma^2/5}$ and its estimate is $\sqrt{2 \times 0.0234/5} = 0.097$.

Confidence intervals for means or for differences between means can now be calculated, or particular pairs of means compared using t-tests. If significance tests are used, it is necessary to avoid making too many comparisons: if every possible pair of means is compared then these comparisons are certainly not all independent of one another and we can violate quite disastrously the level of significance at which we claim to be working. A good rule-of-thumb is not to make more comparisons than there are degrees of freedom for treatments in the analy-sis of variance, but even these comparisons have to be chosen carefully. Some problems in interpreting experimental results are discussed in Clarke (1994) and Clarke and Kempson (1997).

22.6.2 Exercise

Five cars of the same model have slightly different modifications made in their engines; they are then tested with three different fuels A, B, C, the order in which each fuel is used in each car being determined at random. The distance (in kilometres) that each car will travel on 1 litre of fuel is found, by the same test driver driving a standard course. Do the following results indicate that differences exist (i) between cars; (ii) between fuels?

	Car 1	Car 2	Car 3	Car 4	Car 5
Fuel A	11.3	11.7	11.0	11.6	10.6
Fuel B	9.9	10.3	9.2	10.7	9.6
Fuel C	13.8	14.2	11.7	12.4	12.7

22.7 DESIGN OF EXPERIMENTS

In a scientific experiment, the person in charge of it will have some control over the treatments to be applied and the general running of the experiment while it is in progress. In the experiment using six different varieties of wheat, described in the previous section, it was assumed at the planning stage that there would be a systematic trend in fertility through the area of land available. Therefore an arrangement in randomized blocks was adopted to remove this, so that the comparison between varieties was free of this trend. All the remaining variation among the experimental units was then assumed to be random, natural unit-to-unit variability. If this assumption is to be reasonable, all cultivation operations during the experiment must be carried out in exactly the same way throughout the area, all varieties must receive the same fertilizers, herbicides, irrigation or any other factor that may affect their performance. The aim of the experiment is to compare varieties; all other possible systematic differences must either be designed out (like the trend in soil fertility) or controlled.

An important part of planning is therefore to think of all the likely systematic sources of variation that may be present among the experimental units, and design out all the unwanted ones so that a genuine comparison of treatments can be made. Exercise 22.6.2 presents data which were collected in order to compare the three different fuels; but because there were also likely to be differences between cars the experiment was planned so that these could be taken out in the analysis before the fuels were compared. If different test drivers or courses had been used, these could have caused systematic differences; so this was not allowed.

In the investigation into the price of petrol in Section 22.2, the data arose from a situation where there was perhaps sufficient control to describe the study as an experiment rather than a survey. Three areas were identified, city, town and country, and units (filling stations) were chosen at random from those available. Systematic differences were expected between areas and only random variations within them. A one-way analysis of variance could thus be used, with the areas acting as 'treatments'. But if the size of a filling station had a direct relation to the price of its petrol, a properly designed study would have chosen, say, three size classes, and one station in each class would have been chosen in each area. A two-way analysis would then have been needed, to allow for this extra source of variation.

As an example in marketing, suppose that a supermarket chain is investigating the response of its customers to a new, brighter packaging of a product; it is suggested that the position on the shelves (top, middle, lower) also has an effect on the response. Planning would allow for this extra factor by arranging that in some of its stores the newly packaged product was placed on top shelves, in others on middle shelves, and in others on lower shelves. Sales would be compared with those before the change of packaging; therefore an experiment should begin with the old packaging of the product on shelves of various heights. Sometimes a randomized block arrangement is not enough to remove all known or suspected sources of systematic variation, and more complex designs are needed (Clarke and Kempson, 1997).

It is possible to make the blocks in a randomized block layout remove more than one source of variation. In the wheat experiment of Section 22.6, the individual plots would each need to be cut separately and the yield of grain weighed; other measurements besides yield might also be taken. Large-scale harvesting machinery could not be used, and the harvesting process might take more than one day. This could introduce another possible systematic difference, due to the time of harvest. We can remove that by having the whole of one block dealt with first, then moving on to another block and completing that, and so on; any time difference that exists is thus removed by the blocking, as well as any trend in fertility.

Clearly some knowledge of the background to an experiment is needed to make informed guesses of likely sources of variation, and a statistician always needs to carry out the planning and design stage in close collaboration with the people who have commissioned the experiment and will use its results. More discussion, and a wide range of examples, will be found in Clarke and Kempson (1997).

22.8 MINITAB: ANALYSIS OF VARIANCE; F-DISTRIBUTION

MINITAB allows the response variate y, for a one-way analysis of variance, to be entered in one column or over many columns. The menu choice is **ANOVA>One-Way** or **ANOVA>One-Way (Unstacked)**. Let us assume that the data on petrol prices in Section 22.2 are entered in column C1, named 'Price', with corresponding labels in C2, named 'Location'. MINITAB accepts either numbers or text for column entries so the names 'City', etc., could be used as labels but numbers or letters are usually more convenient. The alternative method of entering unstacked data would be to put the prices for each location in a separate column.

To carry out the analysis of variance, choose

Stat>ANOVA>One-Way/
Enter 'Price' in **Response** and 'Location' in **Factor/**

'Factor' is a term to cover anything that is applied to a plot, or describes a plot; the values will be specified by labels as opposed to the response variable that takes numerical values.

The output begins with an analysis of variance table similar to Table 22.1. The F-statistic, which is the ratio of the groups mean square to the residual mean square (see Section 22.4.2), is quoted together with its p-value. This allows a test of the NH that all locations have the same mean petrol price. Under the table are the group means and number of replicates on which they are based. There is also the value of s, the square root of the residual mean square. The quantities R^2, R^2-adj and individual confidence intervals for means should all be ignored.

A useful graph, that indicates the nature of possible group differences, is obtained by selecting **Graphs. . ./ Boxplots of data** from the dialogue box.

A check on the validity of the analysis is usually made by inspecting the residuals. The mathematical model assumes that the random variable element U, of an observation Y, has a normal distribution with mean 0 and constant variance σ^2, and is independent of the random element of any other observation. We take the residuals as proxies for these unknown random elements and test if they have appropriate properties. Checking the assumptions is best done graphically (see Section 24.8). We recommend the selection of **Histogram of residuals, Normal plot of residuals** and **Residuals versus fits** following choice of the option **Graphs. . .** in the dialogue box. It is possible to store the values of the residuals and fits, which appear by default in the next available columns.

Data for a two-way analysis of variance must be entered in three columns. Using a row for each unit plot, there will be entries for response variate y, block label and treatment label. Thus, for Example 22.2, entries would be made, plot by plot, into columns named 'Yield', 'Block' and 'Variety'. To carry out an analysis choose

> **Stat>ANOVA>Two-Way/**
> Enter 'Yield' in **Response**, 'Block' in **Row factor**, 'Variety' in **Column factor**, select **Display means** for 'Variety' and possibly for 'Block',
> Select **Fit additive model/**

We have chosen the *row* and *column* factors to agree with the data table in Example 22.2 but, for MINITAB, the choice is arbitrary. In fact, one would normally put 'Variety' first as being the more important factor. It is important to select **Fit additive model**. This fits a model that is a simple extension of that in Section 22.5 for a one-way analysis. Without the selection, a more elaborate model is fitted.

The output gives an analysis of variance table similar to that in Section 22.6.1. The general features of the output are the same as for the one-way analysis and the same reservations apply. Box plots for treatments can be difficult to interpret when block effects are present. The residuals should be inspected, as above.

22.8.1 *F-distribution*

Use of MINITAB to evaluate the probability functions of F is discussed in Section 18.9.

22.9 EXERCISES ON CHAPTER 22

1. Nine people volunteer to act as 'subjects' in a psychological trial, and they are assigned at random to three groups L, M, N. All the subjects are new to a special task which they have to carry out; each group does the task under different conditions and they achieve the following scores:

$$
\begin{array}{llll}
\text{L} & 2, & 7, & 3 \\
\text{M} & 7, & 9, & 5 \\
\text{N} & 9, & 12, & 9.
\end{array}
$$

(a) Find the total sum of squares S, and the between-groups sum of squares S_G.

(b) Find the sum of squares of deviations within each of the groups L, M, N; check that the sum of these three is equal to $S - S_G$.

(c) What are the degrees of freedom for (i) S; (ii) S_G; (iii) each sum of squares within a group, L or M or N? Find the degrees of freedom for $S - S_G$ and check it by another method.

(d) Construct an analysis of variance table to show these results.

(e) What is the estimate of the variance of an individual score?

(f) Do the three groups appear to have the same mean score?

2. For each of the following experimental designs state the degrees of freedom corresponding to each component source of variation in the analysis of variance table. Also, in each case, state the critical value of F to be used in testing the null hypothesis at the 5% level that there is no difference between treatments (i.e. between groups).

(a) Completely randomized design, 30 observations, 6 treatments.

(b) Completely randomized design, 42 observations, 2 treatments.

(c) Randomized block design, 20 observations, 4 treatments, 5 blocks.

3. In the analysis of variance F-test, why is it that only the area in the upper tail of the F-distribution is of interest?

4. The analysis of a set of data from a completely randomized design led to the following analysis of variance table:

Source	d.f.	Sum of squares	Mean square
Treatments	2	1565.43	782.72
Error (residual)	18	587.14	32.62
Total	20	2152.57	

The three treatment totals (each the sum of 7 observations) were: A, 143; B, 291; C, 220. Interpret these results by carrying out an F-test and t-tests.

5. To compare the effectiveness of five different types of phosphorescent coating for instrument dials of a certain kind, 30 dials were divided at random

into five groups of six dials and each group given a different coating. The dials were illuminated by ultra-violet light for a fixed period of time and the time each glowed after the source of ultra-violet light was extinguished was recorded in minutes. The sums of the glow-times for coatings A, B, C, D and E were 51, 67, 79, 61 and 72 respectively, and the sum of the squares of all 30 glow-times was 3811. Draw up an analysis of variance table and test for differences between the mean glow-times of the five types of coating. In addition, give a 95% confidence interval for the difference between the mean glow-times for coatings A and B.

State carefully the statistical model underlying your analysis.   (*IOS*)

6. In a completely randomized herbicide trial, each of two treatments was applied to eight plots of seedling plants; but two of the plots receiving treatment B suffered mishaps. The total plant growth in each of the remaining plots in the first year was recorded, as follows (metres).

Treatment A (x): 35, 38, 30, 39, 41, 44, 46, 37.
Treatment B (y): 42, 35, 45, 40, 39, 38.

$$\sum x = 310; \sum x^2 = 12\,192; \sum y = 239; \sum y^2 = 9579.$$

(i) Use a t-test to discuss whether there is a real difference between the two treatments.
(ii) Construct an analysis of variance and use this to compare the treatments.
(iii) State how the t-statistic and the pooled variance in (i) are related to the analysis of variance table in (ii).

7. The following data were obtained in a study concerning the oxygen consumption of the limpet *Acmaea scabra* at three different concentrations of seawater. The variable measured, x, is $\mu l\,O_2$/mg dry body mass/min. Each value is taken from one limpet kept in one of three tanks over a period of time.

100% seawater		75% seawater	50% seawater
12.1	9.7	10.4	14.6
8.9	7.2	7.2	11.1
13.6	14.1	6.4	16.9
9.7	8.3	13.3	18.8
$n = 8$		$n = 4$	$n = 4$
$\sum x = 83.6$		$\sum x = 37.3$	$\sum x = 61.4$
$\sum x^2 = 918.30$		$\sum x^2 = 377.85$	$\sum x^2 = 975.42$

Carry out an analysis of variance on these data to test whether mean oxygen consumption of *A. scabra* is the same at all three concentrations of seawater. Give also a table of means with standard errors, and comment on the results.   (*IOS*)

8. The scores in a psychological test of four groups of subjects A, B, C, D, are given below. The groups were all given the same test, but they came from different major subjects in a university. Examine whether there is evidence that this test produces different scores for different subject groups.

A: 13, 14, 22, 21.
B: 17, 19, 30, 15, 19.
C: 14, 27, 22, 24, 32, 21.
D: 36, 24, 30, 26, 34.

[The sum of the squares of observations, $13^2 + 14^2 + \cdots + 26^2 + 34^2$, is 11 480.]

9. The lipid composition of myelin from the central nervous system of four specimens from each of five different species gave the following levels of cholesterol (in micromoles per milligram of lipid):

		Specimen		
Species	1	2	3	4
Man	0.70	0.73	0.68	0.72
Ox	0.58	0.62	0.65	0.56
Rat	0.68	0.64	0.71	0.67
Rabbit	0.63	0.65	0.67	0.60
Frog	0.64	0.66	0.69	0.67

$$\sum x = 13.15, \sum x^2 = 8.6841$$

Analyse the data to discover whether there are real differences in cholesterol level between the species, and if so examine the means to discover where these differences lie.

Describe the assumptions necessary for the validity of your analysis.

10. Twenty chickens have their water consumption measured while they are taking part in an experiment involving five different diets A–E.

Diet	Water consumption (ml/hour)				Total
A	9.2	8.6	9.3	9.1	36.2
B	8.3	8.6	8.7	8.0	33.6
C	7.6	7.3	8.2	7.9	31.0
D	7.2	6.5	7.6	6.3	27.6
E	6.2	7.2	6.8	6.1	26.3
					154.7

Construct an analysis of variance table for water consumption on the different diets. ($\sum x^2 = 1216.41$.)

A is a standard diet; carry out suitable tests to compare the other diets with it, and comment on the results.

The experimenter is interested in the size of the difference in consumption between diets D and E; construct a 95% confidence interval for this.

11. The protein content of three varieties of winter kale was examined in a field trial arranged in five randomized blocks. The crude protein content (as a percentage of the total dry matter) was measured on each plot.

	Block				
Variety	I	II	III	IV	V
Canson	20.8	18.6	19.5	21.3	19.2
Proteor	18.8	18.0	18.6	20.1	18.4
Thousand head	18.7	18.9	19.6	19.0	18.0
	$(\sum x^2 = 5523.21)$				

Examine the data to decide whether there are real differences between the protein contents for the three varieties, and compare the varietal means.

12. Six varieties of corn are planted in five randomized blocks, and the yields (in kilograms) on each plot are as follows. Examine whether there are differences between blocks and/or varieties.

	Block				
Variety	I	II	III	IV	V
A	35	20	26	26	29
B	34	25	28	27	31
C	33	26	25	28	30
D	34	19	26	25	31
E	30	21	24	24	25
F	28	17	23	26	26

13. In order to test the possible existence of differences between observers in a routine laboratory examination by microscope, five slides I–V were prepared, and four observers A–D each counted the number of a certain type of cell which they saw when studying the slides through the microscope. They obtained the following results:

	Preparation number				
Observer	I	II	III	IV	V
A	48	51	44	75	63
B	62	58	57	96	75
C	57	54	55	83	78
D	39	50	44	70	66

Given that the total sum of squares for the above figures is 4077.75, the sum of squares between observers A–D is 839.75, and that between preparations I–V is 3045.00, set up an analysis of variance to discover whether there are significant differences among the observers. Find the standard error of the difference between two observer means, and a 99% confidence interval for the difference between the means found by observers A and B.

Comment on any other feature of the results.

14. Three experimental methods P, Q, R of making a certain component are being examined and compared with a standard method O. The component should be exactly 2 cm long. Experimental results obtained are:

O: 2.04, 2.01, 1.98, 2.00, 2.03, 2.01, 1.96, 1.99, 2.02, 1.97.
P: 2.05, 2.03, 2.00, 2.04, 2.07.
Q: 2.05, 2.08, 2.08, 2.04, 2.03.
R: 2.01, 1.98, 1.95, 1.99, 2.00.

(a) Find a 95% confidence interval for the mean on method O.
(b) Find the standard error of the difference between the mean on method O and the mean on one of the other methods.
(c) Comment on these results and make any other calculations which you would consider useful in writing a report on this experiment.

15. The gains in weight of mice from ten litters are measured when they live in different environmental conditions: four mice of the same sex are used from each litter to test four types of condition. In a fixed period of time, the following weight gains are made (coded in suitable units).

Environment	Litter											Total
	I	II	III	IV	V	VI	VII	VIII	IX	X		
A	8	9	5	7	6	8	10	8	7	12	:	80
B	11	9	10	8	8	8	12	11	10	13	:	100
C	6	8	5	6	5	6	8	7	9	10	:	70
D	7	8	6	6	8	7	10	8	7	8	:	75
Total	32	34	26	27	27	29	40	34	33	43	:	325

Construct an analysis of variance and use it to make a report on the results of the experiment.

22.10 COMPUTING EXERCISES

1. Carry out one-way analyses of variance on the following data sets:
   (i) petrol prices in Section 22.2;
   (ii) children's test scores in Exercise 22.2.3(2).

2. Carry out two-way analyses of variance on the following data sets:
   (i) wheat yields in Example 22.2;
   (ii) cell counts in Exercise 22.9(13).

3. Generate 24 observations for a one-way classification according to the model

$$Y_{ij} = \mu_i + U_{ij}$$

where $i = 1, \ldots, 4$; $j = 1, \ldots, 6$; $\mu_1 = 10, \mu_2 = 14, \mu_3 = 16, \mu_4 = 20$ and U_{ij} is a random variable from a normal distribution with mean 0 and standard deviation 1.8. Analyse the data by an analysis of variance and interpret the results. Note the estimates of the means for the four treatments, and compare with the true values μ_i.
   Repeat 50 times and look at the distribution of the estimates.

4. Generate 24 observations for a two-way classification according to the model

$$Y_{ij} = \mu + \beta_j + \tau_i + U_{ij}$$

where $i = 1, \ldots, 4$; $j = 1, \ldots, 6$; $\mu = 15$; $(\beta_1, \beta_2, \beta_3, \beta_4, \beta_5, \beta_6) = (-8, -4, 0, 1, 3, 8)$; $(\tau_1, \tau_2, \tau_3, \tau_4) = (-5, -1, 1, 5)$. Also U_{ij} is a random variable from a normal distribution with mean 0 and standard deviation 1.8. Analyse the data by an analysis of variance and interpret the results. Note the estimates of the means for the four treatments, and compare with the true values of these means, which are $\mu + \tau_i$.
   Repeat 50 times and look at the distribution of the estimates of the means.

22.11 PROJECTS

1 Estimation of current time

Aim. To test whether the impression a person has of the current time varies with age and sex.

Equipment. An accurate watch.

Collection of data. Ask a person to say, without looking at a watch or clock, what the current time is. The observation recorded is the absolute value of the difference between the person's stated time and the true time. Divide people into four groups: young and old women, young and old men. Choose at least five people at random from each group (but the numbers in the sample need not be the same for all groups).

Analysis. Carry out a one-way analysis of variance of the records of error in stated time. Interpret the differences in group means if any exist.

2 Card sorting

Aim. To test whether background noise has an effect on the time taken to sort a pack of playing cards.

Equipment. Standard pack of 52 playing cards, record player, stop watch.

Collection of data. The variate measured is the time taken to sort a pack of playing cards into 13 piles, one pile for each value of card (Ace, King, ..., 3, 2). The time should not be recorded until the subject has checked the cards in each pile and moved any card that is in the wrong pile. The sorting should be done quickly but not frantically. A subject should practise a number of times before the experiment begins.

The experimental treatments are three types of background noise: (1) quiet; (2) pop music; (3) confusing speech, which might be a colleague slowly reciting a random sequence of card values (e.g. 3, 7, King, Ace, 2, 9, ...) as the subject sorts the cards. Each subject should sort the cards with each type of background noise once; the order of the three treatments must be chosen at random, using an independent randomization for each subject. At least five subjects should be used.

Analysis. Carry out a two-way analysis of variance. Examine whether there are differences between the treatment means, after taking account of subject-to-subject variation.

Note. Readers who are studying a science subject, e.g. biology or geography, may find that practical work produces data for which the methods of this chapter are needed in analysis (as well as the earlier work on significance testing and confidence intervals).

SUMMARY

A possible **linear model** to explain an observation Y_{ij}, for $i = 1, \ldots, k$ and $j = 1, 2, \ldots, n_i$ is $Y_{ij} = \mu_i + U_{ij}$, where μ_i is the mean of a population from which a random sample of n_i observations has been made and U_{ij} is a normally distributed random term with expected value 0 and variance σ^2 (which is the same for each observation). This leads to a **one-way analysis of variance** in which the **total corrected sum of squares** $S = \sum_{i,j} (y_{ij} - \bar{y})^2$ is split into two components, one of which measures differences *between groups* or populations ('treatments') and the other measures random *residual variation*. The **residual mean square** estimates σ^2, and the ratio of between-groups mean square to residual mean square provides an **F-test** of the null hypothesis that there are no differences between groups.

Source of variation	Sum of squares	Degrees of freedom	Mean square
Between groups	$S_G = \sum T_i^2/n_i - G^2/N$	$k-1$	$S_G/(k-1)$
Residual	$S_R = S - S_G$	$N-k$	$S_R/(N-k)$
Total	$S = \sum_i \sum_j y_{ij}^2 - G^2/N$	$N-1$	

Symbols are defined in Section 22.2. The mean $\bar{y}_i$ for the ith group has variance $\hat{\sigma}^2/n_i$, in which $\hat{\sigma}^2$ is the residual mean square, and is normally distributed because of the assumptions in the linear model.

If there is some systematic variation among the experimental units, this should be removed by dividing them into *blocks*, adding a term to the linear model and taking out the between-block variation by an extra row in the analysis of variance. The detail of a **two-way analysis of variance** is given in Section 22.6. Blocks of units, each block as homogeneous as possible within itself, allow *treatments* to be compared as precisely as possible.

Planning an experiment requires (1) random choice of which units receive which treatment, (2) removal of systematic variations among experimental units and (3) controlling the conduct of the experiment to keep $\hat{\sigma}^2$ as low as possible.

23 SIMPLE LINEAR REGRESSION

OVERVIEW

In this chapter we study relationships between pairs (x_i, y_i) of measurements which are either linear or capable of being made linear in the model parameters through using a transformation. Estimation of the parameters, analysis of data and use for prediction are discussed.

23.1 INTRODUCTION

Often the results of an investigation, by experiment or survey, may be conveniently summarized by drawing a straight-line graph. But when the data are plotted, we often find that the points do not all lie *exactly* on a straight line; and sometimes the data form a cloud of points like the examples we considered in Chapter 21 when studying correlation. How can we then decide which straight line to draw? We consider this problem in this chapter.

23.2 FITTING A STRAIGHT LINE WHEN THERE IS ERROR IN ONLY ONE VARIABLE

A simple experiment, that you may have done, is to hang weights on the end of a spring and record how the length of the spring changes when the weights are changed. We shall describe a little more fully how this experiment may be done. The spring under test is suspended freely from one end, and a scale pan attached at the other end. A measuring scale is secured firmly by the side of the spring, so that the zero of the scale coincides with the end-point of the spring when the scale pan is attached. Weights of differing mass are put on the scale pan and the extension, i.e. the increase in length of the spring, due to each weight, is read from the scale. Each weight creates a load or force on the spring so we measure the load in newtons. We suppose five different loadings are used and that the five observations are made in random order to give the first run of the experiment (see Fig. 23.1). The five observations are then repeated, in random order to give the second run. The repetition is desirable because when we repeat an experiment, even under conditions which appear unchanged, we rarely obtain

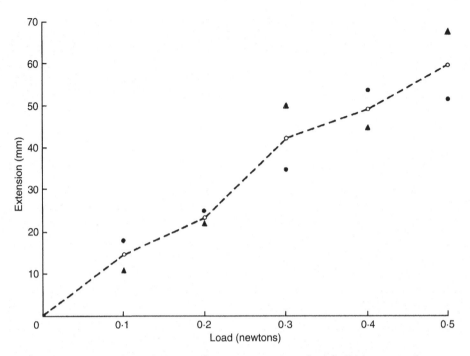

Figure 23.1 *Plot of extension of spring against load. Points are from first run (•), or second run (▲), or means of the two runs (○).*

exactly the same result. We want to be confident that we shall not obtain very different results every time we run the experiment, otherwise there is little point in saying anything general about the results.

The load takes five fixed values, which we decide before beginning the experiment. The extension corresponding to each particular load is subject to measurement error, and also to small changes in the environment, so it may take a range of values. More precisely, this extension has a *distribution*. A reasonable model, for this distribution, is a normal distribution: the conditions set out on p. 352 all seem appropriate. We may therefore regard the two observations for a particular load as being a random sample from a normal distribution.

What is the relation between the extension of the spring and the load applied? We might replace the two observations that we have obtained for each load by their mean, plot these on a graph and then join the means together by a line. We have done this in Fig. 23.1, and have included the origin on the line since the extension will be zero when the load is zero. From the line we can predict the extension that would be obtained for any load that does not exceed the maximum load used. But we might argue that the sudden changes in the direction of the line are not reasonable; usually natural things change smoothly. Would it not be better to draw a *straight line* through the points rather than a zig-zag line? Also, the prediction of the extension corresponding to a load in between any pair of loads used in the experiment is determined by only four experimental points, the two repeat observations at each of the pair of loads. But if the overall relationship really is a straight line, then we ought to determine the line from the whole set of ten points used in the experiment. Predictions of the extension for given loads would then also be based on *all* the points used in the experiment, and we might expect this to make for more accurate predictions. What principle shall we use to determine the line?

We replaced each set of observations at a given load by their mean. Now we have used many times already in this book the sum of squares of deviations about the mean, $\sum_{i=1}^{n}(x_i - \bar{x})^2$. It is quite easy to show that if we take the sum of squares of deviations about any other point it will be greater than this. In other words, if we wish to find a value of c such that $Q = \sum_{i=1}^{n}(x_i - c)^2$ is a minimum, then the appropriate value of c is $\bar{x}$. To show this, we set $dQ/dc = 0$. This gives

$$\frac{dQ}{dc} = \sum_{i=1}^{n}[-2(x_i - c)] = 0,$$

i.e.

$$\sum_{i=1}^{n}(x_i - c) = 0,$$

so that

$$nc = \sum_{i=1}^{n}x_i \quad \text{and} \quad c = \bar{x}.$$

A second differentiation shows that $d^2Q/dc^2 > 0$, and therefore we have obtained a minimum. The reader may see this result without using the calculus if he follows a method similar to that given in Section 23.3, minimizing Q by expanding $\sum_{i=1}^{n}(x_i - c)^2$ in the form $\sum_{i=1}^{n}(x_i - \bar{x} + \bar{x} - c)^2$.

We adopt a similar principle in fitting a line: we draw a line through the points in such a way that the sum of squares of deviations of the points from the line is a minimum (Fig. 23.2). In this we consider the *vertical* deviations only, that is, those corresponding to the extension measurements of the spring in the experiment. The loads x were fixed exactly; they had no variability involved in them.

This method of fitting a line uses the values of all the observations. The resulting line must obviously pass close to the means we drew in Fig. 23.1; we have attempted to draw it, by eye, in Fig. 23.2. We can improve on this attempt if we specify the model algebraically. The basic idea of estimating quantities by minimizing a sum of squares of deviations (or 'errors') was first proposed by Gauss, and is called the *principle of least squares*.

We first assume that the particular values x_i, y_i of two variables X, Y are related by the equation

$$Y_i = \beta x_i + e_i, \tag{23.1}$$

in which β is the slope of the underlying straight-line relationship and e_i is the error or residual of the determination of y_i. This form of straight line, passing through the origin, is best fitted to our experiment, in which X corresponds to load and Y to extension. Note that X is *fixed* equal to x_i, and has no error attached to it: it is a *mathematical variable*, not a random variable, and we can choose the values we give to it. On the other hand, Y is a *random variable*. Besides depending

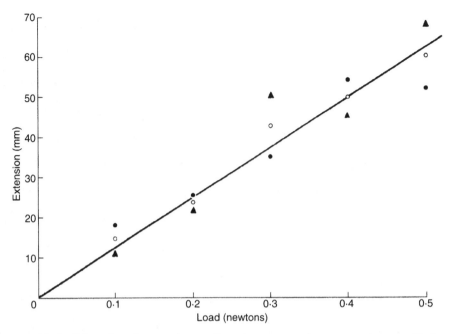

Figure 23.2 *Line fitted to points of spring extension experiment. Points are from first run (●), or second run (▲), or means of the two runs (○).*

on X, there is a random error in determining Y, so that the value y_i corresponding to a given fixed x_i consists of two components, as described in Section 15.7. The first of these is βx_i, accounting for the straight-line relation between y_i and x_i, and we also add e_i, which is the random error or residual in determining y_i. This leads to Equation (23.1) relating the values y_i and x_i observed in the experiment. The residual error terms e_i ($i = 1, 2, \ldots, n$) are assumed to be independent random samples from a distribution with mean 0 and variance σ^2. It therefore follows that each y_i is a random sample from a distribution with mean βx_i and variance σ^2, and that all the y_i are independent of one another.

Figure 23.3 illustrates how a typical point (x_i, y_i) deviates from the fitted line. The deviation or residual is equal to $y_i - \beta x_i$, so that the sum of squares of deviations (residuals) is

$$Q = \sum_{i=1}^{n} (y_i - \beta x_i)^2.$$

We are seeking to minimize Q; we do this by choosing the appropriate value of β, which we will call b. We are allowing the slope of the line to be adjusted to find the best possible fit to the observed set of points (x_i, y_i), so we treat the values of x_i and y_i as constants in Q, with β the only variable element in Q. To find the minimum of Q, we therefore consider

$$\frac{\mathrm{d}Q}{\mathrm{d}\beta} = \sum_{i=1}^{n} [-2x_i(y_i - \beta x_i)].$$

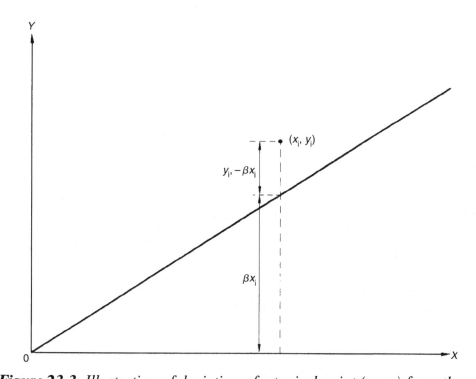

Figure 23.3 *Illustration of deviation of a typical point (x_i, y_i) from the line $y = \beta x$.*

Setting this equal to 0, we have

$$\sum_{i=1}^{n} x_i(y_i - \beta x_i) = 0$$

(after an elementary simplification).

The value of β that satisfies this is b, so we obtain

$$\sum_{i=1}^{n} x_i y_i = b \sum_{i=1}^{n} x_i^2, \quad \text{or} \quad b = \frac{\sum_{i=1}^{n} x_i y_i}{\sum_{i=1}^{n} x_i^2}.$$

(We may differentiate $dQ/d\beta$ again, to check that we do have a minimum of Q when we take $\beta = b$.) The value b, obtained by using the n pairs of observed values x_i and y_i, is the *estimate* of the population parameter β found by the method of least squares.

In the two runs of the spring experiment, the observed results were:

Load, x (newtons)	0.1	0.1	0.2	0.2	0.3	0.3	0.4	0.4	0.5	0.5
Extension, y (mm)	18	11	25	22	35	50	54	45	52	68

We find

$$\sum_{i=1}^{10} x_i y_i = 137.4 \quad \text{and} \quad \sum_{i=1}^{10} x_i^2 = 1.1,$$

so that

$$b = \frac{137.4}{1.1} = 124.91.$$

The estimated line fitting these results is then $y = 124.91x$.

23.3 FITTING A STRAIGHT LINE THAT NEED NOT PASS THROUGH THE ORIGIN

There is not usually any reason why the line that we fit should pass through the origin. More commonly, we want to fit a straight line whose equation is assumed to be

$$y = \alpha + \beta x.$$

The observed y_i will, in this case, be related to the fixed x_i by the equation

$$Y_i = \alpha + \beta x_i + e_i, \tag{23.2}$$

in which the e_i have exactly the same properties as in Section 23.2. The Y_i corresponding to a given x_i will thus be a random variable whose mean is $\alpha + \beta x_i$ and whose variance is σ^2. In order to fit this line, we proceed exactly as in Section 23.2, although this time we have to estimate α as well as β; the formula for the estimate of β becomes a little more complicated as a result.

Let us consider an example. In a chemical process the temperature is fixed at various levels and the yield of the process found at each of the temperatures. The data are summarized in the table, and plotted in Fig. 23.4.

Temperature, x (°C)	50	60	70	80	90
Yield, y	2	2	4	3	5

From past experience of this process, it is known that the response is linear. It is required to estimate the equation of the line.

We shall apply the method of least squares to the model (23.2). From Fig. 23.4, we see that the vertical deviation for a typical point is $(y_i - \alpha - \beta x_i)$. We therefore want to find the values of α and β that minimize

$$Q = \sum (y_i - \alpha - \beta x_i)^2,$$

in which the $\{x_i\}$ and $\{y_i\}$ are the experimental observations, and only α and β can vary.

Since two quantities may vary we cannot use simple differentiation. The minimum can be found using partial differentiation; this is explained in Appendix VII,

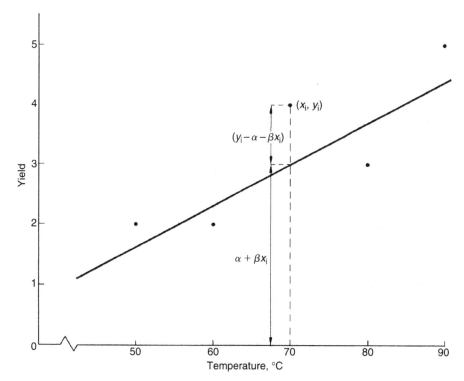

Figure 23.4 *Plot of yield against temperature, showing also the deviation of a typical point (x_i, y_i) from the line $y = \alpha + \beta x$.*

but there is an alternative algebraic method. The algebra is simplified if we use the coding $u_i = x_i - \bar{x}$, $v_i = y_i - \bar{y}$. We shall later make use of the relations $\sum u_i = 0$, $\sum v_i = 0$ that follow from this coding.

We can write

$$y_i - \alpha - \beta x_i \equiv (y_i - \bar{y}) - \beta(x_i - \bar{x}) - (\alpha - \bar{y} + \beta\bar{x})$$
$$\equiv v_i \qquad - \beta u_i \qquad - (\alpha - \bar{y} + \beta\bar{x}).$$

Thus

$$Q = \sum (y_i - \alpha - \beta x_i)^2 = \sum [v_i - \beta u_i - (\alpha - \bar{y} + \beta\bar{x})]^2.$$

The cross-product term in the expansion of this expression is

$$-2\sum (v_i - \beta u_i)(\alpha - \bar{y} + \beta\bar{x}) = -2(\alpha - \bar{y} + \beta\bar{x})\sum (v_i - \beta u_i),$$

since the term $(\alpha - \bar{y} + \beta\bar{x})$ is constant for all i and can be brought in front of $\sum$. The sum is $\sum v_i - \beta \sum u_i$, which is 0 because $\sum u_i = 0 = \sum v_i$. The cross-product is therefore 0, and we find

$$Q = \sum (v_i - \beta u_i)^2 + \sum (\alpha - \bar{y} + \beta\bar{x})^2$$
$$= \sum (v_i - \beta u_i)^2 + n(\alpha - \bar{y} + \beta\bar{x})^2.$$

The two expressions into which we have split Q are positive or zero, since each is proportional to a square or a sum of squares. Q will be a minimum when each of the two expressions attains its minimum value. If we put $\alpha = \bar{y} - \beta\bar{x}$, the second expression $n(\alpha - \bar{y} + \beta\bar{x})^2$ is zero. Being a square, it cannot take a negative value, so this is its minimum value. It is still open to us to choose β as we like, and we choose it to minimize $\sum(v_i - \beta u_i)^2$, the first expression in the partition of Q. But we solved a similar problem on p. 606. It follows from the earlier result that the expression is minimized when we put β equal to the particular value

$$b = \frac{\sum u_i v_i}{\sum u_i^2}.$$

To summarize, Q attains its minimum when we give β the particular value

$$b = \frac{\sum (x_i - \bar{x})(y_i - \bar{y})}{\sum (x_i - \bar{x})^2} \tag{23.3}$$

and we give α the value

$$a = \bar{y} - b\bar{x}. \tag{23.4}$$

The estimated line is therefore given by

$$y = a + bx \tag{23.5}$$

where b and a are derived from the sample values using Equations (23.3) and (23.4). From Equations (23.4) and (23.5) it follows that $(\bar{x}, \bar{y})$ is on the line. The line may therefore be written

$$(y - \bar{y}) = b(x - \bar{x}).$$

This is often the most convenient way to calculate it, and the fact that the line goes through $(\bar{x}, \bar{y})$ is also useful when we come to draw a graph.

For our chemical example we find $\bar{x} = 70$ and $\bar{y} = 3.2$. We tabulate the deviations:

						Total
$x_i - \bar{x}$	-20	-10	0	10	20	0
$y_i - \bar{y}$	-1.2	-1.2	0.8	-0.2	1.8	0
$(x_i - \bar{x})(y_i - \bar{y})$	24	12	0	-2	36	70
$(x_i - \bar{x})^2$	400	100	0	100	400	1000

$$b = \frac{\sum (x_i - \bar{x})(y_i - \bar{x})}{\sum(x_i - \bar{x})^2} = \frac{70}{1000} = 0.07.$$

The equation of the line is therefore

$$(y - 3.2) = 0.07(x - 70)$$

or

$$y = -1.7 + 0.07x.$$

A computer program for fitting a line to data is given in *CCC*, Section 11.2.

23.3.1 Exercises

1. The length L of a metal rod is measured at various temperatures T. It is known that $L = \alpha + \beta T$ but there are measurement errors when determining L. Estimate the equation of the line from the data:

Temperature, t (°C)	20	30	40	50	60
Length, l (mm)	102	107	109	109	113

Mark the data points on a graph and draw in the calculated line.

2. The deviation, in the y direction, of a point (x_i, y_i) from the estimated line is given by $(y_i - a - bx_i)$. Show that the sum of these deviations, over all points in the sample, is zero.

3. Use the facts that (a) the estimated line passes through the mean $(\bar{x}, \bar{y})$ of the data, (b) the sum of deviations about the estimated line is zero, in order to draw an appropriate line fitted by eye to the data of the chemical example given on p. 606. After having done this, draw in the calculated line on the same graph.

4. Fit a line to the spring experiment data on p. 605, but do *not* assume that the line passes through the origin.

5. A small factory employs 12 staff to produce speciality garden furniture. The employees are paid by results. The factory owner believes there is a link between length of experience and the number of (standard) units of furniture produced. Data for a typical month are:

Employee	A	B	C	D	E	F	G	H	I	J	K	L
Experience (months)	36	26	23	16	29	20	21	31	30	12	41	46
Production (units)	116	102	92	86	118	99	83	110	94	71	145	137

Draw a scatter diagram, find the regression of production on experience, and show this on the diagram.

23.4 THE DISTRIBUTIONS OF a AND b

In the example above we made an estimate of the slope β; it was $b = 0.07$. How reliable is this value b? It would be useful to know the standard error of b, and to obtain a confidence interval for β. Another useful item of information would be a confidence interval for a predicted value of y, and for this we require the standard error of a also.

Our first step is to give an adequate specification of the random terms e_i in our model. We require that

(i) each e_i is independent of every other e_j;

and that each e_i is from a distribution that

(ii) has mean zero, i.e. $E[e_i] = 0$;
(iii) has variance σ^2, i.e. $E[e_i^2] = \sigma^2$; and
(iv) is normal.

We can summarize these briefly as *the e_i are mutually independent and all* $\mathcal{N}(0, \sigma^2)$. It follows that the Y_i are mutually independent and each y_i is from an $\mathcal{N}(\alpha + \beta x_i, \sigma^2)$ distribution.

We first find the mean and variance of the random variable b. The formula for b is

$$\frac{\sum_{i=1}^{n} (x_i - \bar{x})(Y_i - \overline{Y})}{\sum_{i=1}^{n} (x_i - \bar{x})^2}.$$

Remembering that each x_i is a fixed number, we see that $\bar{x}$ is a fixed constant, and also that the sum $\sum(x_i - \bar{x})^2$ is a constant number. If therefore we write

$$w_i = \frac{x_i - \bar{x}}{\sum (x_i - \bar{x})^2},$$

the only part of w_i that changes with i is the x_i in the numerator, and this is *not* a *random* variable. We may thus consider $\{w_i\}$ as a set of *weights* determined as soon as the $\{x_i\}$ are known; the w_i are fixed (not random) numbers. Using these weights w_i, we write b as a weighted sum of the random variables $(Y_i - \overline{Y})$:

$$b = \sum w_i(Y_i - \overline{Y}).$$

Now

$$\sum w_i = \frac{1}{\sum (x_i - \bar{x})^2} \cdot \sum (x_i - \bar{x}) = 0,$$

since the first factor on the right is a constant and the second is zero by the definition of $\bar{x}$. Therefore

$$b = \sum w_i(Y_i - \overline{Y}) = \sum w_i Y_i - \sum w_i \overline{Y}$$
$$= \sum w_i Y_i - \overline{Y} \cdot \sum w_i$$
$$= \sum w_i Y_i \qquad \left(\text{since } \sum w_i = 0 \right).$$

Thus b is a weighted sum of the random variables $\{Y_i\}$.

Now since $Y_i = \alpha + \beta x_i + e_i$, and $E[e_i] = 0$, we have $E[Y_i] = \alpha + \beta x_i$. Therefore

$$E[b] = E\left[\sum w_i Y_i \right] = \sum w_i E[Y_i],$$

since each w_i is a constant, and so

$$
\begin{aligned}
E[b] &= \sum w_i(\alpha + \beta x_i) \\
&= \sum w_i \alpha + \sum w_i \beta x_i = \alpha \sum w_i + \beta \sum w_i x_i \\
&= \beta \sum w_i x_i,
\end{aligned}
$$

since $\sum w_i = 0$. Now

$$
\sum w_i x_i = \frac{\sum (x_i - \bar{x}) x_i}{\sum (x_i - \bar{x})^2}.
$$

Because $\bar{x}$ is a constant, as is $\sum (x_i - \bar{x})^2$, we have

$$
\frac{\sum (x_i - \bar{x}) \bar{x}}{\sum (x_i - \bar{x})^2} = \frac{\bar{x}}{\sum (x_i - \bar{x})^2} \cdot \sum (x_i - \bar{x}) = 0,
$$

since the last factor is 0. Thus we can write

$$
\sum w_i x_i = \frac{\sum (x_i - \bar{x}) x_i}{\sum (x_i - \bar{x})^2} - \frac{\sum (x_i - \bar{x}) \bar{x}}{\sum (x_i - \bar{x})^2} = \frac{\sum (x_i - \bar{x})^2}{\sum (x_i - \bar{x})^2} = 1,
$$

leading finally to $E[b] = \beta \sum w_i x_i = \beta$. The mean of b is β, showing that b is an unbiased estimator of β.

We also have

$$
\text{Var}[b] = \text{Var}\left[\sum w_i Y_i\right] = \sum w_i^2 \, \text{Var}[Y_i],
$$

since the w_i are constants and the Y_i are independent.

Thus

$$
\text{Var}[b] = \sum w_i^2 \sigma^2 = \sigma^2 \sum w_i^2.
$$

But

$$
\sum w_i^2 = \sum \left(\frac{x_i - \bar{x}}{\sum (x_i - \bar{x})^2} \right)^2 = \frac{\sum (x_i - \bar{x})^2}{[\sum (x_i - \bar{x})^2]^2} = \frac{1}{\sum (x_i - \bar{x})^2}
$$

and

$$
\text{Var}[b] = \frac{\sigma^2}{\sum (x_i - \bar{x})^2}.
$$

Therefore we need to know or estimate σ^2 in order to find the variance of b. A reasonable quantity on which to base an estimate of σ^2 is the sum of squares of differences between observed and predicted values, i.e. $\sum_i [Y_i - \bar{Y} - b(x_i - \bar{x})]^2$ which may also be written

$$
\sum_i (Y_i - \bar{Y})^2 - b \sum_i (x_i - \bar{x})(Y_i - \bar{Y}).
$$

This is most directly found by the analysis of variance method described in Section 23.9 below; it is the *residual sum of squares*, and the individual differences between observed and predicted values are the *residuals*. The expected value of this residual sum of squares is $(n-2)\sigma^2$, and hence this sum of squares divided by $(n-2)$ is an unbiased estimator of σ^2.

Since b is a linear function of the Y_i which are normally distributed, it follows that b also has a normal distribution. We summarize these results by stating that b has the distribution $N(\beta, \sigma^2/\sum(x_i - \bar{x})^2)$.

Turning to a, which is equal to $\bar{y} - b\bar{x}$, we note that $\bar{y}$ and b are both linear functions of normally distributed variables and $\bar{x}$ is a constant; therefore a also is a linear function of normally distributed variables and so is itself normally distributed. As we remarked at the beginning of this section, Y_i is from a normal distribution whose mean is $\alpha + \beta x_i$; thus $E[Y_i] = \alpha + \beta x_i$ and $E[\bar{Y}] = \alpha + \beta\bar{x}$. Finally

$$E[a] = E[\bar{Y} - b\bar{x}] = E[\bar{Y}] - \bar{x}E[b] = \alpha + \beta\bar{x} - \beta\bar{x} = \alpha,$$

and a (like b) is an unbiased estimator.

The results of this section may be used to find a confidence interval for β, as we show in the following example. This assumes that σ^2 is known; if it has been estimated, we must use the t-distribution instead of the normal in our calculation, and this is dealt with in Example 23.4.

Example 23.1 Confidence interval for slope

A biologist researching into the effects of various drug treatments on animal weight has found that when an animal of initial weight x is given a drug treatment its weight increases by an amount y given by

$$y = \alpha + \beta x + e,$$

where α, β are constants dependent upon the particular drug treatment administered, and e is a random error which is normally distributed with mean zero and standard deviation 0.1. The table shows the values of x and y for six animals given the same drug treatment.

x	1.8	1.8	2.0	2.0	2.2	2.2
y	0.5	0.7	0.8	0.9	1.2	1.3

(i) Calculate the least-squares estimates of α and β.
(ii) Obtain an estimate of the expected increase in the weight of an animal of initial weight 2.1.
(iii) Calculate 90% confidence limits for β. *(Welsh)*

(i) We find $\bar{x} = 2.0$ and $\bar{y} = 0.9$. We express the data as deviations:

							Total
$x_i - \bar{x}$	-0.2	-0.2	0	0	0.2	0.2	0
$y_i - \bar{y}$	-0.4	-0.2	-0.1	0	0.3	0.4	0
$(x_i - \bar{x})(y_i - \bar{y})$	0.08	0.04	0	0	0.06	0.08	0.26
$(x_i - \bar{x})^2$	0.04	0.04	0	0	0.04	0.04	0.16

$$b = \frac{\sum(x_i - \bar{x})(y_i - \bar{y})}{\sum(x_i - \bar{x})^2} = \frac{0.26}{0.16} = 1.625.$$

The equation of the line is $(y - \bar{y}) = b(x - \bar{x})$. Putting in the estimates,

$$(y - 0.9) = 1.625(x - 2.0)$$

or

$$y = -2.35 + 1.625x.$$

The estimates of α and β are therefore -2.35 and 1.625 respectively.

    (ii) The expected increase in the weight of an animal whose initial weight is 2.1 is

$$y = -2.3500 + 1.625(2.1) = -2.3500 + 3.4125 = 1.0625.$$

(iii) $\text{Var}[b] = \dfrac{\sigma^2}{\sum(x_i - \bar{x})^2} = \dfrac{\text{Var}[e]}{\sum(x_i - \bar{x})^2} = \dfrac{(0.1)^2}{0.16} = \dfrac{0.01}{0.16} = \dfrac{1}{16}.$

Since b is normally distributed with mean β and variance $\frac{1}{16}$ the required confidence interval has for its limits

$$b \pm 1.645\sqrt{\tfrac{1}{16}} = 1.625 \pm 1.645(0.25) = 1.625 \pm 0.411.$$

The interval is therefore $(1.214, 2.036)$.

23.4.1 Exercises

1. In the chemical example of Section 23.3, the random error e has standard deviation 0.5. Find a 95% confidence interval for β.

2. A research chemist knows that, for temperatures ranging from 50 to 60°C, a variable y is linearly related to temperature. In order to determine this relationship, he conducts two experiments with the temperature controlled at 50 and 60°C, and obtains y-values of 76 and 63, respectively. Realizing that this method for determining a y-value is subject to error variation, he decided to repeat the two experiments and obtained a y-value of 75 when the temperature was 50°C and a y-value of 66 when the temperature was 60°C. Use the results to determine the equation of the least-squares estimate of the linear relationship connecting y and temperature.

    Given that the method for determining a y-value is subject to an error which is normally distributed with mean zero and variance $\sigma^2 = 2$, derive 95% confidence limits for the true change in the value of y resulting from an increase in temperature of 1°C.   (*Welsh*)

23.5 REGRESSION

The examples of fitting a line that we have discussed are those in which one variable, usually labelled X, takes fixed values. Either it is controlled, or any variation in it is so small that it can be neglected; all the variability in the experiment is a result of random variation in determining the values of Y. In those situations we can estimate the relation between the two variables. This may be of intrinsic interest or we may want it in order to predict Y from X. These examples usually occur only in experiments in the physical sciences.

There are many situations where we are interested in the relation between two variables but we cannot control either of them. Examples are the relation between the height and weight of people, and the relation between the income of a family and the amount it spends on housing. We shall show how it is possible to obtain a line that allows us to predict one variable from another. We begin with a rather artificial example to make the ideas clear.

23.6 PREDICTION OF THE TOTAL SCORE
ON TWO DICE

Suppose we throw two fair dice, one red and one white, but we look only at the red one. How would we predict the total score from the score on the red die? We begin by drawing a diagram (Fig. 23.5) to illustrate the 36 equally likely outcomes from the throw. We represent each outcome by the score on the red die X, and the total score Y. Suppose the score on the red die is 3, what is our prediction for

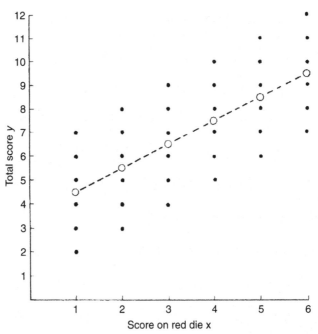

Figure 23.5 *The points (•) show the possible values of the total score y on two dice as the score on the red die x varies. The regression line is obtained by joining the means (○) of each column of points.*

the total score? We notice first that there is not a unique total score when the red die shows 3. There is a distribution of scores: the total score is equally likely to be 4, 5, 6, 7, 8 or 9. If we are required to produce a single value, the best we can do is to choose some single value to represent the distribution and the obvious choice is the mean. Thus if the red die shows 3 we would predict the total score to be 6.5. In fact, this is a total that cannot actually occur, but it is a value that will be close to many of the total scores that are obtained when the red die shows 3. We can repeat this procedure for every value that the red die may show, marking in the mean of the corresponding distribution of the total score (Fig. 23.5). (These distributions of Y for given X-values are said to be the *conditional distributions* of Y; compare this with 'conditional probability', Section 7.7.) If we join up the means we obtain what is called the *regression* of 'total score' on 'score on red die'. It is represented by the equation

$$y = 3.5 + x.$$

It will be seen that if we know the joint distribution of two random variables, we can find the regression of one random variable on the other. We find the conditional distribution of Y for each value of X, find the means of these conditional distributions, and then connect these means, to give the regression of Y on X. From our method of derivation it will be seen that the regression need not be a straight line; it could be a curve, or even consist of disjointed pieces of lines or curves. In fact, we shall restrict ourselves in this chapter to regression lines that are straight lines.

23.6.1 Exercise

The pairs of possible values of the continuous random variables X, Y are uniformly distributed inside the triangle whose vertices are the origin and the points with coordinates $(2, 0)$ and $(2, 2)$. Draw a sketch, mark in the regression of Y on X and give the equation of the regression line. (Note that no calculation is necessary. The mean of Y for each X-value is the centre of the line over which the conditional distribution of Y is spread.)

23.7 PREDICTION OF SON'S HEIGHT FROM FATHER'S HEIGHT

The idea of regression was proposed by Galton (British, 1822–1911). Galton began life as an explorer but when his travels were over he settled to work as a scientist. He was obsessed by the idea of describing phenomena by numbers and is claimed to be the father of modern statistics. One of his major interests was the study of inheritance. It is in these studies that the idea of *regression* arose.

Let us look at one of the problems that interested Galton. How are the heights of a father and his son related? To make the problem more precise we ask the question 'How is a son's height predicted from his father's height?' We must

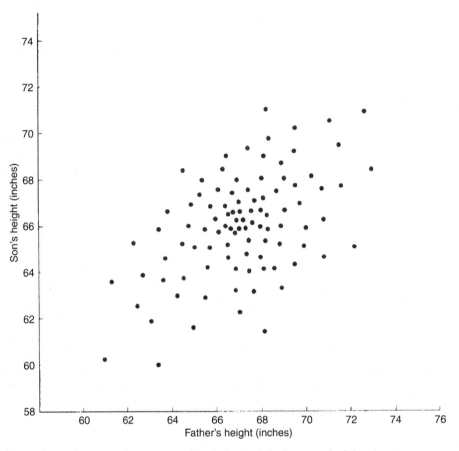

Figure 23.6 *Scatter diagram of heights of father and eldest son.*

begin by collecting data. From a population of families in which there is at least one mature son, we would choose families at random and, in each chosen family, measure the height of the father and of the eldest son. A scatter diagram of the values obtained would give us a picture such as Fig. 23.6. There is obviously a correlation between the two heights; taller sons tend to go with taller fathers.

We want to set up a statistical model of this relation. First we need to specify a joint probability distribution of the two continuous random variables X and Y, in which X corresponds to father's height and Y to son's height. We shall assume that X and Y are jointly normally distributed. This means that when we fix a value x of X then the distribution of Y conditional on this x is normal. In Fig. 23.7 we show the distribution of Y at each of a number of typical X values, $x_1, x_2, \ldots$. The means of the distributions of Y conditional on each of $x_1, x_2, \ldots$ are also shown, and the regression is obtained by finding all such Y-means. The regression relation turns out to be a straight line, as we have indicated in Fig. 23.7.

If we know the regression line, then for fathers of a given height we can predict the *mean* height of their sons. This is the best we can do. The relation between fathers' and sons' heights is a complex one and even if there were no measurement error it could only be fully described by a joint statistical distribution.

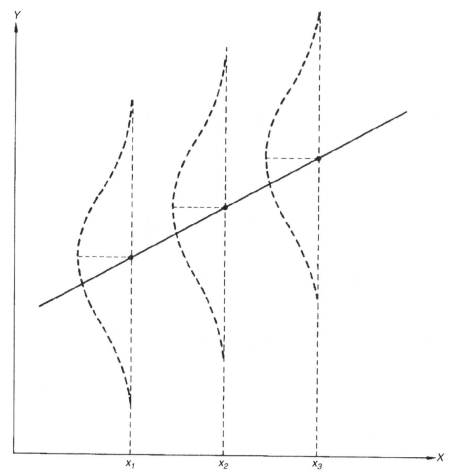

Figure 23.7 *Distribution of Y for chosen values of X when X, Y have a bivariate normal distribution*

But by using the idea of regression we can predict the mean height of sons of fathers of a given height, and this predicted mean height will be close to a high proportion of the sons' heights. These types of predictions are very useful in practice.

If we have a sample of pairs of values (x_i, y_i) the procedure for estimating the regression line is identical with the line-fitting procedure that we used earlier in the chapter. For a given value x_i, there exists a conditional distribution of Y and, as we have remarked, the regression line passes through the mean of this distribution. But we have observed only one single value, y_i, from the conditional distribution corresponding to x_i. We therefore use the principle of least squares, taking account of variation only in Y, and choose the line so that the sum of squares of deviations of the y_i from the line is a minimum. Use of the idea of regression has made our new problem identical with the problem we considered earlier in the chapter. The formulae for the parameters of the regression line are the same as in Section 23.3, and b is called the *regression coefficient*.

Example 23.2 Fitting a regression equation

Find the regression line of son's height on father's height from the data:

Father's height, x (in)	59	61	63	65	67	69	71	73	75
Son's height, y (in)	64	66	67	67	68	69	70	72	72

Predict the mean height of sons whose fathers are 70 inches in height.

The computations are simplified by the codings $u = \frac{1}{2}(x - 67)$, $v = y - 68$.

										Total
u	−4	−3	−2	−1	0	1	2	3	4	0
v	−4	−2	−1	−1	0	1	2	4	4	3
uv	16	6	2	1	0	1	4	12	16	58
u^2	16	9	4	1	0	1	4	9	16	60

$$\bar{u} = 0, \qquad\qquad \bar{x} = 67,$$
$$\bar{v} = \tfrac{3}{9} = 0.33, \qquad\qquad \bar{y} = 68.33.$$

$$\sum (x_i - \bar{x})(y_i - \bar{y}) = 2 \sum (u_i - \bar{u})(v_i - \bar{v})$$

$$= 2 \left(\sum u_i v_i - \frac{(\sum u_i)(\sum v_i)}{9} \right)$$

$$= 2(58 - 0) = 116.$$

$$\sum (x_i - \bar{x})^2 = 4 \sum (u_i - \bar{u})^2 = 4 \left(\sum u_i^2 - \frac{(\sum u_i)^2}{9} \right)$$

$$= 4(60 - 0) = 240.$$

$$b = \frac{\sum (x_i - \bar{x})(y_i - \bar{y})}{\sum (x_i - \bar{x})^2} = \frac{116}{240} = 0.4833.$$

The estimated regression line is given by $(y - \bar{y}) = b(x - \bar{x})$, which is

$$(y - 68.33) = 0.4833(x - 67),$$

or

$$y = 0.4833x + 35.95.$$

When x is 70,

$$y = 0.4833 \times 70 + 35.95 = 33.83 + 35.95 = 69.78.$$

A similar calculation to that in Example 23.2, but based on data from over 1000 families, was carried out by Karl Pearson, a friend of Galton. The regression line Pearson found was $y = 0.516x + 33.73$. Galton noticed that tall fathers tended to have tall sons, but that on the average the sons were not as tall as their fathers. At the other end of the scale, the short fathers had sons who were short, but not so short as their fathers. (There is a trend, in western industrial societies, for average height to increase slightly from one generation to the next, but the effect which Galton noticed is in addition to that.) This led to the regression line having a slope appreciably less than one. Galton said that there was a *regression*, or movement back, towards the mean in height, and in many other measurable human characteristics. He called the line that demonstrated this the regression line and the word regression has been adopted as a technical term in statistics to describe a line obtained by using the procedure that Galton and Pearson used.

23.7.1 Exercises

1. Find the regression of Y on X from the data:

x	5	6	7	8	9
y	2	4	6	7	11

Predict the mean value of Y when X is 10.

2. The following table gives data relating to tests on ten specimens of brass:

Specimen number	1	2	3	4	5	6	7	8	9	10
Hardness, H (in Rockwell units)	57	49	59	45	54	51	49	57	46	55
Tensile strength, T (in 1000 lbf in^{-2})	76	69	83	64	74	73	66	79	65	80

Given that

$$\sum(H-45) = 72, \quad \sum(H-45)^2 = 734,$$
$$\sum(T-64) = 89, \quad \sum(T-64)^2 = 1197,$$

and

$$\sum(H-45)(T-64) = 923,$$

calculate the equation of the line of regression of H on T.

A further specimen of brass has a tensile strength of $75\,000\,\text{lbf in}^{-2}$. What do you predict about its hardness in Rockwell units? Demonstrate that the straight-line method is in fact a reasonable approach to prediction in this case. *(MEI)*

23.8 TWO POSSIBLE REGRESSION LINES

In regression we do not treat both the variables involved in the same way. One variable, which we have called Y, is the predicted variable, and the other variable, X, is the one we use to make the prediction. If we change X and Y in our definition of regression we obtain the regression of X on Y, which will usually be different from the regression of Y on X. (They are the same if and only if X and Y are perfectly correlated.) In some cases prediction in one way may make sense, but it may be nonsense to predict in the reverse way.

Let us alter our notation slightly: write β_{YX} for the regression coefficient of Y on X, and b_{YX} for its estimate. If we also write the regression coefficient of X on Y as β_{XY} then it is estimated by

$$b_{XY} = \frac{\sum (x_i - \bar{x})(y_i - \bar{y})}{\sum (y_i - \bar{y})^2},$$

the positions of x and y simply being interchanged in the calculation. It follows that

$$b_{XY} \cdot b_{YX} = \frac{\sum (x_i - \bar{x})(y_i - \bar{y})}{\sum (y_i - \bar{y})^2} \frac{\sum (x_i - \bar{x})(y_i - \bar{y})}{\sum (x_i - \bar{x})^2} = r^2,$$

where r is the sample correlation coefficient between X and Y.

Example 23.3 Two regression lines

In Utopia the wholesale food price index x and the retail food price index y over twelve years had the following values in order:

x	100	98	96	97	95	92	89	87	88	90	86	88
y	100	97	95	98	94	95	90	89	91	93	89	87

Find the equations of the regression lines: (i) of y on x; (ii) of x on y. Calculate the product-moment coefficient of correlation.

When the wholesale food price index is 84, what is the estimated value of the retail price index?
(*AEB*)

Code both variates by decreasing their values by 92: $u = x - 92, v = y - 92$.

													Total
u	8	6	4	5	3	0	−3	−5	−4	−2	−6	−4	$26 - 24 = 2$
v	8	5	3	6	2	3	−2	−3	−1	1	−3	−5	$28 - 14 = 14$
uv	64	30	12	30	6	0	6	15	4	−2	18	20	203
u^2	64	36	16	25	9	0	9	25	16	4	36	16	256
v^2	64	25	9	36	4	9	4	9	1	1	9	25	196

$$\bar{u} = \tfrac{2}{12} = 0.1667, \quad \bar{x} = 92.1667,$$
$$\bar{v} = \tfrac{14}{12} = 1.1667, \quad \bar{y} = 93.1667.$$

$$\sum (x_i - \bar{x})(y_i - \bar{y}) = \sum (u_i - \bar{u})(v_i - \bar{v})$$

$$= \sum u_i v_i - \frac{\left(\sum u_i\right)\left(\sum v_i\right)}{12} = 203 - \frac{2 \times 14}{12}$$

$$= 203 - \frac{7}{3} = \frac{602}{3}.$$

$$\sum (x_i - \bar{x})^2 = \sum (u_i - \bar{u})^2 = \sum u_i^2 - \frac{\left(\sum u_i\right)^2}{12} = 256 - \frac{4}{12} = \frac{767}{3}.$$

$$\sum (y_i - \bar{y})^2 = \sum (v_i - \bar{v})^2 = \sum v_i^2 - \frac{\left(\sum v_i\right)^2}{12} = 196 - \frac{196}{12} = \frac{539}{3}.$$

(i) The estimated slope of the regression of Y on X is

$$b_{YX} = \frac{\sum (x_i - \bar{x})(y_i - \bar{y})}{\sum (x_i - \bar{x})^2} = \frac{602}{767} = 0.7849.$$

The regression equation is

$$(y - 93.1667) = 0.7849(x - 92.1667)$$
$$y = 20.825 + 0.7849x.$$

(ii)

$$b_{XY} = \frac{\sum (x_i - \bar{x})(y_i - \bar{y})}{\sum (y_i - \bar{y})^2} = \frac{602}{539} = 1.1169.$$

The regression equation of X on Y is

$$(x - 92.1667) = 1.1169(y - 93.1667)$$
$$x = -11.891 + 1.1169y.$$

The correlation coefficient

$$r = \frac{\sum (x_i - \bar{x})(y_i - \bar{y})}{\sqrt{\sum (x_i - \bar{x})^2 \sum (y_i - \bar{y})^2}} = \frac{602}{\sqrt{767 \times 539}} = 0.936.$$

To predict (mean) retail price index y from a given wholesale price index x we use the regression of Y on X. The required value is therefore

$$y = 20.825 + 0.7849(84) = 20.825 + 65.93 = 86.76.$$

A value such as this which is a prediction of y for an x-value outside the range of x-values on which the regression line is based (i.e. an *extrapolation*) should be used with caution. The relation between X and Y might change drastically outside the range that has been investigated.

23.8.1 Exercise

Calculate the regression line of X on Y for:

  (i) the example of Section 23.6, where $X =$ score on red die and $Y =$ total score;

 (ii) the situation described in Exercise 23.6.1.

   Using diagrams, compare each of these with the corresponding regression of Y on X.

23.9 AN ANALYSIS OF VARIANCE FOR REGRESSION

In Section 23.4, a method was suggested for estimating the residual error variance σ^2 when fitting a regression; this used the sum of squares of the differences between each observed value y_i and the corresponding predicted value $\bar{y} + b(x_i - \bar{x})$. Let us call this sum of squares the *residual sum of squares*, and denote it by S_R. Then

$$S_R = \sum [y_i - \bar{y} - b(x_i - \bar{x})]^2 = \sum (y_i - \bar{y})^2 - 2b \sum (y_i - \bar{y})(x_i - \bar{x}) + b^2 \sum (x_i - \bar{x})^2.$$

Since $b = \sum (x_i - \bar{x})(y_i - \bar{y}) / \sum (x_i - \bar{x})^2$, the final term equals

$$b\left[b \sum (x_i - \bar{x})^2 \right] = b\left[\sum (x_i - \bar{x})(y_i - \bar{y}) \right].$$

Therefore $S_R = \sum (y_i - \bar{y})^2 - b \sum (x_i - \bar{x})(y_i - \bar{y})$.

   We may set out the calculation in the form of an analysis of variance.

Source of variation	Sum of squares	Degrees of freedom	Mean square
Linear regression	$S_{LR} = b \sum (x_i - \bar{x})(y_i - \bar{y})$	1	S_{LR}
Residual	$S_R = S - S_{LR}$	$n - 2$	$S_R/(n - 2)$
Total	$S = \sum (y_i - \bar{y})^2$	$n - 1$	

The expected value of S_R is $(n - 2)\sigma^2$, and therefore the residual mean square gives an unbiased estimate of σ^2.

   It is also possible to use an F-test to investigate whether the slope β of the regression line is zero. On the NH that β is zero, the regression mean square has expected value σ^2, and if β is non-zero this expected value is greater than σ^2. Also, on the NH, the ratio of the regression and residual mean squares, $S_{LR}/(S_R/(n - 2))$, has an F-distribution with 1 and $n - 2$ degrees of freedom. Making a one-tail test gives evidence of whether there is a real regression or not.

   Most computer programs include in their output the value of the *coefficient of determination*, R^2; this is the proportion of the total sum of squares S which is accounted for by the regression S_{LR}. It is usually given as a percentage, $100S_{LR}/S$.

A high value, near 100%, is taken to mean that a linear model fits well. However, the value of R^2 can be quite high even when the departures from a linear model are systematic, so that really a curve should have been fitted. The only way of discovering this is to draw the scatter diagram, which should always be included in a regression analysis as a matter of routine. The ratio S_{LR}/S is simply the square of the correlation coefficient r; but capital R^2 is generally used because there is a similar ratio R^2 that is defined in multiple regression.

Example 23.4 Analysis of variance for regression

The following data give the reaction times of 10 men, of various ages, to a visual stimulus (a flashing light) in a psychological experiment; X is age (in years) and Y is reaction time (in milliseconds):

x	37	35	41	43	42	50	49	54	60	65
y	190	197	205	210	218	226	228	230	234	240

(i) Fit a model $Y = \alpha + \beta x + e$ to the data, where α and β are constants and e is a random error with mean 0 and variance σ^2.
(ii) Estimate σ^2.
(iii) Find a 95% confidence interval for β.
(iv) Test whether $\beta = 0$.
(v) Find the coefficient of determination R^2.

From the data, $\sum x_i = 476$, $\sum y_i = 2178$, $\bar{x} = 47.6$, $\bar{y} = 217.8$, $\sum (x_i - \bar{x})^2 = 872.40$, $\sum (x_i - \bar{x})(y_i - \bar{y}) = 1375.20$, $\sum (y_i - \bar{y})^2 = 2505.60$.

(i) Hence $b = \sum (x - \bar{x})(y_i - \bar{y}) / \sum (x_i - \bar{x})^2 = 1375.20/872.40 = 1.5763$. The regression line of Y on X is

$$(y - 217.8) = 1.5763(x - 47.6)$$

i.e.

$$y = 142.8 + 1.576x.$$

(ii) For an analysis of variance,

$$S = \sum (y_i - \bar{y})^2 = 2505.60; \quad S_{LR} = b \sum (x_i - \bar{x})(y_i - \bar{y}) = 2167.78$$

Setting this out in a table, we have:

Source of variation	Sum of squares	Degrees of freedom	Mean square
Linear regression	2167.78	1	2167.78
Residual	337.82	8	42.23
Total	2505.60	9	

The estimate of σ^2, usually denoted by s^2, is the residual mean square, 42.23.

(iii) The estimated variance of b is $s^2 / \sum (x_i - \bar{x})^2$ (see Section 23.4). This is equal to $42.23/872.40 = 0.048\,407$; the standard error of b is $\sqrt{0.048\,407} = 0.220$.

Since we have to use an estimated value of the variance, the multiplier used in the confidence interval must be the t-value with the same degrees of freedom as s^2, namely 8 d.f. The 95% critical value of $t_{(8)}$ is 2.306.

Therefore the 95% confidence interval for β has limits $b \pm t_{(8)} \sqrt{\mathrm{Var}(b)}$, i.e. $1.5763 \pm 2.306 \times 2.220$ i.e. 1.576 ± 0.507 or $(1.069, 2.083)$.

(iv) From the analysis of variance table, the ratio of the regression mean square to the residual mean square is $2167.78/42.23 = 51.3$. On the NH that $\beta = 0$, this is distributed as $F_{(1,8)}$. From Table A7, the 1% critical value of $F_{(1,8)}$ is 11.26; the observed value 51.3 greatly exceeds this and so there is very strong evidence that β is not zero.

There are two further methods of testing the NH that $\beta = 0$, which are equivalent to the F-test. Since zero is not contained in the 95% confidence interval for β, we may reject the NH at the 5% level. If we calculate a 99% confidence interval, using the t-value 3.355 with 8 d.f, we shall also find that this interval does not contain zero; thus we reject the NH that $\beta = 0$, at the 1% level.

The final method is to calculate

$$t = \frac{b - 0}{\text{standard error of } b} = \frac{1.576}{0.220} = 7.16.$$

This is compared with the critical values of $t_{(8)}$ in Table A3. It exceeds even the 0.1% value of $t_{(8)}$ and so provides very strong evidence indeed against the NH.

(v) From the analysis of variance table, we see that

$$R^2 = S_{LR}/S = 2167.78/2505.60 = 0.865.$$

We can say that 86.5% of the variability in reaction time is explained, or accounted for, by the regression on age. Since $S_{LR}/S = r^2$, in simple linear regression, it is often argued that the strength of association between two variates is better assessed using r^2 rather than the correlation coefficient r.

23.9.1 Limits for α, expected values and individual values of Y

We saw in Section 23.3 that $a = \bar{y} - b\bar{x}$; therefore

$$\mathrm{Var}(a) = \mathrm{Var}(\bar{y} - b\bar{x}) = \mathrm{Var}(\bar{y}) + \bar{x}^2\,\mathrm{Var}(b) - 2\bar{x}\,\mathrm{Cov}(\bar{y}, b).$$

Intuitively, there should be no correlation between the average value of the observations, $\bar{y}$, and the regression coefficient b which is the slope of the fitted line. This can be shown by repeating the same type of algebraic argument as we used to establish $\mathrm{Var}(b)$, but we shall not take up space to do so. However, assuming $\text{Cov}\,(\bar{y}, b) = 0$, we have

$$\text{Var}\,(a) = \text{Var}\,(\overline{Y}) + \bar{x}^2\,\text{Var}\,(b) = \frac{\sigma^2}{n} + \frac{\sigma^2 \bar{x}^2}{\sum_{i=1}^{n}(x_i - \bar{x})^2} = \sigma^2 \left\{ \frac{1}{n} + \frac{\bar{x}^2}{\sum_{i=1}^{n}(x_i - \bar{x})^2} \right\}.$$

Since a follows a normal distribution, confidence limits for α can be found as $a + t\sqrt{\text{Var}\,(a)}$, where σ^2 is replaced by s^2 from the analysis of variance and t has the same degrees of freedom as s^2.

In some problems it is sufficient to give confidence limits for β and possibly also for α; but often we wish to know how precisely the predicted value $\hat{Y}$, at a given value of x_i, can be estimated. $\text{Var}\,(\hat{Y}) = \text{Var}\,(a + bx_i)$, and in order to calculate this we require $\text{Cov}\,(a, b)$, which is equal to

$$\text{Cov}\,(\bar{y} - b\bar{x}, b) = \text{Cov}\,(\bar{y}, b) - \bar{x}\,\text{Var}\,(b) = 0 - \bar{x}\,\text{Var}\,(b) = -\sigma^2 \frac{\bar{x}}{\sum_{i=1}^{n}(x_i - \bar{x})^2}.$$

(See Section 21.4, p. 549.) These results give

$$\text{Var}\,(\hat{Y}) = \text{Var}\,(a + bx_i) = \text{Var}\,(a) + x_i^2\,\text{Var}\,(b) + 2x_i\,\text{Cov}\,(a, b)$$

$$= \sigma^2 \left\{ \frac{1}{n} + \frac{(\bar{x}^2 + x_i^2 - 2x_i\bar{x})}{\sum_{i=1}^{n}(x_i - \bar{x})^2} \right\}$$

$$= \sigma^2 \left\{ \frac{1}{n} + \frac{(x_i - \bar{x})^2}{\sum_{i=1}^{n}(x_i - \bar{x})^2} \right\}$$

Since Y follows a normal distribution we can use the same method as above to obtain limits for a predicted expected value (i.e. the mean value) of Y at a given value x_i of X. The further x_i is from $\bar{x}$, the wider these limits will be; this reflects the facts that there is uncertainty in the estimate b but the regression line does pass through the mean. Figure 23.8 shows the situation.

We can also see from Fig. 23.8 how much wider the limits will be if we try to predict where *individual* values of $\hat{Y}$ will lie for a given value of X. This is because the variance of an individual value has an additional component σ^2 to allow for the variation of these values about their mean, so that it becomes

$$\sigma^2 \left\{ 1 + \frac{1}{n} + \frac{(x_i - \bar{x})^2}{\sum_{i=1}^{n}(x_i - \bar{x})^2} \right\}.$$

Occasionally this information is useful as well, but the extra component can dominate the value of the variance quite seriously and lead to much wider limits.

Example 23.4 (continued)

For the data on reaction times (p. 623), estimate α and find the standard error of the estimate. Also find 95% confidence limits for the mean value of Y predicted for a man aged 45. Comment on the results.

We have already found $a = 142.8$ when working out the regression equation in Example 23.4(i). $\text{Var}(a)$ is estimated as

$$s^2(1/10 + 47.6^2/872.40) = 42.23(0.1 + 2.597) = 113.90,$$

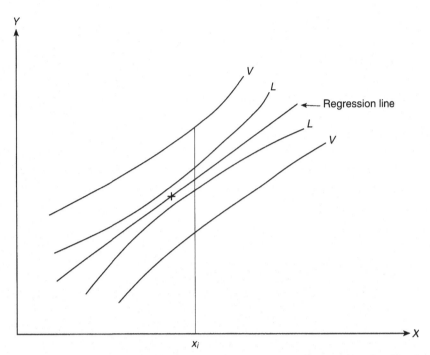

Figure 23.8 *Limits for predicted values of Y and for individual values:*
$+ = \mathrm{mean}(\bar{x}, \bar{y})$; $L =$ *limits for predicted* $\mathrm{E}[Y]$; $V =$ *limits for individual values of Y.*

so the standard error of a is 10.67. In this problem a is really of no interest because it is the value of Y when $X = 0$ which is a very long way away from all the available data. It would be the reaction time of a new-born male child! But there is no evidence at all that a linear (or any other) relation holds below the age of 35 which was the smallest value of X in the set of data.

For age 45, however, x is not far from $\bar{x}$, and well within the range of available data, so we proceed to find

$$\mathrm{Var}(\hat{Y}) = 42.23(0.1 + \{45.0 - 47.6\}^2/872.40) = 4.5502.$$

The t-test will use the critical values for 8 d.f., since s^2 has 8 d.f.; the 95% limits for the predicted (mean) value are therefore $\hat{y} \pm 2.306\sqrt{4.5502}$. At $X = 45$,

$$\hat{y} = 142.8 + (45 \times 1.576) = 213.72.$$

The required 95% limits are therefore 213.72 ± 4.92 or (208.8 to 218.6). These limits are relatively narrow; it is left as an exercise for the reader to show that they get steadily wider for ages further from 47.6 (see Exercise 23.9.2(1)).

If we had been interested in the range of likely values for individual observations of Y at $X = 45$, the extra term in the variance would give

$$\mathrm{Var}(\hat{Y}) = 42.23(1 + 0.1 + \{45.0 - 47.6\}^2/872.40) = 46.7802,$$

completely dominated by this additional term. This leads to much wider limits, which are

$$213.72 \pm 2.306\sqrt{46.7802} = 213.72 \pm 15.77 \text{ or } (198.0 \text{ to } 229.5).$$

23.9.2 Exercises

1. For the data of Example 23.4, calculate the 95% limits for individual values of Y predicted at $X = 35, 40, 47.6, 55, 60$ and show both the regression line and these limits in a diagram (compare Fig. 23.8).

2. For the data of Exercise 23.3.1(5), find the standard errors of a and b. Complete an analysis of variance and comment on the results.

23.10 TRANSFORMATIONS TO MAKE A MODEL 'LINEAR'

The relationship $E[Y \mid x] = \alpha + \beta x$ is called 'linear' because it is a linear function of the *parameters* α and β, not because it is linear in x and y. The parameters can then be estimated very simply in a least-squares form of analysis. If instead we thought the relationship was $E[Y^2 \mid x] = \alpha + \gamma x^2$, that is still linear in its parameters α and γ. We would estimate α and γ by making $y^2 = Y^*$ and $x^2 = X^*$ in the linear regression equation $Y^* = \alpha + \gamma X^*$:

$$\hat{\gamma} = \sum (X^* - \overline{X}^*)(Y^* - \overline{Y}^*) / \sum (X^* - \overline{X}^*)^2 \quad \text{and} \quad \hat{\alpha} = \overline{Y}^* - \hat{\gamma}\overline{X}^*.$$

We have *transformed* the original observations on to new scales of measurement in order to solve the problem. A scatter diagram of the points (x_i^2, y_i^2) would be a useful item of initial data analysis before making any calculations. If a computer program is then used the original values (x_i, y_i) have to be transformed to (x_i^2, y_i^2) at the beginning of the calculations.

In economic models, and also in biological growth models, relationships are frequently curvilinear. One method of dealing with these would obviously be to fit a polynomial $y = \alpha + \beta x + \gamma x^2 + \delta x^3 + \cdots + \varepsilon$ using the methods of *multiple regression* from Chapter 24; but the parameters $\alpha, \beta, \gamma, \ldots$ are usually extremely difficult to interpret. However, some very useful relationships exist involving *logarithmic* and *exponential* functions. If a scatter diagram of the observed data (x_i, y_i) shows a systematic tendency towards a curve, with in addition random deviations from it, there are two quite simple possibilities. Figure 23.9(a) is a situation which may be modelled by $E[Y \mid x] = \alpha + \beta \ln x$, so that a simple linear regression of y on $\ln x$ provides estimates of α and β.

Figure 23.9(b) suggests $y = \alpha \exp(\beta x)$; this is equivalent to $\ln y = \ln \alpha + \beta x$, so that a regression of $\ln y$ on x gives $\ln \alpha$ (and hence α) and β. In the equation $y = \alpha \exp(\beta x)$, α is the value of y when $x = 0$ and β is the constant factor by which Y is multiplied to find the gradient at any point on the curve. This is because

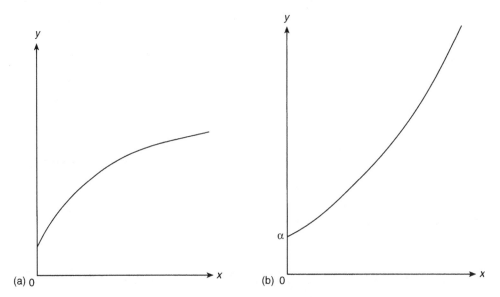

Figure 23.9 *(a) Relation $y = \alpha + \beta \ln x$. (b) Relation $y = \alpha \exp(\beta x)$.*

$dy/dx = \beta y$ has the solution $y = \alpha \exp(\beta x)$, and the rate of change of y is always β times its present size. Biological organisms often grow according to this rule, at a rate proportional to their present size.

A more general curved relation is $E[Y \mid x] = Ax^B$, a 'power law' which can include both of the shapes in Fig. 23.9 according to whether B is less or greater than 1. Example 23.5 shows how A and B can be estimated after the model has been 'linearized'.

Before working through this example we must consider what assumptions can be made about the residual (error) terms in the models. In order to solve for A and B, the relation will be written $\ln Y = \ln A + B \ln x$. If we use the standard least-squares linear regression, followed by tests and confidence interval calculations based on the analysis of variance and the estimated s^2, we are adding a term ε to the *transformed* relation and thus effectively assuming that $\ln Y$ is normally distributed. Had we tried to add a residual term ε to the original power-law relation, so making Y normally distributed, the least-squares equations necessary for estimating the two parameters would have been very awkward to obtain and to handle. In practice when we meet this problem we will work with the linearized model and hope that the necessary assumptions about the properties of residual terms are not unreasonable on the transformed scale of measurement.

Example 23.5 *Estimating the parameters in $Y = AX^B$*

The yield Y (kilograms per plot) of a corn crop in a controlled experiment is measured, and is thought to be related to the amount of irrigation supplied, X (hours per day), by the model $Y = AX^B$. The following results have been obtained.

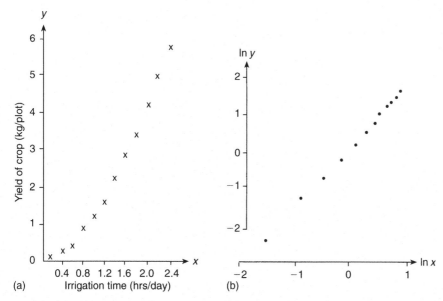

Figure 23.10 (a) Data of Example 23.5. (b) Data of Example 23.5 when plotted on logarithmic scales.

x	0.2	0.4	0.6	0.8	1.0	1.2	1.4	1.6	1.8	2.0	2.2	2.4
y	0.10	0.27	0.48	0.91	1.24	1.67	2.26	2.78	3.43	4.21	5.00	5.80

Plot these on a graph of y against x, and comment on the shape of this graph. Make a suitable transformation to a linear model, and show the results of this on a graph. Estimate A and B.

Figure 23.10(a) shows the data plotted on the original scales of x and y. If we did not have the data below $x = 0.8$ we might be tempted to assume that a linear relation existed between these two measurements without any transformation; however, a careful study would suggest a gentle curve, that is systematic and not random variation about a straight line if one had been fitted.

The relation $Y = AX^B$ becomes $\ln Y = \ln A + B \ln X$ in which $\ln A$ is the intercept on the y-axis and B is the gradient of the fitted line. The transformed values are given in the following table and the graph of $\ln Y$ against $\ln X$ is plotted in Fig. 23.10(b).

$\ln x$	−1.61	−0.92	−0.51	−0.22	0	0.18	0.34	0.47	0.59	0.69	0.79	0.88
$\ln y$	−2.30	−1.31	−0.73	−0.09	0.22	0.51	0.82	1.02	1.23	1.44	1.61	1.76

A linear regression calculation can be done using $\ln y$ as the dependent variable and $\ln x$ as the independent variable; since x was controlled so is $\ln x$ and the usual statistical conditions are satisfied. The results of calculating sums, squares and products of transformed variables are as follows: $\sum \ln x = +0.68$; $\sum \ln y = +4.18$; $n = 12$; $\sum (\ln x)^2 = 6.3386$; $\sum (\ln x)(\ln y) = 10.6903$; if required for an analysis of variance, $\sum (\ln y)^2 = 18.8446$.

The mean value of $\ln x$ is $0.68/12 = 0.0567$, and of $\ln y$ is $4.18/12 = 0.3483$. Finally, the slope B is

$$\frac{10.6903 - (0.68)(4.18)/12}{6.3386 - (0.68)^2/12}$$

using Equation (23.3) with $\ln x$ instead of x and $\ln y$ for y. Hence

$$B = \frac{10.4534}{6.3001} = 1.659,$$

leading to the equation

$$\ln y - 0.3483 = 1.659\{\ln x - 0.0567\}$$

or

$$\ln y = 0.2542 + 1.659 \ln x.$$

The estimate of $\ln A$ is 0.2542, and so $\hat{A} = e^{0.2542} = 1.29$. The original power-law model $Y = AX^B$ is therefore estimated as

$$y = 1.29x^{1.66}.$$

If we require confidence intervals for A or any fitted values of Y, we must calculate them indirectly; intervals for $\ln A$ can be found in the same way as for a in Example 23.4 (continued) and the upper and lower limits then transformed back to the original scale of measurement (see Examples 15.2 and 18.9, pp. 364 and 453). In this power-law relation B takes the place of b in Example 23.4 and so is easy to deal with directly.

There are several useful nonlinear relations that can be treated in a similar way (see, for example, Exercises 23.12(14) and (15)), and a thorough discussion of the problem will be found in Mead *et al.* (2001).

23.11 MINITAB: SIMPLE LINEAR REGRESSION

To simply obtain a graph of the regression line fitted to pairs of data, choose **Graph>Scatterplot** and select **With Regression**, instead of **Simple**, in the dialogue box. To obtain numerical information, as well as a graph, choose

> **Stat>Regression>Fitted Line Plot/**
> Enter names in **Response Y** and **Predictor X**, select **Linear** under **Type of Regression Model/**

The output gives the regression equation and an analysis of variance similar to that we give in Section 23.9. The F-value, used to test the NH '$\beta = 0$', is quoted, together with its p-value. Also quoted are s, the square root of the residual mean square, and R^2, the coefficient of determination or proportion of the total sum of squares accounted for by the regression (Section 23.9). R^2-adj is based on mean squares and may be ignored.

It would be very useful also to have the following table:

	Estimate of Coef	SE of Estimate of Coef	T	p
Constant	a	$\sqrt{\mathrm{Var}\,(a)}$		
x	b	$\sqrt{\mathrm{Var}\,(b)}$		

However, **Fitted Line Plot** does not provide it, though **Stat>Regression> Regression**, as discussed in Section 24.10, does when used on the same data. The column headed T shows the values of the t-statistics on the null hypotheses that $\alpha = 0$ and $\beta = 0$ respectively, and the next column gives the corresponding p-values.

As recommended when discussing ANOVA in Section 22.8, a check should be made of the validity of the model by inspecting the residuals. The graphs **Histogram of residuals, Normal plot of residuals** and **Residuals versus fits** may be selected following choice of the option **Graphs . . .** in the dialogue box.

Choice of **Quadratic** or **Cubic** under **Type of Regression Model** allows fitting a quadratic or cubic equation in the single variable x. If the relation between observed values is curved rather than linear, a quadratic equation often gives an adequate fit. (See also p. 650.)

Choice of **Options** in the dialogue box allows the fitting of a power law as in Example 23.5: the **Transformations Logten of Y** and **Logten of X** may be selected. This will carry out a satisfactory fit but if it is desired to use natural logarithms, as in our example, then one would make use of **Calc>Calculator**, with **Natural log** as the **function** in the **Expression**, to provide two new columns of transformed data to enter into the analysis.

23.12 EXERCISES ON CHAPTER 23

1. The number of colour television licences issued each quarter of the year is shown below for 9 quarters. Plot the least-squares regression line of the logarithm of number of licences against time.

Year	1971	1972				1973			
Quarter	4	1	2	3	4	1	2	3	4
Number of licences (10 000s)	120	153	186	217	265	317	361	405	478

2. The gross weekly earnings of non-manual workers of different ages in 1974 are summarized in the table below. (Age is in years, and earnings is the median weekly earnings in £.) Calculate the least-squares regression line of

earnings on age, writing it in the form $Y = a + bX$. *Without recalculation:* (i) find the values of a and b if all non-manual workers had an increase of £6 per week; and (ii) find the value of b if all non-manual workers had an increase of 6% of their previous earnings.

Age	18	20	22	27	35	45	55
Earnings	15.5	23.2	34.0	44.9	53.1	55.0	51.2

(IOS)

3. Calculate a least-squares regression line of crop yield on applied fertilizer from the records below. Plot the data and the line on a graph and comment on the results.

Crop yields (cwt/acre)	160	168	176	179	183	186	189	186	184
Fertilizer (oz/yd^2)	0	1	2	3	4	5	6	7	8

4. A survey of the pocket money received by children in a primary school was made by choosing at random four children of each of the ages 5, 7, 9 and 11 years. The amounts of pocket money received are given in the following table:

Age (years)	Pocket money (p)			
5	2	8	10	12
7	9	13	14	16
9	9	14	16	21
11	18	19	23	36

Plot these data on a scatter diagram. Plot also the mean pocket money for each age.

Find the regression equation for predicting pocket money from age, and draw it on your diagram. *(Oxford)*

5. Fifteen boys took two exam papers in the same subject and the marks as percentages were as follows, where each boy's marks are in the same column:

Paper I, x	65	73	42	52	84	60	70	79	60	83	57	77	54	66	89
Paper II, y	78	88	60	73	92	77	84	89	70	99	73	88	70	85	89

Given that $\sum xy = 83\,734$, calculate the equation of the line of regression of y on x.

Two boys were each absent for one paper. One scored 63 on paper I, the other scored 81 on paper II. In which case can you use your regression line to estimate the mark that the boy should be allocated for the paper he did not take, and what is that mark? *(Southern)*

6. The data (x_i, y_i), where $x > 0$, $i = 1, \ldots, n$, are thought to conform to a proportional regression model (linear regression through the origin):

$$y_i = \beta x_i + e_i, \quad i = 1, \ldots, n,$$

where $\{e_i\}$ are independent normally distributed error terms with mean zero.

(i) If $\{e_i\}$ have constant known variance σ^2, show that the maximum likelihood (ML) estimate of β minimizes $\sum_{i=1}^{n} e_i^2$ and is given by

$$\hat{\beta}_1 = \sum_{i=1}^{n} x_i y_i / \sum_{i=1}^{n} x_i^2.$$

(ii) If instead the variance of e_i is given by $\sigma^2 x_i, i = 1, \ldots, n$, show that the ML estimate of β minimizes $\sum_{i=1}^{n}(e_i^2/x_i)$ and is given by $\hat{\beta}_2 = \bar{y}/\bar{x}$, where $\bar{x}, \bar{y}$ are the sample mean values of $\{x_i\}$ and $\{y_i\}$ respectively.

(iii) Each time a motorcycle is filled with petrol, a record is kept of the amount of petrol in litres (x) used, and the distance travelled in miles (y) since the previous fill-up. Values of x and y recorded on the last 9 occasions were:

x	4.3	4.9	5.7	6.5	7.2	8.3	8.4	9.6	10.1
y	123	156	183	183	204	234	270	273	324

Plot the data and calculate $\hat{\beta}_1$ and $\hat{\beta}_2$. Which of the models (i) or (ii) do you think better represents the data? (RSS)

7. (i) Explain why the regression line of y on x is not necessarily the same as the regression line of x on y. How do you decide which is the appropriate regression in any particular situation?

(ii) When do the lines coincide and when are they at right angles?

8. From n pairs of values (x_i, y_i), $i = 1, 2, \ldots, n$, the following quantities are calculated:

$$n = 20, \quad \sum x_i = 400, \quad \sum y_i = 220, \quad \sum x_i^2 = 8800,$$

$$\sum x_i y_i = 4300, \quad \sum y_i^2 = 2620.$$

Find the linear regression equations of y on x and x on y. Which would be the more useful if:

(a) x is the age in years and y is the reaction time in milliseconds of 20 people;

(b) x is the cost (£000) and y the floor-space (in 1000 ft^2) of 20 buildings? (Oxford)

9. Briefly state the principle of the method of least squares for fitting a straight line to an observed set of corresponding values of two variables and state what assumptions, if any, need to be made in order to apply the method.

A test was performed to determine the relationship between the chemical content of a particular constituent $(y, \mathrm{g\,l^{-1}})$ in solution and the crystallization temperature $(x,$ kelvin$)$. The results were as follows:

x	0.3	0.4	1.2	2.3	3.1	4.2	5.3
y	3.2	2.4	4.3	5.4	6.6	7.8	8.8

The following quantities have been calculated from these data:

$$\sum x = 16.8, \quad \sum y = 38.5, \quad \sum x^2 = 62.32, \quad \sum xy = 119.36.$$

Assuming a linear relationship $y = \alpha + \beta x$, calculate the least-squares estimates of α and β, each correct to three significant figures. Suppose that the y-determinations are subject to independent random errors each having mean zero and standard deviation $0.5\,\mathrm{g\,l^{-1}}$. Stating clearly any further assumptions you make regarding the data, determine 95% confidence limits for β. *(Welsh)*

10. The weight per cent, W, of nitrous oxide in a mixture of nitrous oxide and nitrogen dioxide at temperature T is given by the relation

$$W = \alpha + \beta T^{-1},$$

where α and β are constants. In an experiment to estimate α and β, the temperature T was carefully controlled at four different values and three determinations of W were made at each temperature value. Letting $x = 1000T^{-1}$, the following calculations were made from the 12 observed values of (x, W):

$$\sum x = 43.2, \quad \sum W = 13.2, \quad \sum x^2 = 161.28, \quad \sum xW = 44.64.$$

(i) Determine the least-squares estimate of the equation connecting W and x.
(ii) Estimate the value of W when $T = 250$.
(iii) Given that determinations of W are subject to a normally distributed random error having mean zero and standard deviation 0.08, calculate 99% confidence limits for β. *(Welsh)*

11. Two variables x and y are known to be such that for all $0 \leqslant x \leqslant a$,

$$y = \alpha + \beta x,$$

where α and β are unknown. An experimental observation of the value of y corresponding to a fixed value of x is subject to a random error of mean zero and variance σ^2. In order to estimate β it is decided to carry out a total of $2n$ experiments with x fixed at the values $x_1, x_2, \ldots, x_{2n}$, respectively and to observe the corresponding values of y. Show that the estimate of β

obtained from an application of the method of least squares to the values $(x_i, y_i), i = 1, 2, \ldots, 2n$, is unbiased.

State, with your reasons, which of the following two sets of $2n$ values of x you would recommend as being the better to use in order to estimate β.

Set 1 $x_1 = x_2 = \cdots = x_n = 0$, $x_{n+1} = x_{n+2} = \cdots = x_{2n} = a$.

Set 2 $x_r = (r-1)a/(2n-1)$, $r = 1, 2, \ldots, 2n$.   (*Welsh*)

12. Set out the analysis of variance for the data in Revision Exercises F, Question 8. Test the null hypothesis that the slope is zero, and find a 95% confidence interval for the true value of β.

13. Construct an analysis of variance for the data in Revision Exercises F, Question 15, for the regression of weight on age. Find an estimate of the variance of the weight measurements y, and use this to obtain a 95% confidence interval for the true value of the regression slope.

14. The weights y (in kilograms) of batches of 10 randomly chosen plants, all taken from the same source, are measured after they have grown for the same length of time in standard conditions of temperature, light and humidity. Each batch has received a carefully controlled amount x of standard nutrients. The ratio $u = x/y$ has been calculated. The results are:

x	1	2	3	4	5	6	7	8	9	10
y	13.8	24.9	33.1	38.2	51.0	60.6	64.4	71.5	76.8	80.3
u	0.0725	0.0803	0.0906	0.1047	0.1176	0.1320	0.1553	0.1958	0.2344	0.2989

(a) Plot y against x and comment on the form of the relationship between them.

(b) The equation $1/y = a/x + b$, where a and b are constants, is proposed as a model for these data. Write this in the form of a linear relation between u and x.

(c) Find estimates of a and b, given the following *uncorrected* sums of squares and products:

$$\sum u = 1.4821, \quad \sum x = 90, \quad \sum u^2 = 0.268\,868\,85,$$

$$\sum ux = 18.3723, \quad \sum x^2 = 1326.$$

Using these estimates of a and b, verify that the fitted values of y for $x = 1$ and 2 are (approximately) 14.25 and 25.02 respectively. Find the remaining fitted values and comment briefly on the adequacy of the model.   (*RSS*)

15. Water uptake of seeds of a particular variety of wild oats was measured every two hours for a period of 24 hours. The uptake, in grams per gram of

dryweight of seed, is shown in the following table:

Time (t)	Water uptake (y)	$w\,(=t/y)$
2	0.69	2.898 55
4	1.50	2.666 67
6	1.43	4.195 80
8	1.64	4.878 05
10	2.01	4.975 12
12	1.96	6.122 45
14	2.28	6.140 35
16	2.10	7.619 05
18	2.02	8.910 89
20	2.33	8.583 69
22	2.32	9.482 76
24	2.42	9.917 35

Theory suggests a 'hyperbolic law' between water uptake and time, $y = at/(b+t)$. Show that this can be written as a linear relationship between $w = t/y$ and t. Plot w against t. By carrying out a suitable linear regression, and transforming back to the original units, estimate a and b. What water uptake is predicted by the fitted hyperbolic law at $t = 16$ hours?　　(*RSS*)
[*Note:* $\sum w = 76.390\,73$; $\sum wt = 1190.460\,22$.]

23.13 COMPUTING EXERCISES

1. The heights, H (in centimetres), and weights, W (in kilograms), of 54 male students are given below.
   (i) Produce a scatter diagram of the data.
   (ii) Find the correlation coefficient between H and W.
   (iii) Find an equation to predict W from H.

H	W	H	W	H	W
176	72	175	65	179	66
172	58	173	61	184	74
177	60	178	68	181	64
174	68	174	64	177	67
167	59	176	66	177	69
191	72	167	64	178	73
190	85	185	76	178	71
176	73	174	55	181	60
173	57	170	60	182	70

(*contd*)

H	W	H	W	H	W
180	76	176	68	183	63
178	69	160	53	184	92
170	74	174	71	176	61
173	63	182	72	170	66
168	50	184	80	174	67
173	64	183	73	160	52
178	71	179	68	172	64
182	66	164	63	191	96
175	67	177	72	174	52

2. The relationship between mean oxygenation rate (y) and the depth of the stream (x) was studied using data from 12 different streams:

y (ppm per day)	x (m)	y (ppm per day)	x (m)
1.44	1.54	0.28	3.46
2.27	0.99	3.36	0.64
0.98	1.34	2.79	0.89
0.50	1.86	1.57	1.38
0.74	1.72	0.46	2.89
1.13	2.17	0.39	1.91

  (i) Plot the data in a scatter diagram.
 (ii) Fit a linear equation relating y to x.
(iii) Fit a linear equation relating $\ln y$ to $\ln x$.
(iv) For each of (ii), (iii) plot the residuals (differences between observed and fitted values) against the fitted values $\hat{y}$.
 (v) Plot the power-law curve $y = Ax^b$, using appropriate estimates of A and b. Compare this with the scatter diagram (i).   (RSS)

3. In the following five sets of 11 data points (x, y) the four sets A, B, C, D have the same x-values.

x	y(A)	y(B)	y(C)	y(D)	x	y(E)
10.0	21.26	22.40	20.70	22.94	8.0	21.20
8.0	20.57	21.84	20.43	19.85	8.0	18.74
13.0	20.15	21.35	25.31	19.65	8.0	21.98
9.0	22.28	22.15	20.56	20.25	8.0	19.70
11.0	21.36	22.27	20.80	22.38	8.0	22.66
14.0	22.30	20.43	21.19	23.84	8.0	19.32
6.0	21.35	20.25	20.21	20.55	8.0	21.62

(*contd*)

x	$y(A)$	$y(B)$	$y(C)$	$y(D)$	x	$y(E)$
4.0	18.81	17.66	19.90	19.53	8.0	20.91
12.0	23.63	21.90	20.95	21.55	8.0	20.13
7.0	18.73	21.15	20.31	19.72	8.0	20.44
5.0	20.01	19.05	20.09	20.19	19.0	23.75

(i) Fit the regression line of y on x for each of the five sets of data.

(ii) By looking at the scatter plots discuss the usefulness of the predicted y-value, when $x = 10$, for each of the data sets.

23.14 PROJECTS

1 Relation between arm length and finger length

Aim. To set up a regression equation to predict arm length of adults from the length of their index finger.

Equipment. Measuring tape.

Collection of data. The index finger (next to the thumb) is measured as accurately as possible. Make sure it is kept straight, define beforehand what is 'length', and keep the same definition throughout the study. (Do not include long nails!) Decide also whether to measure right or left hand. The arm length is now measured for the same subject, using a careful definition, e.g. armpit to wrist bone.

Repeat on 15 to 20 subjects if possible. If at least 10 of each sex are available it will be possible to analyse the results for each sex separately.

Analysis. Plot a scatter diagram. Calculate the regression line, specified in the aim above, and draw it on the diagram. (If the points do not cluster uniformly about the line but suggest a curvilinear relation, use of log x and/or log y may lead to a better fit.)

2 Trisection of lines

Aim. To determine the relation between the absolute error in trisecting a line and the length of the line.

Equipment. Plain sheets of A4-size paper. Ruler.

Collection of data. Draw lines of length 12, 15, 18, 21, 24 or 27 cm on the sheets of paper, one line to a sheet. Repeat each length on three more sheets of paper. Put the sheets in random order and present them one by one to a subject who is asked to mark in a trisection point for each line. Measure the absolute error (i.e. error ignoring the sign) made in each trisection. The whole experiment may be repeated with other subjects.

Analysis. For each subject, plot the errors against the length of the line. Find the equation of the regression line and mark it on the graph.

3 Distance between two places

Aim. To find a relation between the distance 'as the crow flies' between two places in a town, and the distance that would have to be travelled on foot or by car.

Equipment. Large-scale map of a town. Measuring tape or ruler and string.

Collection of data. Choose several pairs of points on the map, using random numbers to locate the points; two random digits give the distance along the horizontal axis from the origin (bottom left-hand corner of map) and two more the distance along the vertical axis. Use about 20 pairs of points. Measure the direct distance between the two members in each pair. Then measure the shortest route on the map that would be possible if walking or driving from one to the other: measure this as accurately as possible using a piece of string.

Analysis. Plot the travelling distance (y) against the direct distance (x) on a graph, and find the equation of the regression line to predict y from x. Draw the line on the graph. (Consider whether the line should be fitted with or without an intercept, or constant, term.)

Note: A similar project could be carried out using a map of a large region or country to find the relation between distance 'as the crow flies' and distance travelled by car.

SUMMARY

When a **straight-line relation** is thought to exist between Y and X, it may be written $Y = \alpha + \beta X + e$, in which α, β are constants (parameters) to be estimated from n pairs of observations (x_i, y_i). The $\{e_i\}$ are independent normally distributed random residual terms with the same variance σ^2.

The estimate of β is

$$b = \frac{\sum (x_i - \bar{x})(y_i - \bar{y})}{\sum (x_i - \bar{x})^2},$$

of α is $\bar{y} - b\bar{x}$, and of σ^2 is $(S - S_{LR})/(n - 2)$, where S is the total sum of squares $\sum (y_i - \bar{y})^2$ and $S_{LR} = b \sum (x_i - \bar{x})(y_i - \bar{y})$ is that part of the sum of squares accounted for by the linear regression. The **coefficient of determination** S_{LR}/S is sometimes used as a measure of how well the observed data fit the line $\hat{y} = a + bx$.

When $X = x_i$, the **fitted** or **predicted** (expected) value of Y is $\hat{y}_i$, and the **residual** is $y_i - \hat{y}_i$. The sum of squares of residuals, divided by its degrees of freedom $(n - 2)$, also gives the estimate $\hat{\sigma}^2$ for use in tests and confidence intervals.

$\text{Var}[b] = \hat{\sigma}^2 / \sum (x_i - \bar{x})^2$, and b is normally distributed; hence a 95% *confidence interval* for β is

$$b - 1.96\sqrt{\text{Var}(b)} \leqslant \beta \leqslant b + 1.96\sqrt{\text{Var}(b)}.$$

Since S_{LR} has 1 d.f., the ratio $S_{LR}/\hat{\sigma}^2$ is distributed $F_{(1, n-2)}$ for a test of the null hypothesis '$\beta = 0$'.

A 95% confidence interval for the *expected* value of Y at $X = x_i$ is found in the same way, using

$$\text{Var}[\hat{y}_i] = \sigma^2 \left\{ \frac{1}{n} + \frac{(x_i - \bar{x})^2}{\sum (x_i - \bar{x}^2)} \right\}.$$

Sound inference can only be made for x_i inside the range of values in the available data.

Methods for simple linear regression can be adapted to other forms of relation which can be made linear in their parameters by a suitable *transformation*, e.g. $Y = AX^B$ becomes $\ln Y = \ln A + B \ln X$.

REVISION EXERCISES F

1. What distribution might be used to describe the following data which show the numbers of minutes in which there were $0, 1, 2, \ldots$ arrivals at a service station? How well does your distribution fit the data?

Arrivals per minute	0	1	2	3	4	5	6	7	8 or more
Number of minutes	4	16	21	23	17	11	6	2	0

2. The number of times a year that a river flooded in a 30-year period is shown in the table below. Estimate the mean and variance of the number of floods per year. Which distribution do you expect will fit these data? Fit this distribution, and test how well it explains the data. Does this suggest anything about weather conditions in the area drained by the river?

Number of floods	0	1	2	3	4
Number of years	8	10	7	4	1

3. An engineer is employed to service a particular item of electrical equipment. Each morning he receives the previous day's incoming repairs and completes them in the day. The daily number of repairs has a Poisson distribution with mean 3. He is paid a flat rate of £8 for up to four items in a day and a bonus of £2 extra for each item in excess of four. Find, working to four decimal places, the probability that he receives some bonus payment on any given day. Write down an expression for the mean amount earned per day and show that this may be simplified, for purposes of calculation, to the form

$$\sum_{r=0}^{\infty} 2rP(r) + 8P(0) + 6P(1) + 4P(2) + 2P(3)$$

(where $P(r)$ is the probability of r repairs in a day). Hence find the mean amount earned per day.

A five-day week is worked both at the collection point for repairs and by the engineer. His employer decides to change to a weekly basis for calculation and to pay bonus for each item in excess of 20 for the week. State the mean weekly number of repairs. Assuming that the number of repairs per week also has a Poisson distribution, use the normal distribution as an approximation to calculate the probability that some bonus payment is obtained for any given week on this new system. (*JMB*)

4. The following data were collected in 100 ten-second counts of particles from a radioactive source:

Count	2600–2629	2630–2659	2660–2689	2690–2719	2720–2749	2750–2779	2780–2809
Frequency	2	2	4	5	8	9	14

Count	2810–2839	2840–2869	2870–2899	2900–2929	2930–2959	2960–2989
Frequency	12	12	8	12	8	4

Represent the data as a histogram.

Calculate the mean and variance for this sample.

The timing of these counts was done using a stop-clock, so that the counts are affected by errors in the timing and are of the form $x + e$ where x is the true count and e is the error due to inaccurate timing. Assuming that e has zero mean and is independent of x, and that x has a Poisson distribution, use your sample mean as an estimator of the population mean of x, and state the corresponding estimate of the variance of x. Hence estimate the variance of e.

Give your opinion as to whether or not the data support the installation of an automatic timing device which would eliminate the error e. (*JMB*)

5. Define the Poisson distribution and derive its mean. Describe how, and under what circumstances, the Poisson distribution may be approximated by a normal distribution.

Misprints in books produced by a particular printer occur at random in such a way that the number of misprints in books of a given length has a Poisson distribution. A certain book has 106 misprints. Calculate an approximate 95% symmetric two-sided confidence interval for the mean number of misprints in books of the same length as this one. (*JMB*)

6. Define the *arithmetic mean*, the median and the *standard deviation* of a probability distribution.

Evaluate the mean and standard deviation for each of the following distributions:

(i) a binomial distribution in which $n = 5$ and $p = \frac{1}{5}$;
(ii) a Poisson distribution with parameter 2;
(iii) a normal distribution for which the probability of a negative value of the random variable is 0.3085 and the mode is 2;
(iv) a rectangular distribution over the range 0 to 2. (*Oxford*)

7. The numbers of goals scored by a football team in 60 matches have been as shown below. Calculate the expected frequencies on a Poisson distribution with the same mean. Obtain approximate 95% confidence limits for the total number of goals likely to be scored in their next 10 matches.

Number of goals	0	1	2	3	4	5
Number of games	13	22	14	5	5	1

8. A group of scientists collected a number of sea urchins from various locations. At each collecting point they noted the mean summer temperature, x, and the average magnesium content, y, of the specimens. The results were:

$x(°C)$	18.1	23.0	17.5	20.2	14.7	13.8	15.1	13.8	24.2
y (% $MgCO_3$)	8.8	9.5	8.9	9.1	8.6	8.3	8.5	8.2	9.5

(a) Find the linear regression line of y on x.
(b) Find the correlation coefficient for these data.
(c) Test whether the correlation coefficient in this sample is consistent with a population correlation coefficient of 0.94.
(d) Comment on the assumptions needed to answer (c).

9. (a) Let X and Y be random variables having finite variances. Show that

$$\text{Var}(X + Y) = \text{Var}(X) + \text{Var}(Y) + 2\,\text{Cov}\,(X, Y).$$

(b) Explain what is meant by an *unbiased estimator* of a parameter.
   Observations X_1 and X_2 have the same mean μ, the same variance σ^2, and correlation coefficient ρ. Show that $\overline{X} = \frac{1}{2}(X_1 + X_2)$ is an unbiased estimator of μ and obtain its variance.
   Under which of the following conditions would $\overline{X}$ estimate μ most precisely: (i) if X_1 and X_2 were uncorrelated; (ii) if X_1 and X_2 were positively correlated; (iii) if X_1 and X_2 were negatively correlated?

10. The joint probability distribution of two discrete random variables R and S is shown in the following table. Find the covariance between R and S, and also their correlation coefficient.

	R		
S	1	2	3
1	3/16	1/8	1/4
2	0	1/8	1/16
3	1/16	1/8	1/16

11. The continuous random variable X has probability density function
$$f(x) = \begin{cases} kx & 0 \leqslant x \leqslant 2, \\ 0 & \text{elsewhere.} \end{cases}$$
Find the correlation coefficient between X and X^2.

12. Continuous independent random variables X_1 and X_2 each have mean μ_X and variance σ_X^2. Another set of continuous random variables, Y_1, Y_2 and Y_3, each have mean μ_Y and variance σ_Y^2; Y_1, Y_2 and Y_3 are all independent of one another but the correlation coefficient between X_i and Y_j ($i=1$ or 2, and $j=1, 2$ or 3) is ρ.
If $\overline{X} = \frac{1}{2}(X_1 + X_2)$, and $\overline{Y} = \frac{1}{3}(Y_1 + Y_2 + Y_3)$ find the correlation coefficient between $\overline{X}$ and $\overline{Y}$.

13. The heights of fathers and their eldest sons are tabulated below:

| Height of father, x (in) | 63 | 68 | 70 | 64 | 66 | 72 | 67 | 71 | 68 | 62 |
| Height of son, y (in) | 65 | 66 | 72 | 66 | 69 | 74 | 69 | 73 | 65 | 66 |

(a) Find the regression lines of y on x and of x on y.
(b) Plot a scatter diagram for the above data and draw and label the two regression lines on the same diagram.

14. On each of 30 items, two measurements are made, x and y. The following summations are given:

$$\sum x = 15, \quad \sum y = -6, \quad \sum xy = 56, \quad \sum x^2 = 61, \quad \sum y^2 = 90.$$

Calculate the product-moment correlation coefficient and obtain the regression lines of y on x and x on y, giving the constants to three significant figures.
If the variable x is replaced by X, where $X = (x-1)/2$, find the correlation coefficient between X and y and the regression lines of y on X and X on y, giving the constants to three significant figures. (JMB)

15. The following data refer to the weights y_i (pounds) and ages x_i (years, to one place of decimals) of 10 randomly chosen primary schoolchildren in a particular geographic area:

| x_i (years) | 7.8 | 6.2 | 7.0 | 6.4 | 6.0 | 8.0 | 5.6 | 8.0 | 5.2 | 7.3 |
| y_i (lb) | 61 | 48 | 52 | 56 | 50 | 60 | 44 | 57 | 38 | 57 |

Construct a scatter diagram of these data and determine the least-squares linear regression line for the regression of weight on age. Use the estimated regression line to predict the weight of a primary schoolchild whose age is 7.0 years from the same geographic area. Explain why it might be inadvisable to use the same relationship to predict the weight of a primary schoolchild of age 10.0 years from this area, illustrating your comments by reference to the scatter diagram. (JMB)

16. X is a discrete random variable with the pdf shown below:

x	0	1	2	3	4	5
$P(X=x)$	0.1	0.2	0.3	0.2	0.1	0.1

Obtain the mean and variance of the random variable $Y = X - 2$.

Find the covariance of X and Y and evaluate the probability that Y is less than 1 conditional on the value of X exceeding 1. *(IOS)*

17. The scores of twelve candidates in an examination in mathematics and physics are:

Candidate	A	B	C	D	E	F	G	H	I	J	K	L
Mathematics, x	68	77	52	78	90	42	33	61	25	46	63	37
Physics, y	70	62	60	71	82	55	48	51	35	62	32	20

Fit straight lines by the method of least squares (the regression lines) making: (i) x; (ii) y the independent variable. A drawing is *not* required.

Find the product-moment coefficient of correlation between the two sets of marks. *(AEB)*

18. Annual sales of electricity (in 1000 GWh) from 1965 to 1973 were:

Year	1965	1966	1967	1968	1969	1970	1971	1972	1973
Sales	151	158	164	176	189	198	203	210	225

Calculate the least-squares regression line of sales on year. How useful would this have been, early in 1974, for predicting sales in the following three years?

19. Random variables U, V, W are uncorrelated and have variances 5, 1, 3 respectively. Find values of a, b, c such that the three variables

$$X = U + V + W, \quad Y = U + aV, \quad Z = U + bV + cW$$

are uncorrelated, and find their variances.

20. A personnel manager is investigating a selection procedure for salesmen in which applicants for posts of salesmen are given an aptitude test and have an interview. The sales records of ten recruits during their first year are noted and compared with their rating in the aptitude test and their interview assessment.
The three rankings are:

Salesman	A	B	C	D	E	F	G	H	I	J
Aptitude test	3	7	6	8	2	4	1	10	5	9
Interview assessment	4	5	3	2	1	7	6	10	9	8
Sales	1	2	3	4	5	6	7	8	9	10

Calculate rank correlation coefficients between: (i) sales and aptitude test; and (ii) sales and interview assessment. Comment on the results. *(IOS)*

21. (a) A manufacturer of lead-covered components suspects that right-hand components are not receiving on average the same thickness of lead covering as left-hand ones. He tells you that two random samples of size eight

gave the following data:

	Mean thickness (cm $\times 10^{-3}$)							
Left-hand components	366	362	376	399	355	332	382	340
Right-hand components	338	346	379	363	312	321	360	301

What is your opinion?

(b) The manufacturer now tells you that only eight pieces of apparatus were dismantled for this test, and that the left-hand and right-hand components had come in pairs from these eight pieces of apparatus:

	Mean thickness (cm $\times 10^{-3}$)							
Piece of apparatus	A	B	C	D	E	F	G	H
Left-hand component	366	362	376	399	355	332	382	340
Right-hand component	338	346	379	363	312	321	360	301

How does this new piece of information alter your opinion?   (*IOS*)

22. Detonators are tested for safety in handling by dropping them on to a metal plate from various heights. Twenty detonators are dropped from each of six heights and the number exploding from each height is recorded in the following table:

Height of drop, x (cm)	25	28	31	34	37	40
Number exploding	3	5	10	15	18	19

The proportions exploding from different heights may be assumed to be given by normal areas corresponding to a mean μ and a standard deviation σ. Use the table of areas under the standard normal curve to find values of $(x - \mu)/\sigma$ for each value of x. Plot these values on a graph and from this graph obtain estimates of the values of μ and σ.

Estimate the proportion that will explode if dropped from a height of 30 cm.   (*O&C*)

23. A certain meteorological event may or may not happen in any given year, years behaving independently of one another in this respect. The probability of an occurrence in a given year is p (constant). Show that the waiting time W between consecutive years in which the event occurs has distribution

$$P(W = s) = (1 - p)^{s-1}p, \quad s = 1, 2, \ldots.$$

Find $E[W]$, Var $[W]$, and a general formula for $P(W \geq t)$.

24. A seaside landlord has four flats which he lets out for a total of 20 weeks each summer, each letting of each flat being for one complete week. Past experience has shown that in $\frac{3}{5}$ of the lettings the tenant desires garage

accommodation, so he builds three garages. Assuming that the flats are fully occupied throughout the season, estimate:

(i) the proportion of weeks in which there are more requests for garages than garages available;

(ii) the probabilities that $0, 1, 2$ or 3 garages are occupied in any week.

If the landlord charges £1 per week per garage, estimate his expected income from the garages during the 20-week letting period. Estimate how much extra income he can expect if he builds four garages instead of three.

(*JMB*)

25. (a) In a game of chance each player throws two unbiased dice and scores the difference between the larger and smaller numbers which arise. Two players compete and one or the other wins if, and only if, he scores at least 4 more than his opponent. Find the probability that neither player wins.

(b) On a test drop a parachutist is equally likely to land at any point on a line joining two points A and B. If Y is the ratio of his distance from A to his distance from B determine the probability density function of Y, and sketch this function. Find the probability that Y exceeds 9. (*JMB*)

26. A fair coin and two dice, one red and one green, are thrown simultaneously. If the coin falls heads the 'score' X is the sum of the scores on both dice, but if it falls tails the 'score' X is the score on the red die alone. Draw up a table showing the probability $P(X = r)$ that X takes the value r for $r = 1, 2, \ldots, 12$. Show that

$$P(5 \leqslant X \leqslant 7) = 27/72.$$

Calculate the mean of X. (*Cambridge*)

27. A sales representative has been assigned to a large town for one month to sell a new type of photocopying machine to business and industrial concerns in the town. Previous experience with such sales campaigns indicates that if the representative were to visit n concerns during the month then the probability that he would sell k machines is given by

$$p(k) = \frac{2}{n+1} \left(1 - \frac{k}{n} \right), \quad k = 0, 1, 2, \ldots, n.$$

Determine the expected number of sales in terms of n.

Suppose that the representative's expenses in making n visits amount to £$\frac{1}{2}n^2$ and that he receives a commission of £60 on each machine he sells. Show that the representative's expected net profit (commission less expenses) is greatest when $n = 20$. For this value of n calculate the probabilities that at the end of the month the representative will: (i) be out of pocket; (ii) have made a profit of at least £100. (*Welsh*)

28. Consider the quadratic equation in x given by

$$x^2 - 2x - y = 0.$$

(a) If the value of y is chosen at random from the interval $(-1, 1)$, calculate:
   (i) the probability that the larger root of the equation will exceed 1.5;
   (ii) the mean and the variance of the larger root.
(b) If, instead, the value of y is taken to be the largest of three randomly chosen values from $(-1, 1)$, calculate the probability that the larger root will exceed 2.
(Welsh)

29. Observations $x_1, x_2, \ldots, x_n$ are taken at times $t_1, t_2, \ldots, t_n$ and it is desired to fit a curve of the form

$$x = ab^t$$

for positive constants a and b. Describe how the theory of linear regression can be used to accomplish this, and explain the assumptions underlying your analysis.
(IOS)

30. Explain the meaning of the terms *null hypothesis, alternative hypothesis* and *level of significance* in tests of hypotheses.

The lifetime of an important electronic component of a certain measuring instrument has an exponential distribution with probability density function

$$f(x) = \frac{1}{\mu} e^{-x/\mu}, \quad x > 0,$$

where μ is the mean of the distribution. The manufacturer specifies a mean lifetime of 120 hr for this component, but in a test of a particular instrument chosen at random, the component failed after only 47.6 hr of use. Is this enough evidence to conclude that the mean lifetime of such components is less than that specified by the manufacturer?
(IOS)

31. (a) Use the definitions of expectation and variance to show that if X is a continuous random variable and a, b are constants, then $E[aX + b] = aE[X] + b$, and $Var[aX + b] = a^2 Var[X]$.
(b) A retailer buys six batches of perishable food at £100 each. In a week, the demand X from his customers for batches of the food has a distribution whose mean and variance are both 4, and each batch he sells is sold for £200. Any batches which are unsold at the end of the week are given away, so he makes a loss of £100 on each unsold batch.
   Let Y be his profit, i.e. $Y = $ income $- 600$ (in pounds).
   (i) Find $E[Y]$ and $Var[Y]$ under the assumption that the probability that X exceeds 6 is zero.
   (ii) Find $E[Y]$ under the assumption that X has a Poisson distribution with mean 4 and also that, if demand exceeds 6 batches, it is not possible to obtain additional batches, so that excess demand cannot be met. (IOS)

32. An indicator (GFR) of a medical condition is difficult to measure, but an alternative measurement (Cr) related to GFR can be made. You have been given the following data and asked to investigate whether there is any transformation that will allow a simple unweighted linear regression to be used to fit a model to the transformed variables. For the given data, plot GFR against 1/Cr; and also plot ln (GFR) against ln (Cr). Comment on the appearance

of each plot and indicate, with reasons, which transformation you would recommend.

Cr	GFR	1/Cr	ln(GFR)	ln(Cr)
0.71	167.0	1.41	5.12	−0.34
0.45	103.0	2.22	4.63	−0.80
1.20	95.0	0.83	4.55	0.18
1.10	80.0	0.91	4.38	0.10
0.99	133.0	1.01	4.89	−0.01
0.80	98.0	1.25	4.58	−0.22
4.35	19.0	0.23	2.94	1.47
0.98	92.0	1.02	4.52	−0.02
0.85	102.0	1.18	4.62	−0.16
1.31	88.0	0.76	4.48	0.27
1.11	68.0	0.90	4.22	0.10
1.77	29.0	0.56	3.37	0.57
8.21	6.2	0.12	1.82	2.11
4.50	7.7	0.22	2.04	1.50
2.47	27.0	0.40	3.30	0.90

(*RSS*)

24 MULTIPLE REGRESSION

OVERVIEW

A close study is made of the regression equation by which a response variate Y is predicted from two explanatory variates X_1 and X_2. This provides the understanding needed to interpret the output of computer analyses which may involve the regression of a response variate on many explanatory variates.

24.1 INTRODUCTION

We saw in Chapter 23 that, given an appropriate set of data, we may derive an equation to predict the expected height of a mature son from the height of his father. The method was based on the linear model

$$Y = \alpha + \beta x + e$$

where the random errors e are mutually independent, each with mean 0 and variance σ^2. These assumptions are adequate for estimation of the unknown parameters α, β and σ; but in order to make significance tests and to set up confidence intervals, we need to make the additional assumption that the random errors are normally distributed. Having found estimates a and b of the parameters in the equation then the mean height of sons whose fathers' height is x is predicted by the mean of the distribution of Y for given x:

$$\hat{Y} = E[Y \mid x] = a + bx. \tag{24.1}$$

We would expect to make a better prediction if we took into account the mother's height as well. If we denote father's height by X_1 and mother's height by X_2 we are now considering a linear model

$$Y = \alpha + \beta_1 x_1 + \beta_2 x_2 + e \tag{24.2}$$

with the same assumptions as before about the random errors.

We have been calling Y the *predicted* variate and the Xs the *predictor* variates. Other terminology we shall use is that Y is the *response* variate and the Xs are *explanatory* variates. In the literature the reader will sometimes find the Xs referred to as *regressor* variates; together they make up a *regression set*. Another usage, that we do not recommend since it may mislead, is to call the Xs independent variates and Y the dependent variate.

A further example we might imagine is the prediction of the time Y taken to run a mile by army recruits. Possible explanatory variates are lung capacity X_1, resting heart rate X_2, weight X_3 and height X_4; in fact, any variate that might be correlated with Y and is measurable. An appropriate model might be

$$Y = \alpha + \beta_1 x_1 + \beta_2 x_2 + \beta_3 x_3 + \beta_4 x_4 + e.$$

Explanatory variates are often intercorrelated; we would expect, for example, weight X_3 and height X_4 to be positively correlated. Because of this, a variate which is correlated with Y may, when added to the model, improve the prediction very little because the information it contains has already been derived from the other explanatory variates that have been used in the model.

A final example is where the yield Y of a crop might be related to the amount X of a fertilizer by a parabolic relation described by the linear statistical model

$$Y = \alpha + \beta_1 x + \beta_2 x^2 + e.$$

Each of the cases above is an example of a linear statistical model. In this context 'linear' refers to the fact that each model equation is linear in the unknown parameters (e.g. α, β_1 and β_2) and the random error e. Because of this, estimation of the unknown parameters is straightforward whereas direct estimation of the parameters in a nonlinear model, as for example:

$$Y = 1/(\alpha + \beta x) + e$$

is difficult (p. 627).

Each of the cases is also an example of multiple linear regression. 'Regression' describes relationships of the form

$$E[Y \mid x] = f(x),$$

or

$$E[Y \mid x_1, x_2, \ldots, x_p] = f(x_1, x_2, \ldots, x_p),$$

involving functions of one or many predictor variates; 'multiple' denotes that there is more than one predictor variate and, in this context, any simple function of X, e.g. X^2, is considered to be a different variate from X. If two random variables, X and Y, have a bivariate normal distribution then the mean of the distribution of Y, conditional on X taking a fixed value x, takes the form of the linear model (24.1) above. Similarly, if Y, X_1 and X_2 have a trivariate normal distribution then the distribution of Y, conditional on X_1 and X_2 taking fixed values x_1 and x_2, takes the form (24.2) above. In each of these cases, an underlying multivariate normal distribution leads directly to a regression relationship which is linear in the unknown parameters and in the explanatory variables. Often we do not know whether the underlying distribution is multivariate normal; none the less, multiple regression models similar to those described above may be worth trying. By fitting and testing the models we may judge how satisfactory they are.

24.2 ESTIMATION OF UNKNOWN PARAMETERS

We consider the model

$$Y = \alpha + \beta_1 x_1 + \beta_2 x_2 + e$$

where we wish to estimate α, β_1 and β_2, given a sample of n sets of values $(y_i, x_{1i}, x_{2i})(i = 1, 2, \ldots, n)$. The algebra of the derivation of the estimates is simpler if we code the variates, as we did when discussing simple linear regression. We set $v_i = y_i - \bar{y}, u_{1i} = x_{1i} - \bar{x}_1, u_{2i} = x_{2i} - \bar{x}_2$. We now write the model

$$V = \alpha' + \beta_1 u_1 + \beta_2 u_2 + e.$$

The constant term changes from α to α'. Again we use the principle of least squares. The quantity to be minimized, with respect to variation in α', β_1 and β_2, is

$$Q = \sum (v_i - \alpha' - \beta_1 u_{1i} - \beta_2 u_{2i})^2.$$

The most convenient method is to use partial differentiation. (Appendix VII gives a brief description for readers who are not familiar with the technique.) On differentiating, setting the derivatives to zero, and replacing the variables with their estimates a', b_1 and b_2, we obtain:

$$\frac{\partial Q}{\partial \alpha'} = -2 \sum (v_i - a' - b_1 u_{1i} - b_2 u_{2i}) = 0, \tag{24.3}$$

$$\frac{\partial Q}{\partial \beta_1} = -2 \sum u_{1i}(v_i - a' - b_1 u_{1i} - b_2 u_{2i}) = 0, \tag{24.4}$$

$$\frac{\partial Q}{\partial \beta_2} = -2 \sum u_{2i}(v_i - a' - b_1 u_{1i} - b_2 u_{2i}) = 0. \tag{24.5}$$

From Equation (24.3), $a' = 0$ since $\sum v_i = \sum u_{1i} = \sum u_{2i} = 0$. Equations (24.4) and (24.5) may be written

$$b_1 \sum u_{1i}^2 + b_2 \sum u_{1i} u_{2i} = \sum u_{1i} v_i,$$

$$b_1 \sum u_{1i} u_{2i} + b_2 \sum u_{2i}^2 = \sum u_{2i} v_i.$$

These equations, like any others derived by the use of least squares, are called the *normal equations* though there is no connection with the normal distribution. The numerical quantities in the equations are corrected sums of squares and products. Using the notations

$$S_{pq} = \sum u_{pi} u_{qi} = \sum (x_{pi} - \bar{x}_p)(x_{qi} - \bar{x}_q) \quad (p = 1, 2; q = 1, 2),$$

and

$$S_{0p} = \sum u_{pi}v_i = \sum (x_{pi} - \bar{x}_p)(y_i - \bar{y}) \quad (p = 1, 2),$$

we may write the normal equations:

$$b_1 S_{11} + b_2 S_{12} = S_{01},$$
$$b_1 S_{12} + b_2 S_{22} = S_{02}.$$

To solve the equations it is first worth calculating the quantity $D = S_{11}S_{22} - S_{12}^2$. Then

$$b_1 = (S_{22}S_{01} - S_{12}S_{02})/D \quad \text{and} \quad b_2 = (S_{11}S_{02} - S_{12}S_{01})/D.$$

The calculations may be illustrated by an example. An artificial example has been chosen to give simple numbers and a simple statistical structure.

Example 24.1 Calculation of regression equation

Dice were thrown to obtain 10 values of each of the following variates: X_1 = value on a twelve-sided die, X_2 = twice the value on a six-sided die, Z = value on a six-sided die, $Y = X_1 + X_2 + Z$. The values obtained were:

											Total
X_1	8	7	11	3	10	7	5	11	8	7	77
X_2	6	12	4	8	12	10	4	2	4	4	66
Y	18	23	21	17	25	18	13	19	14	14	182

Calculate the regression equation of Y on X_1 and X_2. Find the predicted value $\hat{y}$ for each of the 10 observed pairs of values of X_1 and X_2. Calculate the residuals $(\hat{y} - y)$.

The linear model used to generate the data was $Y = 3.5 + x_1 + x_2 + e$. Compare your estimated regression equation with this.

From the table,

$$S_{11} = \sum x_1^2 - \left(\sum x_1\right)^2 \bigg/ n = 651 - 592.9 = 58.1,$$

$$S_{22} = \sum x_2^2 - \left(\sum x_2\right)^2 \bigg/ n = 556 - 435.6 = 120.4,$$

$$S_{12} = \sum x_1 x_2 - \left(\sum x_1\right)\left(\sum x_2\right) \bigg/ n = 492 - 508.2 = -16.2,$$

$$S_{01} = \sum x_1 y - \left(\sum x_1\right)\left(\sum y\right) \bigg/ n = 1447 - 1401.4 = 45.6,$$

$$S_{02} = \sum x_2 y - \left(\sum x_2\right)\left(\sum y\right) \bigg/ n = 1286 - 1201.2 = 84.8$$

The normal equations are:

$$58.1b_1 - 16.2b_2 = 45.6$$
$$-16.2b_1 + 120.4b_2 = 84.8$$

$$D = 58.1 \times 120.4 - (16.2)^2 = 6732.8$$
$$b_1 = (120.4 \times 45.6 + 16.2 \times 84.8)/D = 6864.0/6732.8 = 1.019$$
$$b_2 = (58.1 \times 84.8 + 16.2 \times 45.6)/D = 5665.6/6732.8 = 0.841.$$

The regression equation is

$$(y - 18.2) = 1.019(x_1 - 7.7) + 0.841(x_2 - 6.6)$$

i.e.

$$y = 4.803 + 1.019x_1 + 0.841x_2.$$

Alternatively, a may be found directly as $y - b_1\bar{x}_1 - b_2\bar{x}_2$.

The observed values y, the values $\hat{y}$ predicted from the regression equation, and the residuals $(\hat{y} - y)$, are:

y	18	23	21	17	25	18	13	19	14	14
$\hat{y}$	18.0	22.0	19.4	14.6	25.1	20.3	13.3	17.7	16.3	15.3
$y - \hat{y}$	0	1.0	1.6	2.4	−0.1	−2.3	−0.3	1.3	−2.3	−1.3

The true coefficients of x_1 and x_2 in the underlying model are 1.0, whereas the estimates are 1.019 and 0.841; the estimate of the constant is 4.803 instead of the true value 3.5. For the relatively small sample of 10, the estimates are satisfactory. The residuals demonstrate a random scatter about the plane.

24.2.1 Exercises

1. Three variates were recorded on each of nine units included in a survey. From the data the following sums were calculated:

$$\sum x_1 = 9, \quad \sum x_2 = 18, \quad \sum y = 9,$$
$$\sum x_1^2 = 69, \quad \sum x_2^2 = 112, \quad \sum y^2 = 111,$$
$$\sum x_1 y = 78, \quad \sum x_2 y = 84, \quad \sum x_1 x_2 = 72.$$

Find the regression equation of Y on X_1 and X_2.

2. In a similar method to that used in Example 24.1, data were generated by throwing six-sided dice. $X_1 =$ four times value on die 1, $X_2 =$ square of value on die 1, $Z =$ three times value on die 2; $Y = X_1 + X_2 + Z$. Using the data below, find the regression equation of Y on X_1 and X_2.

										Total
x_1	4	12	16	16	8	20	8	16	8	24	132
x_2	1	9	16	16	4	25	4	16	4	36	131
y	8	36	47	50	18	60	15	44	24	75	377

24.3 INTERPRETATION BY ANALYSIS OF VARIANCE

We introduced an analysis of variance for simple linear regression in Chapter 23. It provided a method of estimating the variability σ^2 of the term e in the linear model and of testing the significance of the regression. An analysis of variance for multiple regression provides us with the same estimate and test and allows us to make further tests on the regression. It is a valuable tool for interpreting the regression.

Using the regression equation we may predict the ith value

$$\hat{y}_i = E[Y \mid x_{1i}, x_{2i}] = \bar{y} + b_1(x_{1i} - \bar{x}_1) + b_2(x_{2i} - \bar{x}_2).$$

Note that

$$\sum \hat{y}_i = \sum \bar{y} + b_1 \sum (x_{1i} - \bar{x}_1) + b_2 \sum (x_{2i} - \bar{x}_2) = n\bar{y}.$$

Thus the mean of the predicted values is the same as the mean of the observed values, and

$$\sum (\hat{y}_i - \bar{y}) = 0.$$

The discrepancy between the observed value y_i and the predicted value $\hat{y}_i$ is the residual $(y_i - \hat{y}_i)$. If we sum the residuals,

$$\sum (y_i - \hat{y}_i) = \sum y_i - \sum \hat{y}_i = n\bar{y} - n\bar{y} = 0,$$

i.e. the mean of the residuals is zero.

The analysis of variance is based on the identity, proved in the next section, that

$$\sum (y_i - \bar{y})^2 = \sum (\hat{y}_i - \bar{y})^2 + \sum (y_i - \hat{y}_i)^2.$$

To express this in words, note that each sum of squares is a corrected sum of squares, i.e. it is a sum of squares of deviations of values from their mean. Remembering that a corrected sum of squares is a measure of variability, we can say: the total variability in y is equal to the variability $\sum (\hat{y}_i - \bar{y})^2$ accounted for, or explained, by the regression plus a residual unexplained variability $\sum (y_i - \hat{y}_i)^2$. The regression sum of squares is usually calculated most conveniently by using the identity (also proved in the next section):

$$\sum (\hat{y}_i - \bar{y})^2 = b_1 \sum (x_{1i} - \bar{x}_1)(y_i - \bar{y}) + b_2 \sum (x_{2i} - \bar{x}_2)(y_i - \bar{y})$$
$$= b_1 S_{01} + b_2 S_{02}.$$

The analysis of variance table is:

Source of variation	Sum of squares	Degrees of freedom	Mean square
Regression	$S_{MR} = b_1 S_{01} + b_2 S_{02}$	2	$S_{MR}/2$
Residual	$S_R = S - S_{MR}$	$n - 3$	$S_R/(n - 3)$
Total	$S = \sum (y_i - \bar{y})^2$	$n - 1$	

The total sum of squares and the regression sum of squares are calculated directly. The residual sum of squares is obtained by taking the difference. The residual mean square $S_R/(n - 3)$ is an unbiased estimator of σ^2, usually denoted by s^2.

The degrees of freedom of the regression sum of squares are two since two parameters, β_1 and β_2, are fitted. By difference, the residual sum of squares has $n - 3$ degrees of freedom. Provided that Y has a normal distribution, or approximately so, we may make an F-test of whether the overall regression is significant. On the null hypothesis that $\beta_1 = \beta_2 = 0$, the ratio of the regression mean square to the residual mean square has an F-distribution with 2 and $n - 3$ degrees of freedom. On the same null hypothesis, the regression mean square has expected value σ^2, whereas if the null hypothesis is *not* true the expected value is greater than σ^2. Therefore a one-tail test is used. As we saw with simple linear regression (p. 622), the ratio of the regression sum of squares to the total sum of squares, S_{MR}/S, is a useful measure of the effectiveness of a regression. The ratio equals R^2, the square of the multiple correlation coefficient. R is a generalization of the simple correlation coefficient r; for our purposes, we may take it as being defined by the last sentence. R^2 is also known as the *coefficient of determination*.

It is informative to partition the regression sum of squares. In the example of regression of son's height Y on father's height X_1 and mother's height X_2 we might find the sum of squares corresponding to the regression of Y on father's height alone and then find the extra sum of squares when mother's height is also included in the regression.

Source of variation	Sum of squares	Degrees of freedom	Mean square
Regression on X_1	S_{01}^2/S_{11}	1	
Extra when X_2 included	$S_{MR} - S_{01}^2/S_{11}$	1	$S_{MR} - S_{01}^2/S_{11}$
Regression on X_1 and X_2	S_{MR}	2	

The significance of the additional component may be assessed by testing the ratio of its mean square to the residual mean square as F with 1 and $n - 3$ degrees of freedom. For example, we may test whether including mother's height, as well as father's height, significantly improves the regression. It is usually worth repeating the partition but carrying out first the regression on X_2 alone; we may find one of the variates explains substantially more of the variation in Y than does the other when considered alone.

Example 24.1 (continued) Analysis of variance for regression

Carry out an analysis of variance for the regression fitted in Example 24.1. Test if both X_1 and X_2 contribute to the regression.

From the original table of data (p. 652), total sum of squares is

$$S = \sum y_i^2 - \left(\sum y_i\right)^2 \Big/ n = 3454 - 3312.4 = 141.6.$$

The regression sum of squares is

$$S_{MR} = b_1 S_{01} + b_2 S_{02} = 1.019 \times 45.6 + 0.841 \times 84.8 = 117.85.$$

The residual sum of squares is

$$S_R = S - S_{MR} = 141.60 - 117.85 = 23.75.$$

We put these sums of squares in the analysis of variance table.

Source of variability	Sum of squares	Degrees of freedom	Mean square	F
Regression	117.85	2	58.925	17.4
Residual	23.75	7	3.393	
Total	141.60	9		

The F-test of the significance of the overall regression is based on the assumption that the distribution of Y is normal. This is obviously not the case here from the way in which Y was derived. None the less, the relevant distributions that underlie the test are those of sums of Y-variates and these are approximately normal.

The F-statistic equals $58.925/3.393 = 17.4$ and is associated with 2 and 7 degrees of freedom. From tables, the 1% point is 9.55. The calculated statistic is much bigger than this so we reject the null hypothesis that $\beta_1 = \beta_2 = 0$, and conclude that there is a significant regression.

We proceed now to partition the regression sum of squares. For a regression on X_1 alone, the regression sum of squares is

$$b \sum (x_1 - \bar{x}_1)(y - \bar{y}) = \left[\sum (x_1 - \bar{x}_1)(y - \bar{y})\right]^2 \Big/ \sum (x_1 - \bar{x}_1)^2$$
$$= (45.6)^2/58.1 = 35.79$$

(see p. 622). Similarly, the sum of squares for the regression on X_2 alone is $(84.8)^2/120.4 = 59.73$. Putting these results in the analysis of variance table, we obtain:

Source of variability	Sum of squares	Degrees of freedom	Mean square	F
Regression on X_1 alone	35.79	1		
Extra when X_2 included	82.06	1	82.06	24.2
Regression on X_2 alone	59.73	1		
Extra when X_1 included	58.12	1	58.12	17.1
Regression on X_1 and X_2	117.85	2		
Residual	23.75	7	3.393	
Total	141.60	9		

The 1% value of F with 1 and 7 degrees of freedom is 12.25. Comparing the calculated F-values in the analysis of variance table with this, we conclude that adding the extra variate, whether it is X_1 or X_2, significantly improves the regression. This agrees with the way the data were constructed. Y was made to be correlated with both X_1 and X_2; moreover, X_1 and X_2 were independent and hence contribute separately to the prediction of Y.

Example 24.2 Analysis of variance for regression (Exercise 24.2.1(2) continued)

After fitting the regression of Y on X_1 and X_2 for the data of Exercise 24.2.1(2) the following analysis of variance may be constructed. Interpret the results.

Source of variability	Sum of squares	Degrees of freedom	Mean square	F
Regression on X_1 alone	4069.68	1		
Extra when X_2 included	0.55	1	0.55	0.05
Regression on X_2 alone	3912.57	1		
Extra when X_1 included	157.66	1	157.66	15.35
Regression on X_1 and X_2	4070.23	2	2035.12	
Residual	71.87	7	10.27	
Total	4142.10	9		

This example presents an interesting contrast with Example 24.1 (continued). In that case the explanatory variates were independent, whereas in this case they are highly correlated (the value of r, calculated from the data, is 0.98). The regression on X_1 and X_2, jointly, is highly significant, explaining $4070.23/4142.10 = 98.3\%$ of the variability in Y; regressions on X_1 and X_2 individually are also significant explaining, respectively, 98.2% and 94.5% of the variability in Y. If to the regression on X_1 alone, X_2 is added, we find there is no significant improvement; the F-statistic is less than one. All the information in the X_2-values has been communicated through X_1. Perhaps rather surprisingly, if X_2 is introduced into the regression set first, then adding X_1 does significantly improve the regression.

The general conclusion is that if we wish to predict Y we can do it most satisfactorily from X_1 alone.

24.3.1 Exercises

1. The plasma lipid level of total cholesterol was recorded for a sample of eight patients before they received drug therapy. The investigator wanted to see how these levels depended upon the weight and the age of the patient. From analyses of the data the following sums of squares were obtained:

Regression on weight alone	0.1486
Regression on age alone	4.0439
Regression on weight and age	4.1139
Total	4.9150

Compile an analysis of variance table and test whether lipid levels depended on weight and/or age, or neither.

2. Data on X_1, X_2 and Y, collected on 11 individuals, yielded the following sums of squares and products:

$$\sum (x_1 - \bar{x}_1)^2 = 55, \qquad \sum (x_2 - \bar{x}_2)^2 = 59, \qquad \sum (y - \bar{y})^2 = 233,$$

$$\sum (x_1 - \bar{x}_1)(y - \bar{y}) = 51, \quad \sum (x_2 - \bar{x}_2)(y - \bar{y}) = 94, \quad \sum (x_1 - \bar{x}_1)(x_2 - \bar{x}_2) = -7.$$

(i) Find the estimates b_1, b_2 of the coefficients in the regression of Y on X_1 and X_2.
(ii) Test whether both X_1 and X_2 should be included in the regression.
(iii) Estimate σ^2.

24.4 MATHEMATICAL DERIVATION OF ANALYSIS OF VARIANCE IDENTITY

Recall that the predicted value is

$$\hat{y} = \bar{y} + b_1(x_1 - \bar{x}_1) + b_2(x_2 - \bar{x}_2) \tag{24.6}$$

and the residual

$$y - \hat{y} = (y - \bar{y}) - b_1(x_1 - \bar{x}_1) - b_2(x_2 - \bar{x}_2). \tag{24.7}$$

24.4.1 The analysis of variance identity

To prove $\sum (y - \bar{y})^2 = \sum (\hat{y} - \bar{y})^2 + \sum (y - \hat{y})^2$.

The normal equations are:

$$b_1 \sum (x_1 - \bar{x}_1)^2 + b_2 \sum (x_1 - \bar{x}_1)(x_2 - \bar{x}_2) = \sum (x_1 - \bar{x}_1)(y - \bar{y}), \quad (24.8)$$

$$b_1 \sum (x_1 - \bar{x}_1)(x_2 - \bar{x}_2) + b_2 \sum (x_2 - \bar{x}_2)^2 = \sum (x_2 - \bar{x}_2)(y - \bar{y}). \quad (24.9)$$

The first normal equation (24.8) may be written

$$\sum (x_1 - \bar{x}_1)[(y - \bar{y}) - b_1(x_1 - \bar{x}_1) - b_2(x_2 - \bar{x}_2)] = 0,$$

i.e.

$$\sum (x_1 - \bar{x}_1)(y - \hat{y}) = 0. \quad (24.10)$$

This shows that the correlation coefficient of X_1 and the residuals, calculated from the sample data, will be zero. Similarly, for X_2 we have

$$\sum (x_2 - \bar{x}_2)(y - \hat{y}) = 0 \quad (24.11)$$

Multiply Equation (24.10) by b_1 and (24.11) by b_2 and add, to give

$$\sum [b_1(x_1 - \bar{x}_1) + b_2(x_2 - \bar{x}_2)](y - \hat{y}) = 0.$$

Using Equation (24.6),

$$\sum (\hat{y} - \bar{y})(y - \hat{y}) = 0. \quad (24.12)$$

This shows that the predicted values are uncorrelated with the corresponding residuals. Finally,

$$\sum (y - \bar{y})^2 = \sum [\hat{y} - \bar{y} + (y - \hat{y})]^2$$
$$= \sum (\hat{y} - \bar{y})^2 + \sum (y - \hat{y})^2,$$

since the cross-product term $2 \sum (\hat{y} - \bar{y})(y - \hat{y}) = 0$ from Equation (24.12). This proves the result.

24.4.2 The regression sum of squares identity

To prove $\sum (\hat{y} - \bar{y})^2 = b_1 \sum (x_1 - \bar{x}_1)(y - \bar{y}) + b_2 \sum (x_2 - \bar{x}_2)(y - \bar{y})$.

Using the normal equations, multiply Equation (24.8) by b_1 and Equation (24.9) by b_2 and add,

$$b_1^2 \sum (x_1 - \bar{x}_1)^2 + 2b_1 b_2 \sum (x_1 - \bar{x}_1)(x_2 - \bar{x}_2) + b_2^2 \sum (x_2 - \bar{x}_2)^2$$
$$= b_1 \sum (x_1 - \bar{x}_1)(y - \bar{y}) + b_2 \sum (x_2 - \bar{x}_2)(y - \bar{y}). \quad (24.13)$$

The left-hand side (LHS) of Equation (24.13) equals

$$\sum [b_1(x_1 - \bar{x}_1) + b_2(x_2 - \bar{x}_2)]^2$$

$$= \sum (\hat{y} - \bar{y})^2, \qquad \text{using Equation (24.6)}$$

$$= \text{LHS of the identity to be proved.}$$

The RHS of Equation (24.13) is the RHS of the required identity, and completes the proof.

24.5 CONFIDENCE INTERVALS AND t-TESTS FOR THE REGRESSION COEFFICIENTS

To find confidence intervals and carry out t-tests we need to know the variances of the regression coefficients. We have seen in simple linear regression (Chapter 23) that the regression coefficient β is estimated by b and that $\text{Var}[b] = \sigma^2/\sum(x - \bar{x})^2$. The variance σ^2 is estimated by the residual mean square in the analysis of variance. When there are two explanatory variates the variances are determined in a similar way. Each of the estimates, b_1 and b_2, is expressed in the form $\sum w_i Y_i$ and then its variance equals $\sigma^2 \sum w_i^2$. We shall not work through the algebra but simply give the results.

The variances involve the denominator D that appeared when solving the normal equations:

$$D = S_{11}S_{22} - S_{12}^2 = \sum (x_1 - \bar{x}_1)^2 \sum (x_2 - \bar{x}_2)^2 - \left[\sum (x_1 - \bar{x}_1)(x_2 - \bar{x}_2) \right]^2.$$

The variances are:

$$\text{Var}[b_1] = \sigma^2 \sum (x_2 - \bar{x}_2)^2/D, \qquad \text{Var}[b_2] = \sigma^2 \sum (x_1 - \bar{x}_1)^2/D.$$

The estimates, b_1 and b_2, may be correlated so we need to note the covariance

$$\text{Cov}[b_1, b_2] = \sigma^2 \sum (x_1 - \bar{x}_1)(x_2 - \bar{x}_2)/D.$$

Example 24.3 t-tests and confidence intervals for regression coefficients

Estimate the variances and covariance of the estimated regression coefficients, b_1 and b_2, calculated in the regression illustrated in Examples 24.1 and 24.1 (continued). Test if both X_1 and X_2 contribute to the regression. Find 95% confidence intervals for β_1 and β_2.

From previous calculations,

$$S_{11} = 58.1, \quad S_{22} = 120.4, \quad S_{12} = -16.2, \quad D = 6732.8.$$
$$b_1 = 1.019, \quad b_2 = 0.841.$$

The variance σ^2 is estimated by the residual mean square, $s^2 = 3.393$. The estimates of the variances and covariances are:

$$\text{Var}[b_1] = s^2 S_{22}/D = 3.393 \times 120.4/6732.8 = 0.060\,676,$$

$$\text{Var}[b_2] = s^2 S_{11}/D = 3.393 \times 58.1/6732.8 = 0.029\,280,$$

$$\text{Cov}[b_1, b_2] = s^2 S_{12}/D = 3.393 \times (-16.2)/6732.8 = -0.008\,164.$$

The values are often set out in a matrix, called a *covariance matrix*:

$$\begin{bmatrix} 0.060\,676 & -0.008\,164 \\ -0.008\,164 & 0.029\,280 \end{bmatrix}.$$

We may make a t-test using the null hypothesis that $\beta_1 = 0$. The t-statistic is

$$b_1/\sqrt{(\text{Var}[b_1])} = 1.019/\sqrt{0.060\,676} = 1.019/0.246 = 4.14,$$

which, on the null hypothesis, follows a t-distribution with 7 degrees of freedom (the same number of degrees of freedom as the residual mean square from which s^2 was obtained). The 1% value of t_7 is 3.499; hence we reject the null hypothesis and conclude that β_1 differs from zero. The test is completely equivalent to the F-test we made in Example 24.1 (continued); note that $t^2 = (4.14)^2 = 17.1 = F$. We have shown that if to a regression on X_2 alone the further explanatory variate X_1 is added then there is a significant improvement in the regression. It is *not* a simple test of whether there is a significant regression on X_1 ignoring X_2.

In a similar way, to test the null hypothesis $\beta_2 = 0$, we calculate the t-statistic

$$b_2/\sqrt{\text{Var}[b_2]} = 0.841/\sqrt{0.029\,280} = 4.91.$$

We conclude that the addition of X_2 improves the regression on X_1.

A confidence interval for β_i $(i = 1, 2)$ is calculated in the standard way by $b_i \pm t\sqrt{\text{Var}[b_i]}$ where t is the value of the t-distribution for the chosen probability level and with the same number of degrees of freedom as the residual mean square from which s^2 is obtained. Hence a 95% confidence interval for β_1 is given by

$$1.019 \pm 2.365 \times 0.246 = 1.019 \pm 0.582 = (0.437,\ 1.601).$$

Similarly, a 95% confidence interval for β_2 is given by

$$0.841 \pm 2.365 \times 0.171 = 0.841 \pm 0.404 = (0.437, 1.245).$$

24.6 REGRESSION WITH MORE THAN TWO EXPLANATORY VARIATES

As the number of explanatory variates increases, both the algebra and arithmetic of multiple regression become much more complex. The algebra is simplified by the use of matrix algebra, but we do not pursue this. The arithmetic is easily handled by computer and we discuss computer analyses in the next section.

The basis for fitting any regression is the set of normal equations derived by the least-squares principle. For a regression with p explanatory variates there will be p equations for the p estimates $b_1, b_2, \ldots, b_p$ of the unknown regression coefficients $\beta_1, \beta_2, \ldots, \beta_p$. The coefficient of b_j ($j = 1, 2, \ldots, p$) in the ith equation is $S_{ij} = \sum(x_i - \bar{x}_i)(x_j - \bar{x}_j)$. The term on the RHS of the ith equation is $S_{0i} = \sum(y_i - \bar{y})(x_i - \bar{x}_i)$. Solving such a set of simultaneous equations is a forbidding task by hand, when p is much above two, but an easy task for a computer.

24.7 REGRESSION BY COMPUTER

We give an example of the output that may be obtained when regression is carried out by computer; our illustration is based on MINITAB, but other statistical packages give similar results.

The data are the prices quoted for double-glazed windows of particular widths and heights. The aim of the analysis is to predict the price for a given width and height.

Width (cm)	66	183	124	239	76	109	196	251	165	81	249	142	170	254
Height (cm)	122	122	122	122	30	58	61	76	86	81	117	61	30	30
Price (£)	66	78	75	90	45	57	73	83	71	59	95	61	53	64

After the data have been keyed into the computer, a simple REGRESS command gives the following output:

PRICE $= 24.8 + 0.138$ WIDTH $+ 0.272$ HEIGHT

Predictor	Coef	St. dev.	t-ratio	p
Constant	24.823	2.909	8.53	0.000
WIDTH	0.13800	0.01288	10.72	0.000
HEIGHT	0.27226	0.02392	11.38	0.000

$s = 3.132$ R-sq $= 95.9\%$

Analysis of Variance

SOURCE	DF	SS	MS	F	p
Regression	2	2515.0	1257.5	128.20	0.000
Error	11	107.9	9.8		
Total	13	2622.9			

SOURCE	DF	SEQSS
WIDTH	1	1244.0
HEIGHT	1	1271.0

The overall regression is highly significant, as may be seen from the analysis of variance table. The regression accounts for 95.9% of the variability in price, as we see from the value of R^2. The p-value for the F-test exceeds the 0.1% level (as do the t-tests above) so a value of zero is given. The t-ratio comes from dividing the coefficient value by its standard deviation; thus for the width coefficient the ratio is $0.13800/0.01288 = 10.72$. The t-tests show that the addition in the regression of either width or height, to the other explanatory variate, significantly improves the regression. The effect on the sums of squares of adding height to the regression on width may be seen in the sub-table which includes SEQSS, sequential sums of squares. The first sum of squares, 1244.0, is the sum of squares for the regression of price on width alone; the second is the extra sum of squares when height is also included in the regression. To derive similar sums of squares when height is introduced first requires another regression command of (i) the multiple regression with width and height in a different order, or (ii) a simple regression on height.

24.8 CHECKING THE FIT OF THE REGRESSION

Examination of the residuals is the major method by which the model is checked, and it is best done graphically. We give a list of the most useful plots.

1. *Dot plot or stem plot of all residuals.* Do the residuals look as if they might be a sample from a normal distribution? Are there any outliers?

2. *Plot of residuals $(y - \hat{y})$ against fitted values $\hat{y}$.* (Fitted values are better than observed values because they are not correlated with the residuals.) The most useful single plot. For the assumptions to be satisfied the plotted points should show random scatter over the plane and exhibit no pattern. A common failing is for the variance to change with the size of the fitted value; a transformation often rectifies the fault.

3. *Plot of residuals against X variates.* If there is a pattern it reveals that the appropriate functional form is not being used for the particular X. For example, an additional variate X^2 may need to be added to the model.

4. *Plot of residuals against time, or space or some other nuisance factor, if relevant.* Any association indicates that the model should be extended to include the associated factor.

5. *Plot of ith residual against $(i+1)$th residual.* Association indicates serial correlation in the data, i.e. successive observations are not independent.

6. *Plot residuals on probability paper or as normal scores.* A test of normality.

24.9 GENERAL STRATEGIES

Multiple regression is the most misused statistical procedure on computers. The major misapprehension by naïve users is to assume it does what they would like it to do rather than work out what it actually does do. It requires a great deal of care and active thought in its use. The computer does not provide ready-made answers.

The usual aims when using multiple regression are *either* prediction (to find an equation that predicts well with as few variates as possible) *or* explanation (to relate the variability in a response variate to a number of explanatory variates). When considering general strategies it is useful to consider separately the cases when (i) the number of explanatory variates is fixed, (ii) a 'best' subset of explanatory variates is to be chosen from those available.

24.9.1 Number of explanatory variates fixed

We list possible stages in the analysis.

1. *Initial data examination.* The aim is to gain understanding of the distribution of the variates involved and of their interrelation; it is also an opportunity to check on data quality, as discussed in Chapter 5. Look at variates individually using dot plots and stem plots. Look at plots of Y against each explanatory variate. Find the correlation matrix, i.e. the correlation coefficients of all pairs of the variates involved. Perhaps plot some of the explanatory variates against each other. *Think* about the nature of the variates and whether a linear relation makes sense. Perhaps transform some of the variates.

2. *Fit the equation.* Obtain estimates of the regression coefficients and of their standard errors. Obtain an analysis of variance and test the overall regression. Look at R^2.

3. *Inspect residuals.* As described in Section 24.8.

4. *Test individual variates.* Although we have been assuming the number of explanatory variates is fixed, there is no point in keeping a variate if it is making no contribution. When there are many explanatory variates then individual variates are best tested by t-tests. Remember that a t-test on β_k, say, is testing whether adding X_k to the set of explanatory variates significantly improves the regression over the fit to all the other variates except X_k. If X_k is found not to be significant and is removed from the regression set, the complete equation has to be recalculated with the smaller set of variates before considering rejecting other variates.

24.9.2 Selection of 'best' subset of explanatory variates

The general criteria are (i) few explanatory variates, (ii) small residual mean square s^2, (iii) large value of R^2 (since R^2 will automatically increase as the number of variates increases, some packages take account of this and use an adjusted R^2).

If there are p explanatory variates then one method would be to look at all possible 2^p regressions as each variate is or is not included in the regression set. The combination of explanatory variables which seems best to meet the criteria above may be called the *best subset* of those variables. It was not often used in the past because it required considerable computer memory and execution time. Instead *stepwise* methods were used, in which variates are either added or removed one at a time; there might be a build-up from few or no variates, or a process of elimination from a complete set. Many statistical packages offer automatic ways of doing this but the analyst should control the process at each stage. Assuming the set is being built up, the question, at least in the early stages, is which variate to introduce next. The possible variates should be plotted against the current residuals; any association suggests the variate in question would be a candidate for introduction. Another indicator is the F-value associated with the extra sum of squares if a particular variate is introduced. As the number in the regression set grows one should also consider rejecting some variates. However, modern packages do offer a *best-subsets* method, and this should be used for preference.

Finally, it should be realized that multiple regression works best when the explanatory variates are uncorrelated. Therefore, at the beginning of an analysis it is worth using the correlation matrix in order to divide the explanatory variates into groups such that variates in a group are highly correlated but variates in different groups are poorly correlated. Then choosing one variate from each group would give a good set of variates to start with.

24.10 MINITAB: MULTIPLE REGRESSION

The example in Section 24.7 gives a set of data, and the output obtained by using the MINITAB multiple regression program. The data were entered into a worksheet with columns headed WIDTH, HEIGHT and PRICE. Choice of

Stat>Regression>Regression/
Enter **Response** PRICE and **Predictors** WIDTH and HEIGHT/

gives the output shown. Always check that a constant term has been fitted. This is the default option; the alternative exists though it is rarely appropriate in practice. It can also be checked by choosing **Options** in the dialogue box and confirming that **Fit Intercept** has been selected. The interpretation of the output is discussed in Section 24.7.

A check should be made that the model is appropriate by inspecting the residuals, as discussed in Section 24.8. The **Graphs...** option should be chosen in the first dialogue box. In the Graphs dialogue box, ensure that **Regular** is selected under **Residuals for Plots**, and select various graphs. **Histogram of residuals**, **Normal plot of residuals** and **Residuals versus fits** might be chosen.

MINITAB aids the pursuit of a 'best' subset of explanatory variables as discussed in Section 24.9.2. The menu choices available are

Stat>Regression>Best subsets/

and

Stat>Regression>Stepwise/.

1. The variance of $\hat{Y}$, the *predicted value* of the mean of Y for given X_1 and X_2, is

$$\frac{\sigma^2}{n} + \sum (x_1 - \bar{x}_1)^2 \operatorname{Var}[b_1] + 2 \sum (x_1 - \bar{x}_1)(x_2 - \bar{x}_2) \operatorname{Cov}[b_1, b_2]$$
$$+ \sum (x_2 - \bar{x}_2)^2 \operatorname{Var}[b_2].$$

For the data of Examples 24.1, 24.1 (continued) and 24.3, find a 95% confidence interval for $\hat{Y}$ when (a) $x_1 = 7.7, x_2 = 6.6$; (b) $x_1 = 12, x_2 = 12$.

2. Calculate least-squares estimates of the parameters of the equation

$$Y_t = \alpha + \beta X_{1t} + \gamma X_{2t} + U_t$$

from the data on 33 observations summarized below:

$$\overline{Y} = 19.5, \quad \overline{X}_1 = 6, \quad \overline{X}_2 = 10.$$

The corrected sums of squares and cross-products are

	Y	X_1	X_2
Y	3	7	4
X_1	7	30	40
X_2	4	40	80

Then

(a) Calculate the coefficient of determination.
(b) Test the hypothesis that $\beta = 0$, stating any assumptions made.
(c) How would you test the hypothesis that $\beta + \gamma = 2$? (*IOS*)

3. A company enrolled 12 sales trainees and on their first day gave them an aptitude test (X_1, marks out of 10) and an interview (X_2, score from 0 to 10). After one month the value of the sales made by each trainee was recorded; for analysis, a variate 'performance' $Y = 10 \ln$ (sales in £s) was used. The data were:

X_1	8	6	10	8	8	10	8	4	9	9	7	9
X_2	3	8	5	9	7	6	0	5	5	2	5	5
Y	26	29	31	36	30	41	22	22	35	36	17	35

Carry out the following analyses in order to investigate if the aptitude test and interview are useful for predicting performance.

(i) Find the regression equation of performance Y on aptitude test mark X_1 and interview score X_2.
(ii) Inspect the residuals by making (a) a dot plot or stem plot of them, (b) a plot of residuals against fitted values.

(iii) Construct an analysis of variance and make F-tests of (a) the overall regression, (b) the addition of each explanatory variate to the regression on the other.

(iv) Calculate the variances of the estimates of the regression coefficients, and make t-tests of whether they differ from zero.

(v) Summarize your findings in the form of a brief report that you, as a statistician, might send to the chief executive.

4. (a) What is the purpose of multiple linear regression?

(b) The number of convictions for drunkenness (from annual figures per 1000 population) Y, the occupancy rate (average number of people per room) X_1, and the unemployment rate (per 100 in the workforce) X_2, were determined for a particular month for 20 local government areas chosen at random throughout the country. Obtain the least-squares regression equation of Y on X_1 and X_2, given the following data:

$$\sum Y = 700, \quad \sum X_1 = 12.6, \quad \sum X_2 = 45,$$

$$\sum (Y - \overline{Y})^2 = 2150, \quad \sum (X_1 - \overline{X}_1)^2 = 0.25, \quad \sum (X_2 - \overline{X}_2)^2 = 37,$$

$$\sum (Y - \overline{Y})(X_1 - \overline{X}_1) = 15.0, \quad \sum (Y - \overline{Y})(X_2 - \overline{X}_2) = 170,$$

$$\sum (X - \overline{X}_1)(X_2 - \overline{X}_2) = 0.5. \qquad (IOS)$$

5. Construct an analysis of variance for the regression in Question 4. Test whether both X_1 and X_2 contribute significantly to the regression. State any assumptions you make in order to carry out the tests.

6. The equation of the regression of Y on X_1 and X_2 is often written

$$y = \alpha + \beta_1 x_1 + \beta_2 x_2.$$

(a) Explain exactly what all the quantities (i.e. $y, \alpha, \beta_1, \ldots$) in the equation represent.

(b) What assumptions are necessary in order that the principle of least squares may be used in order to obtain the equation?

(c) What additional assumptions are necessary in order to test and derive confidence intervals for the regression coefficients?

(d) How would you check whether the assumptions are being satisfied?

7. R. A. Fisher in *Statistical Methods for Research Workers* (11th edition, Oliver and Boyd, 1950) investigated, for 57 rainfall stations in Hertfordshire, the dependence of rainfall on position and altitude. In January 1922 the mean rainfall at these stations was 3.87 inches; their mean longitude was 12.4' W, mean latitude 51°48.5' N and mean altitude 302 feet. In the analysis the variates used, with their units, were: X_1, two minutes of longitude; X_2, one minute of latitude; X_3, 20 feet of altitude; Y, 0.1 inch. The corrected sums of squares

and products are tabulated:

	x_1	x_2	x_3	y
x_1	1934.1	−772.2	924.1	1137.4
x_2		2889.5	119.6	−592.9
x_3			1750.8	891.8
y				1786.6

Solution of the normal equations gave the following estimates of the regression coefficients:

$$b_1 = 0.39624, \quad b_2 = -0.11204, \quad b_3 = 0.30788.$$

(i) Test if the overall regression is significant.
(ii) Calculate the six correlation coefficients for all possible pairs of X_1, X_2, X_3 and Y.
(iii) Fit the regression of Y on each of the possible pairs of X-variates, and find the regression sum of squares in each case.
(iv) Hence test whether all three X-variates contribute significantly to the regression.

8. How (if at all) could simple or multiple linear regression theory be applied to estimating the parameters α, β, γ in the following equations?
   (i) $y = \alpha x_1^{\beta} x_2^{\gamma}$,
   (ii) $y = \alpha e^{(\beta x_1 + \gamma x_2)}$,
   (iii) $y = \alpha + \beta x_1 + x_2^{\gamma}$,
   (iv) $y = \alpha + \beta e^{\gamma x}$,
   (v) $\frac{x}{y} = \alpha + \beta x + \gamma x^2$.

24.12 COMPUTING EXERCISES

1. The data below are selected from a much larger body of data referring to candidates for the General Certificate of Education (GCE) who were being considered for a special award. Here, y denotes the candidate's total mark, out of 1000, in the GCE examination. Of this mark the subjects selected by the candidate account for a maximum of 800; the remainder, with a maximum of 200, is the mark in the compulsory papers – 'General' and 'Use of English' – this mark is shown as x_1, while x_2 denotes the candidate's mark, out of 100, in a compulsory English Language paper taken on a previous occasion.

   Compute the multiple regression of y on x_1 and x_2, and make the necessary tests to enable you to comment intelligently on the extent to which current performance in the compulsory papers may be used to predict aggregate performance in the GCE examination, and on whether previous performance in English Language has any predictive value independently of what has already emerged from the current performance in the compulsory papers.

Candidate	y	x_1	x_2	Candidate	y	x_1	x_2
1	476	111	68	9	645	117	59
2	457	92	46	10	556	94	97
3	540	90	50	11	634	130	57
4	551	107	59	12	637	118	51
5	575	98	50	13	390	91	44
6	698	150	66	14	562	118	61
7	545	118	54	15	560	109	66
8	574	110	51				

(*Cambridge Diploma in Statistics, 1953*)

2. In 1979, the Mathematics Department of Extown University sent a questionnaire to its former students graduating in the years 1970–1978. A high proportion replied and the data from the returns were tabulated:

Ref. number	Sex	Year of graduation	Age at 1 July 1979	Class of degree	Type of work	Salary group	Number of jobs*
1	M	1970	31	1	D	4	2
2	M	1970	31	2b	C	3	2
3	M	1970	32	2b	A	5	1
4	M	1970	33	3	A	3	2
5	M	1970	32	3	A	4	4
6	M	1970	32	Ord.	A	4	1
7	M	1971	30	2a	D	3	3
8	M	1972	29	2a	D	5	2
9	F	1972	30	2a	A	1	3
10	M	1972	29	Ord.	C	3	1
11	M	1972	33	Ord.	A	4	4
12	F	1972	29	Ord.	D	4	2
13	F	1972	29	Ord.	A	3	3
14	M	1972	29	Ord.	C	3	2
15	M	1973	28	2a	A	3	1
16	M	1973	29	Ord.	A	3	2
17	F	1973	29	Ord.	A	3	4
18	M	1973	29	Ord.	D	4	2
19	F	1973	31	Ord.	A	3	1
20	M	1974	27	2a	A	4	2
21	M	1974	28	2a	A	2	1
22	M	1974	28	2b	D	2	2
23	M	1974	29	Ord.	B	5	1
24	F	1974	26	Ord.	A	2	3
25	F	1974	26	Ord.	A	2	2
26	M	1975	27	1	C	3	3
27	M	1975	28	2a	A	4	3
28	F	1975	26	Ord.	D	2	2
29	M	1976	28	2a	C	2	1

(*contd*)

Ref. number	Sex	Year of graduation	Age at 1 July 1979	Class of degree	Type of work	Salary group	Number of jobs*
30	F	1976	25	2a	B	3	1
31	M	1976	26	2a	A	2	1
32	M	1976	26	2a	C	2	1
33	M	1976	27	2a	C	2	1
34	F	1976	25	2a	A	2	1
35	M	1976	25	2b	A	1	1
36	F	1976	25	2b	D	2	1
37	M	1976	25	3	E	1	3
38	M	1976	26	3	A	2	2
39	F	1976	25	Ord.	A	2	1
40	M	1977	24	2a	B	2	2
41	M	1977	24	3	B	1	1
42	F	1977	24	Ord.	E	1	1
43	M	1978	23	1	A	1	2
44	F	1978	24	2a	E	1	1
45	M	1978	23	2a	A	1	1
46	M	1978	29	2a	C	2	1
47	M	1978	27	2b	C	2	1
48	M	1978	23	2b	E	1	1
49	F	1978	23	2b	A	2	1
50	M	1978	23	3	E	1	1

* Number of jobs *since graduation*

Key   *Sex*   *Class of degree*
        M = Male   1 = First-class honours
        F = Female   2a = Upper Second-class honours
                                                 2b = Lower Second-class honours
                                                 3 = Third-class honours
                                                 Ord. = Ordinary

        Type of Work   *Salary group* (£)
        A = Management Services and Computing   1 = 2000–3999
        B = Financial Work   2 = 4000–4999
        C = Research & Development   3 = 5000–5999
        D = Teaching & Lecturing   4 = 6000–6999
        E = Other   5 = 7000–9999
                                                  6 = 10 000 and over

It is required to investigate whether 'salary group' can be predicted by the other variates.

(1) For analysis the following codings are suggested: (a) Sex: code 'M' = +1, 'F' = −1; this has the effect of changing the constant in the regression equation depending on whether the subject is male or female. (b) Year of graduation: subtract 1970 from the values. (c) Class of degree: code the classes 1 to 5 from 'Ordinary' to 'First-class'. (d) Type of work: no obvious coding; do not introduce into regression equation but look at association with residuals.

(2) 'Year of graduation', 'age' and 'number of jobs' are obviously highly correlated. Introduce just one into the initial regression fit; 'age' is probably the best.

(3) Begin by fitting regression of salary on sex, age and degree class. Follow general principles recommended in Section 24.9.1.

(4) Make a table of residuals classified by type of work. Investigate for association.

(5) Test if remaining variates improve regression if admitted to it.

24.13 PROJECT

1 Prediction of weight

Aim. To predict the weight of adults from measurements of their height and waist.

Equipment. Medical scales for weight and height measurement if available but bathroom scales and a straight wall will serve. Measuring tape for waist.

Collection of data. Take measurements of at least 12 subjects if all of same sex; at least 10 of each sex if both sexes can be recorded.

Analysis. Fit a regression of weight on height and waist measurement following the general procedure recommended in Section 24.9.1. Better results may be obtained if all three variates are transformed to logarithms. Analyse men and women separately in the first instance, but later consider including both groups in the same regression using a dummy variate 'male' $= +1$, 'female' $= -1$, as in Exercise 24.12(2).

(If it is impossible to organize people, humbler objects such as potatoes, apples or oranges might be measured, in two directions, and weighed.)

SUMMARY

(1) **Multiple regression** is based on a linear statistical model of the form

$$Y = \alpha + \beta_1 x_1 + \beta_2 x_2 + \cdots + \beta_k x_k + e$$

where the **response variate** Y is related to the particular values $x_1, x_2, \ldots, x_k$ of the **explanatory variates** $X_1, X_2, \ldots, X_k$ such that the expression for Y is *linear* in the unknown parameters $\alpha, \beta_1, \beta_2, \ldots, \beta_k$ and the random error e. The random (residual) errors are mutually independent with mean 0 and variance σ^2.

(2) The mean of Y, for a particular value of $X_1, X_2, \ldots, X_k$, is the **predicted value**

$$\hat{Y} = \mathrm{E}[Y|x_1, x_2, \ldots, x_k] = \alpha + \beta_1 x_1 + \beta_2 x_2 + \cdots + \beta_k x_k.$$

(3) Given a sample of observations $(y_i, x_{1i}, x_{2i}, \ldots, x_{ki})$, $i = 1, 2, \ldots, n$, the regression coefficients $\alpha, \beta_1, \beta_2, \ldots, \beta_k$ are estimated using the **principle of least squares.** This gives rise to the **normal equations:** k equations which are

solved to give the estimates $b_1, b_2, \ldots, b_k$. The constant α is estimated by $a = \bar{y} - b_1 \bar{x}_1 - b_2 \bar{x}_2 - \ldots - b_k \bar{x}_k$.

(4) In order to make tests on, and derive confidence intervals for, the regression coefficients, the additional assumption is made that the residual errors $\{e_i\}$ (and hence the response variates $\{Y_i\}$) are normally distributed.

(5) For the regression of Y on X_1 and X_2,

(i) the **normal equations** are:

$$b_1 \sum (x_1 - \bar{x}_1)^2 + b_2 \sum (x_1 - \bar{x}_1)(x_2 - \bar{x}_2) = \sum (x_1 - \bar{x}_1)(y - \bar{y}),$$

$$b_1 \sum (x_1 - \bar{x}_1)(x_2 - \bar{x}_2) + b_2 \sum (x_2 - \bar{x}_2)^2 = \sum (x_2 - \bar{x}_2)(y - \bar{y}),$$

$$a = \bar{y} - b_1 \bar{x}_1 - b_2 \bar{x}_2.$$

(ii) The **analysis of variance** takes the form:

Source of variation	Sum of squares	Degrees of freedom	Mean square
Regression	S_{MR}	2	$S_{MR}/2$
Residual	S_R	$n-3$	$S_R/(n-3)$
Total	S	$n-1$	

where $S = \sum (y_i - \bar{y})^2$, $S_{MR} = b_1 \sum (x_1 - \bar{x}_1)(y - \bar{y}) + b_2 \sum (x_2 - \bar{x}_2)(y - \bar{y})$ and $S_R = S - S_{MR}$. The significance of the regression is tested by comparing the ratio regression mean square/residual mean square to F with 2 and $n - 3$ degrees of freedom in a one-tail test. The residual mean square provides an estimate s^2 of σ^2.

(iii) The **regression sum of squares** may be partitioned.

Source of variation	Sum of squares	Degrees of freedom	Mean square
Regression on X_1	S_{LR}	1	
Extra when X_2 included	$S_{MR} - S_{LR}$	1	$MS(\text{Extra})$
Regression on X_1 and X_2	S_{MR}	2	

where S_{MR} is as above, and $S_{LR} = [\sum (x_1 - \bar{x}_1)(y - \bar{y})]^2 / \sum (x_1 - \bar{x}_1)^2$.

To test whether the inclusion of X_2 significantly improves the regression, the ratio $MS(\text{Extra})$ to the residual mean square is compared to F with 1 and $n - 3$ degrees of freedom using a one-tail test. A similar test may be made on X_1 by fitting first a regression on X_2.

(iv) **Confidence intervals and *t*-tests:**

$$\text{Var}[b_1] = \sigma^2 \sum (x_2 - \bar{x}_2)^2 / D, \quad \text{Var}[b_2] = \sigma^2 \sum (x_1 - \bar{x}_1)^2 / D,$$

$$\text{Cov}[b_1, b_2] = \sigma^2 \sum (x_1 - \bar{x}_1)(x_2 - \bar{x}_2) / D$$

where $D = \sum (x_1 - \bar{x}_1)^2 \sum (x_2 - \bar{x}_2)^2 - [\sum (x_1 - \bar{x}_1)(x_2 - \bar{x}_2)]^2$.
σ^2 is replaced by its estimate s^2.
  A confidence interval for β_i is given by

$$b_i \pm t \sqrt{\text{Var}[b_i]}$$

where t is the value of the t-distribution with $n - 3$ degrees of freedom for the chosen probability level.
  To test the null hypothesis that $\beta_i = 0$, the ratio $b_i / \sqrt{\text{Var}[b_i]}$ is compared with t with $n - 3$ degrees of freedom.

(6) Most regression analyses are done by computer. Typical output includes:
  (a) the estimated regression equation,
  (b) estimates of the regression coefficients with their standard errors,
  (c) the analysis of variance,
  (d) the proportion of the variability in Y accounted for by the regression.
Many additional analyses are usually offered.
  A check on the assumptions of the model is made by inspection of the residuals $(y_i - \hat{y}_i)$. A particularly useful plot is that of the residuals against the fitted, or predicted, values.

APPENDIX

I THE BINOMIAL SERIES EXPANSION

Theorem

When a and b are any two numbers, and n is a positive integer,

$$(a + b)^n = a^n + \binom{n}{1}a^{n-1}b + \binom{n}{2}a^{n-2}b^2 + \cdots + \binom{n}{n-1}ab^{n-1} + b^n.$$

Proof. Consider a related expression, $(a + c_1)(a + c_2) \cdots (a + c_n)$, and expand this by multiplying out brackets, one by one: first we find that the expression is equal to

$$(a^2 + a(c_1 + c_2) + c_1c_2)(a + c_3) \cdots (a + c_n),$$

which becomes

$$(a^3 + a^2(c_1 + c_2 + c_3) + a(c_1c_3 + c_2c_3 + c_1c_2) + c_1c_2c_3)(a + c_4) \cdots (a + c_n)$$

and eventually

$$a^n + a^{n-1}(c_1 + \cdots + c_n) + a^{n-2}(c_1c_2 + c_1c_3 + \cdots + c_{n-1}c_n)$$
$$+ a^{n-3}(c_1c_2c_3 + \cdots + c_{n-2}c_{n-1}c_n) + \cdots + c_1c_2c_3 \ldots c_n.$$

Let us first count the number of terms in the coefficient of each power of a. The coefficient of a^{n-1} is $c_1 + \cdots + c_n$, consisting of n terms. The coefficient of a^{n-2} is $c_1c_2 + \cdots + c_{n-1}c_n$, in which all the possible pairs c_ic_j, for $i \neq j$, and i, j taking values 1 to n, appear once. There are $\binom{n}{2}$ such pairs. In the same way, the coefficient of a^{n-3} consists of the $\binom{n}{3}$ terms $c_ic_jc_k, i \neq j \neq k$ and i, j, k taking values 1 to n. This argument can be repeated to find out how many terms there are in the coefficient of the general term a^{n-r}, and the answer is $\binom{n}{r}$.

Now replace each c_i by b, so that $c_1 + \cdots + c_n$ becomes $nb, c_1c_2 + \cdots + c_{n-1}c_n$ becomes $\binom{n}{2}b^2, \ldots$, the coefficient of a^{n-r} becomes $\binom{n}{r}b^r$, and the theorem is established so long as n is a positive integer. This is all that is required for studying the binomial distribution.

We include also the statement of a more general theorem, which we do not prove here. If α is not a positive integer, the series will be infinite.

Theorem

When α is any real number, and x is a real number such that $-1 < x < 1$,

$$(1+x)^\alpha = 1 + \frac{\alpha}{1!}x + \frac{\alpha(\alpha-1)}{2!}x^2 + \cdots + \frac{\alpha(\alpha-1)\ldots(\alpha-r+1)}{r!}x^r + \cdots .$$

II THE EXPONENTIAL FUNCTION

Definition

The **exponential function** e^x, where e is a special constant, is that function which is its own derivative. Thus

$$\frac{d}{dx}(e^x) = e^x.$$

It is sometimes convenient to write the exponential function in the form $\exp(x)$.

We suppose that this function can be expressed as a power series in x, with coefficients $a_0, a_1, a_2, \ldots$, in the form $e^x = a_0 + a_1 x + a_2 x^2 + a_3 x^3 + a_4 x^4 + \cdots + a_n x^n + \cdots$, and that this series can be differentiated term by term. Then

$$\frac{d}{dx}(e^x) = a_1 + 2a_2 x + 3a_3 x^2 + 4a_4 x^3 + \cdots + na_n x^{n-1} + \cdots .$$

Because $d(e^x)/dx = e^x$ by definition, we have, on equating the coefficients of the powers of x,

$$a_0 = a_1 \quad a_1 = 2a_2 \quad a_2 = 3a_3$$
$$a_3 = 4a_4 \quad \ldots \quad a_{n-1} = na_n.$$

When $x = 0$, the series is equal to a_0. But any number raised to the power 0 must equal 1, so $a_0 = e^0 = 1$. Hence

$$a_1 = a_0 = 1$$

$$a_2 = \frac{1}{2}a_1 = \frac{1}{2} = \frac{1}{2!}$$

$$a_3 = \frac{1}{3}a_2 = \frac{1}{2 \times 3} = \frac{1}{3!}$$

$$a_4 = \frac{1}{4}a_3 = \frac{1}{2 \times 3 \times 4} = \frac{1}{4!}$$

$$\vdots$$

$$a_n = \frac{1}{n!}$$

This leads us to the following theorem.

Theorem

The series expansion of the exponential function e^x is

$$1 + x + \frac{x^2}{2!} + \frac{x^3}{3!} + \cdots + \frac{x^r}{r!} + \cdots.$$

The constant

$$e = 1 + 1 + \frac{1}{2!} + \frac{1}{3!} + \cdots + \frac{1}{r!} + \cdots.$$

This is an irrational number and equals approximately 2.718.

III DERIVATIVES AND INTEGRALS OF THE EXPONENTIAL FUNCTION

We give a table of derivatives and integrals of exponential functions. The results come from the definition of the exponential function or by substitution.

$f'(x)$	$f(x)$	$\int f(x)\,\mathrm{d}x$
e^x	e^x	e^x
$a\,e^{ax}$	e^{ax}	$\frac{1}{a}\,e^{ax}$
$g'(x)\,e^{g(x)}$	$e^{g(x)}$	$-$

IV INTEGRALS RELATED TO THE NORMAL DISTRIBUTION

(a) Theorem

$$\int_{-\infty}^{\infty} e^{-x^2/2}\,\mathrm{d}x = \sqrt{2\pi} = 2.506\,628.$$

The proof of this result demands more advanced mathematics than we can assume. We therefore content ourselves with a demonstration of the reasonableness of the result by carrying out a numerical integration using Simpson's rule.

We take ordinates at unit intervals from $x = -4$ to $x = 4$ and write the ordinates $y_{-4}, y_{-3}, \ldots, y_0, \ldots, y_3, y_4$. The area under the curve outside the interval $|x| \leqslant 4$ is very small.

The integral approximately equals

$$\frac{1}{3}[(y_{-4} + y_4) + 4(y_{-3} + y_{-1} + y_1 + y_3) + 2(y_{-2} + y_0 + y_2)]$$

$$= \frac{1}{3}[2\exp(-8) + 8\exp(-0.5) + 8\exp(-4.5) + 4\exp(-2) + 2\exp(0)]$$

$$= \frac{1}{3}[2(0.000\,335) + 8(0.606\,531) + 8(0.011\,109) + 4(0.135\,335) + 2]$$

$$= \frac{7.483\,129}{3} = 2.494\,376.$$

(b) Theorem

$$\int_{-\infty}^{\infty} x\,e^{-x^2/2}\,dx = 0.$$

Proof. Consider the positive and negative regions of integration for this integral, which we denote by I:

$$I = \int_{-\infty}^{0} x\,e^{-x^2/2}\,dx + \int_{0}^{\infty} x\,e^{-x^2/2}\,dx.$$

In the integral over the negative region we make the substitution $x = -u$. Hence $dx = -du$ and the integral becomes

$$\int_{\infty}^{0} (-u)e^{-u^2/2}(-du) = \int_{\infty}^{0} u\,e^{-u^2/2}\,du = -\int_{0}^{\infty} u\,e^{-u^2/2}\,du.$$

Thus

$$I = -\int_{0}^{\infty} u\,e^{-u^2/2}\,du + \int_{0}^{\infty} x\,e^{-x^2/2}\,dx = 0.$$

(c) Theorem

$$\int_{-\infty}^{\infty} x^2\,e^{-x^2/2}\,dx = \sqrt{2\pi}.$$

Proof. We write the integral

$$I = \int_{-\infty}^{0} x^2\,e^{-x^2/2}\,dx + \int_{0}^{\infty} x^2\,e^{-x^2/2}\,dx.$$

We make the substitution $x = -u$ in the integral over the negative region to obtain

$$\int_{\infty}^{0} u^2\,e^{-u^2/2}(-du) = -\int_{\infty}^{0} u^2\,e^{-u^2/2}\,du = \int_{0}^{\infty} u^2\,e^{-u^2/2}\,du.$$

Hence

$$I = \int_{0}^{\infty} u^2\,e^{-u^2/2}\,du + \int_{0}^{\infty} x^2\,e^{-x^2/2}\,dx = 2\int_{0}^{\infty} x^2\,e^{-x^2/2}\,dx.$$

We now integrate by parts:

$$\frac{1}{2}I = \int_0^\infty x(xe^{-x^2/2})dx = \left[-x\,e^{-x^2/2)}\right]_0^\infty + \int_0^\infty e^{-x^2/2}\,dx$$

$$= 0 + \int_0^\infty e^{-x^2/2}\,dx = \frac{1}{2}\int_{-\infty}^\infty e^{-x^2/2}\,dx.$$

Hence

$$I = \int_{-\infty}^\infty e^{-x^2/2}\,dx = \sqrt{2\pi}, \quad \text{from (a)}.$$

V THE LIMIT OF $(1 + x/n)^n$ AS $n \to \infty$

Theorem

$$\lim_{n\to\infty}\left(1+\frac{x}{n}\right)^n = e^x.$$

Proof. By the binomial series expansion, we have (for n an integer)

$$\left(1+\frac{x}{n}\right)^n = 1 + n\cdot\frac{x}{n} + \frac{n(n-1)}{2!}\cdot\frac{x^2}{n^2} + \frac{n(n-1)(n-2)}{3!}\cdot\frac{x^3}{n^3} + \cdots$$

$$+ \frac{n(n-1)\cdots(n-r+1)}{r!}\cdot\frac{x^r}{n^r} + \cdots$$

$$= 1 + x + \frac{x^2}{2!}\left(1-\frac{1}{n}\right) + \frac{x^3}{3!}\left(1-\frac{1}{n}\right)\left(1-\frac{2}{n}\right) + \cdots$$

$$+ \frac{x^r}{r!}\left(1-\frac{1}{n}\right)\left(1-\frac{2}{n}\right)\cdots\left(1-\frac{r-1}{n}\right) + \cdots.$$

As $n \to \infty$,

$$\left(1-\frac{1}{n}\right) \to 1, \quad \left(1-\frac{2}{n}\right) \to 1,\ldots,\left(1-\frac{r-1}{n}\right) \to 1,\ldots.$$

Hence

$$\lim_{n\to\infty}\left(1+\frac{x}{n}\right)^n = 1 + x + \frac{x^2}{2!}\cdot 1 + \frac{x^3}{3!}\cdot 1\cdot 1 + \cdots + \frac{x^r}{r!}\cdot 1\cdot 1\ldots 1 + \cdots$$

$$= 1 + x + \frac{x^2}{2!} + \frac{x^3}{3!} + \cdots + \frac{x^r}{r!} + \cdots = e^x.$$

Note also that, replacing x by $-x$ in the series expansion, we have

$$e^{-x} = 1 - x + \frac{x^2}{2!} - \frac{x^3}{3!} + \frac{x^4}{4!} - \cdots + (-1)^n\frac{x^n}{n!} + \cdots,$$

and the limit expression above becomes

$$\lim_{n \to \infty} \left(1 - \frac{x}{n}\right)^n = e^{-x}.$$

VI A DERIVATION OF THE POISSON DISTRIBUTION

Consider the binomial probability

$$P(r) = \binom{n}{r} \pi^r (1 - \pi)^{n-r},$$

in the usual notation (p. 174). For very large sample sizes n, and small probabilities π, write $\lambda = n\pi$. Let $n \to \infty$ and $\pi \to 0$, and λ remain constant. Then

$$P(r) = \frac{n!}{r!(n-r)!} \left(\frac{\lambda}{n}\right)^r \left(1 - \frac{\lambda}{n}\right)^{n-r}$$

$$= \frac{n(n-1)(n-2)\cdots(n-r+1)}{n^r} \cdot \frac{\lambda^r}{r!} \left(1 - \frac{\lambda}{n}\right)^{n-r}$$

$$= 1\left(1 - \frac{1}{n}\right)\left(1 - \frac{2}{n}\right)\cdots\left(1 - \frac{r-1}{n}\right) \cdot \frac{\lambda^r}{r!} \cdot \left(1 - \frac{\lambda}{n}\right)^n \cdot \left(1 - \frac{\lambda}{n}\right)^{-r}.$$

As $n \to \infty$, each of the factors

$$\left(1 - \frac{1}{n}\right), \left(1 - \frac{2}{n}\right), \ldots, \left(1 - \frac{r-1}{n}\right) \to 1,$$

and also

$$\left(1 - \frac{\lambda}{n}\right) \to 1 \text{ so that } \left(1 - \frac{\lambda}{n}\right)^{-r} \to 1 \quad \text{for each integer } r.$$

Finally,

$$\lim_{n \to \infty} \left(1 - \frac{\lambda}{n}\right)^n = e^{-\lambda}$$

as shown above. Hence $P(r) \to e^{-\lambda}\lambda^r/r!$.

VII PARTIAL DIFFERENTIATION

Consider a function containing three variables, such as $z = 2x + 3y + x^2y^2$. It may be represented graphically as the height z vertically above the point (x, y) in a horizontal plane. As (x, y) moves around the plane, z traces out a surface. The rate at which z changes depends on x and y.

If y is held constant, z is restricted to movement parallel to OX; the rate of change of z in this direction is called the *partial derivative* $\partial z/\partial x$ of the function z with respect to x. It is found using all the standard rules of differentiation but treating y as constant. For $z = 2x + 3y + x^2y^2$, $\partial z/\partial x = 2 + 2xy^2$. When x is kept constant, $\partial z/\partial y = 3 + 2x^2y$ measures the rate of change in z as movement in (x, y) is restricted to be parallel to OY. These partial derivatives show the rates of change of z in cross-sections of the surface in three dimensions.

Second partial derivatives may also be defined: $\partial^2 z/\partial x^2$ is the partial derivative of $\partial z/\partial x$ with respect to x and $\partial^2 z/\partial y^2$ is the partial derivative of $\partial z/\partial y$ with respect to y. For the function above,

$$\frac{\partial^2 z}{\partial x^2} = 2y^2 \quad \text{and} \quad \frac{\partial^2 z}{\partial y^2} = 2x^2.$$

Also

$$\frac{\partial^2 z}{\partial y\, \partial x} = \frac{\partial}{\partial y}\left(\frac{\partial z}{\partial x}\right) = \frac{\partial}{\partial y}(2 + 2xy^2) = 4xy$$

in the example above, and

$$\frac{\partial^2 z}{\partial x\, \partial y} = \frac{\partial}{\partial x}\left(\frac{\partial z}{\partial y}\right) = \frac{\partial}{\partial x}(3 + 2x^2y) = 4xy.$$

For many functions, including continuous functions generally used in statistics,

$$\frac{\partial^2 z}{\partial y\, \partial x} = \frac{\partial^2 z}{\partial x\, \partial y};$$

but this is not universally true.

Example
For

$$z = e^{kxy}(\sin ax + \cos by)$$

$$\frac{\partial z}{\partial x} = e^{kxy}(a \cos ax) + (\sin ax + \cos by)ky\, e^{kxy}$$

$$\frac{\partial z}{\partial y} = e^{kxy}(-b \sin by) + (\sin ax + \cos by)kx\, e^{kxy}.$$

Although it is not easy to represent the results graphically, these methods can be extended to functions of any number of variables. For $z = ax^2y^3t^4$, three first-order partial derivatives are

$$\frac{\partial z}{\partial x} = 2axy^3t^4; \quad \frac{\partial z}{\partial y} = 3ax^2y^2t^4; \quad \frac{\partial z}{\partial t} = 4ax^2y^3t^3;$$

second-order derivatives, including those like

$$\frac{\partial^2 z}{\partial x\,\partial t} = \frac{\partial}{\partial x}\left(\frac{\partial z}{\partial t}\right),$$

can be found in the same way as described above.

Turning points (maxima, minima, saddle points) of the surface described by z occur when the first partial derivatives are zero:

$$\frac{\partial z}{\partial x} = \frac{\partial z}{\partial y} = \frac{\partial z}{\partial t} = \ldots = 0.$$

Tests for each type (maximum, etc.) of turning points are given in books on mathematical methods.

VIII SELECTION OF MINITAB MENUS

(Numbers after an item refer to pages where the item is discussed.)

Data Input

Calc>Make Patterned Data   158

Data Manipulation

Calc>Calculator/Partial Sum (and pocket calculator functions)   17, 18, 495, 631
Calc>Standardize   218
Data>Code   158–9
Data>Rank   564
Data>Stack>Columns: Data>Unstack Columns   40
Stat>Tables>Tally Individual Values   18, 159

Plots and Diagrams

Editor>Layout Tool   18
Graph>
> **Bar Chart**   17
> **Boxplot**   79
> **Dotplot**   40–1
> **Histogram**   16–17, 366
> **Matrix Plot**   712
> **Scatterplot**   94, 630
> **Stem-and-Leaf**   18
> **Time Series Plot**   95

Stat>Basic Statistics>Normality Test   366
Stat>Quality Tools>Symmetry Plot   366

Simple Statistics

Stat>Basic Statistics>
> **Display Descriptive Statistics**   40, 257
> **Correlation**   564
> **Graphical Summary**   95

Calc>Column Statistics/Sum, Mean, Median, St. Dev., Range   40
Calc>Row Statistics/Sum, Mean, Median, St. Dev., Range   58

Significance Tests and Confidence Intervals

Stat>Basic Statistics>
> **1-Sample Z**   436–7, 465–6
> **1-Sample t**   436–7, 465–6
> **2-Sample t**   436–7, 465–6
> **Normality Test, Kolmogorov–Smirnov**   495
> **Paired t**   436–7, 465–6
> **1 Proportion**   437, 465–6
> **2 Proportions**   437, 465–6

Stat>Nonparametrics>
> **1-Sample Sign**   299
> **1-Sample Wilcoxon**   299
> **Mann-Whitney**   299

Stat>Tables>
> **Chi-Square Test (Table in Worksheet)**   495
> **Cross-Tabulation and Chi-Square**   495

Sampling

Regression

Probability Distributions

Analysis of Variance

Control Charts

BIBLIOGRAPHY

I PARTICULAR TOPICS

The following suggestions for further reading may be useful for anyone who wishes to pursue any of these topics further. (Numbers after the authors' names refer to chapters.)

Birth and death rates. Anderson, 10; Hill and Hill.
Computing methods. Cooke *et al.*
Demography. Anderson, 9, 10, 11.
Design of experiments. Clarke and Kempson; Clarke, 12, 13; Hill and Hill, for clinical trials.
Exploratory data analysis. Tukey; Du Toit *et al.*
Graphical methods. Du Toit *et al.*
Index numbers. Anderson, 8.
Published statistics. Anderson, 6.
Quality control. Bissell; Wetherill and Brown.
Questionnaires. Moser and Kalton, 12.
Sample surveys. Moser and Kalton; Barnett; also Anderson, 14, 17.
Time series. Anderson, 12, 18.

ANDERSON, A. J. B., 1989 *Interpreting Data: A First Course in Statistics*, Chapman & Hall, London.
BARNETT, V., 2002 *Sample Survey Principles and Methods*, 3rd edn, Arnold, London.
BISSELL, A. F., 1994 *Statistical Methods in SPC and TQM*, Chapman & Hall, London.
CHATFIELD, C., 1995 *Problem Solving*, 2nd edn, Chapman & Hall, London.
CLARKE, G. M., 1994 *Statistics and Experimental Design*, 3rd edn, Arnold, London.
CLARKE, G. M. and KEMPSON, R. E., 1997 *Introduction to the Design and Analysis of Experiments*, Arnold, London.
COOKE, D., CRAVEN, A. H. and CLARKE, G. M., 1990 *Basic Statistical Computing,* 2nd edn, Arnold, London. (*CCC*)
DU TOIT, S. H. C., STEYN, A. G. W. and STUMPF, R. H., 1986 *Graphical Exploratory Data Analysis*, Springer-Verlag, New York.
HILL, A. B. and HILL, I. D., 1991 *Bradford Hill's Principles of Medical Statistics*, Arnold, London.
MEAD, R., CURNOW, R. N. and HASTED, A. M., 2001 *Statistical Methods in Agriculture and Experimental Biology*, 3rd edn, CRC Press, Boca Raton, Fla.
MOSER, C. A. and KALTON, G., 1974 *Survey Methods in Social Investigation*, 2nd edn, Heinemann, London.
SPRENT, P. and SMEETON, N. C., 2001 *Applied Nonparametric Statistical Methods*, 3rd edn, CRC Press, Boca Raton, Fla.
TUKEY, J. W., 1977 *Exploratory Data Analysis*, Addison-Wesley, Reading, Mass.
WETHERILL, G. B. and BROWN, D. W., 1991 *Statistical Process Control*, 3rd edn, Chapman & Hall, London.

II REFERENCE TEXTS

These texts are of undergraduate level in theoretical and applied statistics, and probability.

FREUND, J. E. and WALPOLE, R. E., 1987 *Mathematical Statistics*, 3rd edn, Prentice Hall, Englewood Cliffs, N.J.

MORGAN, B. J. T., 1984 *Elements of Simulation*, Chapman & Hall, London.

MOSTELLER, F., ROURKE, R. E. K. and THOMAS, G. H., 1970 *Probability: A First Course*, 2nd edn, Addison-Wesley, Reading, Mass.

SNEDECOR, G. W. and COCHRAN, W. G., 1980 *Statistical Methods*, 7th edn, Iowa State University Press, Ames, Iowa.

ANSWERS

Chapter 1

1.7 1 (a) Quant., cont. (b) Quant., disc. (c) Qual. (d) Quant., cont. (e) Qual. (f) Quant., disc. (g) Qual. (h) Quant., disc. (i) Qual.

3 (a) $59.5–64.5, 64.5–69.5, \ldots; 62.0, 67.0, \ldots$ (b) 30 years 0 days–34 years 364 days, $\ldots$; 32 years 182 days, $\ldots$ (c) $4, 10\frac{1}{2}, 14\frac{1}{2}, 18\frac{1}{2}, 22\frac{1}{2}, 27\frac{1}{2}$. (d) $0–4\frac{1}{2}$ mins, $4\frac{1}{2}–9\frac{1}{2}, 9\frac{1}{2}–14\frac{1}{2}, 14\frac{1}{2}$–whatever the maximum observed waiting time was. $2\frac{1}{4}, 7, 12$ mins, strictly unknown but say $20'0''$ to allow for a few very large times.

4 Histogram on intervals quoted (0–8, etc.); not entirely satisfactory because data discrete.

5 Histogram; take last interval 15–25 (unless higher figure observed).

6 *Number* 0 1 2 3 4 5 6 7
Frequency 3 3 6 5 5 3 3 2; bar diagram.

7 *Interval* 35–39 40–44 45–49 50–54 55–59 60–64 65–69 70–74
Frequency 1 7 11 4 2 2 2 1 ;
Modal class 45–49.

8 (b) Besides showing summary data as frequencies in intervals, stem-and-leaf also retains original data (and data in rank order too).

9 *Number* 0 1 2 3 4 5
Frequency 3 13 30 33 17 4; bar diagram.

10 *Number of spades* 0 1 2 3 4 5 6 7 8 9 10 11 12 13
Number of hands 2 6 20 28 23 12 4 2 1 0 1 1 0 0 ;
bar diagram.

11 *Height* 26–30 31–35 36–40 41–45 46–50 51–55 56–60 61–65
Frequency 4 5 23 58 61 30 3 3 ;
histogram.

Other groupings possible, e.g. 25–29, 30–34, ...
When using narrower intervals, the histogram often looks less smooth, and it can be harder to see a pattern; using larger intervals loses information because most of the frequency is concentrated in a few intervals.

13 Asian centre figures more variable.

Chapter 2

2.2.1 (a) 21, (b) 21 (c) 12, (d) 11.5.

2.3.1 (a) Median 73, mean 85.3; (b) 175, 175.2; (c) 16, 12.25.

2.3.2 **1** 2628, mean larger, because it is more affected by the few very large observations.

2 Class-centres 17.5, 22.5, 27.5, 32.5, 37.5, 42.5, 50.0, 60.0, 70.0. Mean 41.125, median 39.12.

2.6.1 **2** 2.033.

2.11 **1** Not if 'average' = mean. **2** 10; 10. **3** (i) 13.2, (ii) 8 children of mean age 4.25, 3 adults of mean age 37. **4** Medians 13 (A), 15 (B). **5** (i) 157, (ii) 158.9, (iii) all equal to 165. **6** (a) 196.6, 185, 185, mean high due to one large figure. (b) 41.6, 42, 42, little to choose. (c) 34.1, 34, 34, little to choose. (d) 10, 10, no mode. **7** (a) 53, (b) 21, (c) 6, (d) 6, (e) 18. **8** (a) $\sum_{i=0}^{4} p_i$. (b) $\sum_{i=1}^{4} x_i y_i$. (c) $\sum_{i=1}^{5} i^2$. (d) $\sum_{i=1}^{5} i x_i$. (e) $\sum_{i=1}^{5} 2i$. (f) $\sum_{i=0}^{3} (2i+1) r^i$. **13** (a) 10.94, (b) 1054.5, (c) 0.144. **14** (a) Mean 1.55, median 1.04. (b) 45.6, 45.8. (c) 3.79, 3. **15** (i) 130.4, (ii) 45.2. G.M. not so seriously affected by very high values. **16** (i) Weights of population of 40-year-old males. (ii) Numbers of passengers carried on buses: several buses full (rush-hours), several nearly empty, fewer in-between numbers. (iii) Number of heads in 5 tosses of a fair coin; the tossing of 5 coins is repeated many times. (iv) Number of days, after the first child showed symptoms, to the onset of an infectious disease among a class of children. **17** 16, 17, 19, 14.2. Dot plot suitable. **18** 22.28; mean, because data skew with tail on right. **19** (a) 141.5, (b) 132.4. **20** $a = 10$, $b = 20$; or $a = 16.67$, $b = 6.67$.

Chapter 3

3.10 **1**

Mean	$\frac{1}{2}$	1	$1\frac{1}{2}$	2	$2\frac{1}{2}$	3	$3\frac{1}{2}$	4	$4\frac{1}{2}$	5	$5\frac{1}{2}$	6	$6\frac{1}{2}$	7	$7\frac{1}{2}$	8	$8\frac{1}{2}$
Frequency	1	1	2	2	3	3	4	4	5	4	4	3	3	2	2	1	1;

mean value $4\frac{1}{2}$. **2** (a) room (or store); all rooms (or stores) in school; all rooms (or stores) that are accessible; number of chairs per room (or store); mean number of chairs per sampled room × number of rooms in school. (b) household; all households in town; all who will co-operate; microwave or not; proportion in sample who have microwave. (c) a rectangular unit area of field (often called a *quadrat*), taken to a suitable depth; all the field to this depth; all the field except any parts not accessible; number of worms per quadrat; mean number of worms per quadrat × area of field ÷ area of quadrat. (d) child; all children in the road; all those available; height of a child; sample mean height. **8** (a) No; only first 9 (or 10) letters used. (b) Yes, but wasteful of digits. (c) No; not all letters have same probability of choice. (d) Yes. **10** (a) Not strictly; probability of selection depends on number of entries per page. (b) Yes. **11** (a) Not random. Use electoral roll (the list of those able to vote at elections). (b) Probability depends on family size. Make a list of all children first, with random choice from that. (c) Not random. Use a school list.

Chapter 4

4.3.1 **1** (a) 19 and 5.67; (b) 29 and 9.   **2** 7, 7; 2, 3.25; mean deviation.

4.4.1 **1** (a) 2, 1.41, (b) 18, 4.24, (c) 18, 4.24, (d) 9, 3, (e) 2, 1.41, (f) 0.0417, 0.204.
   2 167, 24.67. M 171.7, 4.22. F 162.3, 1.56.   **3** I 41, 36.2, 205.1, 14.3. II 90,
   33.7, 581.8, 24.1.

4.5.2 **1** 0.62, 0.6644, 0.815.   **2** 1.01, 1.005, 4, variance or s.d.; 0.7, 0, 0, mean.
   3 48.4475, 6.96; 7.5094, 2.74.   **4** 5.9844, 2.43   **5** 10.0, 58.25, 7.63. (a) 17.63,
   (b) 25.26. 19.2%; 4.7%.

4.6.1 62, 46, 75, 14.5; Sept. 46, 12.7, 15.3; Dec. 63, 15.2, 18.5.

4.7.1 **1** 12.7, 22.3; cannot calculate mean and s.d. as 'over 52' is too vague to be
   used.   **2** 1.77, 1.27; yes because the distribution is symmetric.

   4.9 **1** 4.71, 2.52.   **2** (a) 23, 4. (b) 200, 40. (c) 20 004, 4000. (d) 4, 2.   **3** 2.5, 1.43,
   1.20.   **4** 0, 3.045, 2, 6.68, 2.59.   **5** (a) 2.2075, 1.49. (b) 38.06, 6.17. (c) 3.1775,
   1.78.   **6** (a) 3.7264, 1.93. (b) 659.75, 25.7. (c) 0.000 129, 0.0114.   **7** 18, 4.73,
   5.49, 8.   **8** $\sqrt{(N^2 - 1)/12}$.

Chapter 5

5.6 **1** 5-number summaries. M: 50.0, 60.5, 66.5, 72.0, 85.0. F: 33.0, 53.0, 57.0,
   61.5, 68.0. M, on average 9.5 kg heavier than F; F outlier at 33.0. Otherwise
   both fairly symmetrical.   **2** (i) SD about 0.45, round to 1 d.p.; note B more
   variable than others. (iii) Rounded data, mean, median, range are: A, 13.7,
   13.8, 1.0; B, 14.7, 14.5, 1.6; C, 15.1, 15.1, 0.7; D, 14.4, 14.3, 1.1. A differs from B,
   C, D; also C differs from D but neither differs from B.   **3** (i) 0's preferred,
   and possibly 5's; 2's rare. (ii) 5-number summaries: Cash, 11.0, 16.0, 19.5,
   30.0, 40.0; Cheque, 37.0, 52.0, 55.5, 60.5, 70; Card, 43.0, 54.5, 59.0, 65.0, 67.0.
   Cheque and card distributions similar and symmetric; cash much quicker
   and skew.   **4** SD about 0.34, record to 1 d.p., no change.   **5** Frequency
   table suggests rolls aimed to be 60 feet long and were measured to nearest
   half-inch. Outlier 18.542.   **6** Some evidence that X, Y are associated.   **7** 5-
   number summaries: Placebo, 380, 420, 475, 490, 540; Drug, 310, 405, 420, 470,
   600. Outlier 600. Drug reduces respiratory capacity.   **8** Heart rate decreases
   as age increases, rapidly initially but levelling out as adulthood approached.
   9 Evidence of association. Outliers H: 148; W: 33.

Revision Exercises A

1 (i) Bar diagram. (ii) 19.5, 1.02. (iii) 0.975.   **2** (i) 31.33, 7.99. (ii) 35.73, 7.99.
(iii) 34.46, 8.79.   **3** £14, 25.50, 36, 48 approx.; 310/725, i.e. approx. 40%; falls
below first data point, project graph smoothly back to approx. £6.50, but may
be underestimate if few wages far below £10.

4

Number of words	10–	20–	30–	40–	50–	60–	70–	80–	90–
Frequency	15	24	25	13	8	6	4	3	2 ;

mean = 38.1 from table (37.9 from raw data).

5 11.023, 6.06   **6** Take last class as 10, to allow for the few larger dwellings.
4.69, 1.594, 34.0%; 4.92, 1.649, 33.5%.   **7** Median weights decrease as density

x

increases; variability in weights remains about the same. **8** 30.6, 32, 31, 30.46; 30.8, 29, 31, 30.67. **9** 4.35, 4. Two populations of figures, cannot be summarized by a single measure. **10** A 53.09, 35.34, 59.05. B 24.87, 15.59, 21.60. Geom., because dist. is skew; dist. of logs of obs. less skew, and G.M. of original obs. corresponds to A.M. of logs. **12** 0.167 12, 0.167, 0.0022, 0.165 71, 0.002 94. **13** 105.44, 9.91. **14** 34.95, 32.6, 186.46, 13.65, 60.5%. **15** 19y $7\frac{1}{2}$m. (b) 22y 1m, 11m. **16** 21.415, 135.91, 11.66. **17** (b) $10^4/a_1$ etc. (i) $\frac{1}{3}(a_1 + a_2 + a_3)$, $\frac{10\,000}{3}(\frac{1}{a_1} + \frac{1}{a_2} + \frac{1}{a_3})$. (ii) $\sqrt[3]{a_1 a_2 a_3}, 10^4/\sqrt[3]{a_1 a_2 a_3}$. **18** Household should be the sampling unit. (iv) worst because not all households covered, (i) does not give equal probability to households, (iii) best in theory, (ii) easier and cheaper, acceptable method. **19** A, 58.7; B, 54.6; C, 36.1; D, 38.2; E, 56.1; F, 35.8; G, 46.9; H, 54.4. Ranks 1, 3, 7, 6, 2, 8, 5, 4. New marks A, 59.6; B, 56.5; C, 37.6; D, 42.5; E, 57.2; F, 33.3; G, 52.1; H, 55.4. No change in ranks. (a) All pass on first method, F fails on second, (b) C, D, F fail on first method; C, F but not D on second.

Chapter 6

6.2.1 **1** 0.62, **2** (a) 0.65, (b) 0.8. **3** $\frac{1}{10}$. **4** $\frac{1}{8} = 0.125$. **5** (a) $\frac{1}{6}$, (b) $\frac{1}{2}$, (c) $\frac{1}{2}$, since primes are 2, 3, 5. **6** (a) $\frac{1}{36}$, (b) $\frac{1}{2}$, (c) $\frac{1}{3}$. **7** (a) $\frac{1}{4}$, (b) $\frac{7}{100}$, (c) $\frac{3}{100}$. **8** Assuming birthdays random throughout year, ignoring leap years, (a) $\frac{1}{7}$, (b) $\frac{1}{365} = 0.0027$.

6.3.1 **1** (a) $S = \{1, 2, 3, 4, 5, 6\}$. $A = \{2, 4, 6\}$. $B = \{3, 6\}$. $P(A) = \frac{1}{2}$, $P(B) = \frac{1}{3}$. (b) Label coats 1, 2, 3. Outcome specified by listing coats received by 3 guests in order, assuming $(1, 2, 3)$ is outcome in which all guests receive correct coat. $S = \{(1, 2, 3), (1, 3, 2), (2, 1, 3), (2, 3, 1), (3, 1, 2), (3, 2, 1)\}$. $A = \{(1, 2, 3)\}$. $B = \{(1, 3, 2), (2, 1, 3), (3, 2, 1)\}$. $P(A) = \frac{1}{6}$, $P(B) = \frac{1}{2}$. (c) $S = \{T, C, W\}, A = \{T, C, W\}, P(A) = 1$. **2** $S = \{(-2, -2), (-2, -1), (-2, 0), (-2, 1), (-2, 2), (-1, -2), (-1, -1), (-1, 0), (-1, 1), (-1, 2), (0, -2), (0, -1), (0, 0), (0, 1), (0, 2), (1, -2), (1, -1), (1, 0), (1, 1), (1, 2), (2, -2), (2, -1), (2, 0), (2, 1), (2, 2)\}$. $P(A) = \frac{1}{5}, P(B) = \frac{2}{5}, P(C) = \frac{13}{25}$. **3** (a) $\frac{1}{3}$, (b) $\frac{2}{3}$, (c) $\frac{1}{9}$, (d) $\frac{10}{27}$, (e) $\frac{2}{27}$, (f) $\frac{4}{27}$, (g) $\frac{11}{27}$, (h) $\frac{1}{9}$. **4** (a) $\frac{1}{2}$, (b) $\frac{1}{3}$, (c) $\frac{1}{5}$, (d) $\frac{1}{2}$. **5** (a) $\frac{1}{720}$. (b) $\frac{4!3!}{6!} = \frac{1}{5}$. (c) $\frac{1}{10}$. (d) $\frac{2 \times 5!}{6!} = \frac{1}{3}$. **6** (a) 0.65, (b) 0.30, (c) 0.047.

6.4.1 **1** $P(A) = \frac{4}{7}$, $P(B) = \frac{2}{7}$, $P(C) = \frac{1}{7}$. **2** $P(A) = P(B) = \frac{1}{4}$, $P(C) = \frac{1}{2}$. **3** p (because $p + q = 1$).

6.7.1 **1** $26^2 = 676$. **2** 48. **3** $4! = 24$. **4** $26^3 \times 10^3 = 17\,576\,000$. **5** (a) $4! = 24$. (b) $\frac{4!}{2!} = 12$. **6** 6, 36. **7** 6 497 400. **8** $\binom{8}{2} = 28$. **9** (a) 15, (b) 1956. **10** 1980. **11** 120.

6.11 **1** 24, 12, 6. **2** (i) $\frac{5}{18}$, (ii) $\frac{1}{18}$. **3** 0.59, (a) $n \geqslant 23$, (b) $n \geqslant 29$. **4** 4620 (out of total of 13 860). (a) $\frac{2}{45} = 0.044$, (b) 1, (c) $\frac{2320}{3600} = 0.644$. **5** (a) $\frac{1}{16}$, (b) $\frac{3}{8}$, (c) $\frac{1}{16}$. An outcome may be specified by an ordered quadruple, e.g. (1, 2, 2, 4), where the $\frac{1}{4}$ hour in which a person arrives is numbered 1, 2, 3 or 4. **6** (i) $\frac{290}{563} = 0.515$. (ii) $\frac{321}{563} = 0.570$. (iii) $\frac{102}{563} = 0.181$. (iv) $\frac{222}{563} = 0.394$. **7** (a) $\frac{7}{15} = 0.467$, (b) $\frac{1}{5}$. (i) $\frac{22}{35} = 0.629$, (ii) $\frac{33}{91} = 0.363$.

Chapter 7

7.3.1 1 Whole of S; points in A; points in B; {HH}; {HH}; {HH}; use $N(S), N(A)$ etc. to calculate probabilities. **2** (a) No. (b) Yes. (c) No. (d) No. **3** $\frac{5}{36} = 0.139$; $\frac{1}{2}$; 1.

7.5.1 1 (i) $\frac{11}{12}$, (ii) $\frac{2}{9}$, (iii) $\frac{3}{4}$. **3** 0.2581.

7.6.1 1 (a) mut. excl.; (b) mut. excl. and exh.; (c) mut. excl. and exh.; (d) mut. excl. **2** (i) $\frac{3}{8}$, (ii) $\frac{3}{8}$, (iii) $\frac{1}{8} \cdot \frac{7}{8}$.

7.7.1 1 (a) $\frac{2}{5}$, (b) $\frac{1}{2}$. **3** $\frac{1}{45}$. **4** (a) (i) $\frac{1}{7}$, (ii) $\frac{6}{7}$, (iii) $\frac{1}{35}$. (b) $\frac{1}{30}, \frac{1}{12}$. **5** (i) $\frac{51}{290} = 0.176$. (ii) $\frac{123}{244} = 0.504$. (iii) $\frac{369}{479} = 0.770$. (iv) $\frac{36}{194} = 0.186$.

7.8.2 1 (a) Neither, (b) indep., (c) indep., (d) neither, (e) mut. excl. **3** (a) $(\frac{4}{9})^3 = 0.088$, (b) 0.115, (c) 0.115, (d) 0.828. **6** $U = \pm 1, \pm 2, \pm 3, \pm 4, \pm 5, \pm 6$, each with prob. $\frac{1}{12}$.

7.9.1 1 Probs are $\frac{0.8\pi}{0.1 + 0.7\pi}$ and $\frac{0.95\pi}{0.2 + 0.75\pi}$.

7.9.3 1 $\frac{4}{11}$. **2** (a) 0.155. (b) 0.645. **3** (i) $\frac{9}{10}$, (ii) $\frac{1}{5}$, (iii) $\frac{23}{35} = 0.657$, (iv) $\frac{8}{35} = 0.229$. $\frac{3}{16}$. **4** 0.33, 0.39, 0.28, 0.332. **5** (a) (i) $\frac{1}{9}$, (ii) $\frac{2}{9}$, (iii) $\frac{2}{3}$. (b) 0.661. **6** 80, 0.7125.

7.12 2 (i) 0.4, (ii) 0.1, (iii) 0.8. **3** 0.936. **4** (i) 0.32, (ii) 0.5625 $(= \frac{9}{16})$, (iii) 1.48. **5** 0.952, 45. **6** $\frac{5}{7} = 0.714$. **7** (a) (i) 0. (ii) P(A)P(B). (b) (i) 0.79, (ii) 0.34. (c) 0.118. **8** (i) $\frac{1}{32} = 0.031\,25$. (ii) $\frac{13}{20} = 0.65$. (iii) $\frac{17}{125} = 0.136$. (iv) $\frac{24}{125} = 0.192$. (v) $\frac{1}{64} = 0.016$.

Chapter 8

8.2.1 1 $P(R=0) = 0.04$, $P(R=1) = 0.90$, $P(R=2) = 0.05$, $P(R=3) = 0.01$. **2** (a) $P(R=0) = 0.25$, $P(R=1) = 0.5$, $P(R=2) = 0.25$. (b) $P(R=0) = (1-\pi)^2$, $P(R=1) = 2\pi(1-\pi)$, $P(R=2) = \pi^2$. **3** (i) $P(R=i) = \frac{1}{6}$ for $i = 1, 2, 3, 4, 5, 6$. (ii) For example: $P(R=0) = 0.1$, $P(R=1) = 0.2$, $P(R=2) = 0.6$, $P(R=3) = 0.1$ (will vary for different people). (iii) (a) $P(R=0) = 0.410 = P(R=1)$, $P(R=2) = 0.154$, $P(R=3) = 0.026$, $P(R = 4) = 0.002$ (probs to 3 d.p., sum not exactly 1 due to rounding error). (b) $P(R = 0) = (1 - \pi)^4$, $P(R = 1) = 4\pi(1-\pi)^3$, $P(R = 2) = 6\pi^2(1 - \pi)^2$, $P(R=3) = 4\pi^3(1 - \pi)$, $P(R=4) = \pi^4$. (iv) For example: denoting brown, black, fair, red hair by b, c, f, g, $P(R=b) = 0.60$, $P(R=c) = 0.15$, $P(R=f) = 0.20$, $P(R=g) = 0.05$ (will vary for different groups of people).

8.6.1 1 (i) Yes, $n = 3, \pi = \frac{1}{6}$. (ii) As (i). (iii) No, n not constant. (iv) No, prob. of obtaining ace changes as cards dealt. (v) Yes, $n = 30, \pi = \frac{1}{7}$ approx. (vi) No, n is not fixed. (vii) No, the people ($=$ trials) in groups are not independent of one another. **2** (i) $0.205, 0.0067, 1 - (0.8)^5 = 0.672$. **3** No. $1 - (\frac{2}{3})^3 = 0.704$. **4** (i) 0.774, (ii) 0.226. **6** 125, 75, 15, 1; fits fairly well (see Ch. 19).

8.8.1 1 Probabilities for same values as Table 8.3 are 0.756, 0.664, 0.639, 0.632, 0.618, 0.589, 0.545, 0.489, 0.427, 0.364, 0.247, 0.156, 0.092, 0.051, 0.027. This is very strict for π small but not for π between about 5 and 10%. Best is $n = 12$, $r = 0$. **2** (i) 0.9418, (ii) 0.8483. **3** (a) (i) 0.8122, (ii) 0.4114, (iii) 0.0442.

(b) (i) 0.9245, (ii) 0.6769, (iii) 0.2061; (c) (i) 0.5405, (ii) 0.1117, (iii) 0.0013.
(d) (i) 0.9392, (ii) 0.6475, (iii) 0.1227.   **4** 0.1544.

5 $\sum_{r=0}^{4} \binom{100}{r} \pi^r (1-\pi)^{100-r}$; 0.6120; 1.634 (geometric distribution).

8.10.1 1 0.5, 7, 12.5, 1.5.   **2** Two values of Q correspond to each value of R.

8.13 1 (i) $\frac{12}{25}$, (ii) $\frac{13}{25}$.   **2** $P(R=0)=\frac{3}{4}$, $P(R=1)=\frac{1}{4}$.

3
Amount	−5	−4	−3	−2	−1	0	1	2	3	4	5
Probability	$\frac{1}{36}$	$\frac{2}{36}$	$\frac{3}{36}$	$\frac{4}{36}$	$\frac{5}{36}$	$\frac{6}{36}$	$\frac{5}{36}$	$\frac{4}{36}$	$\frac{3}{36}$	$\frac{2}{36}$	$\frac{1}{36}$

4 Examples: (a) Number that shows on one throw, $\kappa=6$. (b) Throw die once. Record $R=0$ when 1 or 2 shows, $R=1$ when 3, 4, 5 or 6. $\pi=\frac{2}{3}$. (c) Throw die four times. Record number of sixes. $n=4$, $\pi=\frac{1}{6}$. (d) Record number of throws required to obtain a six. $\pi=\frac{1}{6}$.   **6** 4.

7 (i)
R	0	1	2
Probability	$\frac{1}{4}$	$\frac{1}{2}$	$\frac{1}{4}$

(ii)
$1/(1+R)$	1	$\frac{1}{2}$	$\frac{1}{3}$
Probability	$\frac{1}{4}$	$\frac{1}{2}$	$\frac{1}{4}$

8 (i) $(\frac{4}{5})^4 = 0.410$. (ii) 0.154. (iii) $1-(\frac{1}{5})^4 = 0.998$.   **9** One.   **10** See Fig. 15.1.   **12** Binomial, $n=6$, $\pi=\frac{1}{2}$. (i) 0.226. (ii) 0.073.   **13** (a) 3. (b) (i) $\frac{1}{5}$, (ii) $\frac{3}{80}=0.038$.   **14** (i) $6(\frac{1}{50})^3 = 3/62\,500$. (ii) $1164/6\,250\,000$. $P(R=r)=(\frac{1}{50})(\frac{49}{50})^{r-1}$; 0.18.   **16** 0.410, 0.410, 0.514, 0.026, 0.002.
17 (a) $n=3$, $\pi=\frac{1}{4}$ gives probs. 0.4219, 0.4219, 0.1406, 0.0156 for 0, 1, 2, 3. Associate runs of 4 digits 0001–4219 with value 0, 4220–8438 with 1, 8439–9844 with 2, 9845–0000 with 3. (b) Use digits in pairs: 01–20 corresponds to $r=0$, 21–60 to $r=1$, 61–80 to $r=2$, 81–90 to $r=3$, 91–97 to $r=4$, 98–00 to $r=5$.   **18** Study 3, accept: $(1-\pi)^3$; study 6, accept: $3\pi(1-\pi)^5$; study 3, inspect: $\pi^2(3-2\pi)$; study 6, inspect: $3\pi(1-\pi)^2\{1-(1-\pi)^3\}$. (i) 0.729, 0.177, 0.028, 0.066. (ii) 0.512, 0.197, 0.104, 0.187.   **19** (i) 0.4401, (ii) 0.7103, (iii) 0.2897.

Chapter 9

9.1.1 1 75p.   **2** 7.   **3** 2.   **4** $\frac{1}{2}(\kappa+1)$.

9.2.3 1 (i) $\frac{1}{2}$, (ii) 1.   **2** $8\frac{1}{3}$.   **3** $n\pi=900$.   **4** 12, 24.   **5** 0, gives very little information.   **6** (a) 5, (b) 2, (c) 20.

9.4.1 1 (i) $\frac{3}{2}$, (ii) 3, (iii) 1.   **2** $E[R]=1$, $E\left[\frac{1}{R+1}\right]=\frac{7}{12}$.

9.6.1 1 $\sigma^2=\frac{1}{2}$.   **2** (a) 0, 4, 2. (b) 1.7, 0.61, 0.78. (c) 11.7, 0.61, 0.78. (d) 0, 1.2, 1.095.   **3** (i) 0.475, 0.689, (ii) 6.944, 2.635. (iii) 90, 9.49. (iv) 4, 2 and 16, 4.
4 5, 2.5, 0.377.   **5** 8, 2.19, 0.126, 0.016.   **6** $\mu=3$, $\sigma=1.64$; (a) 0.35, (b) 0.04.
7 π, $\pi(1-\pi)$.

9.7.2 1 (i) 5μ, $25\sigma^2$. (ii) $-\mu$, σ^2. (iii) $10-\mu$, σ^2. (iv) 0, σ^2. (v) μ/σ, 1. (vi) $\frac{2}{3}\mu-\frac{10}{3}$, $\frac{4}{9}\sigma^2$.   **2** $(R-2)/1.265$.

9.8.3 1 (i) 0.0336, 10.62. (ii) 0.1993, 7.15.   **2** $n(2-p)$. 0.6144. 0.2013. 8.07.

9.11 1 $\frac{1}{15}$, $\frac{8}{3}$, $\frac{14}{9}$, $\frac{3}{5}$.   **2** $\frac{1}{2}(k+1)$, $\frac{1}{12}(k^2-1)$.   **3** £95, £115.   **4** 52.8, 251 mins.
6 When $\sigma^2=0$, i.e. there is only one possible value of R.   **7** $\frac{9}{4}$, $\sqrt{\frac{19}{16}}=1.09$.
8 Binomial, $n=10$, $\pi=\frac{1}{5}$. 2; 0. 4; 10.   **9** 0.322, 50p, 50, 21.   **10** $2\frac{3}{4}$.   **13** $\frac{1}{2}(1+t^2)$,

1, 1. **14** $\frac{\pi t}{1-t(1-\pi)}\cdot\frac{1}{\pi},\frac{1-\pi}{\pi^2}$. **15** 0.2219; 0.8079; 0.1921. 0.2769; 0.5816; 0.4184. Quite well; P(reject) twice as high for $\pi = 0.06$.

Revision Exercises B

1 $\frac{4}{7}$. **2** 10 out of 32. (a) 50%, (b) 10%. **3** $\frac{1}{2}$, 0, 0, $\frac{1}{3}$. No. **4** $a_n = (n-1)(n-2)/2^{n+1}$ for $n \geqslant 3$. **5** $\frac{125}{216} = 0.579$. **6** $\frac{11}{36} = 0.306$. **7** (i) Separate accidents to a person are not independent events. The probability 0.02 is the probability of one *or more* accidents, not just of one accident. (ii) $\frac{2}{3}$ is the mean number of answers. Prob. $= 1 - (\frac{5}{6})^4 = \frac{671}{1296} = 0.518$. **8** (a) $\frac{13}{16} = 0.813$. (b) $\frac{4}{7}, \frac{2}{7}, \frac{1}{7}$. (c) $\frac{3}{7}$. **9** (i) $(\frac{5}{6})^5 = 0.402$. (ii) 0.263. (iii) 0.015. $5^{r-1}/6^r$; 4. $\sum_{(r=1)}^{\infty} rq^{r-1}p = 6; \sum_{(r=1)}^{\infty} r^2q^{r-1}p = 66$; **10** $P(R=r) = 2^{r-1}/3^r$ for $r \geqslant 1$. $P(R=r) = \frac{1}{3}$ for $r = 1, 2, 3$. (i) $\frac{8}{27} = 0.296$, (ii) $\frac{38}{81} = 0.469$. **11** (a) 295/2 666 667 = 0.000 111. (b) $\frac{7}{295} = 0.0237$. (c) £0.003 75. (d) 4.5%. (e) 952. **12** 0.507. **13** $\frac{1}{4}\cdot\mu = -\frac{1}{3}, \sigma^2 = \frac{43}{18} = 2.389$. **14** 0.322, 0.268, 0.224, 0.186, $\frac{1}{8}, \frac{1}{8}, \frac{1}{4}, \frac{1}{2}$. **16** 0.04, 0.022, £761.

17

r	0	1	2	3	4	5
Frequency	16.3	80.2	157.6	154.8	76.1	15.0

18 For $N, 3/2^k + 7/2^{3k} - 9/2^{2k}$ for $k \geqslant 1$. For $M, 7/2^{3k}$ for $k \geq 1$. **19** (i) 0.328, (ii) 0.410, (iii) 0.263. **20** 6, 3. $\frac{11}{26} = 0.423$. **21** (a) 1/120. (b) 7/120. **22** $(1-p)/p^2$ (geometric). **23** (i) Prefer B: long-term (expected) loss $= £5600$. (ii) E[loss] $= £4580$. **24** Accept on kth batch with geometric prob. $P(0)\{P(1)\}^{k-1}$; hence $P(\text{accept}) = P(0)/\{1 - P(1)\}$. $P(\text{reject})$ for any batch is $1 - P(0) - P(1)$; hence required average is $\{1 - P(0) - P(1)\}^{-1}$. 0.1818; 1.68.

25 (i)

$Y(£)$	4000	2000	0	-2000	-4000
$P(Y)$	0.2	0.4	0.2	0.1	0.1

$P(\text{loss}) = 0.2$. (ii) 11.5, 1.45, £1000, 5.8×10^6.

Chapter 10

10.1.1 1 $P(X=1, Y=1) = \frac{1}{6}$; $P(X=1, Y=-1) = \frac{1}{2}$; $P(X=0, Y=1) = \frac{1}{12}$, $P(X=0, Y=-1) = \frac{1}{4}$. **2** No.

3

X:		1	2	3	Y:		-1	0	1
Probability:		0.4	0.4	0.2	*Probability*:		0.35	0.35	0.30
(i) $X+Y$:	0	1	2	3	4				
Probability:	0.20	0.15	0.40	0.20	0.05				
(ii) $X-Y$:	0	1	2	3	4				
Probability:	0.15	0.15	0.45	0.20	0.05				
(iii) XY:	-3	-2	-1	0	1	2	3		
Probability:	0.05	0.10	0.20	0.35	0.15	0.10	0.05		

10.2.1 1 $P(X=0) = \frac{1}{4}$, $P(X=1) = \frac{1}{2}$, $P(X=2) = \frac{1}{4}$, $P(Y=0) = \frac{11}{24}$, $P(Y=4) = \frac{1}{4}$, $P(Y=5) = \frac{7}{24}$. **2** (i) 2μ, (ii) 0, (iii) μ, (iv) μ.

$3 \frac{1}{2}(R+S)$: -2 $-\frac{3}{2}$ -1 $-\frac{1}{2}$ 0 $\frac{1}{2}$ 1 $\frac{3}{2}$ 2

Probability: $\frac{5}{225}$ $\frac{14}{225}$ $\frac{26}{225}$ $\frac{40}{225}$ $\frac{55}{225}$ $\frac{40}{225}$ $\frac{26}{225}$ $\frac{14}{225}$ $\frac{5}{225}$ $E[\frac{1}{2}(R+S)]=0.$

4 μ. **5** (a) μ; (b) μ.

10.3.1 1 (i) XY: -1 0 1

Probability: $\frac{1}{2}$ $\frac{1}{3}$ $\frac{1}{6}$ $E[XY]=-\frac{1}{3}.$

(ii) XY: -4 -2 -1 0 1 2 4

Probability: $\frac{26}{225}$ $\frac{44}{225}$ $\frac{20}{225}$ $\frac{81}{225}$ $\frac{16}{225}$ $\frac{28}{225}$ $\frac{10}{225}$ $E[XY]=-\frac{4}{9}.$

2 X: 0 1 $X+Y$: 0 1 2 XY: 0 1

$Y=0$: $\frac{1}{4}$ $\frac{1}{4}$ *Probability*: $\frac{1}{4}$ $\frac{1}{2}$ $\frac{1}{4}$ *Probability*: $\frac{3}{4}$ $\frac{1}{4}$ $E[X+Y]=1, E[XY]=\frac{1}{4}.$

1: $\frac{1}{4}$ $\frac{1}{4}$

3 XY: 1 2 3 4 5 6 8 9 10 12 15 16 18 20 24 25 30 36

Probability: $\frac{1}{36}$ $\frac{2}{36}$ $\frac{2}{36}$ $\frac{3}{36}$ $\frac{2}{36}$ $\frac{4}{36}$ $\frac{2}{36}$ $\frac{1}{36}$ $\frac{2}{36}$ $\frac{4}{36}$ $\frac{2}{36}$ $\frac{1}{36}$ $\frac{2}{36}$ $\frac{2}{36}$ $\frac{2}{36}$ $\frac{1}{36}$ $\frac{2}{36}$ $\frac{1}{36}$

$E[XY]=12.25.$

10.4.1 1 (i) $4\sigma_R^2 + \sigma_S^2, \sqrt{4\sigma_R^2 + \sigma_S^2}$ (ii) $\frac{9}{16}(\sigma_R^2 + \sigma_S^2), \frac{3}{4}\sqrt{\sigma_R^2 + \sigma_S^2}.$ (iii) $16\sigma_S^2 + \sigma_R^2,$

$\sqrt{16\sigma_S^2 + \sigma_R^2}.$ **2** 2.5, 1.81. **3** $\frac{17}{6}=2.83; \frac{145}{72}=2.01; 1.42.$ **4** $\sigma^2/n, \sigma/\sqrt{n}.$

10.5 1 22.5, 41.25. **2** (i) $\frac{20}{3}$, (ii) $\frac{40}{3}$, (iii) $\frac{80}{3}$, (iv) $\frac{100}{3}$. **3** (ii) $\frac{4}{3}$. (iii) not independent. (iv) $P(Z=0)=\frac{1}{9}; P(Z=1)=P(Z=2)=\frac{4}{9}\cdot\frac{4}{3}.$ **4** $\frac{1}{6}, 1, \frac{2}{3}, \frac{2}{3}, \frac{1}{3}.$
5 $\frac{4}{5}, \frac{1}{5}, \frac{4}{5}.$ **6** $\frac{1}{64}$, 8, 12, 6, 1. **7** $-\frac{3}{38}, \frac{789}{1444}. -\frac{5}{14}, \frac{275}{588}.$ **8** $\frac{1}{10}$. (i) 6, 4. (ii) 0, 4.

U: 1 2 3 4 6 8 9 12 16

Probability: 0.01 0.04 0.06 0.12 0.12 0.16 0.09 0.24 0.16 $\mu=9, \sigma^2=19.$

Chapter 11

11.3.1 1 (i) $\mu=1.5, \sigma^2=0.25.$

(ii) *Sample*: 111 112 121 211 122 212 221 222

$\bar{x}$: 1 4/3 4/3 4/3 5/3 5/3 5/3 2

Probability: 1/8 3/8 3/8 1/8 **3** (i) 4.5, 8.25.

(ii) *Mean*: 0 $\frac{1}{2}$ 1 $1\frac{1}{2}$ 2 $2\frac{1}{2}$ 3 $3\frac{1}{2}$ 4 $4\frac{1}{2}$ 5 $5\frac{1}{2}$ 6 $6\frac{1}{2}$ 7 $7\frac{1}{2}$ 8 $8\frac{1}{2}$ 9

$100 \times$ *Probability*: 1 2 3 4 5 6 7 8 9 10 9 8 7 6 5 4 3 2 1

4.5, 4.125.

11.5.1 1 Median, 3. **3** Larger variance.

11.7 1 $\alpha=1/16.$ $\mu=4.$ $\sigma^2=2.5.$ (i) 4, 10. (ii) 20, 32.5.

2 x: 0 1 2 3 4

$P(x)$: 1/22 10/33 5/11 2/11 1/66. $\frac{20}{11}; \frac{84}{121}.$

3 *Mid-range*: 1 $1\frac{1}{2}$ 2 $2\frac{1}{2}$ 3

Probability: 1/64 9/32 13/32 9/32 1/64, expected value 2. 33/32.

5 $\frac{1}{n}\{A + \frac{1}{2}B(n+1)\} > \frac{1}{2}B$ for all finite n. Variance of mean would increase as n increased.

Chapter 12

12.6 1 (a) List of qualified electors for that village. (b) Individual. (c), (d) Stratified random if possible, otherwise simple random. (e) Age, sex, occupation affect people's leisure activities; also some have jobs that form leisure for others.

4(a)

Days open	Shop type 1	2	3	Total
5	—	$2(54\frac{1}{2})$	—	$2(54\frac{1}{2})$
6	$3(59\frac{2}{3})$	$4(52\frac{1}{2})$	$4(76\frac{1}{2})$	$11(63\frac{2}{11})$
7	$3(66\frac{2}{3})$	—	$2(97\frac{1}{2})$	$5(79)$
	$6(63\frac{1}{6})$	$6(53\frac{1}{6})$	$6(83\frac{1}{2})$	$18(66.61)$

(c) Better to use same enumerators in each town and to select shops at random.

Chapter 13

13.2.1 1 (a) NH: The coin is fair. AH: The coin is biased. (b) NH: There is no difference in taste between cups containing 1 and 2 teaspoonfuls of sugar. AH: There is a difference in taste. (c) NH: There is no difference in life span between rats fed on the two diets. AH: There is a difference. (d) NH: There is no difference between the employees' mean travelling time on the day of the accident and that of previous months. **2** $\frac{1}{32}$. Two-tail test; probability of the given result or a similar extreme result is 6.25%. So not significant at 5% level and NH not rejected. **3** NH: $\pi = \frac{1}{6}$, one-tail test, probability of 3 or more sixes 6.23%, not significant. Evidence is that die has not been successfully loaded.

13.5.3 1 (a) (i), (ii), (v), (vi) 1/70. (iii), (iv) 1/35. (b) (i) {10, 11, 25, 26}. (ii) {10}. (iii) {24, 25, 26}. **2** $T_A = 11$. Yes, at 5% significance level.

13.6.1 1 $T_A = 94$. $U_{BA} = 58$. No difference. **2** (i) $T_+ \leqslant 8$ $(P = 0.0244)$. (ii) $T_+ \leqslant 11$ $(P = 0.0527)$. **3** (a) $T_- = 13$. No evidence of difference. (b) P = 0.344. No evidence of difference.

13.7.1 (a) 1.76%. (b) (i) 7.06%. (ii) 50%.

13.10 1 Probability of results equally, or more, extreme = 34.4%. No reason to reject claim. **2** Specify outcomes by number of correct tastings. At 5%, critical region is {0, 1, 2, 10, 11, 12}; 1%, {0, 1, 11, 12}; 0.1%, {0, 12}. **3** (a) {4, 5}, (b) {5}. **4** Two-tailed test. Probability of given, or more extreme, result = 1.29%. Reject NH that methods are equally effective.

5 One-tailed test. P(7 or more values < 11.0) $= 17.2\%$. No reason to reject NH. **6** (i) 0.938, (ii) 0.0087. (a) 0.0087. (b) 0.062. **7** (i) $p^4 + (1-p)^4$. (ii) $p^7 + 7p^6(1-p) + 7p(1-p)^6 + (1-p)^7$. Procedure 2. **8** Distribution of times symmetrical, so mean $=$ median approx.; unreasonable. **9** Sign test. $n = 25, n_+ = 18$. Sig. at 5%. Mean $= 7.804$. Median $= 6.6$. Distribution very skew. **10** $U_{\text{BA}} = 132$. No evidence of difference. **11** Sign test using 22 preferences. $n_+ = 15$. No evidence of difference. **12** $T_+ = 193$. Evidence of difference. **13** Cannot reasonably assume normality. (i) Sign test, cannot reject NH of no difference. (ii) Wilcoxon signed-rank test gives evidence of difference. Test (ii) has greater power through using magnitudes.

Revision Exercises C

1 Show $E[XZ] = 0 = E[X]E[Z]$. **2** 2.506, 1.51, 0.946. Use $\sigma_x^2 + \sigma_y^2 = \sigma_{x+y}^2$.

3 $(1-p), p(1-p)$

(i)

$Y =$	0	1
$X = 0$	p^2	$p(1-p)$
1	$p(1-p)$	$(1-p)^2$. Equal for all p.

(ii)

$Y =$	0	1
$X = 0$	p^2	$p(1-p)$
1	$(1-p)^2$	$p(1-p)$. Equal for $p = \frac{1}{2}$.

4 (a) $P(Y = 1) = 0.7, P(Y = 9) = 0.3$.

(b)

$X =$	-3	-1	1	3
$Y = 1$	0	0.3	0.4	0
9	0.2	0	0	0.1.

(c) -3.6.

5 No. $2\beta + 4\alpha$. $4\beta(1-\beta) + 16\alpha(1-\alpha-\beta)$.

6 (i)

$X =$	0	1
$Y = 0$	$\frac{1}{6}$	$\frac{5}{18}$
1	0	$\frac{5}{9}$.

(ii) $P(Z = 0) = \frac{1}{6}, P(Z = 1) = \frac{5}{18}, P(Z = 2) = \frac{5}{9}$. $N = Z + 1$. 1.556, 3.858.

7 (b)

s^2	$\frac{1}{2}$	2	$4\frac{1}{2}$	8
Probability	0.4	0.3	0.2	0.1.

$\sigma^2 = 2, E[s^2] = 2.5$. Sampling is without replacement.

8 Var $[Z_2] = \frac{pq}{m+n} < \text{Var}[Z_1] = \frac{1}{4}pq\left(\frac{1}{m} + \frac{1}{n}\right)$; use Z_2.

9 R is geometric. $E\left(\frac{1}{R}\right) = \pi \sum_{r=1}^{\infty}(1-\pi)^{r-1}/r \neq \pi$. **10** $E[S] = \frac{1-\pi}{\pi}$, whereas $E[R] = \frac{1}{\pi}$. Var$[R] = Var[S]$. **11** $k = \frac{4}{3}$. **12** (a) Yes. (b) $\mu^2 + \sigma^2/50$. No; bias $\pi\sigma^2/50$. (c) $\pi(\mu^2 + \sigma^2)$. No; bias $\pi\sigma^2$. **15** $(1-p)^{10}$. 0.865. **16** (b) $(n+3)/2^{n+2}$. **17** (a) 27, 9; 18 and 36. **18** $p\mu_1 + (1-p)\mu_2$; $p\sigma_1^2 + (1-p)\sigma_2^2 + p(1-p)(\mu_1 - \mu_2)^2$; $\mu_1 - \mu_2$; $\sigma_1^2 + \sigma_2^2$. **19** 0.614. **20** 0.617; yes (prob. of this event is 0.149). **21** No. (Prob. $= 0.070$.)

695

22 (i) 2.212×10^{-5}. (ii) 20, 4. **23** (a) 0.799, (b) 0.700. **24** (a) $\frac{1}{2}$; 0, 0, $\frac{1}{6}$, 1. (b) 0. Probs of type II errors are 0, $\frac{1}{2}, \frac{5}{6}$ 1. (c) $1 - \text{prob}$ of type II error. **25** Binomial, $n = 3, \pi = m/6$; $\frac{1}{2}m$, $\frac{1}{12}m(6 - m)$. $\hat{m} = 0, 2, 4, 6$. $P(\hat{m} = 0) = 0.2 = P(\hat{m} = 4), P(\hat{m} = 2) = 0.6. \frac{1}{27}$. Power $= 1$, 0.579, 0.296, 0.125, 0.005, 0 for $m = 0, 1, 2, 3, 5, 6$. **26** The 12 firm results indicate there is a difference. **27** $k = 1/20$. $P(X = 0) = 3/10 = P(X = 3).P(X = 1) = 1/5 = P(X = 2)$. Y same. Not independent: $P(X = x, Y = y) \neq P(X = x)P(Y = y).|x - y|$ takes values 1, 3 each with prob. 3/10, and 2 with prob. 2/5. 2. 3/5. **28** $T_S = 20, U = 5$. Evidence new treatment better. **29** (a) Rank sum, or U. Significance doubtful, borderline 5%, need more data. (b) Use rank sum, allow for ties. No evidence of difference. A much more variable than C, distributions may not be similar, hence best to compare means and variances separately.

Chapter 14

14.1.1 1 1.46. **2** (i) $f(x) = \frac{1}{k}, 0 \leqslant x \leqslant k$. (ii) $F(b) = b/k, 0 \leqslant b \leqslant k$. $M = \frac{1}{2}k$.

14.2.1 1 1, $F(b) = \frac{1}{2}b^2 (0 \leqslant b \leqslant 1)$ or $2b - \frac{1}{2}b^2 - 1(1 \leqslant b \leqslant 2); 1, 1$. **2** Some area under graph is negative (below axis); cannot have negative probabilities. **3** (i) $k = 1/2$. $F(b) = \frac{b^2}{4}, 0 < b < 2$. (ii) No (see question 2). (iii) No. (iv) $k = \frac{3}{14}$. $F(b) = \frac{1}{7}(b^{3/2} - 1), 1 < b < 4$. (v) $k = \frac{1}{2}$. $F(b) = \frac{1}{2}(1 + \sin b)$ $\left(-\frac{\pi}{2} < b < \frac{\pi}{2}\right)$. (vi) $k = 2$. $F(b) = 1 - e^{-2b}(0 < b < \infty)$.

14.5.1 1 $\frac{8}{15} = 0.533$. 0.0489. 0.707. **2** (i) 2, (ii) $\frac{3}{4}$, (iii) $F(b) = b^3 (0 < b < 1)$, (iv) 0.109. **3** $\frac{3}{32}$. Origin moves by transformation $y = x + 2$; no change in scale. $2, 0.894, 2, 0.374$.

14.7.1 1 $F(b) = 1 - e^{-b} (b > 0); \ln 2, \frac{1}{2} - \frac{1}{e} = 0.132$. **2** $k = \lambda$. 100, 10 000. 0.607, 0.368. **3** 2. **4** (i) $\frac{1}{\lambda} + \frac{1}{\mu} + \frac{1}{v}$, (ii) $\frac{1}{\lambda \mu}$, (iii) $\frac{1}{\lambda^2} + \frac{1}{\mu^2} + \frac{1}{v^2}$. (iv) $\frac{1}{\mu^2} + \frac{9}{v^2}$. **5** $\frac{1}{\mu}e^{-t/\mu}$. $P(t > 3000) = 0.135$. 5000 h.

14.8.2 1 $\frac{1}{2}(1 + t), \frac{1}{2}(1 + e^t), \frac{1}{2}$. **2** $\frac{e^{bt} - e^{at}}{(b - a)t}; \frac{1}{2}(a + b); \frac{1}{12}(b - a)^2$.

14.10.1 1 $f(y) = 2y/3a, \sqrt{a} < y < 2\sqrt{a}$; 14 $\sqrt{a}/9$; $0.0802a$; $\frac{7}{12}$. **2** $f(h) = \pi h^2/3$ $(0 < h < (9/\pi)^{1/3} = 1.42)$; (i) 0.349; (ii) 1.06; (iii) 0.075; (iv) 0.422.

14.12 1 1, $\frac{1}{6}$, 0.42. **2** $\sigma = (b - a)/\sqrt{12}$. **3** $\frac{1}{12}, \frac{9}{8}$, 1; 0.167; 0.145. **4** 0, $\frac{1}{3}, \frac{2}{3}$. **6** (i) $1/\pi$, (ii) $\frac{1}{2}$, (iii) $\frac{1}{3}$. **7** (i) $2\lambda/3, \lambda^2/18$. (ii) 16, £3. **8** $k = \frac{6}{5}, c = 3$. 837.5. **10** Mgf is $(1 - \frac{1}{2}t)^{-1}$. **11** $f(y) = \frac{1}{9}e^y$ $(0 < y < \log_e 10 = 2.3026)$. $0.519 = e(e - 1)/9$. 0.171. **12** $3\pi, 3\pi^2$. 0.62.

Chapter 15

15.2.1 (i) 34.1, (ii) 13.6, (iii) 15.7, (iv) 15.7, (v) 0.13, (vi) 0.50.

15.4.1 1 (i) $\frac{1}{\sqrt{8\pi}} \exp\{-(x - 3)^2/8\}$; (ii) $\frac{1}{\sqrt{10\pi}} \exp(-x^2/10)$;
(iii) $\frac{1}{\sqrt{2\pi}} \exp\{-(x + 2)^2/2\}$; (iv) $\frac{1}{\sqrt{20\pi}} \exp\{-\pi(x + 6)^2/20\}$.
2 0, 3. **3** $1/\sqrt{6\pi}$.

15.5.1 **1** (i) 0.8413, (ii) 0.9773, (iii) 0.1360, (iv) 0.6915, (v) 0.1587, (vi) 0.6826. **2** $a = b = 0.95, c = 1, d = 0.4, e = -0.75, f = 0.99$. **3** (a) 0.2524, (b) 0.0478, (c) 0.0913, (d) 0.9522, (e) 0.3413, (f) 0.9044, (g) 0.0085, (h) 0.2486. **4** (i) 0.3830, (ii) 0.0228, (iii) 0.1336, (iv) 0.9938, (v) 0.9785.

15.5.3 **1** 75, 122. **2** 63. **3** 95, 9, 1%, 0.2%. **4** Time in decimal minutes after 5 p.m.: 25.23, 2.90; (i) 18.88, (ii) 13.

15.10 **1** 7.52 a.m. **2** 0.753. **3** 0.224. **4** 0.0098, 0.971, 6.19 m. **5** 0.212, 0.673, 0.115; 0.90p; 0.60p. **6** 30.01, 0.163.

mm	≤29.6	29.7	29.8	29.9	30.0	30.1	30.2	30.3	≥30.4
Frequency	6.7	20.8	53.8	96.8	120.5	104.1	62.3	25.9	9.2

7 (i) 0.0054, (ii) 9, 0.180, (iii) 0.009. **8** Normal approx., using continuity correction (see p. 377) 149.875, 150.0625. **9** Recommend (ii). Min. loss and least cost per article. **10** Answer depends on proportion of packets <140 g that can be accepted. With mean 146 g proportion is 0.001 35; for 148 g, it is 0.000 032.

Chapter 16

16.1.1 **1** (i) 0.013, (ii) 0.186. **2** 54, using sample variance; pop. variance unknown; small sample, so sample variance may not be very good estimate of pop. variance. **3** 166. **4** 0.136.

16.2.1 **1** 0.0013. Large sample, need not be normal, because mean will be approx. normal **2** 11.01 a.m.; 0.922. **3** (i) 0.06, (ii) 0.94, (iii) 0.

16.3.1 **1** (i) 0.342, (ii) 0.656, (iii) 0.0012. **2** (a) 37, sixth form like rest of school; may be epidemics affecting younger groups more. (b) 3 (using normal approx.). **3** 0.848. (1413; 1527). **4** 0.153; 0.284. **5** With 100 questions, probabilities (a) 0.0004, (b) negligibly small. Smallest n for 1% pass probability (a) 52, (b) 27.

16.4.1 **1** 0 (to 3 dec. pl.). **2** (a) 0.0068, (b) 0.0001. **3** 0.6, 0.11.

16.5.1 **1** (a) 0.391, (b) 0.609, (c) about $8.22\frac{1}{2}$. **2** $\mu_1 \pm \mu_2$, $\sigma_1^2 + \sigma_2^2$. (a) (94.4; 105.6). (b) $92\frac{1}{2}$%. (c) 22.1%. **3** (i) 0.924, (ii) 0.865, (iii) 0.789.

16.6.4 **1** (a) 45.10, 54.90; 41.77, 58.23. (b) (i) in control; (ii) outside warning, but inside action; (iii) one outside warning; (iv) both outside warning, opposite directions; stop. **2** Warning 113.0, 127.0; action 108.2, 131.8. All means in control. **3** Warning 109.1, 130.9 for totals; action 101.8, 138.2. Point 12 outside warning, but inside action; 15, 16, 17 outside warning, stop; 18, 19, 20 outside action. Ranges: proceed as in Section 16.6.3, all in control (including 15, which is the highest).

16.6.6 **1** For totals, warning 4, 16, action 0, 20. One isolated 16, take no action; two adjacent 17s, stop. **2** For totals, warning 15.5 and 34.5, action 9.0, 41.0. One 36 outside warning, next almost as large, next below warning. Check variability but no need to stop.

16.8 **1** (i) 1.248×10^{-21}, (ii) 0.996. **2** 0.0008, 0.9994. **3** Using continuity correction, 0.62%, 35.2%; 6, 5; 0.115. **4** 9%. **5** (i) $\mathcal{N}(8, \frac{1}{4})$. (ii) $\mathcal{N}(100, 50)$. (iii) No. Weights not necessarily normal, sample small. (iv) No, because π

too small (see Chapter 20). (v) $\mathcal{N}$ $(340, 360)$. (vi) $\mathcal{N}$ $(206, 288)$.   **6** 19.22.
7 (i) $\frac{3}{4}$, (ii) 0.0925, (iii) 0.0005.   **8** 0.0005.   **9** 0.81.   **10** Every 433 hours
approx.   **11** 0.18, 3500.   **12** (i) 243 mm, 1.78 mm. (ii) 0.253, (iii) 15.33 mm.
13 $M = 63$ approx.; 63.05, 11.28; normal, from the appearance of frequency
table.   **14** (a) 0.0266, (b) 6.504, (c) 64.4%.

Revision Exercises D

1 $c = 1/\sigma$.   **2** Expected value $= \frac{16}{7}$ (number is geometric dist. with $\pi =$
$\frac{7}{16}$).   **3** 0.57.   **4** $\sigma = 1/\lambda$, SIQR $= 0.549/\lambda$.   **5** $x_0 =$ minimum wage. £3167,
£^{2}8 022 222; 0.239.   **6** $f(w) = (2/9\pi)^{1/3} w^{-2/3}$ for $\frac{\pi}{6} \leqslant w \leqslant \frac{4\pi}{3}$.1.96, 1.06.   **7** $\mu_r =$
$\sum_{j=0}^{r} \binom{r}{j} \mu'_{r-j}(-\mu)^j$.   **8** $f(z) = 2z$ for $0 \leqslant z \leqslant 1$. Prob. is $\frac{1}{4}$.   **9** 150, 25, 0.841.
10 a, $\frac{1}{3}(3a^2 + c^2)$; $\frac{1}{9}c^2(6a^2 + c^2)$.   **11** 0.159, 225; 1200, 1225; 1280; 0.0052.
12 0.504.   **13** (i) 0.034, (ii) 0.000 to 3 d.p., (iii) 0.244, (iv) 2.43.   **14** 5.   **16** 1;
$P(X - Y < 0) = 0.76$.   **17** $a = \frac{1}{3}, b = \frac{1}{6}; \frac{1}{2}(-1 + \sqrt{13}) = 1.30$.   **18** 0.253, 0.090;
weights of M, F independent.   **19** 102.7, 5.76, using normal distribution and con-
tinuity correction.   **21** (i) 0.106, (ii) 0.266.   **22** 0, $n/12$; 2638.   **23** $n_1/(n_1 + 2n_2)$.
24 $\frac{5}{16}$.   **25** (a) (i) $P(B) = 0.36 > P(A) = 0.31$. (ii) B mean $= 13.3 > A$ mean $= 13.0$.
(b) A profit $= 26p > B$ profit $= 25p$.   **26** $F(z) = z^2/a^2$ and $f(z) = 2z/a^2$ for
$0 \leqslant z \leqslant a$. $E[Z] = \frac{2}{3}a$.   **27** 89.0%, (i) 84.1%, (ii) 71.5%.   **28** (b) $k = (e^\theta - 1)^{-1}$.
29 0.181, £87.31.   **30** (i) $\mathcal{N}$ $(2000 - 250p, 10000(25p^2 - 32p + 16))$. (ii) 0.
(iii) 16/25. (iv) $2000, $1840. (v) 0.0968, 0.0668. (v) Use (iii), lower prob. of
ruin (lower expectation but also lower variance).

Chapter 17

17.3.2 1 Evidence that chocolates wrapped. One-tail test. $z = 2.0$.   **2** No
evidence of effect of ill-health. Two-tail test. $z = -1.4$.
17.4.1 1 $z = 2.25$. (a) Reject NH '$\mu = 50$'. (b) Do not reject NH. **2** Do not reject
NH '$\mu = 10$'. $z = 1.33$.   **3** Yes; reject NH '$\mu = 64$'. $z = 3.67$.   **4** Time
increased. Reject NH '$\mu = 150$'. $z = 2.68$.
17.8.1 1 Do not reject NH 'mean difference $= 0$'. $z = 0.99$.   **2** Reject NH
'$\mu = 1800$'. $z = -3.16$. Mean of large sample approx. normally dist.
3 NH 'True median $M_0 = 1500$'. AH '$M_0 \neq 1500$'. Find distribution of
sample median M. Calculate $P(M < 1400$ or $M > 1600)$. Reject NH if this
prob. $< 5\%$.   **4** Performance improved. Reject NH '$\mu = 60$'. $z = 1.73$.
No. Pair coached and uncoached.   **5** Yes. $z = 12.01$.   **6** Paired data. No
effect, $z = 1.81$.
17.9.1 1 No. $z = 1.167$.   **2** No sex difference. $z = 1.21$.   **3** Not biased.
$z = -1.73$.   **4** Yes. $z = -2.00$.   **5** No. $z = 1.48$.
17.11.1 1 8.60. Reject manufacturer's claim. $t_{(5)} = 3.54$.   **2** (i) Reject NH.
$z = -1.68$ (one-tail). (ii) Do not reject NH. $t_{(15)} = -1.39$. (i) uses all
the information.
17.12.1 1 Assume diffs. normal; $\bar{d} = 1.2$. Reject NH 'mean diff. $= 0$'; $t_{(7)} = 2.83$.
Pairing removes litter differences.   **2** Yes. $t_{(7)} = 3.01$. $(x - y)$ normally
distributed.

17.13.1 1 Yes. $t_{(23)} = 2.83$. Samples from independent normal distributions. **2** No. $t_{(20)} = 6.01$.

17.17 1 No change. $z = -1.50$. **2** Yes. $z = -5.30$. **3** Yes. $z = 2.45$. 0.037. **4** 188. Two separate samples not significantly less; $z = -1.6$, $z = -1.4$. Combined sample significantly less; $z = -2.12$. **5** Take $P(89 \leqslant X \leqslant 111) = \frac{1}{2}$. Sample is unusual. $z = 2.53$. **6** A stronger. $z = 3.81$. **7** Wheel biased. $z = 4.06$. **8** 0.0062. Mean weight different; $z = -2.74$. **9** Yes. $z = 2.39$. **10** (i) 0.201, (ii) 0.322. 0.137. Yes; $z = 1.77$. No; $z = 3.54$. **11** Paired data. Performance affected. $t_{(9)} = -2.70$. **12** Assume variances of 64, 100 are unbiased estimates of a common variance. First group not better. $t_{(22)} = 1.57$. **13** Yes. $t_{(10)} = 2.38$. Pair rods if possible. **14** $\frac{1}{3}, \frac{4}{9}$. **15** Yes. $t_{(9)} = 2.95$. **16** (i) Inconclusive: $P = 0.055$ in one-tail test. (ii) Strong evidence that it is less: $t_{(9)} = -3.56$. Weights normally distributed (so test of median = test of mean). Uses *all* information. **17** Do not reject NH of no difference: $t_{(5)} = 1.46$ in paired comparison. Differences normal and independent. **18** (b) (i) $T_- = 3.5$, sig. at 1%. Evidence of difference. (ii) Discard only if known reason. Outlier little effect on nonparametric test. **19** Do not reject NH 'mean = 30': $t_{(8)} = -1.77$.

Chapter 18

18.3.1 1 If obs. $= x$, interval is $(x - 3.92, x + 3.92)$. **2** If sample mean $= \bar{x}$, interval is (i) $(\bar{x} - 2.45, \bar{x} + 2.45)$, (ii) $(\bar{x} - 3.22, \bar{x} + 3.22)$, (iii) $(\bar{x} - 4.11, \bar{x} + 4.11)$. **3** $(1017.6, 1022.4)$.

18.4.1 1 $(0.08, 0.16)$. **2** $(7200, 9300)$ approx.

18.5.1 1 $(0.64, 1.50)$. **2** $(\pounds 8630, \pounds 12\,370)$ approx. **3** $(-2.33, +2.80)$. **4** $(-0.47, +0.70)$.

18.7.1 1 $(0.17, 1.2)$. **2** $\bar{x} = 11.38$, $s^2 = 0.4633$; (i) $(10.95, 11.82)$, $(0.233, 1.33)$; (ii) $(10.77, 11.99)$, $(0.190, 1.96)$.

18.8.3 1 Mean $= \frac{1}{2}(\alpha + \beta) = 3.96$. $E[X^2] = \frac{1}{3}(\alpha^2 + \alpha\beta + \beta^2) = 15.976$. Hence (α, β) is $(3.02, 4.90)$. **2** Differentiate $\ln P$ with respect to π; $\hat{\pi} = r/n$. **3** $\bar{x}$.

18.10 1 $(5.27, 7.98)$. **2** 174.61, 7.997. $(173.3, 175.9)$. **3** 0.6, 0.11, $(0.38, 0.82)$. **4** (a) (i) 0.383, (ii) 0.866. (b) $(-0.15, +0.17)$. **5** $\theta, \sqrt{\theta(1-\theta)/n}$. $(0.83, 0.94)$. **6** Neither different. Means sig. diff. at 5% from expected diff. of 3. $(3.03, 3.33)$. **7** No; sig. diff. from expected 50 at 1%. $(0.26, 0.46)$. **8** $(0.17, 4.40)$. Differences assumed normal. **9** $(159.4, 167.9)$. About 64. **10** (a) One-tail $t_{(6)} = 2.04$, sig. at 5%. (c) $(-0.26, +0.89)$ **11** Box 1 $(26.24, 150.6)$. Box 2 $(20.95, 120.3)$. **12** (b) 1/mean $= 0.001\,99$. **13** $E[X] = c\alpha/(\alpha - 1)$, $E[X^2] = c^2\alpha/(\alpha - 2)$. $\tilde{\alpha} = 1 + \sqrt{1 + \bar{x}^2/(m - \bar{x}^2)}$, where $m = \sum x_i^2/n$; then $\tilde{c} = (\tilde{\alpha} - 2)m/(\tilde{\alpha} - 1)\bar{x}$. **14** Observation 28 impossible if $a < 28$. $\hat{a} = 28$. $\tilde{a} = 26$, impossible because of apparent outlier at 28.

Chapter 19

19.2.1 1 $a = 3.84$, $b = 0.05$, $c = 30$, $d = 0.48$, $e = 0.04$, $f = 0$, $g = 1.24$, $h = 30$.
2 (a) $(2.18, 17.53)$. (b) $(2.60, 26.76)$. **3** 0.975. **4** 0.009.

19.4.1 1 No. $X^2 = 2.00$; 1 d.f. **2** Yes. $X^2 = 5.33$; 1 d.f. **3** No. $X^2 = 2.80$; 5 d.f.
4 Not in agreement. $X^2 = 6.81$; 2 d.f.

19.6.1 1 10, 40, 60, 40, 10. $X^2 = 4.37$; 4 d.f. No evidence of bias. **2** Expected
frequencies 2, 36, 270, 1080, 2430, 2916, 1458. Do not reject 3:1 ratio.
$X^2 = 8.18$; 5 d.f. (combine 0, 1). **3** 5.2, 13.9, 16.2, 10.8, 4.5, 1.2, 0.2, 0, 0.
Pool classes $x \geqslant 4$ in test. $X^2 = 5.97$; 4 d.f. Do not reject hypothesis.

19.7.1 1 No. $X^2 = 5.70$; 2 d.f. **2** Reject NH of no relation. $X^2 = 10.65$; 3 d.f.
3 $X^2 = 100.2$; 4 d.f. Very strong evidence of relation.

19.8.1 1 Yes. $X^2 = 3.57$ (3.00 with Yates' correction); 1 d.f. **2** $z = 1.89$ ($= \sqrt{X^2}$ in
Qu.1), not sig. $(0.265, 0.435)$ and $(0.133, 0.317)$. **4** Replace $(ad - bc)$ by
$(|ad - bc| - 0.5 N)$. **5** Yes. $X^2 = 4.84$ (4.15 with Yates' correction); 1 d.f.
$X^2 = 48.4$.

19.9.2 1 Yes. $X^2 = 11.54$; 8 d.f. (Right-hand class is $\geqslant 400$). **2** No evidence
against hypothesis of uniform distribution. $X^2 = 6.35$; 4 d.f. **3** Pool two
extreme left-hand and two extreme right-hand classes. Satisfactory fit.
$X^2 = 4.30$; 2 d.f.

19.11 1 It does not. $X^2 = 9.11$; 3 d.f. **2** Yes. $X^2 = 5.25$ (4.79 with Yates' correction); 1 d.f. **3** Yes. $X^2 = 1.81$; 2 d.f. **4** No. $X^2 = 3.21$ (2.66 with Yates'
correction); 1 d.f. **5** No. $X^2 = 5.20$; 5 d.f. **6** Yes. $X^2 = 1.47$; 3 d.f. **7** Yes.
$X^2 = 7.58$ (5.88 with Yates' correction); 1 d.f. **8** Yes. $X^2 = 1.58$ (0.99 with
Yates' correction); 1 d.f. **9** No evidence of differential colour preference. $X^2 = 4.89$; 2 d.f. No colour preferred. $X^2 = 0.62$; 2 d.f. **10** No.
Ignoring 'not known', $X^2 = 61.9$; 4 d.f. **11** Pool classes 0– and 1–, giving
total of 7 classes. Satisfactory fit. $X^2 = 2.87$; 6 d.f. **12** Expected frequencies 4.9, 21.0, 38.3, 27.3, 8.5. Fit good. $X^2 = 0.11$; 2 d.f. **13** $\lambda = 8/185$.
Expected frequencies 57.3, 29.9, 15.7, 8.1, 4.3, 2.3, 2.4. Satisfactory fit.
$X^2 = 2.40$; 4 d.f. **14** 69.95; 1.26; 141.7; 176.2, 177.5. **15** It is. $X^2 = 1.78$; 4
d.f. **16** $X^2 = 11.35$; 6 d.f. $X^2 = 0.635$; 1 d.f. Cannot reject either hypothesis (strains same; ratio 9:7). **17** Neither significant. Max. diff. in K–S is 10,
or 0.25 in cdf. Need detailed information, not grouped, for precise analysis.

Revision Exercises E

1 5.97, 18.1830. **2** (i) 0.62%, (ii) 4.78%, (iii) 8.27%, (iv) 42.18%. **3** (i) 0.329,
(ii) 0.539. 0.9975. **4** $P(M = 0) = P(M = 2) = p^2(3 - 2p)$, $P(M = 1) = (1 - 2p)$
$(1 + 2p - 2p^2)$. **5** 35. **6** Standard deviation of estimate of π is 0.016. 550.
7 Use difference on each piece; mean = 1.27, variance = 1.269. Test of 'mean = 0'
gives $t_{(9)} = 3.57$, sig. at 1%. Interval $(0.464, 2.076)$. **8** $(0.127, 0.193)$. **9** Test for
binomial with $n = 1000$, $\pi = \frac{1}{3}$, 1-tail, $z = 1.79$, sig. at 5%. Suggest discrimination.
Interval $(0.321, 0.399)$. $n = 15\,300$. **10** $(34.47, 34.73)$. **11** 0.701; 70% conf. int.
is $\bar{x} - 0.3 < \mu < \bar{x} + 0.3$. **12** 16. **13** 80, 4; 0.0009, 0.063. **14** 95%. Yes, at 5%
(1-tail). $(16.99, 17.61)$ min. **15** 9.8059, 0.002 88. $(9.8043, 9.8075)$. 128 (using 95%
level). **16** 129.25, 13.804, 0.8661. $(127.0, 131.5)$. **17** 724.6. $(720.25, 728.95)$.

$t_{(11)} = 2.416$, sig. at 5%, suggests difference. **18** $\mu - \lambda$, $\sigma^2(\frac{1}{16} + \frac{1}{25})$. 0.0059.
19 $(-0.027, 2.967)$. No. 0.397. 467 g. **20** In whole population following diet, mean reduction lies in given range with probability 0.95. Assuming normal distribution, $(1.64, 3.36)$. **21** 8.28; 7.86 and 8.63 (approx.); 0.59. **22** $(2.74, 3.66)$; $(2.71, 3.49)$. **23** As 3×2 table, $\chi^2_{(2)} = 0.71$, no difference. Combining H + P to avoid low expected values, as 2×2 table $\chi^2_{(1)} = 0.57$, no difference. **24** Ratio A : B : C not same for two moths: $\chi^2_{(2)} = 25.49$, sig. at 0.1%. **25** $\chi^2_{(5)} = 9.04$, not sig.
26 (a) $\chi^2_{(11)} = 21.33$, sig. at 5%. (b) $\chi^2_{(1)} = 9.14$, sig. at 1%, for difference between rates; $\chi^2_{(5)} = 8.00$ n.s. in May–Oct., $\chi^2_{(5)} = 3.60$ n.s. in Nov.–April. **27** $\chi^2_{(1)} = 3.49$, n.s. $\chi^2_{(2)} = 6.70$, sig. at 5%. **28** Either gives sig. result. $t_{(9)} = 4.16$. Sign test prob. $= 0.021$. Since the differences between A and B on individuals seem to be normally distributed, parametric rather better because more powerful.
29 Should be paired comparison. **30** (i) Unpaired t-test (exact). (ii) Unpaired large-sample z-test (approx.) **31** (i) $(15.9, 26.1)$. (ii) Claim substantiated (clear from (i)). (iii) Normality. Same variances. **32** (i) Reject hypothesis. $t_{(9)} = -2.36$. (Assume normality). (ii) Do not reject NH. Sign test. **33** $T_- = 12.5$. No evidence of difference. **34** (i) $(18.2, 21.9)$, $(9.47, 34.9)$. (ii) $(17.5, 22.6)$, $(8.06, 45.5)$. **35** $\chi^2_{(12)} = 111.8$. Very strong evidence of relation. **36** (i) $F(X) = X/\theta$ $(0 \leqslant X \leqslant \theta)$. $\theta/2$. $\theta^2/12$. (ii) All observations independent, all must be $< X_{(n)}$. Pdf of $X_{(n)}$ is nx^{n-1}/θ^n. (iii) $E[(n+1)X_{(n)}/n] = \theta$. Variance $\theta^2/\{n(n+2)\}$. Yes. (iv) $\theta^2/3n$. Use estimator based on $X_{(n)}$ which is m.l.

Chapter 20

20.3.1 1 (a) Poisson 0.368, binomial 0.335. (b) 0.368, 0.402. (c) 0.264, 0.263.
2 0.013. **3** 0.0513.
20.6.1 1 (i) Poisson, $\lambda = 15$; should be random and independent. Can be approximated by $\mathcal{N}(15, 15)$. (ii) Poisson, with quite a large mean, so a normal distribution also holds approx. (iii) Binomial, $n = 10$, $\pi = \frac{1}{6}$. (iv) Number of throws in 5 minutes not fixed, so binomial not satisfied; but the number of throws is likely to be large so a normal approximation would hold with $\mu \approx 50$, σ quite small. (v) Discrete; not binomial since max. capacity not constant; numbers small, so no normal approximation possible. (vi) Poisson.
20.7.1 1 (i) 0.736, (ii) 0.677. **2** No. of replacements $= 6$. **3** 3.7, 0.884.
20.8.2 2 $n = $ no. per quadrat, $r = $ no. marked per quadrat.

$$P(r) = \sum_{n=0}^{\infty} \frac{e^{-\lambda}\lambda^n}{n!} \binom{n}{r} \pi^r(1 - \pi)^{n-r} = \frac{e^{-\lambda}\pi^r\lambda^r}{r!} \sum_{n=r}^{\infty} \frac{\lambda^{n-r}(1 - \pi)^{n-r}}{(n - r)!}$$

$$= \exp(-\lambda\pi)\{(\lambda\pi)^r/r!\}.$$

20.9.1 1 (i) 0.19, (ii) 0.018, (iii) 0.264; 2. **2** $\chi^2_{(5)} = 1.43$ not sig., could be Poisson with mean 3; use mean of data (2.9) for λ, reduce d.f. by 1.
3 $\chi^2_{(2)} = 3.17$ n.s., can accept engineer's hypothesis. 0.647.

20.12.1 1 33.   **2** No significant evidence.   **3** Sig. at 5%. Report may be based only on % improvement.   **4** No.

20.13.1 1 (a) (16.64, 19.47), (b) 18.717.   **2** 0.944.

20.15.1 1 1.283, 1.4485. Satisfactory fit.   **2** Reject Poisson hypothesis at 0.1% level.

20.17 1 0.974, 51.8.   **2** No; 0.082; 0.828.   **3** 0.407; 0.366; 0.165; 0.063 (to 3 d.p.). 0.817. 0.052.   **4** $P(r) = \frac{e^{-\lambda}\lambda^r}{r!(1-e^{-\lambda})}$ for $r = 1, 2, 3, \ldots$.   **5** (a)13.5, (b) 14.3, (c) 40.6.   **6** (i) 0.082, (ii) 0.242, 6.16.   **7** 95.24; 136.68; 98.08; 46.92; 16.84; 4.84; 1.40 (for $\geqslant 6$). 57.7.   **8** 2; 16.24; 32.48; 32.48; 21.65; 10.83; 6.32 (for $\geqslant 5$). $\chi^2_{(3)} = 7.69$, fit adequate.   **9** 0.202, 0.271; independence of two machines.   **10** Yes, mean approx. = variance. (5.56, 6.94). Exact method: (5.59, 6.98).   **12** $\bar{r} = 2.1$. $I = 116.67$, n.s. as $\chi^2_{(99)}$, Poisson appears to fit.
13 (i) P = 0.6376. (ii) P = 0.0012. 0.812.   **14** $Var[X] = \lambda$. $\hat{\lambda} = \bar{x}$, unbiased. $\lambda^* = s^2$. Prefer m.l. estimator which has minimum variance property (in large samples). 7.0, 112.0, 7.0, 112.0. Poisson model not satisfactory. Replace 35 by 8: 4.0, 4.0, 4.0, 4.0, model now very satisfactory.

Chapter 21

21.2.1 1 Yes.   **2** No.   **3** Yes.   **4** No.   **5** Yes.
21.3.1 1 −0.455.   **2** (a) r_0, (b) r_0.   **4** 0.744, 0.082.
21.3.2 1 0.935, 0.088, 0.757.
21.4.1 1 $Var[X] - Var[Y]$.   **3** $\sqrt{3/8} = 0.612$ **4** $Var[U] = \sigma_X^2 + \sigma_Y^2 + 2\rho\sigma_X\sigma_Y$.
$Var[V] = \sigma_X^2 + \sigma_Y^2 - 2\rho\sigma_X\sigma_Y \cdot \rho_{UV} = (\sigma_X^2 - \sigma_Y^2)/\sqrt{(\sigma_X^2 + \sigma_Y^2)^2 - 4\rho^2\sigma_X^2\sigma_Y^2}$.
21.5.2 Significant at 0.1%; not sig.; sig. at 0.1%.
21.6.1 (Space permits only indication of answers to these questions.)   **1** Both variables subject to time trends. Nonsense correlation.   **2** Appears to measure a cause-and-effect relationship (though not by itself demonstrating it); wood laid down in first season largely determines crop in second.   **3** Cannot infer causal relation. Correlation may indicate that more intelligent mothers have children later.   **4** No. Compare Figs 21.5 and 21.6. Regard boys and girls as distinct populations.   **5** Spurious correlations of parts with whole.   **6** If there is inflation, increases in both X, Z and a substantial positive correlation. Not very informative.   **7** Populations best regarded as distinct. Possible to pool estimates of correlation *within* populations, to obtain overall measure of relation within, e.g., the whole of the south-east.
21.8.1 1 No. $z = 0.67$.   **2** (−0.331, 0.469).   **3** Not sig. different. $z = 0.96$.
21.9.1 2 0.718.
21.10.1 1 Sig. at 5% level; $t_{(9)} = 3.09$. Similar conclusion: marks are associated. Probability level similar in two tests but not identical.   **2** $r_s = 0.329$, $t_{(10)} = 1.10$; no sig. association.   **3** Yes; longer time associated with better quality. $r_s = 0.636$, $t_{(10)} = 2.61$.   **4** (i) −0.380, not sig. (ii) −0.600, not sig.

21.12 1 -0.218. **2** (i) $\mu_X + \mu_Y$, (ii) $\sigma_X^2 + \sigma_Y^2 + 2\rho\sigma_X\sigma_Y$. 32, 7, 0.352.
3 $\text{Var}[X] - \text{Var}[Y]$. 3, 1. **6** (a) Both sig. diff. from 0; $t_{(18)} = 4.81$ and $t_{(28)} = 8.54$, or use Table A5. (b) No. $z = -1.83$. **7** -0.042. Suggests no association. **8** $r_s = 0.481$. Not a sig. association. $t_{(10)} = 1.73$. **9** (i) 0.171, (ii) $r_s = 0.371$. **10** $r_s = 3/7$. Not sig. **11** (i) Rank $r_s = -0.857$. Sig. association. (ii) Product-moment $r = 0.916$. Sig. association.

Chapter 22

22.2.3 1 Biscuit types $SS = 2.9611$, d.f. $= 2$; Residual $SS = 1.1856$, d.f. $= 24$.
2 Methods $SS = 42.50$, d.f. $= 3$; Residual $SS = 64.22$, d.f. $= 30$.
22.3.1 1 $(33.79, 34.08)$; $(0.01, 0.45)$. **2** $(3.94, 6.06)$; $(5.87, 8.13)$; $(-3.85, -0.15)$.
22.4.3 (i) $F_{(2,24)} = 29.97$, means not equal. (ii) $F_{(3,30)} = 6.62$, means not equal.
22.6.2 Cars $SS = 3.951$, d.f. $= 4$; Fuels $SS = 22.948$, d.f. $= 2$; Residual $SS = 2.445$, d.f. $= 8$. (i) $F_{(4,8)} = 3.23$, not significant. No. (ii) $F_{(2,8)} = 37.50$, significant at 1%. Yes.
22.9 1 (a) $S = 82$, $S_G = 54$. (b) L 14, M 8, N 6. (c) (i) 8. (ii) 2. (iii) 2. d.f. $= 6$. (d) Residual $SS = 28$; d.f. $= 6$. (e) 4.67. (f) No. $F_{(2,6)} = 5.79$.
2 (a) Groups 5; residual 24; $F_{(5,24)} = 2.62$ at 5%. (b) Groups 1; residual 40; $F_{(1,48)} = 4.08$ at 5%. (c) Blocks 4; groups 3; residual 12; $F_{(3,12)} = 3.49$ at 5%. **3** See Section 22.4. **4** $F_{(2,18)} = 24.00$, sig. at 1%; means differ. Means A: 20.43; B: 41.57; C: 31.43. S.E. of diff. of 2 means $= 3.053$. $t_{(18)} = 2.101$ at 5%. All differences significant. **5** Groups $SS = 76$, d.f. $= 4$; Residual $SS = 105$, d.f. $= 25$. $F_{(4,25)} = 4.52$, sig. at 1%, evidence of differences. S.E. of diff. of two means $= 1.183$. Limits $(0.23, 5.10)$. **6** (i) $t_{(12)} < 1$. (ii) Groups $SS = 4.024$, d.f. $= 1$; Residual $SS = 238.33$, d.f. $= 12$; $F_{(1,12)} < 1$. $F_{(1,12)} = t_{(12)}^2$. Same variance estimates by pooling or from residual mean square. **7** Groups $SS = 86.852$, d.f. $= 2$; Residual $SS = 107.637$, d.f. $= 13$. $F_{(2,13)} = 5.24$, sig. at 5%. S.E. of mean A $= 1.02$, B or C $= 1.44$. **8** Groups $SS = 411.67$, d.f. $= 3$; Residual $SS = 488.33$, d.f. $= 16$. $F_{(3,16)} = 4.50$, sig. at 5%. Evidence of difference. **9** Groups $SS = 0.025\,15$, d.f. $= 4$; Residual $SS = 0.012\,825$, d.f. $= 15$. $F_{(4,15)} = 7.36$, sig. at 1%. S.E. of diff. of two means $= 0.021$. **10** Groups $SS = 16.858$, d.f. $= 4$; Residual $SS = 2.9475$, d.f. $= 15$. All except B differ from A at 1% level. $(-0.342, 0.992)$. **11** Groups (varieties) $SS = 3.8253$, d.f. $= 2$; Blocks $SS = 5.5666$, d.f. $= 4$; Residual $SS = 3.4014$, d.f. $= 8$, $F_{(2,8)} = 4.50$, sig. at 5%. Canson $>$ others (using $t_{(8)}$). **12** Varieties $SS = 97.07$, d.f. $= 5$; Blocks $SS = 400.53$, d.f. $= 4$; Residual $SS = 60.27$, d.f. $= 20$. $F_{(4,20)} = 33.22$, sig. at 1% for blocks. $F_{(5,20)} = 6.44$, sig. at 5% for varieties. **13** d.f. observers $= 3$, preparations $= 4$, residual $= 12$. $F_{(3,12)} = 17.49$, sig. at 1% for observers. S.E. diff. of two means $= 2.53$. $(5.7, 21.1)$. **14** (a) $(1.985, 2.017)$. (b) 0.014. (c) $t_{(21)}$, tests on differences between means. **15** Litter $SS = 71.625$, d.f. $= 9$; Environments $SS = 51.875$, d.f. $= 3$; Residual $SS = 30.875$, d.f. $= 27$. $F_{(9,27)} = 6.96$, sig. at 5% for litters, as may be expected. $F_{(3,27)} = 15.12$, sig. at 1% for environments; t-tests show that the sig. result is mainly due to difference between B and others.

Chapter 23

23.3.1 1 $L = 98.4 + 0.24T$. **4** $y = 2.9 + 117.0x$. **5** $P = 49.0 + 2.01E$.

23.4.1 1 $(0.04, 0.10)$. **2** $y = 130.5 - 1.1t$. $(-1.38, -0.82)$.

23.6.1 1 $y = \frac{1}{2}x$.

23.7.1 1 $y = 2.1x - 8.7; 12.3$. **2** $H = 1.389 + 0.697T$. 53.66. Draw a graph.

23.8.1 (i) $x = \frac{1}{2}y$. (ii) $x = 1 + \frac{1}{2}y$.

23.9.2 1 Limits $(181.0, 214.9)$, $(189.7, 222.0)$, $(202.1, 233.5)$, $(213.3, 245.6)$, $(220.4, 254.3)$. **2** 7.771, 0.266. $S_{LR} = 4484.8$. $S = 5270.7$. $F_{(1,10)} = 57.07$ (sig. at 0.1%) shows line is good fit between 10 and 50 months.

23.12 1 $y = 2.110 + 0.0735t$. Origin of t is 1971(4). **2** $y = 10.56 + 0.914x$. (i) 16.56, 0.914. (ii) 0.969. **3** $y = 166.8 + 3.05x$. Curve needed. **4** $y = 2.5x - 5$. **5** $y = 32.05 + 0.726x$. 63 on I $\Rightarrow$ 78 on II. **6** (i) ln $L = $ constant $- (\sum e_i^2)/2\sigma^2$, as a function of β; max. when $\sum e_i^2$ min. (ii) ln $L = $ constant $- (\sum e_i^2/x_i)/2\sigma^2$, as a function of β; max. when $(\sum e_i^2/x_i)$ min. (iii) $\hat{\beta}_1 = 30.034$. $\hat{\beta}_2 = 30.000$. Residuals increase with x; hence prefer model (ii). **7** (i) Exact linear relation. (ii) Uncorrelated x, y. $(\rho = 0)$. **8** $y = 13.5 - 0.125x; x = 25.5 - 0.5y$. (a) y on x, (b) x on y. **9** $a = 2.56, b = 1.23$. Assuming errors in y normally distributed, interval is $(1.02, 1.44)$. **10** (i) $W = 2.9 - 0.5x$. (ii) 0.9. (iii) $(-0.586, -0.414)$. **11** Set 1 gives smaller value of Var(b). **12** $F_{(1,7)} = 169.3$, reject NH. $(0.097, 0.141)$. **13** $s^2 = 10.205$. $(4.21, 9.04)$. **14** Smooth curve, levelling off. $u = a + bx$. $\hat{a} = 0.060\,42, \hat{b} = 0.009\,755$. 33.45, 40.23, 50.44, 57.78, 63.30, 71.07, 76.27, 81.48. Very good. **15** $w = b/a + t/a. w = 1.880 + 0.345t; \hat{a} = 2.90, \hat{b} = 5.45$. 2.16.

Revision Exercises F

1 Poisson, $\lambda = 3$. Fits well. **2** 1.33, 1.26. Poisson, $\lambda = 4/3$. Final class is $\geqslant$ 3 floods. Fits well ($\chi^2 = 0.08$; 2 d.f.). **3** 0.1847, £8.64, 15, 0.0778. **4** 2822.7, 7742.18. $2822.7, 4919.48$, using $(n-1)$ divisor in variance. **5** $(86, 126)$. **6** (i) 1, $\sqrt{4/5}$. (ii) 2, $\sqrt{2}$. (iii) 2, 4. (iv) 1, $1/\sqrt{3}$. **7** 13.4, 20.1, 15.1, 7.5, 2.8, 1.1 (final class is $\geqslant 5$). $(7.4, 22.6)$. **8** (a) $y = 0.119x + 6.70$. (b) 0.980. (c) Yes. **9** (b) Var $(\overline{X}) = \frac{1}{2}\sigma^2(1 + \rho)$; negatively correlated. **10** -0.023, -0.036. **11** $\frac{1}{5}\sqrt{24} = 0.980$. **12** $\rho\sqrt{6}$. **13** (a) $y = 0.811x + 14.1; x = 0.784y + 13.4$. **14** 0.856; $y = 1.10x - 0.751$, $x = 0.664y + 0.633$. 0.856; $y = 2.21X + 0.351, X = 0.332y - 0.184$. **15** $y = 6.63x + 7.58; 54.0$. Extrapolation unreliable. **16** $0.3, 2.01$, $2.01, \frac{3}{7}$. **17** (i) $y = 0.621x + 19.2$; (ii) $x = 0.783y + 13.7, 0.697$. **18** $y = 9.20x + 186$, where x is the year measured from 1969. Some guide, but limited because it is an extrapolation. **19** $a = -5, b = 1, c = -2$; Var $(x) = 9$, Var $(y) = 30$, Var $(z) = 18$. **20** (i) $r_s = 0.248$ n.s. (ii) $r_s = 0.709$. sig. at 5%. Interview better predictor. **21** (a) No evidence of difference: $t_{(14)} = 1.94$. (b) Evidence of difference: $t_{(8)} = 4.14$. **22** z scores are -1.04, -0.67, 0, 0.67, 1.28, 1.64. From line drawn by eye $\mu = 30.8$, $\sigma = 5.2$, 0.44. **23** p^{-1}, $(1-p)p^{-2}$, $(1-p)^{t-1}$. **24** (i) $\frac{81}{625} = 0.130$. (ii) $\frac{16}{625} = 0.026$, $\frac{96}{625} = 0.154$, $\frac{216}{625} = 0.346$, 0.475. £45.4, £2.6. **25** (a) $\frac{5}{6}$. (b) $(1+y)^{-2}$, $\frac{1}{10}$.

26 *r*:

	1	2	3	4	5	6	7	8	9	10	11	12
Probability:	$\frac{6}{72}$	$\frac{7}{72}$	$\frac{8}{72}$	$\frac{9}{72}$	$\frac{10}{72}$	$\frac{11}{72}$	$\frac{6}{72}$	$\frac{5}{72}$	$\frac{4}{72}$	$\frac{3}{72}$	$\frac{2}{72}$	$\frac{1}{72}$.

Mean $= 5.25$.
27 $\frac{1}{3}(n-1)$. (i) $\frac{74}{210} = 0.352$. (ii) $\frac{4}{7} = 0.571$. **28** (a) (i) 0.875, (ii) $\frac{1}{3}(3+2\sqrt{2})$, $\frac{1}{9}$. (b) $\frac{7}{8}$.
29 Transform to $\log x = \log a + t \log b$; regress $\log x$ on t; assume $\log x$ is normal and values independent. **30** $F(47.6) = 0.327$. No. **31** (b) $Y = 200X - 600$. (i) $E[Y] = 200$, $\text{Var}(Y) = 160\,000$. (ii) Prob. of 6 sales $=$ Prob. (demand $\geqslant 6$) in Poisson with mean 4. $E[Y] = 160.9$. **32** Both plots reasonably linear except for the point where $Cr = 0.45$; as this is by some way the lowest Cr, could omit it. Possibly GFR against $1/Cr$ has variance increasing with GFR. Prefer other suggested transformation.

Chapter 24

24.2.1 1 $E[Y|x_1, x_2] = -0.307 + 1.022x_1 + 0.142x_2$.
   2 $E[Y|x_1, x_2] = -6.039 + 3.207x_1 + 0.107x_2$.

24.3.1 1 SS(d.f.) are: extra for A, 3.8953 (1); extra for W, 0.0700 (1); residual, 0.8011 (5). W does not contribute: W alone, $F_{(1,6)} < 1$; extra for W, $F_{(1,5)} < 1$. A is sig. at 1%, $F_{(1,6)} = 27.9$. **2** (i) $b_1 = 1.147$, $b_2 = 1.729$. (ii) Both to be included. Extra for X_2, $F_{(1,8)} = 116.5$, sig. at 0.1%. Extra for X_1, $F_{(1,8)} = 47.8$, sig. at 0.1%. (iii) 1.4912.

24.11 1 (a) (16.8, 19.6). (b) (23.8, 30.4). **2** $E[Y|x_1, x_2] = 18.5 + 0.5x_1 - 0.2x_2$. (a) 0.9. (b) $t_{(30)} = 15.8$, reject NH. (c) t-test, using $\text{Var}(b+c) = \text{Var}(b) + 2\text{Cov}(b,c) + \text{Var}(c)$. **3** (i) $E[Y|x_1, x_2] = 0.021 + 2.985x_1 + 1.219x_2$. (ii) One extreme residual, $Y = 17$, $\hat{Y} = 27$. (iii) Residual $MS = 24.74$ (9 d.f.). (a) $F_{(2,9)} = 6.78$, sig. at 5%. (b) Extra for X_2, $F_{(1,9)} = 3.99$ n.s. Extra for X_1, $F_{(1,9)} = 11.27$, sig. at 1%. (iv) $\text{Var}(b_1) = 0.7909$, $t_{(9)} = 3.36$, sig. at 1%. $\text{Var}(b_2) = 0.3722$, $t_{(9)} = 2.00$ n.s. (v) Aptitude test a useful predictor, explains 42% of variability in performance. Interview not sig., but worth more investigation. **4** (b) $E[Y|x_1, x_2] = -6.65 + 52.22x_1 + 3.89x_2$. **5** Residual $SS = 706$ (17 d.f.). Overall regression sig. at 0.1%, $F_{(2,17)} = 17.4$. Extra for X_2 sig. at 1%, $F_{(1,17)} = 13.1$. Extra for X_1 sig. at 0.1%, $F_{(1,17)} = 16.0$. **7** (i) $F_{(3,53)} = 14.1$, sig. at 0.1%. (ii) Denoting y as 0, $r = (01)\ 0.61$, $(02) -0.26$, $(03)\ 0.50$, $(12) -0.33$, $(13)\ 0.50$, $(23)\ 0.05$. (iii) Regression SS, $(12)\ 676.34$, $(23)\ 602.62$, $(31)\ 761.57$. (iv) Latitude X_2 does not contribute to overall regression although regression on X_2 alone is almost sig. at 5%. **8** (i) Regress $\log y$ on $\log x_1$ and $\log x_2$. (ii) Regress $\log y$ on x_1 and x_2. (iii) Nonlinear model. (iv) Nonlinear model. (v) Regress $1/y$ on $1/x$ and x.

HINTS ON COMPUTING EXERCISES

We have aimed to do everything by menus but there are instances where the menu method is very awkward or gives unsatisfactory results. In such cases we have used session commands. It is in relatively simple procedures such as drawing histograms, tallying continuous variates or calculating means from frequency data that we have found this so. We recommend always **Enable Session Commands**, partly for the reasons above but also because it provides a record of what has been done. It is good practice to give columns informative names. To conserve space we have done this only occasionally in these hints.

Further hints on the exercises will be found on the website.

Chapter 1

1 Assume data in C1.

(a), (b) **Graph>Histogram/** Select **Simple/** Enter C1 as **Graph variable/ Graph>Stem-and-Leaf/** Enter C1 as **Graph variable/**

(c) **Graph>Bar Chart/**Select **Counts of unique values** and **Simple/** Enter C1 as **Categorical variable//**

(d) Although number of words is discrete, a histogram gives a better impression than a bar chart.

2 Enter Male weights in C1, Female in C2. **Graph>Histogram/** Select **Simple**, enter Male and Female as **Graph variables**, and select **Multiple Graphs/** Choose **Overlaid on same graph/** Select **Data View/** Select **Area** and **Bars// Graph>Stem-and-Leaf/** Enter Male and Female as **Graph variables**, do NOT **trim outliers/**

3 The commands

HISTOGRAM C1;
CUTPOINTS 19.45:25.45/0.1.

produce a histogram with class-interval 0.1. Repeat using class-intervals 0.3, 0.5, 0.7 and 0.9. Choose **Editor>Layout Tool** and put histograms side by side for comparison.

4 The commands HISTOGRAM C1; NINTERVAL K. produce a histogram with K intervals. Choose a number of values of K, including the square root of n, to investigate the rule.

Chapter 2

1 Enter data in X and Y. Use **Calc>Calculator**, to derive Z. Enter test name: SET C4 9(A) 8(B) END. **Stat>Basic Statistics>Display Descriptive Statistics/** Enter **Graph variable** X, and **By Variable** C4/. **Graph>Dotplot/** Select **With Groups/** Enter **Graph variable** X and **Categorical variable** C4/.

2 **Stat>Basic Statistics>Display Descriptive Statistics/** Enter C1 in **Variable**, select **Statistics/** Select **Mean** and **Median/**

Chapter 3

1 See hints above, Chapter 1, Exercise 2.

2 **Calc>Random Data>Integer/** Enter 100 **rows of data**, **store** in C1, **min** 0, **max** 9/. **Graph>Bar Chart/**

3 **Calc>Random Data/ Uniform/** Enter 100 **rows of data**, **store** in C1, **Lower endpoint** 0, **Upper** 1/. **Graph>Histogram/**

4 Data in C1. **Calc>Random Data>Sample from Columns/ Sample** 100 **rows** from C1, **store** in C2, select **Sample with Replacement/**. Repeat, putting samples in C3, C4, C5. **Calc>Row Statistics/** Select **Mean**, **Input variables** C2–C5, **store** in C6/.

Chapter 4

1 See hints on exercises in earlier chapters.

2 Enter Sex in C1, Salary in C2. **Stat>Basic Statistics>Display Descriptive Statistics/** Enter **Variable** Salary, and **By Variable** Sex/. Also make diagrams.

3 First part as in hint for Chapter 3, Exercise 4.

Chapter 6

1 Generate trials using **Calc>Random Data>Bernoulli**, 50 columns each with 100 rows. Label and calculate columns: (1) 'No. of H' in 50 trials using **Calc>Row Statistics/ Sum/** (2) 'Total H' to date, using **Calc>Calculator**, with function **Partial Sums** of 'No. of H' (3) 'No. of trials' using **Calc>Make Patterned Data/ Store patterned data in** 'No of trials' **From first value** 50 **To last value** 5000 **In steps of** 50/ (4) 'Proportion H' = 'Total H'/ 'No. of trials'. Plot 'Proportion H' against time using **Graph>Time Series Plot**.

2 Generate trials in C1–C3, 100 rows each, using **Calc>Random Data>Bernoulli**. LET C4 = C1 + C2 + C3, as command, or use **Calc>Row Statistics/ Sum/**. **Stat>Tables>Tally Individual Values/** Select **Counts** and **Percents/**.

Chapter 7

1 Generate 500 Bernoulli trials in C1 with probability of success 0.16667 (**Calc>Random Data**). Put partial sums in C2 (**Calc>Calculator**). Tally the data in C2 putting categories in C3 and counts in C4. (TALLY C2; COUNTS; STORE C3 C4.) Delete C4(1) opposite category 0, and the last entry in C4, because of end effects. The values in C4 then form a random sample from the required distribution.

2 Generate 200 values in C1 chosen from the **Integer** distribution on 1 to 10. Replace by 0, throughout C1, any of the distinct values that occur in the first nine entries in C1 while leaving other values unchanged. Put the result in C2. In a particular example this was done by the menu **Data>Code>Numeric to Numeric/Code data from** C1 **into** C2, **Original values** 1 2 5 6 7 8, **New value** 0/. Looking at the result suggested a second coding of (3 4 10) to 0. Using this left only '9' which first occurred at Observation 41. Hence 41 was a random sample from the required distribution. To generate a sizeable sample by this method is obviously very laborious. The method can be automated by using a MINITAB macro.

Chapter 8

1 Set the digits 0, 1, ..., 8 in C1. **Calc>Probability Distributions>Binomial/** Select **Probability, Input** C1, **Output** C2, $n = 30$, $p = 0.1$. LET C2(10) = 1 − SUM(C2).

2 Generate 100 values in C1 using **Calc>Random Data>Binomial/** with $n = 30$, $p = 0.1$. **Stat>Tables>Tally Individual Variables/** Select **Counts** and **Percentages**.

3 Proceed as recommended above for Chapter 7, Exercise 1. Having obtained the values of 'number of throws to a 6', add successive pairs of values, to give values from the required distribution.

Chapter 9

1 Generate 500 values from **Integer** distribution, in range 1 to 6, and put in C1. **Data>Code>Numeric to Numeric/ Code data from** C1 **into** C2, **Original values** 2:5, **New value** −2/. **Stat>Tables>Tally Individual Variables**.

2 The game may be summarized by a table.

No of throws, x	1	2	3	4	5 with H	5 with T
Probability	0.5	0.25	0.125	0.0625	0.031 25	0.031 25
Ann's gain	1	2	3	2	−5	−62

Required distribution of gain given by table above provided the probabilities for '2' are added. Assume gain in C1, probabilities in C2. Choose random sample from distribution of gains using: **Calc>Random Data>Discrete/** 100 **rows, store in** C3, **Values** in C1, **Probabilities** in C2/. **Calc>Calculator**, with **Partial sums** of C3, put in C4/. Finally, add 600 to every value in C4 and use **Graph>Time Series Plot**.

Chapter 10

1 Label C1 'X', C2 'Y', C3 'Z'. **Calc>Random Data>Binomial/** 500 rows, **store in X**, $n = 4, p = 0.6$/. Repeat for Y. Let $Z = X + Y$.

On new worksheet, label C1 'X', C2 'Y1', C3 'Y2', C4 'Y', C5 'Z'. Generate X as before. Code X values: 0, 1, 2 to 1, and 3, 4 to 0; store in Y1 (**Data>Code>Numeric to Numeric/**). Generate 500 values of **Binomial** with $n = 3, p = 0.4$, and store in Y2. Let $Y = Y1 + Y2$. Let $Z = X + Y$. Choose **Stat>Basic Statistics>Display Descriptive Statistics**.

2 Label C1 'Toss', C2 'Factor', C3 'Score', C4 'Gain'. Generate 500 Bernoulli trials, with $p = 0.5$, in Toss. Code values in Toss, 1 replaced by 2 and 0 by -1, store in Factor. Generate 500 values in Score from **Integer** distribution, min 1, max 6. Let Gain $=$ Factor $*$ Score. Look at distribution and descriptive statistics of Gain.

Use **Calc>Calculator** with function **Partial Sums** of Gain. Plot change over time using **Graph>Time Series Plot**.

Chapter 11

1 Generate 500 values in k columns using **Calc>Random Data>Integer**, with min 0 and max 9. Choose **Calc>Row Statistics** to find the mean for each row of the k columns. Carry out for $k = 1, 4, 9, 16, 25$.

2 Label C1 'R' and C2 'P'. Generate 500 values in R using **Calc>Random Data>Binomial**, with $n = 20$ and $p = 0.1$, say. Let $P = R/20$. P contains random values from the required distribution. A histogram may be constructed and descriptive statistics determined.

Chapter 13

1 (i) See Section 13.9. (ii) Calculate differences. Proceed as in (i) except that Test median is zero and a one-tailed test is used.

2 Set data in two columns. **Graph>Dotplot/** Select **Multiple Y's**, **Simple/**. **Stat>Nonparametrics>Mann-Whitney/**.

3 **Stat>Nonparametrics>1-Sample Wilcoxon/**.

Chapter 14

1 Generate 500 values in C1 using **Calc>Random Data>Exponential**, with mean 0.5. Make partial sums of C1 and put in C2 using **Calc>Calculator**. Find integral part of each value in C2 by using **Expression** ROUND(C2-0.5,0) in **Calculator** and storing in C3. Tally values in C3 using **Stat>Tables>Tally Individual Variables**. In the frequency table, the counts opposite $5, 10, 15, \ldots$ are observations from the required distribution. Do not overlook zero counts.

2 Generate data using **Calc>Random Data>Uniform**.

Chapter 15

1 Set data in C1, -1000 89.5:189.5/20 1000. Find values of cumulative distribution function corresponding to values in C1 using **Calc>Probability Distribution>Normal**, with mean 131.5, s.d. 20. Store in C2. Find differences between successive values of C2 using **Stat>Time Series>Differences** (**Lag** $= 1$ is appropriate). Store in C3. Multiply all values in C3 by 100.

2 (a) Find expected frequencies in a similar way to previous question.
(b) Generate 500 observations in each of 10 columns, finding sum, instead of mean, for each row (see the hint for Chapter 11, Exercise 1, above). Code all values falling in a chosen interval by centre of interval: **Data>Code>Numeric to Numeric/Original** (0.5:10.5), **New** 5.5; **Original** (10.5:20.5), **New** 15.5, etc. May take more than one operation. Suppose coded values are in C20. Tally values using TALLY C20; COUNTS; STORE C21 C22. To give observed frequencies opposite class centres.

3 Generate 160 observations from each of the 3 distributions and put into C1, C2 and C3. Produce histograms from the columns using the option **Labels** to give a title to each graph indicating distribution and no. of observations. Use **Editor>Layout Tool** to put the three histograms side by side. Reduce the no. of observations in each column by choosing **Data>Delete Rows**, and repeat.

Chapter 16

1 Generate 500 values in each of C1–C12 using **Calc>Random Data>Uniform**, with lower endpoint 0, upper 12. **Calc>Row Statistics/Mean**, and store in C20. **Calc>Calculator**, C21 $= (C20 - 6)$. For part (iii) look at the hint to Chapter 15, Exercise 2(b).

2 **Calc>Random Data>Binomial/** with $n = 9$, $p = 0.1$. **Calc>Calculator**, C2 $= (C1 - 0.9)/0.9$. Direct use of **Calc>Histogram** gives unsatisfactory histograms with many empty intervals. The number of intervals should be reduced either by editing the histogram or by using session commands.

Chapter 17

1 **1-Sample t>**Enter **Test mean** 37.659/. **2-Sample t/** Do not pool variance.

2 Generate 1000 values in C1 using **Calc>Random Data>Normal**, with mean 12, s.d. 0.8. LET C2 $= (C1 - 12)/ 0.8$. **Data>Code>Numeric to Numeric/Original** $(-1000:-1.96)$, **New** -1; **Original** $(-1.96:1.96)$, **New** 0; **Original** (1.96, 1000), **New** 1.

Chapter 18

1 (ii) **Data>Code>Numeric to Numeric/Original** 11:20 **New** 1, **Original** 0:10 **New** 0/. **Stat>Basic Statistics>1 Proportion**. Select **Option, Use test and interval based on normal distribution**.

2 Generate 1000 values in each of C1–C10 using **Calc>Random Data>Normal**, with mean 1000, s.d. 2.5. **Calc>Row Statistics>Mean** of C1–C10, store in C11. Use **Calc>Calculator** to make $K1 = (1.96 * 2.5)/3.16228$, $C12 = C11 - K1, C13 = C11 + K1$. Replace one endpoint of any interval that does not contain 1000, using **Data>Code>Numeric to Numeric/Original value** (0: 999.99999) **New** 0, data in C12, store in C22; **Original value** (1000.00001:10000) **New** 0, data in C13, store in C23. No. of zeros in C22 and C23 is no. of intervals not containing 1000.

3 Generate 500 values in each of C1–C15 using **Calc>Random Data>Uniform**, with lower endpoint 0, upper 2. Use **Calc>Row Statistics** to put **Mean**, **Median** and **Max** in C21, C22, C23. Use **Calc>Calculator** to give $C24 = (8 * C23)/15$.

Chapter 19

1 To obtain expected values, proceed as in the hint to Chapter 15, Exercise 1, starting with setting −1000 90:160/10 1000 in C1. Store expected frequencies in E. Set observed values 3 12 47 77 65 32 13 5 in O. Use **Calc>Calculator** to find $C10 = (O - E) ** 2/E$ and $K1 = SUM$ C10. **Calc>Probability Distributions>Chi-Square/** Select **Cumulative Probability**.

2 Generate 500 values in C1 using **Calc>Random Data>Normal**. Use coding in a similar way to Chapter 15, Exercise 2(b), to find observed frequencies. Find expected frequencies as in Chapter 15, Exercise 1. Then proceed as hint immediately preceding.

3 Input data to C1–C3 with 4 rows each. Use **Stat>Tables>Chi-Square Test (Table in Worksheet)**. Alternatively, stack the data and choose **Stat>Tables>Cross Tabulation and Chi-Square**, which gives greater flexibility in the output.

Chapter 20

1 Generate 500 values in C1 using **Calc>Random Data>Poisson**, with **Mean** 1, etc.

2 **Stat>Tables>Individual values**, operating on data in C1, gives observed frequencies. Store in C2. Find mean and store in K1. Set 0:6 in C3. **Calc>Probability Distributions>Poisson/Select Probability**, enter K1 for **Mean**, store in C4. Find probability of values $\geqslant 7$ as complement of sum of C4. These values, when multiplied by 90, are the expected frequencies. Then proceed as in latter part of hint, Chapter 19(1).

3 **Calc>Probability Distributions>Binomial/Select Probability**, enter n and π/. **Calc>Probability Distributions>Poisson/** Select **Probability**, enter **Mean**/.

Chapter 21

1 Name columns 'WT', 'FAT', '%FAT' and 'SEX'. Read in data using, say, 1 for male and 2 for female. Also unstack data using **Data>Unstack columns/Unstack 'WT'**, 'FAT', '%FAT' using **subscripts** in 'SEX', **store after last column in use/**. **Stat>Basic Statistics>Correlation/**Enter 'WT' and '%FAT' as **Variables/** gives r for all data. Repeat on separate data for men and women. Investigate by diagram, **Graph>Scatterplot/** Select **With Groups**, enter 'WT' and '%FAT' in **Variables**, 'SEX' in **Categorical variable/**. Continue to investigate using similar commands.

2 Similar menu choices to previous exercise.

Chapter 22

1 Follow the instructions for one-way analysis in Section 22.8, with (i) prices or (ii) scores as **response y**.

2 Follow the instructions for two-way analysis in Section 22.8 with (i) yields or (ii) counts as **response y**, (i) varieties or (ii) observers as **row factor** and (i) blocks I–V or (ii) preparations as **column factor**.

Chapter 23

1 (i) Use **Graph>Scatterplot** (see Section 5.5). (ii) Use **Stat>Basic Statistics>Correlation** as in Section 21.10. (iii) Use **Stat>Regression>Fitted Line Plot** as in Section 23.11. (Note that **Fitted Line Plot** also gives the information in **Scatterplot**.)

2 (i) Use **Graph>Scatterplot**. (ii) Follow the instructions in Section 23.11 using y and x as variables. (iii) Proceed as in (ii) but find $\ln y$ and $\ln x$ as in Section 23.11.

3 Use **Fitted Line Plot** in each case, including its diagram.

Chapter 24

1 Follow the instructions in Section 24.10 with x_1 and x_2 as **predictors**.

2 Enter salary as **response y** in C1 and $x_1 \ldots x_6$ in C2–C7 as the **predictors** class, age, sex, year, jobs, type. Follow the instructions (1)–(5) in the question, and Section 24.10 for the chosen analysis. **Graph>Matrix Plot** is useful to show which of $x_1 - x_6$ are related to y, and to one another.

Table A1 *Random digits*

12005	84000	51051	92674	76575	35789	04180	75029	32490	39949
98859	09884	45275	09467	93026	32912	13941	23206	62419	67776
26604	95099	93751	00590	93060	64776	83565	69919	51623	27483
82984	65780	94428	30160	86023	52284	62463	70712	40687	92630
70888	14063	96700	83008	17579	71321	62664	51514	92195	46722
77803	61872	86245	68220	66267	01379	11304	01658	82404	46728
35228	49673	53552	51215	45611	83927	00772	99295	72154	24126
69965	74926	63366	47688	14279	42943	98863	86630	53925	22310
89716	61713	30650	49028	20285	37791	69149	41701	42403	64009
68348	85228	97590	90997	83339	95822	72969	14037	32379	96225
33821	41538	86376	71823	16285	92630	89531	59337	05421	17043
63162	18167	32088	41917	60942	63252	83886	54130	31841	04502
03431	44528	41760	68035	33731	43262	12789	40348	15532	95309
99198	35092	63655	23987	31112	88069	58720	41729	18757	96096
75535	45156	49477	10673	48262	78240	94031	06192	75221	13363
98554	52502	11780	04060	56634	58077	02005	80217	65893	78381
89725	00679	28401	79434	00909	22989	31446	76251	17061	66680
49221	37750	26367	44817	09214	82674	65641	14332	58211	49564
31783	96028	69352	78426	94411	38335	22540	37881	10784	84658
61025	72770	13689	21456	48391	00157	61957	11262	12640	17228
10581	30143	89214	52134	76280	77823	61674	96898	90487	43998
51753	56087	71524	64913	81706	33984	90919	86969	75553	87375
96050	08123	28557	04240	33606	10776	64239	81900	74880	92654
93998	95705	73353	26933	66089	25177	62387	34932	62021	34044
70974	45757	31830	09589	31037	91886	51780	21912	16444	52881
25833	71286	76375	43640	92551	46510	68950	60168	26399	04599
55060	28982	92650	71622	36740	05869	17828	29377	01020	90851
29436	79967	34383	85646	04715	80695	39283	50543	26875	94047
80180	08706	17875	72123	69723	52846	71310	72507	25702	33449
40842	32742	44671	72953	54811	39495	05023	61569	60805	26580

(contd)

31481	16208	60372	94367	88977	35393	08681	53325	92547	31622
06045	35097	38319	17264	40640	63022	01496	28439	04197	63858
41446	12336	54072	47198	56085	25215	89943	41153	18496	76869
22301	07404	60943	75921	02932	50090	51949	86415	51919	98125
38199	09042	26771	15881	80204	61281	61610	24501	01935	33256
06273	93282	55034	79777	75241	11762	11274	41685	24117	98311
92201	02587	31599	27987	25678	69736	94487	41653	79550	92949
70782	80894	95413	36338	04237	19954	71137	23584	87069	10407
05245	40934	96832	33415	62058	87179	31542	18174	54711	21882
85607	45719	65640	33241	04852	87636	43840	42242	22092	28975
61175	56493	93453	90267	99471	04519	78694	17115	00371	64703
36079	22448	22686	31272	01245	66265	12670	29560	49346	20049
94688	39732	02785	73373	44876	39888	69352	40488	43849	95406
54047	85793	53994	28605	46114	91174	49646	85123	66246	72392
24997	69553	46802	24331	88523	89026	69776	55460	21984	76677
58817	18317	58358	37063	90744	65601	77282	85925	43361	78952
40830	81825	87707	94714	64180	83492	39137	76232	92009	52966
94454	88221	30380	09953	34841	97815	12056	66094	07407	72757
27342	91018	64227	67603	63903	17313	88745	72451	61949	52481
74162	64984	84195	02337	49352	37786	20017	12481	07796	96277
55135	89786	81282	06410	86600	56549	28268	94715	40449	55914
23710	10833	28936	78630	15868	58827	58707	42688	03162	09539
44221	13093	60948	80173	81905	02559	68076	68417	59856	89401
60153	35928	56904	56391	63031	10917	81241	84050	44236	17841
81975	28513	74788	30647	60238	77031	28893	43525	30920	95417
72177	07774	43400	50468	47951	99475	50534	82409	61221	67633
89791	65378	53609	75192	12980	46429	59055	00220	60366	74283
22027	20739	71805	58619	68863	17715	62202	83159	21504	26514
55957	57627	24052	69693	87768	46443	43484	94488	46063	05525
75080	29073	37102	43620	95419	88526	51255	13706	96545	21285
41146	09986	77034	94207	39412	17739	22258	32945	41330	76710
31265	68613	12868	82946	75980	47114	81368	54747	86798	36950
59956	43733	29735	05443	31254	85275	03763	48739	11081	49940
81755	74610	53964	32865	47419	13292	81889	88900	84116	67995
34604	11342	77046	22569	58028	61191	53234	43983	84692	66186
94088	19798	25396	89752	28658	87582	08228	75960	15916	16757
18762	26079	72976	03233	12093	56659	31445	38213	55499	23751
61828	85552	69667	88964	84312	02078	91508	78999	76176	11799
94724	44409	03514	84530	73463	86319	12902	39116	65488	74392
48158	94298	71559	06531	76381	19781	57515	42274	75305	52205

TABLES

714

(contd)

70403	35908	81568	69574	44311	52191	72364	68880	58891	85188
81888	43309	47024	71212	22659	45093	39622	36131	07971	34954
51015	65209	20753	37890	26636	82332	11244	64872	03892	53269
07370	37778	51004	66389	81507	81677	25029	14773	77256	84701
67159	74475	96734	01410	62677	29354	87497	44005	20054	02046
31725	62743	83282	21429	88893	25479	82374	27000	25835	23433
70018	81532	35723	67758	38665	19602	78029	14571	01153	17416
17378	32479	55575	08509	27804	68850	76844	61631	16190	73383
66921	65761	12748	87280	98471	13679	39731	31664	86856	93563
14338	86192	95534	90708	96728	94873	26248	36704	81386	20191
19795	52147	19995	05701	49411	69386	04673	16357	95928	80781
38239	14674	92628	99014	86202	82223	03855	36677	12769	39373
45454	83549	26116	07766	04554	62626	17645	29750	64346	10103
47770	02076	96110	94395	02177	46334	67060	20444	72212	18548
69071	58342	78798	21776	36415	76543	62379	39579	27507	40249
43050	10241	97342	69774	70440	78092	17229	79994	06689	16317
70021	33726	62364	12923	41687	04077	91319	46977	28684	64710
30995	46330	94666	47311	66039	86338	97476	68144	77233	00666
96099	19925	41976	37198	50891	03966	82313	62789	49932	83549
01090	47392	70665	58300	96531	32880	19482	75930	36672	88467
67366	42813	25092	73451	18786	30436	17354	72165	11983	58676
38455	78612	48404	25830	42510	82650	50284	79953	89398	40674
39866	98018	39053	85401	03342	57554	34452	46263	20795	79672
23520	25411	03569	68579	64688	91549	58802	89324	51917	70996
27118	19535	84009	15953	81913	50393	64334	29635	16694	78352
90558	86484	03273	55219	27949	21473	65524	13104	26491	25001
80889	89030	00953	59867	73574	74606	86154	63400	32908	98812
14989	43548	20444	13850	41286	04618	53980	78430	35310	39771
65393	43012	52164	72837	71446	48970	84940	20035	05283	08078
77739	22912	37037	68732	04426	42222	84355	62851	07763	61883

These random digits were generated on a computer and will be adequate for the exercises in this book.

Table A2 *Values of z, the standard normal variable, from 0.0 by steps of 0.01 to 3.9, showing the cumulative probability up to z (Probability correct to 4 decimal places)*

z	0.00	0.01	0.02	0.03	0.04	0.05	0.06	0.07	0.08	0.09
0.0	.5000	.5040	.5080	.5120	.5160	.5199	.5239	.5279	.5319	.5359
.1	.5398	.5438	.5478	.5517	.5557	.5596	.5636	.5675	.5714	.5753
.2	.5793	.5832	.5871	.5910	.5948	.5987	.6026	.6064	.6103	.6141
.3	.6179	.6217	.6255	.6293	.6331	.6368	.6406	.6443	.6480	.6517
.4	.6554	.6591	.6628	.6664	.6700	.6736	.6772	.6808	.6844	.6879
.5	.6915	.6950	.6985	.7019	.7054	.7088	.7123	.7157	.7190	.7224
.6	.7257	.7291	.7324	.7357	.7389	.7422	.7454	.7486	.7517	.7549
.7	.7580	.7611	.7642	.7673	.7704	.7734	.7764	.7794	.7823	.7852
.8	.7881	.7910	.7939	.7967	.7995	.8023	.8051	.8078	.8106	.8133
.9	.8159	.8186	.8212	.8238	.8264	.8289	.8315	.8340	.8365	.8389
1.0	.8413	.8438	.8461	.8485	.8508	.8531	.8554	.8577	.8599	.8621
.1	.8643	.8665	.8686	.8708	.8729	.8749	.8770	.8790	.8810	.8830
.2	.8849	.8869	.8888	.8907	.8925	.8944	.8962	.8980	.8997	.9015
.3	.9032	.9049	.9066	.9082	.9099	.9115	.9131	.9147	.9162	.9177
.4	.9192	.9207	.9222	.9236	.9251	.9265	.9279	.9292	.9306	.9319
.5	.9332	.9345	.9357	.9370	.9382	.9394	.9406	.9418	.9429	.9441
.6	.9452	.9463	.9474	.9484	.9495	.9505	.9515	.9525	.9535	.9545
.7	.9554	.9564	.9573	.9582	.9591	.9599	.9608	.9616	.9625	.9633
.8	.9641	.9649	.9656	.9664	.9671	.9678	.9686	.9693	.9699	.9706
.9	.9713	.9719	.9726	.9732	.9738	.9744	.9750	.9756	.9761	.9767
2.0	.9772	.9778	.9783	.9788	.9793	.9798	.9803	.9808	.9812	.9817
.1	.9821	.9826	.9830	.9834	.9838	.9842	.9846	.9850	.9854	.9857
.2	.9861	.9864	.9868	.9871	.9875	.9878	.9881	.9884	.9887	.9890
.3	.9893	.9896	.9898	.9901	.9904	.9906	.9909	.9911	.9913	.9916
.4	.9918	.9920	.9922	.9925	.9927	.9929	.9931	.9932	.9934	.9936
.5	.9938	.9940	.9941	.9943	.9945	.9946	.9948	.9949	.9951	.9952
.6	.9953	.9955	.9956	.9957	.9959	.9960	.9961	.9962	.9963	.9964
.7	.9965	.9966	.9967	.9968	.9969	.9970	.9971	.9972	.9973	.9974
.8	.9974	.9975	.9976	.9977	.9977	.9978	.9979	.9979	.9980	.9981
.9	.9981	.9982	.9982	.9983	.9984	.9984	.9985	.9985	.9986	.9986
3.0	.9987	.9987	.9987	.9988	.9988	.9989	.9989	.9989	.9990	.9990
.1	.9990	.9991	.9991	.9991	.9992	.9992	.9992	.9992	.9993	.9993
.2	.9993	.9993	.9994	.9994	.9994	.9994	.9994	.9995	.9995	.9995
.3	.9995	.9995	.9995	.9996	.9996	.9996	.9996	.9996	.9996	.9997
.4	.9997	.9997	.9997	.9997	.9997	.9997	.9997	.9997	.9997	.9998
.5	.9998	.9998	.9998	.9998	.9998	.9998	.9998	.9998	.9998	.9998
.6	.9998	.9998	.9999	.9999	.9999	.9999	.9999	.9999	.9999	.9999
.7	.9999	.9999	.9999	.9999	.9999	.9999	.9999	.9999	.9999	.9999
.8	.9999	.9999	.9999	.9999	.9999	.9999	.9999	.9999	.9999	.9999
.9	1.0000									

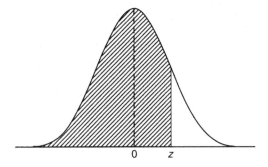

The curve is $\mathcal{N}(0, 1)$, the standard normal variable. The table entry is the shaded area $\Phi(z) = P(Z < z)$. For example, when $z = 1.96$ the shaded area is 0.9750. Critical values of the standard normal distribution will be found in the bottom row of Table A3.

Table A3 *Student's t-distribution*
Values exceeded in two-tail test with probability P

d.f.	P = 0.1	0.05	0.02	0.01	0.002	0.001
1	6.314	12.706	31.821	63.657	318.31	636.62
2	2.920	4.303	6.965	9.925	22.327	31.598
3	2.353	3.182	4.541	5.841	10.214	12.924
4	2.132	2.776	3.747	4.604	7.173	8.610
5	2.015	2.571	3.365	4.032	5.893	6.869
6	1.943	2.447	3.143	3.707	5.208	5.959
7	1.895	2.365	2.998	3.499	4.785	5.408
8	1.860	2.306	2.896	3.355	4.501	5.041
9	1.833	2.262	2.821	3.250	4.297	4.781
10	1.812	2.228	2.764	3.169	4.144	4.587
11	1.796	2.201	2.718	3.106	4.025	4.437
12	1.782	2.179	2.681	3.055	3.930	4.318
13	1.771	2.160	2.650	3.012	3.852	4.221
14	1.761	2.145	2.624	2.977	3.787	4.140
15	1.753	2.131	2.602	2.947	3.733	4.073
16	1.746	2.120	2.583	2.921	3.686	4.015
17	1.740	2.110	2.567	2.898	3.646	3.965
18	1.734	2.101	2.552	2.878	3.610	3.922
19	1.729	2.093	2.539	2.861	3.579	3.883
20	1.725	2.086	2.528	2.845	3.552	3.850
21	1.721	2.080	2.518	2.831	3.527	3.819
22	1.717	2.074	2.508	2.819	3.505	3.792
23	1.714	2.069	2.500	2.807	3.485	3.767
24	1.711	2.064	2.492	2.797	3.467	3.745
25	1.708	2.060	2.485	2.787	3.450	3.725
26	1.706	2.056	2.479	2.779	3.435	3.707
27	1.703	2.052	2.473	2.771	3.421	3.690
28	1.701	2.048	2.467	2.763	3.408	3.674
29	1.699	2.045	2.462	2.756	3.396	3.659
30	1.697	2.042	2.457	2.750	3.385	3.646
40	1.684	2.021	2.423	2.704	3.307	3.551
60	1.671	2.000	2.390	2.660	3.232	3.460
120	1.658	1.980	2.358	2.617	3.160	3.373
∞	1.645	1.960	2.326	2.576	3.090	3.291

The last row of the table (∞) gives values of z, the standard normal variable.

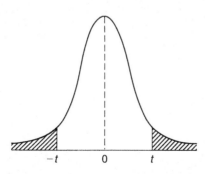

P is the shaded area.
To make a one-tail test, use $P' = \frac{1}{2}P$ as the probability level.

Table A4 *Values of the χ^2 distribution exceeded with probability P*

d.f.	P						
	0.995	0.975	0.050	0.025	0.010	0.005	0.001
1	3.9×10^{-5}	9.8×10^{-4}	3.84	5.02	6.63	7.88	10.83
2	0.010	0.051	5.99	7.38	9.21	10.60	13.82
3	0.072	0.22	7.81	9.35	11.34	12.84	16.27
4	0.21	0.48	9.49	11.14	13.28	14.86	18.47
5	0.41	0.83	11.07	12.83	15.09	16.75	20.52
6	0.68	1.24	12.59	14.45	16.81	18.55	22.46
7	0.99	1.69	14.07	16.01	18.48	20.28	24.32
8	1.34	2.18	15.51	17.53	20.09	21.96	26.13
9	1.73	2.70	16.92	19.02	21.67	23.59	27.88
10	2.16	3.25	18.31	20.48	23.21	25.19	29.59
11	2.60	3.82	19.68	21.92	24.73	26.76	31.26
12	3.07	4.40	21.03	23.34	26.22	28.30	32.91
13	3.57	5.01	22.36	24.74	27.69	29.82	34.53
14	4.07	5.63	23.68	26.12	29.14	31.32	36.12
15	4.60	6.26	25.00	27.49	30.58	32.80	37.70
16	5.14	6.91	26.30	28.85	32.00	34.27	39.25
17	5.70	7.56	27.59	30.19	33.41	35.72	40.79
18	6.26	8.23	28.87	31.53	34.81	37.16	42.31
19	6.84	8.91	30.14	32.85	36.19	38.58	43.82
20	7.43	9.59	31.41	34.17	37.57	40.00	45.32
21	8.03	10.28	32.67	35.48	38.93	41.40	46.80
22	8.64	10.98	33.92	36.78	40.29	42.80	48.27
23	9.26	11.69	35.17	38.08	41.64	44.18	49.73
24	9.89	12.40	36.42	39.36	42.98	45.56	51.18
25	10.52	13.12	37.65	40.65	44.31	46.93	52.62
26	11.16	13.84	38.89	41.92	45.64	48.29	54.05
27	11.81	14.57	40.11	43.19	46.96	49.64	55.48
28	12.46	15.31	41.34	44.46	48.28	50.99	56.89
29	13.12	16.05	42.56	45.72	49.59	52.34	58.30
30	13.79	16.79	43.77	46.98	50.89	53.67	59.70
40	20.71	24.43	55.76	59.34	63.69	66.77	73.40
50	27.99	32.36	67.50	71.42	76.15	79.49	86.66
60	35.53	40.48	79.08	83.30	88.38	91.95	99.61
70	43.28	48.76	90.53	95.02	100.43	104.22	112.32
80	51.17	57.15	101.88	106.63	112.33	116.32	124.84
90	59.20	65.65	113.15	118.14	124.12	128.30	137.21
100	67.33	74.22	124.34	129.56	135.81	140.17	149.45

For degrees of freedom $f > 100$, test $\sqrt{2\chi^2_{(f)}}$ as $\mathcal{N}(\sqrt{2f - 1}, 1)$.

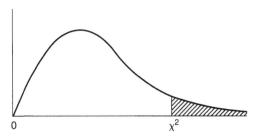

P is the shaded area.

Table A5 *Values of the correlation coefficient, r, based on samples of size n, which differ significantly from 0 at the 5%, 1%, 0.1% levels, using a two-tail test*

n	d.f.	0.05	0.01	0.001
3	1	0.997	0.999	1.000
4	2	.950	.990	0.999
5	3	.878	.959	.991
6	4	.811	.917	.974
7	5	.754	.875	.951
8	6	0.707	0.834	0.925
9	7	.666	.798	.898
10	8	.632	.765	.872
11	9	.602	.735	.847
12	10	.576	.708	.823
13	11	0.553	0.684	0.801
14	12	.532	.661	.780
15	13	.514	.641	.760
16	14	.497	.623	.742
17	15	.482	.606	.725
18	16	0.468	0.590	0.708
19	17	.456	.575	.693
20	18	.444	.561	.679
21	19	.433	.549	.665
22	20	.423	.537	.652
27	25	0.381	0.487	0.597
32	30	.349	.449	.554
37	35	.325	.418	.519
42	40	.304	.393	.490
47	45	.288	.372	.465
52	50	0.273	0.354	0.443
62	60	.250	.325	.408
72	70	.232	.302	.380
82	80	.217	.283	.357
92	90	.205	.267	.338
102	100	.195	.254	.321

Table A6 *Values of Spearman's rank correlation coefficient r_S which differ significantly from 0 at the 5% and 1% levels, using a two-tail test*

Sample size, n	0.05	0.01
5	1.000	–
6	0.886	1.000
7	.786	0.929
8	.738	.881
9	.700	.833
10	.648	.794

For sample sizes greater than 10 use Table A5.

Table A7 Table of F-distribution. Upper 5% points

v_2 \ v_1	1	2	3	4	5	6	7	8	9	10	12	15	20	24	30	40	60	120	∞
1	161.4	199.5	215.7	224.6	230.2	234.0	236.8	238.9	240.5	241.9	243.9	245.9	248.0	249.1	250.1	251.1	252.2	253.3	254.3
2	18.51	19.00	19.16	19.25	19.30	19.33	19.35	19.37	19.38	19.40	19.41	19.43	19.45	19.45	19.46	19.47	19.48	19.49	19.50
3	10.13	9.55	9.28	9.12	9.01	8.94	8.89	8.85	8.81	8.79	8.74	8.70	8.66	8.64	8.62	8.59	8.57	8.55	8.53
4	7.71	6.94	6.59	6.39	6.26	6.16	6.09	6.04	6.00	5.96	5.91	5.86	5.80	5.77	5.75	5.72	5.69	5.66	5.63
5	6.61	5.79	5.41	5.19	5.05	4.95	4.88	4.82	4.77	4.74	4.68	4.62	4.56	4.53	4.50	4.46	4.43	4.40	4.36
6	5.99	5.14	4.76	4.53	4.39	4.28	4.21	4.15	4.10	4.06	4.00	3.94	3.87	3.84	3.81	3.77	3.74	3.70	3.67
7	5.59	4.74	4.35	4.12	3.97	3.87	3.79	3.73	3.68	3.64	3.57	3.51	3.44	3.41	3.38	3.34	3.30	3.27	3.23
8	5.32	4.46	4.07	3.84	3.69	3.58	3.50	3.44	3.39	3.35	3.28	3.22	3.15	3.12	3.08	3.04	3.01	2.97	2.93
9	5.12	4.26	3.86	3.63	3.48	3.37	3.29	3.23	3.18	3.14	3.07	3.01	2.94	2.90	2.86	2.83	2.79	2.75	2.71
10	4.96	4.10	3.71	3.48	3.33	3.22	3.14	3.07	3.02	2.98	2.91	2.85	2.77	2.74	2.70	2.66	2.62	2.58	2.54
11	4.84	3.98	3.59	3.36	3.20	3.09	3.01	2.95	2.90	2.85	2.79	2.72	2.65	2.61	2.57	2.53	2.49	2.45	2.40
12	4.75	3.89	3.49	3.26	3.11	3.00	2.91	2.85	2.80	2.75	2.69	2.62	2.54	2.51	2.47	2.43	2.38	2.34	2.30
13	4.67	3.81	3.41	3.18	3.03	2.92	2.83	2.77	2.71	2.67	2.60	2.53	2.46	2.42	2.38	2.34	2.30	2.25	2.21
14	4.60	3.74	3.34	3.11	2.96	2.85	2.76	2.70	2.65	2.60	2.53	2.46	2.39	2.35	2.31	2.27	2.22	2.18	2.13
15	4.54	3.68	3.29	3.06	2.90	2.79	2.71	2.64	2.59	2.54	2.48	2.40	2.33	2.29	2.25	2.20	2.16	2.11	2.07
16	4.49	3.63	3.24	3.01	2.85	2.74	2.66	2.59	2.54	2.49	2.42	2.35	2.28	2.24	2.19	2.15	2.11	2.06	2.01
17	4.45	3.59	3.20	2.96	2.81	2.70	2.61	2.55	2.49	2.45	2.38	2.31	2.23	2.19	2.15	2.10	2.06	2.01	1.96
18	4.41	3.55	3.16	2.93	2.77	2.66	2.58	2.51	2.46	2.41	2.34	2.27	2.19	2.15	2.11	2.06	2.02	1.97	1.92
19	4.38	3.52	3.13	2.90	2.74	2.63	2.54	2.48	2.42	2.38	2.31	2.23	2.16	2.11	2.07	2.03	1.98	1.93	1.88
20	4.35	3.49	3.10	2.87	2.71	2.60	2.51	2.45	2.39	2.35	2.28	2.20	2.12	2.08	2.04	1.99	1.95	1.90	1.84
21	4.32	3.47	3.07	2.84	2.68	2.57	2.49	2.42	2.37	2.32	2.25	2.18	2.10	2.05	2.01	1.96	1.92	1.87	1.81
22	4.30	3.44	3.05	2.82	2.66	2.55	2.46	2.40	2.34	2.30	2.23	2.15	2.07	2.03	1.98	1.94	1.89	1.84	1.78
23	4.28	3.42	3.03	2.80	2.64	2.53	2.44	2.37	2.32	2.27	2.20	2.13	2.05	2.01	1.96	1.91	1.86	1.81	1.76
24	4.26	3.40	3.01	2.78	2.62	2.51	2.42	2.36	2.30	2.25	2.18	2.11	2.03	1.98	1.94	1.89	1.84	1.79	1.73
25	4.24	3.39	2.99	2.76	2.60	2.49	2.40	2.34	2.28	2.24	2.16	2.09	2.01	1.96	1.92	1.87	1.82	1.77	1.71
26	4.23	3.37	2.98	2.74	2.59	2.47	2.39	2.32	2.27	2.22	2.15	2.07	1.99	1.95	1.90	1.85	1.80	1.75	1.69
27	4.21	3.35	2.96	2.73	2.57	2.46	2.37	2.31	2.25	2.20	2.13	2.06	1.97	1.93	1.88	1.84	1.79	1.73	1.67
28	4.20	3.34	2.95	2.71	2.56	2.45	2.36	2.29	2.24	2.19	2.12	2.04	1.96	1.91	1.87	1.82	1.77	1.71	1.65
29	4.18	3.33	2.93	2.70	2.55	2.43	2.35	2.28	2.22	2.18	2.10	2.03	1.94	1.90	1.85	1.81	1.75	1.70	1.64
30	4.17	3.32	2.92	2.69	2.53	2.42	2.33	2.27	2.21	2.16	2.09	2.01	1.93	1.89	1.84	1.79	1.74	1.68	1.62
40	4.08	3.23	2.84	2.61	2.45	2.34	2.25	2.18	2.12	2.08	2.00	1.92	1.84	1.79	1.74	1.69	1.64	1.58	1.51
60	4.00	3.15	2.76	2.53	2.37	2.25	2.17	2.10	2.04	1.99	1.92	1.84	1.75	1.70	1.65	1.59	1.53	1.47	1.39
120	3.92	3.07	2.68	2.45	2.29	2.17	2.09	2.02	1.96	1.91	1.83	1.75	1.66	1.61	1.55	1.50	1.43	1.35	1.25
∞	3.84	3.00	2.60	2.37	2.21	2.10	2.01	1.94	1.88	1.83	1.75	1.67	1.57	1.52	1.46	1.39	1.32	1.22	1.00

v_1, v_2 are upper, lower d.f. respectively.
Tabulated values are those exceeded with probability 0.05 in a one-tail test.

v_2 \ v_1	1	2	3	4	5	6	7	8	9	10	12	15	20	24	30	40	60	120	∞
1	4052	4999.5	5403	5625	5764	5859	5928	5981	6022	6056	6106	6157	6209	6235	6261	6287	6313	6339	6366
2	98.50	99.00	99.17	99.25	99.30	99.33	99.36	99.37	99.39	99.40	99.42	99.43	99.45	99.46	99.47	99.47	99.48	99.49	99.50
3	34.12	30.82	29.46	28.71	28.24	27.91	27.67	27.49	27.35	27.23	27.05	26.87	26.69	26.60	26.50	26.41	26.32	26.22	26.13
4	21.20	18.00	16.69	15.98	15.52	15.21	14.98	14.80	14.66	14.55	14.37	14.20	14.02	13.93	13.84	13.75	13.65	13.56	13.46
5	16.26	13.27	12.06	11.39	10.97	10.67	10.46	10.29	10.16	10.05	9.89	9.72	9.55	9.47	9.38	9.29	9.20	9.11	9.02
6	13.75	10.92	9.78	9.15	8.75	8.47	8.26	8.10	7.98	7.87	7.72	7.56	7.40	7.31	7.23	7.14	7.06	6.97	6.88
7	12.25	9.55	8.45	7.85	7.46	7.19	6.99	6.84	6.72	6.62	6.47	6.31	6.16	6.07	5.99	5.91	5.82	5.74	5.65
8	11.26	8.65	7.59	7.01	6.63	6.37	6.18	6.03	5.91	5.81	5.67	5.52	5.36	5.28	5.20	5.12	5.03	4.95	4.86
9	10.56	8.02	6.99	6.42	6.06	5.80	5.61	5.47	5.35	5.26	5.11	4.96	4.81	4.73	4.65	4.57	4.48	4.40	4.31
10	10.04	7.56	6.55	5.99	5.64	5.39	5.20	5.06	4.94	4.85	4.71	4.56	4.41	4.33	4.25	4.17	4.08	4.00	3.91
11	9.65	7.21	6.22	5.67	5.32	5.07	4.89	4.74	4.63	4.54	4.40	4.25	4.10	4.02	3.94	3.86	3.78	3.69	3.60
12	9.33	6.93	5.95	5.41	5.06	4.82	4.64	4.50	4.39	4.30	4.16	4.01	3.86	3.78	3.70	3.62	3.54	3.45	3.36
13	9.07	6.70	5.74	5.21	4.86	4.62	4.44	4.30	4.19	4.10	3.96	3.82	3.66	3.59	3.51	3.43	3.34	3.25	3.17
14	8.86	6.51	5.56	5.04	4.69	4.46	4.28	4.14	4.03	3.94	3.80	3.66	3.51	3.43	3.35	3.27	3.18	3.09	3.00
15	8.68	6.36	5.42	4.89	4.56	4.32	4.14	4.00	3.89	3.80	3.67	3.52	3.37	3.29	3.21	3.13	3.05	2.96	2.87
16	8.53	6.23	5.29	4.77	4.44	4.20	4.03	3.89	3.78	3.69	3.55	3.41	3.26	3.18	3.10	3.02	2.93	2.84	2.75
17	8.40	6.11	5.18	4.67	4.34	4.10	3.93	3.79	3.68	3.59	3.46	3.31	3.16	3.08	3.00	2.92	2.83	2.75	2.65
18	8.29	6.01	5.09	4.58	4.25	4.01	3.84	3.71	3.60	3.51	3.37	3.23	3.08	3.00	2.92	2.84	2.75	2.66	2.57
19	8.18	5.93	5.01	4.50	4.17	3.94	3.77	3.63	3.52	3.43	3.30	3.15	3.00	2.92	2.84	2.76	2.67	2.58	2.49
20	8.10	5.85	4.94	4.43	4.10	3.87	3.70	3.56	3.46	3.37	3.23	3.09	2.94	2.86	2.78	2.69	2.61	2.52	2.42
21	8.02	5.78	4.87	4.37	4.04	3.81	3.64	3.51	3.40	3.31	3.17	3.03	2.88	2.80	2.72	2.64	2.55	2.46	2.36
22	7.95	5.72	4.82	4.31	3.99	3.76	3.59	3.45	3.35	3.26	3.12	2.98	2.83	2.75	2.67	2.58	2.50	2.40	2.31
23	7.88	5.66	4.76	4.26	3.94	3.71	3.54	3.41	3.30	3.21	3.07	2.93	2.78	2.70	2.62	2.54	2.45	2.35	2.26
24	7.82	5.61	4.72	4.22	3.90	3.67	3.50	3.36	3.26	3.17	3.03	2.89	2.74	2.66	2.58	2.49	2.40	2.31	2.21
25	7.77	5.57	4.68	4.18	3.85	3.63	3.46	3.32	3.22	3.13	2.99	2.85	2.70	2.62	2.54	2.45	2.36	2.27	2.17
26	7.72	5.53	4.64	4.14	3.82	3.59	3.42	3.29	3.18	3.09	2.96	2.81	2.66	2.58	2.50	2.42	2.33	2.23	2.13
27	7.68	5.49	4.60	4.11	3.78	3.56	3.39	3.26	3.15	3.06	2.93	2.78	2.63	2.55	2.47	2.38	2.29	2.20	2.10
28	7.64	5.45	4.57	4.07	3.75	3.53	3.36	3.23	3.12	3.03	2.90	2.75	2.60	2.52	2.44	2.35	2.26	2.17	2.06
29	7.60	5.42	4.54	4.04	3.73	3.50	3.33	3.20	3.09	3.00	2.87	2.73	2.57	2.49	2.41	2.33	2.23	2.14	2.03
30	7.56	5.39	4.51	4.02	3.70	3.47	3.30	3.17	3.07	2.98	2.84	2.70	2.55	2.47	2.39	2.30	2.21	2.11	2.01
40	7.31	5.18	4.31	3.83	3.51	3.29	3.12	2.99	2.89	2.80	2.66	2.52	2.37	2.29	2.20	2.11	2.02	1.92	1.80
60	7.08	4.98	4.13	3.65	3.34	3.12	2.95	2.82	2.72	2.63	2.50	2.35	2.20	2.12	2.03	1.94	1.84	1.73	1.60
120	6.85	4.79	3.95	3.48	3.17	2.96	2.79	2.66	2.56	2.47	2.34	2.19	2.03	1.95	1.86	1.76	1.66	1.53	1.38
∞	6.63	4.61	3.78	3.32	3.02	2.80	2.64	2.51	2.41	2.32	2.18	2.04	1.88	1.79	1.70	1.59	1.47	1.32	1.00

v_1, v_2 are upper, lower d.f. respectively.
Tabulated values are those exceeded with probability 0.01 in a one-tail test.

Table A8 Common statistical distributions
I. Discrete distributions.

Name, parameters	Possibility space	Probability mass function $f(r)$	Mean, $E[R]$	Variance, $E[(R - E[R])^2]$	Moment generating function, $E[e^{tR}]$
Discrete uniform κ, no. of values in the possibility space (κ a positive integer)	$\{1, 2, \ldots, \kappa\}$	$\dfrac{1}{\kappa}$	$\dfrac{\kappa + 1}{2}$	$\dfrac{\kappa^2 - 1}{12}$	$\dfrac{1 - e^{\kappa t}}{\kappa(e^{-t} - 1)}$
Bernoulli π, probability of success at the single trial ($0 < \pi < 1$)	$\{0, 1\}$	$\pi^r (1 - \pi)^{1-r}$	π	$\pi(1 - \pi)$	$(1 - \pi) + \pi e^t$
Binomial n, no. of trials; π, probability of success at each trial (n a positive integer; $0 < \pi < 1$)	$\{0, 1, 2, \ldots, n\}$	$\dbinom{n}{r} \pi^r (1 - \pi)^{n-r}$	$n\pi$	$n\pi(1 - \pi)$	$\{(1 - \pi) + \pi e^t\}^n$
Geometric π, probability of success at each trial ($0 < \pi < 1$)	$\{1, 2, 3, \ldots\}$	$(1 - \pi)^{r-1}\pi$	$\dfrac{1}{\pi}$	$\dfrac{1 - \pi}{\pi^2}$	$\dfrac{\pi}{1 - (1 - \pi)e^t}$
Poisson λ, average no. of random events in a given interval ($\lambda > 0$)	$\{0, 1, 2, \ldots\}$	$\dfrac{\lambda^r e^{-\lambda}}{r!}$	λ	λ	$\exp\{\lambda(e^t - 1)\}$

The probability generating function $G(u)$ is obtained from the moment generating function by replacing e^t by u.

II. Continuous distributions.

Name, parameters	Possibility space	Probability density function $f(x)$	Mean, $\mathrm{E}[X]$	Variance, $\mathrm{E}[(X - \mathrm{E}[X])^2]$	Moment generating function, $\mathrm{E}[e^{tX}]$
Continuous uniform α, minimum value; β, maximum value	$\alpha < x < \beta$	$\dfrac{1}{\beta - \alpha}$	$\dfrac{\beta + \alpha}{2}$	$\dfrac{(\beta - \alpha)^2}{12}$	$\dfrac{e^{\beta t} - e^{\alpha t}}{(\beta - \alpha)t}$
Exponential λ, average no. of random events in a given interval ($\lambda > 0$)	$x \geqslant 0$	$\lambda e^{-\lambda x}$	$\dfrac{1}{\lambda}$	$\dfrac{1}{\lambda^2}$	$\dfrac{\lambda}{\lambda - t}$
Normal μ, mean; σ^2, variance ($\sigma > 0$)	$-\infty < x < \infty$	$\dfrac{1}{\sigma\sqrt{2\pi}} \exp\left(-\dfrac{(x - \mu)^2}{2\sigma^2}\right)$	μ	σ^2	$\exp\left(\mu t + \dfrac{\sigma^2 t^2}{2}\right)$

Table A9 *Kolmogorov–Smirnov one-sample statistic, D*

*Each table entry is the value of a Kolmogorov–Smirnov one-sample statistic D for sample size n such that its **right-tail** probability P is the value given on the top row.*

n	P = 0.100	0.050	0.020	0.010	n	P = 0.100	0.050	0.020	0.010
1	0.950	0.975	0.990	0.995	21	0.259	0.287	0.321	0.344
2	0.776	0.842	0.900	0.929	22	0.253	0.281	0.314	0.337
3	0.636	0.780	0.785	0.829	23	0.247	0.275	0.307	0.330
4	0.565	0.624	0.689	0.734	24	0.242	0.269	0.301	0.323
5	0.509	0.563	0.627	0.669	25	0.238	0.264	0.295	0.317
6	0.468	0.519	0.577	0.617	26	0.233	0.259	0.290	0.311
7	0.436	0.483	0.538	0.576	27	0.229	0.254	0.284	0.305
8	0.410	0.454	0.507	0.542	28	0.225	0.250	0.279	0.300
9	0.387	0.430	0.480	0.513	29	0.221	0.246	0.275	0.295
10	0.369	0.409	0.457	0.489	30	0.218	0.242	0.270	0.290
11	0.352	0.391	0.437	0.468	31	0.214	0.238	0.266	0.285
12	0.338	0.375	0.419	0.449	32	0.211	0.234	0.262	0.281
13	0.325	0.361	0.404	0.432	33	0.208	0.231	0.258	0.277
14	0.314	0.349	0.390	0.418	34	0.205	0.227	0.254	0.273
15	0.304	0.338	0.377	0.404	35	0.202	0.224	0.251	0.269
16	0.295	0.327	0.366	0.392	36	0.199	0.221	0.247	0.265
17	0.286	0.318	0.355	0.381	37	0.196	0.218	0.244	0.262
18	0.279	0.309	0.346	0.371	38	0.194	0.215	0.241	0.258
19	0.271	0.301	0.337	0.361	39	0.191	0.213	0.238	0.255
20	0.265	0.294	0.329	0.352	40	0.189	0.210	0.235	0.252

For $n > 40$, right-tail critical values based on the asymptotic distribution can be calculated as follows:

0.100	0.050	0.020	0.010
$1.22/\sqrt{n}$	$1.36/\sqrt{n}$	$1.52/\sqrt{n}$	$1.63/\sqrt{n}$

Source: Adapted from L. H. Miller (1956), Table of percentage points of Kolmogorov statistics, *Journal of the American Statistical Association*, **51**, 111–121, with permission.

Table A10 *Factors for obtaining control chart limits for range*

Sample size n	Lower action (99.9%)	Lower warning (95%)	Upper warning (95%)	Upper action (99.9%)
2	0.00	0.04	2.81	4.12
3	0.04	0.18	2.17	2.99
4	0.10	0.29	1.93	2.58
5	0.16	0.37	1.81	2.36
6	0.21	0.42	1.72	2.22
7	0.26	0.46	1.66	2.12
8	0.29	0.50	1.62	2.04
9	0.32	0.52	1.58	1.99
10	0.35	0.54	1.56	1.94

Index

Bold numbers are used to indicate definitions and theorems. *Italic* numbers are used to indicate projects. References to the use of MINITAB are given in the Selection of MINITAB menus on page 681.

INDEX

728